HANDBOOK
OF FERTILITY

HANDBOOK OF FERTILITY

NUTRITION, DIET, LIFESTYLE AND REPRODUCTIVE HEALTH

Edited by

RONALD ROSS WATSON
University of Arizona, Arizona Health Sciences Center, Tucson, AZ, USA

AMSTERDAM • BOSTON • HEIDELBERG • LONDON
NEW YORK • OXFORD • PARIS • SAN DIEGO
SAN FRANCISCO • SINGAPORE • SYDNEY • TOKYO
Academic Press is an Imprint of Elsevier

Academic Press is an imprint of Elsevier
32 Jamestown Road, London NW1 7BY, UK
525 B Street, Suite 1800, San Diego, CA 92101-4495, USA
225 Wyman Street, Waltham, MA 02451, USA
The Boulevard, Langford Lane, Kidlington, Oxford OX5 1GB, UK

Notices
Knowledge and best practice in this field are constantly changing. As new research and experience broaden our understanding, changes in research methods, professional practices, or medical treatment may become necessary.

Practitioners and researchers must always rely on their own experience and knowledge in evaluating and using any information, methods, compounds, or experiments described herein. In using such information or methods they should be mindful of their own safety and the safety of others, including parties for whom they have a professional responsibility.

To the fullest extent of the law, neither the Publisher nor the authors, contributors, or editors, assume any liability for any injury and/or damage to persons or property as a matter of products liability, negligence or otherwise, or from any use or operation of any methods, products, instructions, or ideas contained in the material herein.

Medical Disclaimer:
Medicine is an ever-changing field. Standard safety precautions must be followed, but as new research and clinical experience broaden our knowledge, changes in treatment and drug therapy may become necessary or appropriate. Readers are advised to check the most current product information provided by the manufacturer of each drug to be administered to verify the recommended dose, the method and duration of administrations, and contraindications. It is the responsibility of the treating physician, relying on experience and knowledge of the patient, to determine dosages and the best treatment for each individual patient. Neither the publisher nor the authors assume any liability for any injury and/or damage to persons or property arising from this publication.

British Library Cataloguing-in-Publication Data
A catalogue record for this book is available from the British Library

Library of Congress Cataloging-in-Publication Data
A catalog record for this book is available from the Library of Congress

ISBN: 978-0-12-800872-0

For information on all Academic Press publications
visit our website at http://store.elsevier.com/

Typeset by Thomson Digital

Contents

List of Contributors   xi
Preface   xiii
Acknowledgments   xv

A

OVERVIEW ON FERTILITY

1. Perceptions of Environmental Risks to Fertility

KAREN P. PHILLIPS

Introduction   3
Conclusions   13
References   13

2. Paternal Smoking as a Cause for Transgenerational Damage in the Offspring

DIANA ANDERSON, THOMAS ERNST SCHMID, ADOLF BAUMGARTNER

Introduction   19
Tobacco Smoke   20
Genetic, Epigenetic, and other Transmissible Factors   21
Quality Control of Paternal Germ Cells, Post-fertilization
    Repair of Sperm DNA Damage   23
Conclusions   23
References   24

3. Impact of Cigarette Smoking During Pregnancy on Conception and Fetal Health through Serum Folate Levels

MEHMET YEKTA ONCEL, OMER ERDEVE

Introduction   27
General Knowledge on Folate   27
Smoking and Pregnancy   28
Conclusion   31
References   31

4. The Effect of Maternal Metabolic Health and Diet on the Follicular Fluid Composition and Potential Consequences for Oocyte and Embryo Quality

S.D.M. VALCKX, J.L.M.R. LEROY

Introduction   35
Metabolic Health Related Composition of Ovarian
    Follicular Fluid and its Relation to Oocyte and Embryo
    Quality Parameters   35
Fatty Acids in the Ovarian Follicular Fluid   37
The Effect of Elevated NEFA Concentrations on Oocyte
    and Embryo Quality   39
Conclusions   41
References   41

5. The Effect of Heavy Metals on Preterm Mortality and Morbidity

GÜLCAN TÜRKER

Introduction   45
Heavy Metals and Preterm Mortality and Morbidity   46
Lead   47
Mercury   50
Cadmium   51
Arsenic   52
Conclusions   53
References   53

6. Nitrate, Nitrite, Nitrosatable Drugs, and Congenital Malformations

JEAN D. BRENDER

Introduction   61
Maternal Exposures to Nitrate, Nitrite, and N-Nitroso
    Compounds   61
Nitrosatable Drugs and Birth Defects   62
Nitrate in Drinking Water and Birth Defects   65
Dietary Intake of Nitrates, Nitrites, and Nitrosamines
    in Foods and Birth Defects   69
Joint Exposures of Nitrosatable Drugs and Nitrate/Nitrite
    and Birth Defects   69
Vitamin C, Prenatal Nitrosatable Drug Use, and Birth
    Defects in Offspring   71
Conclusions, Clinical Implications, and Suggestions for
    Further Research   72
References   72

7. The Application of Reproductive Techniques (ART): Worldwide Epidemiology Phenomenon and Treatment Outcomes

GIULIA SCARAVELLI, ROBERTA SPOLETINI

Background   75
Evaluation of ART Procedures Worldwide: The Diverse
    ART Data Collection Systems   75
Reasons for Different ART Utilization   77
ART Application and Availability   78
ART Application/Number of Treatment Cycles   79
ART Principal Outcomes   83
Discussion   85
References   86

8. The Effects of Environmental Hormone Disrupters on Fertility, and a Strategy to Reverse their Impact

FRANK H. COMHAIRE, WIM A.E. DECLEER

Introduction	89
Historic Aspects	89
Methodological Aspect	90
The Estrogen Receptor Binding Assay	90
Application of the ERBA Test to Environmental Samples	90
Xenoestrogens in Flemish Surface Waters	91
Strategy to Reduce Environmental Contamination	92
Effect of Reducing Environmental Contamination of Sperm Quality in Flanders	92
Effect of Antiestrogen Treatment on Semen Quality and Male Fertility	93
Complementary Treatment with a Specific Nutraceutical	93
Female Infertility	95
Estrogen Metabolism	96
Conclusions	96
Appendix	97
References	97

9. Embryo Losses During Nutritional Treatments in Animal Models: Lessons for Humans Embryo Losses and Nutrition in Mammals

CAROLINA VIÑOLES, CECILIA SOSA, ANA MEIKLE, JOSÉ-ALFONSO ABECIA

Human Infertility and Nutrition	99
The Sheep as a Model	100
Embryo Development and Maternal Recognition of Pregnancy	100
Effect of Undernutrition on Embryo Development	101
Lessons for Humans	103
References	103

B

BARIATRIC SURGERY, OBESITY, AND FERTILITY

10. Fertility and Testosterone Improvement in Male Patients After Bariatric Surgery

MICHAELA LUCONI, GIOVANNI CORONA, ENRICO FACCHIANO, MARCELLO LUCCHESE, MARIO MAGGI

Male Obesity and Sex Hormones	109
Reproductive Functions in Obese Men	111
Bariatric Surgery and its Effect on Male Fertility	112
Conclusions	114
References	114

11. Obesity and Reproductive Dysfunction in Men and Women

TOD FULLSTON, LINDA WU, HELENA J. TEEDE, LISA J. MORAN

Introduction	119
Obesity Impacts on Female Fertility	120
Obesity Impacts on Male Fertility	124
Practical Advice and Implications	126
Conclusions	126
References	127

12. Obesity and Gestational Outcomes

AOIFE M. EGAN, MICHAEL C. DENNEDY

Introduction: Obesity in Women of Childbearing Age	133
Obesity and Adverse Maternal Outcomes	133
Obesity and Adverse Neonatal Outcomes	134
Mechanisms of Adverse Outcomes	135
Obesity Management: Prepregnancy	137
Management of Obesity During Pregnancy	138
Conclusions	139
References	139

C

EXERCISE AND LIFESTYLE IN FERTILITY

13. Lifestyle Factors and Reproductive Health

ASHOK AGARWAL, DAMAYANTHI DURAIRAJANAYAGAM

Introduction	145
Effects of Lifestyle Factors on Fertility	145
Other Factors	152
Conclusions and Recommendations	152
References	153

14. Effect of Diet and Physical Activity of Farm Animals on their Health and Reproductive Performance

ANNA WILKANOWSKA, DARIUSZ KOKOSZYŃSKI

Introduction	159
Farm Animal Vigor and Health Status	159
Reproductive Performance of Livestock Species	160
Influence of Diet on the Health and Reproduction Performance of Livestock	162
The Effect of Physical Activity on Health and Reproduction of Livestock Species	166
Conclusions	167
References	167

15. Prenatal Physical Activity and Gestational Weight Gain

JENNIFER L. KRASCHNEWSKI, CYNTHIA H. CHUANG

Introduction	173
Gestational Weight Gain Guidelines	173
Health Care Provider Advice on GWG	174
Physical Activity in Pregnancy	176
GWG Interventions	177
Achieving Healthy GWG	178
References	179

16. Lifestyle Intervention for Resumption of Ovulation in Anovulatory Women with Obesity and Infertility

WALTER K. KUCHENBECKER, ANNEMIEKE HOEK

Introduction	183
Defining Obesity and its Contribution to Anovulation	183
The Role of Lifestyle Intervention for Resumption of Ovulation	184
Implementation of Lifestyle Intervention	187
Conclusions	188
References	188

17. Effects of Lifestyle on Female Reproductive Features and Success: Lessons from Animal Models

ANTONIO GONZALEZ-BULNES, SUSANA ASTIZ

Introduction	191
Animal Models	192
Effects of Diet and Lifestyle on Puberty	194
Effects of Diet and Lifestyle on Fertility	194
Effects of Diet and Lifestyle on Pregnancy and Offspring Development	195
Conclusions	197
Acknowledgments	197
References	197

D

NUTRITION AND REPRODUCTION

18. Green Leafy Vegetables: A Health Promoting Source

MUHAMMAD ATIF RANDHAWA, AMMAR AHMAD KHAN, MUHAMMAD SAMEEM JAVED, MUHAMMAD WASIM SAJID

Introduction	205
Therapeutic Value and Health Benefits of Green Leafy Vegetables	207
Effect of Green Leafy Vegetables on Mother and Fetus Health	214
Conclusions	216
References	216

19. Nutrition, Lifestyle, and Obesity in Urology

LEONARDO REIS, RICARDO MIYAOKA

Obesity and Benign Urological Disease	221
Obesity and Urological Malignancy: Possible Mechanisms	223
Protecting Nutritional Factors	224
Physical Exercise and Weight Loss	224
Key Points	224
References	225

20. Pregnancy with Multiple Micronutrients: Perinatal Mortality Reduction

BATOOL A. HAIDER, ZULFIQAR A. BHUTTA

Introduction	227
Methods	228
Results	228
Discussion	235
Conclusions	235
References	236

21. Nutrition and Hormones: Role in Male Infertility

LAKSHANA SREENIVASAN

Introduction	237
References	239

22. Nutrition and Developmental Programming of Central Nervous System (CNS)

SAYALI CHINTAMANI RANADE

Introduction	241
Developmental Origins of Adult Disease	241
Role of Maternal Nutrition in Early Life Programming	241
Nutritional Deficiencies	242
The Developing Nervous System	242
Probable Mechanism of Action	244
Nutritional Inadequacies and their Neurological Consequences	246
Nutritional Deficiencies and Psychiatric Disorders	247
Early Life Malnutrition and Aging Brain	248
Nutritional Intervention	248
References	249

23. Herbal Supplements in Pregnancy: Effects on Conceptions and Delivery

FABIO FACCHINETTI, GIULIA DANTE, ISABELLA NERI

Traditional Medicine: Introduction	253
Traditional Chinese Medicine	253
Herbal Remedies and Fertility	254
Herbal Remedies During Pregnancy	255
References	259

24. Selenium in Fertility and Reproduction

HITEN D. MISTRY, LESIA O. KURLAK

Selenium Background	261
Selenium and Reproductive Health	264
Selenium and Pregnancy	265
Selenium and Early Childhood	267
Conclusions	268
References	268

25. The Role of Oxidative Stress in Endometriosis

ADITI MULGUND, SEJAL DOSHI, ASHOK AGARWAL

Introduction	273
Background of Disease	273
The Origin of Endometriosis	274
Relationship Between Oxidative Stress and Endometriosis	275
Oxidative Stress Markers in Endometriosis	275
Nitric Oxide and Endometriosis	276
Role of Iron	277
Role of Cytokines	278
Treatment of Endometriosis	278
Conclusions	279
References	280

26. Oxidative Stress in Preeclampsia

ASHOK AGARWAL, EVA TVRDA, ADITI MULGUND

Introduction	283
Definition of Preeclampsia	283
Pathophysiology of Preeclampsia	283
Risk Factors for Preeclampsia	284
Effects on Mother and Baby	284
Treatment of Preeclampsia	284
Preeclampsia-related Oxidative Stress	284
Complications in the Fetus and Newborn	287
Interventions to Overcome Oxidative Stress in Preeclampsia	287
Future Directions	288
Conclusions	288
References	289

E

NUTRITION AND LIFESTYLE IN *IN VITRO* FERTILIZATION

27. The Psychological Management of Infertility

MANDY J. RODRIGUES, FRANCISCO ANTONIO RODRIGUES

Introduction	293
Stress and Infertility	293
The Cycle of Infertility	293
The Consequences of Infertility	294
Infertility and One's Relationship	294
Psychoneuroimmunology	294
Real and Self-induced Stress	295
Time Urgency Perfectionism Stress	295
Adoption and the Impact on Infertility	296
Identifying Chronic Stress	297
Managing One's Stress	298
Other Stress Management Techniques in Infertility	298
Spousal Involvement	299
Conclusions	299
References	299

28. The Role of Body Mass Index on Assisted Reproductive Treatment Outcome

KATHRYN GEBHARDT, DEIRDRE ZANDER-FOX

Introduction: Obesity and Fertility	303
Maternal Obesity and Fertility	303
Paternal Obesity and Fertility	306
Conclusions	309
References	309

29. Health Outcomes of Children Conceived Through Assisted Reproductive Technology

FIONA LANGDON, ABBIE LAING, ROGER HART

Introduction	313
Congenital Malformations	314
Perinatal Complications	315
Imprinting Disorders	316
Cardiovascular and Metabolic Effects	317
Malignancy	317
Respiratory and Allergy Disorder: Asthma, Allergy, and Atopy	318
Endocrine	318
Ophthalmological and Auditory Disorders	318
Growth and Pubertal Development	318
General Health and Well-being	319
Testicular Function	319
Cerebral Palsy	319
Neuromotor Development	320
Cognitive Function	320
School Performance	320
Socio-emotional Development and Psychological Disorders	320
Autism	321
Attention Deficit Disorder	321
Conclusions	321
References	321

30. Influence of Male Hyperinsulinemia on IVF Outcome

EDOLENE BOSMAN, ALETTA DOROTHEA ESTERHUIZEN, FRANCISCO ANTONIO RODRIGUES

Introduction	327
Semen Parameters and Functional Assays in Hyperinsulinemic Patients	328
Results	328
IVF Outcome of Hyperinsulinemic Sperm Donors	330
Influence of Male Hyperinsulinemia on IVF Outcome	332
The Effect of Metformin Therapy and Dietary Supplements on Semen Parameters in Hyperinsulinemic Males	334
References	336

F
NUTRITION, LIFESTYLE, AND MALE FERTILITY

31. Dietary Zinc Deficiency and Testicular Apoptosis

DEEPA KUMARI, NEENA NAIR, RANVEER SINGH BEDWAL

Zinc Biology	341
Zinc Absorption	341
Zinc Deficiency	342
Zinc in Male Reproduction	343
Apoptosis and Apoptotic Genes	343
Zinc and Apoptosis	344
Spermatogenesis and Germ Cell Apoptosis	347
References	348

32. Environmental Factors, Food Intake, and Social Habits in Male Patients and its Relationship to Intracytoplasmic Sperm Injection Outcomes

DANIELA PAES DE ALMEIDA FERREIRA BRAGA, EDSON BORGES JR.

Introduction	355
Weight and Exercise	355
Recreational Substances and Illicit Drugs	357
Occupational and Environmental Factors	359
Caffeine	361
Food Intake	361
Conclusions	362
References	363

33. The Role of Over-the-Counter Supplements in Male Infertility

ALAN SCOTT POLACKWICH JR., EDMUND S. SABANEGH

Introduction	369
Oxidative Stress and Male Infertility	369
Supplements	370
Conclusions	376
References	376

34. Sperm Physiology and Assessment of Spermatogenesis Kinetics *In Vivo*

SANDRO C. ESTEVES, RICARDO MIYAOKA

Introduction	383
The Testis: Structure and Function	383
The Epididymis: Structure and Function	387
Sperm Function	390
Assessment of Spermatogenesis Kinetics *In Vivo*	392
Conclusions	393
References	394

35. Antioxidant Treatment and Prevention of Human Sperm DNA Fragmentation: Role in Health and Fertility

C. ABAD GAIRÍN, J. GUAL FRAU, N. HANNAOUI HADI, A. GARCÍA PEIRÓ

Introduction	397
Antioxidants and General Health	399
Oxidative Stress and DNA Fragmentation Prevention in Infertile Man	400
Antioxidant Treatment	402
Conclusions	405
References	406

36. Epigenetics and its Role in Male Infertility

EVA TVRDA, JAIME GOSALVEZ, ASHOK AGARWAL

Introduction	411
Epigenetics of the Male Gamete	411
DNA Methylation	412
Chromatin Reorganization	413
Histone Modifications	414
Sperm RNAs	414
Male Infertility	415
Epigenetics and Testicular Cancer	416
Nutrition and Epigenetics	417
The Impact of Environment on Epigenetics	417
Epigenetics and ART	418
Current Techniques for the Assessment of Spermatozoa Epigenetics	418
The Future of Epigenetics	419
References	419

Index	**423**

List of Contributors

José-Alfonso Abecia, DVM, MSc, PhD   Universidad de Zaragoza, Zaragoza, Spain

Ashok Agarwal, PhD   Center for Reproductive Medicine, Cleveland Clinic, Cleveland, Ohio, USA

Diana Anderson, PhD   University of Bradford, Medical Sciences, Bradford, West Yorkshire, UK

Susana Astiz, MVD, PhD, Dip ECBHM   INIA, Madrid, Spain

Adolf Baumgartner, PhD   University of Bradford, Medical Sciences, Bradford, West Yorkshire, UK

Ranveer Singh Bedwal, MSc, PhD   Cell Biology Laboratory, Department of Zoology, University of Rajasthan, Jaipur, India

Zulfiqar A. Bhutta, PhD, MBBS, FRCPCH, FAAP   Robert Harding Chair in Global Child Health and Policy, Centre for Global Child Health, Hospital for Sick Children, Toronto, Canada

Edson Borges Jr., MD, PhD   Fertility – Assisted Fertilization Centre; Sapientiae Institute – Educational and Research Centre in Assisted Reproduction, São Paulo, SP – Brazil

Edolene Bosman, MSc   TUPS Stress, Medfem Fertility Clinic, Bryanston, Johannesburg, South Africaa

Jean D. Brender, PhD, RN, FACE   Texas A&M Health Science Center, College Station, Texas, USA

Cynthia H. Chuang, MD, MSc   Penn State College of Medicine, Hershey, Pennsylvania, USA

Frank H. Comhaire, MD, PhD   Fertility-Belgium Centre, Belgium

Giovanni Corona, MD, PhD   University of Florence, Florence; Azienda Usl Bologna Maggiore-Bellaria Hospital, Bologna, Italy

Giulia Dante, MD   Policlinico Hospital – University of Modena and Reggio-Emilia, Modena, Italy

Wim A.E. Decleer, MD   Fertility-Belgium Centre, Belgium

Michael C. Dennedy, MD, PhD   Galway Diabetes Research Centre; National University of Ireland, Galway, Ireland

Sejal Doshi, MD   Northeastern Ohio Medical University, Rootstown, Ohio, USA; Center for Reproductive Medicine, Cleveland Clinic, Cleveland, Ohio, USA

Damayanthi Durairajanayagam, PhD   Department of Physiology, Faculty of Medicine, MARA University of Technology, Selangor, Malaysia

Aoife M. Egan, MB, BCh, BAO   Galway Diabetes Research Centre, Galway, Ireland

Omer Erdeve, MD   Ankara University School of Medicine Children's Hospital, Ankara, Turkey

Aletta Dorothea Esterhuizen, PhD   TUPS Stress, Medfem Fertility Clinic, Bryanston, Johannesburg, South Africa

Sandro C. Esteves, MD, PhD   ANDROFERT, Referral Center for Male Reproduction, Campinas, SP, Brazil

Enrico Facchiano, MD   Azienda Sanitaria di Firenze, Santa Maria Nuova Hospital, Florence, Italy

Fabio Facchinetti, MD   Policlinico Hospital – University of Modena and Reggio-Emilia, Modena, Italy

Daniela Paes de Almeida Ferreira Braga, DVM, MSc   Fertility – Assisted Fertilization Centre; Sapientiae Institute – Educational and Research Centre in Assisted Reproduction, São Paulo, SP – Brazil

J. Gual Frau, MD Bs   Urology Department, Corporació Sanitària Parc Taulí, Institut Universitari Parc Taulí, Sabadell, Spain

Tod Fullston, BSc (Hons), PhD   University of Adelaide, South Australia, Australia

C. Abad Gairín, MD Bs   Urology Department, Corporació Sanitària Parc Taulí, Institut Universitari Parc Taulí, Sabadell, Spain

Kathryn Gebhardt, PhD   Repromed; University of Adelaide, Adelaide, South Australia, Australia

Antonio Gonzalez-Bulnes, MVD, PhD   INIA, Madrid, Spain

Jaime Gosalvez, PhD   Universidad Autonoma de Madrid, Madrid, Spain

N. Hannaoui Hadi, MD Bs   Urology Department, Corporació Sanitària Parc Taulí, Institut Universitari Parc Taulí, Sabadell, Spain

Batool A. Haider, MD, MSc, ScD   Harvard School of Public Health, Boston, Massachusetts, USA

Roger Hart, MD   School of Women's and Infants' Health, University of Western Australia, King Edward Memorial Hospital, Perth, Western Australia, Australia; Fertility Specialists of Western Australia, Bethesda Hospital, Perth, Western Australia, Australia

Annemieke Hoek, MD, PhD   Department of Obstetrics and Gynecology, University Medical Center Groningen, Groningen, The Netherlands

Muhammad Sameem Javed, BSc, MSc   University of Agriculture, Faisalabad, Pakistan

Ammar Ahmad Khan, BSc, MSc   University of Agriculture, Faisalabad, Pakistan

Dariusz Kokoszyński, PhD   Department of Poultry Breeding and Animal Products Evaluation, University of Technology and Life Sciences, Bydgoszcz, Poland

Jennifer L. Kraschnewski, MD, MPH   Penn State College of Medicine, Hershey, Pennsylvania, USA

Walter K. Kuchenbecker, MBChB   Isala Fertility Center, Isala Clinics, Zwolle, The Netherlands

Deepa Kumari, MSc, PhD   Cell Biology Laboratory, Department of Zoology, University of Rajasthan, Jaipur, India

Lesia O. Kurlak, PhD   University of Nottingham, Nottingham, UK

Abbie Laing, MBChB   King Edward Memorial Hospital, Perth, Western Australia, Australia

Fiona Langdon, MBBS   King Edward Memorial Hospital, Perth, Western Australia, Australia

J.L.M.R. Leroy, MSc, PhD-student   Gamete Research Centre, Veterinary Physiology and Biochemistry, Department of Veterinary Sciences, University of Antwerp, Belgium

Marcello Lucchese, MD   Azienda Sanitaria di Firenze, Santa Maria Nuova Hospital, Florence, Italy

Michaela Luconi, PhD   University of Florence, Florence, Italy

Mario Maggi, MD   University of Florence, Florence, Italy

Ana Meikle, DVM, MSc, PhD   Universidad de la República, Montevideo, Uruguay

Hiten D. Mistry, PhD   University of Bern, Berne, Switzerland

Ricardo Miyaoka, MD, PhD   University of Campinas, Campinas SP; ANDROFERT, Referral Center for Male Reproduction, Campinas SP; Sumaré State Hospital, Sumaré SP, Brazil

Lisa J. Moran, BSc (Hons), BND, PhD   University of Adelaide, South Australia; Monash University, Victoria, Australia

Aditi Mulgund, MD   Northeastern Ohio Medical University, Rootstown; Center for Reproductive Medicine, Cleveland Clinic, Cleveland, Ohio, USA

Neena Nair, MSc, PhD   Cell Biology Laboratory, Department of Zoology, University of Rajasthan, Jaipur, India

Isabella Neri, MD   Policlinico Hospital – University of Modena and Reggio-Emilia, Modena, Italy

Mehmet Yekta Oncel, MD   Zekai Tahir Burak Maternity Teaching Hospital, Ankara, Turkey

A. García Peiró, PhD   Centro de Infertilidad Masculina y Análisis de Barcelona (CIMAB), Edifici Eureka, Bellaterra, Spain

Karen P. Phillips, PhD   Interdisciplinary School of Health Sciences, University of Ottawa, Ottawa, Ontario, Canada

Alan Scott Polackwich Jr., MD   Glickman Urological and Kidney Institute, Cleveland Clinic Foundation, Cleveland, Ohio, USA

Sayali Chintamani Ranade, PhD   National Brain Research Centre, Manesar, Haryana, India

Muhammad Atif Randhawa, BSc, MSc, PhD   University of Agriculture, Faisalabad, Pakistan

Leonardo Reis, MD, MSc, PhD   University of Campinas; Pontifical Catholic University of Campinas, Campinas SP, Brazil

Francisco Antonio Rodrigues, MBBCh   TUPS Stress, Medfem Fertility Clinic, Bryanston, Johannesburg, South Africa

Mandy J. Rodrigues, MA   TUPS Stress, Medfem Fertility Clinic, Bryanston, Johannesburg, South Africa

Edmund S. Sabanegh, MD   Glickman Urological and Kidney Institute, Cleveland Clinic Foundation, Cleveland, Ohio, USA

Muhammad Wasim Sajid, BSc, MSc   University of Agriculture, Faisalabad, Pakistan

Giulia Scaravelli, MD, PhD   National Health Institute, Rome, Italy

Thomas Ernst Schmid, PhD   Klinikum rechts der Isar, Technische Universität München, Department of Radiooncology, Germany

Cecilia Sosa, DVM, MSc, PhD   Universidad de Zaragoza, Zaragoza, Spain

Roberta Spoletini, PhD   National Health Institute, Rome, Italy

Lakshana Sreenivasan, BSc   The University of Arizona, Arizona, USA

Helena J. Teede, MBBS, PhD, FRACP   Monash University; Diabetes and Vascular Medicine, Monash Health, Victoria, Australia

Gülcan Türker, MD   Kocaeli University, Kocaeli, Turkey

Eva Tvrda, PhD   Center for Reproductive Medicine, Cleveland Clinic, Cleveland, Ohio, USA; Department of Animal Physiology, Slovak University of Agriculture, Nitra, Slovakia

S.D.M. Valckx, PhD   Gamete Research Centre, Veterinary Physiology and Biochemistry, Department of Veterinary Sciences, University of Antwerp, Belgium

Carolina Viñoles, DVM, MSc, PhD   Instituto Nacional de Investigación Agropecuaria, Tacuarembó, Uruguay

Anna Wilkanowska, MSc   Department of Agricultural, Environmental and Food Sciences, University of Molise, Campobasso, Italy

Linda Wu, BPharmSc, PhD   University of Adelaide, South Australia, Australia

Deirdre Zander-Fox, PhD   Repromed; University of Adelaide, Adelaide, South Australia, Australia

Preface

Many factors affect fertility, positively and negatively. Toxins and contaminants are important in cell functions and the growth of eggs.

SECTION A: OVERVIEW ON FERTILITY

Karen P. Phillips describes the public's perception of environmental risks, especially those that affect the ability to conceive and carry to birth. Diana Anderson, Thomas Ernst Schmid, and Adolf Baumgartner focus on the well-known role of mothers who use tobacco. This leads to transgenerational damage in the offspring. Mehmet Y. Oncel and Omer Erdeve review the role of tobacco smoking on after conception and its damage on fetal health, and they describe a key mechanism – serum folate levels. S. D. M. Valckx and J. L. M. R. Leroy approach this issue in another way by evaluating the effects of maternal metabolic health. Then they evaluate the effects of maternal health as well as diet on follicular fluid composition. Finally, they review the consequences on the embryo as well as the oocyte. Gülcan Türker reviews heavy metals, such as mercury, which are well known major toxins. He describes their actions on preterm mortality and morbidity. Jean D. Brender reviews common materials on foods, like bacon, and known to cause cancer. They review nitrate and nitrite, which are metabolites of foods that mothers consume, and nitrosable drugs for their effects on congenital malformations. Giulia Scaravelli and Roberta Spoletini describe the application of reproductive techniques on worldwide epidemiology phenomenon and treatment outcomes. Frank H. Comhaire and W. A. E. Decleer discuss the effect of environmental hormone disrupters on infertility. Then the authors review strategies to reverse the effects of agents affecting hormones. Finally, Carolina Viñoles, Cecilia Sosa, Ana Meikle, and José-Alfonso Abecia review animal models where experimental studies can be done. They review damage to embryos during nutritional supplementation in animal models. Finally, they review human embryo losses and nutrition and nutritional deficiencies.

SECTION B: BARIATRIC SURGERY, OBESITY, AND FERTILITY

Humans are more frequently obese, which affects male and female fertility. In chapter "Fertility and Testosterone Improvement in Male Patients After Bariatric Surgery" written by Michaela Luconi, Giovanni Corona, Enrico Facchiano, Marcello Lucchese, and Mario Maggi shows that surgery to reduce weight has a positive effect in male fertility via improvements in hormone levels. Tod Fullston, Linda Wu, Helena J. Teede, and Lisa J. Moran broadly review the role of obesity in men as well as women in reproductive dysfunction. Michael Conall Dennedy reviews obesity per se and gestational outcomes.

SECTION C: EXERCISE AND LIFESTYLE IN FERTILITY

Ashok Agarwal and Damayanthi Durairajanayagam review lifestyle factors and reproductive health. Anna Wilkanowska and Dariusz Kokoszyński use animal models wherein animals are not purposely exercised. They review the normal activity of farm animals on their health and reproductive performance. Jennifer L. Kraschnewski describes the data and mechanism of the effects of physical activity on gestational weight gain. Clearly, exercise can change the growth and body mass. Walter K. Kuchenbecker and Annemieke Hoek review lifestyles to modulate ovulation. Lifestyle intervention was able to increase ovulation in anovulatory women, especially those with obesity or infertility. Finally, A. Gonzalez-Bulnes reviews reproduction learned from animal models.

SECTION D: NUTRITION AND REPRODUCTION

In chapter "Green Leafy Vegetables: A Health Promoting Source" written by Muhammad Atif Randhawa, A. Khan, Muhammad Sameem Javed, and Wasim Sajid begin the reviews of food, metabolism, and nutrition on fertility. They found that green leafy vegetables, as with most health situations, are very helpful for the promotion of infants' health. Leonardo Reis and Ricardo Miyaoka found nutrition lifestyle and obesity in urology. Batool A. Haider and Zulfiqar A. Bhutta describe the role of multiple micronutrients in perinatal health and mortality reduction. Similarly, Lakshana Sreenivasan and Ronald R. Watson found that nutrition and hormones have a role in male infertility. S. C. Ranade found that nutrition had effects on the development and programming of neurological systems. Fabio Facchinetti, Giulia Dante, and Isabella Neri review herbs and their components in dietary supplements that affect delivery and conception. Nutrition supplements also affect pregnancy. Hiten D. Mistry and Lesia O. Kurlak review a key nutrient with the potential to provide toxicity with too much consumption. They found that selenium affects reproduction and fertility. It may function in part by antioxidant activity. Finally, Aditi Mulgund, Sejal Doshi, and Ashok Agarwal found a role for antioxidant stress in endometriosis, while Ashok Agarwal, Eva Tvrda, and Aditi Mulgund found that oxidative stress promotes preeclampsia.

SECTION E: NUTRITION AND LIFESTYLE IN *IN VITRO* FERTILIZATION

In vitro fertilization has been shown to be affected by nutrition and lifestyle. Mandy J. Rodrigues and Francisco Antonio Rodrigues review the regulation and management of the psychological effects of infertility. Kathryn Gebhardt and Deidre L. Zander-Fox found that the role of body mass index or obesity is associated with the success or failure on assisted reproductive treatment outcome. Fiona Langdon, Abbie Laing, and Roger Hart review the health outcomes of children conceived through assisted reproductive technology. Edolene Bosman, Aletta Dorothea Esterhuizen, and Francisco Antonio Rodrigues review chronic diseases and their influences on *in vitro*

outcomes. They emphasize hyperinsulinemia on male regulation of *in vitro* fertilization.

SECTION F: NUTRITION, LIFESTYLE, AND MALE FERTILITY

Clearly, the needs of males for fertility are different from females. Deepa Kumari, Neena Nair, and R.S. Bedwal found that zinc deficiency changed testicular apoptosis. Inadequate cell division is a serious inhibition to sperm growth. Edson Borges Jr. and Daniela Braga found that environmental factors, food intake, and social habits modify male patients and their relationship to intracytoplasmic sperm injection outcomes. Alan Polackwich and Edmund S. Sabanegh Jr. review the role of over-the-counter supplements in male infertility as diet and medicines do not have to be new pharmaceutical drugs. Sandro C. Esteves and Ricardo Miyaoka review sperm physiology and assessment of spermatogenesis kinetics *in vivo*, which is critical for understanding the role of lifestyle and food in cell growth for male fertility. Carlos Abad, N. Hannaoui Hadi, and A García Peiró discuss antioxidant treatment and prevention of human sperm DNA fragmentation, specifically its role in health and fertility. Finally, Eva Tvrda, Jaime Gosalvez, and Ashok Agarwal review epigenetics and its role in male infertility.

Acknowledgments

The work of Dr. Watson's editorial assistant, Bethany L. Stevens, in communicating with authors and working on the manuscripts was critical to the successful completion of the book. It is very much appreciated. The encouragement, advice, and support of Jeffrey Rossetti in Elsevier book preparation were very helpful. Support for Ms. Stevens and Dr. Watson was graciously provided by the Natural Health Research Institute (www.naturalhealthresearch.org), a nonprofit health promotion institute. Finally, the work of the librarian at the Arizona Health Science Library, Mari Stoddard, was vital and very helpful in identifying the key researchers who participated in the book.

OVERVIEW ON FERTILITY

1

Perceptions of Environmental Risks to Fertility

Karen P. Phillips, PhD

Interdisciplinary School of Health Sciences, University of Ottawa, Ottawa, Ontario, Canada

INTRODUCTION

Environmental risks, including lifestyle habits, infectious disease, and occupational exposures, are increasingly examined as modifiable infertility risk factors. This chapter will provide an overview of infertility risk factors and explain how risk perception is formed. Perceptions of fertility risks within specific populations will also be examined. Finally, a public health framework is proposed to enhance fertility-specific health promotion and enable mitigation of modifiable risks to fertility.

Infertility – Definitions, Manifestations, and Risk Factors

Infertility is generally defined as the absence of conception after 1 year of regular, unprotected intercourse. Prevalence of current infertility in North America ranges from 11.5% to 15.7% in Canada [1] and 15.5% among US women [2]. The pathophysiology of infertility may include genetic/chromosomal anomalies, congenital or infectious malformations of the reproductive tract, and endocrine disorders (hypogonadism, anovulation) in both men and women. Additionally, female infertility is caused by advanced age, endometriosis, polycystic ovarian syndrome, and diminished ovarian reserve [3]. Infertility manifests as ovulatory dysfunction, amenorrhea, implantation failure/spontaneous abortion, and diminished semen parameters [3]. For about 10–15% of infertile couples, infertility is unexplained, with minimal abnormal findings upon medical investigation [3].

Most modifiable risk factors can be categorized under "environment." Diet, lifestyle habits (cigarette smoking, alcohol use), history of sexually transmitted infections (STI), pharmaceutical/medical treatments, as well as environmental exposures to chemicals, radiation, air pollution, and heavy metals are identified as risk factors to fertility [4–6] (Box 1.1). Environmental exposures and lifestyle habits represent sex-specific risks to fertility based on differences in reproductive physiology and urogenital anatomy. Common lifestyle and environmental determinants of fertility and reproductive health are summarized next.

Lifestyle and Infertility

Body Weight

Obesity (body mass index (BMI) $\geq$ 30 kg/m^2) is increasingly described as a global phenomenon, contributing significantly to morbidity and mortality (cardiovascular disease, diabetes, mobility issues, etc.). For both men and women, obesity contributes to hypogonadism and other perturbations of the endocrine system that may contribute to infertility [7–10]. Obese women have generally poor outcomes with assisted reproductive technologies (ART) [11,12] including fewer metaphase II oocytes retrieved, oocyte spindle defects [13], more canceled cycles [14], decreased fertilization [15] and decreased live birth rates [12, 16–18]. Similarly, obesity in the male partner may be associated with decreased clinical pregnancy [19,20] and live birth rates [10,20]. Obese women also face increased risks during pregnancy, such as gestational diabetes, hypertension, spontaneous abortion, large fetuses for gestational age, and concomitant delivery complications [21,22]. Poor ART outcomes, coupled with concerns of increased pregnancy complications in the obese mother, have prompted suggestions that ART be denied to women with BMI $\geq$ 30–35 kg/m^2 [22]. Critics of such policies argue that greater emphasis should be placed on weight–hip ratio as a predictor of reproductive risk, BMI limits are discriminatory and stigmatizing, and sustained weight loss is difficult to achieve for many individuals [22]. Over half of American men and women surveyed supported BMI limits for

Handbook of Fertility. http://dx.doi.org/10.1016/B978-0-12-800872-0.00001-9

BOX 1.1

MODIFIABLE RISKS TO INFERTILITY

- BMI $\geq$ 30 mg/kg^2
- BMI $<$ 18 mg/kg^2
- Alcohol
- Cigarette smoke
- Recurrent sexually transmitted infections

Less evidence for

- Environmental contaminants (ambient exposure levels)
- Caffeine

ART, citing risks of pregnancy complications in obese patients and possibility of health complications including obesity in offspring [23]. BMI ART treatment limits have not been proposed for obese male patients. As lifestyle modifications are difficult to achieve and sustain, support for weight management should be a component of preconception care [11,24].

Underweight individuals also face fertility challenges and may be less responsive to ART [11,21]. Women with a BMI $<$18 mg/kg^2 are at increased risk for ovulatory dysfunction and pregnancy complications, such as preterm labor. General recommendations for fertility planning include maintenance of healthy body weight, nutritious diet, and moderate exercise [11,21,25].

Tobacco

Tobacco smoking is a leading cause of adverse health effects, including cancer, cardiovascular, and respiratory disease. Although cigarette smoking is declining, 18.1% of American adults [26] and 16% of Canadian adults [27] were smokers in 2012. Globally, tobacco use is increasing rapidly among young girls aged 13–15 years. Despite health promotion efforts, continued use of tobacco by these young girls may herald future increases in global rates of smoking among adult women [28]. Cigarette smoke represents a well-established mixture of reproductive toxicants for both men and women [29,30]. Reproductive pathology may result from vasoconstriction, generation of reactive oxygen species and concomitant oxidative stress, and alterations in endocrine signaling. Male smoking is associated with decreased sperm count, motility, seminal volume, and reduced ART outcomes [31,32] – manifested at a cellular level as sperm DNA damage (e.g., DNA fragmentation, formation of DNA adduct) and pathological changes to the acrosomal membrane and tail [25,33,34].

For women, smoking is associated with earlier age at menopause, premature ovarian failure, infertility, and spontaneous abortion [35]. Female smokers respond poorly to ART gonadotropin stimulation, producing fewer oocytes with low developmental potential, and decreased live birth rates [18,36,37]. Cadmium, nicotine-metabolite cotinine, and polycyclic aromatic hydrocarbons, such as benzo[a]pyrene (B[a]P), are examples of some of the toxic compounds contained in cigarette smoke and have also been detected in follicular fluid [29,38,39]. Accumulation of toxicants in the follicular compartment has the potential to perturb oocyte growth, induce follicular atresia, and dysregulate necessary hormonal support for the initiation of meiotic resumption and ovulation [38,40,41].

Side-stream or second-hand smoke exposure is also detrimental to fertility. Women exposed to side-stream smoke or mainstream smoke face compromised ART outcomes including reductions in implantation and pregnancy rates [39]. As described, obese women tend to respond poorly to ART; however, obese smokers may face even greater reproductive challenges [18]. Given the significant health risks posed by tobacco smoking, it is clearly evident that smoking cessation should be recommended prior to commencement of natural conception or ART for both men and women [42]. Encouragingly, the effects of smoking on oocyte/embryo developmental potential may be transient, as former smokers have an increased pregnancy rate compared with smokers [36].

Alcohol

Heavy alcohol consumption is generally associated with poor fertility outcomes in both men and women, along with significant pregnancy complications and adverse fetal development [21,35]. Both moderate (7–8 drinks/week) and low alcohol consumption (1 drink/week) may be associated with impaired fertility [21]. Alcohol's reproductive toxicity may be due to impaired hypothalamic–pituitary–testicular signaling manifesting as decreased semen quality [35]. Female fertility may also be reduced by alcohol consumption with possible effects on the hypothalamus resulting in decreased luteinizing hormone secretion and anovulation [35]. Alcohol's teratogenic effects on the developing embryo and fetus have prompted general advisements for women attempting to conceive to abstain from alcohol [21,35].

Caffeine

The literature examining effects of caffeine intake on fertility is limited by methodological issues (retrospective studies – recall bias of caffeine intake) and lack of consensus [11,21,25,35]. The effects of caffeine consumption on prolonged time-to-pregnancy may be confounded by other negative lifestyle habits such as smoking [11]. It is accepted that caffeine intake of <200–300 mg per day is unlikely to cause detrimental reproductive outcomes such as spontaneous abortion or fetal development [21]. Women who are trying to conceive or are pregnant are advised to limit caffeine intake to 100–200 mg per day [21].

Sexually Transmitted Infections

STI risk is particularly high in adolescents and young adults. Despite mandatory sexual health education in public schools in Canada [43], increased rates of infection with *Chlamydia trachomatis* and *Neisseria gonorrhoeae* have been reported [44]. In the United States, public school sexual health education varies by state [45] with regional, ethnic, and race disparities evident in STI prevalence [46]. Rates of chlamydial infections have continued to rise in many European countries despite implementation of screening programs [47]. Recurrent STI within high-risk groups, safe sex fatigue, high-risk sexual encounters facilitated by social media, and lack of awareness of long-term consequences of STI, such as cancer and infertility, may explain these trends [48].

STI are caused by viruses, bacteria, and other pathogens. Although viral infections may pose risks for reproductive health, most studies have emphasized bacterial STI as risks to fertility. Although isolated infections of *C. trachomatis* or *N. gonorrhoeae* pose little risk to fertility or perinatal outcomes if promptly treated, for many men and women, STI may be asymptomatic, which may preclude medical interventions [47,49]. Women with a history of bacterial STI may experience inflammatory obstructions of the reproductive tract (pelvic inflammatory disease – PID), including scarring of the fallopian tubes – a major cause of female infertility and risk factor for ectopic pregnancy [50]. PID had been strongly associated with chlamydial and gonorrheal infections; however, rates of PID seem to be decreasing along with related complications such as tubal factor infertility and ectopic pregnancy [47,51–53]. A history of recurrent chlamydial or gonorrheal infections remains a significant risk factor for these reproductive tract pathologies [47,50]; however, improved screening and prompt treatment may be responsible for reducing the rates of these complications even in high-risk groups [52].

The relationship between bacterial STI and male infertility is less clear [54]. Bacterial STI gain entry to the reproductive tract via urogenital passages, producing urethritis, or less commonly, prostatitis or epididymitis [49,55]. Men with chlamydial infections tend to exhibit diminished semen quality along with sperm DNA damage; however, it is unclear whether these deficits significantly impair fertility [56,57]. Further study is warranted regarding the history of recurrent STI and male infertility.

Summary

In addition to the described lifestyle risks (obesity, being underweight, alcohol, smoking, and STI), stress and anxiety prior to ART have also been proposed as infertility risk factors [30,58]. Men and women planning to conceive should optimize fertility by maintaining a healthy body weight, abstaining from excessive alcohol or caffeine intake, and be supported to quit smoking. Sexual health promotion strategies (education, screening, condom distribution) enable early detection and treatment of STI, thereby preventing later fertility issues.

Environmental Disruption of Reproductive Health

Reproductive and developmental health are adversely affected by ionizing radiation, heavy metals, lead, industrial chemicals/solvents, temperature, and contaminated food and water (biological/chemical) [4,6] (Fig. 1.1). It is also biologically plausible, though there is less evidence, that exposures to air pollution [6,59,60], electromagnetic or radiofrequency fields (WiFi [61], cell phones [62,63]), along with social environmental factors, particularly disparities in socioeconomic status (SES) and social stress [64–66], impair human reproduction and development. Endocrine disrupters are perhaps the most widely studied environmental modulators of fertility and reproductive health, with effects demonstrated in wildlife, laboratory animals, and humans [67,68]. Endocrine disrupters encompass a wide range of chemical species, mixtures, and compounds including phytoestrogens, pesticides, industrial chemicals, and by-products [67,69]. Endocrine disrupters have been linked to adverse health effects including infertility, abnormal prenatal and childhood development (spontaneous abortion, male reproductive tract abnormalities and other birth defects, sex ratios, precocious puberty), and reproductive cancers (prostate, breast, ovarian, endometrial, and testicular) [67,71]. It is hypothesized that global declines in men's reproductive health characterized by poor semen quality, urogenital birth defects (cryptorchidism, hypospadias), and testicular cancer, collectively called "testicular dysgenesis syndrome," are caused by exposures to endocrine disrupters [67,70,71].

Endocrine disruption represents one of the most plausible mechanisms by which environmental chemical exposures may adversely affect reproductive capacity [67,69,71]. Molecular mechanisms of these contaminants may include endocrine receptor agonists, antagonists,

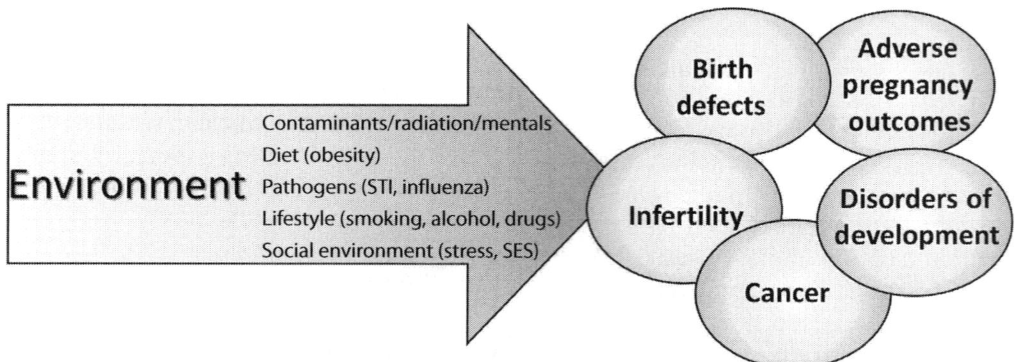

FIGURE 1.1   **Environmental determinants of reproductive health.** Reproductive health may be adversely affected prior to conception, during gestation and delivery, childhood development, or as adult-onset diseases. Lifestyle habits (smoking, alcohol, drugs, stress, obesity, STI) and environmental exposures (endocrine disrupters/contaminants, radiation, pollution, metals, solvents, temperatures, social stressors) have been identified as risk factors for infertility and other adverse reproductive health outcomes [11,21,30]. STI, sexually transmitted infections; SES, socioeconomic status.

enzyme inhibitors, and steroid hormone transport interference (Box 1.2). Perhaps the most widely studied endocrine disrupters are xenoestrogens or chemicals that mimic endogenous estrogens by binding to estrogen receptors and regulating gene expression [67]. Diethylstilbestrol (DES), the phytoestrogens genistein and coumestrol, certain polychlorinated biphenyls (PCBs), nonylphenol, the insecticides methoxychlor and chlordecone, and bisphenol A (BPA) are examples of xenoestrogens [72]. Some chemicals (e.g., the fungicide vinclozolin, dichlorodiphenyl trichloroethane (DDT)-metabolite dichlorodiphenyl dichloroethylene (p,p'-DDE)) exhibit antiandrogenic activity [67]. Other endocrine disrupters are aromatase inhibitors (fungicide fenarimol) and aryl hydrocarbon receptor agonists (dioxins, PCBs, polychlorinated dibenzofurans (PCDFs)), while other chemicals exhibit mixed mechanisms of action (diethylhexyl phthalate (DEHP), methoxychlor-antiandrogen, weak estrogen) [67]. By disrupting endogenous endocrine pathways, endocrine disrupters have the potential to dysregulate the growth and maturation of gametes, disrupt ovulation, embryo and fetal development, and labor and delivery.

Gene–environment interactions, such as epigenetic processes (methylation, histone modifications), represent additional pathways by which environmental chemicals may perturb reproductive function [73–75]. DNA methylation has been recognized as a key event in epigenetic silencing of genes involved in embryonic development, X-chromosome inactivation, genomic imprinting, and chromosome instability. Epigenetic processes may be triggered by stress, diet, or environmental exposures [74,75]. Germ line epigenetic alterations are associated with birth defects, spontaneous abortions, and abnormal

---

BOX 1.2

___

ENVIRONMENTAL PERTURBATION OF FERTILITY: MOLECULAR MECHANISMS

- Endocrine disruption
  - Hormone receptor agonists
  - Hormone receptor antagonists
  - Enzyme inhibitors
  - Alterations in hormone synthesis, transport
- Cell injury
  - Oxidative stress
  - Heat stress
  - Inflammation/scar tissue
- DNA damage/alterations
  - Epigenesis
  - Carcinogenesis
- Cell death
  - Apoptosis
  - Toxicity/necrosis
- Gamete-specific damage
  - Oocyte spindle defects (congression failure)
  - Follicular atresia
  - Inhibited spermatogenesis
  - Sperm-defects (membrane, motility, DNA damage)

fetal developmental programming [74,76,77]. Beyond the disruption of endocrine signaling, epigenetic transformations have the potential to modulate gene expression and contribute to infertility in later generations [78].

Human Exposure

The routes of human exposure to environmental contaminants typically include inhalation, oral (water, diet), and dermal. Ongoing monitoring of the arctic peoples and wildlife is necessitated by bioaccumulation of organic pollutants present in marine mammals and human lipid stores (e.g., adipose tissue, breast milk) [67]. Similarly, consumption of Great Lakes fish may pose a risk due to bioaccumulation of lake pollution [67]. Farmworkers, agricultural workers, and their families also represent populations typically exposed to greater levels of pesticides that are associated with increased risk of birth defects [79], spontaneous abortion [30,80], and prolonged time to pregnancy [81]. In the general public, biomonitoring data do indicate detectable levels of endocrine disrupters and other environmental contaminants in semen (reviewed) [82] and follicular and amniotic fluid [83–90]. The presence of contaminants in reproductive compartments provides biological plausibility for perturbation of gametes and embryo/fetal development; however, the extent to which exposures to environmental contaminants contribute to infertility and reproductive dysfunction remains unclear.

Summary

Environmental hazards, replete with complex scientific names and equivocal evidence of reproductive risks, have nonetheless become social issues, in part because of their putative threats to health and safety [91]. For most individuals, exposure to such hazards is at "environmental" levels; however, the science is unclear regarding the adverse health effects of these low-dose exposures [30,67–69,92]. Occupational exposures are recognized as significant health threats [5,6,30,31], necessitating health and safety workplace standards and education, use of personal protective equipment (PPE), and regulatory oversight. Individuals attempting to conceive should identify any environmental exposures obtained through leisure activities, diet, community, and workplace with their healthcare providers.

Risk Perception

Navigating often complex and conflicting environmental and lifestyle risk information is challenging for infertility patients. Which environmental exposures should be minimized? Which lifestyle choices should be adopted? How can lifestyle modifications improve the success of ART? Perceptions of risk may contribute to anxiety, worry, or stimulate behavior change to diminish risk [93]. Awareness, understanding, and ultimately mitigation of modifiable risks to fertility may enable some patients to regain a sense of control while undergoing treatment for infertility. Infertility prevention may also be optimized by knowledge, awareness, and understanding of infertility risk factors [68,94]. Risk mitigation requires both awareness of environmental determinants of reproductive health and the perception of these factors as potential hazards. Understanding how people perceive risk is fundamental to designing effective environmental health promotion strategies.

Environmental Risk Perception

Risk perception is suggested to be shaped by social context and culture [95], personal ability to influence risk, bias, familiarity with the hazard [91,96], and personal beliefs [97]. The public may appear more concerned about low-probability risks (e.g., nuclear power) than high-probability risks (speeding without seat belts), which has led to the characterization of the public as "scientifically illiterate" and unable to comprehend technical and scientific issues [91]. Most people have difficulties with interpretation of probabilities, are less concerned with inconsistencies in the scientific evidence, and are generally most concerned about public risks that would generate mass-scale death and disability [91]. Mental models or heuristics enable individuals to evaluate situations and form judgments based on prior experience, beliefs, and common knowledge. Although heuristics may assist in rapid problem solving and learning, they may also lead to erroneous conclusions, particularly with technical or scientific risk issues [91,98].

The Social Amplification of Risk Framework [99] provides a model for heightened risk perception within the social context of communications from media, activist organizations, opinion leaders, peers, and public agencies (Fig. 1.2). The public is exposed to increasingly polarized social discourse regarding these often technical environment-related topics, with perceptions of environmental risks suggested to be amplified [98]. In spite of sometimes weak or inconclusive evidence linking environmental exposures to adverse human health effects, risk perceptions are influenced by media's emphasis on health hazards [98]. It is highly evident that controversial health topics are dramatized by communication/information channels (e.g., media, Internet), which serve to amplify risk perceptions [96,99]. Controversial issues are repeated within media cycles such that repetitive risk messaging (volume), controversy, and dramatization of the risk collectively influence social amplification of a particular hazard [99]. Risks to populations symbolically portrayed as "vulnerable" (e.g., developing fetus, infants, and children) [100] and likewise information perceived to have symbolic connotations also contribute to amplification [99]. Models of risk perception enable a

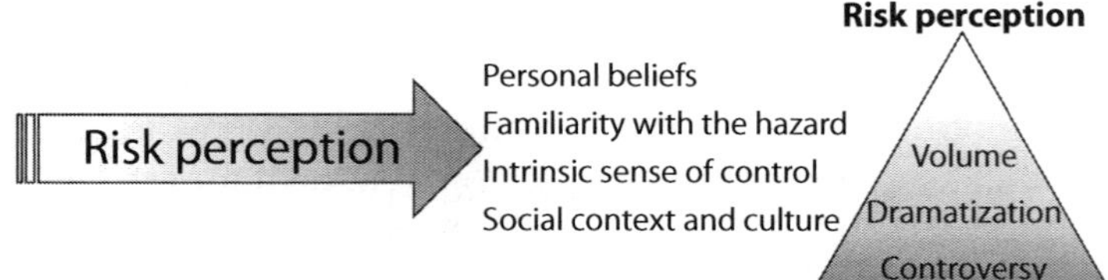

FIGURE 1.2   **Social amplification of risk framework.** Perceptions of risk are shaped by personal beliefs, familiarity/experience with hazard, and personal sense of control [95]. Within the context of society and culture, certain risks may be amplified through repetitive social discourses, dramatizations of potential outcomes, and controversy [99].

greater understanding of health risk information uptake by the public and the variables that ultimately shape the development of individual risk heuristics.

Gender and Risk Perception

Men and women exhibit gendered preferences for occupations, lifestyle habits, leisure activities, and health behaviors; all of which may shape perceptions of environmental influences on fertility, pregnancy, and development. It is well established that men and women perceive risks differently [100–103]. Men and women worry about different types of risks: women express concerns with hazards associated with home and family, particularly hazards that may produce accidents, adverse health, or death, whereas men tend to be more concerned about risks to occupation or career [101]. Women report greater concern about environmental risks and infectious diseases, while men worry about health and safety including occupational accidents [101,104]. Studies also consistently demonstrate that men exhibit lower risk perception compared with women for a range of hazards [101,104]. Gender and race are consistent predictors of risk perception such that women and minorities are more likely to identify hazards as high risk compared with white men [103,105]. It is evident that sociopolitical status, sense of control, power, and economic freedom may collectively diminish concerns about environmental hazards [103,105].

Reproductive health is a societal norm, such that the issue of infertility produces a social stigma, isolation, and altered perceptions of gender roles [93]. In particular, women's social and cultural constructions of motherhood greatly influence their perceptions of risk. Infertile women are willing to assume greater risks than their male partners to conceive [93]. Their risk–benefit compromise will require infertile women to risk potential long-term effects of fertility drugs and ART complications, such as ovarian hyperstimulation syndrome and multiple births, in exchange for the benefit of having a baby [93]. Although women will assume personal risks to conceive, upon becoming a mother and caregiver, risk perception becomes heightened regarding home and family [102,103].

Reproductive risks have been examined in many contexts including DES exposure *in utero*, reproductive risks in the workplace, risks of childbirth, and adverse pregnancy outcomes. Studies from social sciences examine how reproductive risks are perceived, constructed, and influenced by gender, race/ethnicity, culture [101], economics, and the law [106]. During the 1970s through 1990s, many US companies used discriminatory screening practices to exclude presumed fertile women from employment on the basis of reproductive risk due to occupational hazards [106]. Johnson Controls, Inc., a battery manufacturer, excluded fertile women from jobs working with lead, going so far as to require women to produce proof of sterilization [106]. Such "fetal exclusion policies" were perceived by employers to mitigate reproductive risk in the workplace and avoid media stigma and lawsuits. In addition to being discriminatory, such hiring practices implicitly negate reproductive and other health risks to nonfertile women and male workers [106]. Occupational reproductive risk assessments tend to be biased toward women's health, often omitting screening and education for male employees [107]. For occupational exposures where the reproductive risk to men has been proven significant (e.g., the pesticide dibromochloropropane (DBCP)), rather than restrict opportunities for male workers, the pesticide itself was banned, suggesting a gender-biased approach to occupational risk assessment [107]. Ultimately, the *Johnson Controls* 1991 U.S. Supreme Court decision ruled that exclusion of fertile women from jobs involving toxic exposures was discriminatory [106], providing legal protection for equal employment opportunity. However, the concept of fetal protectionism remains pervasive within Western culture. Pregnant women who engage in behaviors perceived as reproductively risky, such as smoking, alcohol consumption, or use of unnecessary pharmaceuticals, are viewed as irresponsible or reckless [108]. Increasingly, the social discourse around pregnancy and environmental risk reduction, such as uptake of "environmentally friendly" foods and products, has become conflated with "good mothering," representing additional social pressure for women to mitigate environmental risks to reproductive health [108].

Race/Ethnicity and Risk Perception

Individuals of low SES and belonging to racial/ethnic minority communities are disproportionately affected by environmental health risks [91,109]. Environmental health perspectives of minority and low-income communities have been assessed in urban US studies. Women living in Manhattan- and Boston-area minority communities express environmental concerns about neighborhood local pollution, lead exposure, garbage, pests (e.g., rodents, roaches), noise levels, and community safety [110,111]. Low SES and minority communities face barriers to health promotion messages due to reduced uptake/access local media and Internet [111]. In North America, influences of SES and race/ethnicity on environmental risk perception may be difficult to delineate. Examination of environmental risks to health in poor, minority communities may require consideration of multiple, interacting risks rather than a focus on single contaminants. This approach better reflects the experiences of many communities exposed to several types of hazards within substandard housing, occupational, and educational facilities [109]. The concept of environmental justice is used to reflect these disparities in home and work environments experienced by poor and minority members of society [109].

Population-Based Environmental Risk Perception Studies

Public risk perception has been a driving force for the development of environmental policy in Australia [100], Canada [112], and the United States [113]. Population-based environmental risk perception studies have been conducted in Canada in the past two decades [114,115], with cigarette smoking, obesity, and unprotected sex perceived as the top three health hazards [115]. Hazards that are exotic, unfamiliar, unfairly distributed, anthropomorphic (e.g., affecting pregnant women or children), or scientifically uncertain were perceived by Australians to pose significant risk [100]. Passive smoking, water scarcity, and sun exposure were identified as the top three environmental risks by Western Australians and rated as "high" risks to health [116]. A US ecological study did not emphasize health hazards, but did report that public ranking of ecological risks was influenced by values, beliefs, and worldviews [117]. Compared with risk professionals, US lay public perceives low-probability, high-consequence risks (e.g., hazardous waste sites, persistent organics, sewage, and radiation) to be of greatest concern [117]. These population studies also report that advanced age and lower SES were associated with higher environmental risk perception [100,115–117].

Summary

Men and women assess reproductive and environmental risks differently, which may ultimately produce gender-specific health behaviors and choices. Minority and low-income communities face significant environmental health challenges due to SES disparities,

neighborhood-level hazards, and heightened risk perception. Understanding the many influences used by men and women to process and assess environmental health risk information is necessary to optimize reproductive-specific health promotion programs.

Risk Perception of Environmental Risks to Fertility

In addition to gender, personal experiences also shape risk perception. Dramatic life experiences may be conflated with hazard identification and produce heightened risk perception [99]. Personal experience with family planning, pregnancy, parenthood, and infertility may influence perception of environmental risks to fertility. Risk perceptions are examined within three specific populations: (1) adolescents and young adults, (2) general public, and (3) occupationally exposed workers.

Preconception: Adolescents and Young Adults

Fertility knowledge and awareness has been examined in university students from Sweden [118–120], Finland [121], Italy [122], the United States [123], the United Kingdom [124], Israel [125], and Canada [126,127], with the general consensus that young adults lack an understanding of the age-related decline in female fertility. University students, typically aged 18–25 years, are generally engaged in pregnancy prevention, with parenthood postponed upon completion of academic studies and achievement of career, economic, and relationship stability [118–120,128]. For most young adults, awareness and knowledge of fertility risk factors will come from health professionals, media, Internet, health promotion/education, and anecdotal information from social networks [129,130].

Lifestyle habits, such as smoking, alcohol consumption, and STI, were identified as risks to female fertility in studies from the United Kingdom [124], Sweden [118], British Columbia, Canada [126], and Ottawa, Canada [127]. About a quarter of British Columbian female undergraduates perceived physical or emotional stress and exposure to cigarette smoke as higher risks to female fertility than age [126]. In a qualitative study, Ottawa male and female undergraduates described the modern culture of stress as contributing to infertility and adverse health outcomes in general [48]. Ottawa male and female undergraduates recognized that fertility risks due to lifestyle factors, such as alcohol and smoking, were related to timing of exposures, dose, and frequency [48]. Secondary school female adolescents from Toronto, Ontario, also exhibit awareness of infertility, with STI, smoking, being overweight or underweight, street drugs, and heavy alcohol consumption identified as risks to fertility [131]. Most female adolescents, but fewer males from low-income, predominantly African-American neighborhoods of San

Francisco identified STI including chlamydia and gonorrhea as risk factors for female infertility [132]. These studies indicate that adolescents and young adults are generally aware of lifestyle risk factors for female infertility.

Few studies have examined awareness and perceptions of risk factors for male infertility. Most Toronto high school boys asserted that their fertility was an important issue, and recognized lifestyle habits as risks for male infertility [131]. Although few were concerned about the possibility of infertility, most Ottawa men identified they wished to become fathers in the future [128]. Turkish male and female university students identified smoking, alcohol, and STIs as risk factors for both male and female infertility; however, these lifestyle risks were perceived to primarily affect women [133]. Male and female Ottawa undergraduates also described lifestyle risks to male fertility with more men than women believing that advanced paternal age was also a risk factor [127].

Very few studies have examined university students' perceptions of environmental hazards as risks to fertility. Environmental exposures (e.g., pollution, chemicals, radiation/X-rays) were considered risk factors for both male and female infertility by Turkish students [133]. Similarly, Ottawa undergraduates believed that environmental exposures to pollutants, pesticides, plastics, and BPA would increase risk of infertility [48]. Negative perceptions of environmental risks were influenced by media reports and anecdotal information, with Ottawa students generally asserting that industrial chemicals contribute to widespread public exposure and adverse health [48].

There is a global emphasis on the promotion of a "healthy lifestyle" including smoking cessation, alcohol moderation, maintenance of healthy body weight, and HIV/STI risk reduction [134]. Although a healthy lifestyle may mitigate some modifiable fertility risk factors, it will not abrogate risks caused by maternal age, hormonal, genetic, or congenital abnormalities. Such confusion is evident in UK students who asserted that healthy lifestyle habits would actually increase female fertility [124]. Ottawa undergraduates similarly believed that adherence to a healthy lifestyle ensured overall good health, including fertility [48]. Despite the existence of knowledge gaps, particularly among boys and men, adolescents and young adults seem to be aware of lifestyle risk factors for infertility. It is also possible that any hazards to adverse health are also assumed to be risk factors for infertility, with little understanding of the mechanisms or fertility-specific risk factors such as maternal age.

General Public: Childless Adults and Parents

Studies examining reproductive risk perception in the general population have recruited participants within broad age ranges, typically with a mean age 28–33 years [135–140]. These slightly older participants will have completed the transition to adulthood with school/college/university matriculation, entrance to workforce, development of relationships, and contemplation of parenthood [141]. Very few studies have specifically examined public perceptions of environmental risks to fertility. Australians living in Queensland perceive pesticides, household chemicals, and animal-borne diseases to be harmful to human reproduction, with women more likely to perceive hazards as risky compared with men [140]. Lead exposure followed by stress was perceived as the most harmful hazard to either male or female reproduction [140]. Differences in environmental risk perception were apparent in older participants (45+ years) who perceived household chemicals/paint, radiation, and lead exposures as more significant hazards to reproductive health compared with participants aged 18–34 years [140]. Obesity and smoking were identified as fertility risks in a second study from Australia [135], and by childless Canadian women who also considered STI as risk factors [139]. Similarly, American women identified alcohol, smoking, history of STI, and being overweight or underweight as risks to female fertility [137]. Stress was perceived by more than 90% of American women as a risk factor for infertility [137]. Childless Canadian men [138], but not male Australians [135], identified smoking as a risk factor for male fertility; however, neither group recognized male obesity as a risk factor.

The public has a general understanding of fertility risk factors; however, there is a pronounced gender gap. Awareness of risk factors for infertility is related to female gender, high human development index nationality, advanced education, employment, and medical consultation for infertility problems [135].

Occupationally Exposed: Agricultural Workers

US agricultural workers, including farmers, pesticide applicators, and workers involved in the manufacture of pesticides, are mostly immigrants from Mexico or Latin America with low SES, poor English proficiency and literacy, and limited access to healthcare [142–144]. Several studies have reported reproductive risk perceptions of workers and family members employed in California, Washington State, Florida, and North Carolina. Pesticides were perceived to be poisonous and dangerous, with risk signified by the strong chemical odor of the pesticide [142,145,146]. Both men and women believed that pesticides caused a range of health risks, including dermal irritation, respiratory conditions, and headaches, as well as adverse reproductive health such as infertility, miscarriage, and birth defects [142,144–148]. Female farmworkers and women in farmworker households conflated pesticide "exposure" with "infection," believing that pesticides cause STI and blood contamination, which could be genetically transmitted by exposed men to the fetus [142,147]. Health and safety practices recommend the use of PPE, frequent hand-washing, and removal of contaminated clothing prior to family interactions to reduce the "take-home" exposure pathway [144]. Workers and family members perceived a lack of control over pesticide exposures and identify barriers to risk mitigation such as lack of workplace facilities, resources, time, and health and safety knowledge gaps [142,144,145].

Summary

Knowledge and awareness of environmental risks to fertility develops over the life-course as young adults transition from pregnancy prevention to family planning. In addition to US agriculture, many industries are associated with occupational contaminant exposures, representing potential risk for workers. Although the perception of reproductive risks may be heightened, workers and their families may lack the resources or voice to mitigate risks of occupational exposures. More risk perception studies are required focusing on male fertility and perceptions related to exposures to environmental contaminants, radiation, and other hazards.

Mitigation of Environmental Risks to Fertility

The evaluation of risk information, characterization of hazards, and assessment of public reaction to environmental health topics are collectively used to improve communication of related risk information [96]. Risk communication is an essential component of health promotion and health education; ultimately intended to influence health behaviors [96]. Health promotion and education remain the most important correlates of fertility knowledge rather than personal fertility or parenting experience [135]. Fertility-specific health promotion will enable the development of self-efficacy [68,94,149], such as lifestyle modifications and early recognition and treatment of fertility-related medical conditions [150]. Evidence-based infertility risk communication targeted at the general public should include information on prevalence, causes, signs and symptoms, consequences, and risk mitigation strategies [150]. Designing such reproductive risk communication programs is not simple. The 2001 American Society for Reproductive Medicine (ASRM) infertility prevention campaign received significant criticism for use of shock tactics, contributing to stigma against people with infertility and the individuals with the four target risk factors (smoking, weight problems, STIs, and advance maternal age) [151,152]. Emphasis on individual factors also fails to communicate the increased likelihood of infertility posed by multiple, interacting risk factors in both the male and female [152].

FertiSTAT, a self-administered, multifactorial tool, provides women with a personalized fertility risk assessment and recommendations, and has the potential to provide effective risk communication [153]. This fertility-specific risk communication tool addresses two prerequisites for behavior modification: education and personalized risk [149,154]. Although FertiSTAT does not include environmental/occupational exposures, women may select individualized risk factors (e.g., age), reproductive factors (e.g., PID and STI), lifestyle factors (alcohol, illegal drugs, caffeine, smoking, stress, BMI > 25, unprotected sex with multiple partners), as well as indicators for male infertility (postpubertal mumps, undescended testicles) to achieve an infertility risk profile [153]. An Australian program – FAST (Fertility ASsessment and advice Targeting lifestyle choices and behaviors) provides infertility patients with individualized counseling and support, which has proved successful in a pilot study [155]. A similar Dutch lifestyle intervention incorporated fertility nurses in the provision of preconception care [156]. Characteristics of successful lifestyle interventions include structured approach, involvement of trained health-care professionals, and a patient-centered and motivational approach to help achieve behavior changes and optimize fertility [11,155,156].

Barriers to Risk Mitigation

Comprehensive sexual and reproductive health education, which includes signs and symptoms of reproductive health problems and consequences of STI, such as cancer and infertility, is essential for adolescents and adults, and yet rarely delivered [43,157]. Fundamental gaps in infertility surveillance, such as prevalence of specific subtypes of infertility especially among different populations (race/ethnicity, SES, occupationally exposed workers), limit the design of effective fertility risk mitigation [150,158]. Infertility studies focus predominantly on women's experiences; however, men's awareness and perceptions of male-factor infertility along with other reproductive issues are additional gaps in sexual health promotion [150,159,160]. Addressing environmental risks is particularly problematic, as the evidence of adverse human health effects is equivocal, with issues such as dose effects and mechanism subject to scientific debate [98]. Environmental risk communication strategies should adopt a life-course approach; targeting adolescents, young adult women and men, along with medical providers [158].

Media is undoubtedly the most significant source of scientific information for the lay-public; however, it is also the source of health risk misinformation [113,130]. Inaccurate environmental health information may prevent risk mitigation and may cause unnecessary anxiety and concern about adverse health outcomes [158].

Interprofessional collaborations with media to provide expert interpretation of emerging fertility science, along with dissemination of targeted evidence-based fertility health promotion, may help reduce fertility-related misinformation.

Finally, mitigation of environmental risks may well depend on family economic resources. As described, agricultural workers' limited resources prevent replacement of damaged PPE, which results in greater occupational pesticide exposures [142,144]. "Environmentally friendly" products (e.g., organic food, toys, and baby products) are generally more expensive, leaving socioeconomically disadvantaged families little option but to purchase cheap products, which may be more hazardous [161]. Inexpensive products will have greater uptake and influence on health, but will also be perceived as safe, due to widespread availability [162].

Public Health Approaches to Infertility

As discussed, many of the modifiable risks to fertility, such as STI, smoking, and obesity, are currently addressed within the public health domain, leading to recommendations that infertility be framed as a public health issue [150,152,163,164]. Due to the over-representation of modifiable fertility risk factors in low-SES communities [150,152], existing community-specific public health approaches may ensure greater uptake of reproductive health promotion programs [150,152]. Infertility/ART-associated stress, grief, depression, and anxiety [165–170] may also be better supported by a public health framework, through dissemination of realistic ART success rates and management of expectations [163]. Building on these recommendations, a public health framework designed to mitigate modifiable risks to fertility is proposed (Fig. 1.3). Informed by advances in reproductive

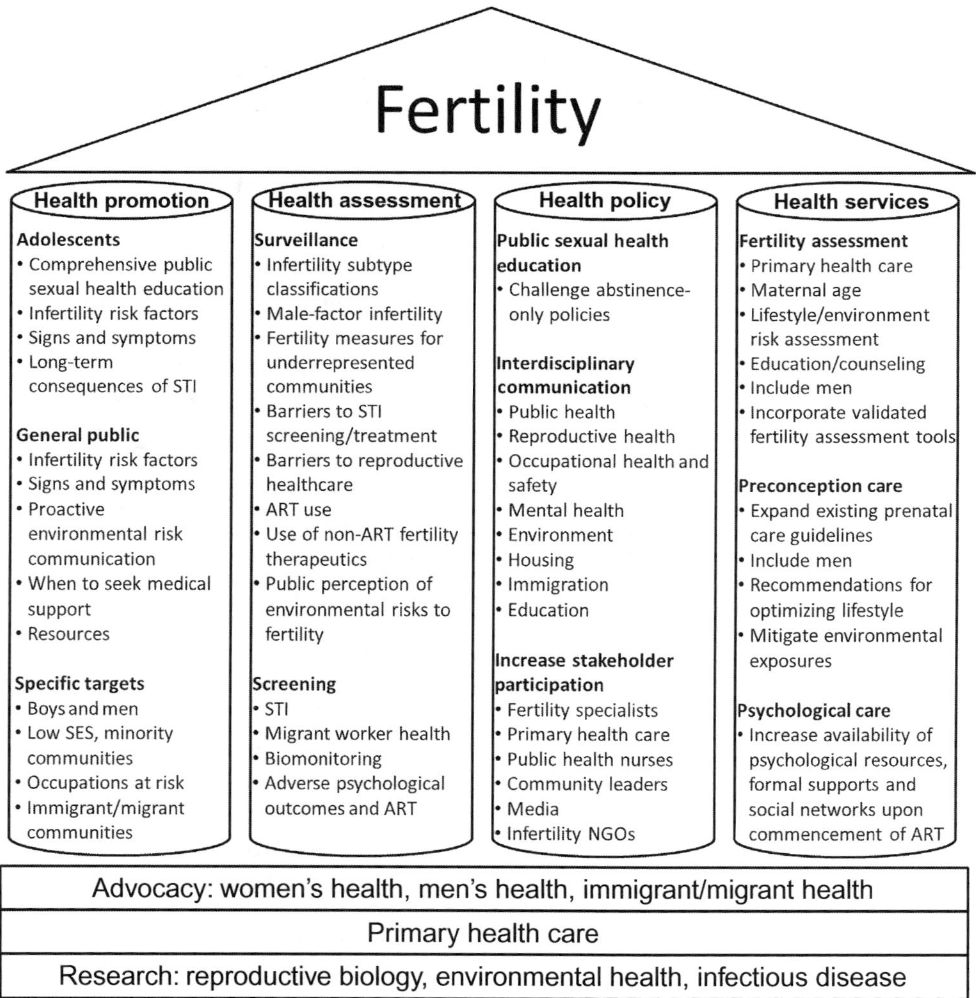

FIGURE 1.3 **Proposed public health framework to mitigate modifiable risks to fertility.** As most modifiable risks to fertility are also public health threats, it is proposed that infertility should be encompassed within the field of public health [150,152,154,163,164]. Building on these recommendations, a fertility-specific public health framework is proposed with four major pillars: (1) health promotion, (2) health assessment, (3) health policy, and (4) health services. STI, sexually transmitted infections; SES, socioeconomic status; NGO, nongovernmental organization; ART, assisted reproductive technologies.

biology, environmental health, and infectious diseases, this framework describes four major pillars: (1) health promotion, (2) health assessment, (3) health policy, and (4) health services. Integral to the framework is the interdisciplinary, intersectoral collaboration, which brings together community groups and primary health-care and mental-health providers, in addition to government agencies (occupational health and safety, environment, housing, immigration, education). Boys, men, low SES, minority, immigrant, and migrant communities are recommended as specific targets for reproductive health promotion, surveillance, and preconception care.

Summary

Infertility is a complex, multifactorial health issue that may be best addressed using public health frameworks for education, health promotion, and prevention. Mitigating environmental risks to fertility will necessitate lifestyle changes, best facilitated by formalized supports such as coaching or counseling. Fertility-specific health promotion programs that provide information about infertility detection, risk factors, and infertility prevention are recommended.

CONCLUSIONS

Environmental risks to fertility include lifestyle factors (e.g., obesity, smoking, alcohol, STI) as well as exposures to contaminants or occupational hazards. Many of these exposures confer additional risks during pregnancy, necessitating both pre- and postconception risk communication strategies. Mitigation of environmental risks to fertility requires education, health promotion, and support to facilitate lifestyle changes. A public health approach that incorporates environmental exposures is suggested as a framework for reproductive health promotion. Whereas lifestyle risk modification is fundamental to disease prevention and primary care, discourses around environmental risks to fertility and reproductive health remain tangential.

References

[1] Bushnik T, Cook JL, Yuzpe AA, Tough S, Collins J. Estimating the prevalence of infertility in Canada. Hum Reprod 2012;27:738–46.

[2] Thoma ME, McLain AC, Louis JF, King RB, Trumble AC, Sundaram R, et al. Prevalence of infertility in the United States as estimated by the current duration approach and a traditional constructed approach. Fertil Steril 2013;99:1324–31.

[3] Adamson GD, Baker VL. Subfertility: causes, treatment and outcome. Best Pract Res Clin Obstet Gynaecol 2003;17:169–85.

[4] Friedler G. Paternal exposures: impact on reproductive and developmental outcome. An overview. Pharmacol Biochem Behav 1996;55:691–700.

[5] Figà-Talamanca I. Occupational risk factors and reproductive health of women. Occup Med (London) 2006;56:521–31.

[6] Jurewicz J, Hanke W, Radwan M, Bonde JP. Environmental factors and semen quality. Int J Occup Med Environ Health 2009;22:305–29.

[7] Phillips KP, Tanphaichitr N. Mechanisms of obesity-induced male infertility. Expert Rev Endocrinol Metabol 2010;5:229–51.

[8] Reese EA. Perspectives on obesity, pregnancy and birth outcomes in the United States: the scope of the problem. Am J Obstet Gynecol 2008;198:23–7.

[9] Brewer CJ, Balen AH. The adverse effects of obesity on conception and implantation. Reproduction 2010;140:347–64.

[10] Colaci DS, Afeiche M, Gaskins AJ, Wright DL, Toth TL, Tanrikut C, et al. Men's body mass index in relation to embryo quality and clinical outcomes in couples undergoing *in vitro* fertilization. Fertil Steril 2012;98:1193–9.

[11] Homan GF, Davies M, Norman R. The impact of lifestyle factors on reproductive performance in the general population and those undergoing infertility treatment: a review. Hum Reprod Update 2007;13:209–23.

[12] Luke B, Brown MB, Stern JE, Missmer SA, Fujimoto VY, Leach R. Female obesity adversely affects assisted reproductive technology (ART) pregnancy and live birth rates. Hum Reprod 2011;26:245–52.

[13] Machtinger R, Combelles CM, Missmer SA, Correia KF, Fox JH, Racowsky C. The association between severe obesity and characteristics of failed fertilized oocytes. Hum Reprod 2012;27:3198–207.

[14] Spandorfer SD, Kump L, Goldschlag D, Brodkin T, Davis OK, Rosenwaks Z. Obesity and *in vitro* fertilization: negative influences on outcome. J Reprod Med 2004;49:973–7.

[15] Bellver J, Ayllon Y, Ferrando M, Melo M, Goyri E, Pellicer A, et al. Female obesity impairs *in vitro* fertilization outcome without affecting embryo quality. Fertil Steril 2010;93:447–54.

[16] Chavarro JE, Ehrlich S, Colaci DS, Wright DL, Toth TL, Petrozza JC, et al. Body mass index and short-term weight change in relation to treatment outcomes in women undergoing assisted reproduction. Fertil Steril 2012;98:109–16.

[17] Moragianni VA, Jones SM, Ryley DA. The effect of body mass index on the outcomes of first assisted reproductive technology cycles. Fertil Steril 2012;98:102–8.

[18] Lintsen AM, Pasker-de Jong PC, de Boer EJ, Burger CW, Jansen CA, Braat DD, et al. Effects of subfertility cause, smoking and body weight on the success rate of IVF. Hum Reprod 2005;20:1867–75.

[19] Merhi ZO, Keltz J, Zapantis A, Younger J, Berger D, Lieman HJ, et al. Male adiposity impairs clinical pregnancy rate by *in vitro* fertilization without affecting day 3 embryo quality. Obesity (Silver Spring) 2013;21:1608–12.

[20] Bakos HW, Henshaw RC, Mitchell M, Lane M. Paternal body mass index is associated with decreased blastocyst development and reduced live birth rates following assisted reproductive technology. Fertil Steril 2011;95:1700–4.

[21] Anderson K, Nisenblat V, Norman R. Lifestyle factors in people seeking infertility treatment – a review. Aust NZ J Obstet Gynaecol 2010;50:8–20.

[22] Pandey S, Maheshwari A, Bhattacharya S. Should access to fertility treatment be determined by female body mass index? Hum Reprod 2010;25:815–20.

[23] Shah DK, Ginsburg ES, Correia KF, Barton SE, Missmer SA. Public opinion regarding utilization of assisted reproductive technology (ART) in obese women. J Women's Health 2013;22:978–82.

[24] Clark AM, Thornley B, Tomlinson L, Galletley C, Norman RJ. Weight loss in obese infertile women results in improvement in reproductive outcome for all forms of fertility treatment. Hum Reprod 1998;13:1502–5.

[25] Sharma R, Beidenharn KR, Fedor JM, Agarwal A. Lifestyle factors and reproductive health: taking control of your fertility. Reprod Biol Endocrinol 2013;11:66.

[26] Centers for Disease Control and Prevention. Adult cigarette smoking in the United States: current estimates [updated February 14, 2014; cited May 26, 2014]. Available from: http://www.cdc.gov/tobacco/data_statistics/fact_sheets/adult_data/cig_smoking/.

[27] Health Canada. Canadian Tobacco Use Monitoring Survey (CTUMS) 2012 [updated October 1, 2013; cited May 26. 2014]. Available from: http://www.hc-sc.gc.ca/hc-ps/tobac-tabac/research-recherche/stat/_ctums-esutc_2012/ann_summary-sommaire-eng.php.

[28] World Health Organization. Prevalence of tobacco use [updated 2014; cited May 26, 2014]. Available from: http://www.who.int/gho/tobacco/use/en/.

[29] Sadeu JC, Foster WG. The cigarette smoke constituent benzo[a]pyrene disrupts metabolic enzyme, and apoptosis pathway member gene expression in ovarian follicles. Reprod Toxicol 2013;40:52–9.

[30] Younglai EV, Holloway AC, Foster WG. Environmental and occupational factors affecting fertility and IVF success. Hum Reprod Update 2005;11:43–57.

[31] Vine MF. Smoking and male reproduction: a review. Int J Androl 1996;19:323–37.

[32] Künzle R, Mueller MD, Hänggi W, Birkhäuser MH, Drescher H, Bersinger NA. Semen quality of male smokers and nonsmokers in infertile couples. Fertil Steril 2003;79:287–91.

[33] Soares S. Cigarette smoking and fertility. Reprod Biol Insights 2009;2:39–46.

[34] Zitzmann M, Rolf C, Nordhoff V, Schräder G, Rickert-Föhring M, Gassner P, et al. Male smokers have a decreased success rate for *in vitro* fertilization and intracytoplasmic sperm injection. Fertil Steril 2003;79:1550–4.

[35] Sadeu JC, Hughes CL, Agarwal S, Foster WG. Alcohol, drugs, caffeine, tobacco, and environmental contaminant exposure: reproductive health consequences and clinical implications. Crit Rev Toxicol 2010;40:633–52.

[36] Van Voorhis BJ, Dawson JD, Stovall DW, Sparks AE, Syrop CH. The effects of smoking on ovarian function and fertility during assisted reproduction cycles. Obstet Gynecol 1996;88:785–91.

[37] Kazemi A, Nasr Esfahani NH, Ahmadi M, Ehsanpour S, Ganji J. Maternal exposure to second-hand smoke and super ovulation outcome for assisted reproduction. Int J Fertil Steril 2009;3:52–5.

[38] Neal MS, Zhu J, Foster WG. Quantification of benzo[a]pyrene and other PAHs in the serum and follicular fluid of smokers versus non-smokers. Reprod Toxicol 2008;25:100–6.

[39] Neal MS, Hughes EG, Holloway AC, Foster WG. Sidestream smoking is equally as damaging as mainstream smoking on IVF outcomes. Hum Reprod 2005;20:2531–5.

[40] Gannon AM, Stämpfli MR, Foster WG. Cigarette smoke exposure leads to follicle loss via an alternative ovarian cell death pathway in a mouse model. Toxicol Sci 2012;125:274–84.

[41] Gannon AM, Stämpfli MR, Foster WG. Cigarette smoke exposure elicits increased autophagy and dysregulation of mitochondrial dynamics in murine granulosa cells. Biol Reprod 2013;88:63.

[42] The Practice Committee of the American Society for Reproductive Medicine. Smoking and infertility: a committee opinion. Fertil Steril 2012;98:1400–6.

[43] Phillips KP, Martinez A. Sexual and reproductive health education: contrasting teachers', health partners' and former students' perspectives. Can J Public Health 2010;101:374–9.

[44] Public Health Agency of Canada. The chief public health officer's report on the state of public health in Canada, 2013. Infectious disease – the never-ending threat. Sexually transmitted infections – a continued public health concern [updated October 23, 2013; cited May 26, 2014]. Available from: http://www.phac-aspc.gc.ca/cphorsphc-respcacsp/2013/sti-its-eng.php.

[45] McClung AA, Perfect MM. Research-based practice sexual health education: social and scientific perspectives and how school psychologists can be involved. NASP Communiqué 2012;40(6):1, 14–16.

[46] Centers for Disease Control and Prevention. Sexually transmitted disease surveillance 2012 [updated January 7, 2014; cited May 26, 2014]. Available from: http://www.cdc.gov/sTD/stats12/toc.htm.

[47] Paavonen J. *Chlamydia trachomatis* infections of the female genital tract: state of the art. Ann Med 2012;44:18–28.

[48] Remes O, Whitten AN, Sabarre KA, Phillips KP. University students' perceptions of environmental risks to infertility. Sex Health 2012;9:377–83.

[49] Bachir BG, Jarvi K. Infectious, inflammatory, and immunologic conditions resulting in male infertility. Urol Clin North Am 2014;41:67–81.

[50] Lareau SM, Beigi RH. Pelvic inflammatory disease and tubo-ovarian abscess. Infect Dis Clin North Am 2008;22:693–708.

[51] Bohm MK, Newman L, Satterwhite CL, Tao G, Weinstock HS. Pelvic inflammatory disease among privately insured women, United States, 2001–2005. Sex Transm Dis 2010;37:131–6.

[52] Rekart ML, Gilbert M, Meza R, Kim PH, Chang M, Money DM, et al. Chlamydia public health programs and the epidemiology of pelvic inflammatory disease and ectopic pregnancy. J Infect Dis 2013;207:30–8.

[53] French CE, Hughes G, Nicholson A, Yung M, Ross JD, Williams T, et al. Estimation of the rate of pelvic inflammatory disease diagnoses: trends in England, 2000–2008. Sex Transm Dis 2011;38:158–62.

[54] Ochsendorf FR. Sexually transmitted infections: impact on male fertility. Andrologia 2008;40:72–5.

[55] Cunningham KA, Beagley KW. Male genital tract chlamydial infection: Implications for pathology and infertility. Biol Reprod 2008;79:180–9.

[56] Pacey AA. Environmental and lifestyle factors associated with sperm DNA damage. Hum Fertil 2010;13:189–93.

[57] Joki-Korpela P, Sahrakorpi N, Halttunen M, Surcel HM, Paavonen J, Tiitinen A. The role of *Chlamydia trachomatis* infection in male infertility. Fertil Steril 2009;91:1448–50.

[58] Lynch CD, Sundaram R, Maisog JM, Sweeney AM, Buck Louis GM. Preconception stress increases the risk of infertility: results from a couple-based prospective cohort study – the LIFE study. Hum Reprod 2014;29:1067–75.

[59] Veras MM, Caldini EG, Dolhnikoff M, Saldiva PH. Air pollution and effects on reproductive-system functions globally with particular emphasis on the Brazilian population. J Toxicol Environ Health B Crit Rev 2010;13:1–15.

[60] Somers CM. Ambient air pollution exposure and damage to male gametes: human studies and *in situ* 'sentinel' animal experiments. Syst Biol Reprod Med 2011;57:63–71.

[61] Carpenter DO, Sage C. Setting prudent public health policy for electromagnetic field exposures. Rev Environ Health 2008;23:91–117.

[62] Kesari KK, Kumar S, Nirala J, Siddiqui MH, Behari J. Biophysical evaluation of radiofrequency electromagnetic field effects on male reproductive pattern. Cell Biochem Biophys 2013;65:85–96.

[63] Merhi ZO. Challenging cell phone impact on reproduction: a review. J Assist Reprod Genet 2012;29:293–7.

[64] Yen IH, Syme SL. The social environment and health: a discussion of the epidemiologic literature. Annu Rev Public Health 1999;20:287–308.

[65] Brunton PJ. Effects of maternal exposure to social stress during pregnancy: consequences for mother and offspring. Reproduction 2013;146:R175–89.

[66] Eskenazi B, Marks AR, Catalano R, Bruckner T, Toniolo PG. Low birthweight in New York City and upstate New York following the events of September 11th. Hum Reprod 2007;22:3013–20.

[67] WHO – World Health Organization. Global assessment of the state-of-the-science of endocrine disruptors. In: Damstra T, Barlow S, Bergman A, Kavlock R, Van Der Kraak G, editors. International Programme on Chemical Safety; 2002 [cited May 26, 2014].

Available from: http://www.who.int/ipcs/publications/new_issues/endocrine_disruptors/en/.

[68] Foster WG, Neal MS, Han MS, Dominguez MM. Environmental contaminants and human infertility: hypothesis or cause for concern? J Toxicol Environ Health B Crit Rev 2008;11:162–76.

[69] Phillips KP, Foster WG. Key developments in endocrine disrupter research and human health. J Toxicol Environ Health B Crit Rev 2008;11:322–44.

[70] Skakkebaek NE, Rajpert-De Meyts E, Main KM. Testicular dysgenesis syndrome: an increasingly common developmental disorder with environmental aspects. Hum Reprod 2001;16:972–8.

[71] National Research Council. Hormonally active agents in the environment. Washington, DC: The National Academies Press; 1999.

[72] Kuiper GG, Lemmen JG, Carlsson B, Corton JC, Safe SH, van der Saag PT, et al. Interaction of estrogenic chemicals and phytoestrogens with estrogen receptor beta. Endocrinology 1998;139:4252–63.

[73] Wadhwa PD. Psychoneuroendocrine processes in human pregnancy influence fetal development and health. Psychoneuroendocrinology 2005;8:724–43.

[74] Anway MD, Skinner MK. Epigenetic transgenerational actions of endocrine disruptors. Endocrinology 2006;27:868–79.

[75] Chang HS, Anway MD, Rekow SS, Skinner MK. Transgenerational epigenetic imprinting of the male germline by endocrine disruptor exposure during gonadal sex determination. Endocrinology 2006;147:5524–41.

[76] Barton TS, Robaire B, Hales BF. Epigenetic programming in the preimplantation rat embryo is disrupted by chronic paternal cyclophosphamide exposure. Proc Natl Acad Sci USA 2005;102(22):7865–70.

[77] Newbold RR, Padilla–Banks E, Snyder RJ, Phillips TM, Jefferson WN. Developmental exposure to endocrine disruptors and the obesity epidemic. Reprod Toxicol 2007;23:290–6.

[78] Anway MD, Skinner MK. Epigenetic programming of the germ line: effects of endocrine disruptors on the development of transgenerational disease. Reprod Biomed Online 2008;16:23–5.

[79] Weselak M, Arbuckle TE, Wigle DT, Walker MC, Krewski D. Pre- and post-conception pesticide exposure and the risk of birth defects in an Ontario farm population. Reprod Toxicol 2008;25:472–80.

[80] Arbuckle TE, Lin Z, Mery LS. An exploratory analysis of the effect of pesticide exposure on the risk of spontaneous abortion in an Ontario farm population. Environ Health Perspect 2001;109:851–7.

[81] Curtis KM, Savitz DA, Weinberg CR, Arbuckle TE. The effect of pesticide exposure on time to pregnancy. Epidemiology 1999;10:112–7.

[82] Phillips KP, Tanphaichitr N. Human exposure to endocrine disrupters and semen quality. J Toxicol Environ Health B Crit Rev 2008;11:188–220.

[83] Jarrell JF, Villeneuve D, Franklin C, Bartlett S, Wrixon W, Kohut J, et al. Contamination of human ovarian follicular fluid and serum by chlorinated organic compounds in three Canadian cities. CMAJ 1993;148:1321–7.

[84] Younglai EV, Foster WG, Hughes EG, Trim K, Jarrell JF. Levels of environmental contaminants in human follicular fluid, serum, and seminal plasma of couples undergoing in vitro fertilization. Arch Environ Contam Toxicol 2002;43:121–6.

[85] Dominguez MA, Petre MA, Neal MS, Foster WG. Bisphenol A concentration-dependently increases human granulosa-lutein cell matrix metalloproteinase-9 (MMP-9) enzyme output. Reprod Toxicol 2008;25:420–5.

[86] Ikezuki Y, Tsutsumi O, Takai Y, Kamei Y, Taketani Y. Determination of bisphenol A concentrations in human biological fluids reveals significant early prenatal exposure. Hum Reprod 2002;17:2839–41.

[87] Tsutsumi O. Assessment of human contamination of estrogenic endocrine-disrupting chemicals and their risk for human reproduction. J Steroid Biochem Mol Biol 2005;93:325–30.

[88] Iguchi T, Sato T. Endocrine disruption and developmental abnormalities of female reproduction. Am Zool 2000;40:402–11.

[89] Weiss JM, Bauer O, Blüthgen A, Ludwig AK, Vollersen E, Kaisi M, et al. Distribution of persistent organochlorine contaminants in infertile patients from Tanzania and Germany. J Assist Reprod Genet 2006;23(9-10):393–9.

[90] Foster WG, Neal MS, Younglai EV. Ovarian toxicity of environmental toxicants. Immunol Endocrin Metabol Agents Med Chem 2006;6:37–43.

[91] Cvetkovich G, Earle TC. Environmental hazards and the public. J Social Issues 1992;48:1–20.

[92] Phillips KP, Foster WG, Leiss W, Sahni V, Karyakina N, Turner MC, Kacew S, Krewski D. Assessing and managing risks arising from exposure to endocrine-active chemicals. J Toxicol Environ Health B Crit Rev 2008;11:351–72.

[93] Becker G, Natchtigall RD. 'Born to be a mother': the cultural construction of risk in infertility treatment in the U.S. Soc Sci Med 1994;39:507–18.

[94] Tough S, Tofflemire K, Benzies K, Fraser-Lee N, Newburn-Cook C. Factors influencing childbearing decisions and knowledge of perinatal risks among Canadian men and women. Matern Child Health J 2007;11:189–98.

[95] Steg L, Sievers I. Cultural theory and individual perceptions of environmental risks. Environ Behav 2000;32:250–69.

[96] Slovic P, Fischhoff B, Lichtenstein S. Why study risk perception? Risk Anal 1982;2:83–93.

[97] Helgeson J, van der Linden S, Chabay I. The role of knowledge, learning and mental models in perceptions of climate change related risks. In: Wals Arjen EJ, Corcoran Peter Blaze, editors. Learning for sustainability in times of accelerating change. Wageningen, The Netherlands: Wageningen Academic Publishers; 2012. p. 329–46. ISBN 9789086867578.

[98] Tyshenko MG, Phillips KP, Mehta M, Poirier R, Leiss W. Risk communication of endocrine-disrupting chemicals: improving knowledge translation and transfer. J Toxicol Environ Health B Crit Rev 2008;11:345–50.

[99] Kasperson RE, Renn O, Slovic P, Brown HS, Emel J, Goble R, et al. The social amplification of risk. A conceptual framework. Risk Anal 1988;8:177–87.

[100] Starr G, Langley A, Taylor A. Environmental health risk perception in Australia. A research report to the Commonwealth Department of Health and Aged Care. Centre for Population Studies in Epidemiology, South Australian Department of Human Services; 2000 [cited May 26, 2014]. Available from: http://www.health.gov.au/internet/main/publishing.nsf/Content/health-pubhlth-publicat-document-metadata-envrisk.htm.

[101] Gustafson PE. Gender differences in risk perception: theoretical and methodological perspectives. Risk Anal 1998;18:805–11.

[102] Slovic P. Trust, emotion, sex, politics, and science: surveying the risk-assessment battlefield. Risk Anal 1999;19:689–701.

[103] Finucaine ML, Slovic P, Mertz CK, Flynn J, Satterfield TA. Gender, race, and perceived risk: the 'white male' effect. Health Risk Soc 2000;2:159–72.

[104] Fischer GW, Morgan G, Fischoff B, Nair I, Lave LB. What risks are people concerned about? Risk Anal 1991;11:303–14.

[105] Flynn J, Slovic P, Mertz CK. Gender, race, and perception of environmental health risks. Risk Anal 1994 Dec;14:1101–8.

[106] Draper E. Reproductive hazards and fetal exclusion policies after Johnson Controls. Stan L Pol'y Rev 2001;12:117–42.

[107] Hoeksma NG. Regulating risk: reproductive toxins in the workplace in the post-Johnson Controls era. S Cal Rev Law Wom Stud 2005;14:289–313.

[108] Kukla R. The ethics and cultural politics of reproductive risk warnings: a case study of California's Proposition 65. Health Risk Soc 2010;12:323–34.

[109] Evans GW, Kantrowitz E. Socioeconomic status and health: the potential role of environmental risk exposure. Ann Rev Pub Health 2002;23:303–31.

[110] Evans DT, Fullilove MT, Green L, Levison MAT Awareness of environmental risks and protective actions among minority women in Northern Manhattan. Environ Health Perspect 2002;110:271–5.

[111] Taylor-Clark K, Koh H, Viswanath K. Perceptions of environmental health risks and communication barriers among low-SEP and racial/ethnic minority communities. J Health Care Poor Underserved 2007;18:165–83.

[112] Leiss W. In the chamber of risk: understanding risk controversies. Montreal and Kingston, Canada: McGill-Queens University Press; 2001.

[113] Morrone M. From cancer to diarrhea: the moving target of public concern about environmental health risks. Environ Health Insights 2011;5:87–96.

[114] Krewski D, Slovic P, Barlett S, Flynn J, Mertz CK. Health risk perception in Canada I: Rating hazards, sources of information and responsibility for health protection. Hum Ecol Risk Assess 1995;1:117–32.

[115] Krewski D, Lemyre L, Turner MC, Lee JEC, Dallaire C, Bouchard L, et al. Public perception of population health risks in Canada: health hazards and sources of information. Hum Ecol Risk Assess 2006;12:626–44.

[116] Government of Western Australis, Department of Health. Community survey of perceived environmental health risks in Western Australia; 2009 [updated 2014; cited May 26, 2014]. Available from: http://www.public.health.wa.gov.au/3/931/1/community_surve.pm.

[117] Slimak MW, Dietz T. Personal values, beliefs, and ecological risk perception. Risk Anal 2006;26:1689–705.

[118] Tydén T, Svanberg AS, Karlström PO, Lihoff L, Lampic C. Female university students' attitudes to future motherhood and their understanding about fertility. Eur J Contracept Reprod Health Care 2006;11:181–9.

[119] Lampic C, Svanberg AS, Karlström P, Tydén T. Fertility awareness, intentions concerning childbearing, and attitudes towards parenthood among female and male academics. Hum Reprod 2006;21:558–64.

[120] Svanberg AS, Lampic C, Karlström PO, Tydén T. Attitudes toward parenthood and awareness of fertility among postgraduate students in Sweden. Gender Med 2006;3:187–95.

[121] Virtala A, Vilska S, Huttunen T, Kunttu K. Childbearing, the desire to have children, and awareness about the impact of age on female fertility among Finnish university students. Eur J Contracept Reprod Health Care 2011;16:108–15.

[122] Rovei V, Gennarelli G, Lantieri T, Casano S, Revelli A, Massobrio M. Family planning, fertility awareness and knowledge about Italian legislation on assisted reproduction among Italian academic students. Reprod BioMed Online 2010;20:873–9.

[123] Peterson BD, Pirritano M, Tucker L, Lampic C. Fertility awareness and parenting attitudes among American male and female undergraduate university students. Hum Reprod 2012;27:1375–82.

[124] Bunting L, Boivin J. Knowledge about infertility risk factors, fertility myths and illusory benefits of healthy habits in young people. Hum Reprod 2008;23:1858–64.

[125] Hashiloni-Dolev Y, Kaplan A, Shkedi-Rafid S. The fertility myth: Israeli students' knowledge regarding age-related fertility decline and late pregnancies in an era of assisted reproduction technology. Hum Reprod 2011;26:3045–53.

[126] Bretherick KL, Fairbrother N, Avila L, Harbord SH, Robinson WP. Fertility and aging: do reproductive-aged Canadian women know what they need to know? Fertil Steril 2010;93:2162–8.

[127] Sabarre K-A, Khan Z, Whitten AN, Remes O, Phillips KP. A qualitative study of Ottawa university students' awareness, knowledge and perceptions of infertility, infertility risk factors and assisted reproductive technologies (ART). Reprod Health 2013;10:41.

[128] Whitten AN, Remes O, Sabarre K-A, Khan Z, Phillips KP. Canadian university students' perceptions of future personal infertility. Open J Obstet Gynecol 2013;3:561–8.

[129] Slauson-Blevins KS, McQuillan J, Greil AL. Online and in-person health-seeking for infertility. Soc Sci Med 2013;99:110–5.

[130] Lampi E. What do friends and the media tell us? How different information channels affect women's risk perceptions of age-related female infertility. J Risk Res 2011;14:365–80.

[131] Quach S, Librach C. Infertility knowledge and attitudes in urban high school students. Fertil Steril 2008;90:2099–106.

[132] Trent M, Millstein SG, Ellen JM. Gender-based differences in fertility beliefs and knowledge among adolescents from high sexually transmitted disease-prevalence communities. J Adolesc Health 2006;38:282–7.

[133] Gungor I, Rathfisch G, Kizilkaya Beji N, Yarar M, Karamanoglu F. Risk-taking behaviours and beliefs about fertility in university students. J Clin Nurs 2013;22(23–24):3418–27.

[134] World Health Organization. A healthy lifestyle [updated 2014; cited May 26, 2014]. Available from: http://www.euro.who.int/en/health-topics/disease-prevention/nutrition/a-healthy-lifestyle.

[135] Bunting L, Tsibulsky I, Boivin J. Fertility knowledge and beliefs about fertility treatment: findings from the international fertility decision-making study. Hum Reprod 2013;28:385–97.

[136] Hammarberg K, Setter T, Norman RJ, Holden CA, Michelmore J, Johnson L. Knowledge about factors that influence fertility among Australians of reproductive age: a population-based survey. Fertil Steril 2013;99:502–7.

[137] Lundsberg LS, Pal L, Gariepy AM, Xu X, Chu MC, Illuzzi JL. Knowledge, attitudes, and practices regarding conception and fertility: a population-based survey among reproductive-age United States women. Fertil Steril 2014;101:767–74.

[138] Daniluk JC, Koert E. The other side of the fertility coin: a comparison of childless men's and women's knowledge of fertility and assisted reproductive technology. Fertil Steril 2013;99:839–46.

[139] Daniluk JC, Koert E, Cheung A. Childless women's knowledge of fertility and assisted human reproduction: identifying the gaps. Fertil Steril 2012;97:420–6.

[140] Shepherd A, Jepson R, Watterson A, Evans JMM. Risk perceptions of environmental hazards and human reproduction: a community-based survey. ISRN Public Health 2012 [cited May 26, 2014]. Available from: http://www.hindawi.com/journals/isrn.public.health/2012/748080/.

[141] Rumbaut RG. Young adults in the United States: a profile. Research network working paper no. 4; 2004 [cited May 26, 2014]. Available from: http://papers.ssrn.com/sol3/papers.cfm?abstract_id=1887827.

[142] Rao P, Quandt SA, Doran AM, Snively BM, Arcury TA. Pesticides in the homes of farmworkers: Latino mothers' perceptions of risk to their children's health. Health Educ Behav 2007;34:335–53.

[143] Remoundou K, Brennan M, Hart A, Frewer LJ. Pesticide risk perceptions, knowledge, and attitudes of operators, workers, and residents: a review of the literature. Hum Ecol Risk Assess 2014;20:1113–38.

[144] Arcury TA, Quandt SA, Russell GB. Pesticide safety among farmworkers: perceived risk and perceived control as factors reflecting environmental justice. Environ Health Perspect 2002;110:233–40.

[145] Strong LL, Starks HE, Meischke H, Thompson B. Perspectives of mothers in farmworker households on reducing the take-home pathway of pesticide exposure. Health Educ Behav 2009;36:915–29.

[146] Elmore RC, Arcury TA. Pesticide exposure beliefs among Latino farmworkers in North Carolina's Christmas tree industry. Am J Ind Med 2001;40:153–60.

[147] Flocks J, Kelley M, Economos J, McCauley L. Female farmworkers' perceptions of pesticide exposure and pregnancy health. J Immigr Minor Health 2012;14:626–32.

[148] Cabrera NL, Leckie JO. Pesticide risk communication, risk perception, and self-protective behaviors among farmworkers in California's Salinas Valley. Behav Sci 2009;31:258–72.

[149] Bandura A. Self-efficacy: toward a unifying theory of behavioral change. Psychol Rev 1977;84:191–215.

[150] Macaluso M, Wright-Schnapp TJ, Chandra A, Johnson R, Satterwhite CL, Pulver A, et al. A public health focus on infertility prevention, detection, and management. Fertil Steril 2010;93:16. e1–16.e10.

[151] Soules MR. The story behind the American Society for reproductive medicine's prevention of infertility campaign. Fertil Steril 2003;80:295–9.

[152] Lemoine ME, Ravitsky V. Toward a public health approach to infertility: the ethical dimensions of infertility prevention. Pub Health Ethics 2013;6:287–301.

[153] Bunting L, Boivin J. Development and preliminary validation of the fertility status awareness tool: FertiSTAT. Hum Reprod 2010;25:1722–33.

[154] Boivin J, Bunting L, Gameiro S. Cassandra's prophecy: a psychological perspective. Why we need to do more than just tell women. Reprod BioMed Online 2013;27:11–4.

[155] Homan G, Litt J, Norman RJ. The FAST study: Fertility ASsessment and advice Targeting lifestyle choices and behaviours: a pilot study. Hum Reprod 2012;27:2396–404.

[156] Ockhuijsen HD, Gamel CJ, van den Hoogen A, Macklon NS. Integrating pre-conceptional care into an IVF programme. J Adv Nurs 2012;68:1156–65.

[157] Lyon M, Rogers J, Summers D. Abstinence-only education policies and programs: a position paper of the Society for Adolescent Medicine. J Adolesc Health 2006;38:83–7.

[158] Grason HA, Misra DP. Reducing exposure to environmental toxicants before birth: moving from risk perception to risk reduction. Pub Health Rep 2009;124:629–41.

[159] Sternberg P, Hubley J. Evaluating men's involvement as a strategy in sexual and reproductive health promotion. Health Prom Int 2004;19:389–96.

[160] Saewyc EM. What about the boys? The importance of including boys and young men in sexual and reproductive health research. J Adolesc Health 2012;51:1–2.

[161] Schmidt CW. Face to face with toy safety: understanding an unexpected threat. Environ Health Perspect 2008;116:A70–6.

[162] Cohen DA, Scribner RA, Farley TA. A structural model of health behavior: a pragmatic approach to explain and influence health behaviors at the population level. Prev Med 2000;30:146–54.

[163] Deonandan R. The public health implications of assisted reproductive technologies. Chron Dis Can 2010;30:119–24.

[164] Fidler AT, Bernstein J. Infertility: from a personal to a public health problem. Pub Health Rep 1999;114:494–511.

[165] Casey Jacob M, McQuillan J, Greil AL. Psychological distress by type of fertility barrier. Hum Reprod 2007;22:885–94.

[166] Ramezanzadeh F, Aghssa MM, Abedinia N, Zayeri F, Khanafshar N, Shariat M, et al. A survey of relationship between anxiety, depression and duration of infertility. BMC Women's Health 2004;4:9.

[167] Franco JG Jr, Razera Baruffi RL, Mauri AL, Petersen CG, Felipe V, Garbellini E. Psychological evaluation test for infertile couples. J Assist Reprod Genet 2002;19:269–73.

[168] Hjelmstedt A1, Andersson L, Skoog-Svanberg A, Bergh T, Boivin J, Collins A. Gender differences in psychological reactions to infertility among couples seeking IVF- and ICSI-treatment. Acta Obstet Gynecol Scand 1999;78:42–8.

[169] Lechner L, Bolman C, van Dalen A. Definite involuntary childlessness: associations between coping, social support and psychological distress. Hum Reprod 2007;22:288–94.

[170] Daniluk JC. Reconstructing their lives: a longitudinal, qualitative analysis of the transition to biological childlessness for infertile couples. J Couns Dev 2001;79:439–49.

2

Paternal Smoking as a Cause for Transgenerational Damage in the Offspring

Diana Anderson, PhD, Thomas Ernst Schmid, PhD**, Adolf Baumgartner, PhD**

*University of Bradford, Medical Sciences, Bradford, West Yorkshire, UK
**Klinikum rechts der Isar, Technische Universität München, Department of Radiooncology, Germany

INTRODUCTION

According to the World Health Organization (WHO) the tobacco epidemic is the biggest public health threat that the world has ever faced with more than one billion smokers worldwide. Every year, smoking kills nearly 6 million people with 10% being deaths of nonsmokers who were exposed to secondhand smoke [1]. This review will focus on a possibly even more serious threat to the next generation as recent scientific findings suggest that smoking fathers are able to transmit transgenerational damage caused by tobacco smoke to their offspring.

Three critical windows of exposure have been proposed when the offspring is being predominantly susceptible [2]: (1) preconceptional exposure of maternal and paternal gametes, (2) prenatal *in utero* exposure of the fetus, and (3) early postnatal exposures of the newborn, for example, through environment/lactation [3–5]. Post-conception exposures to the embryo and newborn have been and are progressively being investigated focusing on environmental and lifestyle contaminants as well as demographic factors in various mother–childbirth cohorts. Such exposures frequently trigger adaptive responses of the fetus resulting in genetic and/or epigenetic changes, hence subsequently leading to transformed physiological conditions [6,7]. DNA damage, impaired DNA repair, and mutations following DNA damage due to exposures to medical, environmental, occupational, and lifestyle genotoxins are a major threat to the integrity of the genome. As a consequence, these factors can result in genetic and epigenetic changes in the unexposed offspring via exposures of male or female germ cells before conception [8–11].

The role of preconceptional exposures in the induction of transgenerational changes via stable chromosomal alterations or epigenetic modifications has been largely neglected in population studies despite a significant theoretical basis corroborated by data from animal models [12–18]. Compared to animal studies, the evaluation of preconceptional exposures for the F1 offspring in human populations is highly challenging and can be merely achieved by reconstruction using detailed questionnaires. While laboratory experiments on animals in a controlled environment allow for relatively clear analyses of transgenerational effects in the F1 and further generations, it will always be difficult in the human population since different exposures occur at the same time in unknown concentrations, combinations, and periods of time – in addition to variations of demographic factors such as gender, age, race, and lifestyle [16,19]. Hence, transgenerational effects have so far only been demonstrated in animal models; however, direct evidence for the existence of transgenerational DNA damage in humans due to the induction of DNA damage in sperm has recently been confirmed for smoking fathers [20].

Human germ cell toxicants or mutagens inducing transgenerational alterations in gene expression, DNA damage, or mutations in the unexposed offspring have not been reliably identified until very recently despite a profound knowledge of the mechanisms, functions, and consequences of many human (somatic) toxicants. The reasons for this discrepancy may seem rather simple: first, individuals in human populations do not live in a controlled environment, and second, it is rather difficult to separate preconceptional from gestational exposures as well as to choose experimental approaches to

Handbook of Fertility. http://dx.doi.org/10.1016/B978-0-12-800872-0.00002-0

detect transmissible genetic or epigenetic alterations in humans. Human and animal studies have contributed to our current understanding of the effects transmitted through the male. Studies in the 1980s to 1990s suggested that genetic damage after radiation and chemical exposure might be transmitted to the offspring [21–28]. With the increasing understanding of spermatogenesis and sperm maturation – this includes histone retention and modification, protamine incorporation into the chromatin, DNA methylation, and the importance of spermatozoal RNA transcripts for the epigenetic state of sperm – heritable studies became viewed in a different light. Molecular approaches have shown that DNA damage can be transmitted to babies from smoking fathers [20], and minisatellite mutations were found in the germ line of exposed fathers postChernobyl radioactive exposure [18,29]. In epidemiological studies, it is possible to clarify if damage is transmitted to the sons after exposure of the fathers. Paternally transmitted damage to the offspring is now recognized as a complex issue with genetic as well as epigenetic components.

TOBACCO SMOKE

The mainstream smoke of modern cigarette smoking as the industrialized form of tobacco use can be defined as an aerosol containing liquid droplets (particulate phase) in a blend of gases and semivolatile substances. The WHO considers tobacco smoking as a pronounced health threat for humans because of the strong biological and toxicological *in vitro* and *in vivo* effects of tobacco smoke [30]. In 2001, Hoffmann et al. identified in cigarette smoke among 4800 mainstream compounds, 69 carcinogens including cocarcinogens and tumor promoters [31]. This is supported by the International Agency for Research on Cancer (IARC) of the WHO, which lists a similar number of carcinogens in tobacco smoke [1]. These compounds are mainly classed as polycyclic aromatic hydrocarbons such as benzo[*a*]pyrene, heterocyclic hydrocarbons, *N*-nitrosamines, aromatic amines, aldehydes, phenolic compounds, and various other organic and inorganic substances. All of them show sufficient carcinogenicity in animals and some are even classed to induce cancer in humans [31–33].

In 2008, 35% of men and 30% of women were smokers of reproductive age in the United States and among this exposed cohort cigarette smoke is significantly correlated with adverse reproductive outcomes, for example, spontaneous abortion, birth defects, and severe genetic diseases [34,35]. The IARC also correlated preconceptional maternal and paternal tobacco smoke exposure with an increased risk for developing childhood cancer in the unexposed offspring [36] and studies like the European Union Framework Program "Newborns

and Genotoxic Exposure Risk" (NewGeneris) examined the possible roles of periconceptional and gestational exposures to the offspring including lifestyle toxins such as smoking [7]. Exposing germ cells to cigarette smoke leads to an increase in oxidative damage, DNA strand breaks, DNA adducts, chromosomal aberrations, and consequently to mutations [37,38]; hence, due to evidence from demographic and rodent studies, cigarette smoke has been recently considered as the first identified germ cell mutagen [16,39,40].

Nicotine, a strong parasympathomimetic alkaloid found in tobacco and tobacco smoke, has also been linked to human carcinogenesis and teratogenicity. Growing human fetal cells *in vitro* in the presence of nicotine lead to significant direct genotoxic effects, in particular to a significant increase in numerical abnormalities, which might give rise to aneuploidies in the offspring *in vivo* [41,42]. Robbins and colleagues also linked lifestyle exposures, such as cigarette smoke, to male reproductive health. In addition to a negatively affected semen quality, the frequency of disomic sperm had significantly increased in sperm of smokers when compared to those of nonsmokers. Such results already caused concerns in view of induced damage that could be transmitted to the offspring causing transgenerational effects [43]. However, despite the known detrimental effects of cigarette smoke on spermatozoa when analyzing the effects of male and female cigarette smoking on the outcomes of *in vitro* fertilization (IVF), such as semen parameters, oocyte quality, fertilization rate, transfer day embryo scores, and pregnancy, men did not have a significant negative effect on outcomes of IVF whether their partners were smokers or nonsmokers [44].

For cigarette smoke exposure, the genotoxin and human carcinogen benzo[*a*]pyrene is considered an important biomarker as transcriptome analysis revealed that gene expression patterns are significantly altered in human sperm [45]. Cigarette smoke and many other lifestyle and environmental toxins can affect the function of microRNAs (miRNAs) as negative regulators leading to posttranscriptional silencing or suppression of target gene expression [46,47]. In an *in vitro* human model using HepG2 cells [48], the response to the exposure to benzo[*a*]pyrene was able to induce the expression of several miRNAs and alter mRNA levels; hence, these mRNAs themselves were targets of altered miRNAs. The potential of miRNAs, such as miR-29b, miR-26a-1, and miR-122, as novel players in the benzo[*a*] pyrene response was identified to be associated with genotoxicity, DNA damage response, DNA repair, cell cycle arrest, and apoptotic signaling. In adult rats, an increase of miR-29b is generally linked to a reduction in DNA methyl transferase expression and a decrease of anti-apoptotic protein Mcl-1 resulting in increased adult germ cell apoptosis leading to epigenetic changes and a higher risk of infertility [49]. Oxidation as well as inflammation reactions may also play a crucial role in miRNA alterations [46]. Not only toxins, such as benzo[*a*]pyrene, found in cigarette smoke show the potential to have a significant male-mediated impact on the offspring, but also factors like paternal food deprivation can cause altered physiological parameters in male mice impairing spermatogenesis [50]. Subsequently, this can lead to a decrease of the average serum glucose, corticosterone, and insulin-like growth factor 1 levels in the offspring demonstrating a transgenerational male-mediated effect regarding parameters of metabolism and growth in the filial generation [51].

There is also evidence that paternal smoking is associated with adverse reproductive outcome [35,52]. Additionally, according to the IARC there is sufficient proof to link both preconceptional paternal and maternal as well as gestational tobacco smoke consumption with an increased risk for childhood cancer in the unexposed offspring [36] due to unrepaired damage in critical genes such as tumor suppressors [4,53].

GENETIC, EPIGENETIC, AND OTHER TRANSMISSIBLE FACTORS

The human diploid genome is continuously changing, which gives rise to an average of 70 *de novo* mutations per generation, being mostly paternal in origin [54]. An increased number of mutations are conveyed mainly by the father rather than the mother to their common children. Generally, the father's age is the main factor for the number of new mutations in the offspring. This age effect contributes two additional mutations per year or a doubling of the paternal mutation rate every 16.5 years [55]. Hence, it is essential for any transgenerational impact study not to disregard the father's age and to treat the paternal mutation rate as a time-dependent variable. Detrimental environmental impacts by genotoxins found in cigarette smoke may then even increase the rate of new paternal mutations, the expression patterns of regulatory small noncoding RNAs, as well as epigenetic modifications.

The implications of transgenerational destabilization of the filial genome will crucially influence genetic risk estimates and the understanding of childhood cancer etiology. The analysis of nonexposed F1 offspring showed significantly increased transgenerational changes in mutation rates due to mutations and DNA damage introduced by sperm from irradiated male mice [56]. It is well known that in mice cigarette smoke is considered a germ cell mutagen causing permanent, irreversible changes in genetic composition and can persist in future generations [57]; however, also paternal exposure to second-hand smoke only may have reproductive consequences

as it induces tandem repeat mutations in sperm under conditions that may not result in somatic genetic damage [39].

Paternal smoking is known to induce oxidative DNA damage, strand breaks, chromosomal aberrations, mutations, and alteration in gene expression. The baseline frequency of DNA strand breaks in spermatozoal DNA increases by approximately 10% in smokers [20,37,38,45,58,59]. In mice, several mechanistic studies using the polycyclic aromatic hydrocarbon benzo[*a*]pyrene, a major component of cigarette smoke, showed that the induced DNA damage in spermatogonial stem cells was repaired, whereas DNA damage induced in spermiogenesis has not been repaired. Benzo[*a*]pyrene-7,8-diol-9,10-epoxide adducts were found instead, which may then be transmitted with the sperm to the offspring [15,60,61]. In humans, smokers show a significant increase in benzo[*a*]pyrene DNA adducts in sperm as well as increased DNA damage in mature spermatozoa due to exposure to oxidative stress [58,59]. Benzo[*a*]pyrene has been found to be able to cross the blood–testis barrier inducing sperm DNA damage, which is then transmitted to the genome of the unborn offspring [59,62]. Unrepaired sperm DNA damage, that is, DNA adducts, originating from paternal exposure to cigarette smoke relatively close to conception could contribute to the genetic instability in the offspring [20].

Despite being largely transcriptionally inactive due to chromatin remodeling, mRNA profiles from ejaculated sperm of cigarette smokers revealed increased expression of the germ cell specific gene protamine 2 (PRM2), 5.4-fold upregulation of the germ cell specific transcription factor SALF, and a 7.4-fold downregulation of the zinc finger encoding gene TRIM26 [63]. Subsequent analysis of transcription factor networks suggested an inhibition of apoptosis in smokers [45]. An unbalanced PRM2/PRM1 ratio and altered levels of protamine may lead to infertility and increased DNA damage [64]. Results from animal models suggest that fertilization of DNA-fragmented spermatozoa using intracytoplasmic sperm injection (ICSI) leads to immediate adverse effects regarding gene transcription and methylation as well as long-term pathologies in the embryo, fetus, and the offspring [65]. These findings demonstrate that sperm that might potentially fertilize an oocyte can carry damaged DNA as well as altered mRNA profiles into the next generation. Maternal smoking or environmental cigarette smoke exposure additionally contributes to the paternal impact. Active maternal smoking during gestation was identified as a well-documented cause of lower birth weight and congenital malformations as well as fetal mortality and morbidity [35,66].

Although there are multiple routes through which parents can influence their offspring, recent studies of environmentally induced epigenetic variation have highlighted the role of nongenomic pathways [67]. Epigenetics is the study of heritable changes in gene expression that are predominantly caused by modifications of the nuclear chromatin, rather than by changes in the underlying DNA sequence [68]. Histone modifications are the other major epigenetic modification, which consist in reversible posttranslational modifications of the residues at N-terminal tails of histones. Histone modifications include acetylation, methylation, ubiquitylation, and phosphorylation. For example, acetylation of the K9 and K14 lysines of the tail of histone H3 by histone acetyltransferases is highly correlated with transcriptional competence. These two main epigenetic mechanisms, DNA methylation and histone modifications, are known to have direct effects on controlling gene expression.

In order for environmentally induced epigenetic modifications to be transmitted to the offspring, they have to escape two major phases of DNA epigenetic reprogramming when the epigenome (i.e., genome-wide methylation patterns) are subjected to extensive demethylation and remethylation [67]. During embryogenesis, DNA methylation is a highly dynamic trait [69]. Immediately after fertilization, the zygote genome undergoes global depletion of DNA methylation [70]. Postimplantation, the DNA methylation patterns are reestablished again and are similar to those found in adult somatic cells [71]. Some classes of genes within the male germ line can retain their altered methylation states across multiple generations despite these waves of epigenetic reprogramming during development. The best example is genes that are imprinted, whereby one copy from the mother or father is highly methylated and silenced. Retrotransposable elements and imprinted genes appear to be sensitive to environmental exposures and capable of retaining epigenetic marks [72].

Smoking is an exposure strongly associated with DNA methylation in a distinct set of loci, which not only clearly distinguish between current and never smokers, but may also reflect the cumulative amount smoked, and time since quitting in former smokers [73]. A human epidemiological study indicated that early paternal smoking is associated with greater body mass index in male, but not female, offspring [74]. Smoking is one of the most powerful lifestyle modifiers of DNA methylation [75]. It is known that subsets of cytosine methylation and chromatin structure patterns in sperm are transmissible throughout subsequent generations [11], suggesting that adverse DNA methylation can influence the regulation of genes responsible for maintaining genome stability.

The exposure of male mice to environmental toxins like cigarette smoke can induce changes in the epigenome of germ cells. Male mice that were exposed to steel plant air (comprised of various pollutants) have hyper-methylated DNA in sperm compared to control animals even following removal from the exposure [76].

DNA damage and deficiencies in DNA repair are a major threat to the integrity of the genome and result in epigenetic changes in the unexposed offspring via exposures of male or female germ cells before conception [11]. These effects presumably arise as a result of differential methylation of imprinted and imprinted-like genes in the male germ line [77].

In humans, the most widely studied epigenetic modification is the methylation of cytosine residues at the carbon 5 position (5mC) within the CpG dinucleotides mediated by a family of DNA methyltransferases (DNMTs). Cigarette smoke may modulate it through DNA damage and subsequent recruitment of DNMTs. DNMT1 is the most abundant DNMT and is considered to be the key maintenance methyltransferase in mammals. Carcinogens in cigarette smoke can damage DNA by causing double-stranded breaks, as shown in mouse embryonic stem cells exposed to cigarette-smoke condensate [78]. The DNA repair recruits DNMT1, which methylates CpGs adjacent to the repaired nucleotides [79]. Another study showed that nicotine downregulates DNMT1 mRNA in mouse brain neurons [80]. It is also possible that smoking also changes DNA methylation patterns via hypoxia, because carbon monoxide can bind to hemoglobin [81]. The hypoxia-inducible factor (HIF)-1α, which has a major role in cell adaptation to hypoxia, is mainly regulated at posttranslational levels. This leads to an upregulation of methionine adenosyltransferase 2A, which is important for the DNA methylation processes [82].

Epidemiological and laboratory studies suggest that paternal toxicological exposures like cigarette smoke may lead to variations in offspring development. However, it is very important to consider the interplay between maternal and paternal influences in future studies when exploring the role of epigenetic variation within the germ line as a mediator of these effects.

It has been shown that functional information can also be transmitted by spermatozoal RNA to the offspring; during fertilization, not only the sperm's DNA content but also the remaining spermatozoal RNA content are transferred into the fertilized egg consequently modulating gene expression and epigenetic programming in the embryo [83–85]. Paternal smoking on the other hand significantly affects the spermatozoal mRNA content [45,58] altering gene expression patterns during the very early stages of cell division (up to the 8-cell stage) of mouse embryos from oocytes fertilized *in vitro* with a spermatozoon that was derived from mice treated intraperitoneally with glycidamide [86] or with benzo[*a*]pyrene [87], a product metabolite and a product, respectively, of cigarette smoke. Thus, it is very plausible that spermatozoal RNA content transferred to the fertilized egg is affected by paternal smoking consequently influencing the zygote in humans as well. However, further studies are needed to confirm this hypothesis that paternal RNAs affected by smoking can play a crucial role in modulating genetics and epigenetics in the offspring.

Besides paternal genetic and epigenetic factors that contribute to transgenerational DNA damage and in turn compromise the genome integrity of the progeny, there are other not yet well understood paternal factors, which will additionally increase the risk of serious disease in the offspring. Bromfield et al. published in 2014 results on mice revealing that paternal seminal fluid plays an important role in conception and the disease state of the offspring by affecting gene expression in the female reproductive tract, in particular, the embryotrophic factors Lif, Csf2, Il6, and Egf as well as the oviductal apoptosis-inducing factor Trail [88]. Hence, paternal seminal fluid composition affects the offspring's health, most prominently the health of males, either by influencing sperm integrity or by altering embryo-specific signals in the female reproductive tract, and reveals that an optimal periconceptional environment is crucial for embryonic health. A suboptimal environment during conception could lead to metabolic diseases such as hypertension or adiposity [88].

QUALITY CONTROL OF PATERNAL GERM CELLS, POST-FERTILIZATION REPAIR OF SPERM DNA DAMAGE

During spermatogenesis the number and quality of the germ cells are tightly controlled and regulated to ensure the production of undamaged sperm. Apoptosis plays an essential role in this process of quality control defining individual fertility as well as reproductive success [89]. Recently, it was found that the extracellular matrix (ECM) metalloproteinase inducer basigin or CD147 is expressed in germ cells of different developmental testicular stages while migrating from the basal toward the adluminal compartment. Hence, knock-out mice deficient of CD147 are infertile, demonstrating its crucial role in spermatogenesis where CD147 seems to control spermatocyte-specific survival/apoptosis [90]. Studying the fertilizing ability of DNA-damaged spermatozoa in animals, Ahmadi and Ng found that their fertilizing capacity was unaffected – regardless of the degree of damage [91]. However, embryonic development significantly declined with increased sperm DNA damage. Conclusively, the zygotes were proficient to repair sperm DNA damage only if it was damaged less than 8%. Beyond this threshold incomplete repair of sperm DNA damage highly likely resulted in retardation of embryonic development and increased early pregnancy loss [91], for example, by blocking blastocyst formation and subsequently inducing apoptosis [92]. Once spermatozoa escape this level of control, parentally transmitted

genetic defects may lead to birth defects, genetic diseases, pregnancy loss, and infant mortality. Such defects include *de novo* mutations, chromosomal structural aberrations, and numerical abnormalities that are predominantly transmitted by sperm. Germ cells during late spermatogenesis are repair-deficient and when exposed to genotoxins they can accumulate DNA damage, which then persists in the fertilizing sperm. Hence, after fertilization the health of the offspring solely hinges on the DNA repair capacity and efficiency of the zygote [93,94]. Nonetheless, in haploid sperm male-driven *de novo* mutations, such as polymorphisms, being transmittable to the next generation may also represent a physiological mechanism contributing to genetic diversity and evolution [95].

CONCLUSIONS

There is no doubt that the WHO is correct to call the tobacco epidemic the biggest public health threat we face today. In 2004, Demarini already called tobacco smoke the most extreme systemic human mutagen and since then it had even been suggested that smoking also has to be considered a human germ cell mutagen [1,16,38]. Now, it has been shown for the first time by Laubenthal et al. that transgenerational DNA alterations also occur in humans and may be converted to permanent mutations in the F1 offspring of fathers who were exposed to cigarette smoke around conception [20]. Even though the cohort Laubenthal et al. used was rather small, the study yielded enough statistical power to draw the conclusion. Maternal cigarette smoke has already been identified as a significant predictor for DNA damage in newborns, but not their mothers, who are known to have lower susceptibilities to toxicants. Furthermore, many mother–child birth cohort studies showed evidence for the impact of maternal cigarette smoking on the health of unborn offspring inducing DNA damage and mutations [5,19,96–98]; however, none of these studies adjusted for a possible paternal contribution, but exclusively focused on the maternal impact [12,20]. Paternal contributions to childhood disease, such as for epigenetic inheritance through the male germ line, is often not considered at all [99]; hence, it is not surprising that an association between smoking by boys in midchildhood and obesity in adolescent boys of the next generation has been found only recently [100]. Thus, it is imperative for future health studies to rethink basic practices in birth cohort studies and also include the father so as not to neglect a possible paternal contribution to transgenerational damage caused by the parental impact – this includes the mother as well as the father. Although it may still take time for the broad public to become fully aware of the scope of transgenerational mutations and their meaning for the filial generation as well as the entire population, these findings will have immediate and dramatic implications for public health decision makers. The notion of transgenerational consequences needs to be put forward in the public's mind in order to protect against any toxicant exposure, but specifically from smoking before and around conception.

References

[1] WHO-FS339. WHO Fact Sheet 339 - Tobacco, http://www.who.int/mediacentre/factsheets/fs339/en/; 2013 [accessed 27.05.14].

[2] Selevan SG, Kimmel CA, Mendola P. Identifying critical windows of exposure for children's health. Environ Health Perspect 2000;108(Suppl 3):451–5.

[3] Anderson LM, Diwan BA, Fear NT, Roman E. Critical windows of exposure for children's health: cancer in human epidemiological studies and neoplasms in experimental animal models. Environ Health Perspect 2000;108(Suppl 3):573–94.

[4] Wild CP, Kleinjans J. Children and increased susceptibility to environmental carcinogens: evidence or empathy? Cancer Epidemiol, Biomarkers Prev 2003;12(12):1389–94.

[5] Demarini DM, Preston RJ. Smoking while pregnant: transplacental mutagenesis of the fetus by tobacco smoke. JAMA 2005;293(10):1264–5.

[6] Gluckman PD, Hanson MA, Cooper C, Thornburg KL. Effect of *in utero* and early-life conditions on adult health and disease. N Engl J Med 2008;359(1):61–73.

[7] Merlo DF, Wild CP, Kogevinas M, Kyrtopoulos S, Kleinjans J, NewGeneris C. NewGeneris: a European study on maternal diet during pregnancy and child health. Cancer Epidemiol, Biomarkers Prev 2009;18(1):5–10.

[8] Anway MD, Cupp AS, Uzumcu M, Skinner MK. Epigenetic transgenerational actions of endocrine disruptors and male fertility. Science 2005;308(5727):1466–9.

[9] Dubrova YE, Plumb M, Gutierrez B, Boulton E, Jeffreys AJ. Transgenerational mutation by radiation. Nature 2000;405(6782):37.

[10] Ng SF, Lin RC, Laybutt DR, Barres R, Owens JA, Morris MJ. Chronic high-fat diet in fathers programs beta-cell dysfunction in female rat offspring. Nature 2010;467(7318):963–6.

[11] Youngson NA, Whitelaw E. Transgenerational epigenetic effects. Annu Rev Genomic Hum Genet 2008;9:233–57.

[12] Anderson D. Male-mediated developmental toxicity. Toxicol Appl Pharmacol 2005;207(2 Suppl):506–13.

[13] Carone BR, Fauquier L, Habib N, Shea JM, Hart CE, Li C, et al. Paternally induced transgenerational environmental reprogramming of metabolic gene expression in mammals. Cell 2010;143(7):1084–96.

[14] Ng HK, Novakovic B, Hiendleder S, Craig JM, Roberts CT, Saffery R. Distinct patterns of gene-specific methylation in mammalian placentas: implications for placental evolution and function. Placenta 2010;31(4):259–68.

[15] Olsen AK, Andreassen A, Singh R, Wiger R, Duale N, Farmer PB, et al. Environmental exposure of the mouse germ line: DNA adducts in spermatozoa and formation of *de novo* mutations during spermatogenesis. PloS One 2010;5(6):e11349.

[16] Singer TM, Yauk CL. Germ cell mutagens: risk assessment challenges in the 21st century. Environ Mol Mutagen 2010;51(8–9):919–28.

[17] Glen CD, Smith AG, Dubrova YE. Single-molecule PCR analysis of germ line mutation induction by anticancer drugs in mice. Cancer Res 2008;68(10):3630–6.

[18] Dubrova YE. Radiation-induced transgenerational instability. Oncogene 2003;22(45):7087–93.

[19] Bennett LM, Wang Y, Ramsey MJ, Harger GF, Bigbee WL, Tucker JD. Cigarette smoking during pregnancy: chromosome translocations and phenotypic susceptibility in mothers and newborns. Mutat Res 2010;696(1):81–8.

[20] Laubenthal J, Zlobinskaya O, Poterlowicz K, Baumgartner A, Gdula MR, Fthenou E, et al. Cigarette smoke-induced transgenerational alterations in genome stability in cord blood of human F1 offspring. FASEB J 2012;26(10):3946–56.

[21] Anderson D, Edwards AJ, Brinkworth MH, Hughes JA. Male-mediated F1 effects in mice exposed to 1,3-butadiene. Toxicology 1996;113(1–3):120–7.

[22] Anderson D, Hughes JA, Edwards AJ, Brinkworth MH. A comparison of male-mediated effects in rats and mice exposed to 1,3-butadiene. Mutat Res 1998;397(1):77–84.

[23] Brinkworth MH, Anderson D, Hughes JA, Jackson LI, Yu TW, Nieschlag E. Genetic effects of 1,3-butadiene on the mouse testis. Mutat Res 1998;397(1):67–75.

[24] Edwards AJ, Anderson D, Brinkworth MH, Myers B, Parry JM. An investigation of male-mediated F1 effects in mice treated acutely and sub-chronically with urethane. Teratog, Carcinog, Mutagen 1999;19(2):87–103.

[25] Francis AJ, Anderson D, Evans JG, Jenkinson PC, Godbert P. Tumours and malformations in the adult offspring of cyclophosphamide-treated and control male rats – preliminary communication. Mutat Res 1990;229(2):239–46.

[26] Jenkinson PC, Anderson D. Malformed foetuses and karyotype abnormalities in the offspring of cyclophosphamide and allyl alcohol-treated male rats. Mutat Res 1990;229(2):173–84.

[27] Jenkinson PC, Anderson D, Gangolli SD. Increased incidence of abnormal foetuses in the offspring of cyclophosphamide-treated male mice. Mutat Res 1987;188(1):57–62.

[28] Trasler JM, Hales BF, Robaire B. Paternal cyclophosphamide treatment of rats causes fetal loss and malformations without affecting male fertility. Nature 1985;316(6024):144–6.

[29] Dubrova YE, Nesterov VN, Krouchinsky NG, Ostapenko VA, Neumann R, Neil DL, et al. Human minisatellite mutation rate after the Chernobyl accident. Nature 1996;380(6576):683–6.

[30] Thielen A, Klus H, Muller L. Tobacco smoke: unraveling a controversial subject. Exp Toxicol Pathol 2008;60(2–3):141–56.

[31] Hoffmann D, Hoffmann I, El-Bayoumy K. The less harmful cigarette: a controversial issue. A tribute to Ernst L. Wynder. Chem Res Toxicol 2001;14(7):767–90.

[32] Dempsey R, Coggins CR, Roemer E. Toxicological assessment of cigarette ingredients. Regul Toxicol Pharmacol 2011;61(1):119–28.

[33] IARC. Tobacco smoke and involuntary smoking. Lyon, France: International Agency for Research on Cancer; 2004.

[34] PCASRM. Practice Committee of American Society for Reproductive Medicine – smoking and infertility. Fertil Steril 2008;90(5 Suppl):S254–9.

[35] Soares SR, Melo MA. Cigarette smoking and reproductive function. Curr Opin Obstet Gynecol 2008;20(3):281–91.

[36] Secretan B, Straif K, Baan R, Grosse Y, El Ghissassi F, Bouvard V, et al. A review of human carcinogens – Part E: tobacco, areca nut, alcohol, coal smoke, and salted fish. Lancet Oncol 2009;10(11):1033–4.

[37] Aitken RJ, Koopman P, Lewis SE. Seeds of concern. Nature 2004;432(7013):48–52.

[38] Demarini DM. Genotoxicity of tobacco smoke and tobacco smoke condensate: a review. Mutat Res 2004;567(2–3):447–74.

[39] Marchetti F, Rowan-Carroll A, Williams A, Polyzos A, Berndt-Weis ML, Yauk CL. Sidestream tobacco smoke is a male germ cell mutagen. Proc Natl Acad Sci U S A 2011;108(31):12811–4.

[40] Demarini DM. Declaring the existence of human germ-cell mutagens. Environ Mol Mutagen 2012;53(3):166–72.

[41] Demirhan O, Demir C, Tunc E, Inandıklıoğlu N, Sütcü E, Sadıkoğlu N, et al. The genotoxic effect of nicotine on chromosomes of human fetal cells: the first report described as an important study. Inhal Toxicol 2011;23(13):829–34.

[42] Davis DL, Friedler G, Mattison D, Morris R. Male-mediated teratogenesis and other reproductive effects: biologic and epidemiologic findings and a plea for clinical research. Reprod Toxicol 1992;6(4):289–92.

[43] Robbins WA, Elashoff DA, Xun L, Jia J, Li N, Wu G, et al. Effect of lifestyle exposures on sperm aneuploidy. Cytogenet Genome Res 2005;111(3–4):371–7.

[44] Cinar O, Dilbaz S, Terzioglu F, Karahalil B, Yücel C, Turk R, et al. Does cigarette smoking really have detrimental effects on outcomes of IVF? Eur J Obstet Gynecol Reprod Biol 2014;174:106–10.

[45] Linschooten JO, Van Schooten FJ, Baumgartner A, Cemeli E, Van Delft J, Anderson D, et al. Use of spermatozoal mRNA profiles to study gene-environment interactions in human germ cells. Mutat Res 2009;667(1–2):70–6.

[46] Hou L, Wang D, Baccarelli A. Environmental chemicals and microRNAs. Mutat Res 2011;714(1–2):105–12.

[47] Sonkoly E, Pivarcsi A. MicroRNAs in inflammation and response to injuries induced by environmental pollution. Mutat Res 2011;717(1–2):46–53.

[48] Lizarraga D, Gaj S, Brauers KJ, Timmermans L, Kleinjans JC, van Delft JH. Benzo[a]pyrene-induced changes in microRNA-mRNA networks. Chem Res Toxicol 2012;25(4):838–49.

[49] Meunier L, Siddeek B, Vega A, Lakhdari N, Inoubli L, Bellon RP, et al. Perinatal programming of adult rat germ cell death after exposure to xenoestrogens: role of microRNA miR-29 family in the down-regulation of DNA methyltransferases and Mcl-1. Endocrinology 2012;153(4):1936–47.

[50] Brinkworth MH, Anderson D, McLean AE. Effects of dietary imbalances on spermatogenesis in CD-1 mice and CD rats. Food Chem Toxicol 1992;30(1):29–35.

[51] Anderson LM, Riffle L, Wilson R, Travlos GS, Lubomirski MS, Alvord WG. Preconceptional fasting of fathers alters serum glucose in offspring of mice. Nutrition 2006;22(3):327–31.

[52] Venners SA, Wang X, Chen C, Wang L, Chen D, Guang W, et al. Paternal smoking and pregnancy loss: a prospective study using a biomarker of pregnancy. Am J Epidemiol 2004;159(10):993–1001.

[53] Hanahan D, Weinberg RA. Hallmarks of cancer: the next generation. Cell 2011;144(5):646–74.

[54] Keightley PD. Rates and fitness consequences of new mutations in humans. Genetics 2012;190(2):295–304.

[55] Kong A, Frigge ML, Masson G, Besenbacher S, Sulem P, Magnusson G, et al. Rate of de novo mutations and the importance of father's age to disease risk. Nature 2012;488(7412):471–5.

[56] Barber RC, Hickenbotham P, Hatch T, Kelly D, Topchiy N, Almeida GM, et al. Radiation-induced transgenerational alterations in genome stability and DNA damage. Oncogene 2006;25(56):7336–42.

[57] Yauk CL, Berndt ML, Williams A, Rowan-Carroll A, Douglas GR, Stampfli MR. Mainstream tobacco smoke causes paternal germline DNA mutation. Cancer Res 2007;67(11):5103–6.

[58] Linschooten JO, Laubenthal J, Cemeli E, Baumgartner A, Anderson D, Sipinen VE, et al. Incomplete protection of genetic integrity of mature spermatozoa against oxidative stress. Reprod Toxicol 2011;32(1):106–11.

[59] Sipinen V, Laubenthal J, Baumgartner A, Cemeli E, Linschooten JO, Godschalk RW, et al. In vitro evaluation of baseline and induced DNA damage in human sperm exposed to benzo[a]pyrene or its metabolite benzo[a]pyrene-7,8-diol-9,10-epoxide, using the comet assay. Mutagenesis 2010;25(4):417–25.

[60] Verhofstad N, van Oostrom CT, van Benthem J, van Schooten FJ, van Steeg H, Godschalk RW. DNA adduct kinetics in reproductive tissues of DNA repair proficient and deficient male mice after oral exposure to benzo(a)pyrene. Environ Mol Mutagen 2010;51(2):123–9.

[61] Verhofstad N, van Oostrom CT, Zwart E, Maas LM, van Benthem J, van Schooten FJ, et al. Evaluation of benzo(a)pyrene-induced gene mutations in male germ cells. Toxicol Sci 2011;119(1):218–23.

[62] Zenzes MT, Puy LA, Bielecki R, Reed TE. Detection of benzo[a]pyrene diol epoxide-DNA adducts in embryos from smoking couples: evidence for transmission by spermatozoa. Mol Hum Reprod 1999;5(2):125–31.

[63] Linschooten JO, Verhofstad N, Gutzkow K, Olsen AK, Yauk C, Oligschläger Y, et al. Paternal lifestyle as a potential source of germline mutations transmitted to offspring. FASEB J 2013;27(7):2873–9.

[64] Oliva R. Protamines and male infertility. Hum Reprod Update 2006;12(4):417–35.

[65] Fernandez-Gonzalez R, Moreira PN, Perez-Crespo M, Sánchez-Martín M, Ramirez MA, Pericuesta E, et al. Long-term effects of mouse intracytoplasmic sperm injection with DNA-fragmented sperm on health and behavior of adult offspring. Biol Reprod 2008;78(4):761–72.

[66] Leonardi-Bee J, Britton J, Venn A. Secondhand smoke and adverse fetal outcomes in nonsmoking pregnant women: a meta-analysis. Pediatrics 2011;127(4):734–41.

[67] Curley JP, Jensen CL, Franks B, Champagne FA. Variation in maternal and anxiety-like behavior associated with discrete patterns of oxytocin and vasopressin 1a receptor density in the lateral septum. Horm Behav 2012;61(3):454–61.

[68] Schagdarsurengin U, Paradowska A, Steger K. Analysing the sperm epigenome: roles in early embryogenesis and assisted reproduction. Nat Rev Urol 2012;9(11):609–19.

[69] Talens RP, Boomsma DI, Tobi EW, Kremer D, Jukema JW, Willemsen G, et al. Variation, patterns, and temporal stability of DNA methylation: considerations for epigenetic epidemiology. FASEB J 2010;24(9):3135–44.

[70] Lee KW, Pausova Z. Cigarette smoking and DNA methylation. Front Genet 2013;4:132.

[71] Smith ZD, Meissner A. DNA methylation: roles in mammalian development. Nat Rev Genet 2013;14(3):204–20.

[72] Lane N, Dean W, Erhardt S, Hajkova P, Surani A, Walter J, et al. Resistance of IAPs to methylation reprogramming may provide a mechanism for epigenetic inheritance in the mouse. Genesis 2003;35(2):88–93.

[73] Elliott HR, Tillin T, McArdle WL, Ho K, Duggirala A, Frayling TM, et al. Differences in smoking associated DNA methylation patterns in South Asians and Europeans. Clin Epigenet 2014;6(1):4.

[74] Pembrey ME, Bygren LO, Kaati G, Edvinsson S, Northstone K, Sjöström M, et al. Sex-specific, male-line transgenerational responses in humans. Eur J Hum Genet 2006;14(2):159–66.

[75] Breitling LP, Yang R, Korn B, Burwinkel B, Brenner H. Tobacco-smoking-related differential DNA methylation: 27K discovery and replication. Am J Hum Genet 2011;88(4):450–7.

[76] Yauk C, Polyzos A, Rowan-Carroll A, Somers CM, Godschalk RW, Van Schooten FJ, et al. Germ-line mutations, DNA damage, and global hypermethylation in mice exposed to particulate air pollution in an urban/industrial location. Proc Natl Acad Sci U S A 2008;105(2):605–10.

[77] Anway MD, Skinner MK. Epigenetic programming of the germ line: effects of endocrine disruptors on the development of transgenerational disease. Reprod BioMed Online 2008;16(1):23–5.

[78] Huang J, Okuka M, Lu W, Tsibris JC, McLean MP, Keefe DL, et al. Telomere shortening and DNA damage of embryonic stem cells induced by cigarette smoke. Reprod Toxicol 2013;35:89–95.

[79] Cuozzo C, Porcellini A, Angrisano T, Morano A, Lee B, Di Pardo A, et al. DNA damage, homology-directed repair, and DNA methylation. PLoS Genet 2007;3(7):e110.

[80] Satta R, Maloku E, Zhubi A, Pibiri F, Hajos M, Costa E, et al. Nicotine decreases DNA methyltransferase 1 expression and glutamic acid decarboxylase 67 promoter methylation in GABAergic interneurons. Proc Natl Acad Sci U S A 2008;105(42):16356–61.

[81] Olson KR. Carbon monoxide poisoning: mechanisms, presentation, and controversies in management. J Emerg Med 1984;1(3):233–43.

[82] Liu Q, Liu L, Zhao Y, Zhang J, Wang D, Chen J, et al. Hypoxia induces genomic DNA demethylation through the activation of HIF-1alpha and transcriptional upregulation of MAT2A in hepatoma cells. Mol Cancer Ther 2011;10(6):1113–23.

[83] Chandler VL, Stam M. Chromatin conversations: mechanisms and implications of paramutation. Nat Rev Genet 2004;5(7):532–44.

[84] Rassoulzadegan M, Grandjean V, Gounon P, Vincent S, Gillot I, Cuzin F. RNA-mediated non-Mendelian inheritance of an epigenetic change in the mouse. Nature 2006;441(7092):469–74.

[85] Bernstein E, Allis CD. RNA meets chromatin. Genes Dev 2005;19(14):1635–55.

[86] Brevik A, Rusnakova V, Duale N, Slagsvold HH, Olsen AK, Storeng R, et al. Preconceptional paternal glycidamide exposure affects embryonic gene expression: single embryo gene expression study following *in vitro* fertilization. Reprod Toxicol 2011;32(4):463–71.

[87] Brevik A, Lindeman B, Brunborg G, Duale N. Paternal benzo[*a*] pyrene exposure modulates microRNA expression patterns in the developing mouse embryo. Int J Cell Biol 2012;2012:407431.

[88] Bromfield JJ, Schjenken JE, Chin PY, Care AS, Jasper MJ, Robertson SA. Maternal tract factors contribute to paternal seminal fluid impact on metabolic phenotype in offspring. Proc Natl Acad Sci U S A 2014;111(6):2200–5.

[89] Aitken RJ, Findlay JK, Hutt KJ, Kerr JB. Apoptosis in the germ line. Reproduction 2011;141(2):139–50.

[90] Chen H, Lam Fok K, Jiang X, Chan HC. New insights into germ cell migration and survival/apoptosis in spermatogenesis: lessons from CD147. Spermatogenesis 2012;2(4):264–72.

[91] Ahmadi A, Ng SC. Fertilizing ability of DNA-damaged spermatozoa. J Exp Zool 1999;284(6):696–704.

[92] Fatehi AN, Bevers MM, Schoevers E, Roelen BA, Colenbrander B, Gadella BM. DNA damage in bovine sperm does not block fertilization and early embryonic development but induces apoptosis after the first cleavages. J Androl 2006;27(2):176–88.

[93] Olsen AK, Lindeman B, Wiger R, Duale N, Brunborg G. How do male germ cells handle DNA damage? Toxicol Appl Pharmacol 2005;207(2 Suppl):521–31.

[94] Marchetti F, Essers J, Kanaar R, Wyrobek AJ. Disruption of maternal DNA repair increases sperm-derived chromosomal aberrations. Proc Natl Acad Sci U S A 2007;104(45):17725–9.

[95] Gregoire MC, Massonneau J, Simard O, Gouraud A, Brazeau MA, Arguin M, et al. Male-driven *de novo* mutations in haploid germ cells. Mol Hum Reprod 2013;19(8):495–9.

[96] Perera FP, Tang D, Tu YH, Cruz LA, Borjas M, Bernert T, et al. Biomarkers in maternal and newborn blood indicate heightened fetal susceptibility to procarcinogenic DNA damage. Environ Health Perspect 2004;112(10):1133–6.

[97] de la Chica RA, Ribas I, Giraldo J, Egozcue J, Fuster C. Chromosomal instability in amniocytes from fetuses of mothers who smoke. JAMA 2005;293(10):1212–22.

[98] Keohavong P, Xi L, Day RD, Zhang L, Grant SG, Day BW, et al. HPRT gene alterations in umbilical cord blood T-lymphocytes in newborns of mothers exposed to tobacco smoke during pregnancy. Mutat Res 2005;572(1-2):156–66.

[99] Soubry A, Hoyo C, Jirtle RL, Murphy SK. A paternal environmental legacy: evidence for epigenetic inheritance through the male germ line. BioEssays 2014;36(4):359–71.

[100] Northstone K, Golding J, Davey Smith G, Miller LL, Pembrey M. Prepubertal start of father's smoking and increased body fat in his sons: further characterisation of paternal transgenerational responses. Eur J Hum Genet 2014;.

Impact of Cigarette Smoking During Pregnancy on Conception and Fetal Health through Serum Folate Levels

Mehmet Yekta Oncel, MD, Omer Erdeve, MD***

*Zekai Tahir Burak Maternity Teaching Hospital, Ankara, Turkey
**Ankara University School of Medicine Children's Hospital, Ankara, Turkey

INTRODUCTION

Folic acid is an essential micronutrient for fetal development and growth. The demand for folic acid is high especially in preterm infants, particularly during the period of rapid growth [1]. Maternal diet deficient in folate results in decreased neurogenesis and increased apoptosis in fetal mouse brain [2]. In humans, maternal dietary supplementation with folic acid in the peri-conceptional period significantly reduces the risk of neural tube defects [3]. Smoking during pregnancy is associated with an increased risk of maternal and fetal complications [4]. The growth retarding effects of prenatal smoke exposure are well documented [5], resulting directly in genetic damage in the fetus [6]. It has been postulated that the formation of carboxyhemoglobin and nicotine-induced vasoconstriction contribute to fetal hypoxia, which has been implicated in the development of fetal complications [7]. Smoking is associated with decreased maternal levels of folate as well as elevated levels of total homocysteine, both of which are in turn associated with adverse pregnancy outcomes [8].

The main point of this review is to discuss the possible effects of maternal smoking during pregnancy on conception and neonatal serum folate levels.

GENERAL KNOWLEDGE ON FOLATE

Folate (both natural food folate and synthetic folic acid) is a 1-carbon source critical for the replication of deoxyribonucleic acid (DNA) and ribonucleic acid (RNA) during the cell division and methylation of DNA, histones, and other proteins [9]. Folate plays an important role in several physiological functions, including normal cell division, especially for cell systems with a high cell division rate (e.g., blood cells and mucosa of the gut). Clinical folate deficiency is associated with weight loss and slow growth in children, and megaloblastic anemia in severe cases [10,11].

The consumption of folate must be regular, as human beings are unable to synthesize folate. Green vegetables provide one of the most important sources of folate. Spinach, salad, asparagus, tomatoes, cucumbers, whole-grain products, liver, and some tropical fruits are also rich in folate whereas only trace amounts are found in meat and fish. Naturally occurring food folate is rather unstable under exposure to oxygen, light, and higher temperatures. Peas and spinach, for example, stored for 5 days, lose around 50% of their folate content. Also, substantial losses of folate can occur due to food processing or cooking [12].

Several factors such as diminished hepatic stores, rapid growth, increased erythropoiesis, use of antibiotics, use of anticonvulsants, and the potential for malabsorption put the newborns at risk for developing folate deficiency [13]. In humans, maternal dietary supplementation with folic acid in the peri-conceptional period significantly reduces the risk of neural tube defects. Therefore, it is recommended that women should take a supplement of 400 μg/day from at least 4 weeks prior to conception up to at least 8–12 weeks of pregnancy [3,14].

Handbook of Fertility. http://dx.doi.org/10.1016/B978-0-12-800872-0.00003-2

SMOKING AND PREGNANCY

Cigarette smoke contains over 1000 different compounds including carbon monoxide, hydrogen cyanide, carcinogens, and trace elements such as lead, nickel, and cadmium. The two main compounds suspected of causing harmful effects on the developing fetus during pregnancy are carbon monoxide and nicotine. Carbon monoxide has a higher affinity for hemoglobin than oxygen, quickly forming the compound carboxyhemoglobin, which is unable to carry oxygen [15]. The formation of this molecule leads to a potential for decreased oxygen delivery to the fetus and fetal hypoxia. Nicotine is generally regarded as the pharmacologically active ingredient in tobacco responsible for the majority of its effects. It has both cardiovascular and central nervous system effects. Nicotine activates the adrenergic system through the release of catecholamines from the adrenal medulla, autonomic ganglia, and neuromuscular junctions [15,16]. The effects of cigarettes on pregnant women and developing fetuses are numerous with a wide range of sequelae that will remain with the fetuses for the rest of their lives [17].

Maternal smoking is a risk factor for adverse outcomes for the fetus and infants. Smoking and high homocysteine concentrations are individually associated with venous thromboembolism, miscarriage, stillbirth, abruptio placentae, decreased fetal growth, and fetal anomalies [17–20]. Smoking is reported to increase homocysteine concentrations in the nonpregnant state and may do so during pregnancy [21,22]. If it does, it may be one of the pathologic mechanisms by which smoking leads to poor obstetric outcomes. It is therefore important for the health of women who smoke and for their developing fetuses to establish whether homocysteine concentrations increase during pregnancy and whether folate concentrations decrease. These may possibly represent a modifiable risk factor in pregnant women who smoke, through increased folate supplementation. Potentially harmful effects of smoking during pregnancy arising in the perinatal period are summarized in Table 3.1.

Maternal smoking doubles the risk of delivering a low birth weight infant (<2500 g), and significantly affects fetal growth and birth weight more than other factors such as the mother's weight, height, number of previous pregnancies and their outcomes, or the gender of the infant [17]. The growth retarding effects of prenatal smoke exposure are well documented, resulting directly in genetic damage in the fetus [23,24]. However, the exact mechanisms by which fetal complications develop remain largely unknown. It is postulated that the formation of carboxyhemoglobin and nicotine-induced vasoconstriction contribute to fetal hypoxia, which is implicated in the development of fetal complications [25].

Placental abruption, the premature detachment of the normally implanted placenta, accounts for 15–25% of all

TABLE 3.1   Potentially Harmful Effects of Smoking During Pregnancy Arising in the Perinatal Period

1. Delivery complications
   a. Placenta previa
   b. Abruptio placenta
   c. Premature rupture of membranes
   d. Preterm labor

2. Issues about the baby
   a. Low birth weight
   b. Intrauterine growth retardation
   c. Low Apgar scores
   d. Polycythemia

3. Teratogenesis
   a. Orofacial anomalies
   b. Club foot
   c. Optic nerve hypoplasia
   d. Omphalocele
   e. Gastroschisis
   f. The number of fingers anomalies
   g. Cryptorchidism
   h. Microcephaly
   i. Hydrocephalus

perinatal mortality due to complications such as preterm delivery, fetal distress, maternal coagulopathy, and ischemic injury to other organs. Placenta previa, the implantation of the placenta in the lower part of the lower uterine segment in advance of the fetal presenting part, can be complicated by prematurity, placenta accreta, vasa previa, and hemorrhage among others [26,27]. There is a 33% increase in perinatal (after 20 weeks gestation) and neonatal (in the first 28 days of age) mortality in smoking women [26]. This increase occurs independently of the decrease in birth weight. While the mean length of gestation is only slightly shorter in pregnant smokers, the proportion of preterm births (less than 37 weeks gestation) increases significantly [26]. Analysis of the Ontario perinatal mortality study showed that maternal smoking increased the perinatal death risk for mothers smoking less than one pack per day by 20% and by 35% for those mothers smoking more than one pack per day. Little further information exists on the other specific causes for increased perinatal mortality [28].

Studies on the relation between smoking during pregnancy and the incidence of congenital malformations are inconclusive. Prior retrospective studies were flawed by small sample size and failure to clearly define specific malformations. Although three larger prospective studies found no increase in the prevalence of congenital malformations, a few malformations were found to be associated with maternal smoking. Data from the Kaiser Permanente Birth Defects Study and the Collaborative Perinatal Project were together used to assess the relationship between maternal smoking and congenital malformations [29,30]. Effects of smoking during pregnancy

TABLE 3.2   Effects of Smoking During Pregnancy that Arise in the Newborn and Nursling Periods

1. Early abandonment of breastfeeding
2. Sudden infant death syndrome
3. Infantile colic
4. Craniosynostosis
5. Retinal vascular abnormalities
6. Carcinogenesis

that arise in the newborn and nursling periods are summarized in Table 3.2.

Several studies suggest that smoking may be associated with decreased fertility among both women and men. In men, smoking can cause decreased sperm motility whereas abstinence from smoking leads to return of motility. In women, numerous studies confirmed a decreased fertility among women who smoke, documenting adverse effects on several crucial processes such as ovulation, tubal transport, and implantation. Nicotine is a potent vasoconstrictor reducing uterine and placental blood flow. These properties may account for the increase in spontaneous abortions observed in smoking women [30,31]. A study reported an association between cocaine and tobacco use and spontaneous abortions among pregnant women [32].

Maternal Smoking and Neonatal Serum Folate Levels

Smoking is associated with decreased levels of folate as well as elevated levels of total homocysteine both of which are in turn associated with adverse pregnancy outcomes [8,33,34]. Smoking increases homocysteine concentrations in nonpregnant women, and similar elevations can be expected to occur during pregnancy [33]. Considering the implications on both maternal and fetal health, the effect of smoking on homocysteine and folate levels, as modifiable factors associated with poor pregnancy outcomes, as well as the potential benefits of folate supplementation warrant further investigation [35].

The decrease in folate concentrations observed in pregnant women who smoke has important clinical implications, as lower levels of folate are believed to be responsible for higher rates of miscarriage, stillbirth, abruptio placentae, and fetal anomalies [35]. Only few studies have looked for a link between smoking and reductions in serum folic acid levels in pregnant women and in infants born to them [35–37].

McDonald et al. [35] showed that pregnant women who smoke had lower folate levels than pregnant women who do not smoke during early gestation. Furthermore, they observed that as MTHFR activity decreased, there was a further decrease in folate levels with tobacco exposure, which illustrates an important gene/environment interaction. Folate concentrations were lower in

the homozygous mutant form (677TT) of MTHFR than in the heterozygous wild-type/mutant (677CT) and homozygous wild-type (677CC) forms. Within each category of MTHFR, pregnant women who smoked had lower folate levels than pregnant women who did not smoke in their study.

The lower folate levels in pregnant women who smoke may be due to increased metabolism or changes in redox status. It has been suggested that low folate levels may increase chromosome fragility and result in rearrangements of certain genes [38]. This can explain, in part, the increase in congenital anomalies seen in settings of low folate (such as neural tube defects) and the decrease in anomalies (such as neural tube defects and congenital cardiac defects) with folate supplementation [39,40]. Many of the poor obstetric outcomes that are associated with increased levels of homocysteine are due to low folate levels because the two are inextricably linked through the homocysteine metabolism pathway [35,37].

Pregnant women who smoke may benefit from higher doses of peri-conceptional folic acid, particularly if they have the homozygous mutant form (677TT) of MTHFR. The increased incidence of adverse obstetric outcomes in pregnant women who smoke may be, in part, due to lower folate concentrations that are mediated possibly by low MTHFR enzymatic activity. From a public health point of view, it may possibly be cost effective to screen women preconceptionally to determine their MTHFR status, particularly if they are pregnant women who smoke. In the nonpregnant population, it has been shown that increased folate supplementation raises the folic acid concentrations of nonpregnant women who smoke [35]. One study noted that, at the current recommended dietary allowance for folate, nonpregnant smokers had 42% lower plasma folate concentrations than nonsmokers; at three times the recommended daily allowance, smokers and nonsmokers had the same plasma folate concentrations [41].

In our previous study on 108 term infants, we found that maternal smoking was associated with significantly lower levels of serum folate in blood samples obtained from cord blood and 1 month after delivery, compared to infants born to nonsmokers ($p < 0.01$). Term babies born to mothers who smoke had significantly lower serum folate levels compared to those born to mothers who do not smoke, both at birth and 1 month after delivery [42]. We observed a similar result in another study, where more preterm infants born to smokers during pregnancy had significantly lower levels of folate in all three samples obtained (on days 14 and 28 postnatally, and at 36 weeks postmenstrual age or just before discharge; samples A, B, and C, respectively) compared to those born to nonsmokers [43]. This may be due to the lost effect of maternal status beyond the first month of life. The higher mean serum folate levels in preterm infants compared to previous studies are related to total parenteral nutrition, fortified human breast milk, and preterm formula, which were not used in past or in term infants.

In a study by Pagan et al. [44] on 196 pregnant women, investigators observed lower levels of folate in smokers compared to nonsmokers both at 18 weeks and at 30 weeks of gestation. They also reported a negative correlation between serum homocysteine and serum folate levels. Our results are consistent with findings in this study with regards to serum folate levels. In another study by Relton et al. [45], smoking was found to be associated with significant reductions in maternal but not neonatal erythrocyte folate levels. Similar reductions in the serum folate levels of pregnant smokers have also been reported in other studies [35,46]. In fact, this effect of smoking has even been documented in normal adult populations as well [47,48].

Reductions in folate levels of smoking pregnant women may be attributed to several factors. Cigarette smoking contributes to oxidative stress and may alter the ability of cells to metabolize and ultimately store folate [49]. Furthermore, women who smoke are less likely to adhere to a diet rich in folate or to show compliance to the use of folic acid supplements [50]. However, in a recent study from Canada, no significant difference in folate intake (including the contribution of folic acid supplements) between smokers and nonsmokers was reported [35].

In a study comparing maternal and neonatal serum folate levels (both at birth and 1 month after delivery) of smoker and nonsmoker mothers, investigators failed to show a statistically significant difference between groups [36]. In another study, while cigarette smoking was found to be associated with lower levels of maternal erythrocyte folate levels, a similar association was not observed with neonatal erythrocyte folate levels [45]. In contrast, Oncel et al. [42] managed to demonstrate significant reductions in serum folate levels of neonates born to smoking mothers, both at birth (cord blood) and 1 month after delivery. Oncel et al. [42] suggest that folic acid supplementation should be discussed for infants born to women who smoked during pregnancy, and/or continued to smoke after delivery. Folic acid supplements are routinely prescribed to expecting mothers during the first trimester. Being a water-soluble vitamin, folic acid supplementation may be continued throughout pregnancy in smoking mothers to prevent the development of fetal and neonatal folate deficiency. However, there is a dire need for more extensive studies to evaluate the short-term and long-term benefits of maternal and neonatal folic acid supplementation.

Neonatal Serum Folate Levels in Preterm Infants

In the 1980s, with the discovery of significantly low levels of serum folate in the first 2–3 months of life in preterm infants, routine folic acid supplementation was recommended to prevent the development of folate deficiency [51–54]. Recent recommendations for folic acid intake suggest 25–50 µg/kg/day in oral intake and 56 µg/kg/day in parenteral nutrition [55], or 35–100 µg/kg/day considering the need in erythropoietin treatment [56]. Human breast milk contains folic acid at a concentration of 15 µg/100 mL, and when mixed with breast milk fortifiers, concentrations may reach up to 65 µg/100 mL. On the other hand, preterm formulas usually contain around 35 µg/100 mL of folic acid [43,55].

However, with the availability of new parenteral nutrition products and preterm formulas containing folic acid, additional folic acid supplementation has become a source of controversy. Micronutrient support of mothers' milk and the development of modern preterm formulas for preterm infants have decreased the need for folic acid supplementation, although the practice of folic acid supplementation remains commonplace. The higher mean serum folate levels in preterm infants in the study by Oncel et al. [43] might be related to total parenteral nutrition, fortified human breast milk, and preterm formulas, which were not used in our previous study with term infants. None of the patients developed folic acid deficiency during the study period, and based on these results it may be concluded that fortified human breast milk and preterm formula provide sufficient folic acid for the increased demand of growing preterm infants.

Oncel et al. [43] demonstrated that there is no risk of folate deficiency in preterm infants in the experimental population with a high rate of folate supplementation during pregnancy and in the experimental nutritional support associating parenteral nutrition providing a high level of folate intake intravenously and the use of supplemented human milk fortifier and preterm formula. In addition, serum folate concentration tends to decrease slowly when exclusive human breast milk with low folate intake was provided to the preterm infants. These data suggest that the half-life of serum folate is high in preterm infants.

Roberts et al. [57] reported higher levels of red cell folate as well as hemoglobin in patients receiving folic acid supplementation, whereas Kendall et al. [58] observed no discernible difference between infants receiving folic acid supplementation and those taking placebo. None of the preterm infants developed folate deficiency despite not receiving any added folic acid supplementation. Jyothi et al. [59] studied red cell folate and plasma homocysteine in preterm infants and reported that folate deficiency did not occur in their series. They also concluded that modern preterm formulas and breast milk fortifiers provide sufficient folic acid required for growing premature infants. Oncel et al. [43] suggested that preterm infants receiving total parenteral nutrition with high folate content had no risk of folate deficiency during the first 2 months of age. However, preterm infants fed orally from birth with human breast milk or preterm formula with a low folic acid content could be at risk of folate deficiency specially when mothers are smokers and/or do not receive folic acid supplementation during pregnancy.

Smoking During Pregnancy May Associate with Asthma or Recurrent Wheezing in Children Through Serum Folate Levels

The currently available evidence regarding an association between folate in pregnancy and childhood asthma or wheeze is conflicting. Robison et al. [60], whose objective was to evaluate the interactive effects of maternal smoking and prematurity upon the development of early childhood wheezing, found that maternal smoking *in utero* and prematurity appear to have a joint effect on recurrent wheezing and an interactive effect on the number of episodes of early childhood wheezing.

Matsui and Matsui [61] found that serum folate levels were inversely associated with total IgE levels, atopy, and wheeze. On the other hand, Magdelijns et al. [62] did not confirm any meaningful association between folic acid supplement use during pregnancy and atopic diseases, but they showed that higher intracellular folic acid levels in pregnancy was associated with a small decreased risk for developing asthma at age 6 to 7 years. In contrast, Bekkers et al. [63] showed that maternal folic acid use was associated with wheeze at 1 year, but not with wheeze at later ages. In conclusion, maternal smoking has an adverse effect on the development of wheeze and asthma in term and preterm infants. Low serum folate levels due to maternal smoking might aggravate this association. Although studies suggest that the infants of unsupplemented mothers with a smoking habit who have lower serum folate level should be followed for recurrent wheezing or asthma closely, last meta-analysis show that analysis of the published literature to date is difficult to summarize because of heterogeneity in folate and folic acid exposure windows during pregnancy, a variety of asthma and allergy-associated outcomes, and the limited number of studies. This meta-analysis does not support the association of folic acid supplement use in the first trimester with risk of asthma in childhood [64]. We think that maternal smoking habit and its relation to serum folate level should be included in such analysis [65]. On the other hand, we believe that further studies investigating the relation between maternal smoking, folic acid supplementation, serum folate levels, and recurrent wheezing or asthma in infants are needed.

CONCLUSION

Folic acid is essential for intrauterine and extrauterine life. Its deficiency may result in undesirable events such as infertility, neural tube defects, impaired neurogenesis, and megaloblastic anemia. Recent developments in follow up of pregnant women and their supplementation has prevented major developmental defects and all attention is focused on preterm infants in the perinatal period. The folic acid supplementation through total parenteral nutrition and new formulas results in higher neonatal serum levels. Studies show that there is no need for folic acid supplementation in preterm infants whose mothers are supplemented and who receive total parenteral nutrition in the neonatal intensive care unit. The major risk group seems to be infants born to smoking mothers, and these babies should be monitored for any deficiency.

References

[1] Dewey KG. Nutrition, growth, and complementary feeding of the breastfed infant. Pediatr Clin North Am 2001;48:87–104.

[2] Craciunescu CN, Brown EC, Mar MH, Albright CD, Nadeau MR, Zeisel SH. Folic acid deficiency during late gestation decreases progenitor cell proliferation and increases apoptosis in fetal mouse brain. J Nutr 2004;134:162–6.

[3] Moyers S, Bailey LB. Fetal malformations and folate metabolism: review of recent evidence. Nutr Rev 2001;59:215–24.

[4] Himmelberger D, Brown BW, Cohen EN. Cigarette smoking during pregnancy and the occurrence of spontaneous abortion and congenital abnormality. Am J Epidemiol 1978;108:470–9.

[5] Naeye RL. Influence of maternal cigarette smoking during pregnancy on fetal and childhood growth. Obstet Gynecol 1981;57: 18–21.

[6] Jalili T, Murthy GG, Schiestl RH. Cigarette smoke induces DNA deletions in the mouse embryo. Cancer Res 1998;58:2633–8.

[7] Haustein KO. Cigarette smoking, nicotine and pregnancy. Int J Clin Pharmacol Ther 1999;37:417–27.

[8] Frenkel EP, Yardley DA. Clinical and laboratory features and sequelae of deficiency of folic acid (folate) and vitamin B12 (cobalamin) in pregnancy and gynecology. Hematol Oncol Clin North Am 2000;14:1079–100.

[9] Rucker RB, Stites T. New perspectives on function of vitamins. Nutrition 1994;10:507–13.

[10] Beitz R, Mensink GB, Fischer B, Thamm M. Vitamins – dietary intake and intake from dietary supplements in Germany. Eur J Clin Nutr 2002;56:539–45.

[11] de Bree A, van Dusseldorp M, Brouwer IA, van het Hof KH, Steegers-Theunissen RP. Folate intake in Europe: recommended, actual and desired intake. Eur J Clin Nutr 1997;51:643–60.

[12] Ros G, Bernal MJ, Olivares AB, Periago MJ, Martínez C. Funcionalidad de los folatos en la dieta. Form Cont Nutr 2002;5:223–37.

[13] Orzalesi M, Colarizi P. Critical vitamins for low birthweight infants. Acta Paediatr Scand Suppl 1982;296:104–9.

[14] Craciunescu CN, Johnson AR, Zeisel SH. Dietary choline reverses some, but not all, effects of folate deficiency on neurogenesis and apoptosis in fetal mouse brain. J Nutr 2010;140:1162–6.

[15] Luck W, Nau H. Nicotine and cotinine concentrations in serum and milk of nursing smokers. Br J Clin Pharmacol 1984;18:9–15.

[16] Koren G. Fetal toxicology of environmental tobacco smoke. Curr Opin Pediatr 1995;7:128–31.

[17] Roquer JM, Figueras J, Botet F, Jiménez R. Influence on fetal growth of exposure to tobacco smoke during pregnancy. Acta Paediatr 1995;84:118–21.

[18] Stewart MJ, Gillis A, Brosky B, Johnston G, Kirkland S, Leigh G, et al. Smoking among disadvantaged women: causes and cessation. Can J Nurs Res 1996;28:41–60.

[19] McBride C, Pirie P, Curry S. Postpartum relapse to smoking: a prospective study. Health Educ Res 1992;7:381–90.

[20] Mullen P, Quinn V, Ershoff D. Maintenance of non-smoking postpartum by women who stopped smoking during pregnancy. Am J Public Health 1990;80:992–4.

[21] Ellison L, Morrison H, de Groh M, Villeneuve P. Health consequences of smoking among Canadian smokers: an update. Chronic Dis Can 1999;20:36.

[22] Onken C, Kranzler H. Pharmacotherapies to enhance smoking cessation during pregnancy. Drug Alcohol Rev 2003;22:191–202.

[23] Naeye RL. Influence of maternal cigarette smoking during pregnancy on fetal and childhood growth. Obstet Gynecol 1981;57:18–21.

[24] Jalili T, Murthy GG, Schiestl RH. Cigarette smoke induces DNA deletions in the mouse embryo. Cancer Res 1998;58:2633–8.

[25] Haustein KO. Cigarette smoking, nicotine and pregnancy. Int J Clin Pharmacol Ther 1999;37:417–27.

[26] Walsh RD. Effects of maternal smoking on adverse pregnancy outcomes: examination of the criteria of causation. Hum Biol 1994;66:1059–92.

[27] Virk J, Zhang J, Olsen J. Medical abortion and the risk of subsequent adverse pregnancy outcomes. N Engl J Med 2007;357:648–53.

[28] Meyer MB, Tonascia JA, Buck C. The interrelationship of maternal smoking and increased perinatal mortality with other risk factors. Further analysis of the Ontario Perinatal Mortality Study, 1960–1961. Am J Epidemiol 1974;100:443–52.

[29] Steyn K, de Wet T, Saloojee Y, Nel H, Yach D. The influence of maternal cigarette smoking, snuff use and passive smoking on pregnancy outcomes: the Birth to Ten Study. Paediatr Perinat Epidemiol 2006;20:90–9.

[30] Shiono PH, Klebanoff MA, Berendes HW. Congenital malformations and maternal smoking during pregnancy. Teratology 1986;34:65–71.

[31] England LJ, Levine RJ, Mills JL, Klebanoff MA, Yu KF, Cnattingius S. Adverse pregnancy outcomes in snuff users. Am J Obstet Gynecol 2003;189:939–43.

[32] Ness RB, Grisso JA, Hirschinger N, Markovic N, Shaw LM, Day NL, Kline J. Cocaine and tobacco use and the risk of spontaneous abortion. N Engl J Med 1999;340:333–9.

[33] Nygard O, Vollset SE, Refsum H. Total plasma homocysteine and cardiovascular risk profile. The Hordaland Homocysteine Study. JAMA 1995;274:1526–33.

[34] Sanchez SE, Zhang C, Malinow M, Ware-Jauregui S, Larrabure G, Williams MA. Plasma folate, vitamin B12, and homocyst(e)ine concentrations in preeclamptic and normotensive Peruvian women. Am J Epidemiol 2001;153:474–80.

[35] McDonald SD, Perkins SL, Jodouin CA, Walker MC. Folate levels in pregnant women who smoke: an important gene/environment interaction. Am J Obstet Gynecol 2002;187:620–5.

[36] Hay G, Clausen T, Whitelaw A, Trygg K, Johnston C, Henriksen T, Refsum H. Maternal folate and cobalamin status predicts vitamin status in newborns and 6-month-old infants. J Nutr 2010;140:557–64.

[37] Sobczak A, Wardas W, Zielinska-Danch W, Pawlicki K. The influence of smoking on plasma homocysteine and cysteine levels in passive and active smokers. Clin Chem Lab Med 2004;42:408–14.

[38] Chen ATL, Reidy JA, Annest JL, Welty TK, Zhou H. Increased chromosome fragility as a consequence of blood folate levels, smoking status, and coffee consumption. Environ Mol Mutagen 1989;13:319–24.

[39] Czeizel AE, Dudas I. Prevention of the first occurrence of neural-tube defects by periconceptional vitamin supplementation. N Engl J Med 1992;327:1832–5.

[40] Shaw GM, O'Malley CD, Wasserman CR, Tolarova MM, Lammer EJ. Maternal periconceptional use of multivitamins and reduced risk for conotruncal heart defects and limb deficiencies among offspring. Am J Med Genet 1995;59:536–45.

[41] Piyathilake CJ, Macaluso M, Hine RJ, Richards EW, Krumkieck CL. Local and systemic effects of cigarette smoking on folate and vitamin B12. Am J Clin Nutr 1994;60:559–66.

[42] Oncel MY, Ozdemir R, Erdeve O, Dilmen U. Influence of maternal cigarette smoking during pregnancy on neonatal serum folate levels. Eur J Nutr 2012;51:385–7.

[43] Oncel MY, Calisici E, Ozdemir R, Yurttutan S, Erdeve O, Karahan S, Dilmen U. Is folic acid supplementation really necessary in preterm infants ≤32 weeks of gestation? J Pediatr Gastroenterol Nutr 2014;58:190–4.

[44] Pagan K, Hou J, Goldenberg RL, Cliver SP, Tamura T. Effect of smoking on serum concentrations of total homocysteine and B vitamins in mid-pregnancy. Clin Chim Acta 2001;306:103–9.

[45] Relton CL, Pearce MS, Parker L. The influence of erythrocyte folate and serum vitamin B12 status on birth weight. Br J Nutr 2005;93:593–9.

[46] van Wersch JW, Janssen Y, Zandvoort JA. Folic acid, vitamin B12, and homocysteine in smoking and non-smoking pregnant women. Eur J Obstet Gynecol Reprod Biol 2002;103:18–21.

[47] Cogswell ME, Weisberg P, Spong C. Cigarette smoking, alcohol use and adverse pregnancy outcomes: implications for micronutrient supplementation. J Nutr 2003;133:1722–31.

[48] Mannino DM, Mulinare J, Ford ES, Schwartz J. Tobacco exposure and decreased serum and red blood cell folate levels: data from the Third National Health and Nutrition Examination Survey. Nicotine Tob Res 2003;5:397–9.

[49] Alberg A. The influence of cigarette smoking on circulating concentrations of antioxidant micronutrients. Toxicology 2002;15:121–37.

[50] Ortega RM, Requejo AM, Lopez-Sobaler AM, Navia B, Mena MC, Basabe B, et al. Smoking and passive smoking as conditioners of folate status in young women. J Am Coll Nutr 2004;23:365–71.

[51] Ek J, Behnecke L, Halvorsen KS, Magnus E. Plasma and red cell folate values and folate requirements in formula-fed premature infants. Eur J Pediatr 1984;142:78–82.

[52] Vanier TM, Tyas JF. Folic acid status in premature infants. Arch Dis Child 1967;42:57–61.

[53] Strelling MK, Blackledge DG, Goodall HB. Diagnosis and management of folate deficiency in low birthweight infants. Arch Dis Child 1979;54:271–7.

[54] Hoffbrand AV. Folate deficiency in premature infants. Arch Dis Child 1970;45:441–4.

[55] Tsang RC, Lucas A, Uauy R. Nutritional needs of the preterm infant: scientific basis and practical guidelines. New York: Caduceus Medical Publishers Inc; 2005. 56-65.

[56] Agostoni C, Buonocore G, Carnielli VP, De Curtis M, Darmaun D, Decsi T, ESPGHAN Committee on Nutrition. et al. Enteral nutrient supply for preterm infants: commentary from the European Society of Paediatric Gastroenterology, Hepatology and Nutrition Committee on Nutrition. J Pediatr Gastroenterol Nutr 2010;50:85–91.

[57] Roberts PM, Arrowsmith DE, Lloyd AV, Monk-Jones ME. Effect of folic acid treatment on premature infants. Arch Dis Child 1972;47:631–4.

[58] Kendall AC, Jones EE, Wilson CI, Shinton NK, Elwood PC. Folic acid in low birthweight infants. Arch Dis Child 1974;49:736–8.

[59] Jyothi S, Misra I, Morris G, Benton A, Griffin D, Allen S. Red cell folate and plasma homocysteine in preterm infants. Neonatology 2007;92:264–8.

[60] Robison RG, Kumar R, Arguelles LM, Hong X, Wang G, Apollon S, et al. Maternal smoking during pregnancy, prematurity and recurrent wheezing in early childhood. Pediatr Pulmonol 2012;47:666–73.

[61] Matsui EC, Matsui W. Higher serum folate levels are associated with a lower risk of atopy and wheeze. J Allergy Clin Immunol 2009;123:1253–9.

[62] Magdelijns FJ, Mommers M, Penders J, Smits L, Thijs C. Folic acid use in pregnancy and the development of atopy, asthma, and lung function in childhood. Pediatrics 2011;128:135–44.

[63] Bekkers MB, Elstgeest LE, Scholtens S, Haveman-Nies A, de Jongste JC, Kerkhof M, et al. Maternal use of folic acid supplements during pregnancy and childhood respiratory health and atopy: the PIAMA birth cohort study. Eur Respir J 2012;39:1468–74.

[64] Crider KS, Cordero AM, Qi YP, Mulinare J, Dowling NF, Berry RJ. Prenatal folic acid and risk of asthma in children: a systematic review and meta-analysis. Am J Clin Nutr 2013;98:1272–81.

[65] Oncel MY, Erdeve O, Ozdemir R, Dilmen U. Does maternal smoking during pregnancy associate recurrent wheezing in infants by altering neonatal serum folate status? Pediatr Pulmonol 2013;48:627–8.

4

The Effect of Maternal Metabolic Health and Diet on the Follicular Fluid Composition and Potential Consequences for Oocyte and Embryo Quality

S.D.M. Valckx, PhD, J.L.M.R. Leroy, MSc, PhD-student

Gamete Research Centre, Veterinary Physiology and Biochemistry, Department of Veterinary Sciences, University of Antwerp, Belgium

INTRODUCTION

Obesity and other maternal metabolic disorders are risk factors for several chronic diseases such as type II diabetes, musculo-skeletal disorders, and cardiovascular disease [1]. Obesity has also been associated with an increased risk of reproductive failure [2,3], with changes in the menstrual pattern, anovulatory cycles, delayed conception, and multiple complications during pregnancy [4]. Besides the lower chance for a spontaneous pregnancy, obesity also impacts on the overall success as well as on the costs of fertility treatments [5].

Overweight and obese women display changes in their serum composition, for example, high serum cholesterol and nonesterified fatty acid (NEFA) concentrations. They also suffer from hyperglycemia and insulin resistance [6]. We previously showed that such serum metabolic changes are reflected in the follicular fluid in high-producing dairy cows [7,8] and in women undergoing assisted reproductive treatment (ART) [9]. Knowing that both the oocyte and the embryo are very vulnerable to changes in their microenvironment [10,11], recent research has been focusing on the follicular microenvironment and its potential effect on oocyte and embryo quality [9,12–14]. Alterations in the composition of the follicular fluid, due to poor metabolic health or through diet, might impair oocyte and cumulus cell quality [8,12,13]. This may in turn hamper embryo quality, for example, by altering

embryo metabolism or gene expression patterns in the preimplantation embryo [15] that could ultimately result in a reduced implantation rate and subsequent developmental abnormalities [16,17].

METABOLIC HEALTH RELATED COMPOSITION OF OVARIAN FOLLICULAR FLUID AND ITS RELATION TO OOCYTE AND EMBRYO QUALITY PARAMETERS

The first information about the effect of metabolic health on the follicular microenvironment and potential consequences for oocyte and embryo quality, is provided by the model of the high-producing dairy cow in negative energy balance [7,8]. High-producing dairy cows typically encounter a state of negative energy balance due to an inadequate appetite in early lactation while milk yield is already peaking. This period is characterized by a massive lipolysis, upregulated ketogenesis, and reduced peripheral insulin sensitivity [18]. Repeated sampling of serum and follicular fluid over time in such cows during both negative and positive energy balance has shown that metabolic changes in the blood profile are reflected in the follicular fluid composition [7]. Besides the metabolic similarities between the pathogenesis of obesity in women and the negative energy balance in

Handbook of Fertility. http://dx.doi.org/10.1016/B978-0-12-800872-0.00004-4

the cow (state of massive lipolysis), there is much resemblance between bovine and human physiological processes involved in follicular growth, oocyte maturation, and early embryonic growth [19]. Only recently, the relation between serum and follicular fluid composition could also be substantiated in women undergoing ART. In this study, we specifically investigated the body mass index (BMI)-related metabolic composition of human serum and preovulatory follicular fluid and associated both metabolic compositions with oocyte quality parameters [9]. It was shown that the serum composition of important metabolic parameters (standard biochemical, fat metabolism related, carbohydrate metabolism related, and hormonal parameters) was partly reflected in the follicular fluid. However, the follicular fluid kept its own characteristics, due to the modulating role of the blood–follicular fluid barrier and/or a substantial contribution of oocyte, granulosa, and cumulus cell metabolism [20,21]. It has previously been suggested that the ovary has the capacity to buffer, to some extent, against major fluctuations in the serum composition, which could play a protective role for the oocyte. For example, the cumulus–oocyte complex within the bovine follicle is protected from extremely low glucose and very high NEFA concentrations in the serum [7].

Biochemical Parameters in the Follicular Fluid

Total Protein

Total protein concentrations in the follicular fluid of women undergoing ART were negatively associated with the number of embryos generated after ovum pick up [9], suggesting a potential harmful effect of intrafollicular total protein concentrations on oocyte developmental competence. In agreement, Iwata et al. showed that oocytes from cows with higher follicular fluid concentrations of total protein and lower concentrations of albumin were impaired in their developmental capacity [22]. Furthermore, high dietary protein levels have been associated with inferior reproductive performance in dairy cattle in most studies [23–25]. The underlying pathogenesis behind this reproductive failure is the potential toxicity of the direct by-products of protein catabolism (ammonia and urea) for the oocyte and the embryo [25].

Urea

Excessive protein and amino acid deamination and detoxification in the liver can result in elevated systemic urea concentrations [25,26]. Importantly, changes in serum urea concentrations are reflected in the follicular fluid composition [7,9] and are potentially hazardous for bovine oocyte developmental competence and subsequent *in vitro* hatching of blastocysts [27,28]. Interestingly, Fahey et al. [29] showed that embryo quality was reduced in donor ewes fed dietary urea. However, the diet of the embryo recipient ewes did not affect embryo

survival [29]. So, it was hypothesized that the adverse effect of urea is likely to be caused by alterations in the follicular microenvironment or the oviduct, rather than a changed uterine environment [25,29].

C-reactive Protein

A strong correlation exists between serum and follicular fluid C-reactive protein (CRP) levels in women undergoing ART, and higher BMI was associated with elevated CRP concentrations in both serum and follicular fluid [9]. Elevated follicular fluid and serum CRP concentrations have previously been associated with chronic anovulation and no conception in women undergoing fertility treatment [30,31]. High CRP levels are characteristic for increased inflammation [32] and are proposed to be involved in oxidative stress and reactive oxygen species (ROS) mechanisms, which can be hazardous for oocyte developmental competence and subsequent embryo quality [33,34].

Follicular Fluid Parameters Related to Lipid Metabolism

Cholesterol

Nutritionally induced hypercholesterolemic conditions during early embryo development have been shown to affect bovine oocyte developmental competence and embryo quality in terms of cell number, apoptotic cell index, and gene expression patterns of metabolically and developmentally important genes [35]. Long-term maternal hypercholesterolemic conditions in the rabbit may even result in growth retardation [36]. HDL cholesterol and its surrogate marker Apolipoprotein A1 (Apo A1) in the serum of women undergoing ART were associated with BMI, as opposed to follicular fluid concentrations that did not associate with BMI [9]. This suggests that factors other than BMI regulate intrafollicular cholesterol concentrations. Interestingly, Apo A1 concentrations in the follicular fluid were negatively associated with embryo quality [9], suggesting that Apo A1 in the follicular fluid could have a negative impact on the oocyte's developmental capacity.

Carnitine

Carnitine concentrations were higher in the follicular fluid of women undergoing ART compared to the serum, and they correlated well between serum and follicular fluid samples [9]. Carnitine is an essential cofactor for carnitine palmitoyl transferase 1, which allows the entry of fatty acids into the mitochondrial matrix for the purpose of beta-oxidation [37]. Increased carnitine levels have been proposed to increase beta-oxidation and improve murine oocyte developmental competence [38]. Moreover, high carnitine concentrations have an antioxidant function, possibly protecting the cumulus-oocyte complex from excess oxidative stress, caused by other parameters present in the follicular fluid [38].

The specific effects of triglyceride and nonesterified fatty acid (NEFA) concentrations on oocyte and embryo quality will be discussed in Sections "Fatty Acids in the Ovarian Follicular Fluid" and "The Effect of Elevated NEFA Concentrations on Oocyte and Embryo Quality," respectively.

Follicular Fluid Parameters Related to Carbohydrate Metabolism

Glucose plays a crucial role in the acquisition of oocyte developmental competence [39]. The oocyte itself has low glycolytic rates and relies on glucose metabolites, such as pyruvate and lactate, provided by the surrounding cumulus cells, for energy substrate provision [39]. Glucose and lactate concentrations did not correlate between serum and follicular samples in women undergoing ART, but lactate concentrations were higher in follicular fluid than in serum [9]. In contrast, hypoglycemic conditions have been shown to be reflected in the bovine follicular microenvironment [7]. Furthermore, adding glucose to the *in vitro* maturation medium of bovine oocytes improved cumulus expansion, nuclear maturation, embryo cleavage rate, and blastocyst development [40,41]. On the other hand, oocyte maturation with glucose concentrations similar to those in the follicular fluid of cows with clinical ketosis (1.4 mM) significantly impaired developmental competence [42]. Mouse models also revealed that oocytes from diabetic (hyperglycemic) mice have increased concentrations of ROS, lower levels of glutathione, and an impaired mitochondrial function (reviewed by Amaral et al. [43] and Grindler and Moley [44]).

Hormonal Parameters in the Follicular Fluid

Insulin

Insulin concentrations are higher in both serum and follicular fluid of obese women compared to normal weight women [9], which might be implicated in the onset of insulin resistance. Under normal circumstances, insulin promotes the follicular response to gonadotropins [45]. However, it can also directly influence the maturing oocyte, because insulin has a growth/anti-apoptotic effect and has the metabolic ability to increase glucose uptake in the cumulus–oocyte complex [46]. It exerts its effect in various cell types and is involved in multiple (steroidogenic) pathways within the follicular unit, with both elevated and decreased levels potentially altering oocyte maturation and quality [47–50].

Insulin-Like Growth Factor 1

Together with insulin, the IGF system plays an important role in follicle growth and development by directly influencing ovarian cell proliferation and steroidogenesis [51–53]. Changes in serum IGF-1 concentrations are mimicked in the follicular fluid [9]. Additionally, IGF-1 influences the oocyte's developmental competence [54]; for example, early preantral mouse follicles, cultured in IGF-1 rich conditions, give rise to oocytes that develop into blastocysts with higher cell numbers, at a higher rate [55]. Furthermore, differences in follicular fluid IGF binding proteins (IGFBPs) and the interaction with locally produced IGF-II may affect the IGF-I bioavailability, and this has been associated with pathways resulting in anovulation [54,56]. Substantiating, the IGFBP especially, and not the IGF-1 concentrations as such, change due to an altered energy balance [26,57,58].

Taken together, these data emphasize that it is not one compound in the follicular fluid that may be seen as a marker for oocyte quality. There are a multitude of factors present, which can interact with each other, and it is the interplay between all those factors that finally determines the effect on oocyte developmental competence.

FATTY ACIDS IN THE OVARIAN FOLLICULAR FLUID

Fatty acids are important compounds in the microenvironment of the ovarian follicle. Besides their role as a cellular energy source, they have important biological functions in cell membrane biogenesis [59,60] and signaling [61]. In addition, they act as precursors for steroids and prostaglandins, which are essential for normal reproductive function [62]. In the follicular fluid, fatty acids are present in an esterified form (triglycerides, cholesterol esters, and phospholipids) or as NEFAs, mainly bound to albumin [63].

Changes in the serum fatty acid composition can be induced in two major ways: through dietary fatty acid supplementation or through an altered metabolism.

Fatty Acid Changes Through Diet

Dietary fatty acids or fat feeding can affect reproductive function by changing ovarian activity, follicular growth, corpus luteum function, and the uterine environment (for review, see Ref. [79]). They furthermore can change the concentration of precursors for the synthesis of reproductive hormones such as steroids and prostaglandins [62]. In this regard, it has been shown that supplementing dietary fat increases the size as well as the estradiol production of the bovine preovulatory follicle [64,65], most likely via the induction of high cholesterol concentrations in the follicular fluid and plasma [25]. Also, it has been shown that alterations in dietary fatty acid intake are reflected in the fatty acid profile of bovine follicular fluid [66,67]. For example, polyunsaturated fatty acid (PUFA) concentrations in bovine follicular fluid correlate well with PUFA concentrations in the diet [68]. Interestingly, acute dietary restriction has been described to increase NEFA concentrations in bovine follicular fluid [58,69,70]. However, a large number of studies indicate that such dietary and/or metabolic fatty acid changes can have repercussions for reproductive function in general and for oocyte and embryo quality more specifically.

Yi et al. [71] showed that a diet rich in n-3 polyunsaturated fatty acids reduced ovulation rate and litter size in mice. In cows, a diet with a high n-3/n-6 ratio has been shown to increase linolenic acid and estradiol levels in the follicle and improve embryo cleavage rate [67]. Feeding linoleic acid to heifers did not affect embryo production, but significantly decreased cryotolerance [72]. Furthermore, high dietary n-3 intake of women in the month before ART was associated with improved embryo morphology [73], even though high maternal dietary n-3 PUFA supplementation periconception reduced normal oocyte development, disturbed mitochondrial metabolism, and adversely affected the morphology of the embryo in the mouse [74]. Furthermore, *in vitro* bovine oocyte maturation in the presence of n-3 PUFA improved nuclear oocyte maturation, while the presence of n-6 PUFA reduced the resumption of meiosis [75,76]. In agreement, Bilby et al. [77] found a lower oocyte developmental competence when cows were fed linoleic acid (C18:2, n-6) rich diets. Studies in sheep revealed that diets rich in fish oil (n-3 PUFA) have a favorable effect on oocyte quality [78], which was associated with changes in the fatty acid composition of the oocyte, resulting in a better membrane integrity following chilling.

A list of important studies about the effects of dietary fat on oocyte and embryo quality, along with their main findings, can be found in a review by Leroy et al. [79].

In conclusion, numerous animal studies, mostly in cattle, sheep, and rodents, have demonstrated the effect of diet-induced fatty acid changes on ovarian physiology in general and on the quality of the oocyte more specifically [66,67,78,80]. Dietary-induced fatty acid changes or obesogenic diets during oocyte maturation or early embryo development affect oocyte and embryo quality [81,82], and these effects may even have consequences for the offspring's health [71,83]. However, results of these studies are not always in agreement and difficult to compare as not only the quantity of the fatty acid offered, but also the physiological status, the type of fatty acid (ratios of n-3, -6, or -9 fatty acids, number of double bonds, carbon chain length), and the duration of treatment may determine the final fatty acid concentrations in the follicular fluid [61].

Interestingly, Zachut et al. [67] found that even though the fatty acid composition of plasma and follicular fluid were similar in dairy cows, it was different from that found in granulosa cells and in the cumulus–oocyte complex. On the contrary, Matorras et al. [84] indicated that the predominant fatty acids present within unfertilized human oocytes were stearic acid, palmitic acid, oleic acid, myristic acid, and linoleic acid, which is similar to the patterns we found in the follicular fluid [85], suggesting a close relationship between the intracellular and extracellular fatty acid composition of the oocyte. In agreement, Adamiak et al. [68] and Rooke et al. [86] found that a change in follicular fluid fatty acid concentrations was also reflected in the fatty acid content and profile of the cumulus–oocyte complex [68,86].

Fatty Acid Changes Through Adverse Metabolic Health

The metabolic health of an individual, as indicated for example by body weight, can affect the fatty acid composition in both serum and follicular fluid [9,12,13]. In this regard, attention has been paid to the serum and follicular fluid composition of obese women and of women with polycystic ovarian syndrome (PCOS) [12,87], but also to cows in a negative energy balance postpartum [7,8]. The common underlying mechanism causing alterations in the fatty acid profile is the massive mobilization of body fats (increased lipolysis). It has been shown that triglyceride and NEFA concentrations are increased in the follicular fluid of obese women subjected to ovum pickup [14], while such elevated NEFA and/or triglyceride concentrations were negatively associated with human cumulus–oocyte complex morphology [13] and affected both murine oocyte maturation [14] and bovine embryo quality [15,88]. Jungheim et al. [83] furthermore suggested, by using a murine model, that maternal diet-induced obesity may affect fertility at the level of the oocyte and preimplantation embryo, and that these effects may contribute to the early development of metabolic problems in the offspring, such as glucose intolerance and increased cholesterol and body fat. More specifically, Igosheva et al. [81] found that maternal diet-induced obesity prior to conception causes an alteration in oocyte and embryo mitochondrial function and poses this as a probable mechanism of obesity-associated reproductive and developmental failure. They describe an altered mitochondrial distribution, hyperpolarization of the mitochondrial membrane, oxidized redox state, and oxidative stress in both oocytes and embryos, associated with significant developmental impairment. In addition, lipid accumulation, endoplasmic reticulum stress, and apoptosis were described in ovarian cells, due to diet-induced maternal obesity [82].

Most research on fatty acids has focused on the total fat fraction, rather than investigating fatty acid differences in different lipid fraction. This is important though, because the follicular cells, the oocyte, and the early embryo may react differently on fatty acids from different fractions. First, triglycerides, stored in lipid droplets, are proposed to be an important energy source

[89]. We previously showed that triglyceride concentrations within the follicular fluid were higher in overweight, compared to normal weight women (trend) [85]. Furthermore, *in vitro* oocytes and embryos are able to accumulate fatty acids from their environment [90] and store them as lipid droplets. An excessive lipid accumulation is, however, known to impair the quality of bovine embryos by increasing their sensitivity to, for example, chilling and cryopreservation [91]. Also, an increased lipid accumulation has been associated with suboptimal mitochondrial function and changes in the relative abundance of developmentally important genes related to stress response in bovine blastocysts [25,91,92].

Second, cholesterol within the follicle is required for steroidogenesis, in addition to membrane synthesis. Cholesterol can be derived from either cellular *de novo* synthesis from acetate or from the uptake of plasma lipoprotein cholesterol [93]. The fatty acids freed from a cholesterol particle can act as precursors for reproductive hormones, such as prostaglandins [62]. We previously found that only HDL cholesterol is present within the follicular fluid of the preovulatory follicle of women undergoing ART [9]. BMI did not affect this follicular fluid HDL cholesterol concentration [9,12,85].

Third, fatty acids in the phospholipid fraction are an important compound of the plasma membrane. It is generally accepted that the quality of an oocyte and embryo is closely related to their fatty acid composition, in particular the composition of the phospholipid fraction [94]. Phospholipids are cleaved from the plasma membrane by phospholipases, which in turn increases the availability of fatty acids for further processing to reproductive hormones [62]. Furthermore, fatty acids regulate membrane stability, since saturated fatty acids decrease and unsaturated fatty acids increase membrane fluidity [95].

Remarkably, even though the NEFA fraction in human preovulatory follicular fluid contained only a small fraction of the total amount of fatty acids in the follicular fluid, only this fraction presented with a great BMI-related variability [85]. This states the NEFA fraction as a potential link between maternal metabolic disorders and reduced oocyte and embryo quality. Part 3 of this chapter will focus solely on the effect of elevated NEFA concentrations on oocyte developmental competence and subsequent embryo quality.

THE EFFECT OF ELEVATED NEFA CONCENTRATIONS ON OOCYTE AND EMBRYO QUALITY

Elevated NEFA concentrations have been proposed as an important link between maternal metabolic disorders and fertility failure. In addition, elevated NEFA concentrations are also involved in the pathogenesis of metabolic disorders, as they for example induce insulin resistance [96], which could contribute to the pathogenesis of type II diabetes or metabolic syndrome. It has been shown that elevated NEFA concentrations are reflected in the follicular fluid (bovine: Leroy et al. [7]; human: Valckx et al. [9]) and that obese women present with higher follicular fluid NEFA concentrations, compared to overweight and normal weight women [85]. However, both the NEFA concentration and the NEFA composition differed between serum and follicular fluid samples in the cow [8]. A possible explanation for this is that in the presence of high NEFA concentrations, a substantial portion of NEFA in serum is deposited into low-density lipoproteins. Saturated fatty acids are preferentially bound to LDL, while unsaturated fatty acids prefer to be bound to albumin [97]. The absence of LDL particles in the follicular fluid [9] could thus account for the differences between serum and follicular fluid NEFA composition.

Elevated NEFA Concentrations: Consequences for Cumulus, Granulosa, and Theca Cells

In most of the following studies, physiological and pathological concentrations of palmitic acid (PA, C16:0), oleic acid (OA, C18:1), and stearic acid (SA, C18:0) were used, because they are the most predominant fatty acids present in both the serum and follicular fluid [8,85]. Frequently, a representative mix of the three NEFAs was made to increase the resemblance to the *in vivo* situation (referred to as "HIGH NEFA").

It was previously shown that lipolysis-linked elevated NEFA concentrations alter the steroidogenic profile and the viability of both granulosa and theca cells [98,99]. More specifically, in bovine granulosa cells, HIGH NEFA concentrations inhibited cell proliferation, which was partly due to the induction of apoptosis as determined by an annexin V-FITC/propidium iodide staining of the granulosa cells. Furthermore, estradiol production was stimulated by HIGH NEFA concentrations [98]. The authors suggested that apoptotic bovine granulosa cells can maintain steroidogenesis as long as the steroidogenic organelles remain intact [100]. A study in immortalized rat granulosa cells also showed that these organelles cluster during the process of apoptosis [101], which could make it look like these follicles have a higher steroidogenic capacity, even though it is caused by an apoptotic event. However, Jorritsma et al. could not detect an effect of elevated OA concentrations on progesterone production by *in vitro* cultured bovine granulosa cells, even though they did document a negative effect on cell proliferation [102]. It was also shown that saturated fatty acids (PA and SA) induce apoptosis in human granulosa cells [103]. Interestingly, arachidonic acid protected granulosa cells from palmitic acid and stearic acid induced apoptosis [103].

Bovine *in vitro* theca cell number was reduced due to exposure to HIGH NEFA concentrations. Furthermore, these HIGH NEFA concentrations reduced progesterone production, an effect most likely induced by the cytotoxic effect of elevated NEFAs, as the authors showed an increased percentage of early apoptotic and late apoptotic/necrotic cells [99].

When bovine *in vitro* maturing cumulus–oocyte complexes were exposed to elevated SA and PA concentrations during the crucial final phase of maturation (24 h), cumulus cell expansion was adversely affected and a significantly higher rate of apoptosis and even necrosis could be detected [8]. Furthermore, HIGH NEFA exposure during maturation caused remarkable differences in gene expression profiles in the cumulus cells surrounding the oocyte [88]. In this regard, cumulus cells that were exposed to HIGH NEFA concentrations presented with a decreased expression level of genes encoding enzymes involved in REDOX regulation: *GADH*, *GPX1*, *GGPD*, and *LDHA* [88]. Because cumulus cells are involved in safeguarding the oocyte from damage induced by ROS, likely by providing glutathione to the oocyte [104], this altered gene expression profile could lead to an impaired oocyte quality.

Elevated NEFA Concentrations: Consequences for the Oocyte

The effect of elevated NEFA concentrations on oocyte and embryo quality has been studied extensively *in vitro*. Aardema et al. [90] showed that bovine oocytes actively take up and metabolize NEFAs out of their environment (store them in triglycerides, phospholipids, and metabolize them), but that only a small fraction stayed in the NEFA fraction within the oocyte. Also, PA and SA had a negative effect on the amount of fat stored in intracellular lipid droplets. Additionally, there was a detrimental effect on oocyte developmental competence. On the contrary, OA increased fat storage in lipid droplets and improved oocyte developmental competence. Remarkably, the negative effects caused by PA and SA could be counteracted by the addition of OA to the treatment [90]. Similar results were described in pancreatic islet cells, where OA was described to reduce palmitate-induced lipotoxicity, possibly by promoting triglyceride formation as a cytoprotective mechanism [105]. These studies highlight the importance of fatty acid ratios and concentrations in maturing oocytes.

Gene expression analysis in bovine oocytes after HIGH NEFA exposure showed that several genes related to REDOX regulation were differentially expressed [88]. Furthermore, oocyte maturation under elevated SA concentrations resulted in a trend for decreased GSH content, compared to oocytes exposed to HIGH NEFA concentrations [106]. The authors hypothesize that this reduced GSH content in high SA oocytes is

the result of more GSH depletion in order to neutralize excessive ROS levels [106]. However, it is also possible that these oocytes were not provided with sufficient GSH by their cumulus cells to begin with, as it was previously shown that high SA exposed cumulus cells show a significantly downregulated expression of *GPX1* [88]. Altogether, this strongly suggests that HIGH NEFA concentrations alter the REDOX status in cumulus–oocyte complexes, which could affect early embryo physiology and quality.

Elevated NEFA Concentrations: Consequences for the Resultant Embryo

Alterations in the oocyte's metabolism, due to exposure to NEFAs during final oocyte maturation, could directly extend to the embryo as well. The quiet embryo hypothesis states that preimplantation embryo survival is best with a relatively low level of metabolism [107]. It was previously shown that exposure of bovine oocytes to HIGH NEFA concentrations during the crucial final phase of maturation indeed alters the metabolism of the resultant embryos [15], which could be indicative for reduced embryo quality.

Bovine blastocysts, originating from HIGH NEFA exposed oocytes, presented with distinct differences in gene expression patterns. By using qRT-PCR, genes involved in *de novo* DNA methylation, metabolism, fatty acid synthesis, mitochondrial biogenesis, oxidative stress, and growth were identified as contributors to the effect of elevated NEFA concentrations during maturation on blastocyst quality [15,88]. Furthermore, microarray approaches revealed that pathways involved in lipid metabolism, carbohydrate metabolism, and cell death were major contributors [106]. These genomewide transcription studies also revealed pathways involved in the development of insulin resistance (protein kinase C). Additionally, HIGH NEFA concentrations increased the overall turnover of amino acids [15], which has been proven to be a valuable quality marker for embryo quality [108] and in this case, is predictive for a lower viability of NEFA embryos, compared to control counterparts. In these embryos, oxygen, pyruvate, and glucose consumption was decreased, while lactate consumption steeply increased [15]. The consumption of oxygen is an indicator of global metabolic activity [109]. In contrast to the findings on amino acid metabolism, these data thus indicate a reduced oxidative metabolism in embryos originating from HIGH NEFA exposed oocytes. Together with the data on pyruvate and glucose consumption, this suggests a drop in the rate of oxidative phosphorylation [15]. Interestingly and in accordance with the study of Aardema et al., blastocysts originating from HIGH NEFA exposed oocytes present with an increased intracellular lipid content, compared to blastocysts originating from oocytes only exposed to elevated SA concentrations [106].

It does not seem surprising that such metabolic changes in preimplantation embryos may have repercussions for the metabolic health of the offspring. In this regard, a study of Jungheim et al. [110] showed that preimplantation exposure of mouse embryos to elevated PA concentrations result in fetal growth restriction followed by a catch-up growth in the offspring. This study emphasizes the ultimate importance of the maternal microenvironment in the earliest stages of life, for the normal and healthy development of the offspring.

CONCLUSIONS

Changes in the serum composition, through diet or an adverse metabolic health, are reflected in the microenvironment of the ovarian preovulatory follicle. Such changes may endanger the oocyte developmental competence by altering, for example, granulosa and cumulus cell function. An impaired oocyte quality may lie at the basis of potential carry-over effects to the embryo, where an altered embryo metabolism, development, and overall quality have been described. Ultimately, these changes in the oocyte's microenvironment can even have consequences for the offspring's health. In conclusion, this chapter clearly states the importance of the follicular microenvironment for optimal oocyte and embryo development, and describes dietary and metabolic mechanisms through which the follicular fluid composition may be altered.

References

[1] WHO. Obesity and overweight. Fact Sheet No. 311, http://www.who.int/mediacentre/factsheets/fs311/en/index.html; 2013.

[2] Lash MM, Armstrong A. Impact of obesity on women's health. Fertil Steril 2009;91(5):1712–6.

[3] Pasquali R, Patton L, Gambineri A. Obesity and infertility. Curr Opin Endocrinol Diabetes Obes 2007;14(6):482–7.

[4] Sebire NJ, Jolly M, Harris JP, Wadsworth J, Joffe M, Beard RW, et al. Maternal obesity and pregnancy outcome: a study of 287 213 pregnancies in London. Int J Obesity 2001;25(8):1175–82.

[5] Koning AMH, Kuchenbecker WKH, Groen H, Hoek A, Land JA, Khan KS, et al. Economic consequences of overweight and obesity in infertility: a framework for evaluating the costs and outcomes of fertility care. Hum Reprod Update 2010;16(3):246–54.

[6] Lebovitz HE. Insulin resistance – a common link between type 2 diabetes and cardiovascular disease. Diabetes Obes Metab 2006; 8(3):237–49.

[7] Leroy JLMR, Vanholder T, Delanghe JR, Opsomer G, Van Soom A, Bols PEJ, et al. Metabolic changes in follicular fluid of the dominant follicle in high-yielding dairy cows early post partum. Theriogenology 2004;62(6):1131–43.

[8] Leroy JLMR, Vanholder T, Mateusen B, Christophe A, Opsomer G, de Kruif A, et al. Non-esterified fatty acids in follicular fluid of dairy cows and their effect on developmental capacity of bovine oocytes *in vitro*. Reproduction 2005;130(4):485–95.

[9] Valckx SD, De Pauw I, De Neubourg D, Inion I, Berth M, Fransen E, et al. BMI-related metabolic composition of the follicular fluid of women undergoing assisted reproductive treatment and the consequences for oocyte and embryo quality. Hum Reprod 2012;27(12):3531–9.

[10] Gilchrist RB, Ritter LJ, Armstrong DT. Oocyte-somatic cell interactions during follicle development in mammals. Anim Reprod Sci 2004;82–83:431–46.

[11] Leroy JLMR, Rizos D, Sturmey R, Bossaert P, Gutierrez-Adan A, Van Hoeck V, et al. Intrafollicular conditions as a major link between maternal metabolism and oocyte quality: a focus on dairy cow fertility. Reprod Fert Develop 2012;24(1):1–12.

[12] Robker RL, Akison LK, Bennett BD, Thrupp PN, Chura LR, Russell DL, et al. Obese women exhibit differences in ovarian metabolites, hormones, and gene expression compared with moderate-weight women. J Clin Endocrinol Metab 2009;94(5):1533–40.

[13] Jungheim ES, Macones GA, Odem RR, Patterson BW, Lanzendorf SE, Ratts VS, et al. Associations between free fatty acids, cumulus oocyte complex morphology and ovarian function during *in vitro* fertilization. Fertil Steril 2011;95(6):1970–4.

[14] Yang X, Wu LL, Chura LR, Liang X, Lane M, Norman RJ, et al. Exposure to lipid-rich follicular fluid is associated with endoplasmic reticulum stress and impaired oocyte maturation in cumulus–oocyte complexes. Fertil Steril 2012;97(6):1438–43.

[15] Van Hoeck V, Sturmey RG, Bermejo-Alvarez P, Rizos D, Gutierrez-Adan A, Leese HJ, et al. Elevated non-esterified fatty acid concentrations during bovine oocyte maturation compromise early embryo physiology. PloS One 2011;6(8):e23183.

[16] Robker RL. Evidence that obesity alters the quality of oocytes and embryos. Pathophysiology 2008;15(2):115–21.

[17] Brewer CJ, Balen AH. The adverse effects of obesity on conception and implantation. Reproduction 2010;140(3):347–64.

[18] Leroy JLMR, Vanholder T, Van Knegsel ATM, Garcia-Ispierto I, Bols PEJ. Nutrient prioritization in dairy cows early postpartum: mismatch between metabolism and fertility? Reprod Domest Anim 2008;43:96–103.

[19] Campbell BK, Souza C, Gong J, Webb R, Kendall N, Marsters P, et al. Domestic ruminants as models for the elucidation of the mechanisms controlling ovarian follicle development in humans. Reprod Suppl 2003;61:429–43.

[20] Webb R, Garnsworthy PC, Gong JG, Armstrong DG. Control of follicular growth: local interactions and nutritional influences. J Anim Sci 2004;82(E-Suppl):E63–74.

[21] Rodgers RJ, Irving-Rodgers HF. Formation of the ovarian follicular antrum and follicular fluid. Biol Reprod 2010;82(6):1021–9.

[22] Iwata H, Tanaka H, Kanke T, Sakaguchi Y, Shibano K, Kuwayama T, et al. Follicle growth and oocyte developmental competence in cows with liver damage. Reprod Domest Anim 2010;45(5):888–95.

[23] Butler WR. Energy balance relationships with follicular development, ovulation and fertility in postpartum dairy cows. Livest Prod Sci 2003;83(2–3):211–8.

[24] Melendez P, Donovan A, Hernandez J, Bartolome J, Risco CA, Staples C, et al. Milk, plasma, and blood urea nitrogen concentrations, dietary protein, and fertility in dairy cattle. J Am Vet Med Assoc 2003;223(5):628–34.

[25] Leroy JL, Van Soom A, Opsomer G, Goovaerts IG, Bols PE. Reduced fertility in high-yielding dairy cows: are the oocyte and embryo in danger? Part II. Mechanisms linking nutrition and reduced oocyte and embryo quality in high-yielding dairy cows. Reprod Domest Anim 2008;43(5):623–32.

[26] Leroy JLMR, Opsomer G, Van Soom A, Goovaerts IGF, Bols P. Reduced fertility in high-yielding dairy cows: are the oocyte and embryo in danger? Part I. The importance of negative energy balance and altered corpus luteum function to the reduction of oocyte and embryo quality in high-yielding dairy cows. Reprod Domest Anim 2008;43(5):612–22.

[27] Ferreira FA, Gomez RGG, Joaquim DC, Watanabe YF, Paula LADE, Binelli M, et al. Short-term urea feeding decreases *in vitro* hatching of bovine blastocysts. Theriogenology 2011;76(2):312–9.

[28] De Wit AAC, Cesar MLF, Kruip TAM. Effect of urea during *in vitro* maturation on nuclear maturation and embryo development of bovine cumulus–oocyte complexes. J Dairy Sci 2001;84(8):1800–4.

[29] Fahey J, Boland MP, O'Callaghan D. The effects of dietary urea on embryo development in superovulated donor ewes and on early embryo survival and development in recipient ewes. Anim Sci 2001;72:395–400.

[30] Lamaita RM, Pontes A, Belo AV, Caetano JP, Andrade SP, Candido EB, et al. Inflammatory response patterns in ICSI patients: a comparative study between chronic anovulating and normally ovulating women. Reprod Sci 2012;19(7):704–11.

[31] Levin I, Gamzu R, Mashiach R, Lessing JB, Amit A, Almog B. Higher C-reactive protein levels during IVF stimulation are associated with ART failure. J Reprod Immunol 2007;75(2):141–4.

[32] Das UN. Is obesity an inflammatory condition? Nutrition 2001;17(11–12):953–66.

[33] Combelles CMH, Gupta S, Agarwal A. Could oxidative stress influence the *in-vitro* maturation of oocytes? Reprod BioMed Online 2009;18(6):864–80.

[34] Agarwal A, Gupta S, Sekhon L, Shah R. Redox considerations in female reproductive function and assisted reproduction: from molecular mechanisms to health implications. Antioxid Redox Signal 2008;10(8):1375–403.

[35] Leroy JLMR, Van Hoeck V, Clemente M, Rizos D, Gutierrez-Adan A, Van Soom A, et al. The effect of nutritionally induced hyperlipidaemia on *in vitro* bovine embryo quality. Hum Reprod 2010;25(3):768–78.

[36] Picone O, Laigre P, Fortun-Lamothe L, Archilla C, Peynot N, Ponter AA, et al. Hyperlipidic hypercholesterolemic diet in prepubertal rabbits affects gene expression in the embryo, restricts fetal growth and increases offspring susceptibility to obesity. Theriogenology 2011;75(2):287–99.

[37] Sutton-McDowall ML, Feil D, Robker RL, Thompson JG, Dunning KR. Utilization of endogenous fatty acid stores for energy production in bovine preimplantation embryos. Theriogenology 2012;77(8):1632–41.

[38] Dunning KR, Akison LK, Russell DL, Norman RJ, Robker RL. Increased beta-oxidation and improved oocyte developmental competence in response to l-carnitine during ovarian *in vitro* follicle development in mice. Biol Reprod 2011;85(3):548–55.

[39] Sutton-McDowall ML, Gilchrist RB, Thompson JG. The pivotal role of glucose metabolism in determining oocyte developmental competence. Reproduction 2010;139(4):685–95.

[40] Sutton-McDowall ML, Gilchrist RB, Thompson JG. Cumulus expansion and glucose utilisation by bovine cumulus–oocyte complexes during *in vitro* maturation: the influence of glucosamine and follicle-stimulating hormone. Reproduction 2004;128(3):313–9.

[41] Bilodeau-Goeseels S. Effects of culture media and energy sources on the inhibition of nuclear maturation in bovine oocytes. Theriogenology 2006;66(2):297–306.

[42] Leroy JLMR, Vanholder T, Opsomer G, Van Soom A, de Kruif A. The *in vitro* development of bovine oocytes after maturation in glucose and beta-hydroxybutyrate concentrations associated with negative energy balance in dairy cows. Reprod Domest Anim 2006;41(2):119–23.

[43] Amaral S, Oliveira PJ, Ramalho-Santos J. Diabetes and the impairment of reproductive function: possible role of mitochondria and reactive oxygen species. Curr Diabetes Rev 2008;4(1):46–54.

[44] Grindler NM, Moley KH. Maternal obesity, infertility and mitochondrial dysfunction: potential mechanisms emerging from mouse model systems. Mol Hum Reprod 2013;19(8):486–94.

[45] Frajblat M, Butler WR. Metabolic effects of insulin and IGF-I in bovine ovarian follicle wall culture. Biol Reprod 2000;62:222.

[46] Purcell SH, Moley KH. The impact of obesity on egg quality. J Assist Reprod Genet 2011;28(6):517–24.

[47] Garnsworthy PC, Fouladi-Nashta AA, Mann GE, Sinclair KD, Webb R. Effect of dietary-induced changes in plasma insulin concentrations during the early post partum period on pregnancy rate in dairy cows. Reproduction 2009;137(4):759–68.

[48] Sanchez F, Romero S, Smitz J. Oocyte and cumulus cell transcripts from cultured mouse follicles are induced to deviate from normal *in vivo* conditions by combinations of insulin, follicle-stimulating hormone, and human chorionic gonadotropin. Biol Reprod 2011;85(3):565–74.

[49] Vlaisavljevic V, Kovac V, Sajko MC. Impact of insulin resistance on the developmental potential of immature oocytes retrieved from human chorionic gonadotropin-primed women with polycystic ovary syndrome undergoing *in vitro* maturation. Fertil Steril 2009;91(3):957–9.

[50] Adamiak SJ, Mackie K, Watt RG, Webb R, Sinclair KD. Impact of nutrition on oocyte quality: cumulative effects of body composition and diet leading to hyperinsulinemia in cattle. Biol Reprod 2005;73(5):918–26.

[51] Spicer LJ, Echternkamp SE. The ovarian insulin and insulin-like growth-factor system with an emphasis on domestic-animals. Domest Anim Endocrinol 1995;12(3):223–45.

[52] Gong JG. Influence of metabolic hormones and nutrition on ovarian follicle development in cattle: practical implications. Domest Anim Endocrinol 2002;23(1–2):229–41.

[53] Lucy MC. Regulation of ovarian follicular growth by somatotropin and insulin-like growth factors in cattle. J Dairy Sci 2000;83(7):1635–47.

[54] Silva JRV, Figueiredo JR, van den Hurk R. Involvement of growth hormone (GH) and insulin-like growth factor (IGF) system in ovarian folliculogenesis. Theriogenology 2009;71(8):1193–208.

[55] Demeestere I, Gervy C, Centner J, Devreker F, Englert Y, Delbaere A. Effect of insulin-like growth factor-I during preantral follicular culture on steroidogenesis, *in vitro* oocyte maturation, and embryo development in mice. Biol Reprod 2004;70(6):1664–9.

[56] Amato G, Izzo A, Tucker AT, Bellastella A. Lack of insulin-like growth factor binding protein-3 variation after follicle-stimulating hormone stimulation in women with polycystic ovary syndrome undergoing *in vitro* fertilization. Fertil Steril 1999;72(3):454–7.

[57] Spicer LJ, Crowe MA, Prendiville DJ, Goulding D, Enright WJ. Systemic but not intraovarian concentrations of insulin-like growth factor-I are affected by short-term fasting. Biol Reprod 1992;46(5):920–5.

[58] Comin A, Gerin D, Cappa A, Marchi V, Renaville R, Motta M, et al. The effect of an acute energy deficit on the hormone profile of dominant follicles in dairy cows. Theriogenology 2002;58(5):899–910.

[59] Renaville B, Bacciu N, Comin A, Motta M, Poli I, Vanini G, et al. Plasma and follicular fluid fatty acid profiles in dairy cows. Reprod Domest Anim 2010;45(1):118–21.

[60] Sturmey RG, Reis A, Leese HJ, McEvoy TG. Role of fatty acids in energy provision during oocyte maturation and early embryo development. Reprod Domest Anim 2009;44(Suppl 3):50–8.

[61] McKeegan PJ, Sturmey RG. The role of fatty acids in oocyte and early embryo development. Reprod Fert Develop 2012;24(1):59–67.

[62] Mattos R, Staples CR, Thatcher WW. Effects of dietary fatty acids on reproduction in ruminants. Rev Reprod 2000;5(1):38–45.

[63] Hughes J, Kwong WY, Li DF, Salter AM, Lea RG, Sinclair KD. Effects of omega-3 and -6 polyunsaturated fatty acids on ovine follicular cell steroidogenesis, embryo development and molecular markers of fatty acid metabolism. Reproduction 2011;141(1):105–18.

[64] Beam SW, Butler WR. Energy balance and ovarian follicle development prior to the first ovulation postpartum in dairy cows receiving three levels of dietary fat. Biol Reprod 1997;56(1):133–42.

[65] Moallem U, Katz M, Lehrer H, Livshitz L, Yakoby S. Role of peripartum dietary propylene glycol or protected fats on metabolism

and early postpartum ovarian follicles. J Dairy Sci 2007;90(3): 1243–54.

[66] Wonnacott KE, Kwong WY, Hughes J, Salter AM, Lea RG, Garnsworthy PC, et al. Dietary omega-3 and -6 polyunsaturated fatty acids affect the composition and development of sheep granulosa cells, oocytes and embryos. Reproduction 2010;139(1):57–69.

[67] Zachut M, Dekel I, Lehrer H, Arieli A, Arav A, Livshitz L, et al. Effects of dietary fats differing in n-6:n-3 ratio fed to high-yielding dairy cows on fatty acid composition of ovarian compartments, follicular status, and oocyte quality. J Dairy Sci 2010;93(2):529–45.

[68] Adamiak S, Ewen M, Rooke J, Webb R, Sinclair KD. Diet and fatty acid composition of bovine plasma, granulosa cells, and cumulus–oocyte complexes. Reprod Fert Develop 2005;17(1–2):200–1.

[69] Jorritsma R, de Groot MW, Vos PLAM, Kruip TAM, Wensing T, Noordhuizen JPTM. Acute fasting in heifers as a model for assessing the relationship between plasma and follicular fluid NEFA concentrations. Theriogenology 2003;60(1):151–61.

[70] Leroy JL, Van Soom A, Opsomer G, Bols PE. The consequences of metabolic changes in high-yielding dairy cows on oocyte and embryo quality. Animal 2008;2(8):1120–7.

[71] Yi D, Zeng S, Guo Y. A diet rich in n-3 polyunsaturated fatty acids reduced prostaglandin biosynthesis, ovulation rate, and litter size in mice. Theriogenology 2012;78(1):28–38.

[72] Guardieiro MM, Machado GM, Bastos MR, Mourao GB, Carrijo LHD, Dode MAN, et al. A diet enriched in linoleic acid compromises the cryotolerance of embryos from superovulated beef heifers. Reprod Fert Develop 2014;26(4):511–20.

[73] Hammiche F, Vujkovic M, Wijburg W, de Vries JHM, Macklon NS, Laven JSE, et al. Increased preconception omega-3 polyunsaturated fatty acid intake improves embryo morphology. Fertil Steril 2011;95(5):1820–3.

[74] Wakefield SL, Lane M, Schulz SJ, Hebart ML, Thompson JG, Mitchell M. Maternal supply of omega-3 polyunsaturated fatty acids alter mechanisms involved in oocyte and early embryo development in the mouse. Am J Physiol Endocrinol Metab 2008;294(2):E425–34.

[75] Marei WF, Wathes DC, Fouladi-Nashta AA. The effect of linolenic acid on bovine oocyte maturation and development. Biol Reprod 2009;81(6):1064–72.

[76] Marei WF, Wathes DC, Fouladi-Nashta AA. Impact of linoleic acid on bovine oocyte maturation and embryo development. Reproduction 2010;139(6):979–88.

[77] Bilby TR, Block J, do Amaral BC, Sa Filho O, Silvestre FT, Hansen PJ, et al. Effects of dietary unsaturated fatty acids on oocyte quality and follicular development in lactating dairy cows in summer. J Dairy Sci 2006;89(10):3891–903.

[78] Zeron Y, Sklan D, Arav A. Effect of polyunsaturated fatty acid supplementation on biophysical parameters and chilling sensitivity of ewe oocytes. Mol Reprod Dev 2002;61(2):271–8.

[79] Leroy J, Sturmey R, Van Hoeck V, De Bie J, McKeegan P, Bols P. Dietary fat supplementation and the consequences for oocyte and embryo quality: hype or significant benefit for dairy cow reproduction? Reprod Domest Anim 2014;49(3):353–61.

[80] Petit HV, Dewhurst RJ, Scollan ND, Proulx JG, Khalid M, Haresign W, et al. Milk production and composition, ovarian function, and prostaglandin secretion of dairy cows fed omega-3 fats. J Dairy Sci 2002;85(4):889–99.

[81] Igosheva N, Abramov AY, Poston L, Eckert JJ, Fleming TP, Duchen MR, et al. Maternal diet-induced obesity alters mitochondrial activity and redox status in mouse oocytes and zygotes. PloS One 2010;5(4):e10074.

[82] Wu LLY, Dunning KR, Yang X, Russell DL, Lane M, Norman RJ, et al. High-fat diet causes lipotoxicity responses in cumulus–oocyte complexes and decreased fertilization rates. Endocrinology 2010;151(11):5438–45.

[83] Jungheim ES, Schoeller EL, Marquard KL, Louden ED, Schaffer JE, Moley KH. Diet-induced obesity model: abnormal oocytes and persistent growth abnormalities in the offspring. Endocrinology 2010;151(8):4039–46.

[84] Matorras R, Ruiz JI, Mendoza R, Ruiz N, Sanjurjo P, Rodriguez-Escudero FJ. Fatty acid composition of fertilization-failed human oocytes. Hum Reprod 1998;13(8):2227–30.

[85] Valckx SD, Arias-Alvarez M, De Pauw I, Fievez V, Vlaeminck B, Fransen E, et al. Fatty acid composition of the follicular fluid of normal weight, overweight and obese women undergoing assisted reproductive treatment: a descriptive cross-sectional study. Reprod Biol Endocrinol 2014;12(1):13.

[86] Rooke J, Ewen M, Reis A, McEvoy TG. Comparison of the fatty acid contents of bovine follicular fluid, granulose cells and cumulus enclosed or denuded oocytes. Reprod Domest Anim 2006;41:306.

[87] Arya BK, Ul Haq A, Chaudhury K. Oocyte quality reflected by follicular fluid analysis in poly cystic ovary syndrome (PCOS): a hypothesis based on intermediates of energy metabolism. Med Hypotheses 2012;78(4):475–8.

[88] Van Hoeck V, Leroy JLMR, Arias-Alvarez M, Rizos D, Gutierrez-Adan A, Schnorbusch K, et al. Oocyte developmental failure in response to elevated nonesterified fatty acid concentrations: mechanistic insights. Reproduction 2013;145:33–44.

[89] Ferguson EM, Leese HJ. A potential role for triglyceride as an energy source during bovine oocyte maturation and early embryo development. Mol Reprod Dev 2006;73(9):1195–201.

[90] Aardema H, Vos PL, Lolicato F, Roelen BA, Knijn HM, Vaandrager AB, et al. Oleic acid prevents detrimental effects of saturated fatty acids on bovine oocyte developmental competence. Biol Reprod 2011;85(1):62–9.

[91] Abe H, Yamashita S, Satoh T, Hoshi H. Accumulation of cytoplasmic lipid droplets in bovine embryos and cryotolerance of embryos developed in different culture systems using serum-free or serum-containing media. Mol Reprod Dev 2002;61(1):57–66.

[92] Rizos D, Gutierrez-Adan A, Perez-Garnelo S, de la Fuente J, Boland MP, Lonergan P. Bovine embryo culture in the presence or absence of serum: implications for blastocyst development, cryotolerance, and messenger RNA expression. Biol Reprod 2003;68(1):236–43.

[93] Grummer RR, Carroll DJ. A review of lipoprotein cholesterol-metabolism – importance to ovarian function. J Anim Sci 1988;66(12):3160–73.

[94] McEvoy TG, Coull GD, Broadbent PJ, Hutchinson JSM, Speake BK. Fatty acid composition of lipids in immature cattle, pig and sheep oocytes with intact zona pellucida. J Reprod Fertil 2000;118(1):163–70.

[95] Macdonald RC, Macdonald RI. Membrane-surface pressure can account for differential activities of membrane-penetrating molecules. J Biol Chem 1988;263(21):10052–5.

[96] Zhang LY, Keung W, Samokhvalov V, Wang W, Lopaschuk GD. Role of fatty acid uptake and fatty acid beta-oxidation in mediating insulin resistance in heart and skeletal muscle. Biochim Biophys Acta, Mol Cell Biol Lipids 2010;1801(1):1–22.

[97] Chung BH, Tallis GA, Simon BH, Segrest JP, Henkin Y. Lipolysis-induced partitioning of free fatty-acids to lipoproteins – effect on the biological properties of free fatty-acids. J Lipid Res 1995;36(9):1956–70.

[98] Vanholder T, Leroy JL, Soom AV, Opsomer G, Maes D, Coryn M, et al. Effect of non-esterified fatty acids on bovine granulosa cell steroidogenesis and proliferation in vitro. Anim Reprod Sci 2005;87(1–2):33–44.

[99] Vanholder T, Leroy JL, Van Soom A, Maes D, Coryn A, Fiers T, et al. Effect of non-esterified fatty acids on bovine theca cell steroidogenesis and proliferation *in vitro*. Anim Reprod Sci 2006;92(1–2):51–63.

[100] Amsterdam A, Dantes A, Selvaraj N, Aharoni D. Apoptosis in steroidogenic cells: structure–function analysis. Steroids 1997;62(1):207–11.

[101] Kerental I, Suh BS, Dantes A, Lindner S, Oren M, Amsterdam A. Involvement of P53 expression in cAMP-mediated apoptosis in immortalized granulosa-cells. Exp Cell Res 1995;218(1):283–95.

[102] Jorritsma R, Cesar ML, Hermans JT, Kruitwagen CLJJ, Vos PLAM, Kruip TAM. Effects of non-esterified fatty acids on bovine granulosa cells and developmental potential of oocytes *in vitro*. Anim Reprod Sci 2004;81(3–4):225–35.

[103] Mu YM, Yanase T, Nishi Y, Tanaka A, Saito M, Jin CH, et al. Saturated FFAs, palmitic acid and stearic acid, induce apoptosis in human granulosa cells. Endocrinology 2001;142(8):3590–7.

[104] Geshi M, Takenouchi N, Yamauchi N, Nagai T. Effects of sodium pyruvate in nonserum maturation medium on maturation, fertilization, and subsequent development of bovine oocytes with or without cumulus cells. Biol Reprod 2000;63(6):1730–4.

[105] Cnop M, Hannaert JC, Hoorens A, Eizirik DL, Pipeleers DG. Inverse relationship between cytotoxicity of free fatty acids in pancreatic islet cells and cellular triglyceride accumulation. Diabetes 2001;50(8):1771–7.

[106] Van Hoeck V, Rizos D, Gutierrez-Adan A, Pintelon I, Jorssen E, Dufort I, et al. Interaction between differential gene expression profile and phenotype in bovine blastocysts originating from oocytes exposed to elevated non-esterified fatty acid concentrations. Reprod Fertil Dev 2013;. doi: 10.1071/RD13263 [Epub ahead of print].

[107] Leese HJ. Quiet please, do not disturb: a hypothesis of embryo metabolism and viability. Bioessays 2002;24(9):845–9.

[108] Sturmey RG, Brison DR, Leese HJ. Assessing embryo viability by measurement of amino acid turnover. Reprod BioMed Online 2008;17(4):486–96.

[109] Leese HJ, Baumann CG, Brison DR, McEvoy TG, Sturmey RG. Metabolism of the viable mammalian embryo: quietness revisited. Mol Hum Reprod 2008;14(12):667–72.

[110] Jungheim ES, Louden ED, Chi MM, Frolova AI, Riley JK, Moley KH. Preimplantation exposure of mouse embryos to palmitic acid results in fetal growth restriction followed by catch-up growth in the offspring. Biol Reprod 2011;85(4):678–83.

The Effect of Heavy Metals on Preterm Mortality and Morbidity

Gülcan Türker, MD

Kocaeli University, Kocaeli, Turkey

INTRODUCTION

Prematurity is the leading cause of neonatal morbidity and mortality. Although the mortality of premature infants has decreased in developed countries over the past decades, low-income and middle-income countries still have a higher neonatal mortality [1–4]. Approximately 15 million babies are born preterm; 5% of whom are under 28 weeks' gestation according to the World Health Organization (WHO). An estimated 2–8 million neonates are born preterm or small-for-gestational-age (SGA). Of the 1.2 million preterm births that occur in high-income regions, more than 0.5 million (42%) are in the United States [3]. Each year, an estimated 1.1 million neonates die from complications of preterm birth as estimated in 2010 [1–5]. Preterm birth is now the second most common cause of death in children under 5 years of age globally after pneumonia, and being born preterm also increases a baby's risk of death due to other causes. Preterm birth is a risk factor in over 50% of all neonatal deaths [3]. Every year an estimated 4 million babies die in the neonatal period [3]. A similar number are stillborn, and 0.5 million mothers die from pregnancy-related causes. Three-quarters of neonatal deaths happen in the first week – the highest risk of death is on the first day of life. Almost all (99%) neonatal deaths arise in low-income and middle-income countries, yet most epidemiological and other research focus on the 1% of deaths in rich countries. In addition, preterm birth is the leading risk factor for 393,000 deaths due to neonatal infections and contributes to long-term growth impairment and significant long-term morbidity such as cognitive, visual, and learning impairments [2,4]. Perinatal mortality is an indicator of the health of the mother and child and may reflect the conditions of reproductive health, which are related to socioeconomic status, quality of antenatal care, nutrition, and delivery factors. Some of the risk factors identified for prematurity or mortality are related to the mother and complications that occur during pregnancy, labor, and delivery, which may affect both the fetus and the newborn. These include maternal hypertension, placenta praevia, premature placental detachment, other abnormalities of the placenta, respiratory distress syndrome (RDS), cardiovascular disorders specific to the perinatal period (such as patent ductus arteriosus, PDA), infections, intrauterine growth restriction (IUGR), and low birth weight (LBW) [6,7]. Some lifestyle factors that contribute to spontaneous preterm birth include stress and excessive physical work or long times spent standing [8]. Smoking and excessive alcohol consumption as well as periodontal disease also has been associated with increased risk of preterm birth [9]. Also, nowadays a change in lifestyle has led to the development of new risk factors for preterm birth and preterm mortality with the worldwide increase of industrial pollution and anthropogenic or natural combustion activities. The environmental pollutants that cause preterm birth are environmental contaminants such as heavy metals, organic hydrocarbons, and pesticides. The influence of these pollutants on public health has been increasingly acknowledged, especially during the period of growth and development. Prenatal life is the most sensitive stage of human development due to the high degree of fetal cellular division and differentiation. The fetus is highly susceptible to teratogens, typically at low exposure levels that do not harm the mother due to differences compared to the adult in many biochemical pathways [10]. Also, if prenatal exposure occurs during organogenesis, permanent structural changes might occur. Alternatively, exposures after completion of organogenesis might result in functional consequences. More recently, studies have indicated that exposure to air pollution may be associated with LBW and preterm birth [11]. In addition, urban air pollution has also been directly implicated in perinatal

Handbook of Fertility. http://dx.doi.org/10.1016/B978-0-12-800872-0.00005-6

mortality [12]. The association between death in the neonatal period and air pollution was identified by Lipfert et al. and Hajat et al. [13,14]. An association between maternal exposure to ambient air pollution and very low birth weight (VLBW) (<1500 g) delivery has been demonstrated [15]. VLBW delivery, which is a high risk for serious morbidity and death of the infant, almost always is associated with delivery before 32 completed weeks of gestation and is often accompanied by IUGR [16].

Also, *in utero* exposure to heavy metals has been studied extensively over the last few decades; the threats posed on pregnancy outcomes and/or adverse developmental effects at levels lower than international guidelines have been observed [17–19]. Mattison suggested that impacts of environmental exposures on pregnancy outcome or development may have no thresholds, and the only reasonable approach is to keep environmental exposures for all individuals as low as possible [20]. Therefore, in this chapter the risk of preterm mortality and the effect of heavy metals on preterm mortality will be revised.

HEAVY METALS AND PRETERM MORTALITY AND MORBIDITY

Fetal exposure to environmental factors occurs through the amniotic fluid, the placenta, and the umbilical cord. During pregnancy, the placenta acts as a selective barrier by allowing nutrients and oxygen to pass to the fetus, and preventing toxic compounds from crossing through [21–23]. However, the placental barrier is not completely impermeable to the passage of harmful substances, such as drugs or toxic agents [22].

Several studies have used either *in vitro* or *in vivo* placental models to investigate the effects of environmental exposure on fetal growth [21,24,25]. Evaluation of *in utero* exposure to environmental pollutants has been mainly achieved by using umbilical cord and/or maternal blood samples [26–28]. As a noninvasive matrix, placenta has been used in biomonitoring studies as a dual biomarker for toxic metals to assess maternal and fetal health [29,30]. Also, the other biomarker for toxic metals is meconium, which can be obtained easily and noninvasively, and is representative of a long duration of fetal exposure to pesticides or metals during gestation [31–35]. Few studies have investigated the effect of heavy metals on prematurity or growth or preterm mortality [21,36–40]. Osman et al. suggested that the levels of lead in cord blood were a negative predictor of birth weight and length [21]. Also, we found higher heavy metals (Pb, Cd) levels in preterm newborns, especially with unknown etiology, than term newborns, and we suggested that these metals may affect gestational age at the fetal stage [36].

Heavy metals, such as arsenic (As), cadmium (Cd), mercury (Hg), and lead (Pb), have no beneficial effects in humans [41]. The general population is exposed to heavy metals at several concentrations either voluntarily through supplementation or involuntarily through intake of contaminated food and water or contact with contaminated soil, dust or air, alcoholic drinks, and tobacco [41–46]. Cigarette smoke contains about 30 metals, of which Cd, As, and Pb have the highest concentrations, and exposure to these toxic metals is associated with many serious diseases [45].

In 2009, Prasher et al. defined heavy metals as chemical elements with a specific gravity that is at least five times the specific gravity of water [44]. Although there is no clear definition of "heavy metal" density, heavy metals can be defined as chemical compounds having a specific density of more than 5 g/cm [3,44].

Hazardous waste sites provide a list of all hazards according to their prevalence and severity of their toxicity assessed by the U.S. Agency for Toxic Substances and Disease Registry (ATSDR). The first, second, third, and sixth hazards on the list are heavy metals: Pb, Hg, As, and Cd, respectively [46].

In small quantities, certain heavy metals are essential nutrients for a healthy life, belonging to the so-called "essential elements or trace minerals" group such as zinc (Zn), iron (Fe), manganese (Mn), and copper (Cu) [41–43]. They are absorbed through food along with proteins and complex carbohydrates and are necessary to maintain the normal functions of many enzymes and transcription factors [41–45]. Trace elements exist in all tissues and cells, and the dynamic balance of elements depends on their intestinal absorption and excretion by the intestine and kidneys [43]. Essential trace element levels change with acute stress and trauma; blood Zn and Fe concentrations often reduce during inflammation and infection [45,46]. At the initiation of the stress response, kidneys increase the excretion of trace elements. Preterm infants show acutely low blood levels of Zn, selenium (Se), and Fe, with low correlation with the degree of prematurity as patients with severe trauma [47,48]. Trace element levels of the body are susceptible to pathological conditions and large amounts of trace elements are needed. Also toxic levels of trace elements are harmful for humans and the high levels of trace elements and heavy metals appear to interfere with the other trace element metabolisms by altering their gastrointestinal and urinary excretion or absorption. Individuals will absorb more lead through their food if their diets are deficient in Ca, Fe, or Zn. During pregnancy, the placenta behaves as a very active transporter of essential elements (Ca, Cu, Zn, and Fe) and toxic elements (Cd, Hg, Nickel) to the developing fetus [22]. For example, elevated blood Pb levels during pregnancy is associated with increased plasma Cu and Fe but a decreased plasma Zn. Extra-lead is released from maternal skeleton during pregnancy due to mobilization of bone Pb stores, particularly in women who smoke or women whose Ca intake is low. Complications of pregnancy, such as gestational hypertension, malaria, and LBW deliveries, may be induced by elevated prenatal blood Pb, which alters trace element metabolism [49]. Pb and/or Cd exposure cause deficiency of antioxidant trace elements like Zn, Cu, Mn, Se, and Fe by promoting their urinary loss, which could result in further oxidative damage [50–52]. Replacing trace element deficiencies or detoxifying after toxic metal or trace elements exposure decreased mortality and morbidity. For example, replacing Fe, Zn, and folic acid during pregnancy has been shown to reduce the number of LBW and SGA babies, preterm delivery, and maternal anemia [53,54]. Replacement of Cu and Se in preterm infant during early postnatal management may decrease the ROP risk factors and may prevent diseases due to trace element deficiencies [55].

Although the effect of a high-level exposure to the heavy metals are well known for animals and humans, the data on the effects of the environmental exposure on the preterm mortality and placental fetal-maternal compartment are limited. Also, despite adults, the fetus is highly susceptible to subtle teratogens at low exposure levels that do not significantly compromise the clinical status of the pregnant woman. The number of live births did not markedly affect the levels of Cd and Pb but was negatively associated with Hg [56,57]. If exposure occurs during the organogenesis, heavy metals could produce permanent structural and anatomical changes. Also, if the exposure occurs after the end of the organogenesis, it might result in functional consequences. The immune, respiratory, and central nervous systems are also susceptible to postnatal exposures because they are immature at birth and characterized by a prolonged period of postnatal maturation [22,42,58].

LEAD

Lead is a widespread environmental pollutant associated with adverse health effects in humans. Sixty percent of lead is used for the manufacture of batteries (automobile batteries, in particular), while the remainder is used in the production of pigments, glazes, solder, plastics, cable sheathing, ammunition, weights, gasoline additive, and a variety of other products. Such industries continue to pose a significant risk to workers, as well as surrounding communities. Water with a lower pH can increase the impact of lead exposures. Lead in soil tends to concentrate in root vegetables and leafy green vegetables. Other more uncommon sources of lead exposure also continue to be sporadically found, such as improperly glazed ceramics, lead crystal, imported candies, certain herbal folk remedies, and vinyl plastic toys and

tobacco smoking [58]. Over the last few decades, however, Pb emissions in developed countries have decreased markedly due to the introduction of unleaded petrol. Subsequently, blood Pb levels in the general population have decreased [58]. Although remarkable reductions in Pb exposure of the general population have been achieved in most of North America and Europe, adults and children worldwide continue to be exposed through a variety of environmental media and informal sector occupations [59].

Lead and Placenta

Lead was found to easily cross the placental barrier by means of passive diffusion [60]. Similarly, a positive correlation was observed in the majority of the studies between placental and cord blood levels Pb, Hg, and Cd [61].

Pb may alter Ca-mediated cellular processes in syncytiotrophoblasts. Pb may precipitate, along with Ca, in the microvilli around the trophoblast [62,63]. Pb storage observed in syncytiotrophoblast cells may be related to a reduced cytochrome oxidase activity [64]. Pb can persist in bone for decades after exposure, which in turn serves as an internal source of exposure [65,66]. Endogenous Pb exposure is an important independent predictor of adverse health outcomes, such as cognitive decline [67,68], cardiovascular disease [69], and decreased fetal growth [70–72].

During pregnancy, these internal stores of Pb mobilize to a marked degree, partitioning into red blood cells (~99%) and plasma (~1%). The pattern of association between plasma and whole blood Pb remains largely unclear and whole blood Pb levels are not an accurate reflection of plasma Pb levels [73,74]. Lead levels in children continue to be a health hazard as the current limit of 10 μg/dl is considered to be too high with the WHO estimate of 40% of children having blood levels greater than 5 μg/dl. Some authors have suggested that a new limit should be set at 2 μg/dl [44,45]. Also, the length of gestation and intrauterine growth decrease inversely with the maternal blood Pb levels, thus Pb increases the risk of prematurity and SGA [75–77]. Thereby, Pb increases the risk of mortality and morbidity due to increased prematurity and SGA [1–5,75–77].

The Effect of Heavy Metals on Hypothalamic-Pituitary-Adrenal Axis

Preterm delivery is a complex condition with a multifactorial etiology. The mechanisms of effect for Pb exposure are unclear on preterm birth. However, it was shown that Pb may alter the hypothalamic-pituitary-adrenal (HPA) axis. Activation of the fetal HPA axis starts parturition through the fetal hypothalamus and/or the placenta increase in secretion of corticotropin-releasing hormone (CRH) ultimately leading to uterine contraction, cervical maturating, and decidual/fetal membrane activation [78]. Altering the pathway of CRH release during pregnancy has been suggested as one possible mechanism that can lead to preterm birth [79–82]. Increased maternal stress is a known risk factor in preterm delivery, which is supposed to act through the maternal/fetal HPA pathway by altering CRH release [83]. Lead exposure may play an important role by increasing the overall baseline level of corticosterone in rats and increasing the response to acute stressors [84,85]. Therefore, we may suggest that increased Pb exposure early in gestation may alter corticotrophin-releasing hormone release alone or together with increased maternal stress responses.

The Effect of Heavy Metals on Oxidative Stress

Heavy metals, such as in preterm mortality and morbidity, can damage human health through an oxidative cell stress (e.g., Cd, Cr, Pb, As), neurological or renal damage (e.g., Pb, Hg), DNA injury (e.g., As, Cr, Cb), or altered glucose (e.g., As) [20–26,84–98]. However, the exact mechanisms of toxicity for these heavy metals are not fully understood. It was shown that Pb exposure raises superoxide and hydrogen peroxide in human endothelial and vascular smooth muscle cells. Pb may play a role in enzymatic inhibition or impaired antioxidant metabolism and in induction of oxidative stress [98,99]. The toxic effects of heavy metals is not only in the induction of oxidative stress, but may also be associated with substitute diverse polyvalent cations (Ca, Zn, and Mg) [98] that function as charge carriers, intermediaries in catalyzed reactions, or as structural elements in the maintenance of protein conformation. Indeed, metals accumulate in cellular organelles and interfere with their function [98–100].

Oxidative stress, defined as an imbalance between cellular reactive oxygen species (ROS) and their control by antioxidants, is one of the most important toxic effects of Pb [101–105]. An oxidative stress occurs at birth in all newborns as a result of the hyperoxic environment due to the transition from the hypoxic intrauterine environment to extrauterine life. Free radical sources (such as inflammation, hydroxyl radicals, hypoxia–hyperoxia, and ischemia), glutamate, and high free Fe release increase oxidative stress during the perinatal period [106–110]. In addition, preterm babies have reduced antioxidant defenses that are not able to counteract the harmful effects of ROS leading to the so called "free-radicals-related disease" of newborns including bronchopulmonary dysplasia (BPD), reperfusion injury after hypoxia, intraventricular hemorrhage (IVH), necrotizing enterocolitis

(NEC), retinopathy of prematurity (ROP), increased ferritin levels due to more packed red cell transfusion, and infections [106–110]. Also, pro-inflammatory cytokines, vascular growth factors, and lipid peroxidation increase in these diseases [109,110]. Free-radicals-related diseases, RDS, PDA, acute renal failure, perinatal hypoxia, and infections are responsible for preterm morbidity and mortality. Also, it is known that Pb may cause a reduction in the activity of antioxidant enzymes and in glutathione content because of the metal's high affinity to sulfhydryl groups and metal cofactors [101]. The activity of antioxidant enzymes and glutathione content in preterm are less than that in term infant [111,112]. Also, preterm infants are more susceptible to oxidative stress than term infants and adults [106–112]. From a mechanistic point of view, as Shachar et al. hypothesized about preterm birth in their review, Pb seems to play a role in the pathogenesis of free-radicals-related diseases and mortality with enzymatic inhibition and in impaired antioxidant metabolism, and affects oxidative stress in preterm infants [102].

In addition to decreasing antioxidant defense, Pb may interfere with membrane integrity and fatty-acid composition, contributing to the production of ROS [113]. Polymorphisms in genes related to oxidative stress have been observed to modulate the effects of heavy metals exposure in medical conditions such as cardiovascular disease and cancer [69,84,99,100,113]. Although there are many findings showing that heavy metals exposure causes preterm birth, no published study has explored the potential interactive effects between oxidative stress genes and heavy metal exposure on risk susceptibility to preterm birth [36–40,70,73,76–79]. However, in light of the information that shows the effects of heavy metals on polymorphisms in genes related to oxidative stress, Shachar et al. noted that lead seems to play a role in the pathogenesis of preterm birth [98–108].

Also, heavy metals affect regulation of detoxification genes, which in turn modulate oxidative stress. Detoxification processes involve two major phases: phase I, entailing the polarization of toxic compounds, and phase II. Cytochrome P450 1A1(CYP1A1) is a phase I hepatic and extra hepatic enzyme whose activity is modulated by heavy metals (Pb, Cd, Cu, chromium, vanadium, and especially Fe) via regulating the aryl hydrocarbon receptor signaling pathway of CYP1A1 [113].

Glutathione S-transferase theta 1 (GSTT1) is a phase II detoxification enzyme catalyzing the conjugation of glutathione to heavy metals and other toxicants [114–122]. For example, the selenoenzymes thioredoxin reductase and glutathione peroxidase were inhibited via interaction with heavy metals especially Hg [115]. CYP1A1 and GSTT1 are highly polymorphic genes and the polymorphisms are associated with differential susceptibility to the adverse effects of heavy metals [114–122]. Deletions in glutathione S-transferases mu 1 (GSTM1) and GSTT1 were found to be associated with higher Hg levels [114–122]. Polymorphisms in CYP1A1, GSTM1, and GSTT1 have been associated with differential risk for preterm birth [120,121]. Nukui et al. found that women and fetuses exposed to cigarette smoke during pregnancy and those who were characterized by a GSTT1 null genotype had an elevated risk for preterm birth [121]. Also, Tsai et al. showed the association between Cd exposure, high-risk CYP1A1 and GSTT1 genotypes polymorphisms, and preterm birth in maternal cigarette smoking [122].

Human paraoxonases (PONs) are a family of proteins that provide protection from physiological oxidative stress and inflammatory processes. In particular, para-oxonase 1 (PON1) is a high-density lipoprotein-binding enzyme exhibiting calcium-dependent enzyme esterase, lactonase, and peroxidase activity and hydrolyzes toxic oxidized lipids and protects against cardiovascular diseases [123,124]. There is a link between genetic polymorphisms of PON1 and hyperlipidemia and increased lipid oxidation [123–126]. Lüersen et al. showed that Pb exposure is associated with decreased serum PON1 activity and PON1 R192 genotypes [127]. The low-antioxidant PON1 R192 allele is associated with increased tumor necrosis factor (TNF)-alpha, a pro-inflammatory cytokine known to be involved in the initiation process of atherosclerosis [127]. Metal ions, particularly Pb, Cu, and Hg, may bind to free sulfhydryl groups of the PON1 enzyme thus reducing its hydrolytic activity and its antioxidant function [128].

Subjects who are homozygous for the Q192R PON1 allele are more susceptible to Pb toxicity, or subjects who have the 192R alloform have higher levels of As and Pb blood levels [128,129]. Based on the relationship of heavy metals with atherosclerosis and of PON1 with lipoproteins, Chen et al. hypothesized that the variants in the gene of PON1 enzyme may be related to vascular endothelial damage, placental insufficiency, and thrombosis, and thus could lead to adverse pregnancy [130]. Later, Ryckman et al. showed maternal and fetal genetic associations of PTGER3 and PON1 with preterm birth [131]. The PON1 gene has three polymorphisms that are considered to be strong determinants of PON1 levels (especially Q192R and L55M in the coding region and C-108T in the promoter region) [132]. Infante-Rivard showed that an increased risk of SGA birth is associated with these variants of the PON1 gene [132]. An increased risk of preterm birth was found among infants with PON1 Q192R genotype compared with other genotypes. Baker et al. found lower mid-gestation PON1 activity in women who later had spontaneous preterm birth [133]. Lee et al. found an increased risk for preterm birth among women with elevated body mass index (BMI > 30) and PON1 Q192R allele [134]. An increased concentration of products of lipid peroxidation,

TNF-alpha, and vascular growth factors were demonstrated in postnatal free-radicals-related diseases as hypoxic ischemic encephalopathy, BPD, NEC, ROP, PVL, and so on were demonstrated in premature infants and higher levels were found in those who subsequently developed BPD [106–110]. A positive correlation was found between the duration of oxygen and ventilation therapy and lipid peroxidation by Inder et al. [135]. Therefore, it may be hypothesized that increased lipid peroxidation, TNF-alpha as pro-inflammatory cytokines and reduced PON1 antioxidant activity due to Pb, Cd, Hg, may promote a pro-oxidant situation and enhance the risk for preterm birth and free-radicals-related diseases, such as BPD, NEC, ROP, and sepsis, and mortality [120–137].

Heme-oxygenase (HO) is the enzyme of the microsomal heme degradation pathway that yields biliverdin, carbon monoxide, and iron [138]. Also, the expression of the HO 1 (HO-1) related gene, which contributes to the maintenance and development of a healthy full-term pregnancy, can be induced by heavy metals, especially Cd [139]. The promoter region of the HO-1 gene contains a Cd response element, the binding site for Cd, which is a potent inducer of HO-1 transcription [139]. The polymorphism of the human HO-1 gene promoter composes of the single nucleotide polymorphism and dinucleotide polymorphism, which may contribute to the fine-tuning of the transcription. Long alleles have been found to be associated with susceptibility to smoking-induced emphysema or coronary artery disease, while they may be linked to resistance to cerebral malaria [135–139]. However, the effects of heavy metals on HO-1 gene polymorphisms have not been studied in association with preterm birth or preterm mortality and morbidity. Despite this, it may be possible that Cd exposure and susceptibility are linked to preterm birth through the binding of Cd to the Cd response element of the HO-1 gene [139–141].

Furthermore, the toxic effects of heavy metals, apart from inducing oxidative stress, can also be attributed to their ability to substitute diverse polyvalent cations (Ca, Zn, and Hg) that function as charge carriers, intermediaries in catalyzed reactions, or as structural elements in the maintenance of protein conformation. Indeed, metals accumulate in cellular organelles and interfere with their function. For example, it has been observed that Pb accumulation in mitochondria induces several changes such as inhibition of Ca^2 uptake, reduction of the transmembrane potential, oxidation of pyridine nucleotides, and a fast release of accumulated Ca^{2+} [142].

The Other Effects of Lead

Moreover, metals bind to proteins [143] and inhibit a large number of enzymes including the mitochondrial ones [144]. Nucleic-acid-binding proteins are also involved, while it has been shown that metals can also bind to DNA, affecting the expression of genes. Finally, some metals interfere with various voltage- and ligand-gated ionic channels exerting neurotoxic effects. For instance, Pb affects the N-methyl-D-aspartic acid (NMDA) receptor, subtypes of voltage- and Ca-gated K channels, cholinergic receptors, and voltage-gated Ca channels [144]. We found an association between heavy metals (Pb and Cd) levels and mortality, in addition to non-surviving preterm infants, with low Apgar scores as in the study of Al-Saleh et al. [37,38]. These ionic channels also play a role in the pathogenesis of hypoxic ischemic encephalopathy. Therefore, it may be suggested that the relationship of ionic channels activity due to Pb, Cd, and Hg may promote perinatal hypoxia and enhance the risk for mortality.

MERCURY

Mercury is present in the environment due to both natural (evaporation of earth's surface and volcanic eruptions) and anthropogenic (emissions from coal-burning power stations and incinerators) sources. Different chemical forms of Hg are present, such as elemental Hg (Hg0), inorganic (divalent and monovalent cationic forms: Hg^{2+} and Hg^{+}), and organic (i.e., methylmercury (MeHg) and ethylmercury (EtHg)) Hg compounds [145].

Metallic Hg is used in thermometers, dental amalgams, and some batteries. Metallic Hg is a liquid. It is not hazardous if ingested. Although metallic Hg is not significantly absorbed in this form, if aerosolized, metallic mercury is well absorbed by the lungs. Divalent and monovalent cationic forms of mercury are encountered in some chemical, metal-processing, electrical-equipment, automotive, and building industries, and in medical and dental services. These cationic forms, inorganic and organic compounds combined with other chemicals, can readily be absorbed through ingestion. All three forms of mercury are toxic to various degrees [146].

MeHg is actively transferred from the placenta to the fetus, reaching the fetal brain as its cysteine conjugate via the neutral amino acid carrier system as well as a great retention in the fetus brain [146–148]. Its high entry in the developing brain is related to the lack of functional blood–brain barrier (BBB) [148]. Maternal exposure to MeHg during pregnancy causes neurological deficits in their offspring [137,145–148]. MeHg affects cell differentiation, migration, and synaptogenesis [147,148]. Therefore, exposure to MeHg during early fetal development leads to subtle brain injury at levels much lower than those affecting the mature brain [147].

EtHg exposure occurs only through thimerosal-containing vaccines widely used in pediatric populations of third-world countries. Ethylmercury is eliminated more rapidly and bioaccumulates much less than methylmercury [149]. Multiple studies have concluded that there is no association between prenatal or perinatal exposure to thimerosal and autism spectrum disorders [150,151]. However, psychomotor development index was negatively affected by EtHg. Age of talking was negatively affected by Hg in hair [152].

Hg0 is absorbed mainly through the respiratory tract. Then it is oxidized mainly by erythrocyte catalase to mercurous (Hg^{+}) and mercuric (Hg^{2+}) ions. These ions are toxic to several organs, such as the central nervous system (CNS), but particularly the kidneys. Conversely, a certain amount of blood Hg0 passes through the BBB, reaching the CNS. Data on the distribution of brain Hg after Hg0 exposure are limited [137,145,146,153–157]. In an experimental study, Hg was found in both glial cells and neurons mainly in the cortical areas and in the fiber systems by Warfvinge et al. [158].

The interactions of MeHg with sulfhydryl-containing proteins (i.e., neurotransmitter receptors, transporters, antioxidant enzymes, etc.), as well as with nonprotein thiols (i.e., glutathione, cysteine), are crucial events in mediating MeHg's neurotoxicity [159–161]. By direct interaction with thiols, as well as indirect mechanisms, MeHg can modify the oxidation state of the –SH groups on proteins, modulating their functions [161]. Consequently, the activities of several SH-containing proteins whose roles are decisive for proper homeostasis of neuronal and glial cells (creatine kinase [162], GSH reductase [163], Ca^{2+}-ATPase [164], thioredoxin reductase [165], choline acetyltransferase and enolase [166]) are disturbed after MeHg exposure. Altered protein function has been posited as a causative factor in MeHg-induced neurotoxicity and neurodegeneration [167]. Moreover, several *in vitro* and *in vivo* studies have shown that MeHg exposure causes GSH depletion [168–172]. Ni et al. compared the responses of astrocytes and microglia following MeHg exposure, specifically addressing the effects of MeHg on cell viability, ROS generation, and GSH levels, as well as Hg uptake and the expression of NF-E2-related factor 2 (Nrf2). They showed that microglia are more sensitive to MeHg than astrocytes, a finding that is consistent with their higher Hg uptake and lower basal GSH levels. Microglia also demonstrated higher ROS generation compared with astrocytes. Nrf2 and its downstream genes were upregulated in microglia with much faster kinetics than in astrocytes. Therefore, Ni et al. suggested that microglia and astrocytes each exhibit a distinct sensitivity to MeHg, resulting in their differential temporal adaptive responses; these unique sensitivities appear to be dependent on the cellular thiol status of the particular cell type [169]. Franco et al. noted that the exposure of lactating mice to MeHg causes significant impairments in motor performance in the offspring, which may be related to a decrease in the cerebellar thiol status, and the increased GSH levels and glutathione reductase activity, observed only in the cerebellums of MeHg-exposed dams, could represent compensatory pathophysiologic responses to the oxidative effects of MeHg toward endogenous GSH [170].

GSH has a crucial role in maintaining redox homeostasis [161]. Drescher et al. recently showed a gene–environment interaction between the rs705379 polymorphism and MeHg exposure on PON1 activity levels in the aboriginal population [173]. Therefore, we suggested that antioxidant depletion, increased lipid oxidation, and increased oxidative stress due to exposure of MeHg seem to play a role in the pathogenesis of preterm birth, mortality, morbidity, and PVL due to free-radicals or inflammatory-related diseases [106–110,168–172].

Also, low-level Hg is toxic to the fetal and infant nervous system. Mothers exposed to Hg in the 1955 disaster in Minamata Bay, Japan, gave birth to infants with mental retardation, retention of primitive reflexes, cerebellar symptoms, and other abnormalities. Recent research in the Faroe Islands has demonstrated that even at much lower levels, it is associated with decrements in motor function, language, memory, and neural transmission in the offspring [174]

Organic mercury, the form of mercury bioconcentrated in fish and whale meat, readily crosses the placenta and appears in breast milk and meconium. Organic mercury can also be excreted via the biliary route, likely complexed to GSH, as a GSH mercaptide [174,175].

Matrix metalloproteinases are also thought to modulate the effects of heavy metals especially Hg. Matrix metalloproteinases hydrolyze extracellular matrix components in a variety of processes ranging from implantation and embryogenesis to inflammation, injury healing, metastasis, and angiogenesis [176]. Matrix metalloproteinases expression may be promoted by inflammatory cytokines, growth factors, hormones, extracellular matrix components, and ROS. BPD is characterized by a tissue remodeling, and is divided into different phases ending in the chronic phase with an increased number of fibroblasts and fibrotic areas. Matrix metalloproteins are important in regulating fibrotic processes; they degrade extracellular matrix proteins and fibrillar collagen. The balance between matrix metalloproteins and their inhibitors is of primary importance in the development and regulation of fibrosis. Oxidative stress increases both matrix metalloproteins and their inhibitors [112], so it may lead to lung damage by increasing collagenase activity, causing disruption of the extracellular matrix. Also, matrix metalloproteins expression in addition to inflammatory cytokines, growth factors, and ROS may play a role in pathogenesis and postnatal angiogenesis of BPD and ROP [174,175]. Conversely, matrix metalloproteins secreting cells also produce tissue inhibitors of metalloproteinases [174,175]. Metalloproteinases, whose production is enhanced by prostaglandins, play an important role in promoting uterine contractions and preterm birth by stimulating cervical ripening and membrane activation [178]. In particular, metalloproteinases degrade extracellular matrix proteins, such as collagens and fibronectins, leading to weakening of membranes and their subsequent rupture [179]. In a recent study by Sundrani et al. [180], placental metalloprotein 1 and metalloprotein 9 levels were significantly increased in women with spontaneous preterm birth compared to those delivering at term, suggesting that placental metalloprotein 1/metalloprotein 9 levels may be involved in the initiation of parturition [181]. Polymorphisms in metalloproteinases and tissue inhibitors of metalloproteinases have been found to modulate the risks for PPROM and preterm birth [181,182]. Several polymorphisms in metalloproteinases and tissue inhibitors of metalloproteinases have been found to modulate the effects of heavy metal exposure (especially Hg) on the risk of cardiovascular diseases [177]. Also, the single nucleotide polymorphisms in these genes are associated with differential risk of preterm birth [183,184]. Several other oxidative stress-related genes have been associated independently with heavy metals exposure and adverse pregnancy outcomes. Early-life exposures to this metal have been associated with long-lasting and enduring neurobehavioral and neurochemical deficits [185]. Al-Saleh et al. showed that Hg in the breast milk of mothers and in the urine of infants affected the excretion of urinary 8-hydroxy-2′-deoxyguanosine (8-OH 2-dG) and malondialdehyde (MDA), which are biomarkers of oxidative stress [186]. Also, they showed the association of high levels of heavy metals (Hg and Cd) with low Apgar score [38]. They suggested that exposure to Hg-induced oxidative stress in the newborn or fetus may play a role in pathogenesis, particularly during neurodevelopment [186].

Therefore, it may be suggested that the effect of Hg on glial cell reactivity and Hg-induced oxidative stress via matrix metalloproteinases or PON1 modulating the effects of heavy metals seem also to play a role in the pathogenesis of preterm birth, perinatal hypoxia, BPD, preterm brain inflammation, and mortality.

CADMIUM

Cadmium is frequently used in electroplating, pigments, paints, welding, and Ni–Cd batteries. Workers in these occupations are exposed to Cd at significantly higher levels than the general population [187]. The general population is exposed to Cd via drinking water, food, and cigarette smoking [188]. Alcoholic beverages, including wine, can be contaminated with metals in concentrations exceeding the tolerable limits and causing toxic effects, particularly in heavy drinkers [46].

Cd, at a high dose, is a potent teratogen in rodents [189,190]. When administered to mice during gestation, Cd-induced malformations of the neural tube, craniofacial region, limbs, trunk, viscera, and axial skeleton in fetuses are seen [190–196]. In addition, maternal Cd exposure during pregnancy-induced fetal growth restriction [197,198], fetal death and birth defects [197,198], sudden infant death [199], and preterm birth [37–39,200]. In contrast to Pb and Hg, the placenta acts as a barrier for Cd and very little (10%) is transferred to the fetus [201]. Wang et al. showed that Cd passed to fetuses whose mothers were exposed to Cd during the late pregnancy period [202]. Lower IQ test, crown–heel lengths, Apgar 5-min scores, birth weights, and SGA births below the 10th percentile and placental thickness were influenced by Cd levels in the umbilical cord [38,203]. Also, prenatal Cd exposure may have a detrimental effect on head circumference at birth and child growth in the first 3 years of life [204]. Cd caused a significant increase in the quantity of micronucleated polychromatic erythrocytes and micronucleated normochromatic erythrocytes in blood cells of both the mothers and their fetuses. An increased level of oxidation was also found, measured by a rise in the levels of the adduct 8-hydroxy-2'-deoxyguanosine. Also, in a dose-dependent manner, Cd caused external, visceral, and skeletal malformations and increased the level of DNA oxidation [205,206].

Cd induces oxidative stress via similar mechanisms to Hg or Pb [90,93,94,102,103]. Also, Samuel et al. showed that Cd exposure increases lipid peroxidation and H_2O_2 concentration [207]. They showed that specific activities of superoxide dismutase, catalase, glutathione peroxidase, glutathione reductase, glutathione-S-transferase, and serum testosterone, estradiol, and progesterone were significantly decreased. Cd reduced the hematocrit, hemoglobin content, and mean corpuscular volume (MCV) levels. However, the number of erythrocytes increased, which has been demonstrated by Samuel et al. [207].

Inhibition of Zn-dependent enzymes, and metabolic changes in the fetus and postnatal hepatic tissues have also been reported on Cd exposure [208]. Pillai et al. showed that early developmental exposure to Pb and Cd both alone and in combination can suppress the hepatic xenobiotic metabolizing enzyme activities in the liver of phase I generation (NADPH- and NADH-cytochrome *c* reductase) and phase II hepatic xenobiotic- and steroid-metabolizing enzymes *in vivo* [209].

It has been demonstrated in rodents that the hypothalamo-pituitary-liver axis is influenced by hormonal signaling during development [210].

Lipid peroxidation indicates free radical toxicity in the liver of Pb- and Cd-treated female and male rats. Reduced glutathione is a major cellular reductant and plays a very important role against the deleterious effects of hydroxyl radicals [211–215].

Maternal Cd exposure during pregnancy induces endoplasmic reticulum (ER) stress in the placenta. Alpha-phenyl-*N*-*t*-butylnitrone, a free-radical spin-trapping agent and antioxidant, significantly alleviated Cd-induced placental ER stress [202].

Thus, the data generated from these studies suggest that prenatal exposure to Cd may play a role via the effect on pathogenesis of related oxidative stress in preterm birth, morbidity, and mortality [211–215].

ARSENIC

Not only millions of people worldwide are exposed to elevated concentrations mainly via drinking water but also via industrial emissions, food, and air [216]. Exposure to As leads to an accumulation of As in tissues, such as the skin, hair, and nails, resulting in various clinical symptoms such as hyperpigmentation and keratosis. Arsenic is a potent human carcinogen and general toxicant but little is known about its effects on maternal and fetal health. Arsenic is metabolized via methylation and reduction reactions, methylarsonic acid and dimethylarsinic acid being the main metabolites excreted in urine [217,218]. Inorganic As and its methylated metabolites easily pass the placenta and both experimental and epidemiologic studies have shown increased risk of impaired fetal growth, and fetal and infant mortality [219–224]. Also, experimental studies have shown associations between fetal exposure and neurotoxicity and behavioral changes [224]. Recent studies have reported that high-level exposure to arsenic *in utero* is associated with increased infant mortality, LBW, and birth defects [225,226]. Also, another recent study has reported that *in utero* exposure to low levels of As may affect the epigenome [227]. Engström et al. showed that the effect of the AS3MT polymorphisms is of importance for risk assessment of As and they thought that a low methylation efficiency is associated with increased risk for As toxicity [228]. Arsenic inhibits DNA repair enzymes and antioxidant-related enzymes (e.g., thioredoxin reductase as Hg) [229, 230]. Altered protein function enzyme inhibition and oxidative stress, as well as epigenetic effects mainly via DNA hypomethylation, have been posited as a causative factor in As-induced neurotoxicity, endocrine effects (most classes of steroid hormones), immune suppression, neurotoxicity, and interaction with enzymes critical for fetal development and programming IUGR, spontaneous abortions, premature delivery, malformations, birth defects, learning and behavior deficits, and neonatal deaths [231–236]. Two to three times more spontaneous abortion and stillbirth have been reported among women especially those who were exposed to high As concentrations in their drinking water (>50 μg/l) [231,237].

Communities with water As levels greater than 5 µg/l should consider a program to document As levels in the population [237]. The fetus and infant are probably partly protected by the increased methylation of As during pregnancy and lactation; the infant is also protected by low As excretion in breast milk [216]. A couple of studies reported increased blood pressure and anemia during pregnancy. Susceptibility to As is dependent on the biomethylation, which occurs via one-carbon metabolism [238–240].

CONCLUSIONS

Heavy metals affect fetal development and programming IUGR, spontaneous abortions, premature delivery, malformations, birth defects, learning and behavior deficits, and neonatal deaths and preterm morbidity through increasing oxidative stress, lipid peroxidation, and TNF-alpha such as pro-inflammatory cytokines and reducing PON1 antioxidant activity. Also, heavy metals may be modulated by gene polymorphisms associated with these activities. They induce DNA biomethylation, interfere in the conjugation of detoxification or catalyzing enzymes by binding to these enzymes, or interfere with the trace elements metabolism by altering their gastrointestinal absorption and urinary excretion. Also, the impacts of environmental exposures on pregnancy outcome or development may have no thresholds. Therefore, the approach is to keep environmental exposures for all individuals as low as possible.

References

[1] UNICEF. Low Birthweight: Country, Regional and Global Estimates. New York: WHO; 2004.

[2] Lee ACC, Katz J, Blencowe H, Cousens S, Kozuki N, Vogel JP, et al. Born too small: national and regional estimates of term and preterm small-for-gestational-age in 138 low-income and middle-income countries in 2010. Lancet Glob Health 2013;1(1):e26–36.

[3] Blencowe H, Cousens S, Chou D, Oestergaard M, Say L, Moller AB, et al. Born too soon: the global epidemiology of 15 million preterm births. Reprod Health 2013;10(Suppl 1):S2.

[4] Kramer MS, Demissie K, Yang H, Platt RW, Sauve R, Liston R, et al. The contribution of mild and moderate preterm birth to infant mortality. JAMA 2000;284:843–9.

[5] Katz J, Lee AC, Kozuki N, Lawn JE, Cousens S, Blencowe H, et al. Mortality risk in preterm and small-for-gestational-age infants in low-income and middle-income countries: a pooled country analysis. Lancet 2013;382(9890):417–25.

[6] Jackson DJ, Lang JM, Ganiats TG. Epidemiological issues in perinatal outcomes research. Paediatr Perinat Epidemiol 1999;13:392–404.

[7] Beck S, Wojdyla D, Say L, Betran AP, Merialdi M, Requejo JH, et al. The worldwide incidence of preterm birth: a systematic review of maternal mortality and morbidity. Bull World Health Organ 2010;88:31–8.

[8] Stoll BJ, Hansen NI, Bell EF, Shankaran S, Laptook AR, Walsh MC, et al. Neonatal outcomes of extremely preterm infants from the NICHD Neonatal Research Network. Pediatrics 2010;126:443–56.

[9] Muglia LJ, Katz M. The enigma of spontaneous preterm birth. N Engl J Med 2010;362:529–35.

[10] Gravett MG, Rubens CE, Nunes TM. GAPPS Review Group. Global report on preterm birth and stillbirth (2 of 7): discovery science. BMC Pregnancy Childbirth 2010;10(Suppl 1):S5.

[11] Stoll BJ, Hansen NI, Bell EF, Shankaran S, Laptook AR, Walsh MC, et al. Neonatal outcomes of extremely preterm infants from the NICHD Neonatal Research Network. Pediatrics 2010;126:443–56.

[12] Wu J, Ren C, Delfino RJ, Chung J, Wilhelm M, Ritz B. Association between local traffic-generated air pollution and preeclampsia and preterm delivery in the south coast air basin of California. Environ Health Perspect 2009;117:1773–9.

[13] Lipfert FW, Zhang J, Wyzga RE. Infant mortality and air pollution: a comprehensive analysis of U.S. data for 1990. J Air Waste Manag Assoc 2000;50:1350–66.

[14] Hajat S, Armstrong B, Wilkinson P, Busby A, Dolk H. Outdoor air pollution and infant mortality: analysis of daily time-series data in 10 English cities. J Epidemiol Community Health 2007;61:719–22.

[15] Rogers JF, Dunlop AL. Air pollution and very low birth weight infants: a target population? Pediatrics 2006;118(1):156–64.

[16] Pedersen M, Giorgis-Allemand L, Bernard C, Aguilera I, Andersen AM, Ballester F, et al. Ambient air pollution and low birthweight: European cohort study (ESCAPE). Lancet Respir Med 2013;1(9):695–704.

[17] Cantonwine D, Hu H, Sánchez BN, Lamadrid-Figueroa H, Smith D, Ettinger AS, et al. Critical windows of fetal lead exposure: adverse impacts on length of gestation and risk of premature delivery. J Occup Environ Med 2010;52(11):1106–11.

[18] Sanin LH, Gonzalez-Cossio T, Romieu I, Peterson KE, Ruíz S, Palazuelos E, et al. Effect of maternal lead burden on infant weight and weight gain at one month of age among breastfed infants. Pediatrics 2001;107:1016–23.

[19] Jelliffe-Pawlowski LL, Miles SQ, Courtney JG, Materna B, Charlton V. Effect of magnitude and timing of maternal pregnancy blood lead (Pb) levels on birth outcomes. J Perinatol 2006;26:154–62.

[20] Mattison DR. Environmental exposures and development. Curr Opin Pediatr 2010;22(2):208–18.

[21] Osman K, Akesson A, Berglund M, Bremme K, Schütz A, Ask K, et al. Toxic and essential elements in placentas of Swedish women. Clin Biochem 2000;33:131–8.

[22] Caserta D, Graziona A, Lo Monte G, Bordi G, Moscarini M. Heavy metals and placental fetal-maternal barrier: a mini-review on the major concerns. Eur Rev Med Pharmacol Sci 2013;17:2198–206.

[23] Wier PJ, Miller RK, Maulik D, DiSant'Agnese PA. Toxicity of cadmium in the perfused human placenta. Toxicol Appl Pharmacol 1990;105:156–71.

[24] Amaya E, Gil F, Freire C, Olmedo P, Fernández-Rodríguez M, Fernández MF, et al. Placental on centrations of heavy metals in a mother–child cohort. Environ Res 2013;120:63–70.

[25] Zdravkovica T, Genbaceva O, Mc Mastera MT, Fisher SJ. The adverse effects of maternal smoking on the human placenta: a review. Placenta 2005;26(Suppl A):S81–6.

[26] Torres-Sánchez LE, Berkowitz G, López-Carrillo L, Torres-Arreola L, Ríos C, López-Cervantes M. Intrauterine lead exposure and preterm birth. Environ Res 1999;81(4):297–301.

[27] Raghunath R, Tripathi RM, Sastry VN, Krishnamoorthy TM. Heavy metals in maternal and cord blood. Sci Total Environ 2000;250(1-3):135–41.

[28] Sakamoto M, Yasutake A, Domingo JL, Chan HM, Kubota M, Murata K. Relationships between trace element concentrations in chorionic tissue of placenta and umbilical cord tissue: potential use as indicators for prenatal exposure. Environ Int 2013;60:106–11.

[29] Leino O, Kiviranta H, Karjalainen AK, Kronberg-Kippilä C, Sinkko H, Larsen EH, et al. Pollutant concentrations in placenta. Food Chem Toxicol 2013;54:59–69.

[30] Iyengar GV, Rapp A. Human placenta as a 'dual' biomarker for monitoring fetal and maternal environment with special reference to potentially toxic trace elements. Part 3: toxic trace elements in placenta and placenta as a biomarker for these elements. Sci Total Environ 2001;280(1-3):221–38.

[31] Turker G, Ergen K, Karakoç Y, Arisoy AE, Barutcu UB. Concentrations of toxic metals and trace elements in the meconium of newborns from an industrial city. Biol Neonate 2006;89:244–50.

[32] Ostrea EM, Morales V, Ngoumgna E, Prescilla R, Tan E, Hernandez E, et al. Prevalence of fetal exposure to environmental toxins as determined by meconium analysis. Neurotoxicology 2002;23:329–39.

[33] Friel JK, Matthew JD, Andrews WL, Skinner CT. Trace elements in meconium from preterm and full-term infants. Biol Neonate 1989;55:214–7.

[34] Haram-Mourobet S, Harper RG, Wapnir RA. Mineral composition of meconium: effect of prematurity. J Am Coll Nutr 1998;17:356–60.

[35] Ostrea EM Jr, Bielawski DM, Posecion NC Jr, Corrion M, Villanueva-Uy E, Jin Y, et al. A comparison of infant hair, cord blood and meconium analysis to detect fetal exposure to environmental pesticides. Environ Res 2008;106(2):277–83.

[36] Özsoy G, Turker G, Özdemir S, Gökalp AS, Barutçu B. The effect of heavy metals and trace elements in the meconium on preterm delivery of unknown etiology. Turkiye Klinikleri J Med Sci 2012;32(4):925–31.

[37] Turker G, Özsoy G, Özdemir S, Barutçu B, Gökalp AS. Effect of heavy metals in the meconium on preterm mortality: preliminary study. Pediatr Int 2013;55:30–4.

[38] Al-Saleh I, Shinwari N, Mashhour A, Rabah A. Birth outcome measures and maternal exposure to heavy metals (lead, cadmium and mercury) in Saudi Arabian population. Int J Hyg Environ Health 2014;217(2-3):205–18.

[39] Cantonwine D, Hu H, Sánchez BN, Lamadrid-Figueroa H, Smith D, Ettinger AS, et al. Critical windows of fetal lead exposure: adverse impacts on length of gestation and risk of premature delivery. J Occup Environ Med 2010;52(11):1106–11.

[40] Vigeh M, Yokoyama K, Seyedaghamiri Z, Shinohara A, Matsukawa T, Chiba M, et al. Blood lead at currently acceptable levels may cause preterm labour. Occup Environ Med 2011;68(3):231–4.

[41] Llobet JM, Falcó G, Casas C, Teixidó A, Domingo JL. Concentrations of arsenic, cadmium, mercury, and lead in common foods and estimated daily intake by children, adolescents, adults, and seniors of Catalonia. Spain. J Agric Food Chem 2003;51:838–42.

[42] Järup L. Hazards of heavy metal contamination. Br Med Bull 2003;68:167–82.

[43] Fraga CG. Relevance, essentiality and toxicity of trace elements in human health. Mol Aspects Med 2005;26:235–44.

[44] Prasher D. Heavy metals and noise exposure: health effects. Noise Health 2009;11:141–4.

[45] Agency for Toxic Substances and Disease Registry (ATSDR). Public health statement for cadmium, September 2008 [displayed January 3, 2012]. Available from: http://www.atsdr.cdc.gov/phs/phs.asp?id=46&tid=15.

[46] Tariba B, Pizent A, Kljaković-Gašpić Z. Determination of lead in Croatian wines by graphite furnace atomic absorption spectrometry. Arh Hig Rada Toksikol 2011;62:25–31.

[47] Black RE. Micronutrients in pregnancy. Br J Nutr 2001;85:S107–93.

[48] Wang BH, Yu XJ, Wang D, Qi XM, Wang HP, Yang TT, et al. Alterations of trace elements (Zn, Se, Cu, Fe) and related metalloenzymes in rabbit blood after severe trauma. J Trace Elem Med Biol 2007;21:102–7.

[49] Saaka M, Oosthuizen J, Beatty S. Effect of joint iron and zinc supplementation on malarial infection and anaemia. East Afr J Public Health 2009;6(1):55–62.

[50] Wang G, Lai X, Yu X, Wang D, Xu X. Altered levels of trace elements in acute lung injury after severe trauma. Biol Trace Elem Res 2012;147:28–35.

[51] Ugwuja EI, Ejikeme B, Obuna JA. Impacts of elevated prenatal blood lead on trace element status and pregnancy outcomes in occupationally non-exposed women. Int J Ocup Environ Med 2011;2(3):143–56.

[52] Wang L, Zhou X, Yang D, Wang Z. Effects of lead and/or cadmium on the distribution patterns of some essential trace elements in immature female rats. Hum Exp Toxicol 2011;30(12):1914–23.

[53] Shah PS, Ohlsson A. Effects of prenatal multimicronutrient supplementation on pregnancy outcomes: a meta-analysis. CMAJ 2009;180(12):E99–E108.

[54] Ladipo OA. Nutrition in pregnancy: mineral and vitamin supplements. Am J Clin Nutr 2000;72(Suppl):280S–90S.

[55] Yang H, Ding Y, Chen L. Effect of trace elements on retinopathy of prematurity. J Huazhong Univ Sci Technolog Med Sci 2007;27(5):590–2.

[56] Hansen S, Nieboer E, Sandanger TM, Wilsgaard T, Thomassen Y, Veyhe AS, et al. Changes in maternal blood concentrations of selected essential and toxic elements during and after pregnancy. J Environ Monit 2011;13:2143–52.

[57] Jain RB. Effect of pregnancy on the levels of blood cadmium, lead, and mercury for females aged 17–39 years old: data from national health and nutrition examination survey 2003–2010. J Toxicol Environ Health A 2013;76(1):58–69.

[58] Hu H. Human health and heavy metals exposure. In: McCally M, editor. Life Support: The Environment and Human Health. MIT Press; 2002. p. 1–10.

[59] Ahmed K, Ayana G, Engidawork E. Lead exposure study among workers in lead acid battery repair units of transport service enterprises, Addis Ababa, Ethiopia: a cross-sectional study. J Occup Med Toxicol 2008;3:30.

[60] Goyer RA. Transplacental transport of lead. Environ Health Perspect 1990;89:101–5.

[61] Esteban-Vasallo MD, Aragones N, Pollan M, Lopezabente G, Perez-Gomez B. Mercury, cadmium, and lead levels in human placenta: a systematic review. Environ Health Perspect 2012;120:1369–77.

[62] Al-Saleh I, Shinwari N, Mashhour A, Rabah A. Birth outcome measures and maternal exposure to heavy metals (lead, cadmium and mercury) in Saudi Arabian population. Int J Hyg Environ Health 2014;217(2-3):205–18.

[63] Al-Saleh I, Shinwari N, Mashhour A, Mohamed Gel D, Rabah A. Heavy metals (lead, cadmium and mercury) in maternal, cord blood and placenta of healthy women. Int J Hyg Envir Health 2011;214:79–101.

[64] Reichrtova E, Dorociak F, Palkovicova L. Sites of Pb and Ni accumulation in the placental tissue. Hum Exp Toxicol 1998;17:176–81.

[65] Hu H, Shih R, Rothenberg S, Schwartz BS. The epidemiology of lead toxicity in adults: measuring dose and consideration of other methodologic issues. Environ Health Perspect 2007;115:455–62.

[66] Hu H, Rabinowitz M, Smith D. Bone lead as a biological marker in epidemiologic studies of chronic toxicity: conceptual paradigms. Environ Health Perspect 1998;106:1–8.

[67] Tellez-Rojo MM, Bellinger DC, Arroyo-Quiroz C, Lamadrid-Figueroa H, Mercado-García A, Schnaas-Arrieta L, et al. Longitudinal associations between blood lead concentrations lower than 10 microg/dL and neurobehavioral development in environmentally exposed children in Mexico City. Pediatrics 2006;118:e323–30.

[68] Lanphear BP, Hornung R, Khoury J, Yolton K, Baghurst P, Bellinger DC, et al. Low-level environmental lead exposure and children's intellectual function: an international pooled analysis. Environ Health Perspect 2005;113:894–9.

[69] Navas-Acien A, Guallar E, Silbergeld EK, Rothenberg SJ. Lead exposure and cardiovascular disease a systematic review. Environ Health Perspect 2007;115:472–82.

[70] Bellinger D, Leviton A, Rabinowitz M, Allred E, Needleman H, Schoenbaum S. Weight gain and maturity in fetuses exposed to low levels of lead. Environ Res 1991;54:151–8.

[71] Gonzalez-Cossio T, Peterson KE, Sanin LH, Fishbein E, Palazuelos E, Aro A, et al. Decrease in birth weight in relation to maternal bone-lead burden. Pediatrics 1997;100:856–62.

[72] Sanin LH, Gonzalez-Cossio T, Romieu I, Peterson KE, Ruíz S, Palazuelos E, et al. Effect of maternal lead burden on infant weight and weight gain at one month of age among breastfed infants. Pediatrics 2001;107:1016–23.

[73] Lamadrid-Figueroa H, Tellez-Rojo MM, Hernandez-Cadena L, Mercado-García A, Smith D, Solano-González M, et al. Biological markers of fetal lead exposure at each stage of pregnancy. J Toxicol Environ Health A 2006;69:1781–96.

[74] Smith D, Hernandez-Avila M, Tellez-Rojo MM, Mercado A, Hu H. The relationship between lead in plasma and whole blood in women. Environ Health Perspect 2002;110:263–8.

[75] Cantonwine D, Hu H, Sánchez BN, Lamadrid-Figueroa H, Smith D, Ettinger AS, et al. Critical windows of fetal lead exposure: adverse impacts on length of gestation and risk of premature delivery. J Occup Environ Med 2010;52(11):1106–11.

[76] Jelliffe-Pawlowski LL, Miles SQ, Courtney JG, Materna B, Charlton V. Effect of magnitude and timing of maternal pregnancy blood lead (Pb) levels on birth outcomes. J Perinatol 2006;26:154–62.

[77] Andrews KW, Savitz DA, Hertz-Picciotto I. Prenatal lead exposure in relation to gestational age and birth weight: a review of epidemiologic studies. Am J Ind Med 1994;26(1):13–32.

[78] Institute of Medicine (US) Committee on Understanding Premature Birth and Assuring Healthy Outcomes. In: Behrman RE, Butler AS, editors. Preterm Birth: Causes, Consequences, and Prevention. Washington (DC): National Academies Press (US); 2007.

[79] Hobel CJ, Arora CP, Korst LM. Corticotrophin-releasing hormone and CRH-binding protein. Differences between patients at risk for preterm birth and hypertension. Ann New York Acad Sci 1999;897:54–65.

[80] Leung TN, Chung TK, Madsen G, Lam C.W., Lam P.K., Walters W.A., et al. Analysis of mid-trimester corticotrophin-releasing hormone and alpha-fetoprotein concentrations for predicting pre-eclampsia. Hum Reprod 2000;15:1813–8.

[81] McLean M, Smith R. Corticotropin-releasing hormone in human pregnancy and parturition. Trends Endocrinol Metab 1999;10:174–8.

[82] McLean M, Smith R. Corticotrophin-releasing hormone and human parturition. Reproduction 2001;121:493–501.

[83] Wadhwa PD, Glynn L, Hobel CJ, Garite TJ, Porto M, Chicz-DeMet A, et al. Behavioral perinatology: biobehavioral processes in human fetal development. Regul Pept 2002;108:149–57.

[84] Rahman I, MacNee W. Oxidative stress and regulation of glutathione in lung inflammation. Eur Respir J 2000;16(3):534–54.

[85] Rastogi SK, Gupta BN, Husain T, Chandra H, Mathur N, Pangtey BS, et al. A cross-sectional study of pulmonary function among workers exposed to multimetals in the glass bangle industry. Am J Ind Med 1991;20:391–9.

[86] Costa M, Yan Y, Zhao D, Salnikow K. Molecular mechanisms of nickel carcinogenesis: gene silencing by nickel delivery to the nucleus and gene activation/inactivation by nickel-induced cell signaling. J Environ Monit 2003;5:222–3.

[87] Damek-Poprawa M, Sawicka-Kapusta K. Damage to the liver, kidney, and testis with reference to burden of heavy metals in yellow-necked mice from areas around steelworks and zinc smelters in Poland. Toxicology 2003;186(1–2):1–10.

[88] Ewan KB, Pamphlett R. Increased inorganic mercury in spinal motor neurons following chelating agents. Neurotoxicology 1996;17(2):343–9.

[89] Garza A, Vega R, Soto E. Cellular mechanisms of lead neurotoxicity. Med Sci Monit 2006;12:RA57–65.

[90] Ghio AJ, Huang YC. Exposure to concentrated ambient particles (CAPs): a review. Inhal Toxicol 2004;16:53–9.

[91] Goering PL. Lead-protein interactions as a basis for lead toxicity. Neurotoxicology 1993;14(2–3):45–60.

[92] Hertog MG, Hollman PC. Potential health effects of the dietary flavonol quercetin. Eur J Clin Nutr 1996;50:63–71.

[93] Huang YC, Ghio AJ. Vascular effects of ambient pollutant particles and metals. Curr Vasc Pharmacol 2006;4:199–203.

[94] Kelly FJ. Dietary antioxidants and environmental stress. Proc Nutr Soc 2004;63:579–85.

[95] Loghman-Adham M. Renal effects of environmental and occupational lead exposure. Environ Health Perspect 1997;105:928–38.

[96] Ratnaike RN. Acute and chronic arsenic toxicity. Postgrad Med J 2003;79(933):391–6.

[97] Rahman A, Vahter M, Smith AH, Nermell B, Yunus M, El Arifeen S, et al. Arsenic exposure during pregnancy and size at birth: a prospective cohort study in Bangladesh. Am J Epidemiol 2009; 169(3):304–12.

[98] Cory-Slechta DA, Virgolini MB, Rossi-George A, Thiruchelvam M, Lisek R, Weston D. Lifetime consequences of combined maternal lead and stress. Basic Clin Pharmacol Toxicol 2008; 102:218–27.

[99] Cory-Slechta DA, Virgolini MB, Thiruchelvam M, Weston DD, Bauter MR. Maternal stress modulates the effects of developmental lead exposure. Environ Health Perspect 2004;112:717–30.

[100] Ni Z, Hou S, Barton CH, Vaziri ND. Lead exposure raises superoxide and hydrogen peroxide in human endothelial and vascular smooth muscle cells. Kidney Int 2004;66(6):2329–36.

[101] Schober SE, Mirel LB, Graubard BI, Brody DJ, Flegal KM. Blood lead levels and death from all causes, cardiovascular disease, and cancer: results from the NHANES III mortality study. Environ Health Perspect 2006;114(10):1538–41.

[102] Shachar BZ, Carmichael SL, Stevenson DK, Shaw GM. Could genetic polymorphisms related to oxidative stress modulate effects of heavy metals for risk of human preterm birth? Reprod Toxicol 2013;42:24–6.

[103] Jomova K, Valko M. Advances in metal-induced oxidative stress and human disease. Toxicology 2011;283(2/3):65–87.

[104] Burton GJ, Jauniaux E. Oxidative stress. Best Pract Res Clin Obstet Gynaecol 2011;25(3):287–99.

[105] Xia D, Yu X, Liao S, Shao Q, Mou H, Ma W. Protective effect of Smilax glabra extract against lead-induced oxidative stress in rats. J Ethnopharmacol 2010;130(2):414–20.

[106] Clerici G, Slavescu C, Fiengo S, Kanninen TT, Romanelli M, Biondi R, et al. Oxidative stress in pathological pregnancies. J Obstet Gynaecol 2012;32(2):124–7.

[107] Karacay O, Sepici-Dincel A, Karcaaltincaba D, Sahin D, Yalvaç S, Akyol M, et al. A quantitative evaluation of total antioxidant status and oxidative stress markers in preeclampsia and gestational diabetic patients in 24–36 weeks of gestation. Diabetes Res Clin Pract 2010;89(3):231–8.

[108] Perrone S, Tataranno MT, Buonocore G. Oxidative stress and bronchopulmonary dysplasia. J Clin Neonatol 2012;1(3):109–14.

[109] Kaindl Angela M, Géraldine Favrais, Pierre Gressens. Molecular mechanisms involved in injury to the preterm brain. J Child Neurol 2009;24(9):1112–8.

[110] Perrone S, Negro S, Tataranno ML, Buonocore G. Oxidative stress and antioxidant strategies in newborns. J Matern Fetal Neonatal Med 2010;23(Suppl 3):63–5.

[111] Saugstad OD. Bronchopulmonary dysplasia-oxidative stress and antioxidants. Semin Neonatol 2003;8:39–49.

[112] Bhandari V. Postnatal inflammation in the pathogenesis of bronchopulmonary dysplasia. Birth Defects Res A Clin Mol Teratol 2014 Mar;100(3):189–201.

[113] Malekirad AA, Oryan S, Fani A, Babapor V, Hashemi M, Baeeri M, et al. Study on clinical and biochemical toxicity biomarkers in a zinc-lead mine workers. Toxicol Ind Health 2010; 26(6):331–7.

[114] Anwar-Mohamed A, Elbekai RH, El-Kadi AO. Regulation of CYP1A1 by heavy metals and consequences for drug metabolism. Expert Opin Drug Metab Toxicol 2009;5(5):501–21.

[115] Branco V, Canario J, Lu J, Holmgren A, Carvalho C. Mercury and selenium interaction *in vivo*: effects on thioredoxin reductase and glutathione peroxidase. Free Radic Biol Med 2012;52:781–93.

[116] Franco JL, Posser T, Dunkley PR, Dickson PW, Mattos JJ, Martins R, et al. Methylmercury neurotoxicity is associated with inhibition of the antioxidant enzyme glutathione peroxidase. Free Radic Biol Med 2009;47:449–57.

[117] Goodrich JM, Wang Y, Gillespie B, Werner R, Franzblau A, Basu N. Glutathione enzyme and selenoprotein polymorphisms associate with mercury biomarker levels in Michigan dental professionals. Toxicol Appl Pharmacol 2011;257(2):301–8.

[118] Chung CJ, Huang CY, Pu YS, Shiue HS, Su CT, Hsueh YM. The effect of cigarette smoke and arsenic exposure on urothelial carcinoma risk is modified by glutathione S-transferase M1 gene null genotype. Toxicol Appl Pharmacol 2013;266(2):254–9.

[119] Gundacker C, Komarnicki G, Jagiello P, Gencikova A, Dahmen N, Wittmann KJ, et al. Glutathione-S-transferase polymorphism, metallothionein expression, and mercury levels among students in Austria. Sci Total Environ 2007;385(1-3):37–47.

[120] Suh YJ, Kim YJ, Park H, Park EA, Ha EH. Oxidative stress-related gene interactions with preterm delivery in Korean women. Am J Obstet Gynecol 2008;198(5):541. e1–e7.

[121] Nukui T, Day RD, Sims CS, Ness RB, Romkes M. Maternal/newborn GSTT1 null genotype contributes to risk of preterm: low birthweight infants. Pharmacogenetics 2004;14(9):569–76.

[122] Tsai HJ, Liu X, Mestan K, Yu Y, Zhang S, Fang Y, et al. Maternal cigarette smoking, metabolic gene polymorphisms, and preterm delivery: new insights on GxE interactions and pathogenic pathways. Hum Genet 2008;123(4):359–69.

[123] Rosenberg O, Rosenblat M, Coleman R, Shih DM, Aviram M. Paraoxonase (PON1) deficiency is associated with increased macrophage oxidative stress: studies in PON1-knockout mice. Free Radic Biol Med 2003;34(6):774–84.

[124] Li WF, Sun CW, Cheng TJ, Chang KH, Chen CJ, Wang SL. Risk of carotid atherosclerosis is associated with low serum paraoxonase (PON1) activity among arsenic exposed residents in southwestern Taiwan. Toxicol Appl Pharmacol 2009;236(2):246–53.

[125] Min J, Park H, Park B, Kim YJ, Park J, Lee H, et al. Paraoxonase gene polymorphism and vitamin levels during pregnancy: relationship with maternal oxidative stress and neonatal birthweights. Reprod Toxicol 2006;22(3):418–24.

[126] Gonzalvo MC, Gil F, Hernandez AF, Villanueva E, Pla A. Inhibition of paraoxonase activity in human liver microsomes by exposure to EDTA: metals and mercurials. Chem Biol Interact 1997;105(3):169–79.

[127] Lüersen K, Schmelzer C, Boesch-Saadatmandi C, Kohl C, Rimbach G, Döring F. Paraoxonase 1 polymorphism Q192R affects the pro-inflammatory cytokine TNF-alpha in healthy males. BMC Res Notes 2011;4.

[128] Li WF, Pan MH, Chung MC, Ho CK, Chuang HY. Lead exposure is associated with decreased serum paraoxonase 1 (PON1) activity and genotypes. Environ Health Perspect 2006;114(8):1233–6.

[129] Hernandez AF, Gil F, Leno E, Lopez O, Rodrigo L, Pla A. Interaction between human serum esterases and environmental metal compounds. Neurotoxicology 2009;30(4):628–35.

[130] Chen D, Hu Y, Chen C, Yang F, Fang Z, Wang L, et al. Polymorphisms of the paraoxonase gene and risk of preterm delivery. Epidemiology 2004;15(4):466–70.

[131] Ryckman KK, Morken NH, White MJ, Velez DR, Menon R, Fortunato SJ, et al. Maternal and fetal genetic associations of PTGER3 and PON1 with preterm birth. PloS One 2010;5(2):e9040.

[132] Infante-Rivard C. Genetic association between single nucleotide polymorphisms in the paraoxonase 1 (PON1) gene and small-for-gestational-age birth in related and unrelated subjects. Am J Epidemiol 2010;171:999–1006.

[133] Baker AM, Haeri S, Klein RL, Boggess K. Association of midgestation paraoxonase1 activity and pregnancies complicated by preterm birth. Am J Obstet Gynecol 2010;203(3):246. e1–e4.

[134] Lee BE, Park H, Park EA, Gwak H, Ha EH, Pang MG, et al. Paraoxonase1 gene and glutathione S-transferase mu 1 gene interaction with preterm delivery in Korean women. Am J Obstet Gynecol 2010;203(6):569. e1–e7.

[135] Inder TE, Darlow BA, Sluis KB, Winterbourn CC, Graham P, Sanderson KJ, et al. The correlation of elevated levels of an index of lipid peroxidation (MDA-TBA) with adverse outcome in the very low birthweight infant. Acta Paediatr 1996;85:1116–22.

[136] Samuel JB, Stanley JA, Princess RA, Shanthi P, Sebastian MS. Gestational cadmium exposure-induced ovotoxicity delays puberty through oxidative stress and impaired steroid hormone levels. J Med Toxicol 2011;7:195–204.

[137] Farina M, Avila DS, da Rocha JB, Aschner M. Metals, oxidative stress and neurodegeneration: a focus on iron, manganese and mercury. Neurochem Int 2013;62:575–94.

[138] Bainbridge SA, Smith GN. HO in pregnancy. Free Radic Biol Med 2005;38(8):979–88.

[139] Takeda K, Ishizawa S, Sato M, Yoshida T, Shibahara S. Identification of a cis-acting element that is responsible for cadmium-mediated induction of the human heme oxygenase gene. J Biol Chem 1994;269:22858–67.

[140] Yamada N, Yamaya M, Okinaga S, Nakayama K, Sekizawa K, Shibahara S, et al. Microsatellite polymorphism in the heme oxygenase-1 gene promoter is associated with susceptibility to emphysema. Am J Hum Gen 2000;66:187–95.

[141] Exner M, Schillinger M, Minar E, Mlekusch W, Schlerka G, Haumer M, et al. Heme oxygenase-1 gene promoter microsatellite polymorphism is associated with restenosis after percutaneous transluminal angioplasty. J Endovasc Ther 2001;5:433–40.

[142] Chavez E, Jay D, Bravo C. The mechanism of lead-induced mitochondrial Ca^{2+} efflux. J Bioenerg Biomembr 1987;19:285–95.

[143] Goering PL. Lead-protein interactions as a basis for lead toxicity. Neurotoxicology 1993;14(2-3):45–60.

[144] Rossi E, Taketani S, Garcia-Webb P. Lead and the terminal mitochondrial enzymes of haem biosynthesis. Biomed Chromatogr 1993;7(1):1–6.

[145] Clarkson TW, Magos L. The toxicology of mercury and its chemical compounds. Crit Rev Toxicol 2006;36:609–62.

[146] Kajiwara Y, Yasutake A, Adachi T, Hirayama K. Methylmercury transport across the placenta via neutral amino acid carrier. Arch Toxicol 1996;70:310–4.

[147] Watanabe C, Yoshida K, Kasanuma Y, Kun Y, Satoh H. *In utero* methylmercury exposure differentially affects the activities of selenoenzymes in the fetal mouse brain. Environ Res 1999;80:208–14.

[148] Costa LG, Aschner M, Vitalone A, Syversen T, Soldin OP. Developmental neuropathology of environmental agents. Annu Rev Pharmacol Toxicol 2004;44:87–110.

[149] Rodrigues JL, Serpeloni JM, Batista BL, Souza SS, Barbosa F Jr. Identification and distribution of mercury species in rat tissues following administration of thimerosal or methylmercury. Arch Toxicol 2010;84(11):891–6.

[150] Yau VM, Green PG, Alaimo CP, Yoshida CK, Lutsky M, Windham GC, et al. Prenatal and neonatal peripheral blood mercury levels and autism spectrum disorders. Environ Res 2014;133:294–303.

[151] Hertz-Picciotto I, Green PG, Delwiche L, Hansen R, Walker C, Pessah IN. Blood mercury concentrations in CHARGE Study children with and without autism. Environ Health Perspect 2010;118(1):161–6.

[152] Marques RC, Bernardi JVC, Dórea JG, de Fatima R, Moreira M, Malm O. Perinatal multiple exposure to neurotoxic (lead,

methylmercury, ethylmercury, and aluminum) substances and neurodevelopment at six and 24 months of age. Environ Pollut 2014;187:e130–5.

[153] Grandjean P, Weihe P, White RF, Debes F, Araki S, Yokoyama K, et al. Cognitive deficit in 7-year-old children with prenatal exposure to methylmercury. Neurotoxicol Teratol 1997;19:417–28.

[154] Llop S, Guxens M, Murcia M, Lertxundi A, Ramon R, Riaño I, et al. Prenatal exposure to mercury and infant neurodevelopment in a multicenter cohort in Spain: study of potential modifiers. Am J Epidemiol 2012;175(5):451–65.

[155] Theunissen PT, Pennings JL, Robinson JF, Claessen SM, Kleinjans JC, Piersma AH. Time-response evaluation by transcriptomics of methylmercury effects on neural differentiation of murine embryonic stem cells. Toxicol Sci 2011;122(2):437–47.

[156] Zimmer B, Schildknecht S, Kuegler PB, Tanavde V, Kadereit S, Leist M. Sensitivity of dopaminergic neuron differentiation from stem cells to chronic low-dose methylmercury exposure. Toxicol Sci 2011;121:357–67.

[157] Grandjean P, Landrigan PJ. Neurobehavioural effects of developmental toxicity. Lancet 2014;13(3):330–8.

[158] Warfvinge K, Hua J, Logdberg B. Mercury distribution in cortical areas and fiber systems of the neonatal and maternal adult cerebrum after exposure of pregnant squirrel monkeys to mercury vapor. Environ Res 1994;67:196–208.

[159] Clarkson TW, Magos L, Myers GJ. The toxicology of mercury – current exposures and clinical manifestations. N Engl J Med 2003;349(18):1731–7.

[160] Sumi D. Biological effects of and responses to exposure to electrophilic environmental chemicals. J Health Sci 2008;54:267–72.

[161] Kim SO, Merchant K, Nudelman R, Beyer WF JJr, Keng T, DeAngelo J, et al. OxyR: a molecular code for redox-related signaling. Cell 2002;109:383–96.

[162] Glaser V, Leipnitz G, Straliotto MR, Oliveira J, dos Santos VV, Wannmacher CM, et al. Oxidative stress mediated inhibition of brain creatine kinase activity by methylmercury. Neurotoxicology 2010;31:454–60.

[163] Stringari J, Nunes AK, Franco JL, Bohrer D, Garcia SC, Dafre AL, et al. Prenatal methylmercury exposure hampers glutathione antioxidant system ontogenesis and causes long-lasting oxidative stress in the mouse brain. Toxicol Appl Pharmacol 2008;227:147–54.

[164] Freitas AJ, Rocha JB, Wolosker H, Souza DO. Effects of Hg^{2+} and CH_3Hg^+ on Ca^{2+} fluxes in rat brain microsomes. Brain Res 1996;738:257–64.

[165] Branco V, Canario J, Lu J, Holmgren A, Carvalho C. Mercury and selenium interaction *in vivo*: effects on thioredoxin reductase and glutathione peroxidase. Free Radic Biol Med 2012;52:781–93.

[166] Kung MP, Kostyniak P, Olson J, Malone M, Roth JA. Studies of the *in vitro* effect of methylmercury chloride on rat brain neurotransmitter enzymes. J Appl Toxicol 1987;7(2):119–21.

[167] Farina M, Rocha JB, Aschner M. Mechanisms of methylmercury-induced neurotoxicity: evidence from experimental studies. Life Sci 2011;89:555–63.

[168] Franco JL, Braga HC, Stringari J, Missau FC, Posser T, Mendes BG, et al. Mercurial-induced hydrogen peroxide generation in mouse brain mitochondria: protective effects of quercetin. Chem Res Toxicol 2007;20:1919–26.

[169] Ni M, Li X, Yin Z, Sidoryk-Wegrzynowicz M, Jiang H, Farina M, et al. Comparative study on the response of rat primary astrocytes and microglia to methylmercury toxicity. Glia 2011;59:810–20.

[170] Franco JL, Teixeira A, Meotti FC, Ribas CM, Stringari J, Garcia Pomblum SC, et al. Cerebellar thiol status and motor deficit after lactational exposure to methylmercury. Environ Res 2006;102:22–8.

[171] Stringari J, Nunes AK, Franco JL, Bohrer D, Garcia SC, Dafre AL, et al. Prenatal methylmercury exposure hampers glutathione antioxidant system ontogenesis and causes long-lasting oxidative stress in the mouse brain. Toxicol Appl Pharmacol 2008;227:147–54.

[172] Dringen R, Pawlowski PG, Hirrlinger J. Peroxide detoxification by brain cells. J Neurosci Res 2005;79:157–65.

[173] Drescher O, Dewailly E, Diorio C, Ouellet N, Sidi EA, Abdous B, et al. Methylmercury exposure, *PON1* gene variants and serum paraoxonase activity in Eastern James Bay Cree adults. J Expo Sci Environ Epidemiol 2014;.

[174] Budtz-Jørgensen E, Grandjean P, Weihe P. Separation of risks and benefits of seafood intake. Environ Health Perspect 2007;115:323–7.

[175] Ballatori N, Gatmaitan Z, Truong AT. Impaired biliary excretion and whole body elimination of methylmercury in rats with congenital defect in biliary glutathione excretion. Hepatology 1995;22:1469–73.

[176] Hernandez-Perez M, Mahalingam M. Matrix metalloproteinases in health and disease: insights from dermatopathology. Am J Dermatopathol 2012;34(6):565–79.

[177] Pepper MS. Role of the matrix metalloproteinase and plasminogen activator-plasmin systems in angiogenesis. Arterioscler Thromb Vasc Biol 2001;21(7):1104–17.

[178] Xu P, Alfaidy N, Challis JR. Expression of matrix metalloproteinase (MMP)-2 and MMP-9 in human placenta and fetal membranes in relation to preterm and term labor. J Clin Endocrinol Metab 2002;87(3):1353–61.

[179] Vu TD, Yun F, Placido J, Reznik SE. Placental matrix metalloproteinase-1 expression is increased in labor. Reprod Sci 2008;15(4):420–4.

[180] Sundrani DP, Chavan-Gautam PM, Pisal HR, Mehendale SS, Joshi SR. Matrixmetalloproteinase-1 and -9 in human placenta during spontaneous vaginal delivery and caesarean sectioning in preterm pregnancy. PLoS One 2012;7(1):e29855.

[181] Maymon E, Romero R, Pacora P, Gervasi MT, Bianco K, Ghezzi F, et al. Evidence for the participation of interstitial collagenase (matrix metalloproteinase 1) in preterm premature rupture of membranes. Am J Obstet Gynecol 2000;183(4):914–20.

[182] Cockle JV, Gopichandran N, Walker JJ, Levene MI, Orsi NM. Matrix metalloproteinases and their tissue inhibitors in preterm perinatal complications. Reprod Sci 2007;14(7):629–45.

[183] Jacob-Ferreira AL, Lacchini R, Gerlach RF, Passos CJ, Barbosa F Jr, Tanus-Santos JE. A common matrix metalloproteinase (MMP)-2 polymorphism affects plasma MMP-2 levels in subjects environmentally exposed to mercury. Sci Total Environ 2011;409(20):4242–6.

[184] Jacob-Ferreira AL, Passos CJ, Gerlach RF, Barbosa F Jr, Tanus-Santos JE. A functional matrix metalloproteinase (MMP)-9 polymorphism modifies plasma MMP-9 levels in subjects environmentally exposed to mercury. Sci Total Environ 2010;408(19):4085–92.

[185] Yorifuji T, Debes F, Weihe P, Grandjean P. Prenatal exposure to lead and cognitive deficit in 7- and 14-year-old children in the presence of concomitant exposure to similar molar concentration of methylmercury. Neurotoxicol Teratol 2011;33:205–11.

[186] Al-Saleh I, Abduljabbar M, Al-Rouqi R, Elkhatib R, Alshabbaheen A. Mercury (Hg) exposure in breast-Fed infants and their mothers and the evidence of oxidative Stress. Biol Trace Elem Res 2013;153:145–54.

[187] Beveridge R, Pintos J, Parent ME, Asselin J, Siemiatycki J. Lung cancer risk associated with occupational exposure to nickel, chromium VI, and cadmium in two population-based case–control studies in Montreal. Am J Ind Med 2010;53(5):476–85.

[188] Honda R, Swaddiwudhipong W, Nishijo M, Mahasakpan P, Teeyakasem W, Ruangyuttikarn W, et al. Cadmium induced renal dysfunction among residents of rice farming area downstream from a zinc-mineralized belt in Thailand. Toxicol Lett 2010;198(1):26–32.

A. OVERVIEW ON FERTILITY

[189] Barr M Jr. The teratogenicity of cadmium chloride in two stocks of Wistar rats. Teratology 1973;7(3):237–42.

[190] Thompson J, Bannigan J. Cadmium: toxic effects on the reproductive system and the embryo. Reprod Toxicol 2008;25(3):304–15.

[191] Hovland DN Jr, Machado AF, Scott WJ Jr, Collins MD. Differential sensitivity of the SWV and C57BL/6mouse strains to the teratogenic action of single administrations of cadmium given throughout the period of anterior neuropore closure. Teratology 1999;60(1):13–21.

[192] Paniagua-Castro N, Escalona-Cardoso G, Chamorro-Cevallos G. Glycine reduces cadmium-induced teratogenic damage in mice. Reprod Toxicol 2007;23(1):92–7.

[193] Robinson JF, Yu X, Hong S, Griffith WC, Beyer R, Kim E, et al. Cadmium-induced differential toxicogenomic response in resistant and sensitive mouse strains undergoing neurulation. Toxicol Sci 2009;107(1):206–19.

[194] Scott WJ Jr, Schreiner CM, Goetz JA, Robbins D, Bell SM. Cadmium-induced postaxial forelimb ectrodactyly: association with altered sonic hedgehog signaling. Reprod Toxicol 2005;19(4):479–85.

[195] Ahokas RA, Dilts PV Jr, LaHaye EB. Cadmium-induced fetal growth retardation: protective effect of excess dietary zinc. Am J Obstet Gynecol 1980;136(2):216–21.

[196] Ji YL, Wang H, Liu P, Zhao XF, Zhang Y, Wang Q, et al. Effects of maternal cadmium exposure during late pregnant period on testicular steroidogenesis in male offspring. Toxicol Lett 2011;205(1):69–78.

[197] Nishijo M, Tawara K, Honda R, Nakagawa H, Tanebe K, Saito S. Relationship between newborn size and mother's blood cadmium levels. Arch Environ Health 2004;59(1):22–5.

[198] Watson ED, Cross JC. Development of structures and transport functions in the mouse placenta. Physiology (Bethesda) 2005;20:180–93.

[199] Wigle DT, Arbuckle TE, Turner MC, Bérubé A, Yang Q, Liu S, et al. Epidemiologic evidence of relationships between reproductive and child health outcomes and environmental chemical contaminants. J Toxicol Environ Health B Crit Rev 2008;11(5–6):373–517.

[200] Laudanski T, Sipowicz P, Modzelewski J, Bolinski J, Szamatowicz J, Razniewska G, et al. Influence of high lead and cadmium soil content on human reproductive outcome. Int J Gynecol Obstet 1991;366:309–15.

[201] Osman K, Akesson A, Berglund M, Bremme K, Schütz A, Ask K, et al. Toxic and essential elements in placentas of Swedish women. Clin Biochem 2000;33(2):131–8.

[202] Wang Z, Wang H, Xu ZM, Ji YL, Chen YH, Zhang ZH, et al. Cadmium-induced teratogenicity: association with ROS-mediated endoplasmic reticulum stress in placenta. Toxicol Appl Pharmacol 2012;259(2):236–47.

[203] Llanos MN, Ronco AM. Fetal growth restriction is related to placental levels of cadmium, lead and arsenic but not with antioxidant activities. Reprod Toxicol 2009;27(1):88–92.

[204] Lin CM, Doyle P, Wang D, Hwang YH, Chen PC. Does prenatal cadmium exposure affect fetal and child growth? Occup Environ Med 2011;68:641–6.

[205] Argüelles-Velázquez N, Alvarez-González I, Madrigal-Bujaidar E, Chamorro-Cevallos G. Amelioration of cadmium-produced teratogenicity and genotoxicity in mice given Arthrospira maxima (Spirulina) treatment. Evid Based Complement Alternat Med 2013;2013:604535.

[206] Kampa Marilena, Castanas Elias. Human health effects of air pollution. Environ Pollut 2008;151(2):362–7.

[207] Samuel JB, Stanley JA, Princess RA, Shanthi P, Sebastian MS. Gestational cadmium exposure-induced ovotoxicity delays puberty through oxidative stress and impaired steroid hormone levels. J Med Toxicol 2011;7(3):195–204.

[208] Samarawickrama GP, Webb M. The acute toxicity and teratogenicity of cadmium in the pregnant rat. J Appl Toxicol 1981;1:264–9.

[209] Pillai P, Patel R, Pandya C, Gupta S. Sex-specific effects of gestational and lactational coexposure to lead and cadmium on hepatic phase I and phase II xenobiotic/steroid-metabolizing enzymes and antioxidant status. J Med Toxicol 2011;7(3):195–204.

[210] Legraverend C, Agneta M, Wells T, Robinson I, Gustafsson JA. Hepatic steroid hydroxylating enzymes are controlled by the sexually dimorphic pattern of growth hormone secretion in normal and dwarf rats. FASEB J 1992;6:711–8.

[211] Valko M, Morris H, Cronin MT. Metals, toxicity and oxidative stress. Curr Med Chem 2005;12:1161–208.

[212] Ercal N, Gurer-Orhan H, Aykin-Burns N. Toxic metals and oxidative stress part I: mechanisms involved in metal-induced oxidative damage. Curr Top Med Chem 2001;1:529–39.

[213] Gobe G, Crane D. Mitochondria, reactive oxygen species and cadmium toxicity in the kidney. Toxicol Lett 2010;198:49–55.

[214] Roberts RA, Smith RA, Safe S, Szabo C, Tjalkens RB, Robertson FM. Toxicological and pathophysiological roles of reactive oxygen and nitrogen species. Toxicology 2010;276:85–94.

[215] Gundacker C, Fröhlich S, Graf-Rohrmeister K, Eibenberger B, Jessenig V, Gicic D, et al. Perinatal lead and mercury exposure in Austria. Sci Total Environ 2010;408:5744–9.

[216] National Research Council. Arsenic in Drinking Water: 2001 Update. Washington, DC: National Academy Press; 2001.

[217] Vahter M. Methylation of inorganic arsenic in different mammalian species and population groups. Sci Prog 1999;82(pt 1):69–88.

[218] Vahter M. Health effects of early life exposure to arsenic. Basic Clin Pharmacol Toxicol 2008;102(2):204–11.

[219] Concha G, Vogler G, Lezcano D, Nermell B, Vahter M. Exposure to inorganic arsenic metabolites during early human development. Toxicol Sci 1998;44:185–90.

[220] Rahman A, Vahter M, Ekstrom EC, Rahman M, Golam Mustafa AH, Wahed MA, et al. Association of arsenic exposure during pregnancy with fetal loss and infant death: a cohort study in Bangladesh. Am J Epidemiol 2007;165:1389–96.

[221] Golub MS, Macintosh MS, Baumrind N. Developmental and reproductive toxicity of inorganic arsenic: animal studies and human concerns. J Toxicol Environ Health B Crit Rev 1998;1:199–241.

[222] Ahmad SA, Sayed MH, Barua S, Khan MH, Faruquee MH, Jalil A, et al. Arsenic in drinking water and pregnancy outcomes. Environ Health Perspect 2001;109:629–31.

[223] Milton AH, Smith W, Rahman B, Hasan Z, Kulsum U, Dear K, et al. Chronic arsenic exposure and adverse pregnancy outcomes in Bangladesh. Epidemiology 2005;16:82–6.

[224] Wang A, Holladay SD, Wolf DC, Ahmed SA, Robertson JL. Reproductive and developmental toxicity of arsenic in rodents: a review. Int J Toxicol 2006;25:319–31.

[225] Rahman A, Vahter M, Smith AH, Nermell B, Yunus M, El Arifeen S, et al. Arsenic exposure during pregnancy and size at birth: a prospective cohort study in Bangladesh. Am J Epidemiol 2009;169(3):301–4.

[226] Rahman A, Persson LÅ, Nermell B, El Arifeen S, Ekström EC, Smith AH, et al. Arsenic exposure and risk of spontaneous abortion, stillbirth, and infant mortality. Epidemiology 2010;21(6):797–804.

[227] Koestler DC, Avissar-Whiting M, Houseman EA, Karagas MR, Marsit CJ. Differential DNA methylation in umbilical cord blood of infants exposed to low levels of arsenic *in utero*. Environ Health Perspect 2013;121(8):971–7.

[228] Engström K, Vahter M, Mlakar SJ, Concha G, Nermell B, Raqib R, et al. Polymorphisms in arsenic (+III oxidation state) methyltransferase (AS3MT) predict gene expression of AS3MT as well as arsenic metabolism. Environ Health Perspect 2011;119(2):182–8.

[229] Hartwig A, Blessing H, Schwerdtle T, Walter I. Modulation of DNA repair processes by arsenic and selenium compounds. Toxicology 2003;193:161–9.

[230] Lin S, Del Razo LM, Styblo M, Wang C, Cullen WR, Thomas DJ. Arsenicals inhibit thioredoxin reductase in cultured rat hepatocytes. Chem Res Toxicol 2001;14:305–11.

[231] von Ehrenstein OS, Guha Mazumder DN, Hira-Smith M, Ghosh N, Yuan Y, Windham G, et al. Pregnancy outcomes, infant mortality, and arsenic in drinking water in West Bengal, India. Am J Epidemiol 2006;163:662–9.

[232] Hopenhayn-Rich C, Browning SR, Hertz-Picciotto I, Ferreccio C, Peralta C, Gibb H. Chronic arsenic exposure and risk of infant mortality in two areas of Chile. Environ Health Perspect 2000;108:667–73.

[233] Rahman M, Vahter M, Wahed MA, Sohel N, Yunus M, Streatfield PK, et al. Prevalence of arsenic exposure and skin lesions. A population based survey in Matlab, Bangladesh. J Epidemiol Community Health 2006;60:242–8.

[234] Wilcox AJ, Weinberg CR, O'Connor JF, Baird DD, Schlatterer JP, Canfield RE, et al. Incidence of early loss of pregnancy. N Engl J Med 1988;319:189–94.

[235] Simpson JL, Mills JL, Holmes LB, Ober CL, Aarons J, Jovanovic L, et al. Low fetal loss rates after ultrasound-proved viability in early pregnancy. JAMA 1987;258:2555–7.

[236] Sweeney AM, LaPorte RE. Advances in early fetal loss research: importance for risk assessment. Environ Health Perspect 1991;90:165–9.

[237] World Health Organization. Guidelines for drinking-water quality, vol.1. 3rd ed. Recommendations. Geneva. Switzerland: World Health Organization; 2004. http://www.who.int/water_sanitation_health/dwq/GDWQ2004web.pdf.

[238] Rahman A, Persson LÅ, Nermell B, El Arifeen S, Ekström EC, Smith AH, et al. Arsenic exposure and risk of spontaneous abortion, stillbirth, and infant mortality. Epidemiology 2010;21(6):797–804.

[239] Chou WC, Chung YT, Chen HY, Wang CJ, Ying TH, Chuang CY, et al. Maternal arsenic exposure and DNA damage biomarkers, and the associations with birth outcomes in a general population from Taiwan. PloS One 2014;9(2):e86398.

[240] Vahter M. Mechanisms of arsenic biotransformation. Toxicology 2002;181–2. 211-217.

A. OVERVIEW ON FERTILITY

6

Nitrate, Nitrite, Nitrosatable Drugs, and Congenital Malformations

Jean D. Brender, PhD, RN, FACE

Texas A&M Health Science Center, College Station, Texas, USA

INTRODUCTION

Congenital malformations remain the greatest contributor to infant mortality in the United States [1] and other developed countries [2]. In 2010, 5107 US children under 1 year of age died from birth defects, accounting for 20.8% of infant deaths [3]. During the same year, approximately 394,000 discharges of newborns from short-stay hospitals were attributed to congenital malformations [4]. About 3% of births in the United States have major structural or genetic birth defects [5]; yet, causes are unknown for an estimated 65 to 75% of these defects [6].

Numerous epidemiologic studies have explored the potential risk factors of birth defects, including prenatal use of medications, patterns of maternal dietary intake, and maternal exposures to environmental contaminants. This chapter will focus on the potential separate and combined effects of nitrosatable drugs, dietary intake of nitrate and nitrite, and exposure to nitrate in drinking water on the risk of congenital malformations. N-nitroso compounds are known to cause congenital malformations in animal models [7,8], and these compounds can be formed in the human stomach after ingestion of nitrosatable compounds in conjunction with nitrate or nitrite [9,10]. In this chapter, the following will be discussed: (1) sources of exposure to nitrate, nitrite, and N-nitroso compounds; (2) the characteristics of nitrosatable drugs and their potential role in birth defects; (3) the role that endogenous formation of N-nitroso compounds might have in the risk of birth defects; and (4) the experimental and epidemiologic evidence regarding vitamin C as an inhibitor of the formation of endogenous N-nitroso compounds.

MATERNAL EXPOSURES TO NITRATE, NITRITE, AND N-NITROSO COMPOUNDS

Various dietary components contribute to exogenous exposure to nitrate, nitrite, and N-nitroso compounds, although drinking water sources may also be a major contributor of daily nitrate intake if levels are high in the water supplies [11,12]. Estimates of daily nitrate intake from the dietary sources range from 31 mg in Norway to 245 mg in Italy [13], with vegetables contributing the largest proportion of daily intake [14,15]. In a study using control mothers from the National Birth Defects Prevention Study (NBDPS), one of the largest population-based, case-control studies conducted on the etiology of birth defects, Griesenbeck et al. [16] observed a median intake of 40.5 mg/day of nitrate from dietary sources. In this study population, vegetables contributed an estimated 50 to 75% of daily intake of nitrate from the diet, depending on the mother's race/ethnicity. In an earlier report, the National Academy of Sciences (NAS) [17] concluded that 87% of dietary nitrate was from vegetables. Nitrate content varies considerably in vegetables with asparagus, broccoli, carrots, peppers, and tomatoes containing less than 50 mg/100 g, to celery, lettuce, radishes, and spinach containing more than 250 mg/100 g nitrate (fresh weight) [18,19].

Nitrate is the most widespread chemical contaminant in aquifers around the world [20]. Generally, water nitrate contributes only a small percentage of total daily nitrate intake unless levels approach or exceed the allowable maximum contaminant level (MCL) [17]. The U.S. Environmental Protection Agency (EPA) has set the MCL for nitrate in public water supplies at 10 mg/L nitrate-nitrogen (45 mg/L total nitrate) primarily to protect

Handbook of Fertility. http://dx.doi.org/10.1016/B978-0-12-800872-0.00006-8

against methemoglobinemia in infants ("blue-baby syndrome") [21]. Agricultural populations, who often obtain their drinking water from private wells, are more likely to be exposed to elevated levels of nitrate in drinking water than urban populations with an estimated 22% of domestic wells in US agricultural areas exceeding 10 mg/L nitrate-nitrogen [22].

Cured meats and baked goods and cereals contribute most of the dietary nitrite intake with drinking water being a negligible source [17]. However, the endogenous conversion of nitrate to nitrite is also a significant source of exposure to nitrite; an estimated 5% of ingested nitrate in food and water are converted to nitrite in the saliva [23]. Once ingested and absorbed, approximately 25% of nitrate is secreted in saliva [24], where about 20% is converted to nitrite by bacteria in the mouth [25]. The National Academy of Sciences estimated that 47% of dietary nitrite was from meat, with cured meat providing most of this contribution (39%), 34% from baked goods and cereals, 16% from vegetables, and 3% from other sources [17]. In contrast, relative contributions of nitrite in the NBDPS participants' diets included 61% from meat, 12% from grain products, 11% from vegetables, and 16% from other foods [16].

Humans are exposed to *N*-nitroso compounds from exogenous sources and through endogenous formation. Exogenous sources include cured meats and smoked fish [12], beer [26], tobacco products and tobacco smoke [27,28], cosmetics [29,30], and occupational exposures in rubber or rocket fuel factories and leather tanneries [17]. Endogenous formation of nitrosamines contributes 40 to 75% of exposure to these compounds in humans [31], and the formation depends on precursors such as nitrate, nitrite, and secondary/tertiary amines and amides. After several decades of research on nitrosation and *N*-nitroso compounds, Lijinsky concluded that "any circumstance in which a nitrosating agent and a nitrosatable amino compound come in contact is a potential source of *N*-nitroso compounds" [32].

Extensive experimental evidence indicates that *N*-nitroso compounds can be formed *in vivo* via the reaction of nitrosatable amines or amides and nitrosating agents, such as nitrite, in an acidic environment as found in the stomach [33]. Endogenous formation of these compounds can also occur through cell-mediated nitrosation with stimulated macrophages [34] and with some bacterial strains of *Alcaligenes*, *Bacillus*, *Escherichia*, *Klebsiella*, *Neisseria*, *Proteus*, and *Pseudomonas* that are capable of catalyzing the nitrosation of secondary amines [31]. A variety of drugs contribute nitrosatable amines or amides in the endogenous formation of *N*-nitroso compounds. In experiments with simulated gastric conditions, the combination of drugs containing secondary or tertiary amines or amides with nitrite have yielded a range of *N*-nitroso compounds depending on the chemical

structure of the drug [10,35,36]. These drugs will be discussed in more detail in the next section of this chapter. Other significant sources of nitrosatable amino compounds include certain foods and beverages (fish, pork, cereals, spices, coffee, tea, beer, wine), cosmetics, tobacco products, and agricultural chemicals [17].

NITROSATABLE DRUGS AND BIRTH DEFECTS

Some drugs are capable of reacting with nitrite in an acidic environment, such as in the stomach, to form *N*-nitroso compounds [37]. These drugs include secondary amines, tertiary amines, and amides. Early epidemiologic studies of maternal exposure to nitrosatable drugs and birth defects in offspring focused on these drugs as a single group [38–42], but more recent studies have examined the separate associations of amides and secondary or tertiary amines [43,44]. Prenatal exposure to these drugs in relation to childhood cancer has also been investigated [45–48]. Estimates of prevalence of nitrosatable drug exposure during pregnancy have varied across these studies. Olshan and Faustman [42] published one of the first studies on nitrosatable drug use during pregnancy in which they identified a total of 13 drug components deemed to have nitrosation potential. Among 51,228 pregnancies followed between 1959 and 1965 as part of the U.S. National Collaborative Perinatal Project, 3.5% of mothers used nitrosatable drugs during the first four months of pregnancy and 11.8% took these drugs anytime during pregnancy. In a population-based case-control study of childhood brain cancer that included study participants in eight U.S. states, an estimated 5% of control mothers took nitrosatable drugs anytime during pregnancy; medications taken included 24 different nitrosatable components [46].

Three subsequent case-control studies of the relation between prenatal nitrosatable drug exposure and childhood cancer used a broader classification scheme for nitrosatable drugs. Classification with respect to nitrosatability was based on published experimental data or chemical structure [45,47,48]. Prevalence of any nitrosatable drug use among control mothers anytime during pregnancy ranged from 14.9 to 18.4%, but prevalence of use during the first trimester was not reported. Two of these studies also reported prevalence of nitrosatable intake by molecular structure including 7.5–9% for secondary amines, 10–13% for tertiary amines, and 2.5–3% for amides [45,48].

Using data from the NBDPS for births occurring from 1997 to 2005, Brender et al. [49] studied the prevalence and patterns of nitrosatable drug use among study control-mothers who were representative of the base populations in their respective states of residence. In this project, the investigators classified drugs reported taken according to their nitrosatability using the following procedures. First, all orally administered prescription and nonprescription medications reported by control participants and the active drug components in each were identified, including orally inhaled medications (e.g., albuterol, epinephrine) that might be swallowed. The resulting group of active ingredients was cross-referenced against previously compiled lists of nitrosatable medicinal compounds [37,48] and categorized based on the presence of amine (secondary and tertiary) and amide functional groups in their chemical structure. Finally, each component was categorized by its primary indication or therapeutic use (e.g., analgesic, anti-infective) and pharmacologic class (e.g., opioid, macrolide). In this study population, approximately 24% of the control mothers took any nitrosatable drug during the first trimester, including 12.4, 12.2, and 7.6% who took secondary amines, tertiary amines, and amides, respectively [49].

Table 6.1 shows the 10 most commonly taken nitrosatable drugs by the NBDPS control mothers. Five tertiary amines (promethiazine, dextromethorphan, diphenydramine, chlorpheniramine, doxylamine, and caffeine [also classified as an amide]), three secondary amines (pseudoephedrine, albuterol, and fluoxetine), and two amides (amoxicillin and caffeine [also a tertiary amine]) were included on this list. These drugs represented a wide variety of therapeutic indications, including asthma, antidepressant, anti-infective, anti-emetic, antihistamine, cough suppressant, decongestant, and stimulant medications. It is noteworthy that four of the five most commonly taken drugs classified as nitrosatable tertiary amines are also available over the counter. The number of identified nitrosatable drugs in this study was much higher than

other studies on the relation between prenatal nitrosatable drug exposure and birth defects with 24 secondary amine components, 51 tertiary amine components, and 22 amide components detected [49].

Between 1989 and 2014, seven epidemiologic studies were published on the relation between maternal exposure to nitrosatable drugs and birth defects in offspring. Table 6.2 summarizes the characteristics and major findings of each of these studies. Olshan and Faustman [42] studied the association between ingestion of nitrosatable drugs during the first four months and anytime during pregnancy and adverse reproductive outcomes including congenital malformations; low birth weight; 5-min Apgar score; fetal, neonatal, and infant death; neurological abnormalities at the end of seven years; and childhood tumors among offspring of women in the Collaborative Perinatal Project of the National Institute of Neurological and Communicative Disorders and Stroke. Relative to women not taking these drugs, women who took nitrosatable drugs during the first four months of pregnancy were more likely to have infants with eye (relative risk [RR] 2.71, 95% CI 1.43, 5.15), musculoskeletal (RR 1.33, 95% CI 1.05, 1.70), or gastrointestinal abnormalities (RR 1.68, 95% CI 1.19, 2.35); hydrocephaly (RR 1.99, 95% CI 0.47, 8.49); spina bifida (RR 4.58, 95% CI 0.66, 31.62); or children who subsequently developed tumors (RR 2.29, 95% CI 0.99, 5.26). Because of the small number of women exposed and the relatively rare occurrence of specific malformations, some of the data did not yield stable risk estimates for individual defects, and the 95% confidence intervals (CI) for several of the risk estimates included 1.0.

Gardner et al. [40] studied the effect of prenatal drug exposures on the risk for craniosynostosis in offspring

TABLE 6.1 Ten Most Commonly Used Nitrosatable Drug Components During the First Trimester of Pregnancy, NBDPS Control Women, 1997–2005

Rank	Nitrosatable drug component	Type of molecular structure	Number	%*
1	Pseudoephedrine	Secondary amine	515	7.7
2	Amoxicillin	Amide	271	4.0
3	Promethiazine	Tertiary amine	207	3.1
4	Albuterol	Secondary amine	178	2.7
5	Dextromethorphan	Tertiary amine	139	2.1
6	Diphenydramine	Tertiary amine	115	1.7
7	Chlorpheniramine	Tertiary amine	104	1.6
8	Doxylamine	Tertiary amine	82	1.2
9	Caffeine	Amide/tertiary amine	64	1.0
10	Fluoxetine	Secondary amine	53	0.8

** Complete information available for 6678 women for any secondary amines, 6708 women for any tertiary amines, and 6701 women for any amides.*
Adapted from Ref. [49].

TABLE 6.2 Epidemiologic Studies of Prenatal Nitrosatable Drug Exposure and Birth Defects in Offspring

First author, year, country	Study design regional description	Years of outcome ascertainment	Exposure description	Birth defects included	Summary of findings
Brender, 2004, USA [38]	Population-based case-control study 14 Texas counties bordering Mexico	1995–2000	All medications reported taken three months before and after conception classified with respect to nitrosatability based on literature	NTDs	Adjusted OR for NTDs in relation to any nitrosatable drug exposure during the periconceptional period: 2.7 (95% CI 1.4, 5.3)
Brender, 2011, USA [43]	Population-based case-control study NBDPS in 10 states	1997–2005	All medications reported taken one month before and after conception classified with respect to nitrosatability based on literature and further classified by molecular structure	NTDs, grouped and specific phenotypes	OR for any neural tube defect with any nitrosatable drug exposure: 1.31 (95% CI 1.12, 1.55) and stronger association noted for tertiary amines (OR 1.60, 95% CI 1.31, 1.95). Strongest association noted between anencephaly and tertiary amine drug exposure (OR 1.96, 95% CI 1.40, 2.73)
Brender, 2012, USA [44]	Population-based case-control study NBDPS in 10 states	1997–2005	All medications reported taken the first trimester classified with respect to nitrosatability based on literature and further classified by molecular structure	Limb deficiencies, oral clefts, and congenital heart malformations	Limb deficiencies significantly associated with prenatal exposure to secondary amines Selected congenital heart defects associated with secondary and tertiary amine and amide drugs
Croen, 2001, USA [39]	Population-based case-control study California	1989–1991	Drugs taken in early pregnancy classified according to potential for nitrosatability based on method of Olshan and Faustman [42]	NTDs	OR of NTDs in relation to any nitrosatable drug exposure during periconceptional period: 1.2 (95% CI 0.77, 1.8)
Gardner, 1998, USA [40]	Population-based case-control study Colorado	1986–1989	Used Olshan and Faustman [42] list of 13 nitrosatable drugs; study participants reported taking three of these drugs	Craniosynostosis	OR of 1.87 (95% CI 1.08, 3.24) for sagittal/lambdoid suture synostosis with prenatal exposure to nitrosatable drugs chlorpheniramine, chlordiazepoxide, and nitrofurantoin
Kallen, 2005, Sweden and France [41]	Two case-control studies; all deliveries in Sweden and Central-East France	1995–2002 (Sweden); 1980–2002 (France)	Examined exposure to the three nitrosatable drugs identified by Gardner et al. [40] (chlorpheniramine, chlordiazepoxide, and nitrofurantoin)	Craniostenosis	With Swedish data, risk ratio of 3.4 (95% CI 1.10, 7.94) in conjunction with three nitrosatable drugs; no such exposure occurred in French study population
Olshan, 1989, USA [42]	Prospective cohort study National Collaborative Perinatal Project in which women were recruited in 12 study centers	1959–1965	Examined exposure to 13 nitrosatable drugs identified in the literature	Any congenital malformations	RRs increased for several birth defects with exposure to drugs within first four months of pregnancy, including spina bifida and eye, musculoskeletal, and gastrointestinal defects

including exposures to nitrosatable drugs. Women who took any nitrosatable drug (limited to three, including chlorpheniramine, chlordiazepoxide, and nitrofurantoin), as classified by Olshan and Faustman [42], were more likely to give birth to infants with sagittal/lambdoid suture synostosis (odds ratio [OR] 1.87, 95% CI 1.08, 3.24) than women who did not report taking these drugs. However, when the investigators examined all potentially nitrosatable drugs taken by the women, no association was seen between the larger group of compounds and any type of craniosynostosis. In a study of Swedish women and their offspring, Kallen and Robert-Gnansia [41] also noted an association (OR 3.4, 95% CI 1.1, 7.9) between craniostenosis and maternal exposure to the three nitrosatable drugs that Gardner et al. identified in their study.

In a study published in 2001, Croen et al. [39] used the Olshan and Faustman list to classify women's exposure to nitrosatable drugs in a study of neural tube defects (NTDs). No association (OR 1.2, 95% CI 0.77, 1.8) was found between periconceptional exposure with these drugs and NTDs in the offspring of California women giving birth during 1989–1991. On the other hand, Brender et al. [38] found a significant positive association between the reported use of drugs with potential for nitrosation and NTDs in offspring among Mexican-American women who resided in one of the 14 Texas counties bordering Mexico (OR 2.7, 95% CI 1.4, 5.3).

None of these studies examined nitrosatable drugs by their molecular structures (e.g., secondary or tertiary amines, amides) in relation to birth defects, and most focused on the list of 13 nitrosatable drugs that Olshan and Faustman compiled from drugs reported taken by US women who gave birth during 1959–1965, over 45 years ago. Using more recent data from the NBDPS that included women giving birth during 1997 through 2005, Brender et al. [43,44] examined the relation between nitrosatable drug use in early pregnancy and selected birth defects in offspring. As described earlier, nitrosatable drugs were classified with respect to their molecular structure as well as for therapeutic indications and pharmacologic class. For NTDs, the investigators defined the exposure window of interest for nitrosatable drugs as one month prior to one month postconception. Women with NTD-affected pregnancies were more likely (19.5%) than control-mothers of babies with no major birth defects (16.9%) to report taking drugs classified as nitrosatable during this period (adjusted OR 1.31, 95% CI 1.12, 1.55) [43]. All ORs in this study were adjusted for maternal race/ethnicity, educational level, and folic acid supplementation around the time of conception. Furthermore, stronger associations were observed for tertiary amine drugs for NTDs overall (adjusted OR 1.60, 95% CI 1.31, 1.95), and specifically for anencephaly (adjusted OR 1.96, 95% CI 1.40, 2.73) and spina bifida (adjusted OR 1.48, 95% CI 1.15, 1.91).

Associations between medications and pregnancy outcomes might be due to a confounding effect from the condition that is being treated; for example, women with diabetes are more likely to be treated with antidiabetic medications and diabetes is associated with some types of birth defects. This type of confounding is known as confounding by indication and might be considered potentially present if drugs associated with the pregnancy outcome represented relatively few indications. However, in the nitrosatable drug and NTD study with NBDPS participants, NTDs were associated with drugs classified as tertiary amines across a broad range of indications, such as analgesic opioids, antidiabetic drugs, anti-emetic phenothiazines, anti-epileptics, antihistamines, anti-infective macrolides, cough medications, gastrointestinal histamine 2 blockers, and stimulants [43].

Brender et al. [44] also examined prenatal nitrosatable drug exposure and limb deficiencies, oral clefts, and congenital heart defects in offspring. In this study, the critical exposure window was defined as the first trimester of pregnancy. The following associations were noted between prenatal nitrosatable drug exposure and birth defects: amides with cleft palate (OR [adjusted for study center, maternal age, race/ethnicity, education, folic acid supplementation, and smoking] 1.27, 95% CI 1.00, 1.62); secondary amines with longitudinal and transverse limb deficiencies (ORs respectively 1.46 [95% CI 1.04, 2.04] and 1.42 [95% CI 1.07, 1.90] that were adjusted for study center, maternal age, race/ethnicity, education, and multivitamin supplementation); tertiary amines with hypoplastic left heart syndrome (OR 1.50, 95% CI 1.10, 2.04) and single ventricle (OR 1.61, 95% CI 1.06, 2.45 [all ORs for congenital heart defects were adjusted for study center, maternal race/ethnicity, education, smoking, and multivitamin use]); secondary amines with atrioventricular septal defects (OR 1.97, 95% CI 1.19, 3.26); and amides with hypoplastic left heart syndrome (OR 1.49, 95% CI 1.02, 2.17), single ventricle (OR 1.84, 95% CI 1.15, 2.95), and septal defects (OR 1.24, 95% CI 1.04, 1.49). Within the secondary and tertiary amine categories, a wide range of drugs with different indications were associated with the various birth defects, although the only nitrosatable amides associated with these defects to any appreciable degree were sulfonamide anti-infectives.

NITRATE IN DRINKING WATER AND BIRTH DEFECTS

Findings from several epidemiologic studies suggest an effect of prenatal exposure to nitrates in drinking water on the risk of congenital malformations in offspring (Table 6.3). Scragg et al. [50] noted that women who lived in Mount Gambier, South Australia, were more likely than women residing in other areas to give birth to stillborns

TABLE 6.3   Nitrate in Maternal Drinking Water Sources and Birth Defects in Offspring

First author, year, country	Study design regional description	Years of outcome ascertainment	Exposure description*	Birth defects included	Summary of findings
Arbuckle, 1988 Canada [53]	Population-based, case-control study New Brunswick	1973–1983	Collected and analyzed a water sample at maternal residence at time of index birth (as nitrate)	Congenital malformations of CNS	OR of 2.3 (95% CI 0.73, 7.3) for CNS malformations with exposure to nitrate 26 mg/L relative to baseline of 0.1 mg/L
Aschengrau, 1993 USA [54]	Hospital case-control study Massachusetts	1977–1980	Matched maternal residence during pregnancy or outcome to results of tap water sample taken in respective city as part of routine analysis	Congenital malformations combined	Overall, congenital anomalies were not associated with detectable levels of water nitrate (0.2–4.5 mg/L)
Brender, 2004 USA [38]	Population-based, case-control study Texas counties along Texas–Mexico border	1995–2000	Usual periconceptional drinking-water source tested for nitrates (as nitrate)	NTDs	OR of 1.9 (95% CI 0.8, 4.6) for NTDs if water nitrates ≥ 3.52 mg/L. Increased water nitrate associated with spina bifida (OR 7.8) but not with anencephaly (OR 1.0). Drinking water nitrate modified association between nitrosatable drug exposure and NTDs
Brender, 2013 USA [55]	Population-based, case-control study Iowa and Texas	1997–2005	Linked maternal addresses to public water utilities; estimated nitrate intake from bottled water with bottled water survey and testing; modeled and predicted nitrate levels in private drinking water wells (Texas only)	NTDs, oral clefts, limb deficiencies, and congenital heart defects	With the lowest tertile of nitrate intake as the referent group, mothers of babies with spina bifida were 2.0 times more likely (95% CI: 1.3, 3.2) to ingest ≥ 5 mg nitrate daily from drinking water than controls; during 1 month preconception through the first trimester, mothers of limb deficiency, cleft palate, and cleft lip cases were, respectively, 1.8 (95% CI: 1.1, 3.1), 1.9 (95% CI: 1.2, 3.1), and 1.8 (95% CI: 1.1, 3.1) times more likely than control mothers to ingest ≥ 5.42 mg of nitrate daily from drinking water
Cedergren, 2002 Sweden [56]	Population-based, retrospective cohort study Infants born in Ostergotland County	1982–1996	Linked address at periconception or early pregnancy to water supplies	Any congenital heart defect	Weak positive association (OR 1.2, 95% CI 0.97, 1.4) between water nitrate ≥ 2 mg/L and heart malformations
Croen, 2001 USA [39]	Population-based, case-control study California live births, stillbirths, and terminations	1989–1991	Periconceptional addresses linked to water utilities and nitrate databases (as nitrate)	NTDs	Exposure to water nitrates >45 mg/L associated with anencephaly (OR 4.0, 95% CI 1.0, 15.4) but not with spina bifida. Increased risk of anencephaly with water nitrate levels below U.S. EPA MCL among groundwater drinkers only (5–15 mg/L: OR 2.1, 16–35 mg/L: OR 2.3, and 36-67 mg/L: OR 6.9)

TABLE 6.3   Nitrate in Maternal Drinking Water Sources and Birth Defects in Offspring *(cont.)*

First author, year, country	Study design regional description	Years of outcome ascertainment	Exposure description*	Birth defects included	Summary of findings
Ericson, 1988 Sweden [57]	Population-based, case-control study All deliveries in Sweden	1976–1977	Earliest known maternal address linked to water nitrate results	NTDs	Average drinking water nitrate similar between NTD cases and comparison infants
Holtby, 2014 Canada [58]	Population-based, case-control study Deliveries in Kings County, Nova Scotia	1988–2006	Addresses at delivery linked to public water systems and a sample of rural wells selected for routine monitoring	Congenital malformations combined	Adjusted ORs of 1.65 and 1.66, respectively, for nitrate-nitrogen levels in drinking water 1–5.56 mg/L and >5.56 mg/L nitrate-nitrogen relative to < 1 mg/L; for births 1998–2006, ORs 2.44 (95% CI 1.05, 5.66) and 2.25 (0.92, 5.52) for same levels respectively
Scragg, 1982 Dorsch, 1984 Australia [50,51]	Population-based, case-control study Mount Gambier region South Australia	1951–1979	Address at delivery linked to sources of water and data on nitrates	Congenital malformations combined and malformations by system	Elevated ORs for any congenital malformation (2.8, 95% CI 1.6, 5.1), defects of CNS (3.5, 95% CI 1.1, 14.6), musculoskeletal system (2.9, 95% CI 1.2, 8.0) if primarily drank groundwater. Elevated ORs for congenital malformations associated with nitrate levels ≥ 5 mg/L relative to nitrate levels < 5 mg/L
Zierler, 1988 USA [59]	Population-based, case-control study All live born deliveries in Massachusetts	1980–1983	Address at the date of conception matched to water nitrate samples of respective water utilities that were drawn on the date closest to the conception date	Congenital heart disease	No association noted between water nitrates higher than minimum detection limit and congenital heart disease

** Nitrate units are specified as reported in publications. Some articles did not specify type of nitrate.*
Adapted (and updated) from Ref. [52].

and liveborns with congenital malformations. Examination of the source of drinking water revealed that women who drank groundwater from the Mount Gambier region were 2.8 times more likely (95% CI 1.6, 5.1) to give birth to offspring with congenital malformations than women who drank rainwater; central nervous system (CNS) defects showed the greatest groundwater-related increase (OR 3.5, 95% CI 1.1, 14.6). The groundwater in this area was found to have higher levels of nitrates (~15 mg/L as nitrate) than rainwater (< 2 mg/L). A positive association was also noted between prenatal groundwater consumption and defects of the musculoskeletal system (OR 2.9), NTDs (OR 3.5), and oral clefts (OR 4.0) [51]. Relative to women who drank water with less than 5 mg/L nitrate, women who consumed water with estimated nitrate levels of 5 to 15 mg/L were 2.6 times more likely to have offspring with malformations, and women who consumed water with nitrates greater than 15 mg/L were 4.1 times more likely to have offspring with malformations. The

investigators used the maternal address of residence at the time of hospital admission of the index birth to assign nitrate exposure to the source of drinking water instead of addresses during the periconceptional period. Using the address at delivery likely introduced some misclassification of exposure because a proportion of women may have moved during their pregnancies.

A study of water nitrates and CNS defects conducted by Arbuckle et al. [53] among women in New Brunswick, Canada, also used addresses at birth rather than those during the periconceptional period. However, the investigators obtained water samples from 77% of the residences rather than relying on public water supply data. Therefore, they were able to examine the relation between private well water and CNS defects as well as that from water from public water supplies. Results indicated that relative to 0.1 mg/L, women who drank water from private wells at 26 mg/L or more were 2.3 times (95% CI 0.73, 7.29) more likely to give birth to babies with CNS defects, while a negative association was found between water nitrates in spring or public water and these defects. In another study conducted in Canada in the area of Kings County, Nova Scotia, Holtby et al. [58] noted a nonsignificant positive association between nitrate-nitrogen levels in drinking water greater than 1 mg/L (> 4.43 mg/L as nitrate) and all congenital malformations combined in offspring. However, when they stratified the analyses by the periods of pre- and postfortification of grain products with folic acid, higher nitrate in drinking water (> 1 mg/L) was significantly associated with congenital malformations in offspring after fortification (OR for 1–5.56 mg/L 2.44 [95% CI 1.05, 5.66] and OR for greater than 5.56 mg/L 2.25 [95% CI 0.92, 5.52]).

In a study of community drinking water and late adverse pregnancy outcomes, Aschengrau et al. [54] found no increased risk of congenital malformations among women who drank groundwater (relative to women who drank surface water) or whose drinking water supplies had measurable levels of nitrate greater than 0.1 mg/L (relative to women whose water supplies had 0.1 mg/L or less of nitrate). The highest measured nitrate in the public drinking water supplies was reported as 4.5 mg/L, considerably lower than that found in the Australian and Canadian studies.

Several studies have been published that focused on the relation between prenatal exposure to nitrate in drinking water and congenital heart defects. In a retrospective cohort study conducted in Sweden, Cedergren et al. [56] found a weakly positive association between maternal prenatal consumption of water nitrates 2.0 mg/L or greater and congenital cardiac defects in offspring (OR 1.18, 95% CI 0.97, 1.44). On the other hand, Zierler et al. [59] noted no association between nitrate levels higher than the minimum detection limit in public water supplies and congenital heart defects in offspring (OR 1.08, 95% CI 0.72, 1.62) in a case-control study conducted in Massachusetts.

Four published studies have reported on the relation between drinking water nitrates and NTDs. Among women giving birth in Sweden, the average water nitrate did not differ appreciably between women who gave birth to offspring with NTDs (average nitrate: 4.9 mg/L) and control women (5.1 mg/L) [57]. In a study conducted among California residents, Croen et al. [39] estimated drinking water nitrates through linkage of residential addresses during the periconceptional period to water supply companies and accompanying nitrate test results during the most relevant period during periconception. Women whose public water supplies had nitrates above 45 mg/L relative to 45 mg/L or less were four times more likely (95% CI 1.0, 15.4) to have offspring with anencephaly; no association was seen between water nitrate above the MCL and spina bifida in

offspring. Furthermore, increased risks for anencephalic offspring were observed with nitrate levels below the MCL for groundwater drinkers. Relative to women who consumed water with nitrate levels less than 5 mg/L, a dose-response was noted between increasing water nitrate and risk for anencephaly (5–15 mg/L: OR 2.1; 16–35 mg/L: OR 2.3; and 36–67 mg/L: OR 6.9). In a study conducted among Mexican-American women, drinking water nitrate was measured in the usual periconceptional drinking water source if available and accessible [38]. Women whose drinking water nitrates measured 3.5 mg/L or greater (median level of control women's drinking water sources) were 1.9 times more likely (CI 0.8–4.6) to have an NTD-affected pregnancy relative to women with drinking water nitrates below 3.5 mg/L.

Using data from Iowa and Texas participants in the NBDPS, Brender et al. [55] examined the daily intake of nitrate from drinking water and selected birth defects in the offspring, including NTDs, oral clefts, limb deficiencies, and congenital heart malformations. In this study, the investigators estimated daily intake of nitrate from all sources of drinking water reported by the participants, including public water utilities, bottled water, and private wells (Texas only and estimated through modeling of groundwater flow and nitrate transport). Daily intake of nitrate from drinking water (milligrams per day) was calculated for two time periods: for analyses involving NTDs, the period of one month prior to one month post-conception and for the other defects, the period of one month prior to conception through the first trimester. Daily intake was based on the reported sources of drinking water with respective nitrate concentrations and quantity reported consumed at home and at work, use of water filters and type, reported consumption of tea and coffee, and any reported changes in water consumption or source of water during the specified exposure periods. With the lowest tertile of nitrate (less than 0.91 mg/day) around conception serving as the referent group (tertiles based on the control-women's distribution), mothers of spina bifida cases were two times more likely (95% CI 1.3, 3.2) to ingest 5 mg or more per day of nitrate from drinking water than control mothers. During the period of one month prior to conception through the first trimester, mothers of limb deficiency, cleft palate, and cleft lip cases were, respectively, 1.8 (95% CI 1.1, 3.1), 1.9 (95% CI 1.2, 3.1), and 1.8 (95% CI 1.1, 3.1) times more likely than control mothers to ingest 5.42 mg or more of nitrate daily than control mothers (vs. less than 1 mg/day).

In summary, a number of published studies indicate that a higher intake of nitrate from drinking water sources during pregnancy may be associated with birth defects in offspring, especially NTDs. One of the issues in drinking water studies is the potential presence of other contaminants. While some of these studies also examined other water contaminants, none examined exposures to mixtures of contaminants. In a study that evaluated chemical mixtures in U.S. groundwater including 383 public wells distributed across 35 states, nitrate was found to occur with other contaminants in drinking water, especially in conjunction with pesticides, arsenic and other trace metals, and water disinfection by-products [60].

DIETARY INTAKE OF NITRATES, NITRITES, AND NITROSAMINES IN FOODS AND BIRTH DEFECTS

While nitrate intake from drinking water sources has been associated with birth defects, the estimated higher dietary intake of nitrate has not been observed to be associated with selected birth defects, including NTDs, oral clefts, or limb deficiencies. Studies conducted to date have estimated dietary nitrate, nitrite, and nitrosamine intake based on responses to food frequency questionnaires and the published literature of nitrate, nitrite, and nitrosamine content of the various foods included on these questionnaires [38,39,61]. Such an approach is subject to potential measurement error, although it is likely that misclassification of intake is nondifferential with respect to case-control status.

Using the 100-item Block food frequency questionnaire [62], Croen et al. [39] estimated daily intake of nitrate, nitrite, and N-nitroso compounds from the diets in 538 case-mothers with NTD-affected pregnancies and 539 control-mothers of nonmalformed babies. With estimated intake divided into quartiles, no positive associations were noted between estimated higher intake of any of these compounds and NTDs in offspring. All point estimates for the ORs were at or below 1.0 for the second, third, and fourth quartiles of intake relative to the lowest quartiles of intake.

In the Texas Neural Tube Defect Project, 184 case-women and 225 control-women were questioned about their food intake during the period three months prior through three months postconception with a 98-item food frequency questionnaire that was specifically designed for the Mexican-American population living in Texas counties bordering Mexico [38]. Nitrite and total nitrite (5% nitrate intake + nitrite intake) were estimated and categorized into tertiles based on the control-women's distributions. With the lowest tertiles of intake as the referent categories, the ORs for the upper tertiles of nitrite intake were 0.8 and 0.9 and for total nitrite, 0.9 and 0.8.

Using data from the NBDPS, Huber et al. [61] examined maternal dietary intake of nitrates, nitrites, and nitrosamines in a sample of 6544 mothers of cases with NTDs, oral clefts, or limb deficiencies and 6807 control mothers. Daily consumption of nitrate, nitrite, and nitrosamines

was estimated from a 58-item food frequency questionnaire based on the Willett Food Frequency Questionnaire [63,64]. Overall, NTDs, oral clefts, and limb deficiencies were not significantly associated with estimated dietary intake of these compounds. A few positive associations were noted, including daily intake of nitrite from animal sources and cleft lip with or without cleft palate (adjusted OR and CI for fourth quartile compared with the lowest quartile of intake: 1.24, 95% CI 1.05, 1.48), animal nitrite with cleft lip only (fourth quartile OR 1.32, 95% CI 1.01, 1.72), and total nitrite intake and intercalary limb deficiencies (fourth quartile OR 4.70, 95% CI 1.23, 17.93). In contrast to previous studies, the investigators also estimated the potential impact of measurement error using the Simulation Extrapolation (SIMEX) algorithm [65] and hypothetically varying the amount of measurement error included in the model from no error to a multiplicative factor of 1.6 in increments of 0.10. The investigators did not find any evidence based on these simulations that even the highest levels of measurement error made any substantial differences in the results.

JOINT EXPOSURES OF NITROSATABLE DRUGS AND NITRATE/NITRITE AND BIRTH DEFECTS

Previous publications of studies on the relation between birth defects in offspring and prenatal exposure to drinking water nitrate and nitrosatable drugs have hypothesized that observed positive associations might be due to the endogenous formation of N-nitroso compounds and their subsequent teratogenic effects [39,42,51]. Results from animal studies, however, have indicated that nitrite and some nitrosatable compounds were teratogenic only in combination with one another, and no excess incidence in malformations were noted when each was administered separately [66]. Recent published studies in humans also suggest that a higher intake of nitrate and nitrite increases positive associations between prenatal exposure to nitrosatable drugs and several types of congenital malformations in offspring, including NTDs, cleft lip alone, cleft palate alone, conotruncal heart defects, single ventricle, and atrioventricular septal defects [38,43,44,55]. Table 6.4 summarizes the characteristics and major findings of these studies. In each of the studies, reported medications during the periconceptional period and/or the first trimester were classified with respect to their nitrosatability based on the available literature, and dietary intake of nitrite and total nitrite were divided into low, moderate, and high intake based on the control-women's distributions. One study also included daily intake from drinking water in the estimates of total nitrite [55]. Stronger associations between nitrosatable drug exposure and NTDs among

TABLE 6.4   Epidemiologic Studies of the Interaction Effects of Prenatal Nitrosatable Drug Exposure with Nitrite Intake in Relation to Birth Defects in Offspring

First author, year, country	Study design regional description	Birth defects included	Description of exposure assessment including how interaction assessed	Summary of findings
Brender, 2004, USA [38]	Population-based case-control study 14 Texas counties bordering Mexico	NTDs	Nitrosatable drug use during the periconceptional period stratified by tertiles of dietary nitrite and total nitrite intake based on the control-women's distributions of intake	Positive associations between prenatal exposure to nitrosatable drug use and NTDs observed only in offspring of women whose intake of nitrite/total nitrite were in the second or third tertiles. Case-women within the highest tertile of total nitrite intake were 7.5 times more likely than control-women to report use of drugs classified as nitrosatable (95% CI 1.8, 45.4)
Brender, 2011, USA [43]	Population-based case-control study NBDPS in 10 states	NTDs, grouped and specific phenotypes	Nitrosatable drug use (any type, secondary amines, tertiary amines) during the period ± 1 month of conception stratified by tertiles of dietary nitrite and total nitrite based on control-women's distributions of intake; additive and multiplicative interaction assessed	Trends of stronger positive associations between secondary and tertiary amine drug exposures and NTDs among women with higher nitrite and total nitrite intake observed for both anencephaly and spina bifida in offspring. Strongest associations and trend noted between tertiary amines and anencephaly stratified by total nitrite intake: ORs associated with tertiary amines from the lowest tertile to the highest tertile of total nitrite intake were 1.16 (95% CI 0.59, 2.29); 2.19 (95% CI 1.25, 3.86); and 2.51 (95% CI 1.45, 4.37)
Brender, 2012, USA [44]	Population-based case-control study NBDPS in 10 states	Limb deficiencies, oral clefts congenital heart malformations	Nitrosatable drug use (any type, secondary amines, tertiary amines, amides) during the first trimester stratified by tertiles of dietary nitrite and total nitrite based on control-women's distributions of intake; additive and multiplicative interaction assessed	Positive associations between tertiary amines and cleft lip alone and between amides and cleft palate alone noted only among women with the highest tertile of total nitrite intake (ORs of 1.46 [95% CI 0.93, 2.27] and 1.57 [95% CI 1.01, 2.54], respectively) and significant additive interaction was present for both sets of associations; significant additive and multiplicative interaction present between the combination of higher total nitrite intake with tertiary amine and amide drug exposures in relation to conotruncal heart defects; association between amide exposure and single ventricle heart defects strongest in offspring of women in the highest tertile of total nitrite exposure (OR 3.30, 95% CI 1.59, 6.84); strongest association between secondary amine drug exposure and atrioventricular septal defects noted in offspring of women with highest intake of total nitrite (OR 3.30, 95% CI 1.44, 7.58)
Brender, 2013, USA [55]	Population-based, case-control study Iowa and Texas	NTDs, limb deficiencies, oral clefts congenital heart malformations	Nitrosatable drug use (any type) stratified by nitrate intake from drinking water and total nitrite based on control-women's distributions of intake; total nitrite included contributions of nitrate from food and water and nitrite from food	Associations between nitrosatable drug use and selected birth defects in offspring were not stronger among women with higher daily intake of nitrate from drinking water. A pattern of stronger associations between nitrosatable drug exposure and some birth defects were noted for women in the upper two tertiles of total nitrite intake that included contributions from drinking water nitrate; this pattern was noted for the associations between nitrosatable drug exposure and NTDs, cleft lip alone, cleft palate alone, limb deficiencies, atrioventricular septal defects, and single ventricle. Associations between prenatal nitrosatable drug exposure and birth defects in the upper two tertiles of total nitrite intake were particularly strong for cleft palate alone (OR 2.51, 95% CI 1.24, 5.06); atrioventricular septal defect (OR 5.10, 95% CI 1.40, 18.6), and single ventricle (OR 3.25, 95% CI 1.13, 9.31)

women with higher nitrite intake were observed in the Texas Neural Tube Defect Study and the NBDPS populations. The NBDPS had sufficient sample size to examine these effects by the molecular structure of the nitrosatable drugs, and associations between secondary and tertiary amines with anencephaly and spina bifida were strengthened in the presence of estimated higher intake of dietary nitrite. The strongest trend was noted between tertiary amines and anencephaly stratified by total nitrite intake; ORs associated with tertiary amines from the lowest tertile to the highest tertile of total nitrite intake were 1.16 (95% CI 0.59, 2.29), 2.19 (95% CI 1.25, 3.86), and 2.51 (95% CI 1.45, 4.37).

In contrast to the positive associations noted between nitrosatable drug use and birth defects with higher maternal intake of total dietary nitrite, stronger associations between nitrosatable drug exposure and birth defects were not observed with higher daily intake of nitrate from drinking water sources among Iowa and Texas NBDPS participants [55]. On the other hand, stronger associations were noted between these types of drugs and selected birth defects in offspring of this study population in the upper two tertiles of total nitrite intake that included contributions of nitrate from dietary and drinking water sources and dietary nitrite. Associations between nitrosatable drugs and birth defects in the upper two tertiles of total nitrite intake were particularly strong for cleft palate alone (OR 2.51, 95% CI 1.24, 5.06), atrioventricular septal defect (OR 5.10, 95% CI 1.40, 18.6), and single ventricle (OR 3.25, 95% CI 1.13, 9.31). The lack of synergy between nitrosatable drug use and higher intake of nitrate from drinking water in relation to birth defects might be explained by the relatively low contribution of drinking water nitrate intake to daily nitrate intake. In the Iowa and Texas study populations, median contribution of nitrate per day from drinking water was approximately 6%. The World Health Organization recently concluded that the contribution of drinking water to daily nitrate intake is usually less than 14% [67]. As discussed earlier, drinking water sources of nitrate contribute a small percentage of total daily nitrate intake unless levels in the water supply approach or exceed the allowable MCL as set by the U.S. Environmental Protection Agency.

VITAMIN C, PRENATAL NITROSATABLE DRUG USE, AND BIRTH DEFECTS IN OFFSPRING

A number of compounds, including vitamin C, vitamin E (nonesterified forms), and polyphenols (antioxidants found in a variety of foods and beverages) [68], have been noted to inhibit *in vivo* nitrosation when administered with a nitrosatable compound [31,69]. Vitamin C is the most well-studied with respect to its role as a nitrosation

inhibitor, so this chapter will focus on the impact of this vitamin. Vitamin C (ascorbic acid) has been shown to inhibit the formation of N-nitroso compounds through a rapid reduction of nitrous acid to nitric oxide and the production of dehydroascorbic acid [70].

Several experiments in human volunteers have demonstrated the inhibition of nitrosation when vitamin C was given in sufficient doses concomitantly with nitrate and nitrosatable precursors, including proline (a nitrosatable compound) [71] and a fish meal rich in amines [72]. Based on their experimental research in humans with nitrate and proline in conjunction with vitamin C, Mirvish et al. [71] concluded that 120 mg of ascorbic acid taken with each meal would significantly reduce endogenous nitrosamine formation when nitrate and nitrosatable compounds were ingested together. Helser et al. [73] noted a significant inhibition of endogenous formation of N-nitroso compounds when ascorbic acid was given up to 5 h before a nitrosatable compound, although the investigators observed the greatest reduction of nitrosation with concurrent administration of ascorbic acid with proline.

Some evidence from epidemiologic studies supports the potential beneficial effects of vitamin C in weakening associations between prenatal exposures to nitrosatable compounds and precursors and birth defects in offspring [39,43,74]. Croen et al. [39] noted lower ORs for the association between higher water nitrate in groundwater and anencephaly among women who consumed at least 103 mg/day of vitamin C (ORs associated with higher water nitrate ranged from 1.8 to 5.3) than among women who consumed less than 103 mg/day of this vitamin (ORs ranged from 4.0 to 18.0).

With respect to nitrosatable drug exposure, Brender et al. [43] observed diminished associations between prenatal exposure to drugs classified as tertiary amines and anencephaly in offspring (OR 1.52) of women in the NBDPS population who reported taking a daily supplement with vitamin C around the time of conception compared with women who took a supplement less than daily (OR 2.77) or not at all during this period (OR 2.11). While this pattern with supplementation was not noted in conjunction with nitrosatable drugs and spina bifida, a higher dietary intake of vitamin C ($\geq$85 mg/day) lessened the association between tertiary amines and spina bifida (OR 1.15) compared with that (OR 1.81) for offspring of women with a lower dietary intake of vitamin C. Shinde et al. [74] also noted lower ORs for selected limb deficiency, oral cleft, and heart defects in relation to nitrosatable drugs if women reported daily supplementation and/or higher dietary vitamin C intake. This pattern was particularly pronounced for longitudinal limb deficiencies in conjunction with tertiary amines (dietary intake of vitamin C $\geq$85 mg per day/daily supplementation versus dietary intake <85 mg/day/no supplementation: ORs, respectively 0.68 and 1.90), transverse limb deficiencies in conjunction with secondary amines (ORs, respectively 1.23 and 4.16), amides with cleft lip with cleft palate (ORs, respectively 1.08 and 2.46), atrial septal defect (not otherwise specified) with tertiary amines (ORs, respectively 1.65 and 2.44), and atrial septal defect with amides (ORs, respectively 0.79 and 4.43). None of these studies had information on specific doses of vitamin C in supplements or the timing of supplementation in relation to exposures to nitrosatable precursors.

CONCLUSIONS, CLINICAL IMPLICATIONS, AND SUGGESTIONS FOR FURTHER RESEARCH

The findings of several epidemiologic studies indicate that nitrosatable drugs may function as potential teratogens in humans, particularly in the presence of higher dietary intake of nitrate and nitrite. The finding that a high proportion of the nitrosatable drugs taken by the NBDPS population were preparations that were available without a prescription underscores the importance of cautioning women in their childbearing years, especially those who are pregnant, to check with their healthcare providers before taking OTC preparations. On the other hand, higher estimated dietary intakes of nitrate, nitrite, and nitrosamines have not been associated with birth defects in several US populations, unless coupled with nitrosatable compounds such as certain drugs.

Although studies in the past have attributed associations between higher levels of nitrate in drinking water and birth defects to the endogenous formation of N-nitroso compounds, the levels of nitrate associated with these defects were relatively low, suggesting that concomitant pollutants or other factors associated with these sources of water might be responsible for the associations. Since some of the highest nitrate levels in drinking water have been detected in private wells, women of childbearing age who obtain their drinking water from these types of wells should consider having their drinking water tested for nitrate and other routinely monitored contaminants that are tested in public water systems. Alternately, they might consider switching to purified bottle water if pregnant or trying to conceive.

While more research is indicated on the potential beneficial effects of vitamin C in preventing birth defects associated with nitrosatable drugs, the results of the few epidemiologic studies that examined the impact of vitamin C on nitrosatable drugs and birth defects appear promising. Based on results from experimental studies, fruits or vegetables (or their juices) that contain vitamin C might need to be consumed shortly before or concomitantly with drugs that are nitrosatable. Further research is indicated on the timing of supplementation and ingestion of foods high in vitamin C in relation to exposure to nitrosatable compounds and pregnancy outcomes.

References

[1] Petrini J, Damus K, Russell R, Poschman K, Davidoff MJ, Mattison D. Contribution of birth defects to infant mortality in the United States. Teratology 2002;66(Suppl 1):S3–6.

[2] Rosano A, Botto LD, Botting B, Mastroiacovo. Infant mortality and congenital anomalies from 1950 to 1994: an international perspective. J Epidemiol Community Health 2000;54:660–6.

[3] Heron M. Deaths, Leading causes for 2010. National vital statistics reports;, vol 62 no 6. National Center for Health Statistics: Hyattsville, MD; 2013.

[4] Centers for Disease Control and Prevention. Number of all-listed diagnoses for sick newborn infants by sex and selected diagnostic categories. 2014. http://www.cdc.gov/nchs/nhds_tables.htm#number (accessed Feb 13, 2014).

[5] Rynn L, Cragan J, Correa A. Update on overall prevalence of major birth defects – Atlanta, Georgia 1978–2005. MMWR Morb Mortal Wkly Rep 2008;57:1–5.

[6] Brent RL. Environmental causes of human congenital malformations: the pediatrician's role in dealing with these complex clinical problems caused by a multiplicity of environmental and genetic factors. Pediatrics 2004;113:957–68.

[7] Inouye M, Murakami U. Teratogenic effect of N-methyl-N-nitro-N-nitrosoguanidine in mice. Teratology 1978;18:263–8.

[8] Nagao T, Morita Y, Ishizuka Y, Wada A, Mizutani M. Induction of fetal malformations after treatment of mouse embryos with methylnitrosourea at the preimplantation stages. Teratog Carcinog Mutagen 1991;11:1–10.

[9] Lijinsky W, Keefer L, Conrad E, Van de Bogart R. Nitrosation of tertiary amines and some biologic implications. J Natl Cancer Inst 1972;49:1239–49.

[10] Gillatt PN, Palmer RC, Smith PL, Walters CL, Reed PI. Susceptibilities of drugs to nitrosation under simulated gastric conditions. Food Chem Toxicol 1985;23:849–55.

[11] Pennington JA. Dietary exposure models for nitrates and nitrites. Food Control 1998;9:385–95.

[12] Lijinsky W. N-nitroso compounds in the diet. Mutat Res 1999;443:129–38.

[13] Gangolli SD, van den Brandt PA, Feron VJ, Janzowsky C, Koeman JH, Speijers GJ, et al. Nitrate, nitrite, and N-nitroso compounds. Eur J Pharmacol 1994;292:1–38.

[14] Jakszyn P, Ibanez R, Pera G, Garcia-Closas R, Agudo A, Amiano P, et al. Food Content of Potential Carcinogens. Barcelona: Catalan Institute of Oncology; 2004.

[15] Walters CL. Nitrate and nitrite in foods. In: Hill M, editor. Nitrates and Nitrites in Food and Water. New York: Ellis Horwood; 1991. p. 93–112.

[16] Griesenbeck JS, Brender JD, Sharkey JR, Steck MD, Huber JC Jr, Rene AA, et al. Maternal characteristics associated with the dietary intake of nitrates, nitrites, and nitrosamines in women of child-bearing age. Environ Health 2010;9:10.

[17] National Academy of Sciences. The Health Effects of Nitrate, Nitrite, and N-nitroso compounds. Washington, D.C: National Academy of Sciences; 1981.

[18] Hord NG, Tang Y, Bryan NS. Food sources of nitrates and nitrites: the physiologic context for potential health benefits. Am J Clin Nutr 2009;90:1–10.

[19] Santamaria P. Nitrate in vegetables: toxicity, content, intake and EC regulation. J Sci Food Agric 2006;86:10–7.

[20] Spalding RF, Exner ME. Occurrence of nitrate in groundwater – a review. J Environ Qual 1993;22:392–402.

[21] U.S. Environmental Protection Agency. Current drinking water standards. Available from: http://www.epa.gov/safewater/mcl.html. 2005 [accessed 21.12.05].

[22] Ward MH, deKok TM, Levallois P, Brender J, Gulis G, Nolan BT, et al. Workgroup report: drinking-water nitrate and health— recent findings and research needs. Environ Health Perspect 2005;113:1607–14.

[23] Choi BC. N-nitroso compounds and human cancer. Am J Epidemiol 1985;121:737–43.

[24] Mensinga TT, Speijers GJA, Meulenbelt J. Health implications of exposure to environmental nitrogenous compounds. Toxicol Rev 2003;22:41–51.

[25] Spiegelhalder B, Eisenbrand G, Preussmann R. Influence of dietary nitrate on nitrite content of human saliva: possible relevance to in vivo formation of N-nitroso compounds. Food Cosmet Toxicol 1976;14:545–8.

[26] Tricker AR, Preussmann R. Volatile and nonvolatile nitrosamines in beer. J Cancer Res Clin Oncol 1991;117:130–2.

[27] Stepanov I, Hecht SS. Tobacco-specific nitrosamines and their pyridine-N-glucuronides in the urine of smokers and smokeless tobacco users. Cancer Epidemiol Biomarkers Prev 2005;14:885–91.

[28] Bernert JT, Jain RB, Pirkle JL, Wang L, Miller BB, Sampson EJ. Urinary tobacco-specific nitrosamines and 4-aminobiphenyl hemoglobin adducts measured in smokers of either regular or light cigarettes. Nicotine Tob Res 2005;7:729–38.

[29] Havery DC, Chou HJ. N-nitrosamines in cosmetics products. Cosmet Toiletries 1994;109:53–61.

[30] Walters KA, Brain KR, Dressler WE, Green DM, Howes D, James VJ, et al. Percutaneous penetration of N-nitroso-N-methyldodecylamine through human skin in vitro: application from cosmetic vehicles. Food Chem Toxicol 1997;35:705–12.

[31] Tricker AR. N-nitroso compounds and man: sources of exposure, endogenous formation and occurrence in body fluids. Eur J Cancer Prev 1997;6:226–68.

[32] Lijinsky W. Occurrence, formation, and detection of N-nitroso compounds. Chemistry and Biology of N-nitroso Compounds. New York: Cambridge University Press; 1992. p. 19–54.

[33] Preussmann R. Occurrence and exposure to N-nitroso compounds and precursors. In: O'Neill IK, Von Borstel RC, Miller CT, Long J, Bartsch H, editors. N-nitroso compounds: occurrence, biological effects, and relevance to human cancer. Lyon: International Agency for Research on Cancer; 1984. p. 3–15.

[34] Miwa M, Stuehr DJ, Marletta MA, Wishnok JS, Tannenbaum SR. N-nitrosamine formation by macrophages. IARC Sci Publ 1987;84:340–4.

[35] Brambilla G, Cajelli E, Finollo R, Maura A, Pino A, Robbiano L. Formation of DNA-damaging nitroso compounds by interaction of drugs with nitrite. A preliminary screening for detecting potentially hazardous drugs. J Toxicol Environ Health 1985;15:1–24.

[36] Ozhan G, Alpertunga B. Genotoxic activities of drug-nitrite interaction products. Drug Chem Toxicol 2003;26:295–308.

[37] Brambilla G, Martelli A. Genotoxic and carcinogenic risk to humans of drug-nitrite interaction products. Mutat Res 2007;635:17–52.

[38] Brender JD, Olive JM, Felkner M, Suarez L, Marckwardt W, Hendricks KA. Dietary nitrites and nitrates, nitrosatable drugs, and neural tube defects. Epidemiology 2004;15:330–6.

[39] Croen LA, Todoroff K, Shaw GM. Maternal exposure to nitrate from drinking water and diet and risk for neural tube defects. Am J Epidemiol 2001;153:325–31.

[40] Gardner JS, Guyard-Goileau B, Alderman BW, Fernbach SK, Greene C, Mangione EJ. Maternal exposure to prescription and non-prescription pharmaceuticals or drugs of abuse and craniosynostosis. Int J Epidemiol 1998;27:64–7.

[41] Kallen B, Robert-Gnansia E. Maternal drug use, fertility problems, and infant craniostenosis. Cleft Palate Craniofac J 2005;42:589–93.

[42] Olshan AF, Faustman EM. Nitrosatable drug exposure during pregnancy and adverse pregnancy outcome. Int J Epidemiol 1989;18:891–9.

[43] Brender JD, Werler MM, Kelley KE, Vuong AM, Shinde MU, Zheng Q, et al. Nitrosatable drug exposure during early pregnancy and neural tube defects in offspring. Am J Epidemiol 2011;174:1286–95.

[44] Brender JD, Werler MM, Shinde MU, Vuong AM, Kelley KE, Huber JC Jr, et al. Nitrosatable drug exposure during the first trimester of pregnancy and selected congenital malformations. Birth Defects Res A Clin Mol Teratol 2012;94:701–13.

[45] Cardy AH, Little J, McKean-Cowdin R, Lijinsky W, Choi NW, Cordier S, et al. Maternal medication use and the risk of brain tumors in the offspring: the SEARCH international case-control study. Int J Cancer 2006;118:1302–8.

[46] Carozza SE, Olshan AF, Faustman EM, Gula MJ, Kolonel LN, Austin DF, et al. Maternal exposure to N-nitrosatable drugs as a risk factor for childhood brain tumors. Int J Epidemiol 1995;24:308–12.

[47] Cook MN, Olshan AF, Guess H, Savitz DA, Poole C, Blatt J, et al. Maternal medication use and neuroblastoma in offspring. Am J Epidemiol 2004;159:721–31.

[48] McKean-Cowdin R, Pogoda JM, Lijinsky W, Holly EA, Mueller BA, Preston-Martin S. Maternal prenatal exposure to nitrosatable drugs and childhood brain tumours. Int J Epidemiol 2003;32:211–7.

[49] Brender JD, Kelley KE, Werler MM, Langlois PH, Suarez L, Canfield MA, et al. Prevalence and patterns of nitrosatable drug use among U.S. women during early pregnancy. Birth Defects Res A Clin Mol Teratol 2011;91:258–64.

[50] Scragg RK, Dorsch MM, McMichael AJ, Baghurst PA. Birth defects and household water supply. Med J Aust 1982;2:577–9.

[51] Dorsch MM, Scragg RK, McMichael AJ, Baghurst PA, Dyer KF. Congenital malformations and maternal drinking water supply in rural South Australia: a case-control study. Am J Epidemiol 1984;119:473–86.

[52] Ward MH, Brender JD. Drinking Water Nitrate and Health. In: Nriagu JO, editor. Encyclopedia of Environmental Health, vol. 2. Burlington: Elsevier; 2011. p. 167–78.

[53] Arbuckle TE, Sherman GJ, Corey PN, Walters D, Lo B. Water nitrates and CNS defects: a population-based case-control study. Arch Environ Health 1988;43:162–7.

[54] Aschengrau A, Zierler S, Cohen A. Quality of community drinking water and the occurrence of late adverse pregnancy outcomes. Arch Environ Health 1993;48:105–13.

[55] Brender JD, Weyer PJ, Romitti PA, Mohanty BP, Shinde MU, Vuong AM, et al. Prenatal nitrate intake from drinking water and selected birth defects in offspring of participants in the National Birth Defects Prevention Study. Environ Health Perspect 2013;121:1083–9.

[56] Cedergren MI, Selbing AJ, Lofman O, Kallen BA. Chlorination byproducts and nitrate in drinking water and risk for congenital cardiac malformations. Environ Res 2002;89:124–30.

[57] Ericson A, Kallen B, Lofkvist E. Environmental factors in the etiology of neural tube defects: a negative study. Environ Res 1988;45:38–47.

[58] Holtby CE, Guernsey JR, Allen AC, VanLeeuwen JA, Allen VM, Gordon RJ. A population-based case-control study of drinking water nitrate and congenital anomalies using geographic information systems (GIS) to develop individual-level exposure estimates. Int J Environ Res Public Health 2014;11:1803–23.

[59] Zierler S, Theodore M, Cohen A, Rothman KJ. Chemical quality of maternal drinking water and congenital heart disease. Int J Epidemiol 1988;17:589–94.

[60] Toccalino PL, Norman JE, Scott JC. Chemical mixtures in untreated water from public-supply wells in the U. S. – occurrence, composition, and potential toxicity. Sci Total Environ 2012;431:262–70.

A. OVERVIEW ON FERTILITY

[61] Huber JC Jr, Brender JD, Zheng Q, Sharkey JR, Vuong AM, Shinde MU, et al. Maternal dietary intake of nitrates, nitrites and nitrosamines and selected birth defects in offspring: a case-control study. Nutr J 2013;12:34.

[62] Block G, Hartman AM, Dresser CM, Carroll MD, Gannon J, Gardner L, et al. A data-based approach to diet questionnaire design and testing. Am J Epidemiol 1986;124:453–69.

[63] Willett WC, Sampson L, Stampfer MJ, Rosner B, Bain C, Witschi J, et al. Reproducibility and validity of a semiquantitative food frequency questionnaire. Am J Epidemiol 1985;122:51–65.

[64] Willett WC, Reynolds RD, Cottrell-Hoehner S, Sampson L, Browne ML. Validation of a semi-quantitative food frequency questionnaire: comparison with a 1-year diet record. J Am Diet Assoc 1987;87:43–7.

[65] Carroll RJ, Rupert D, Stefanski LA, Crainiceanu CM. Measurement Error in Nonlinear Models: A Modern Perspective. 2nd ed New York: Chapman-Hall/CRC; 2006.

[66] Teramoto S, Saito R, Shirasu Y. Teratogenic effects of combined administration of ethylenethiourea and nitrite in mice. Teratology 1980;21:71–8.

[67] World Health Organization. Nitrate and Nitrite in Drinking-water. WHO/SDE/WSH/07.01/16/Rev/1. Geneva: World Health Organization, 2011. Available from: http://www.who.int/water_sanitation_health/dwq/chemicals/nitratenitrite2ndadd.pdf (accessed 09.08.13).

[68] Perez-Jimenez J, Neveu V, Vos F, Scalbert A. Identification of the 100 richest sources of polyphenols: an application of the phenol-explorer database. Eur J Clin Nutr 2010;64(Suppl 3):S112–220.

[69] Mirvish SS. Inhibition by vitamins C and E of *in vivo* nitrosation and vitamin C occurrence in the stomach. Eur J Cancer Prev 1996;5(Suppl 1):131–6.

[70] Bartsch H, Ohshima H, Pignatelli B. Inhibitors of endogenous nitrosation. Mechanisms and implications in human cancer prevention. Mutat Res 1988;202:307–24.

[71] Mirvish SS, Grandjean AC, Reimers KF, Connelly BJ, Chen SC, Morris CR, et al. Effect of ascorbic acid dose taken with a meal on nitrosoproline excretion in subjects ingesting nitrate and proline. Nutr Cancer 1998;31:106–10.

[72] Vermeer IT, Moonen EJ, Dallinga JW, Kleinjans JC, van Maanen JM. Effect of ascorbic acid and green tea on endogenous formation of *N*-nitrosodimethylamine and *N*-nitrosopiperidine in humans. Mutat Res 1999;428:353–61.

[73] Helser MA, Hotchkiss JH, Roe DA. Temporal influence of ascorbic acid dose on the endogenous formation of *N*-nitrosoproline and *N*-nitrosothiazolidine-4-carboxylic acid in humans. J Nutr Biochem 1991;2:268–73.

[74] Shinde MU, Vuong AM, Brender JD, Werler MM, Kelley KE, Huber JC Jr, et al. Prenatal exposure to nitrosatable drugs, vitamin C, and risk of selected birth defects. Birth Defects Res A Clin Mol Teratol 2013;97:515–31.

7

The Application of Reproductive Techniques (ART): Worldwide Epidemiology Phenomenon and Treatment Outcomes

Giulia Scaravelli, MD, PhD, Roberta Spoletini, PhD

National Health Institute, Rome, Italy

BACKGROUND

Infertility is a global health problem with a mean prevalence around the world that can vary from 4% to 17% in developed nations [1]. However, in some parts of Sub-Saharan Africa [2] and some regions of China, where access to treatments is very limited or even absent, the prevalence could raise 30–40% or 20%, respectively [3]. There are many possible causes for infertility. In developed countries, one of the main problems is the advanced age of women, while in poorer countries there is a high prevalence in both genders of sexually transmitted diseases. Additionally, in these countries, postabortion or postdelivery infections for women, due to limited access to treatment, represent a major cause of infertility leading to tubal occlusion necessitating IVF (*in vitro* fertilization) [2]. ART treatments over more than 35 years have steadily improved – in efficacy and safety – with less aggressive ovarian stimulation and with a better embryo culture system, more efficient oocyte and embryo cryopreservation techniques [4–7], and the trend in reducing multiple pregnancies and deliveries. Currently, these procedures are considered good and efficacious solutions for infertility and for those at risk in transmitting genetic diseases.

EVALUATION OF ART PROCEDURES WORLDWIDE: THE DIVERSE ART DATA COLLECTION SYSTEMS

ART Data Collection Systems

After the first "test tube baby" Louise Brown was born in 1978 in the United Kingdom, ART procedures began to spread from Australia to Europe. It has become necessary to record all the treatments to evaluate the real efficacy and safety of their application. Any country wishing to perform ART has to establish its own national data collection system. The first country to do so was Germany in 1982, followed by France in 1984, the United Kingdom in 1990, and then by many others in the subsequent years. The European Society of Human Reproduction and Embryology (ESHRE), under the proposal of Karl Nygren [8], promoted the idea in collecting all the European data. A European register was then established in 1999, beginning with a few countries; today, 39 countries are participating. During the 1990s, national data collection also started in the United States, Canada, and Australia. Currently, international data collection under the endorsement of the International Committee Monitoring Assisted Reproductive Technologies (ICMART) collects data Worldwide.

Handbook of Fertility. http://dx.doi.org/10.1016/B978-0-12-800872-0.00007-X

Europe

In Europe, data collection is performed by the European IVF monitoring (EIM), a consortium created by the ESRHE in 1999. It has now expanded to include data from 39 countries: Albania, Austria, Belgium, Bosnia, Bulgaria, Croatia, Cyprus, Czech Republic, Denmark, Estonia, Finland, France, Germany, Greece, Hungary, Iceland, Ireland, Italy, Kazakhstan, Latvia, Lithuania, Macedonia, Moldova, Montenegro, Norway, Poland, Portugal, Romania, Russia, Serbia, Slovakia, Slovenia, Spain, Sweden, Switzerland, the Netherlands, Turkey, Ukraine, and the United Kingdom. In the EIM there are some countries that send their collected data through national registers, while others send their data either directly from ART clinics or from reproductive scientific societies [9].

This data collection could be performed on a voluntary or mandatory basis. The last data available from EIM were that of the 2010 activity. At this time in the EIM, there were 31 countries participating in data collection, of which 16 countries sent their data through national registers. The percentage of coverage of clinics reporting data for these 16 countries was 100%, and for 14 of these the data collection was mandatory. The proportion of coverage of the EIM considering all the countries reporting data was 83%. Ten countries (32%) collect their data on ART treatments cycle by cycle, while the other 21 collect only summary data [10].

International Committee for Monitoring Assisted Reproductive Technologies, World Report

Data on the number of ART cycles performed worldwide were first collected in 1989 by the International Working Group for Registers on Assisted Reproduction (IWGROAR). Since 2001 the ICMART has continued this activity of collecting data from 56 countries in seven regions (Asia, Australia and New Zealand, Europe, Latin America, Middle East, North America, and Southern Africa). Of these countries, 32 are included in EIM and 12 belong to the Latin American Network for Reproductive Medicine (REDLARA). Egypt, India, Japan, South Korea, Australia, and New Zealand send their data to ICMART.

All data were provided directly from national or from regional ART registries and analyzed in Sweden by the Uppsala Clinical Research Center (UCR), Uppsala University.

ICMART collaborates with the following national organizations: American Society for Reproductive Medicine (ASRM), ESHRE, Fertility Society of Australia (FSA), Japan Society for Reproductive Medicine (JSRM), REDLARA, Middle East Fertility Society (MEFS), and the Society for Assisted Reproductive Technology (SART) [9].

United States

The United States collects their national data from American Society for Reproductive Medicine (ASRM) and with Society for Assisted Reproductive Technology (SART) in collaboration with the Centers for Disease Control and Prevention (CDC) Atlanta. Data collection is performed cycle by cycle from 451 IVF clinics; the last report was for 2011 activity. The data collection coverage is estimated to be about 80% of all ART activity in the United States [11].

Latin America

The Latin American Registry of Assisted Reproduction (RLA) is part of REDLARA. Its main responsibility is to collect, analyze, and disseminate the results of ART procedures performed in centers that have been certified by REDLARA. The registration is voluntary, and having the center certified by an independent body is a condition in the data system. The REDLARA register collects data from 16 countries: Argentina, Bolivia, Brazil, Chile, Colombia, Costa Rica, Ecuador, Guatemala, Mexico, Nicaragua, Paraguay, Panama, Peru, the Dominican Republic, Uruguay, and Venezuela. It is also the first international register that collects data cycle by cycle [12].

Canada

The Canadian Assisted Reproductive Technologies Register (CARTR) was first established in 1999 for the collection of treatment data from Canadian fertility centers that were using ART. The IVF Directors Group of the Canadian Fertility and Andrology Society (CFAS) manages the CARTR program, which is financially supported by participating ART centers. Data collection is voluntary and made on a cycle-by-cycle basis. In 2012, 32 of 33 centers operating in Canada submitted data to CARTR; its coverage of data collection being nearly 97% [13].

Australia and New Zealand

The Fertility Society of Australia (FSA) is the leading body representing scientists, doctors, researchers, nurses, consumers, and counselors in reproductive medicine in Australia and New Zealand.

Each year the FSA holds a Scientific Meeting attracting experts in reproductive health from around the world to present research, discuss new technologies and treatments, and send their data to ICMART [14].

Japan

In Japan, the National Registry of Japan collects cycle-by-cycle data of ART treatments. All centers conducting ART treatment are required to register online. In addition, registering the data of individual ART treatments is linked to a governmental grant for the patient. Should a facility not provide the data, the patients do not receive the grant. Recorded data include patient's age, residence, any government subsidies for ART treatment, cause of infertility, treatment protocols, postprocedural complications, and obstetric outcomes. Accuracy of the data in registered institutions is audited by the government annually. A summary of the registered data is annually disclosed on the *Journal of Obstetrics and Gynecology* (JSOG) website [15].

Middle East Society

The Middle East Fertility Society is a 21-year-old professional society whose mission is to improve fertility care of couples in the Middle East through the transfer of medical knowledge and to promote scientific research. The society has become a vehicle of knowledge dissemination and a platform for research collaboration in the whole region (http://www.mefs.org/).

South Africa (SARA)

All assisted reproductive technology centers in South Africa were invited to join the register. Participant centers voluntarily submitted information from 2009 on the number of ART cycles, embryo transfers and clinical pregnancies, age of female partners or egg donors, and type of fertilization techniques utilized. Data were anonymized, pooled, and analyzed [16].

India

In order to regulate the ART practices and treatment by different centers in India, the Union Health Ministry and the Indian Society of Assisted Reproduction (ISAR) have set up a registry called the National ART Registry of India (NARI). The aim of the registry is to provide appropriate help and assistance to all those who are involved in infertility problems in the country (http://www.isarindia.net/).

Israel

Israel sends their national data to ICMART. In Israel, a strong health reproductive policy facilitates access to ART treatments. The government pays for all the ART treatments. The number of cycles offered with respect to the population is the highest in the world, approximately 3988 cycles per million inhabitants [9]. In 2012, the Ministry of Health changed the rules for ART access in order to better comply with the efficacy of the techniques. The number of cycles for women aged 42 years is limited to three; thereafter, the technique is known to be ineffective, while access to women aged 39 years is permitted in order to avoid further delays [17].

REASONS FOR DIFFERENT ART UTILIZATION

As ART is expensive and involves sophisticated procedures, its application is strictly related to the wealth of a nation or depends on the priorities of different national health systems. In Europe for instance, almost all of the countries offer ART procedures totally or partially covered within the national health scheme. This is one of the reasons why Europe is the region that performs the most cycles in the world [9,18], while in the United States, these treatments are paid for by the infertile couples. In developing countries, like Africa or India where problems of poverty and overpopulation are present, the health priorities over many years have been concentrated on different issues and not on infertility treatments. Despite the diffusion of sexually transmitted diseases and secondary fertility problems, infertility treatments and particularly IVF were unaffordable for the vast majority of the population. For the last 10 years, the attention and the sensibility of Western countries to the problems of infertility, especially the problems that women in poor developing countries have had to face, has grown constantly. As ART treatments were considered a too-expensive solution, with a low success rate per treatment cycle, a group of Belgian researchers recently developed a simplified laboratory culture system, a new "low-cost IVF," which enables the application of these treatments for underresourced societies. Within the "Walking Egg Project" [19,20], a project aimed to improve universal access to infertility care worldwide, they have proposed a simple laboratory equipment and low-cost pharmacological ovarian stimulation to reduce the cost of the procedure to a minimum of €200 per cycle. The social stigma of infertility, where women who are unable to give birth are excluded from the community and social life in poor countries, not only justifies the policy of ART application, but makes it necessary to fulfill one of the Millennium Development Goals announced by the United Nations, namely to reach universal access to reproductive health care by 2015.

Different legislation regarding the application of ART, the different religious perspectives related to it, especially on some particular procedures like preimplantation genetic diagnosis (PGD), preimplantation genetic screening (PGS) oocyte, sperm, embryo donation, and surrogacy, have determined that those procedures are partially or totally banned in some countries, or are permitted in others. Moreover, the costs of those techniques could vary a lot from one country to another.

The expense factor, as in the United States, where a single egg donor cycle may cost up to $30,000, plus the different legal frames, has prompted the diffusion of cross-border reproductive care [21].

This phenomenon has induced many infertile couples to travel around the world to undergo the treatment they need. The likely reasons could include financial considerations, or to undergo procedures that are not permitted in their countries [22,23]. These "journeys" tend to create a certain amount of discomfort and suffering for the infertile couples, and sometimes they are not always treated in the most appropriate way in the foreign country. In other situations the huge discrepancy existing between rich developed countries and developing ones, where there are less stringent rules on reproductive procedures, is a major problem [24]. This permits, like in the case of egg donation or surrogacy, the exploitation of poorer uneducated women that are not aware of the health risks they take in undergoing these procedures for financial gain. A national law was introduced in India to legalize commercial surrogacy with the intention of rendering this procedure more secure for the participants. However, it remains highly debatable if the safety and rights of surrogate mothers are actually defended [25–29].

Surrogate mothers and oocyte donors in poor countries, like India or Thailand, are often recruited among very young women, which without the application of medical and ethical rules are utilized in these procedures despite their age [30].

Different Religious Perspective Related to ART

Catholic Religion

The Catholic religion is the only one that totally bans the use of ART. For Catholics, the human being can only be conceived through sexual intercourse between married couples. Furthermore, homologous, and intrauterine insemination is also forbidden. On the other hand, the Protestant and the Anglican Church have more liberal attitudes toward ART treatments as they are only opposed to IVF with gamete donation and surrogacy [31].

Muslim Religion

In the Muslim religion, the mentioned techniques can be applied, but there are big differences in their application among the different Islamic countries. In Iran, all the reproductive technologies are accepted including surrogacy and the use of donor eggs and embryos. Other Islamic countries, predominantly Sunni, permit ART treatments only with the use of the couple's gametes. In the Emirates, ART procedures are permitted with the use of the couple's gametes, where surrogacy could be accepted if applied within the same family, from one wife to another, married to the same man. In Turkey, the Higher Council of Religious Affairs, the highest decision-making body (issuing Fatwa) and consultant to the presidency of religious affairs, decided that ART procedures are permitted, utilizing the gametes belonging to the couples. Also, PGD and PGS are allowed if necessary from a medical point of view [32].

Judaism

With the commandment "Be fruitful and multiply," the application of ART is permitted in the Jewish law, with the utilization of the couple's own gametes. Insemination with donor spermatozoa is accepted only by a portion of the Jewish population in Israel. Some Rabbinical Authorities permit sperm donation only if the donor is a non-Jew. Oocyte donation, or embryo donation are permitted under certain conditions; Jewish law dictates that the mother determines the religious status of the child and that the woman receiving the egg must be the mother. If the recipient is Jewish, then the child is Jewish. Surrogacy could be practiced, and the infant born after this procedure belongs to the man who produced spermatozoa for the conception and to the woman who gave birth [32].

ART APPLICATION AND AVAILABILITY

ART application around the world varies according to different health policies present in the countries. The policies may encompass cost of treatments, reimbursement, or government regulations regarding access of women permitted to undergo treatment; number of procedures offered per couple; the age limits of women who can have the procedure; kind of procedures allowed, like gamete donation, or surrogacy; and availability of advanced technological services. One indicator of availability of services offering ART is the number of initiated cycles per million inhabitants; another more precise indicator is the number of initiated cycles per million women of a fertile age (15–45 years). Both those indicators are utilized by the EIM consortium and by ICMART to compare the treatments that different countries offer.

The number of cycles performed worldwide could be considered in two different ways: at a regional level, for North America, South America, Europe, and Asia, Australia, and New Zealand, or alternatively at the country level [9].

If the data at a regional level for the year 2008 are analyzed, Australia and New Zealand are the regions that offered the most cycles, of 2145 cycles per million inhabitants. This was followed by Europe offering 947 cycles per million inhabitants, then North America with 440 cycles, the Middle East with 1320 cycles (including Israel), Asia (excluding China) offered 184 cycles (China offering an additional 129 cycles), and Latin America with just 72 cycles per million (Fig. 7.1).

However, these figures just reflect past data, which have already changed; more recent data are available only for some countries.

Due to different data sources and data collection years, the comparisons are inevitably made with outdated data (Fig. 7.2).

Considering ART treatments at the country level for 2006, Israel was the leading country offering 3988 cycles per million inhabitants, due to its strong health reproductive policy that covers all the ART treatments for all the women up to the age of 45 years. The next biggest contributors were some of the European countries, with a range of cycles offered from 2095 in Denmark to

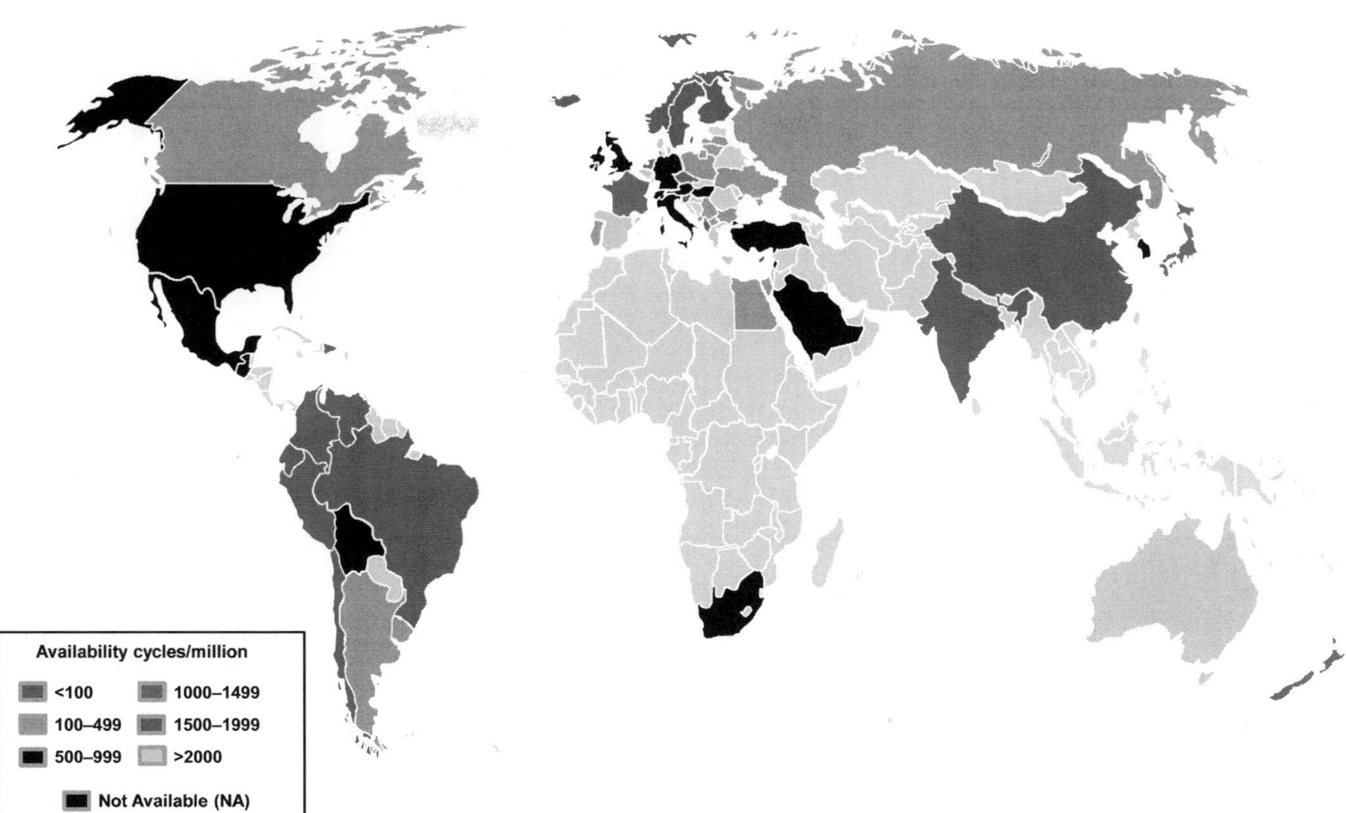

FIGURE 7.1 Total estimated number of cycles in the individual countries divided by its population – according to data collected from ICMART in 2006.

1339 cycles in the Czech Republic. Other contributors for 2006 included Australia with 2166 offered cycles, the United States with 529 cycles offered, and Canada with 356 cycles offered. All the Latin American countries offered from 225 cycles in Argentina to the minimum of 11 cycles offered in the Dominican Republic. In China, they offered just 75 cycles per million inhabitants. In the Middle East, Egypt offered 346 cycles [9] (Table 7.1).

ART APPLICATION/NUMBER OF TREATMENT CYCLES

The vast majority of ART cycles performed around the world at a regional level are in the European countries, where in 2010, 991 clinics from 31 countries performed 550,296 treatment cycles. The ART availability was expressed as the number of initiated cycles per million inhabitants, with an average offer of 1221 cycles per million inhabitants. The highest offer of cycles occurs in the Nordic countries (Iceland 2667, Denmark 2893, Sweden 1943, Norway 1926, and Finland 1772). When considering ART availability expressed as the number of cycles related to the number of women of reproductive age (15–45 years), the average offer was of 6258. A wide range of values occurred; 2710 cycles were offered in Hungary,

3183 in Montenegro, 17,669 in Denmark, and 14,494 in Belgium. Seventeen percent of IVF cycles and 17% of intracytoplasmic sperm injection (ICSI) cycles were administered in women of ≥40 years, with a wide range of values for IVF cycles that varied from 6 to 7% reported by Poland and Kazakhstan, to 30% or 29% reported by Greece and Italy, respectively. The leading countries for number of cycles performed in Europe were France (56,492 aspirations), Italy (52,661), Germany (51,720), the United Kingdom (44,642), and Spain (32,503). The proportion of clinics reporting data was 83% of all clinics practicing ART as reported in the last of publication in the European registers by ESHRE published in Human Reproduction 2014. The proportion of births after ART in those countries, all the clinics reporting to a national register, was 3%. Values ranged from 6% in Denmark to 2% in Italy [10].

The second country level contributor in the world for number of ART treatment cycles is Japan, which alone exceeded the number of cycles applied in the whole of the United States performing 213,800 treatment cycles in 2009 and 242,161 treatment cycles in 2010, collected from 591 clinics. Japan is also the country in which the treated women are the oldest, with more than 36% of cycles being administered in women >40 years. The overall percentage of single-embryo transfer (SET) was 73%. The

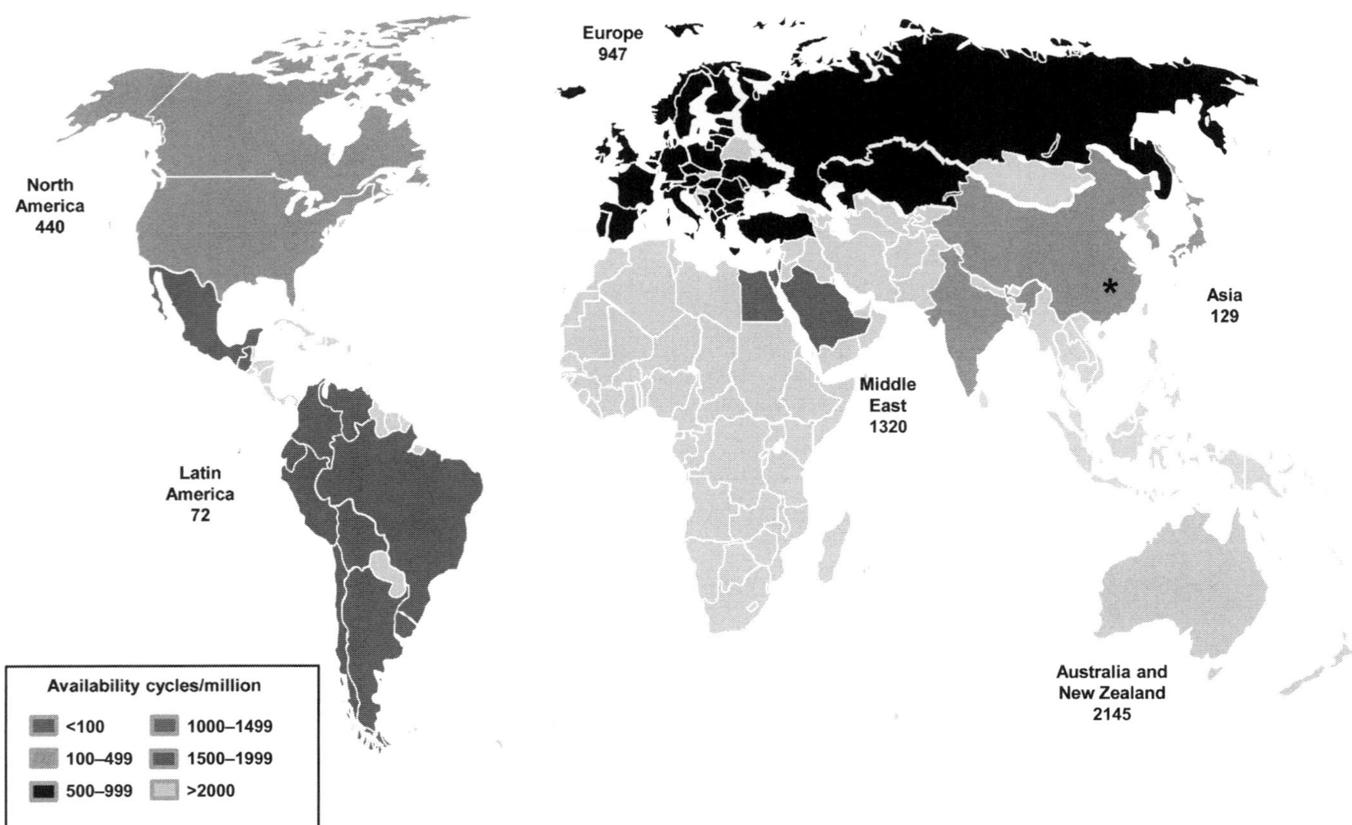

FIGURE 7.2   Total estimated number of cycles in 2008 in the region divided by its population in 2008 (United Nations Statistics Division) – according to data collected from ICMART2008 (ESHRE, Istanbul 2012). In some regions, such as in Latin America, the data expressed by the indicators are representative of all the countries in the region. In other areas, such as the Middle East, Asia or Europe, there are far more diverse average indicator values.

TABLE 7.1   Number of Cycles per Million, the Number of Babies Born per Country and per Region

	Report data and estimation (bold) for year 2006		
Country and region name	Estimated or reported overall total number of cycles	Availability cycles/million	Total babies estimated or reported from all clinics
India	43,877	40	11,228
Japan	140,194	1100	22,808
South Korea	44,090	903	1,096
China*	98,772	75	NA
Asia	*326,933*	*126*	*>35,132*
Australia	43,891	2166	9,097
New Zealand	4,286	1051	3,118
Australia and New Zealand	*48,177*	*1979*	*12,215*
Albania	141	39	78
Austria	5,177	632	1,511
Belgium	22,730	2190	4,214
Bulgaria	2,601	352	709
Cyprus	1,432	1826	391
Czech Republic	13,707	1339	3,661
Denmark	12,566	2305	2,742
Finland	8,749	1672	1,819

TABLE 7.1   Number of Cycles per Million, the Number of Babies Born per Country and per Region *(cont.)*

	Report data and estimation (bold) for year 2006		
Country and region name	Estimated or reported overall total number of cycles	Availability cycles/million	Total babies estimated or reported from all clinics
France	65,680	1079	13,448
Germany	54,695	664	10,846
Greece	**22,061**	**2064**	**6,533**
Hungary	**6,560**	**657**	**1,804**
Iceland	530	1770	149
Ireland	**3,771**	**928**	**919**
Italy	37,771	650	5,368
Latvia	280	123	**98**
Lithuania	**717**	**200**	**48**
Macedonia	911	444	227
Montenegro	245	NA	57
The Netherlands	17,770	1078	4,448
Norway	7,115	1543	1,660
Poland	**11,701**	**304**	**3,420**
Portugal	**4,275**	**403**	**1,088**
Russia	**23,858**	**167**	**6,040**
Serbia	**1,447**	**134**	**679**
Slovenia	2,804	1395	671
Spain	**84,649**	**2095**	**19,311**
Sweden	14,895	1,652	3,420
Switzerland	7,109	945	1,577
Turkey	37,468	532	5,364
Ukraine	**6,125**	**131**	**1,862**
UK	43,949	725	12,631
Europe	*523,489*	*701*	*116,793*
Argentina	**8,994**	**225**	**2,874**
Bolivia	NA	NA	NA
Brazil	12,871	68	3,878
Chile	1,418	88	436
Colombia	**1,671**	**38**	**501**
Dominican Republic	102	11	23
Ecuador	326	24	104
Guatemala	NA	NA	NA
Mexico	NA	NA	NA
Peru	1,236	44	485
Uruguay	**660**	**192**	**176**
Venezuela	909	35	278
Latin America	*>28,187*	*77*	*>8,755*

(Continued)

TABLE 7.1   Number of Cycles per Million, the Number of Babies Born per Country and per Region *(cont.)*

	Report data and estimation (bold) for year 2006		
Country and region name	Estimated or reported overall total number of cycles	Availability cycles/million	Total babies estimated or reported from all clinics
Egypt	**27,306**	**346**	**10,025**
Lebanon	NA	NA	NA
Saudi Arabia	NA	NA	NA
Middle East	*>27,306*	*346*	*>10,025*
Israel	25,333	3988	**5,893**
Canada	11,779	356	3,890
USA	**157,868**	**529**	**63,965**
North America	*169,647*	*512*	*67,855*
South Africa	*NA*	*NA*	*NA*

*Data from China were obtained from an estimation of the average annual values of cycles – Q.JAO 2014. Numbers in bold are estimated; numbers in italics indicate totals for macrogeographic areas.

proportion of births after ART was 3% of all live births in Japan. The number of initiated cycles per million of females of reproductive age (15–45 years), data only available for 2008, was 7831 treatment cycles, comparable with that of other European countries, like Belgium 13,069 or Denmark 12,712 [15].

United States

The third contributor at country level is the United States with 151,923 treatment cycles reported in 2011, from 451 clinics and with 22% of cycles applied to women >40 years. The proportion of births after ART of the total number of births was 1% [11].

China

China at the country level is the fourth contributor in terms of number of cycles performed, with an average number of cycles per year of 98,771,714; last data provided in 2011. In China, the Ministry of Health approved in 2011 the establishment of 178 reproductive centers, plus 13 sperm banks, located in 10 provinces, with the exception of Tibet. At the end of 2012 there were 358 organizations authorized to perform ART treatments. From the data of a recent national survey conducted in 2011, 691,402 ART cycles from 2005 to 2011 were performed in China; 57% (393,538) of them were IVF cycles, 24% (168,498) were ICSI cycles, 18% (124,501) were frozen embryo replacement cycles, and the remainder were IVM and PGD cycles. The average number of cycles per year was 98,771,714. Oocyte donation in China is permitted, but it is strictly regulated and is only available for patients who underwent IVF cycles. Data reported in China on ART application are not included in the ICMART data collection.

China only introduced ART in 1988, due to a variety of social, cultural, and political factors, and where the one-child policy has led to restrictions in the number of IVF treatments [3].

Latin America

In Latin America, 37,853 cycles were performed in 2010; the data were collected by 140 centers from 13 countries. The mean age of women treated was 36 years, while in 23% of cases, ART was applied in women ≥40 years. The access to ART procedures in Latin America was expressed as the total number of initiated cycles per million women of reproductive age (15–45 years) and was 305 cycles [33].

Canada

In 2012, 25,782, ART cycles were performed in Canada, reported from 32 of the 33 centers operating in this territory [13].

South Korea

In South Korea, a national data survey on ART activity performed in 2009 reported that 27,049 cycles of ART treatments were performed, collected by 74 centers; IVF and ICSI cases totaled to 22,049 (79%) [34].

Australia and New Zealand

There were 66,347 ART treatment cycles performed in Australia and New Zealand in 2011 on 34,490 women.

The 92% of cycles (61,158) were applied in Australia, while 8% (5189) were in New Zealand; 95% were nondonor cycles, 65 fresh cycles, 36 were thaw cycles.

In 2010, in 1116 IUI and 5483 IVF cycles, ART was used by 4% of women who gave birth in Australia [14].

South Africa (SARA)

Data were reported from 14 centers, with 4923 cycles performed in 2010; the mean number of cycles per million inhabitants was 98 [16].

ART PRINCIPAL OUTCOMES

ICMART Data, Year 2006 (56 Countries Participating)

A total of 1,050,300 initiated cycles resulted in the following [9]:

- An estimated 256,668 births;
- Overall pregnancy rates (PRs) and delivery rates (DRs) per aspiration for IVF were 31 and 23%, respectively, and for ICSI were 30 and 20%, respectively. The PRs and DRs for frozen embryo transfers (FETs) were 26 and 18%, respectively;
- Multiple DRs per PR were 22% for twins and 2% for triplets following fresh IVF/ICSI, and 16% for twins and 1% for triplets for FETs;
- The overall, proportion of women aged ≥40 years having ART treatment was 16% compared with 15% in 2004 and 16% in 2005;
- The highest cumulative DR (40%) was reported by the United States, even though 19% of the women were ≥40 years old;
- Access to ART varied from 11 to 3988 cycles per million population;
- ICSI comprised 66% of all initiated cycles, FET 27%;
- Single embryo transfer (SET) was applied in 21% of all initiated cycles;
- Perinatal mortality rate was 25 per 1000 births for fresh IVF/ICSI and 18 per 1000 for FETs.

EIM Data, Year 2010 (31 Countries Participating)

- A total of 550,296 ART cycles were performed;
- 120,634 births, 94,609 (78%) were born after IVF/ICSI fresh cycles, 17,689 (15%) after frozen embryo replacement (FER), 7302 (6%) after embryo donation (ED), and 1034 (1%) after cycles in which PGD was performed.

There were significant national variations in clinical outcomes.

- On average, pregnancy rates per aspiration were 29% (+0.3% compared with the year 2009) and 28.8%

(28.7% in 2009) for IVF and ICSI, respectively, and 20% per thawing for FER (−1%).

The highest percentages of women aged ≥40 years were found in Greece, Italy, and Switzerland, whereas the highest percentages of women aged ≤34 years were found in Kazakhstan, Poland, and Ukraine.

- The total proportion of SETs was 26% (24% in 2009 and 22% in 2008).
- Access to ART was on average 1221 cycles per million inhabitants.
- The average number of births out of the general population was 3% with a range that varied from 2% (Italy) to 6% (Denmark) [10].

USA Data, Year 2011

- 151,923 ART cycles (excluding embryo banking cycles) were performed at 451 reporting clinics in the United States during 2011.
- This resulted in 47,818 live births (deliveries of one or more living infants) and 61,610 infants.
- CDC estimates that ART accounts for slightly more than 1% of total US births.
- The percentage of transfers using elective SET (eSET) for women by age group varied from 12% (age 35), to 7% (ages 35–37), to 2% (ages 38–40), to 1% (ages 41–42), to 0.4 % (ages 43–44), and 1% (age >44).

More recent data are available. According to CDC's 2012 preliminary ART Fertility Clinic Success Rates Report, 176,275 ART cycles were performed at 456 reporting clinics in the United States during 2012. This resulted in 51,294 live births (deliveries of one or more living infants) and 65,179 infants. Although the use of ART is still relatively rare when compared to the potential demand, its use has doubled over the past decade. Today, over 1% of all infants born in the United States every year are conceived using ART [11].

Latin America Data, Year 2011 (12 Countries, 145 ART Centers)

The total number of ART procedures was 41,232, resulting in 11,469 live births.

- The clinical pregnancy rate per aspiration for IVF, ICSI, and FET was 30, 35, and 35%, respectively; the delivery rate per aspiration in IVF/ICSI cycles together was 21% (23% in 2010), and the cumulative delivery rate was 28%.
- The proportion of IVF/ICSI cycles performed in women aged 35–39 years was 38% of cycles.
- Thirty-three percent of cycles were applied in women ≥40 years. The mean number of transferred

embryos in IVF/ICSI decreased from 2.4 (2010) to 2.2 in the actual report.
- The proportion of eSET dropped from 4% in 2010 to 3% in 2011 [33].

Japan Data, Year 2010

- The total number of treatment cycles was 242,161 in 2010. There was an annual increase of approximately 30,000 cycles from year 2007 (161,164 in 2007, 190,613 in 2008, and 213,800 in 2009). The number of fresh cycles was 158,391, while the frozen cycles were 83,770. The number of frozen cycles approximately doubled in 2010 compared with that in 2007.
- The number of ART treatments administered to patients aged ≥40 years was 36%.
- The total pregnancy rate per embryo transfer was 22% for fresh cycles and 34% for frozen cycles.
- The single embryo transfer rate was 73%, while the multiple pregnancy rate was 5%.
- The number of infants born after ART was 28,945. From this number, 9934 were born after fresh cycles, while 19,011 were born after frozen cycles. This was 3% of all live births in Japan (1,071,304) [15].

Australia and New Zealand Data, Year 2011

Of the 66,347 initiated cycles, 23% resulted in a clinical pregnancy, and 18% in live deliveries (the birth of at least one live born baby).

- There were 12,443 live births following ART treatments (11,148 in Australia and 1295 in New Zealand). Almost 3/4 of the live births (76%) were full-term singletons of normal birth weight.
- For women aged <30 years, the live delivery rate was 27% for autologous fresh cycles and 23% for autologous thaw cycles.
- For women aged >44 years, the live delivery rate was 1 and 5% for autologous fresh and thaw cycles, respectively.
- The percentage of SET on the total number of cycles was 21.
- In Australia, 301,810 children were born in 2011; 4% of them were born by ART [14].

India Data, Year 2009

- The success rate of fresh embryo transfers was 37% in 2009.
- Clinical pregnancy rates in frozen embryo transfer cycles was 29% per transfer in 2009.
- Number of donor egg cycles was 2130.
- The majority of those cycles were performed in women who were <45 years old. The percentage of donor egg recipients >55 years was 2%.

- Clinical pregnancies were reported in 920 cycles.
- Clinical pregnancy rate was 42% per embryo transfer in 2007. Pregnancy rates in 2009 were higher by 3% when compared to the previous 2 years [35].

South Africa Data, Year 2010

From the 14 participating data collection centers, a total of 4923 aspirations and 4319 embryo transfers were reported, resulting in 1595 clinical pregnancies. The clinical pregnancy rate (CPR) per aspiration and per embryo transfer was 32 and 37%, respectively. Fertilization was achieved by ICSI in 2/3 of the cycles. The number of births was unavailable [16].

China

The data available are reported from the first national survey conducted in China in 2011 that considered two periods, the first included data on ART cycles from 1981 to 2004 (24 years), with a total of 130,844 cycles performed. The second considered ART cycles from 2005 to 2011 (7 years) with 691,402 cycles performed. The survey showed a fivefold increase in the total ART cycles. The compared two periods showed the highest increase with PGD procedures (11-fold) and the lowest with IVM (fourfold). Total cycles (IVF, ICSI, IVM, PGD, and FET) from the period 2005–2011 were 691,402, with a mean number of cycles per year of 98,771,714. The birth defect rates reported from 2005 to 2011 were 1.22% for ICSI and 1.19% for IVF. There were no significant differences in birth defects rates in the two-period comparison.

In another retrospective study conducted on ART cycles performed from 2004 to 2008, data from seven clinics were reported. The results on fresh cycles included 32,754 aspiration cycles and 29,724 embryo transfers, with 7096 births from IVF and 3103 from ICSI. The percentage of births per aspiration cycle was 31%, and multiple pregnancy rate was 33% [3].

Cryopreserved cycles reported were 10,347 embryo transfer cycles and 2761 births; the percentage of births per cycle was 27%, and multiple pregnancy rate was 25%. For all the births, birth defect rates were 1.58 babies born after ICSI procedures, and 1.11 after IVF. The age of women treated was 31.41 ± 4.07 (standard deviation) [36].

Canada Data, Year 2011

A total of 23,997 cycles were reported to CARTR (Canadian), resulting in 7030 clinical pregnancies and 5329 deliveries:
- 5276 live births and 6381 infants.
- 14,866 IVF/ICSI cycles using the women's own oocytes.

- Clinical pregnancy rate per cycle started was between 31 and 38% per embryo transfer; the live birth rate was 24%.
- The number of ART cycles performed in Canada increased by 30% in 2011 compared with the previous year.
- For IVF/ICSI cycles, the multiple birth rate was reduced by 3.3% compared with 2010; consequently, the clinical pregnancy and live birth rates were also lower. In donor oocyte and FET cycles, the multiple birth rates were also lower than in 2010, while the clinical pregnancy and live birth rates remained about the same [37].

DISCUSSION

As previously observed, there are some factors that can explain the differences in the results reported by the different regions or countries:

- Application of different rules or laws can determine different ART procedures and consequently different outcomes. In some European countries, such as Austria, Germany, or Italy, restrictive laws were in force (Italian law changed in June 2014). Hence, a mandatory change in the application of ART procedures has been determined, that is, predetermined number of embryos to be transferred, banning of PGD or PGS utilization, embryo cryopreservation, and gamete donation. Other factors to be taken into consideration include the mean age of women treated. Maternal age is one of the most important factors strictly related to the outcomes of ART procedures. In Europe, like in Greece or in Italy, the mean age of women treated is much higher than in all the other countries negatively affecting outcomes.
- Data reporting inclusion criteria for the cycle started (cycle start recorded when ovarian stimulation begins, or only when the stimulation has success and there is an aspiration of oocytes); this could affect the percentages of pregnancies obtained per cycle started.
- Embryo transfer policies selected single embryo transfer (sSET) versus others. The possibility to select the best embryos to be transferred, and the decision to transfer just one, could affect the outcomes [18].
- Cryopreservation techniques and policies. The efficacy of the new techniques of cryopreservation of embryos and oocytes has improved the possibility of obtaining a pregnancy, supplementing fresh cycles with frozen ones, resulting in better "cumulative pregnancy rates." In those countries where embryo cryopreservation is still banned, totally or partially, the outcomes are adversely affected.

The application and the outcomes of ART are widely different among countries. Different rules and different policies are already evident; these could severely interfere with the diffusion of these techniques and on their access. Once a technique is implemented there is one major factor worldwide affecting outcomes, namely, the age of the population of women. There is no effective procedure in reversing the biological clock. Even the application of donor oocyte procedures to bypass age-related oocyte competence obviates the possibility in achieving a pregnancy and healthy birth, which ultimately is still age dependent. Additionally, ART outcomes are different and differently considered in diverse settings. In Europe, the average percentage of pregnancies obtained per cycle is 29%, while in the United States it is 36%. The application goal of ART is to obtain a single healthy baby; therefore, the European mindset for "best practice" is to transfer one embryo (with a 25.7% proportion of SET) or a maximum of two embryos, minimizing the possibility of obtaining a pregnancy, but maximizing the possibility of a single healthy baby. While in the United States, healthy twins (very low proportion of SET from 7% to 12%) are considered a gold standard; this could be influenced by the different reimbursement policies. In Europe [10], half of all ART procedures are covered by the national health schemes. However, in the United States, all the treatments have to be paid by the couples, with only some states that have insurance coverage for ART procedures. This means that for infertile couples in the United States undergoing ART treatments they would invariably consider the possibility of having twins as being more financially advantageous, unlike the negative views held by European couples toward multiplicity.

The differences in terms of efficacy and safety of the application of those techniques among different countries and different continents, like Europe versus the United States, or Australia versus Japan or India, are still debated and dependent on many variables.

Many ART procedure policies could be ameliorated to improve safety and reduce costs, beginning with policies on the number of embryos to be transferred. In all the Nordic European countries (notably, Sweden, Denmark, Norway, Belgium, and Holland), the gold standard encouraged nowadays is to transfer just one selected embryo, where this procedure accounts for nearly 80–90% of all the cycles performed, while in the United States and Latin America, the trend is still to consider the transfer of two embryos or more. These considerations have led to the observation that because multiple pregnancies and multiple deliveries represent a very high risk for the health of mothers and of the newborns, which also represents a great cost to society, embracing a good policy of the sSET in many more countries will reduce the burden of multiplicity, and ultimately result in financial and safety benefits. The financial gains could be used to partially or totally fund government ART treatment access, so that it could become a positive factor in any national health system [38]. Moreover, the diffusion of "low-cost IVF" in poor-resource countries will represent a possible solution to improve the access to treatments to couples, especially young women who face infertility problems due to the lack of preventive measures for sexually transmitted diseases or postdelivery infections.

References

[1] Boivin J, Bunting L, Collins JA, Nygren KG. International estimates of infertility prevalence and treatment-seeking: potential need and demand for infertility medical care. Hum Reprod 2007;22(6):1506–12.

[2] Leke RJ, Oduma JA, Bassol-Mayagoitia S, Bacha AM, Grigor KM. Regional and geographical variations in infertility: effects of environmental, cultural, and socioeconomic factors. Environ Health Perspect 1993;101(Suppl 2):73–80.

[3] Qiao J, Feng HL. Assisted reproductive technology in China: compliance and non-compliance. Translational Pediatrics 2014; 3(2): 91–97. http://www thetp org/article/view/3545/4408 [accessed 07.01.14]. Available from: http://www.thetp.org/article/view/3545/4408.

[4] Cobo A, Kuwayama M, Perez S, Ruiz A, Pellicer A, Remohi J. Comparison of concomitant outcome achieved with fresh and cryopreserved donor oocytes vitrified by the Cryotop method. Fertil Steril 2008;89(6):1657–64.

[5] Levi Setti PE, Porcu E, Patrizio P, Vigiliano V, de Luca R, d'Aloja P, et al. Human oocyte cryopreservation with slow freezing versus vitrification. Results from the National Italian Registry data, 2007–2011. Fertil Steril 2014; 102 (1): 90–5.

[6] Scaravelli G, Vigiliano V, Mayorga JM, Bolli S, de Luca R, d'Aloja P. Analysis of oocyte cryopreservation in assisted reproduction: the Italian National Register data from 2005 to 2007. Reprod Biomed Online 2010;21(4):496–500.

[7] Sole M, Santalo J, Boada M, Clua E, Rodriguez I, Martinez F, et al. How does vitrification affect oocyte viability in oocyte donation cycles? A prospective study to compare outcomes achieved with fresh versus vitrified sibling oocytes. Hum Reprod 2013;28(8):2087–92.

[8] Brown S. ESHRE: The first 21 years. Bruxells: ESHRE; 2014.

[9] Mansour R, Ishihara O, Adamson GD, Dyer S, de MJ, Nygren KG, et al. International Committee for Monitoring Assisted Reproductive Technologies world report: Assisted Reproductive Technology 2006. Hum Reprod 2014;29(7):1536–51.

[10] Kupka MS, Ferraretti AP, De Mouzon J, Erb K, D'Hooghe T, Castilla JA, et al. Assisted reproductive technology in Europe, 2010: results generated from European registers by ESHRE +. Hum Reprod 2014;29(10):2099–113.

[11] Centers for Disease Control and Prevention, American Society for Reproductive Medicine Society for Assisted Reproductive Technology. 2011 Assisted Reproductive Technology National Summary Report. 2013.

[12] Zegers-Hochschild F. Registro Latinoamericano de Reproducción Asistida – Primer Registro Multinacional Caso a Caso. 5-1-2013. Ref Type: Data File.

[13] Gunby J CARTR Annual report - 2012 Assisted reproductive technologies (ART) in Canada: 2012 results from the Canadian ART Register http://www.cfas.ca/index.php?option=com_content&view=article&id=1365%3Acartr-annual-report-2012&catid=1012%3Acartr&Itemid=668.

[14] Macaldowie A, Wang YW, Chambers GM, Sullivan EA. Assisted reproductive technology in Australia and New Zeland 2011. http://npesu.unsw.edu.au/; 2013. Available from: http://npesu.unsw.edu.au/.

[15] Takeshima K, Saito H, Nakaza A, Kuwahara A, Ishihara O, Irahara M, et al. Efficacy, safety, and trends in assisted reproductive technology in Japan-analysis of four-year data from the national registry system. J Assist Reprod Genet 2014;31(4):477–84.

[16] The Southern African Society of Reproductive Medicine & Gynaecological Endoscopy. SARA 2010 – South African Register of Assisted Reproductive Techniques. http://www.fertilitysa.orgza/ARTDataMonitoring/SARAReport2010.pdf; 2014. Available from: http://www.fertilitysa.org.za/ARTDataMonitoring/SARAReport2010.pdf.

[17] Birenbaum-Carmeli D. 'Cheaper than a newcomer': on the social production of IVF policy in Israel. Sociol Health Illn 2004;26(7):897–924.

[18] Adamson GD. Economics in the development of embryo transfer policies; Transforming Reproductive Medicine Worldwide, Proceedings of The International Federation of Fertility and Sterility and The 69th Annual Meeting of the American Society for Reproductive Medicine, Boston, Massachusetts, 2013, pp. 196–203.

[19] Ombelet W. The Walking Egg Project: universal access to infertility care – from dream to reality. Facts Views Vis Obgyn 2013;5(2):161–75.

[20] Van BJ, Ombelet W, Klerkx E, Janssen M, Dhont N, Nargund G, et al. First births with a simplified culture system for clinical IVF and embryo transfer. Reprod Biomed Online 2014;28(3):310–20.

[21] Inhorn MC, Patrizio P. Rethinking reproductive "tourism" as reproductive "exile". Fertil Steril 2009;92(3):904–6.

[22] Ferraretti AP, Pennings G, Gianaroli L, Natali F, Magli MC. Cross-border reproductive care: a phenomenon expressing the controversial aspects of reproductive technologies. Reprod Biomed Online 2010;20(2):261–6.

[23] Shenfield F, de MJ, Pennings G, Ferraretti AP, Andersen AN, De Wert G, et al. Cross border reproductive care in six European countries. Hum Reprod 2010;25(6):1361–8.

[24] Whittaker A. Cross-border assisted reproduction care in Asia: implications for access, equity and regulations. Reprod Health Matters 2011;19(37):107–16.

[25] Anand Kumar TC. Ethical aspects of assisted reproduction – an Indian viewpoint. Reprod Biomed Online 2007;14(suppl 1):140–2. C 15-11-2006.

[26] Indian Council of Medical Research. National Registry of Assisted Reproductive Technology (ART) Clinics and Banks in India. http://www.icmartivf.org/toolbox/toolbox-main.html; 2010. Available from: http://www.icmartivf.org/toolbox/toolbox-main.html.

[27] Pande A. Transnational commercial surrogacy in India: gifts for global sisters? Reprod Biomed Online 2011;23(5):618–25.

[28] Pillai SM. A national public health plan for infertility in India. Health Sci 2013; 2(1): JS003. http://healthsciences.ac.in/jan-mar-13/downloads/3National_public_health_plan.pdf. Available from: http://healthsciences.ac.in/jan-mar-13/downloads/3.National_public_health_plan.pdf.

[29] Shetty P. India's unregulated surrogacy industry. Lancet 2012;380(9854):1633–4.

[30] Deonandan R, Green S, van BA. Ethical concerns for maternal surrogacy and reproductive tourism. J Med Ethics 2012;38(12):742–5.

[31] Ford NM. A Catholic ethical approach to human reproductive technology. Reprod Biomed Online 2008;17(Suppl 3):39–48.

[32] Schenker JG. Assisted reproductive practice: religious perspectives. Reprod Biomed Online 2005;10(3):310–9.

[33] Zegers-Hochschild F, Schwarze JE, Crosby J, Musri C, Borges de Souza M. Assisted reproductive technologies (ART) in Latin America: The Latin American Registry, 2011. JBRA Assist Reprod 2013; 17(4):216–21.

[34] Choi YM, Chun SS, Han HD, Hwang JH, Hwang KJ, Kang IS, et al. Current status of assisted reproductive technology in Korea, 2009. Obstet Gynecol Sci 2013;56(6):353–61.

[35] Malhotra N, Shah D, Pai R, Pai HD, Bankar M. Assisted reproductive technology in India: a 3 year retrospective data analysis. J Hum Reprod Sci 2013;6(4):235–40.

[36] Yan J, Huang G, Sun Y, Zhao X, Chen S, Zou S, et al. Birth defects after assisted reproductive technologies in China: analysis of 15,405 offspring in seven centers (2004 to 2008). Fertil Steril 2011;95(1):458–60.

[37] Joanne Gunby. CARTR Annual Report 2011 Assisted reproductive technologies (ART) in Canada: 2011 results from the Canadian ART Register. http://www.cfas.ca/index php?option=com_content&view=article&id=1291%3Acartr-annual-report-2011&catid=1012%3Acartr&Itemid=668; 2014.

[38] Chambers GM, Sullivan EA, Ishihara O, Chapman MG, Adamson GD. The economic impact of assisted reproductive technology: a review of selected developed countries. Fertil Steril 2009;91(6):2281–94.

8

The Effects of Environmental Hormone Disrupters on Fertility, and a Strategy to Reverse their Impact

Frank H. Comhaire, MD, PhD, Wim A.E. Decleer, MD

Fertility-Belgium Centre, Belgium

INTRODUCTION

Since the groundbreaking publication by Carlsen et al. [1] much research has been devoted to the study of external factors influencing human fertility. It has become evident that hormone-disrupting, man-made chemicals are involved in the decline of sperm quality and the increased prevalence of congenital diseases of the male and female reproductive genital tract associated with testicular dysgenesis [2].

The governments in several countries, including Flanders (Belgium) have taken measures to reduce the dissemination of hormone disrupters into the environment, which has resulted in at least partial restoration of semen quality [3]. Also, medication and nutraceuticals have become available to counteract the deleterious effects of environmental factors and to improve fertility.

The present chapter will report on observations on semen quality in the Flemish population, on methods to detect environmental pollution with xenoestrogens, and on the effects of antiestrogen and nutraceutical treatment on male and female fertility.

HISTORIC ASPECTS

During the previous century, industrial and agricultural developments took place at a fast pace. The rapid evolution of chemistry and the availability of large amounts of raw materials on the one hand, and the increased buying power of large parts of the world population on the other hand, have resulted in the creation of many thousands of novel chemical agents as well as enormous amounts of waste that have never existed before. Many of the new substances have proven their usefulness, and some have served in attaining a better quality and longer span of human life. However, little attention was given to the possibly unfavorable side effects of these agents and of the generated waste on human health and the condition of wildlife.

It is only recently that concerns have been raised about the fact that large portions of the world population may suffer from particular diseases resulting from exposure to environmental toxicants through the inhaled air, and the ingested food and water. Epidemiological studies have documented an increased prevalence of diseases such as allergy, cancer, and reproductive failure. The simultaneous rise in the prevalence of couple infertility due to poor semen quality, and of cancer of the female breast, as well as of testicular and prostate cancers, has focused the attention on the possible deleterious hormonal effects of certain environmental pollutants [2].

Using several *test methods*, varying from cell culture techniques to entire animal models, it was found that particular chemical agents can deregulate the physiological balance of hormones in the body of humans and animals. Hormones are messenger molecules that are indispensable for the regulation of numerous processes in the body, including fertility, growth, and fetal development. Chemical compounds, called hormone disrupters, interfere with the normal hormonal status and the feed-back system through mechanisms such as the inhibition of hormone synthesis, the redirection of hormone metabolism, the binding to hormone receptors on target cells mimicking hormone effects, and so on. Hormone-disrupting effects were documented at agent concentrations found in

Handbook of Fertility. http://dx.doi.org/10.1016/B978-0-12-800872-0.00008-1

the environment, though these levels are several hundred times lower than those causing toxic effects (no observed adverse effect level, NOAEL).

The hormone disrupters that have raised the most concerns interfere with the sex hormones by exerting an estrogen- or androgen-like effect (xenoestrogens and xenoandrogens), or antagonizing these hormones (antiestrogens and antiandrogens). Human exposure to xenoestrogens has been held responsible for the increased prevalence of breast cancer, endometriosis, and ovulation disturbances (e.g., polycystic ovarian disease) in the female, and testicular maldescent (cryptorchidism), hypospadias (incomplete closure of the urethra in the penis), prostate cancer, testicular cancer, and poor semen quality in men. Important differences in the incidence of some of these diseases have been registered in different regions, which were sometimes only a few dozens of kilometers apart. Also, abnormal semen quality occurred more frequently in regions that presented the highest prevalence of testicular cancer. In addition, trends over time have been observed in these conditions with progressive deterioration over the last five or six decades [1].

Parallel changes have been observed in wildlife, some of which could unambiguously be related to local contamination of the environment with hormone-disrupting chemicals. The changes observed in wildlife animals and in humans could be reproduced in laboratory animals by exposing the latter to particular agents. It appeared that prenatal exposure to xenoestrogenic substances in particular caused abnormalities of the anatomy and function of the reproductive organs of male and female offspring. The combined, albeit indirect, evidence gathered from these studies strongly suggests that environmental hormone disrupters play a pivotal role in the deterioration of human reproductive health.

METHODOLOGICAL ASPECT

The first challenge in solving the problem of hormone disrupters consists of developing adequate, reliable, reproducible, and relevant assays for their detection. Originally, whole animal tests were implemented, including the use of fish, rats, and mice. Although these *in vivo* test systems elucidate the integrated effect in the body elicited by a particular compound, or a mixture of compounds, and account for their bioaccumulation and metabolic conversions in the body, these tests are not suitable for large-scale screening due to high cost, modest responsiveness and moderate reproducibility, and the complex endpoint measurements, which increase the risk for false negative outcomes. Furthermore, the use of numerous laboratory animals per test is ethically contestable. In contrast, the outcome of *in*

vitro tests using isolated organs, cells, or molecules, is – in general – mechanistically more straightforward to interpret. These assays generate a quicker and more sensitive response and are much cheaper than *in vivo* tests. Consequently, *in vitro* and *in vivo* bioassays are being considered complementary rather than substitutes for one another.

The urgent request for information on the possible hormone-like effects of thousands of chemicals has intensified the research for reliable, cost-effective screening tools. Several types of *in vitro* assays have been developed in the past, including those using cultured human breast cancer cells or genetically manipulated yeast [4]. The results of different assays may sometimes diverge and this has caused concerns about their reliability among scientists and the public. Furthermore, hormonal effects may differ between the different cells and tissues. The role of xenoestrogens in the alleged decrease of sperm quality has been a controversial scientific subject in the past years, and the assays that were available suffered from a lack of relevance for fertility.

THE ESTROGEN RECEPTOR BINDING ASSAY

The first step in the development of the estrogen receptor binding assay (ERBA) was the creation of a yeast that contains the "binding pocket" of the human estrogen receptor (TER) incorporated into its genome. This binding pocket was attached to a solid surface by means of a glutathion-S-transferase (GST) tag (Figure 8.1).

Samples suspected of containing xenoestrogens are incubated *in vitro* together with radioactively labeled estradiol and the purified TER. Both classes of compounds compete for the free binding place on the TER. After the competition, the receptors are bound to a solid surface via an antibody against the GST tag, and the amount of bound radioactive estradiol is assessed by means of a luminescence measurement. If the amount of bound labeled estradiol is low, the amount of estrogen-like substances in the test sample will be high, and vice versa. This test is rapid, highly reproducible, and absolutely specific in detecting and quantifying substances that bind to the human estrogen receptor. Since it does not include the use of living cells, the test is not subject to toxic effects.

APPLICATION OF THE ERBA TEST TO ENVIRONMENTAL SAMPLES

The novel ERBA has been applied to surface water samples, and the results were compared to those of the yeast test and a human breast cancer cell test (Figure 8.2). The results of the three assays were similar in water

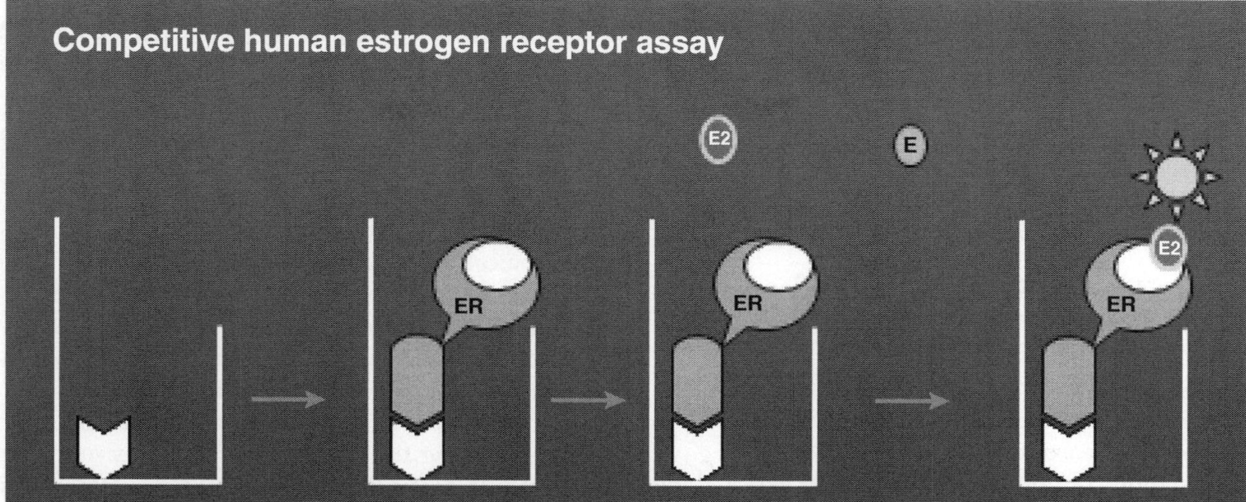

FIGURE 8.1   Principle of the ERBA. ER, human estrogen receptor; dark gray oval (red oval in the web version), radioactively labeled estradiol 17-beta; light gray oval (yellow oval in the web version), xeno-estrogen. In the mixture of the environmental sample or blood serum with radioactively labeled estradiol, competition will take place for binding to the estrogen receptor (ER) that is fixed to the surface of the wells by GST white (green in the web version). After removal of the mixture, radioactivity is measured in the wells. A higher amount of xenoestrogen in the environmental sample will result in a lower binding of the radioactively labeled estradiol to the well.

samples displaying low or moderate estrogen activity, and which were virtually devoid of toxicity. However, the ERBA revealed very high estrogen-like activity in several samples that generated either a negative or only discretely positive reaction in the yeast and breast carcinoma cell assays. Using theoretical calculations and models, it could be proven that the latter assays had yielded incorrectly low (false negative) results in samples that were heavily contaminated by estrogen-like substances and other environmental toxicants.

The test could assess the total estrogen activity in human blood. By distracting the concentration of estradiol from the total estrogen activity measured by the ERBA test, it was possible to estimate the level of internal contamination by xenoestrogens.

XENOESTROGENS IN FLEMISH SURFACE WATERS

Samples of surface water of the upper Schelde river were tested by means of three methods, namely, the *in vitro* breast cancer cell test, the test using genetically modified yeast, and the ERBA (Figure 8.2).

Samples 1 and 2 were only slightly contaminated by estrogen activity (1 and 10 pg/mL estradiol equivalents), and the results of the three tests are concordant. In samples 3 and 4 the estrogen activity is high (approximately 100 pg/mL), which was detected by the ERBA test, but missed by the yeast test, whereas the breast cancer cell test does not show any response. Complementary chemical analysis proved that the extremely high level

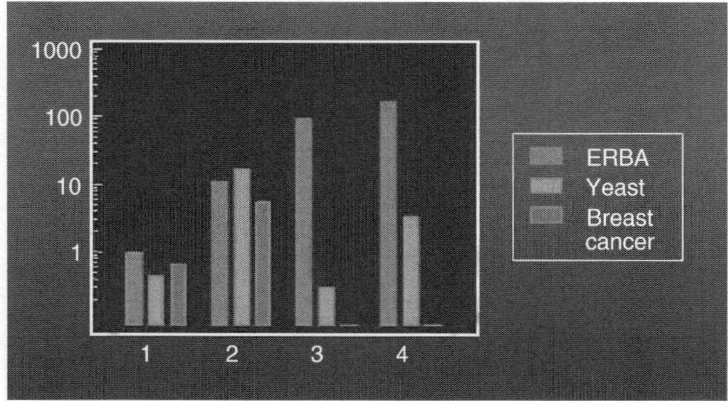

FIGURE 8.2   The histogram represents the level of estrogenic activity (in pg/mL of estradiol-equivalents; note the logarithmic scale on the vertical axis) in four samples of Flemish surface waters taken from the Schelde river.

of contamination in samples 3 and 4 was mainly due to the presence of methylparabene.

The contaminated surface water will affect the vegetables and animals that are ultimately consumed by humans and exert their hormone disrupting effect. Hence, measures had to be taken to reduce, or preferentially eliminate, the sources of xenoestrogen contamination.

STRATEGY TO REDUCE ENVIRONMENTAL CONTAMINATION

Figure 8.3 identifies the steps to be taken in a strategy to reduce human and animal contamination leading to malfunction or disease.

These include continuous monitoring of external environmental samples and of internal exposure (in blood and urine) using biomarkers and by means of screening tests as described earlier. Subsequent chemical analysis should identify the specific agents involved in the hormonal effect. Based on these analyses, the source of the contamination will be detected. Risk analysis is performed and recommendations are formulated that must be implemented by policy-makers.

EFFECT OF REDUCING ENVIRONMENTAL CONTAMINATION OF SPERM QUALITY IN FLANDERS

The strategy depicted in Figure 8.3 has been implemented in Flanders from 1995 onward and aimed at reducing environmental contamination with hormone disrupters.

Biological monitoring was performed by detailed semen analysis of representative samples of the young adult population. Sperm motility, being the most severely affected variable, was measured using a fully validated, computer-assisted system of image analysis [5].

As can be seen in Figure 8.4 important changes have occurred during the observation and intervention periods, with dramatic decrease of environmental pollution by dioxins and important reduction of heavy metals. The contamination with pesticides has also decreased though less dramatically. The least effect was noticed regarding the contamination by polyaromatic hydrocarbons (PAHs) because of the intense industrial activity in the region.

Whereas a reduction to half of the proportion of spermatozoa with grade "a" (rapid linear progressive)

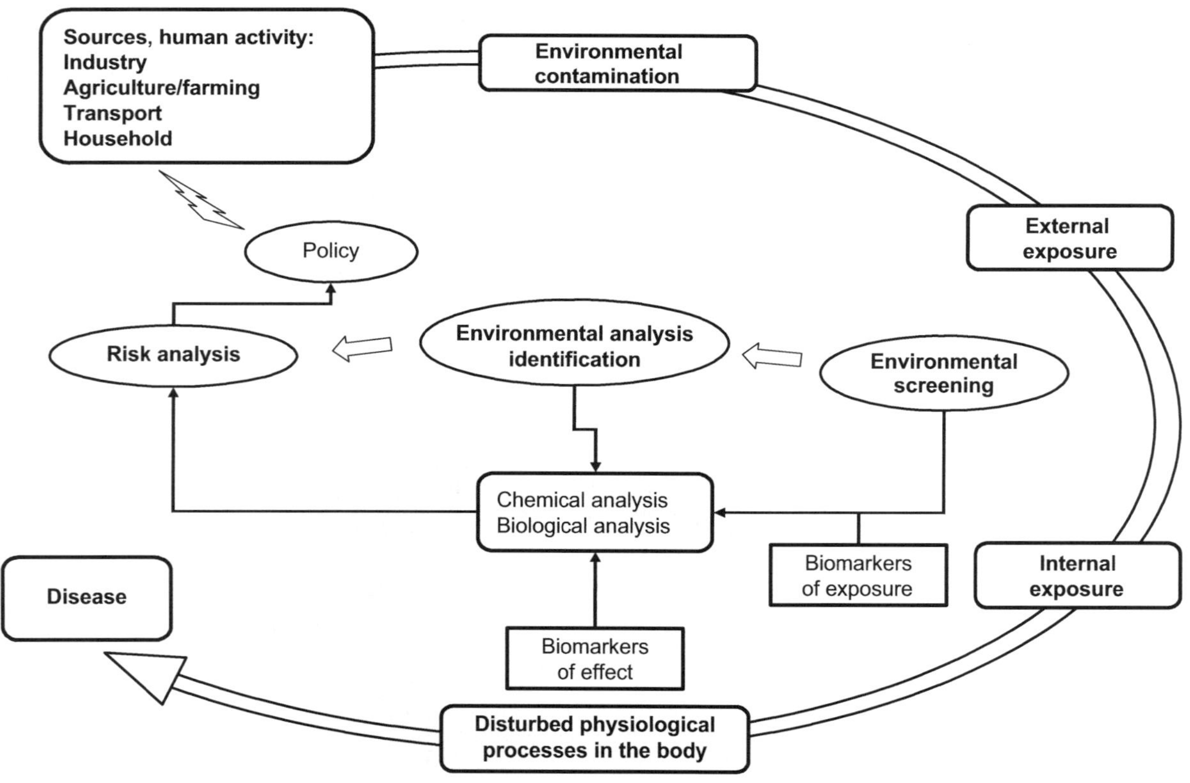

FIGURE 8.3   This flow diagram depicts the strategy to be implemented for eliminating hormone-disrupting agents from the environment and from living organisms.

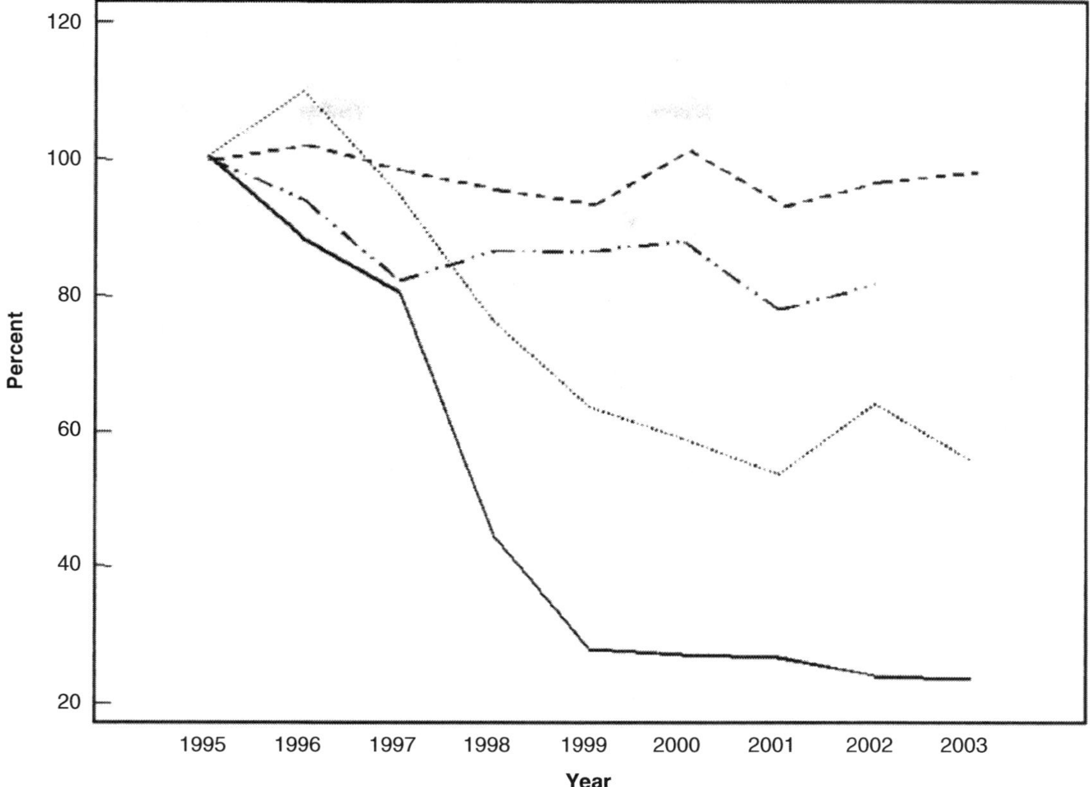

FIGURE 8.4 Proportional changes (vertical axis, initial values at 100%) of the concentration in environmental samples during the time period from 1995 to 2003 (horizontal axis) of: polyaromatic hydrocarbons (PAHs, dashed line), pesticides (dashed and dotted line), heavy metals (dotted line), and dioxins (continuous line).

motility has occurred in the time period between 1977 and 1993, the inverse trend was observed from 1997 to 2004 following the striking reduction of environmental contamination by dioxins (Figure 8.5). Full recovery to the initial value recorded in 1978 has not been achieved, possibly because of the persisting environmental contamination with PAHs and the high concentrations of pesticides.

EFFECT OF ANTIESTROGEN TREATMENT ON SEMEN QUALITY AND MALE FERTILITY

Xenoestrogen contamination has been identified as a major source of deficient sperm quality. Indeed, these agents exert a deleterious influence at both the hypothalamo-pituitary levels, and on testicular spermatogenesis (Figure 8.6).

In 1976 we introduced the treatment of idiopathic oligozoospermia with the selective antiestrogen Tamoxifen [6]. In contrast to Clomiphene citrate, which is a racemic mixture of two isomers exerting both antiestrogenic and intrinsic estrogenic activity, Tamoxifen exclusively inhibits the binding to the cellular estrogen receptor. This effect involves not only the binding of the natural estrogens estradiol 17-beta and its metabolites, but equally the binding of xenoestrogens. Tamoxifen treatment increases the secretion of LHRH and of the gonadotropins LH and FSH, with doubling of the testosterone concentration in blood within one month of intake, and the 2.5-fold increase of sperm concentration within 3–6 months. When given to patients with idiopathic oligozoospermia and sperm concentration below 10 million/mL and LH and FSH levels not elevated, Tamoxifen treatment results in doubling of the pregnancy rate compared to placebo, and a number needed to treat (NNT) of 3.9 (CI: 2.9–7.8)[7].

COMPLEMENTARY TREATMENT WITH A SPECIFIC NUTRACEUTICAL

Aside from their hormone-disrupting effects, environmental factors, unhealthy life style, and unbalanced nutrition impair the fertilizing capacity of spermatozoa through mechanisms such as oxidative stress, inflammation, and epigenetic changes.

In order to be able to fertilize, spermatozoa must have a membrane with optimal fluidity, a normal DNA, and an efficient energy production.

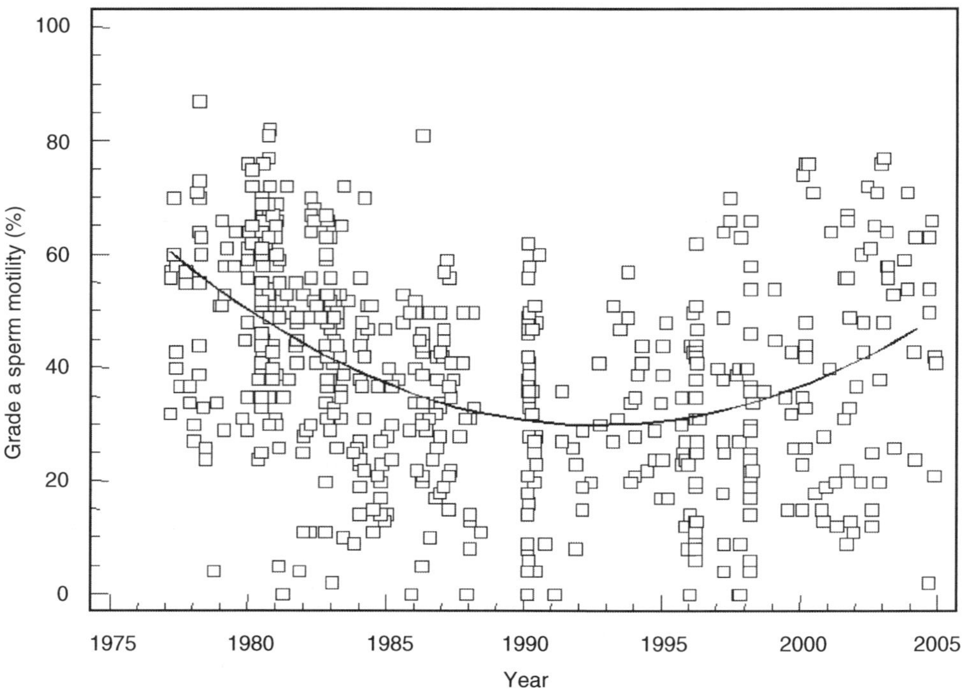

FIGURE 8.5   Correlation between grade "a" sperm motility and year at semen analysis. Evolution of grade "a" (rapid linear progressive) sperm motility (in absolute percentage-value on the vertical axis) during the observation and intervention periods.

The phospholipid composition of the membrane should contain a high proportion of long-chain omega-3 polyunsaturated fatty acids (PUFAs, docosahexaenoic acid and eicosapentaenoic acid) (Figure 8.7)[8]. However, these long-chain PUFAs cannot pass through the tight junctions that create the blood–testis barrier. In fact, we observed an inverse correlation between sperm motility and the nutritional intake of long-chain PUFAs, but a positive correlation with the intake of short-chain PUFAs. The short-chain alfa-linolenic acid serves as a substrate,

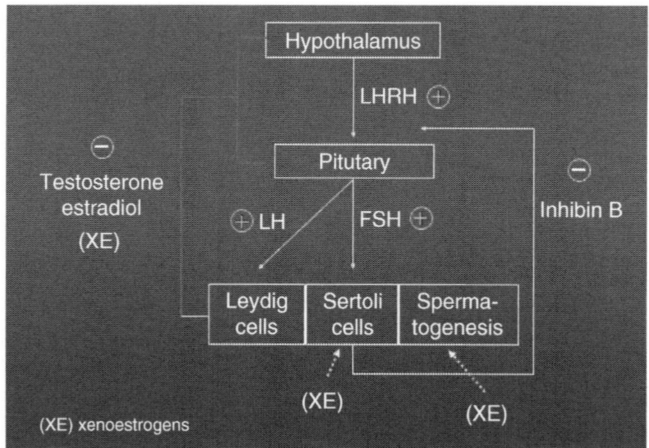

FIGURE 8.6   Xenoestrogens (XE) inhibit the release of LHRH and of the gonadotropins LH and FSH by the hypothalamo-pituitary unit, impair Sertoli cell function, and damage spermatogenesis resulting in lower sperm production (oligozoospermia), and poor sperm quality (astheno- and teratozoospermia).

and is metabolized in the seminiferous tubules into the long-chain PUFAs through the effect of the desaturase and elongase enzymes. The optimal function of these enzymes requires the availability of the cofactors zinc and vitamin B6 (pyridoxine). The food supplementing nutraceutical Qualisperm provides these cofactors, as well as linseed (flaxseed) oil that is rich in alfa-linolenic acid.

Oxidative stress caused by the excess of mainly reactive oxygen species (ROS) impairs DNA by oxidizing guanine in 8-hydroxy-2'-deoxyguanosine (8-OH 2-dG). During cell replication, the latter will combine with thymine instead of cytosine, causing transition mutagenesis. The 8-OH 2-dG content of spermatozoa of a large proportion of infertile patients is increased, which may exert a deleterious effect on embryos, causing either early pregnancy loss, or possibly an increased risk of malignant tumors in the offspring. Antioxidant treatment using the retinoid Astaxanthin (biomass of the *Haematococcus pluvialis*) normalizes the DNA by reducing the amount of oxidized guanine (Figure 8.8)[9].

Oxidative stress also inhibits the production of energy in the form of adenosine triphosphate (ATP) by the mitochondria situated at the midpiece of the spermatozoa. The oxidative overload is reduced by the antioxidant oxido-reductase Ubiquinone Q10.

The transfer of the substrate for the Krebs cycle, namely, the long-chain PUFAs, from the cell cytoplasm into the mitochondria is facilitated by acetyl-carnitine.

Chronic inflammation is common in patients with infertility and causes elevated levels of interleukins (mostly

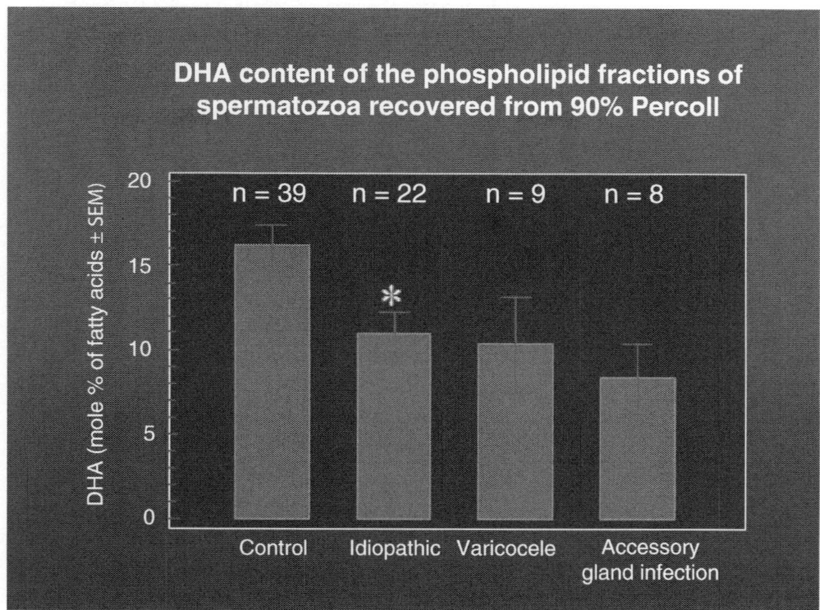

FIGURE 8.7 Proportion of docosahexaenoic acid (DHA, on the vertical axis, in percent of total fatty acids) present in the phospholipids of the membrane of spermatozoa of fertile controls, and of infertile patients with varicocele, or with idiopathic oligozoospermia, or with male accessory gland infection. The bars show the mean values with standard error of the mean. The values of all infertile groups are significantly lower than that of the spermatozoa of fertile controls.

interleukins 1, 2, and 6) and activation of the nuclear factor kappa B (Nf-kB). Also, the production of inflammatory prostaglandins is increased through activation of the cyclo-oxygenase enzymes 1 and 2 (COX 1&2). The extract of the bark of the (Mediterranean) pine tree is rich in proanthocyanidins (PACs) that inhibit both COX enzymes and reduces the activation of the Nf-kB.

The nutraceutical Qualisperm (Nutriphyt Ltd, Oostkamp, Belgium) contains the substances just enumerated, together with the vitamins B9 (folic acid) and B12 (cyanocobalamin) (formulation in appendix).

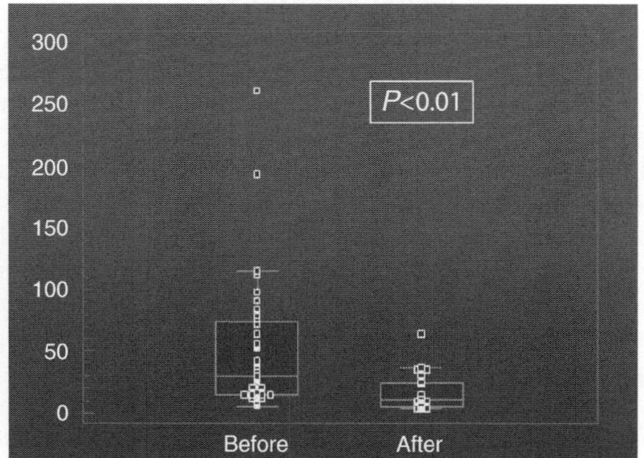

FIGURE 8.8 Box and whisker plots with dots of the 8-OH 2-dG content of spermatozoa (in femtomole, on the vertical axis) of infertile patients, before and after 8 weeks treatment with the antioxidant Astaxanthin.

Given together with linseed oil, the complementary nutraceutical was proven to significantly increase sperm quality (Figure 8.9), and to enhance the probability of spontaneous pregnancy after varicocele treatment as well as the ongoing pregnancy rate after IVF (NNT: 4)[10].

FEMALE INFERTILITY

Less information is available regarding the external influences on female fertility. The epigenomics of the oocytes generated in the female fetus are influenced by the nutritional state and lifestyle of the mother during pregnancy. This explains the transgenerational transmission of certain diseases, which are not explained by genomic alterations [11].

The nutritional intake of long-chain omega-3 PUFA influences the make-up of the oocyte membrane, and the structure of the phospholipid bilayer will affect the capacity to fuse with the sperm membrane. Similarly, oxidative damage to the oocyte membrane reduces its fusogenic capacity.

An elevated level of homocysteine in blood is correlated with increased homocysteine concentration in follicular fluid [12]. The latter may influence the methylation of DNA and induce epigenomic changes that exert an unfavorable effect on the future health of the offspring. Clearly, optimal intake of vitamin B should be aimed for.

Endometriosis, pelvic inflammatory disease, and also the polycystic ovary syndrome cause increased ROS [13]

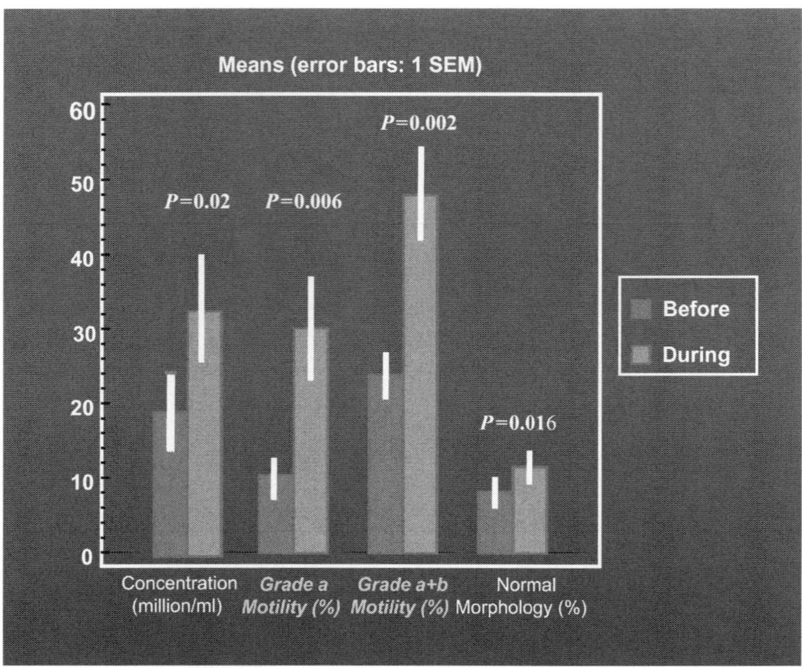

FIGURE 8.9 Histogram (mean and SD) of sperm quality characteristics before and after 3 months treatment of infertile patients with complementary Qualisperm plus linseed oil (Linusit®). On the vertical axis are the concentration (in million/mL), and the percentage of spermatozoa with grade "a" motility, total progressive motility (grades "a" plus "b" added), and a proportion of spermatozoa with normal morphology. Statistical significance has been calculated using paired t-tests.

and inflammation. Interleukins have been found elevated in the peritoneal fluid of these patients [14]. Food supplementation with anti-inflammatory substances, such as pine bark extract, and the antioxidant Astaxanthin are, therefore, recommended in these patients.

ESTROGEN METABOLISM

Estrogens are metabolized into several end-products through the interaction of different enzymes (Figure 8.10). Some of these metabolites exert a strong estrogenic effect, such as the 16-alfa-OH-estrogens. Conversion into the latter metabolite can be corrected by supplying supplementary soy isoflavones (Dadzein and Genistein), or the extract of linseed (*Linum usitatissimum*), or of *Lepidium meyenii* (MACA). In addition, the latter displays an adaptogenic effect since it increases the production of the Heat Shock Protein 70 (HSP 70), reducing the negative influence of stress on the protein configuration. These plant extracts redirect the hydroxylase activity away from the 16-alfa toward the 2- and 4-hydroxy metabolites [15].

Xenoestrogens do influence the hydroxylases favoring the 16-alfa pathway, and ROS promote the metabolism of the *S*-quinones to *O*-quinones. The relative deficiency of GST may drive metabolism into the direction of the metabolite 3,4-*O*-quinone, which is a depurinating derivative that causes mutagenesis. Therefore, food supplementation with antioxidants, including glutathione, is recommended in female infertility.

During the process of embryogenesis cell division takes place, which requires the availability of energy that is delivered as ATP by the mitochondria. Energy production will be enhanced by delivering supplementary long-chain omega-3 PUFA, and acetyl-carnitine, and the oxido-reductase Ubiquinone Q10.

CONCLUSIONS

External factors play an important role in the pathogenesis of male infertility. Hormone disrupters, xenoestrogens in particular, deregulate the hypothalamo-pituitary-testicular axis and cause direct damage to spermatogenesis. The elimination of these man-made agents from the environment improves sperm quality and fertility. Antiestrogen treatment with Tamoxifen and complementary food supplementation with the nutraceutical Qualisperm further enhance the fertilizing capacity of spermatozoa.

As far as female fertility is concerned oxidative damage and inflammation caused by endometrioses, or pelvic inflammatory disease, or polycystic ovary syndrome should be counteracted. The unfavorable effect of xenoestrogens on estrogen metabolism should

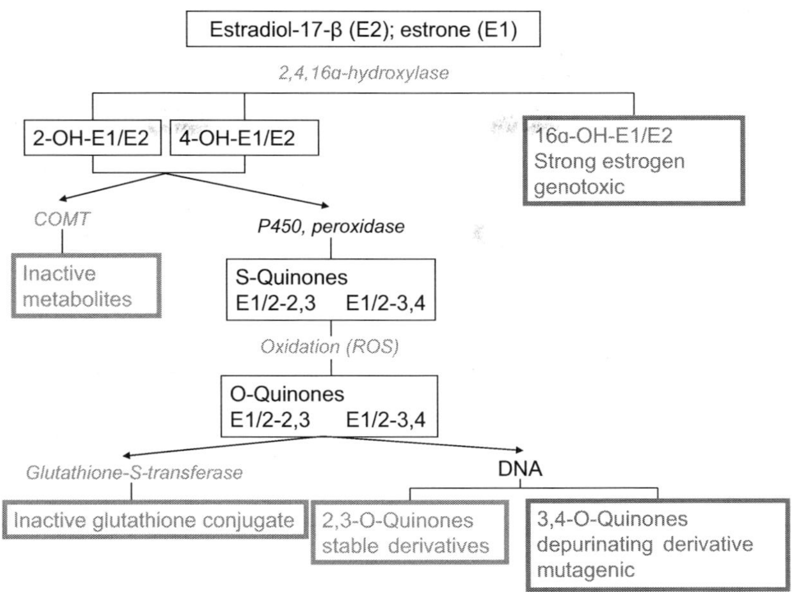

FIGURE 8.10   Steps of the metabolism of estrogens (estradiol 17-beta and estrone).

be corrected using particular plant extracts, antioxidants and an anti-inflammatory agent. Also, epigenomic changes caused by excess homocysteine in follicular fluid should be corrected by vitamin B supplementation.

This approach was proven to result in higher probability of spontaneous conception, and to increase the ongoing pregnancy rate after assisted reproduction among couples consulting professionals about infertility.

APPENDIX

Formulation of the nutraceutical food supplement (Qualisperm/Improve®, Nutriphyt Ltd. Oostkamp, Belgium) for male and female infertility:

Lepidium meyenii 125 mg; *Pinus maritima* 50 mg; acetyl-carnitine 50 mg; coenzyme Ubiquinone Q10 12.5 mg; Astaxanthin 4 mg; zinc-bisglycinate 3.75 mg; vitamin B6 1.5 mg; vitamin B9 0.1 mg; vitamin B12 0.75 μg.

References

[1] Carlsen E, Giwercman A, Kleiding N, Skakkebaek NE. Evidence for decreasing quality of semen during past 50 years. BMJ 1992;305:609–13.

[2] Sharpe RM, Skakkebaek NE. Are oestrogens involved in falling sperm counts and disorders of the male reproductive tract? Lancet 1993;341:1392–5.

[3] Comhaire FH, Mahmoud AM, Schoonjans F. Sperm quality, birth rates and the environment in Flanders (Belgium). Reprod Toxicol 2007;23:133–7.

[4] Eertmans F, Dhooge W, Stuyvaert S, Comhaire F. Endocrine disruptors: effects on male fertility and screening tools for their assessment. Toxicol In Vitro 2003;17:515–24.

[5] Hinting A, Schoonjans F, Comhaire F. Validation of a single-step procedure for the objective assessment of sperm motility characteristics. Int J Androl 1988;11:277–87.

[6] Comhaire F, Vermeulen A. Treatment of oligozoospermia with Tamoxifen. Int Congr Androl 1976;. Abstract P9: 35.

[7] Adamopoulos DA, Pappa A, Billa E, Nicopoulou S, Koukkou E, Michopoulos J. Effectiveness of combined tamoxifen citrate and testosterone undecanoate treatment in men with idiopathic oligozoospermia. Fertil Steril 2003;80:914–20.

[8] Zalata AA, Christophe AB, Depuydt CE, Schoonjans F, Comhaire F. The fatty acid composition of phospholipids of spermatozoa from infertile patients. Mol Hum Reprod 1998;4:111–8.

[9] Comhaire FH, Christophe AB, Zalata AA, Dhooge WS, Mahmoud AM, Depuydt CE. The effects of combined conventional treatment, oral antioxidants and essential fatty acids on sperm biology in subfertile men. Prostaglandins, Leukot Essent Fatty Acids 2000;63:159–65.

[10] Comhaire F, Decleer W. Comparing the effectiveness of infertility treatments by numbers needed to treat (NNT). Andrologia 2012;44:401–4.

[11] Greally JM, Jacobs MN. *In vitro* and *in vivo* testing methods of epigenomic endpoints for evaluating endocrine disrupters. ALTEX 2013;30:445–71.

[12] Ocal P, Ersoylu B, Cepni I, Guralp O, Atakul N, Irez T, Idil M. The association between homocysteine in the follicular fluid with embryo quality and pregnancy rate in assisted reproductive techniques. J Assist Reprod Genet 2012;29:299–304.

[13] Karuputhula NB, Chattopadhyay R, Chakravaty B, Chaudhury K. Oxidative status in granulosa cells of infertile women undergoing IVF. Syst Biol Reprod Med 2013;59:91–8.

[14] Othman Eel-D, Hornung D, Salem HT, Khalifa EA, El-Metwally TH, Al-Hendry A. Serum cytokines as biomarkers for nonsurgical prediction of endometriosis. Eur J Obstet Gynecol Reprod Biol 2008;137:240–6.

[15] Brooks JD, Ward WE, Lewis JE, Hilditch J, Nickell L, Wong E, Thompson LU. Supplementation with flaxseed alters estrogen metabolism in postmenopausal women to a greater extent than does supplementation with an equal amount of soy. Am J Clin Nutr 2004;79:318–25.

Embryo Losses During Nutritional Treatments in Animal Models: Lessons for Humans Embryo Losses and Nutrition in Mammals

Carolina Viñoles, DVM, MSc, PhD,
Cecilia Sosa, DVM, MSc, PhD**, Ana Meikle, DVM, MSc, PhD†,
José-Alfonso Abecia, DVM, MSc, PhD***

*Instituto Nacional de Investigación Agropecuaria, Tacuarembó, Uruguay
**Universidad de Zaragoza, Zaragoza, Spain
†Universidad de la República, Montevideo, Uruguay

HUMAN INFERTILITY AND NUTRITION

Nutrition plays an important role in infertility in women of reproductive age [1,2]. Malnutrition is a major problem in developing countries, and obesity and eating disorders are increasingly common in developing as well as in developed countries [2,3]. The reproductive axis is closely linked to nutritional status, especially undernutrition in the female, and inhibitory pathways involving detectors in the hindbrain suppress ovulation in subjects with severe weight loss [3]. Recovery may occur after minimal reacquisition of weight because an energy balance is more important than any body fat mass [4]. Anorexia nervosa and bulimia nervosa affect up to 5% of women of reproductive age causing amenorrhea, infertility and, in those who conceive, an increased likelihood of miscarriage and preterm births [5–7]. On the other hand, obesity can affect reproduction through fat cell metabolism, steroids, and secretion of proteins such as leptin and adiponectin. Moreover, changes induced on homeostatic factors, such as pancreatic secretion of insulin, androgen synthesis by the ovary, and sex hormone-binding globulin production by the liver, can also be affected [3]. The World Health Organization estimates that in 2014, 300 million women in the world were obese, and when they become mothers, are much more likely to have obese children, especially if they develop gestational diabetes [8]. Obesity-associated anovulation may lead to infertility and to a higher risk of miscarriage [8]. Management of anovulation produced by obesity involves diet and exercise as well as standard approaches to ovulation induction [9]. Most obese women conceive without assistance, although their pregnancies have increased rates of pregnancy-associated hypertension, gestational diabetes, large babies, cesarean section, and perinatal mortality and morbidity [3].

Modern societal pressures and expectations over the past several decades have resulted in the tendency for couples to delay conception [10]. This delay in childbearing is associated to a gradual decrease in fertility, beginning approximately at the age of 32 years, and decreasing more rapidly after the age of 37 [11]. The International Committee for Monitoring Assisted Reproductive Technology's (ICMART) estimated that in 2004, 237,809 babies were born through assisted reproductive technologies (ART) worldwide [12]. In spite of the accumulated knowledge in the use of ART, the statistics from 2011 show an average of 31% live birth rate from transfers in the United States [13]. The success rate of ART is largely dictated by sperm quality on the male side, and oocyte quality and uterine receptivity on the female side of the equation [11,14,15].

Handbook of Fertility. http://dx.doi.org/10.1016/B978-0-12-800872-0.00009-3

Experimental studies using animal models have shown the impact of a short-term differential nutrition on oocyte quality and uterine receptivity in adults, and the long-term effect of both undernutrition and obesity in young mothers on their own health and the health and reproductive outcome of their progeny, results that may be used as lessons to better understand and treat human infertility.

THE SHEEP AS A MODEL

Advances in medical research have shown that choosing the adequate animal model to test specific hypotheses is a true challenge. While most of the published research has been performed using rodents due to their management facility and the low economical costs, the use of sheep and its contribution to biomedical research has been outstanding due mainly to its physiology, which is comparable to humans.

Some comparative advantages are that sheep are genetically closer to humans than rodents. Large animals have a much longer lifespan than small mammals, which may be of particular interest in long-term studies. Besides, the timing of physiological processes in sheep is more similar to humans: gestation and fetal development, number of fetuses, attainment of puberty, and development or aging. In particular, sheep (*Ovis aries*) are an attractive species for scientific procedures: for medical, veterinary, and fundamental biological research; they are docile, rarely show aggression, and are gregarious [16]. Besides, their size makes them easy to handle and to perform invasive surgery, they are relatively inexpensive to maintain and are available in large numbers in most countries, and they present lesser ethical concerns than domestic pets and nonhuman primates [17]. Large volumes of sampling material (blood, urine, feces), and with greater frequency, can be collected in sheep [18]. Taking into consideration the rapid advances of the era of physiological genomics, sheep provide available tissue for different (DNA, mRNA, protein) molecular analysis, which facilitates the determination of several endpoints [19]. While genetic improvement due to animal selection through decades has enhanced sheep production, the phenotypic data registered worldwide provide an interesting platform for genomic-wide association studies (selection assisted by molecular markers). Moreover, sheep are an important productive species that in many countries sustain family economy.

Sheep have been shown to be particularly suitable for studies in the areas of endocrinology, reproduction, pregnancy, and fetal development. This has been due to the success of the translation of experimental work in the sheep to the antenatal and neonatal intensive care in human. The sheep model contribution to the understanding of the mechanisms of fetal programming, including underlying timing of normal birth, the effects of the use of glucocorticoids (for maturing the lungs before birth), and/or inadequate intrauterine environment to fetal development, and consequences in adult health is unquestionable.

On the other hand, a great body of research has been conducted in the last decades to study the associations between the metabolic status of sheep and their reproductive performance. Thus, this review is focused on the effects of nutrition on embryo development and uterine health in sheep.

EMBRYO DEVELOPMENT AND MATERNAL RECOGNITION OF PREGNANCY

Mammalian embryos follow, in general terms, the same stages of development from fertilization to implantation. During the preimplantation period of pregnancy, the early ovine embryo undergoes multiple divisions and enters the uterus as a morula around day 5 postfertilization. By day 6, the blastocyst is formed and it hatches from the zona pellucida on days 8–9. From that moment on, the trophoblast elongates, passing from tubular to filamentous shape between day 12 and day 16. Conceptus elongation signals the onset of implantation, although firm adhesion to the endometrium does not occur until day 16 [20,21].

Nevertheless, the fate of the embryo is decided before implantation. The establishment of pregnancy depends on a successful impairment of the conceptus and the lysis of the corpus luteum (CL). Cyclic uterine physiology is driven by sexual steroid hormones, estrogens, and progesterone, acting through their intracellular receptor (ER and PR, respectively) [22]. The luteolytic mechanism, which takes place around day 14 of the ovine estrous cycle, involves the participation of progesterone, estrogens, and oxytocin acting through their uterine receptors (PR, ERα, and OXTR, respectively) and a countercurrent transfer of prostaglandin F2α from the uterus to the ovary [23]. Along with elongation, mononuclear trophectoderm cells synthesize interferon tau (IFN-τ) that will act in a paracrine manner to ultimately prevent the development of the endometrial luteolytic mechanism [24–26]. Maintenance of the CL, and therefore, the secretion of progesterone, is crucial to support the growing needs of pregnancy. Indeed, progesterone concentrations on day 12 after artificial insemination are higher in pregnant than in nonpregnant ewes [27]. In addition, IFN-τ will also stimulate several endometrial genes to assure the uterine receptivity to implantation [26,28]. Recent works have also provided evidence on the contribution of prostaglandins and cortisol in modulating conceptus development and uterine physiology [29,30].

Human embryos do not elongate and secrete IFN-τ and the physiology of implantation is different in ruminants and human (epitheliochorial vs. hemochorial placentation, respectively). Nevertheless, during the preimplantation period embryos are free-living or loosely adhered to the uterine lumen in all eutherian mammals, and the successful establishment of pregnancy depends on the adequate synchrony between uterine and embryo development. The preimplantation period is the stage where more embryo losses are observed in ruminants and women over 40 years of age, in comparison with fertilization failure and late embryo mortality [31–33]. Until implantation, the conceptus develops unattached or loosely attached in the lumen of the reproductive tract and is supported by histotrophic secretions [34,35]. Although it is accepted that the preimplantation embryo is autonomous to a certain extent, it is also undoubted that the environment where it develops determines its phenotypic marks and will affect its future development [36,37]. This dependence on extrinsic factors is greater in the later stages of the preimplantation period, which confers the uterine lumen, rather than the oviducts – a crucial role in embryo development [36]. The histotroph is the combination of substances derived from transport, synthesis, and secretion by the endometrial glands [34,38]. Although recent efforts have shed light over some of its components, its complex nature is yet to be elucidated [39,40]. A critical landmark in the establishment of pregnancy is a correct and synchronous dialogue between the conceptus and the mother. Sexual steroid hormones drive the differentiation of the secretory epithelia of the reproductive tract [41] and also the timing and composition of uterine secretions. It has been found that the rate of secretion and the composition of the fluids of the reproductive tract of cows was altered following infusion of progesterone [42]. It has also been recently reported that the protein fraction of the uterine luminal fluid of cows varies according to the moment of early of pregnancy [43]. To sum up, changes in the conceptus must be accompanied by structural and functional adaptation of the uterus to sustain pregnancy. Adequate uterine secretions will assure timely growth of the trophoblast, and the continuous progesterone exposure will prepare the uterus to sustain pregnancy.

EFFECT OF UNDERNUTRITION ON EMBRYO DEVELOPMENT

The effect of different levels of dietary energy on embryo survival in sheep has been described in the literature since the 1980s [44–47]. Surprisingly, reduced rates of embryo survival have been reported consistently in overfed and underfed ewes. Overnutrition in early pregnancy (up to day 14 after mating) can increase embryo losses due to an increase in progesterone clearance that leads to reduction in the circulating concentrations of progesterone [48,49]. However, due to annual pasture cycles and the availability of food, the most probable scenario in the majority of the farming systems is undernutrition, and thus, this is the topic where more information has been generated in the past 25 years. An increased rate of ova wastage in ewes, which had a restricted food intake (0.5 times maintenance requirements (M)) compared with adequate feeding (1.5 M) during the 14 days prior to mating, until slaughter 11 days after mating has been observed [46]. Other authors observed a delay in the development of embryos collected from undernourished ewes 8 days after mating [50]. Using ewes slaughtered on day 15 of pregnancy, significantly less embryos were collected from restricted ewes reaching the stage of elongated blastocysts, although no differences in the number of blastocysts were recorded on day 9 of pregnancy [51]. This was consistent with previous reports [50,52,53] that indicated at that moment that the effect of undernutrition on pregnancy rate is only expressed from the second week of pregnancy [51]. This set of preliminary experiments led to the conclusion that the effects of undernutrition on embryo survival were not necessarily mediated through changes in ovarian function or progesterone delivery to the uterus, but may involve changes in the uterine environment resulting from different patterns of embryo–maternal signals.

Undernutrition decreases oocyte quality, embryonic development, and pregnancy rates [54] and it is well known that feeding below the requirements for maintenance increases peripheral plasma progesterone concentrations [47,55]. However, other authors [56] demonstrated a reduction of progesterone concentration in the endometrium consistent with a lower endometrial abundance of progesterone receptors [57] and the impairment of early embryonic development [53]. A differential sensitivity of the reproductive tract to steroids may result in a distinct environment for the embryo, through different growth factor concentrations for instance, as these are mainly regulated by steroids [58]. Both insulin-like growth factors (IGF) I and II are crucial for embryo development [58], and are implicated as regulators of preimplantation and placental development [59]. Despite their relevant role, reports on the effects of undernutrition on the expression of these growth factors along the reproductive tract during the early luteal phase are almost nonexistent. Recently, it has been shown [60] that undernutrition decreases IGF-I and IGF-II mRNA expression in an organ-dependent manner as the effects were observed only in the uterus and oviduct, respectively. Moreover, oviducts presented a greater mRNA expression of IGF-II than the uterus, while the opposite occurred for IGF-I mRNA expression [60]. This is consistent with the different suggested roles of these factors in the specific organs: IGF-II has a more predominant role at these very early stages (day 5) of embryo development [37] and enhances the metabolic pathways necessary for the motility of the oviductal wall [58], while IGF-I plays a crucial role preparing a uterine environment capable of sustaining embryonic survival and implantation [61].

The development of embryos during the preimplantation period is affected, in part, by various hormones, growth factors, and their receptors [37] such as plasma nonesterified fatty acid (NEFA), insulin, IGF-I, and adipokines such as leptin and adiponectin. They have been proposed as indices of metabolic status as well as metabolic signals to the reproductive system, so that they are believed to carry the "metabolic message" to the reproductive system [62]. Thus, exposure of maturing bovine oocytes to elevated NEFA concentrations has a negative impact on fertility, not only through a reduction in oocyte developmental capacity but through compromised early embryo quality, viability, and metabolism [63]. We have previously found that IGF-I and insulin, as well as glucose, presented fluctuations in their plasma concentrations that coincided with key events in the ovine estrous cycle and suggested that the reproductive system could also modulate the metabolic milieu [64,65]. IGF-I is produced locally in many organs of the body, including the reproductive tract of cyclic and pregnant ewes [61]. In the uterus, it stimulates cell proliferation and differentiation of the early embryo and of endometrial cells [37]. Moreover, the adipose tissue secretes leptin, and in ruminants the amount secreted directly reflects the mass of adipose tissue. This hormone serves as a signal of body fat reserves to the central nervous system, acting to decrease appetite and increase energy expenditure, thereby maintaining adiposity around a set point [66]. Besides, leptin has been shown to affect ovarian function [67] and there is accumulating evidence that leptin may be directly involved in preimplantation embryonic development [68], so this cytokine hormone is another candidate for linking the effects of undernutrition on uterine and embryonic function. Reinforcing this concept, it has been found that ewes fed at maintenance level for the first 15 days after mating had less embryo losses than ewes fed at half maintenance levels [27]. However, when the supplemental feed was interrupted, a marked decrease in leptin concentrations was associated to greater embryo losses from day 17 to day 30 of pregnancy [27].

The effect of nutrition on oocyte quality and embryo development is more controversial in ewes that received hormonal stimulation of follicle development. In the context of multiple ovulation and embryo transfer (MOET) programs, it has been reported [69] that undernutrition of donor ewes has a negative effect on oocyte quality, which results in lower rates of cleavage and blastocyst formation, concluding that these specific animals may require a specific nutritional management to provide good quality oocytes for ART. Thus, nutrition of donor animals seems to be a key component affecting the development of oocytes and the preimplantation embryo. Moreover, other authors [70] showed that in the superovulated ewe

the recommended high level of feeding before mating increased ovulation rate, improving the response to the superstimulatory treatment. Related to these results, it has been observed [71] that consumption of the 0.5 M diet compared with either the maintenance or the 1.5 M diet during the periconceptional period (18 days before until 6 days after the day of ovulation) increased the total number of cells in the blastocyst and that this increase occurred entirely within the trophectoderm lineage of the embryo. In contrast, others [72] reported that the cleavage rates of oocytes were negatively affected by an *ad libitum* diet compared with a 0.5 M diet. Undernutrition (0.5 M) of superovulated donor ewes during the period of superovulation and early embryonic development resulted in an increased number of ovulations, but reduced total and viable number of embryos. These effects might be mediated by the disruption of endocrine homeostasis, oviduct environment, and/or oocyte quality. These inconsistencies may be related to the nutritional history of the ewes prior to the short-term intervention and their age among other factors, suggesting that this information should be carefully considered when ewes are selected for MOET programs [73].

The effect of undernutrition on embryo development can also be considered in the medium to long term, through changes in the "developmental programming," described as the programming of various body systems and processes by a stressor of the maternal system during pregnancy or during the neonatal period [74]. One of these stressors is maternal nutrition, especially around conception, which is known as the "periconceptional period." It has been reviewed that maternal periconceptional undernutrition is related to altered development and an enlarged danger of adverse neurodevelopmental and metabolic consequences in childhood and later life, suggesting that environmental signals acting during early development may result in epigenetic changes, which may play a role in the relationship between early life vulnerability and adult phenotype [75]. Maternal undernutrition around conception has numerous effects on several aspects of the physiology of offspring. At the reproductive level, it causes a delay in fetal ovarian development in sheep [76], significantly reduces the number of lambs born per ewe, and increases the birth body weight of the offspring and the oocyte population of 30- and 60-day-old ewe lambs [73]. Since females are born with the follicle population that they will expend during their lifetime, a decrease promoted by inadequate nutrition during gestation may have long-term consequences for their reproductive outcome [77]. Not only ovulation rate and embryo survival have been shown to decrease with inadequate nutrition in early life [78], but also the size of the follicle population is associated to the beginning of the reproductive cycles and fertility at first mating in female lambs [79].

LESSONS FOR HUMANS

Education and enhanced awareness of the effect of nutrition and age on fertility are essential in counseling patients who desire pregnancy. The impact of nutrition on oocyte quality and uterine receptivity must be considered in the periconceptional period. Abrupt changes in the diet must be avoided during the first month of gestation to reduce early embryo losses. An appropriate diet, considering the nutritional history (nutritional disorders, obesity) of the patient must be elaborated, in order to increase the chances of success after ART, especially when superovulation treatments are being used. Moreover, a planned pregnancy offers an opportunity to address weight control prior to conception. At the very least, avoiding excessive weight gain during pregnancy may prevent excessive weight retention postpartum. Finally, based on the concept of *in utero* programming, these lifestyle measures may not only have short- and long-term benefits for the mother but also for her offspring.

References

[1] Negro-Vilar A. Stress and other environmental factors affecting fertility in men and women: overview. Environ Health Perspect 1993;101:59–64.

[2] Catalano PM. Obesity, insulin resistance, and pregnancy outcome. Reproduction 2010;140:365–71.

[3] Baird DT, Cnattingius S, Collins J, Evers J, Glasier A, Heitmann B, et al. Nutrition and reproduction in women. Hum Reprod Update 2006;12:193–207.

[4] Loucks A. Energy availability, not body fatness, regulates reproductive function in women. Exerc Sport Sci Rev 2003;31:144–8.

[5] Blais MA, Becker AE, Burwell RA, Flores AT, Nussbaum KM, Greenwood DN, et al. Pregnancy: outcome and impact on symptomatology in a cohort of eating-disordered women. Int J Eat Disord 2000;27:140–9.

[6] Crow SJ, Thuras P, Keel PK, Mitchell JE. Long-term menstrual and reproductive function in patients with bulimia nervosa. Am J Psychiatry 2002;159:1048–50.

[7] Moutquin JM. Socio-economic and psychosocial factors in the management and prevention of preterm labour. BJOG 2003;110:56–60.

[8] Pasquali R, Petrusi C, Gerghini S, Cacciari M, Gambineri A. Obesity and reproductive disorders in women. Hum Reprod Update 2003;9:359–72.

[9] Haddock CK, Poston WSC, Dill PL, Foreyt JP, Ericsson M. Pharmacotherapy for obesity: a quantitative analysis of four decades of published randomized clinical trials. Int J Obes 2002;26:262–73.

[10] Kovac JR, Addai J, Smith RP, Coward RM, Lamb DJ, Lipshultz LI. The effects of advanced paternal age on fertility (Review). Asian J Androl 2013;15:723–8.

[11] Practice Committee of the American Society for Reproductive Medicine. Committee opinion no. 589: female age-related fertility decline. Obstet Gynecol 2014;123:719–21.

[12] Sullivan EA, Zegers-Hochschild F, Mansour R, Ishihara O, de Mouzon J, Nygren KG, et al. International Committee for Monitoring Assisted Reproductive Technologies (ICMART) world report: assisted reproductive technology. Hum Reprod 2013;28:1375–90.

[13] CDC. Assisted reproductive technology national summary. Centers Dis Control Prev 2011:5. http://www.cdc.gov/art/.

[14] Maroulis GB. Effect of aging on fertility and pregnancy. Endocrinologist 1995;5:364–70.

[15] Castelbaum AJ, Lessey BA. Infertility and implantation defects (Review). Infertil Reprod Med Clin North Am 2001;12:427–46.

[16] Arney DR. Sheep behaviour, needs, housing and care. Scand J Lab Anim Sci 2009;36:69–73.

[17] Mageed M, Berner D, Jülke H, Hohaus C, Brehm W, Gerlach K. Morphometrical dimensions of the sheep thoracolumbar vertebrae as seen on digitised CT images. Lab Anim Res 2013;29:138–47.

[18] Arney DR. Welfare of large animals in scientific research. Scand J Lab Anim Sci 2009;36:97–101.

[19] Lie S, Morrison JL, Williams-Wyss O, Suter CM, Humphreys DT, Ozanne SE, et al. Periconceptional undernutrition programs changes in insulin-signaling molecules and microRNAs in skeletal muscle in singleton and twin fetal sheep 1. Biol Reprod 2014;90:1–10.

[20] Guillomot M. Cellular interactions during implantation in domestic ruminants. J Reprod Fertil 1995;49:39–51.

[21] Wintenberger-Torres S, Flechon JE. Ultrastructural evolution of the trophoblast cells of the pre-implantation sheep blastocyst from day 8 to day 18. J Anat 1974;118:143–53.

[22] Geisert RD, Morgan GL, Short EJ, Zavy MT. Endocrine events associated with endometrial function and conceptus development in cattle. Reprod Fertil Dev 1992;4:301–5.

[23] McCracken JA, Custer EE, Lamsa JC. Luteolysis: a neuroendocrine mediated event. Physiol Rev 1999;79:263–323.

[24] Hansen TR, Austin KJ, Perry DJ, Pru JK, Teixeira MG, Johnson GA. Mechanism of action of interferon-τ in the uterus during early pregnancy. J Reprod Fertil 1999;54:329–39.

[25] Spencer TE, Bazer FW. Conceptus signals for establishment and maintenance of pregnancy. Reprod Biol Endocrinol 2004;2:49.

[26] Spencer TE, Johnson GA, Bazer FW, Burghardt RC, Palmarini M. Pregnancy recognition and conceptus implantation in domestic ruminants: roles of progesterone, interferons and endogenous retroviruses. Reprod Fertil Dev 2007;19:65–78.

[27] Viñoles C, Glover KMM, Paganoni BL, Milton JTB. Embryo losses in sheep during short-term nutritional supplementation. Reprod Fertil Dev 2012;24:1040–7.

[28] Spencer TE, Gray A, Johnson GA, Taylor KM, Gertler A, Gootwine E, et al. Effects of recombinant ovine interferon tau, placental lactogen, and growth hormone on the ovine uterus. Biol Reprod 1999;61:1409–18.

[29] Dorniak P, Bazer FW, Spencer TE. Prostaglandins regulate conceptus elongation and mediate effects of interferon τ on the ovine uterine endometrium. Biol Reprod 2011;84:1119–27.

[30] Dorniak P, Bazer FW, Spencer TE. Physiology and endocrinology symposium: biological role of interferon τ in endometrial function and conceptus elongation. J Anim Sci 2013;91:1627–38.

[31] Ubaldi F, Rienzi L, Baroni E, Ferrero S, Iacobelli M, Minasi MG, et al. Implantation in patients over 40 and raising FSH levels – a review (Review). Placenta 2003;24:S34–8.

[32] Diskin MG, Morris DG. Embryonic and early foetal losses in cattle and other ruminants. Reprod Domest Anim 2008;43:260–7.

[33] Diskin MG, Parr MH, Morris DG. Embryo death in cattle: an update. Reprod Fertil Dev 2012;24:244–51.

[34] Ashworth CJ. Maternal and conceptus factors affecting histotrophic nutrition and survival of embryos. Livest Prod Sci 1995;44:99–105.

[35] Gray CA, Burghardt RC, Johnson GA, Bazer FW, Spencer TE. Evidence that absence of endometrial gland secretions in uterine gland knockout ewes compromises conceptus survival and elongation. Reproduction 2002;124:289–300.

[36] Leese HJ. The effects of advanced paternal age on fertility (Review). Hum Reprod Update 1995;1:63–72.

[37] Kaye PL. Preimplantation growth factor physiology. Rev Reprod 1997;2:121–7.

[38] Dorniak P, Welsh TH Jr, Bazer FW, Spencer TE. Cortisol and interferon τ regulation of endometrial function and conceptus development in female sheep. Endocrinology 2013;154:931–41.

[39] Bazer FW, Wu G, Johnson GA, Kim J, Song G. Uterine histotroph and conceptus development: select nutrients and secreted phosphoprotein 1 affect mechanistic target of rapamycin cell signaling in ewes. Biol Reprod 2011;1107:1094–107.

[40] Bazer FW, Kim J, Ka H, Johnson GA, Wu G, Song G. Select nutrients in the uterine lumen of sheep and pigs affect conceptus development. J Reprod Dev 2012;58:180–8.

[41] Gray CA, Burghardt RC, Johnson GA, Bazer FW, Spencer TE. Developmental biology of uterine glands. Biol Reprod 2001;65:1311–23.

[42] Hugentobler SA, Sreenan JM, Humpherson PG, Leese HJ, Diskin MG, Morris DG. Effects of changes in the concentration of systemic progesterone on ions, amino acids and energy substrates in cattle oviduct and uterine fluid and blood. Reprod Fertil Dev 2010;22:684–94.

[43] Forde N, McGettigan PA, Mehta JP, O'Hara L, Mamo S, Bazer FW, et al. Proteomic analysis of uterine fluid during the pre-implantation period of pregnancy in cattle. Reproduction 2014;147:575–87.

[44] Brien FD, Cumming IA, Clarke IJ, Cocks CS. Role of plasma progesterone concentration in early-pregnancy of the ewe. Aust J Exp Agric 1981;21:562–5.

[45] Parr RA, Davis IF, Fairclough RJ, Miles MA. Overfeeding during early pregnancy reduces peripheral progesterone concentration and pregnancy rate in sheep. J Reprod Fertil 1987;80:317–20.

[46] Rhind SM, McMillen S, Wetherill GZ, McKelvey WAC, Gunn RG. Effects of low levels of food intake before and/or after mating on gonadotropin and progesterone profiles in Greyface ewes. Anim Prod 1989;49:267–73.

[47] Abecia JA, Rhind SM, Goddard PJ, McMillen SR, Ahmadi S, Elston DA. Jugular and ovarian venous profiles of progesterone and associated endometrial progesterone concentrations in pregnant and non-pregnant ewes. Anim Sci 1996;63:229–34.

[48] Parr RA, Davis IF, Squires TJ, Miles MA. Influence of lupin feeding before and after joining on plasma progesterone and fertility in Merino ewes. Proc Aust Soc Anim Prod 1992;19:185–7.

[49] Parr RA, Davis IF, Miles MA, Squires TJ. Liver blood flow and metabolic clearance rate of progesterone in sheep. Res Vet Sci 1993;55:311–6.

[50] Abecia JA, Lozano JM, Forcada F, Zarazaga L. Effect of level of dietary energy and protein on embryo survival and progesterone production on day eight of pregnancy in *Rasa Aragonesa* ewes. Anim Reprod Sci 1997;48:209–18.

[51] Abecia JA, Forcada F, Lozano JM. A preliminary report on the effect of dietary energy on prostaglandin F2 alpha production *in vitro*, interferon-τ synthesis by the conceptus, endometrial progesterone concentration on days 9 to 15 of pregnancy and associated rates of embryo wastage in ewe. Theriogenology 1999;52:1203–13.

[52] Parr RA, Cumming IA, Clarke IJ. Effects of maternal nutrition and plasma progesterone concentrations on survival and growth of the sheep embryo in early gestation. J Agric Res 1982;98:39–46.

[53] Abecia JA, Rhind SM, Bramley TA, McMillen SR. Steroid production and LH receptor concentrations of ovarian follicles and corpora lutea and associated rates of ova wastage in ewes given high and low levels of food intake before and after mating. Anim Sci 1995;61:57–62.

[54] Abecia A, Sosa C, Forcada F, Meikle A. The effect of undernutrition on the establishment of pregnancy in the ewe. Reprod Nutr Dev 2006;45:367–78.

[55] Rhind SM, Leslie LD, Gunn RG, Doney JM, Plasma FSH. LH, prolactin and progesterone profiles of Cheviot ewes with different levels of intake before and after mating, and associated effects on reproductive performance. Anim Reprod Sci 1985;8:301–13.

[56] Lozano JM, Abecia JA, Forcada F, Zarazaga L, Alfaro B. Effect of undernutrition on the distribution of progesterone in the uterus of ewes during the luteal phase of the estrous cycle. Theriogenology 1998;49:539–46.

[57] Sosa C, Lozano JM, Viñoles C, Acuña S, Abecia JA, Forcada F, et al. Effect of plane of nutrition on endometrial sex steroid receptor in ewes. Anim Reprod Sci 2004;84:337–48.

[58] Wathes DC, Reynolds TS, Robinson RS, Stevenson KR. Role of the insulin-like growth factor system in uterine function and placental development in ruminants. J Dairy Sci 1998;81:1778–89.

[59] Kaye PL, Bell KL, Beebe L, Dunglison GF, Gardner HG, Harvey MB. Insulin and the insulin-like growth factors (IGFs) in preimplantation development. Reprod Fertil Dev 1992;4:373–86.

[60] De Brun V, Abecia JA, Carriquiry M, Forcada F. Undernutrition and laterality of the corpus luteum affects gene. Spanish J Agric Res 2013;11:989–96.

[61] Stevenson KR, Gilmour RS, Wathes DC. Localization of insulin-like growth factor-I (IGF-I) and -II messenger ribonucleic acid and type 1 IGF receptors in the ovine uterus during the estrous cycle and early pregnancy. Endocrinology 1994;134:1655–64.

[62] Blache D, Tellam RL, Chagas LM, Blackberry MA, Vercoe PE, Martin GB. Level of nutrition affects leptin concentrations in plasma and cerebrospinal fluid in sheep. J Endocrinol 2000;165:625–37.

[63] Van Hoeck V, Sturmey RG, Bermejo-Alvarez P, Rizos D, Gutierrez-Adan RG, Leese HJ, et al. Elevated non-esterified fatty acid concentrations during bovine oocyte maturation compromise early embryo physiology. PLoS One 2011;6:e23183.

[64] Sosa C, Abecia JA, Forcada A, Viñoles C, Tasende C, Valares JA, et al. Effect of undernutrition on uterine progesterone and oestrogen receptors and on endocrine profiles during the ovine oestrous cycle. Reprod Fertil Dev 2006;18:447–58.

[65] Sosa C, Abecia A, Carriquiry M, Vázquez MI, Fernández-Foren A, Talmon C, et al. Effect of undernutrition on the uterine environment during maternal recognition of pregnancy in sheep. Reprod Fertil Dev 2009;21:869–81.

[66] Chilliard Y, Delavaud C, Bonnet M. Leptin expression in ruminants: nutritional and physiological regulations in relation with energy metabolism. Domest Anim Endocrinol 2005;29:3–22.

[67] Smith GD, Jackson LM, Foster DL. Leptin regulation of reproductive function and fertility. Theriogenology 2002;57:73–86.

[68] Herrid M, Nguyen VL, Hinch G, McFarlane JR. Leptin has concentration and stage-dependent effects on embryonic development *in vitro*. Reproduction 2006;132:247–56.

[69] Borowczyk E, Caton JS, Redmer DA, Bilski JJ, Weigl RM, Vonnahme KA, et al. Effects of plane of nutrition on *in vitro* fertilization and early embryonic development in sheep. J Anim Sci 2006;84:1593–9.

[70] McEvoy TG, Robinson JJ, Aitken RP, Findlay PA, Palmer RM, Robertson IS. Dietary-induced suppression of preovulatory progesterone concentrations in superovulated ewes impairs the subsequent *in-vivo* and *in-vitro* development of their ova. Anim Reprod Sci 1995;39:89–107.

[71] Kakar MA, Maddocks S, Lorimer MF, Kleemann DO, Rudiger SR, Hartwich KM, et al. The effect of peri-conception nutrition on embryo quality in the superovulated ewe. Theriogenology 2005;64:1090–103.

[72] Lozano JM, Lonergan P, Boland MP, O'Callaghan D. Influence of nutrition on the effectiveness of superovulation programmes in ewes: effect on oocyte quality and post-fertilization development. Reproduction 2003;125:543–53.

[73] Abecia JA, Casao A, Lobón S, Meikle A, Forcada F, Sosa C. The effect of periconceptional undernutrition of sheep on the cognitive/emotional response and oocyte quality of offspring at 30 days of age. J Dev Orig Health Dis 2014;5:79–87.

[74] Reynolds LP, Borowicz PP, Caton JS, Vonnahme KA, Luther JS, Hammer CJ, et al. Developmental programming: the concept,

large animal models, and the key role of uteroplacental vascular development. J Anim Sci 2010;88:E61–72.

[75] Bloomfield FH. Epigenetic modifications may play a role in the developmental consequences of early life events. J Neurodev Disord 2011;3:348–55.

[76] Rae MT, Palassio S, Kyle CE, Brooks AN, Lea RG, Miller DW, et al. Effect of maternal undernutrition during pregnancy on early ovarian development and subsequent follicular development in sheep fetuses. Reproduction 2001;122:915–22.

[77] Viñoles C, Paganoni BL, McNatty KP, Heath DA, Thompson AN, Glover KM, et al. Follicle development, endocrine profiles and ovulation rate in adult Merino ewes: effects of early nutrition (pre- and post-natal) and supplementation with lupin grain. Reproduction 2014;147:101–10.

[78] Gunn RG, Sim DA, Hunter EA. Effects of nutrition *in utero* and in early life on the subsequent lifetime reproductive performance of Scottish Blackface ewes in two management systems. Anim Sci 1995;60:223–30.

[79] Lahoz B, Alabart JL, Monniaux D, Mermillod P, Folch J. Anti-Müllerian hormone plasma concentration in prepubertal ewe lambs as a predictor of their fertility at a young age. BMC Vet Res 2012;8:118.

BARIATRIC SURGERY, OBESITY, AND FERTILITY

10

Fertility and Testosterone Improvement in Male Patients After Bariatric Surgery

Michaela Luconi, PhD, Giovanni Corona, MD, PhD*,**,*
*Enrico Facchiano, MD†, Marcello Lucchese, MD†, Mario Maggi, MD**

*University of Florence, Florence, Italy
**Azienda Usl Bologna Maggiore-Bellaria Hospital, Bologna, Italy
†Azienda Sanitaria di Firenze, Santa Maria Nuova Hospital, Florence, Italy

MALE OBESITY AND SEX HORMONES

Obesity, a metabolic condition characterized by an increase in dysfunctional adipose tissue, is exponentially expanding worldwide, becoming more and more not only a medical and health problem, but also a social and economic burden. In fact, this complex pathology affects all social classes and spreads over both developed and developing countries [1]. There are, nowadays, about 1.5 billion people worldwide who are overweight, with 500 million of them being obese with a male:female ratio of 2:3, compared with 850 million who are chronically underweight [1]. Europe is reporting increasing proportions of men and women in their reproductive years who are overweight and obese [2]. These proportions become even higher with aging. Accordingly, the worldwide, age-standardized prevalence of obesity was 9.8% in men in 2008, almost doubling over a 30-year period [3]. Even more worrying, nearly 43 million children under the age of 5 were overweight in 2010. The worldwide global change of body mass index (BMI) was calculated to be around 0.4 kg/m^2 per decade for men [2]. Finally, transmission of obesity tract by paternal inheritage has recently been suggested [4,5]. Excess body weight is a crucial risk factor for mortality and morbidity, mostly due to the cardiovascular and metabolic pathologies often associated, such as type 2 diabetes mellitus (T2DM), metabolic syndrome (MetS), hypertension, cardiovascular diseases (CVD), and cancer, overall resulting in about 3 million deaths every year worldwide [6–10].

Besides these well-known associated morbidities, obesity was recently also associated with a high prevalence of reproductive disorders. In particular, several papers reported that obese males often present altered levels of sex steroid hormones (decreased total (TT) and free (FT) testosterone and increased estradiol (E2)) with low levels of sex steroid-binding globulin (SHBG) [11–22]. Such alterations are accompanied by normal or reduced levels of gonadotropins (luteinizing hormone (LH) and follicular stimulating hormone (FSH)), defining a clear hypogonadotropic hypogonadism characterized by a central dysfunction (Fig. 10.1) [23–29]. Indeed, although LH pulse frequency is not altered in morbid obese, its levels

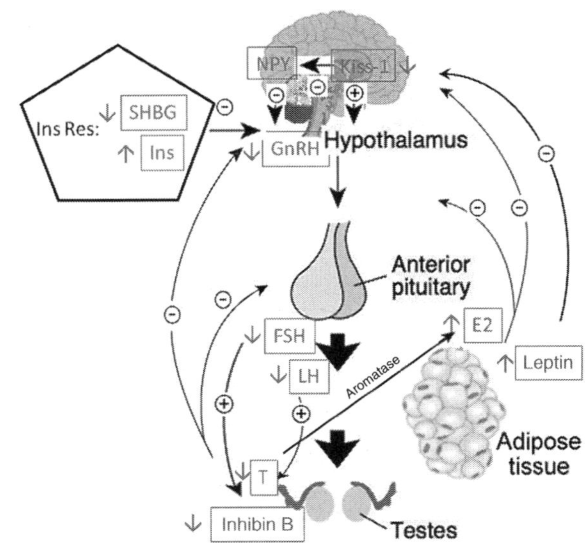

FIGURE 10.1 **The hypothalamus-pituitary-testis axis and its dysregulation in obesity.** (−) indicates a negative and (+) a positive feedback. Insulin (Ins); insulin-resistance (Ins Res); Neuropeptide Y (NPY); Kisspeptin 1 (Kiss-1).

Handbook of Fertility. http://dx.doi.org/10.1016/B978-0-12-800872-0.00010-X

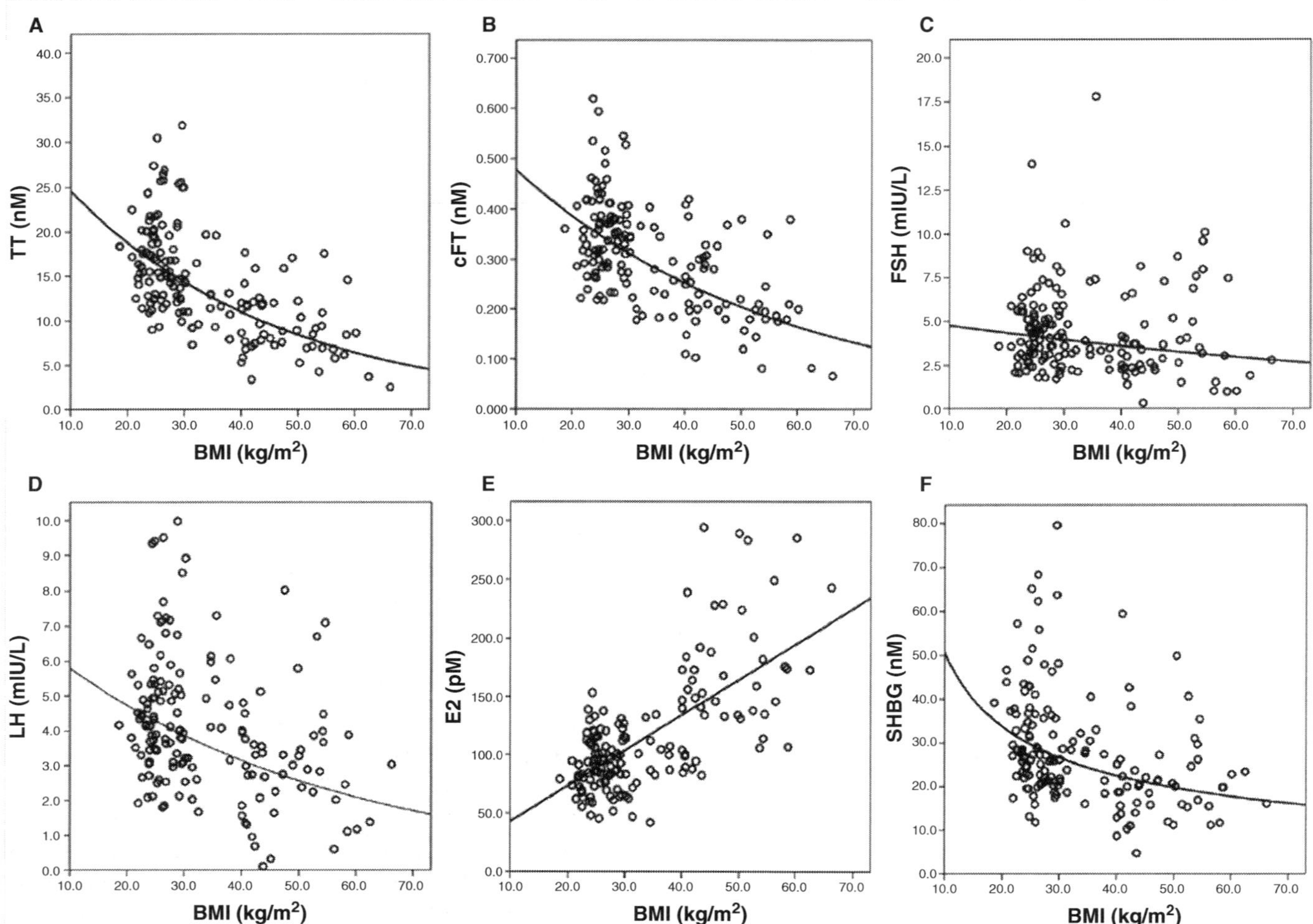

FIGURE 10.2 **Correlation models between BMI and sex steroid hormones in males.** The best fitting regression curves ($P < 0.001$) are shown for relationship between BMI and TT (A, exponential model), cFT (B, exponential model), FSH (C, exponential model), LH (D, exponential model), E2 (E, linear model) and SHBG (F, power model) in a cross-sectional cohort of subjects composed by 161 males from a general population enriched with morbidly obese subjects (median[interquartile]: BMI = 29.2[24.8–41.9] kg/m^2), age = 45.0[42.4–48.0]). Sex hormones were evaluated immunoenzymatically. All statistical analyses were performed on SPSS 18.0 for Windows (Statistical Package for the Social Sciences, Chicago, USA). With permission of Elsevier [20].

and pulse amplitude are significantly reduced compared to normal-weight controls [30,31]. The mechanisms involved in the obesity-related derangement of the gonadotropin axis (Fig. 10.1) include excess leptin [32–34] and higher PRL levels [35], and the increased peripheral conversion of androgens into estrogens through aromatase highly expressed in adipose tissue, which represses LH secretion by a central negative feedback [27,36,37]. In line with this view, we previously reported that body weight loss, obtained through either lifestyle or bariatric interventions, is associated with a fall in estrogen levels and with a concomitant rise in gonadotropins and T [13]. However, other obesity-associated factors, besides estrogens, have been suggested as a link between increased visceral fatness and hypogonadotropic hypogonadism, including glucose intolerance and/or hyperglycemia [23,38]. The reduced SHBG, mainly due to hepatic dysfunction and associated hyperinsulinemia observed in T2DM comorbidity, further contributes to the negative

feedback of estrogens [39]. Finally, reduction in inhibin B [40], which is a marker of Sertoli functions associated with spermatogenic activity in the testis, may contribute to the axis derangements and alterations in testicular functions observed in obesity.

Our group recently explored the correlation models between BMI and sex hormones in a cross-sectional survey performed on a cohort of male subjects ($n = 166$, BMI, mean ± SD = 39.9 ± 11.2 kg/m^2) including $n = 56$ morbidly obese patients (BMI > 40 kg/m^2, mean ± SD = 48.2 ± 6.9 kg/m^2) referred to the bariatric surgery unit of Florence University Hospital of Careggi [20]. We showed that the best fitting models describing these relations were the exponential decay for TT, cFT, gonadotropins, a negative power model for SHBG, and a positive linear regression for E2 (Fig. 10.2). These findings suggest that some compensatory effects may occur above extreme BMI, which regulate androgen and gonadotropin decrease, while E2 may increase continuously.

REPRODUCTIVE FUNCTIONS IN OBESE MEN

Obesity may adversely affect male reproduction by different mechanisms such as endocrine, inflammatory, thermal, genetic, psychological, and sexual mechanisms [10,41]. Infertility is defined as the inability of a couple to obtain a baby after at least one 1 year of unprotected sexual intercourse [42] and can be due to alterations at different steps in the reproductive process.

Different studies demonstrated a strict positive association between BMI and infertility in obese men with different OR showing the maximal negative effects for BMI > 35 kg/m^2 [43,44]. In general, noncontrolled studies conducted on male partners in infertile populations attending andrology clinics often showed an increase in obesity incidence or higher weight in males belonging to the infertile groups [45]. The incidence of obesity is three times higher in men with male factor subfertility [46]. Morbidly obese men had a higher rate of seeking babies by assisted reproductive techniques, which resulted in a decreased blastocyst development and reduced live birth rates associated with paternal BMI [47–49].

Semen Parameters

Semen analysis is a pivotal evaluation in the diagnosis of male infertility, since a decrease in sperm number/concentration, motility, and normal morphology is associated with a reduction in fertility [50]. Notably, the secondary hypogonadism observed in male obesity may not necessarily be associated with a significant worsening of the semen quality.

Impaired semen quality in male obesity is still controversial. In fact, the two current systemic reviews associated with meta-analyses assessing the impact of BMI on semen parameters lead to conflicting results based on the different studies analyzed. In fact, although initially no association between alterations in BMI and any semen parameters considered (motility, viability, sperm number, volume) was found, based on 31 studies taken into consideration by the older meta-analysis [51], the following update review comprising 21 studies highlighted a positive and significant OR of 2.04 [1.59–2.62] of developing oligozoospermia and asthenozoospermia in morbidly obese men [52]. However, compared to the previous one, the more recent review excluded several papers ($n = 19$) from the meta-analysis due to the inability of analyzing the data, probably introducing a big bias in the analysis. We recently reported studies on 222 men seeking medical care for couple infertility. After adjusting for age and testosterone levels, higher BMI was significantly related to higher prostate volume and several CDU features of the prostate, including macrocalcifications, inhomogeneity, higher arterial peak systolic velocity (the latter adjusted also for blood pressure), but not with abnormalities of testis, epididymis, and seminal vesicles. Furthermore, higher BMI and BMI class were significantly related to higher seminal IL-8 (sIL8), a reliable surrogate marker of prostate inflammatory diseases, even after adjustment for age. Conversely, no associations among BMI, clinical symptoms of prostatitis, or semen parameters were observed [53]. In a further study of males of infertile couples, the presence of MetS was positively associated with prostate enlargement, biochemical (sIL8), and ultrasound-derived signs of prostate inflammation, but not with prostate-related symptoms, suggesting that MetS is a trigger for a subclinical, early-onset form of benign prostatic hyperplasia [54]. In contrast, in another population-based prospective cohort of 501 couples attempting to conceive (LIFE study), higher waistline and, to a lower extent, BMI were associated with impaired sperm count [55]. It was suggested that a reduction in sperm volume, probably due to alterations in semen vesicles and prostate in obesity, may be responsible for the observed decrease in the total sperm number [55]. Although no association between BMI or waist circumference with semen parameters was found, a worse semen quality was confirmed in a recent study comparing semen parameters of 23 morbidly obese (mean ± SD BMI = 44.3 ± 5.9 kg/m^2) and 25 age-matched lean (BMI = 24.2 ± 3.0 kg/m^2) subjects [56]. Variability in semen assessment methods between the different studies analyzed as well as significant differences in the cohorts of subjects may justify these conflicting results. In particular, sperm morphology evaluation has been complicated during the last 15 years by differences in reference populations [42].

Sperm Functional Assay

Conventional analysis of sperm parameters evaluated by semen analysis is insufficient to predict sperm fertilization potential. In fact, the functional ability of spermatozoa to undergo pivotal processes required for fertilization, such as acrosome reaction (AR) or zona pellucida (ZP) binding, cannot be predicted on the basis of the routine semen analysis [57]. Sperm DNA fragmentation is another parameter that affects embryo development after fertilization.

ZP binding ability of spermatozoa has been demonstrated to be not affected by obesity [58]. Conversely, AR in response to the physiological stimulus of progesterone [59–61], which is a predictive parameter for fertilization outcome in *in vitro* fertilization (IVF) [60,61] is significantly reduced in morbidly obese men [56]. Notably, the spontaneous AR occurs at higher levels in spermatozoa of morbidly obese men versus lean men, probably contributing to interfere with the fertilization potential of spermatozoa, since AR must occur in a precise timing when spermatozoa reach the oocyte and not before [56].

Studies evaluating the correlation between obesity and DNA fragmentation in spermatozoa of obese men are complicated by the different techniques used to assess chromatin integrity (terminal uridine nick-end labeling or tunnel assay, sperm chromatin structure assay, or SCSA and comet) and the use of flow cytometry or microscopy analysis. Despite this variability in the techniques to assess the DNA fragmentation index (DFI), almost all studies reported its significant increase in obese and overweight compared to lean subjects attending infertility centers [62–65], suggesting an association between obesity and damage in sperm DNA structure, which finally can interfere with fertility potential. However, a significant correlation between BMI and DFI was not found in all studies [55].

Sexual Quality of Life and Erectile Function

Erectile dysfunctions (EDs) and psychological distress interfering with reproductive functions in obese males have been extensively investigated [9,10,66–72]. The studies are made complex by the different definitions used, the different endpoints, and the variety of assessment methods, of which, only some have been more rigorously developed and validated, such as the Brief Sexual Function Inventory (BSFI) [73], Erection Quality Scale (EQS) [74], FSFI [75,76], International Index of Erectile Function (IIEF) standard and abridged versions (IIEF-5; SHIM) [77–79], and structured interview on erectile dysfunction (SIEDY) [80–84]. Despite this heterogeneity, the majority of these studies reported a higher occurrence of ED in obese men than healthy lean men. Indexes of dissatisfaction with sexual quality of life correlated with BMI and weight in the European Male Aging study, a population-based survey on more than 3400 men aged 40–80 years [72]. In men reporting symptoms of ED, overweight and obesity are represented in 79% of the subjects [85]. Moreover, ED in obesity is associated with an increased risk of incident cardiovascular disease, as demonstrated in a consecutive series of 1687 patients attending andrological clinics for ED [9]. For a review of population- and non-population-based studies of obesity and sexual functioning, see Refs [10] and [86].

BARIATRIC SURGERY AND ITS EFFECT ON MALE FERTILITY

Although male gender accounts for only 20% of patients undergoing bariatric surgery, it represents 30% of the obese population eligible for surgery. Moreover, male obesity is characterized by higher BMI and more severe comorbidities than the average woman seeking surgery [87]. Gender differences in social pressures to achieve normal weight may account for this under-representation in men [88,89].

Bariatric surgery has traditionally been divided into three types of procedures: (1) restrictive, producing weight loss by limiting food intake (adjustable gastric band (AGB) and gastric sleeve (GS)); (2) malabsorptive, inducing weight loss totally by interference with digestion and absorption (intestinal bypass); and (3) mixed, limiting intake and producing malabsorption (Roux-en-Y gastric bypass (RYGB) or duodenal switch). Despite this apparently clear classification, the mechanisms of action by which weight loss is achieved and obesity-associated comorbidities reduced and even remitted, remain unclear [90]. Their specific indications are reviewed in the Guidelines of the Society of American Gastrointestinal and Endoscopic Surgeons [91].

Severe obesity virtually affects every system of the body with a broad expression of comorbidities which, when present, can support the indication for bariatric surgery even in subjects with BMI <40 kg/m^2 [91]. These comorbidities include hypertension, T2DM, renal failure, lipid dysmetabolism, immunoincompetence, asthma, gastroesophageal reflux disease, chronic obstructive pulmonary disease, sleep apnea, cardiac failure, atherosclerosis, arthritis of the weight bearing joints, infertility, skin breakdown, and an increased prevalence of cancer. They generally respond favorably to bariatric surgery, often with total and permanent remission, in particular for T2DM and CVD [92–94]. Thus, along with the main goal of bariatric surgery, which is to decrease the excess in body weight, another pivotal endpoint is emerging, which is reducing obesity-related comorbidities and therefore fulfilling the new concept of "metabolic surgery."

As highlighted by subsequent review and meta-analysis appearing in the literature, very few prospective controlled study trials have been performed to assess the effect of bariatric surgery on obesity and comorbidities, in particular with fertility recovery as the primary endpoint. An additional point of interest is to dissect the effects of massive weight loss from some additional effects linked to bariatric surgery, which should be investigated by studies comparing the effects of nonsurgical and surgically induced weight loss in morbidly obese men. Interestingly, we demonstrated that bariatric surgery induces androgen recovery, which is higher than that expected on the basis of weight loss only, suggesting that the occurrence of additional effects may be associated with surgery [20].

Effects on Sex Steroid Hormone Levels

The effects of bariatric surgery on sex steroid hormone levels and obesity-associated hypogonadotropic hypogonadism are the most extensively demonstrated, as shown by a systematic review and meta-analysis of the literature by our group [13]. Of the 24 studies performed on obese

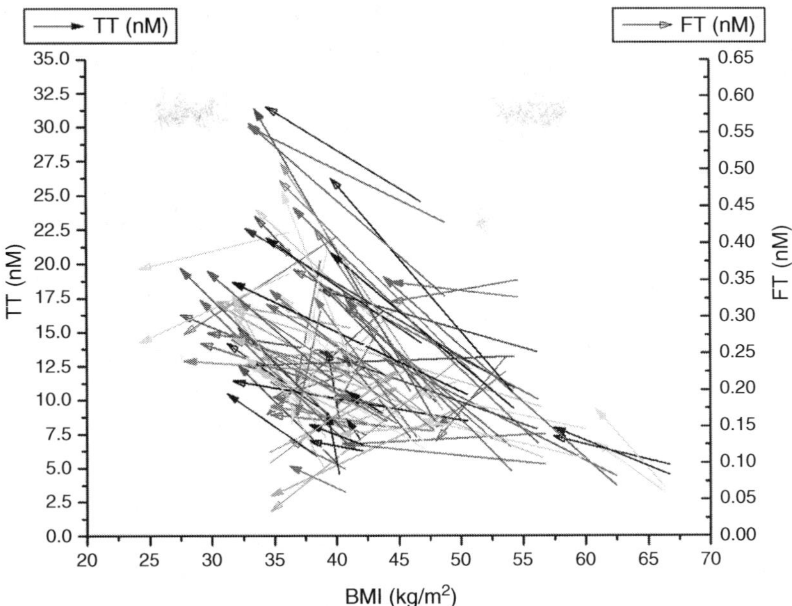

FIGURE 10.3 **Vectorial representation of variations in androgens (TT and FT) and BMI in morbidly obese men subjected to bariatric surgery.** TT and calculated FT were evaluated immunoenzymatically before and at 6 months from bariatric surgery (gastric bypass) in $n = 55$ morbidly obese subjects (BMI 46.6 ± 17.5 kg/m², age 43 ± 12 years). The vectorial lines (arrows) with the same color represent the variation at 6 month from baseline of BMI (bottom axis), TT (left axis) or calculated FT (right axis) in the same patient.

males selected for the meta-analysis, 22 evaluated the effect of diet or bariatric surgery, whereas two directly compared the two procedures. A low caloric diet and bariatric surgery are associated with a significant increase in circulating SHBG and TT levels, with bariatric surgery being more effective (TT fold increase: 8.73 (6.51–10.95) vs. 2.87 (1.68–4.07) for bariatric surgery and the low-calorie diet, respectively; both $P < 0.0001$ vs. baseline). Androgen rise is greater in patients losing more weight as well as in younger, nondiabetic subjects [19] with a greater degree of obesity. Multiple regression analysis shows that the degree of body weight loss is the best determinant of TT rise ($B = 2.50 ± 0.98$, $P = 0.029$). In addition to the TT and FT increase, starting even at 6 months from surgery (Fig. 10.3), a significant increase in SHBG and FT have also been reported at different times of follow-up [13,19–22,66,95,96] as well as in gonadotropins, with a concomitant decrease in estradiol levels. When considering the recent meta-analysis by Corona et al. [13], however, no effect of E2 decrease on TT was observed after weight loss, and when a variation of E2 was introduced in a multiple regression model along with other covariates, Δ-BMI but not Δ-E2 resulted as the best predictor of testosterone increase after surgery. Hence, other fat-associated factors besides estrogens can be speculated to mediate the weight reduction-induced improvement in testosterone levels. The gonadotropin amelioration observed after surgery may contribute to positively affect the pituitary-gonadal axis, contributing to testosterone level improvement (Fig. 10.1). Notably, in a cohort of 76 morbidly obese men undergoing bariatric surgery compared to 27

nonoperated subjects, in parallel with the observed amelioration of TT and cFT levels, a significant increase was observed also in osteocalcin (OCN = 10.4 ± 10.3 ng/mL, $P < 0.001$) and undercarboxylated osteocalcin (ucOCN = 5.4 ± 3.7 ng/mL, $P < 0.001$) in operated but not in nonoperated patients [95]. This osteoblast-produced hormone OCN and its testis-specific undercarboxylated isoform (ucOCN) have been demonstrated to be pivotal for bone-controlled induction of testosterone production in the Leydig cells [97]. OCN recovery observed after bariatric surgery was significantly associated with cFT improvement independently of BMI variation and age in hypogonadal morbidly obese males, suggesting a contribution of the bone–testis axis.

A consistent variability in the variation in sex hormone levels after bariatric surgery is reported among the different studies, which can be only partially explained by the differences in the time of follow-up, type of operation, and ethnicity. Previous works reported that BMI [13,20] and age [19] were among the best predictors of androgen recovery after bariatric surgery. However, such cohorts of morbidly obese patients were not stratified for hypogonadism at baseline. Interestingly, when a cohort of 55 morbidly obese men observed at 6 and 9 months after different types of bariatric surgery was stratified for strict hypogonadism (TT < 8.0 nM), the increase in androgen levels (TT and cFT) was statistically significant in hypogonadal subjects only, whereas, the decrease of estradiol levels was important in eugonadal only [96]. Moreover, estradiol levels at baseline were significantly higher in eugonadal versus hypogonadal men, further confirming that estrogens cannot affect either hypogonadism or the androgen recovery observed after surgery. Thus, hypogonadism before surgery predicts a better recovery in androgen levels as well as waist circumference decrease, with no significant differences in BMI reduction between eugonadal and hypogonadal subjects [96].

Effects on Sperm Parameters

No longitudinal studies assessing the effect of weight loss induced by bariatric surgery on DNA fragmentation levels have been performed yet, probably due to the scarce compliance of male subjects in performing multiple semen analysis during the postoperative follow-up. Similarly, only few anecdotal studies have reported the negative effect of bariatric surgery on semen quality in a few patients at least after 6 months and 1 year after surgery. However, in two of three patients, the quality increased at 2 years resulting in IVF pregnancy [98]. Conversely, the only prospective study performed on 20 morbidly obese men randomized for gastric bypass or nonintervention in which semen analysis was performed at baseline and after 24 months revealed no significant variation in sperm parameters, despite a significant improvement in hormone profile and sexual functions [41]. It is interesting to see the discrepancy occurring between the rapid beneficial effect of surgery on sex-hormone levels and concurrently the negative effect on sperm production. However, as already highlighted, the alteration in androgen production characterizing hypogonadism does not consequently result in a reduced quality of semen and spermatozoa. Similarly, recovery from hypogonadism does not mean a consequent improvement in sperm quality, maybe due to the different timing in the two functions of the testis, namely, steroidogenesis and spermatogenesis, the latter probably requiring longer intervals to recover.

Effects on *In Vitro* Fertilization

Despite some studies assessing the impact of female bariatric surgery on obesity-related infertility and IVF outcome [99], nothing has been done for obese male partners losing weight by bariatric surgery. In the female, despite the improvement in menstrual irregularity and increased ovulation rate, which lead to increased pregnancy rates and reduced risks of obesity-related pregnancy complications such as gestational diabetes and hypertensive disorders, pregnancies after bariatric surgery may be considered a high risk due to the concerns for vitamin deficiencies and gastrointestinal symptoms related to the surgery, and thus requires a strict follow-up [100].

Effects on Sex Quality of Life and Erectile Function

A significant amelioration of sex quality of life and ED after bariatric surgery has been described [21,101–103], although only one study has been designed as a prospective randomized long-term controlled trial assessing sexual functions and in particular ED after massive weight loss obtained surgically and nonsurgically [66]. In the group of 10 morbidly obese men undergoing gastric bypass, a significant increase in TT level and amelioration of ED evaluated by IIEF-5 questionnaire score was observed at 24 months of follow-up ($P = 0.0349$ and 0.0469, respectively), as well as a significant difference in TT, FT, and IIEF-5 score at 24 months compared to the nonoperated control group ($n = 10$) ($P = 0.0224$, 0.0043, and 0.0149, respectively).

Whether such amelioration of sex quality of life and ED are directly due to bariatric surgery or indirectly by the obtained massive weight loss, which improves body image, depression, and other psychogenic factors in these subjects, has still to be clarified.

CONCLUSIONS

Despite conflicting results about altered semen parameters and reduced fertility in male morbidly obese subjects, which deserve further controlled clinical studies, hypogonadism represents a risky associated comorbidity. The Endocrine Society guidelines and the Third International Consultation on Sexual Medicine underline that lifestyle modifications should be encouraged in hypogonadal patients with obesity, T2DM, and MetS. Since lifestyle modifications often fail to obtain a significant and stable weight loss, particularly in the case of morbid obesity, bariatric surgery can provide a more effective weight loss that can finally result in a more effective reduction of comorbidities. The evidence that symptomatic hypogonadism in morbidly obese patients can regress after bariatric surgery should lead us to consider it as an additional indication to bariatric surgery even in male patients with a BMI <40 kg/m^2, as for other comorbidities such as diabetes.

References

[1] World Health Organization Report. Obesity and overweight. Fact sheet No. 311 Updated March 2011.

[2] Sassi F. Obesity and the Economics of Prevention: Fit Not Fat. Paris: OECD; 2010.

[3] Finucane MM, Stevens GA, Cowan MJ, Danaei G, Lin JK, Paciorek CJ, et al. Global Burden of Metabolic Risk Factors of Chronic Diseases Collaborating Group (Body Mass Index). National, regional, and global trends in body-mass index since 1980: systematic analysis of health examination surveys and epidemiological studies with 960 country-years and 9•1 million participants. Lancet 2011;377:557–67.

[4] Soubry A, Schildkraut JM, Murtha A, Wang F, Huang Z, Bernal A, et al. Paternal obesity is associated with IGF2 hypomethylation in newborns: results from a Newborn Epigenetics Study (NEST) cohort. BMC Med 2013;11:29.

[5] Fullston T1, Ohlsson Teague EM, Palmer NO, DeBlasio MJ, Mitchell M, Corbett M, et al. Paternal obesity initiates metabolic disturbances in two generations of mice with incomplete penetrance to the F2 generation and alters the transcriptional profile of testis and sperm microRNA content. FASEB J 2013;27:4226–43.

[6] Ezzati M, Lopez AD, Rodgers A, Vander Hoorn S, Murray CJ. The Comparative Risk Assessment Collaborating Group. Selected major risk factors and global and regional burden of disease. Lancet 2002;360:1347–60.

[7] Prospective Studies Collaboration. Body-mass index and cause specific mortality in 900 000 adults: collaborative analyses of 57 prospective studies. Lancet 2009;373:1083–96.

[8] World Health Organization. Global health risks: mortality and burden of disease attributable to selected major risks. Geneva: World Health Organization; 2009.

[9] Corona G, Monami M, Boddi V, Balzi D, Melani C, Federico N, et al. Is obesity a further cardiovascular risk factor in patients with erectile dysfunction? J Sex Med 2010;7:2538–46.

[10] Corona G, Rastrelli G, Filippi S, Vignozzi L, Mannucci E, Maggi M. Erectile dysfunction and central obesity: an Italian perspective. Asian J Androl 2014;16:581–91.

[11] Corona G, Rastrelli G, Maggi M. Diagnosis and treatment of late-onset hypogonadism: systematic review and meta-analysis of TRT outcomes. Best Pract Res Clin Endocrinol Metab 2013;27:557–79.

[12] Corona G, Vignozzi L, Sforza A, Maggi M. Risks and benefits of late onset hypogonadism treatment: an expert opinion. World J Men's Health 2013;31:103–25.

[13] Corona G, Rastrelli G, Monami M, Saad F, Luconi M, Lucchese M, et al. Body weight loss reverts obesity-associated hypogonadotropic hypogonadism: a systematic review and meta-analysis. Eur J Endocrinol 2013;168:829–43.

[14] Corona G, Rastrelli G, Monami M, Melani C, Balzi D, Sforza A, et al. Body mass index regulates hypogonadism-associated CV risk: results from a cohort of subjects with erectile dysfunction. J Sex Med 2011;8:2098–105.

[15] Corona G, Monami M, Rastrelli G, Aversa A, Sforza A, Lenzi A, et al. Type 2 diabetes mellitus and testosterone: a meta-analysis study. Int J Androl 2011;34:528–40.

[16] Corona G, Monami M, Rastrelli G, Aversa A, Tishova Y, Saad F, et al. Testosterone and metabolic syndrome: a meta-analysis study. J Sex Med 2011;8:272–83.

[17] Corona G, Forti G, Maggi M. Why can patients with erectile dysfunction be considered lucky? The association with testosterone deficiency and metabolic syndrome. Aging Male 2008;11:193–9.

[18] Corona G, Mannucci E, Forti G, Maggi M. Following the common association between testosterone deficiency and diabetes mellitus, can testosterone be regarded as a new therapy for diabetes? Int J Androl 2009;32:431–41.

[19] Facchiano E, Scaringi S, Veltri M, Samavat J, Maggi M, Forti G, et al. Age as a predictive factor of testosterone improvement in male patients after bariatric surgery: preliminary results of a monocentric prospective study. Obes Surg 2013;23:167–72.

[20] Luconi M, Samavat J, Seghieri G, Iannuzzi G, Lucchese M, Rotella C, et al. Determinants of testosterone recovery after bariatric surgery: is it only a matter of reduction of body mass index? Fertil Steril 2013;99:1872–9.

[21] Hammoud A, Gibson M, Hunt SC, Adams TD, Carrell DT, Kolotkin RL, et al. Effect of Roux-en-Y gastric bypass surgery on the sex steroids and quality of life in obese men. J Clin Endocrinol Metab 2009;94:1329–32.

[22] Pellitero S, Olaizola I, Alastrue A, Martínez E, Granada ML, Balibrea JM, et al. Hypogonadotropic hypogonadism in morbidly obese males is reversed after bariatric surgery. Obes Surg 2012;22:1835–42.

[23] Morelli A, Sarchielli E, Comeglio P, Filippi S, Vignozzi L, Marini M, et al. Metabolic syndrome induces inflammation and impairs gonadotropin-releasing hormone neurons in the preoptic area of the hypothalamus in rabbits. Mol Cell Endocrinol 2014;382: 107–19.

[24] Dandona P, Dhindsa S. Update: hypogonadotropic hypogonadism in type 2 diabetes and obesity. J Clin Endocrinol Metab 2011;96:2643–51.

[25] Araujo AB, Dixon JM, Suarez EA, Murad MH, Guey LT, Wittert GA. Clinical review: endogenous testosterone and mortality in men: a systematic review and meta-analysis. J Clin Endocrinol Metab 2011;96:3007–19.

[26] Corona G, Rastrelli G, Forti G, Maggi M. Update in testosterone therapy for men. J Sex Med 2011;8:639–54.

[27] Corona G, Rastrelli G, Morelli A, Vignozzi L, Mannucci E, Maggi M. Hypogonadism and metabolic syndrome. J Endocrinol Invest 2011;34:557–67.

[28] Corona G, Rastrelli G, Vignozzi L, Mannucci E, Maggi M. How to recognize late-onset hypogonadism in men with sexual dysfunction. Asian J Androl 2012;14:251–9.

[29] Corona G, Rastrelli G, Vignozzi L, Maggi M. Emerging medication for the treatment of male hypogonadism. Expert Opin Emerg Drugs 2012;17:239–59.

[30] Giagulli VA, Kaufman JM, Vermeulen A. Pathogenesis of the decreased androgen levels in obese men. J Clin Endocrinol Metab 1994;79:997–1000.

[31] Vermeulen A, Kaufman JM, Deslypere JP, Thomas G. Attenuated luteinizing hormone (LH) pulse amplitude but normal LH pulse frequency, and its relation to plasma androgens in hypogonadism of obese men. J Clin Endocrinol Metab 1993;76:1140–6.

[32] Isidori AM, Caprio M, Strollo F, Moretti C, Frajese G, Isidori A, et al. Leptin and androgens in male obesity: evidence for leptin contribution to reduced androgen levels. J Clin Endocrinol Metab 1999;84:3673–80.

[33] Hausman GJ, Barb CR. Adipose tissue and the reproductive axis: biological aspects. Endocr Dev 2010;19:31–44.

[34] George JT, Millar RP, Anderson RA. Hypothesis: kisspeptin mediates male hypogonadism in obesity and type 2 diabetes. Neuroendocrinology 2010;91:302–7.

[35] Bartke A, Matt KS, Steger RW, Clayton RN, Chandrashekar V, Smith MS. Role of prolactin in the regulation of sensitivity of the hypothalamicpituitary system to steroid feedback. Adv Exp Med Biol 1987;219:153–75.

[36] Buvat J, Maggi M, Guay A, Torres LO. Testosterone deficiency in men: systematic review and standard operating procedures for diagnosis and treatment. J Sex Med 2013 Jan;10(1):245–84.

[37] Gooren L, Spinder T, Spijkstra JJ, van Kessel H, Smals A, Rao BR, et al. Sex steroids and pulsatile luteinizing hormone release in men. Studies in estrogen-treated agonadal subjects and eugonadal subjects treated with a novel nonsteroidal antiandrogen. J Clin Endocrinol Metab 1987;64:763–70.

[38] Morelli A, Comeglio P, Sarchielli E, Cellai I, Vignozzi L, Vannelli GB, et al. Negative effects of high glucose exposure in human gonadotropin-releasing hormone neurons. Int J Endocrinol 2013;2013. 684659.

[39] Stellato RK, Feldman HA, Hamdy O, Horton ES, McKinlay JB. Testosterone, sex hormone-binding globulin, and the development of type 2 diabetes in middle-aged men: prospective results from the Massachusetts male aging study. Diabetes Care 2000;23:490–4.

[40] Teerds KJ, de Rooij DG, Keijer J. Functional relationship between obesity and male reproduction: from humans to animal models. Hum Reprod Update 2011;17:667–83.

[41] Reis LO, Dias FG. Male fertility, obesity, and bariatric surgery. Reprod Sci 2012;19:778–85.

[42] World Health Organization. WHO Laboratory Manual for the Examination and Processing of Human Semen. 5th ed Switzerland: WHO Press; 2010.

[43] Pasquali R, Patton L, Gambineri A. Obesity and infertility. Curr Opin Endocrinol Diabetes Obes 2007;14:482–7.

[44] Nguyen RH, Wilcox AJ, Skjaerven R, Baird DD. Men's body mass index and infertility. Hum Reprod 2007;22:2488–93.

[45] Hammoud AO, Gibson M, Peterson CM, Meikle AW, Carrell DT. Impact of male obesity on infertility: a critical review of the current literature. Fertil Steril 2008;90:897–904.

[46] Magnusdottir EV1, Thorsteinsson T, Thorsteinsdottir S, Heimisdottir M, Olafsdottir K. Persistent organochlorines, sedentary occupation, obesity and human male subfertility. Hum Reprod 2005;20:208–15.

[47] Pauli EM1, Legro RS, Demers LM, Kunselman AR, Dodson WC, Lee PA. Diminished paternity and gonadal function with increasing obesity in men. Fertil Steril 2008;90:346–51.

[48] Bakos HW, Henshaw RC, Mitchell M, Lane M. Paternal body mass index is associated with decreased blastocyst development and reduced live birth rates following assisted reproductive technology. Fertil Steril 2011;95:1700–4.

[49] Colaci DS, Afeiche M, Gaskins AJ, Wright DL, Toth TL, Tanrikut C, et al. Men's body mass index in relation to embryo quality and clinical outcomes in couples undergoing *in vitro* fertilization. Fertil Steril 2012;98:1193–9.

[50] Bonde JP, Ernst E, Jensen TK, Hjollund NH, Kolstad H, Henriksen TB, et al. Relation between semen quality and fertility: a population-based study of 430 first-pregnancy planners. Lancet 1998;352:1172–7.

[51] MacDonald AA, Herbison GP, Showell M, Farquhar CM. The impact of body mass index on semen parameters and reproductive hormones in human males: a systematic review with meta-analysis. Hum Reprod Update 2010;16:293–311.

[52] Sermondade N, Faure C, Fezeu L, Shayeb AG, Bonde JP, Jensen TK, et al. BMI in relation to sperm count: an updated systematic review and collaborative meta-analysis. Hum Reprod Update 2013;19:221–31.

[53] Lotti F, Corona G, Colpi GM, Filimberti E, Degli Innocenti S, Mancini M, et al. Elevated body mass index correlates with higher seminal plasma interleukin 8 levels and ultrasonographic abnormalities of the prostate in men attending an andrology clinic for infertility. J Endocrinol Invest 2011;34:e336–42.

[54] Lotti F, Corona G, Degli Innocenti S, Filimberti E, Scognamiglio V, Vignozzi L, et al. Seminal, ultrasound and psychobiological parameters correlate with metabolic syndrome in male members of infertile couples. Andrology 2013;1:229–39.

[55] Eisenberg ML, Kim S, Chen Z, Sundaram R, Schisterman EF, Buck Louis GM. The relationship between male BMI and waist circumference on semen quality: data from the LIFE study. Hum Reprod 2014;29:193–200.

[56] Samavat J, Natali I, Degl'Innocenti S, Filimberti E, Cantini G, Di Franco A, et al. Acrosome reaction is impaired in spermatozoa of obese males: a preliminary study. Fertil Steril 2014;102:1274–81.

[57] Cooper TG, Noonan E, von Eckardstein S, Auger J, Baker HW, Behre HM, et al. World Health Organization reference values for human semen characteristics. Hum Reprod Update 2010;16:231–45.

[58] Sermondade N, Dupont C, Faure C, Boubaya M, Cédrin-Durnerin I, Chavatte-Palmer P, et al. Body mass index is not associated with sperm-zona pellucida binding ability in subfertile males. Asian J Androl 2013;15:626–9.

[59] Baldi E, Luconi M, Krausz C, Forti G. Progesterone and spermatozoa: a long-lasting liaison comes to definition. Hum Reprod 2011;26:2933–4.

[60] Krausz C, Bonaccorsi L, Luconi M, Fuzzi B, Criscuoli L, Pellegrini S, et al. Intracellular calcium increase and acrosome reaction in response to progesterone in human spermatozoa are correlated with *in-vitro* fertilization. Hum Reprod 1995;10:120–4.

[61] Krausz C, Bonaccorsi L, Maggio P, Luconi M, Criscuoli L, Fuzzi B, et al. Two functional assays of sperm responsiveness to progesterone and their predictive values in *in-vitro* fertilization. Hum Reprod 1996;11:1661–7.

[62] Chavarro JE, Ehrlich S, Colaci DS, Wright DL, Toth TL, Petrozza JC, et al. Body mass index and short-term weight change in relation to treatment outcomes in women undergoing assisted reproduction. Fertil Steril 2012;98:109–16.

[63] Dupont C, Faure C, Sermondade N, Boubaya M, Eustache F, Clément P, et al. Obesity leads to higher risk of sperm DNA damage in infertile patients. Asian J Androl 2013;15:622–5.

[64] Fariello RM, Pariz JR, Spaine DM, Cedenho AP, Bertolla RP, Fraietta R. Association between obesity and alteration of sperm DNA integrity and mitochondrial activity. BJU Int 2012;110:863–7.

[65] Kort HI, Massey JB, Elsner CW, Mitchell-Leef D, Shapiro DB, Witt MA, et al. Impact of body mass index values on sperm quantity and quality. J Androl 2006;27:450–2.

[66] Reis LO, Favaro WJ, Barreiro GC, de Oliveira LC, Chaim EA, Fregonesi A, et al. Erectile dysfunction and hormonal imbalance in morbidly obese male is reversed after gastric bypass surgery: a prospective randomized controlled trial. Int J Androl 2010;33:736–44.

[67] Pasquali R, Casimirri F, Cantobelli S, Melchionda N, Morselli Labate AM, Fabbri R, et al. Effect of obesity and body fat distribution on sex hormones and insulin in men. Metabolism 1991;40:101–4.

[68] Corona G, Lee DM, Forti G, O'Connor DB, Maggi M, O'Neill TW, EMAS Study Group. et al. Age-related changes in general and sexual health in middle-aged and older men: results from the European Male Ageing Study (EMAS). J Sex Med 2010;7:1362–80.

[69] Corona G, Mannucci E, Fisher AD, Lotti F, Petrone L, Balercia G, et al. Low levels of androgens in men with erectile dysfunction and obesity. J Sex Med 2008;5:2454–63.

[70] Corona G, Mondaini N, Ungar A, Razzoli E, Rossi A, Fusco F. Phosphodiesterase type 5 (PDE5) inhibitors in erectile dysfunction: the proper drug for the proper patient. J Sex Med 2011;8:3418–32.

[71] Corona G, Razzoli E, Forti G, Maggi M. The use of phosphodiesterase 5 inhibitors with concomitant medications. J Endocrinol Invest 2008;31:799–808.

[72] Han TS, Tajar A, O'Neill TW, Jiang M, Bartfai G, Boonen S, et al. Impaired quality of life and sexual function in overweight and obese men: the European Male Ageing Study. Eur J Endocrinol 2011;164:1003–11.

[73] O'Leary MP, Fowler FJ, Lenderking WR, Barber B, Sagnier PP, Guess HA, et al. A brief male sexual function inventory for urology. Urology 1995;46:697–706.

[74] Wincze J, Rosen R, Carson C, Korenman S, Niederberger C, Sadovsky R, et al. Erection quality scale: initial scale development and validation. Urology 2004;64:351–6.

[75] Rosen R, Brown C, Heiman J, Leiblum S, Meston C, Shabsigh R, et al. The Female Sexual Function Index (FSFI): a multidimensional self-report instrument for the assessment of female sexual function. J Sex Marital Ther 2000;26:191–208.

[76] Isidori AM, Pozza C, Esposito K, Giugliano D, Morano S, Vignozzi L, et al. Development and validation of a 6-item version of the female sexual function index (FSFI) as a diagnostic tool for female sexual dysfunction. J Sex Med 2010;7:1139–46.

[77] Rosen R, Carson C, Korenman S, Niederberger C, Sadovsky R, McLeod L, et al. Erection quality scale: initial scale development and validation. Urology 2004;64:351–6.

[78] Rosen RC, Riley A, Wagner G, Osterloh IH, Kirkpatrick J, Mishra A. The international index of erectile function (IIEF): a multidimensional scale for assessment of erectile dysfunction. Urology 1997;49:822–30.

[79] Rosen RC, Cappelleri JC, Smith MD, Lipsky J, Peña BM. Development and evaluation of an abridged, 5-item version of the International Index of Erectile Function (IIEF-5) as a diagnostic tool for erectile dysfunction. Int J Impot Res 1999;11:319–26.

[80] Petrone L, Mannucci E, Corona G, Bartolini M, Forti G, Giommi R, et al. Structured interview on erectile dysfunction (SIEDY): a new, multidimensional instrument for quantification of pathogenetic issues on erectile dysfunction. Int J Impot Res 2003;15: 210–20.

[81] Corona G, Boddi V, Balercia G, Rastrelli G, De Vita G, Sforza A, et al. The effect of statin therapy on testosterone levels in subjects consulting for erectile dysfunction. J Sex Med 2010;7:1547–56.

[82] Boddi V, Corona G, Fisher AD, Mannucci E, Ricca V, Sforza A, et al. "It takes two to tango": the relational domain in a cohort of subjects with erectile dysfunction (ED). J Sex Med 2012;9: 3126–36.

[83] Corona G, Ricca V, Bandini E, Rastrelli G, Casale H, Jannini EA, et al. SIEDY scale 3, a new instrument to detect psychological component in subjects with erectile dysfunction. J Sex Med 2012;9: 2017–26.

[84] Corona G, Fagioli G, Mannucci E, Romeo A, Rossi M, Lotti F, et al. Penile Doppler ultrasound in patients with erectile dysfunction (ED): role of peak systolic velocity measured in the flaccid state in predicting arteriogenic ED and silent coronary artery disease. J Sex Med 2008;5:2623–34.

[85] Feldman HA, Johannes CB, Derby CA, Kleinman KP, Mohr BA, Araujo AB, et al. Erectile dysfunction and coronary risk factors: prospective results from the Massachusetts male aging study. Prev Med 2000;30:328–38.

[86] Kolotkin RL, Zunker C, Østbye T. Sexual functioning and obesity: a review. Obesity (Silver Spring) 2012;20:2325–33.

[87] Kolotkin RL, Crosby RD, Pendleton R, Strong M, Gress RE, Adams T. Health related quality of life in patients seeking gastric bypass surgery vs. nontreatment-seeking controls. Obes Surg 2003;13:371–7.

[88] White MA, O'Neil PM, Kolotkin RL, Byrne TK. Gender, race, and obesity-related quality of life at extreme levels of obesity. Obes Res 2004;12:949–55.

[89] Livingston EH, Huerta S, Arthur D, Lee S, De Shields S, Heber D. Male gender is a predictor of morbidity and age a predictor of mortality for patients undergoing gastric bypass surgery. Ann Surg 2002;236:576–82.

[90] Buchwald H, Avidor Y, Braunwald E, Jensen MD, Pories W, Fahrbach K, et al. Bariatric surgery: a systematic review and meta-analysis. JAMA 2004;292:1724–37.

[91] American Society for Bariatric Surgery. Society of American Gastrointestinal Endoscopic Surgeons. Guidelines for laparoscopic and open surgical treatment of morbid obesity. Obes Surg 2000;10:378–9.

[92] Pories WJ. Bariatric surgery: risks and rewards. J Clin Endocrinol Metab 2008;93:S89–96.

[93] Adams TD, Davidson LE, Litwin SE, Kolotkin RL, LaMonte MJ, Pendleton RC, et al. Health benefits of gastric bypass surgery after 6 years. JAMA 2012;308:1122–31.

[94] Ricci C, Gaeta M, Rausa E, Macchitella Y, Bonavina L. Early impact of bariatric surgery on type II diabetes, hypertension, and hyperlipidemia: a systematic review, meta-analysis and meta-regression on 6,587 patients. Obes Surg 2014;24:522–8.

[95] Samavat J, Facchiano E, Cantini G, Di Franco A, Alpigiano G, Poli G, et al. Osteocalcin increase after bariatric surgery predicts androgen recovery in hypogonadal obese males. Int J Obes (Lond) 2014;38:357–63.

[96] Samavat J, Facchiano E, Lucchese M, Forti G, Mannucci E, Maggi M, et al. Hypogonadism as an additional indication for bariatric surgery in male morbid obesity? Eur J Endocrinol 2014;171:555–60.

[97] Oury F, Ferron M, Huizhen W, Confavreux C, Xu L, Lacombe JS, et al. Osteocalcin regulates murine and human fertility through a pancreas-bone-testis axis. J Clin Invest 2013;123:2421–33.

[98] Sermondade N, Massin N, Boitrelle F, Pfeffer J, Eustache F, Sifer C, et al. Sperm parameters and male fertility after bariatric surgery: three case series. Reprod BioMed Online 2012;24:206–10.

[99] Doblado MA, Lewkowksi BM, Odem RR, Jungheim ES. *In vitro* fertilization after bariatric surgery. Fertil Steril 2010;94:2812–4.

[100] Tan O, Carr BR. The impact of bariatric surgery on obesity-related infertility and *in vitro* fertilization outcomes. Semin Reprod Med 2012;30:517–28.

[101] Dallal RM, Chernoff A, O'Leary MP, Smith JA, Braverman JD, Quebbemann BB. Sexual dysfunction is common in the morbidly obese male and improves after gastric bypass surgery. J Am Coll Surg 2008;207:859–64.

[102] Rosenblatt A, Faintuch J, Cecconello I. Sexual hormones and erectile function more than 6 years after bariatric surgery. Surg Obes Relat Dis 2013;9:636–40.

[103] Mora M, Aranda GB, de Hollanda A, Flores L, Puig-Domingo M, Vidal J. Weight loss is a major contributor to improved sexual function after bariatric surgery. Surg Endosc 2013;27:3197–204.

B. BARIATRIC SURGERY, OBESITY, AND FERTILITY

11

Obesity and Reproductive Dysfunction in Men and Women

Tod Fullston, BSc (Hons), PhD, Linda Wu, BPharmSc, PhD*,*
*Helena J. Teede, MBBS, PhD, FRACP**,†,*
Lisa J. Moran, BSc (Hons), BND, PhD,***

*University of Adelaide, South Australia, Australia
**Monash University, Victoria, Australia
†Diabetes and Vascular Medicine, Monash Health, Victoria, Australia

INTRODUCTION

Obesity is an increasing global health problem that is becoming an epidemic worldwide. There are 2.1 billion adults classified as overweight or obese with 671 million of these classified as obese [1]. This epidemic is heightened in developed countries whereby in an archetypal westernized society, such as the United States, overweight (body mass index (BMI) 25–29.9 kg/m^2) and obesity (BMI $\geq$ 30 kg/m^2) are increasingly prevalent and are estimated to affect 66% of adults [2]. Overweight and obesity are associated with a variety of clinical complications, including an elevated risk of metabolic conditions such as impaired glucose tolerance (IGT), type 2 diabetes mellitus (DM2), metabolic syndrome, and cardiovascular disease (CVD). They are also associated with an increased prevalence of other disorders including sleep apnea, cancers, and worsened psychological health [3]. In women, being overweight or obese is also associated with worsened reproductive health and infertility [4,5]. Specifically, in reproductive-age women, the prevalence of overweight and obesity is estimated at 30 and 22%, respectively, in Australia [6] with a 2.5× increase from 1980 to 2000 [7]. US data (2009–2010) similarly report that 31.9% of women of reproductive age (20–39 years) are obese [8]. Given the adverse effects of overweight and obesity on reproductive function, a significant proportion of women of reproductive age are at increased risk of adverse effects of adiposity including reduced fertility and pregnancy complications.

Recently, it has emerged that male obesity also impairs fertility, with a recent meta-analysis concluding that male obesity reduced sperm counts by increasing oligospermia and azoospermia [9]. This may be due to altered sex hormone concentrations (decreased sex hormone-binding globulin (SHBG)/testosterone; increased estrogen) that can be associated with male obesity, directly impairing spermatogenesis and the molecular composition and function of spermatozoa [9]. Reproductive dysfunction due to a male being overweight or obese is further evidenced by an enrichment of overweight/obese men attending fertility clinics [10], increased time to conception [11,12], decreased pregnancy rates, and increased pregnancy loss in couples undergoing assisted reproduction [10,13–17]. This appears to be partly due to reduced sperm–egg binding and impaired embryo development to the blastocyst stage, therefore resulting in reductions to both fertilization and implantation/live births due to paternal obesity at the time of assisted reproduction [10,15,16,18]. There are multiple reports of what impact being overweight or obese has on the sperm parameters that are routinely tested by primary care physicians to assess male fertility (i.e., motility, morphology, and count), with conflicting outcomes. Despite this there is a growing body of evidence that male obesity impairs sperm at the molecular level when more sensitive measures of sperm quality are deployed (e.g., increased ROS concentrations, altered chromatin structure/composition, increased DNA damage, and decreased mitochondrial function) [19–21]. Such molecular perturbations to sperm are independently associated with reduced pregnancy and increased miscarriage during assisted reproduction [22].

**Handbook of Fertility. http://dx.doi.org/10.1016/B978-0-12-800872-0.00011-1

OBESITY IMPACTS ON FEMALE FERTILITY

Basic Science Research

There is a high prevalence of obese women in the infertile population, and numerous studies have highlighted the link between obesity and female infertility. Animal studies have allowed the careful dissection of the impact of obesity on each aspect of the female reproductive system and found that obesity-induced female infertility is associated with increased risk of abnormal estrous cyclicity, ovulatory dysfunction, poor oocyte quality and embryo development, pregnancy complications, as well as diminished offspring health.

Monogenic leptin signaling defects in mouse ($Ob^{-/-}$ and $Db^{-/-}$ mouse) models are widely used as models of obesity. $Ob^{-/-}$ mouse contains a single base pair deletion in the leptin coding region and $Db^{-/-}$ mouse contains a mutation in the leptin receptor. Both mouse strains have impaired leptin signaling in the hypothalamus that leads to persistent hyperphagia and obesity, DM2, and consequent insulin resistance with hyperinsulinemia. These mice are widely used as a setting in which to study the impact of obesity combined with DM2 on fertility [25–27].

Diet-induced obese (DIO) mouse models are often utilized in obesity research as a nonleptin-deficient polygenic mouse model of obesity. Polygenic obesity models typically use C57BL/6 inbred mice, particularly C57BL/6J and C57BL/6N, which are the most widely used for DIO models because they exhibit metabolic abnormalities similar to human metabolic syndrome when fed with a high fat diet (HFD) over a prolonged period of time, that is, glucose intolerance and insulin resistance. The fat composition in the food and how many weeks an animal is kept on the diet depends on research goals (i.e., obesity rate, insulin resistance, etc.). For instance, a short-term HFD feeding leads to obesity and dyslipidemia but not hyperinsulinemia [28]. Many studies began to feed mice at about 6–8 weeks of age for 4–16 weeks. Most HFD feeds contain 22–60% total calories from fat, whereas control diet (CD) feed contains 3.8–6% total calories from fat.

Similar to humans, diet-induced obesity in female mice is known to alter metabolic profile, including excess body fat deposition, increased serum glucose, free fatty acid (FFA), triglycerides, and leptin and reduced adiponectin levels [28–31]. The hypothalamus responds to metabolic, circadian, and endocrine signals to synthesize and release gonadotropin-releasing hormone (GnRH), which stimulates the pituitary gland to synthesize and secrete gonadotropin hormones, luteinizing hormone (LH), and follicle-stimulating hormone (FSH). The normal intrinsic pulsatile secretory patterns of GnRH, FSH, and LH act synergistically in reproduction. The circulating gonadotropins act upon receptors in the ovary to regulate steroidogenesis, folliculogenesis, and gametogenesis. Alterations in the levels of these circulating molecules are thought to be directly involved in the regulation of fertility at every level of the hypothalamus-pituitary-gonadal (HPG) axis [32] (Fig. 11.1). This section will summarize the molecular mechanisms by which

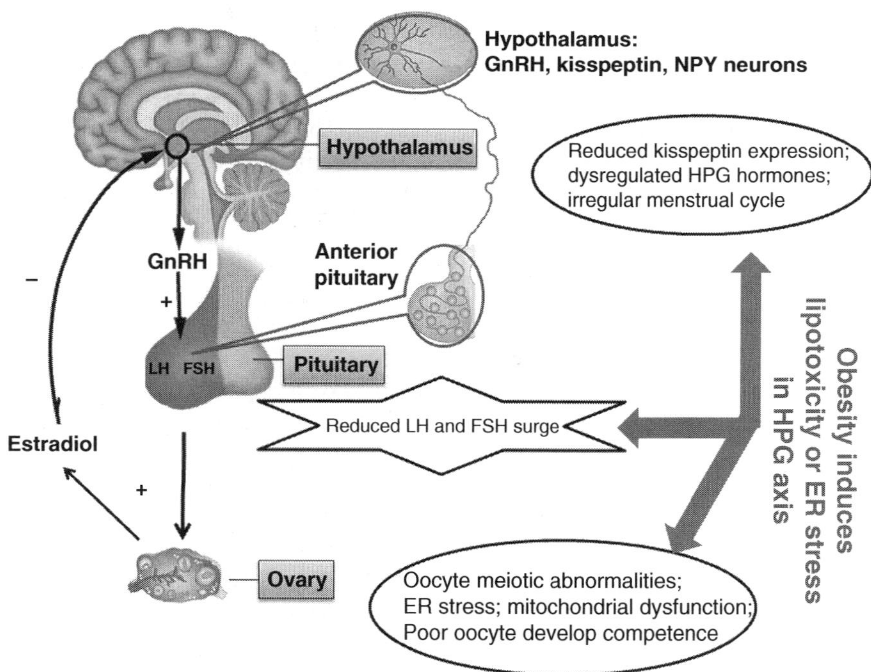

FIGURE 11.1   The effects obesity can have on the hypothalamic gonadal (HPG) axis.

obesity impacts the female HPG axis as elucidated by animal models of obesity.

Effects of Obesity on the Hypothalamus and Pituitary

Obesity is associated with increased circulating triglycerides and FFA, which are released from adipose tissue and result in lipotoxicity in nonadipose tissues. When exposed to this high lipid environment, nonadipose tissue cells accumulate excess lipids resulting in high intracellular FFA, which trigger the intercellular stress machinery, namely, mitochondrial ROS production, endoplasmic reticulum (ER) stress, and the unfolded protein response (UPR) [33,34]. Obesity-induced lipotoxicity has been demonstrated in a variety of tissues including the hypothalamus. Reversal of ER stress by using ER stress inhibitors reinstates normal protein folding in the hypothalamus and thereby improves leptin resistance and insulin sensitivity in DIO mice and in leptin receptor-deficient (*db/db*) obese mice [35,36].

HFD-induced obese female C57BL/6 mice exhibit prolonged estrus and decreased proestrus [37]. Furthermore, ovaries from obese mice at diestrus had reduced number of corpora lutea yet similar numbers of primary/secondary follicles and preantral/antral follicles, indicating a defect in natural ovulation in DIO mice. The female obese mice were found to exhibit reduced LH and FSH surge on the afternoon of proestrus; however, the pituitary GnRH response was exaggerated after stimulation with exogenous GnRH [37], which strongly suggests that dysregulation of HPG hormones is initiated at the hypothalamic level by impaired GnRH production or secretion. Kisspeptin is essential for activation of GnRH neurons, and DIO DBA/2J mice were found to have decreased kisspeptin expression in both the arcuate and rostral periventricular region of the third ventricle and reduced number of kisspeptin-positive neurons in the rostral periventricular region of the third ventricle in the hypothalamus [38]. Thus, there is strong evidence that obesity causes hypothalamic abnormalities, but whether lipotoxicity and ER stress are directly responsible for the loss of these critical reproductive protein hormones in the hypothalamus remains to be determined.

Effects of Obesity on Ovary and Oocyte Quality

Maternal obesity has been shown to be associated with decreased oocyte quality, namely, the ability of the oocyte to form a healthy embryo when fertilized. Animal models have been used to address obesity-related pathology in oocyte quality at the molecular level. Using a "cafeteria diet" to induce obesity in rodents is intended to be a diet model designed to closely reflect Western dietary habits. Female rats fed a cafeteria diet after weaning became obese and insulin-resistant in adulthood, which negatively affected female reproduction by reducing the number of oocytes and preantral follicles, as well as the thickness of the follicular layer [39].

Sixteen weeks of HFD feeding induced obesity with hyperinsulinemia and dyslipidemia in female mice, which also exhibited defects to ovarian function that resulted in poor oocyte quality, subsequently reduced blastocyst survival rates, and abnormal embryonic cellular differentiation into the inner cell mass (ICM) and trophectoderm (TE) lineages. Rosiglitazone is a peroxisome proliferator-activated receptor gamma (PPARγ) transcription factor agonist, which acts as an insulin-sensitizing agent. When treated with rosiglitazone for 4 days before mating, DIO female mice had normalized plasma triglyceride, glucose, and insulin concentrations, as well as normalizing early embryo development rates and ICM:TE ratio, similar to those of embryos from females fed a control diet [31]. This indicates that oocyte quality from obese/hyperinsulinemic females can be improved by a treatment that resensitizes these animals to insulin prior to conception [31]. A 4 week HFD regimen results in increased blood lipid, glucose but not insulin levels in association with increased lipid accumulation and induction of ER stress pathway genes (*Atf4* and *Grp78*) in the cumulus–oocyte complex, decreased mitochondrial activity in oocytes, and increased incidence of apoptosis in both cumulus and granulosa cells. The obese mice also exhibited reduced *in vivo* fertilization rates compared to mice fed control diet [28]. These responses are characteristic biomarkers of lipotoxicity and subsequent *in vitro* studies have demonstrated that reversing these lipid-induced defects in the cumulus–oocyte complex with an ER stress inhibitor improves embryo development [40].

Mitochondrial dysfunction is consistently observed in oocytes from different DIO mouse models. Live cell dynamic fluorescence imaging is widely used to detect mitochondrial status in single oocytes and there are a variety of fluorescent probes to measure specific aspects of mitochondrial function. The Mito Trackers and JC1 (5,5′,6,6′-tetrachloro-1,1′,3,3′-tetraethylbenzimidazolylcarbocyanine iodide; Invitrogen) have been extensively used to study mitochondrial dynamics and function in oocytes; however, these different detection methods often give contradictory results. For example, by JC1 staining, oocytes from mice fed an HFD (22% fat) for 4 weeks exhibited decreased mitochondrial membrane potential [28], but by using tetramethyl rhodamine methyl ester (TMRM) oocytes from obesogenic diet (20% animal lard) showed greater mitochondria activity [29]. Thus, comparisons using these staining methods should be combined with analysis of other mitochondria parameters such as localization, morphology by electron microscopy, ATP energy production, and mtDNA content. Using transmission electron microscopy to examine mitochondrial morphology in GV stage oocytes from HFD and control mice showed that mitochondria in oocytes as well as cumulus cells from HFD mouse have higher rates of abnormal mitochondrial morphology, including disarrayed cristae, decreased density of the matrix, increased swelling, and vacuolization [41]. Surprisingly, the abnormal morphology of the mitochondria did not significantly compromise oocyte metabolism, as evidenced by lower citrate levels but normal ATP production in oocytes from HFD mice. Although it must be noted that mtDNA copy number in oocytes for HFD mice were higher than lean control [29,41].

The ovulated oocytes from DIO female mice were found to have meiotic abnormalities, which manifest as significant spindle or chromosome misalignment defects [41], while fertilization rates were found to not be different from obese mothers. However, both *in vitro* embryo culture and *in vivo* embryo development show that the oocytes from obese female mice exhibited slower development throughout the entirety of preimplantation embryo development and produce fewer blastocysts. Thus, embryonic loss in obese mothers is at least in part due to the poor quality of the oocyte.

Effect of Obesity on Fetal Development and Offspring Programming

Maternal obesity affects not only preconception ovarian function and early embryonic development, but also the growth and development of offspring. Increased chemokine mRNA (*CCL2*, *CCL5*, *CCL7*, and *CxCL10*) and their related regulators (*TLR2*, *CD14*, and *Ccr1*) were found in obese rat uterus, which were associated with ectopic lipid accumulation and expression of lipid metabolic genes. The maternal obesity-caused uterine alterations may link to increased predisposition of offspring to obesity later in life [42]. Obese female mice (HFD fed for 16 weeks) had smaller fetuses on embryonic day 14.5, in association with increased placental *IGF-2R* mRNA, a maternal gene known to decrease embryonic growth and birth weight [43]. The obese mothers also delivered smaller pups; however, at 13 weeks old, the pups delivered from obese mice were significantly larger and exhibited glucose intolerance and increased cholesterol and body fat suggesting the development of an adult metabolic-type syndrome [44]. Even when the blastocysts from obese mother mice were transferred to pseudopregnant nonobese control recipients, to exclude the effect of the obese uterine environment, the pups still exhibited growth retardation and brain developmental

abnormalities indicating the defects responsible for altered postnatal growth are established prior to the blastocyst stage [41].

Sheep studies demonstrate similar results. For example, embryos from donor ewes fed a high-energy diet for 5 months were transferred to nonobese recipients, which resulted in the birth of female lambs with greater total fat mass. A relatively short period of dietary restriction (1 month) in the periconceptional period of DIO ewes caused weight loss and normalized the body fat mass and key adipogenic, lipogenic, and adipokine genes in fat depots of female offspring [45]. Another similar maternal weight loss study during the periconceptional period in the sheep found that maternal obesity during the periconceptional period and dietary restriction in either obese or normal weight mother ewes resulted in marked changes in the abundance of molecules in insulin signaling, glucose transport, and glycogen synthesis pathways in offspring skeletal muscle. Some negative effects of maternal obesity during the periconceptional period on glucose uptake and glycogen synthesis were ablated by dietary restriction, which might provide evidence for dietary restriction to be recommended for overweight or obese women who are seeking to become pregnant [46].

In summary, the mechanisms by which obesity impairs female fertility are not fully known and are likely to be systematic. Many current animal models of obesity-associated female infertility provide helpful tools to investigate the pathologies derived from different components of the HPG axis. This section highlights that obesity-induced lipotoxicity or ER stress affects the HPG axis, and these effects can have lasting impacts on the next generation. Suitable animal models and basic science studies are fundamental to testing novel hypotheses of female obesity-induced reproductive dysfunction.

Human Research

Overweight and Obesity, Early Menarche, Menstrual and Ovulatory Dysfunction and Polycystic Ovary Syndrome

Overweight and obesity contribute to reproductive abnormalities and infertility in women across the lifespan (Fig. 11.2). In early reproductive life, elevated adiposity is associated with early pubertal development and early menarche [48,49] although the causality of

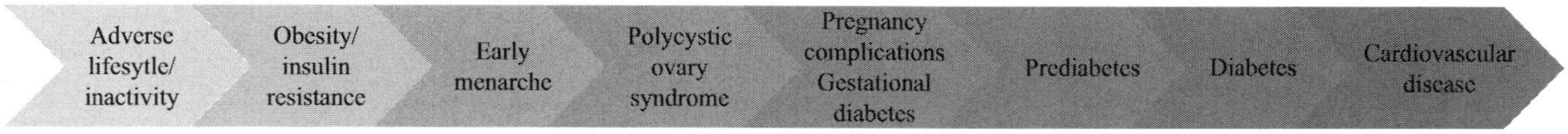

FIGURE 11.2   The continuum of potential lifestyle-related reproductive and metabolic implications across the lifespan in women (adapted from Ref. [47]).

this association is unclear. While the rate of weight gain during early childhood is a key determinant of pubertal timing in girls [50], early menarche is also associated with an elevated risk of adult obesity, DM2, and CVD [51–53].

In reproductive-aged women, overweight and obesity are associated with menstrual irregularity and ovulatory dysfunction [5,54]. The association between being overweight or obese and menstrual and ovulatory abnormalities in women is partially related to polycystic ovary syndrome (PCOS). This is a common endocrine condition that affects up to 18% of women of reproductive age [55]. PCOS is the most common cause of anovulatory infertility in women and is associated with the reproductive diagnostic features of clinical or biochemical hyperandrogenism, anovulation or menstrual disturbances, or the presence of polycystic ovaries on ultrasound [56]. In addition, PCOS is associated with worsened metabolic and psychological health, specifically elevated risk factors and prevalence of IGT, DM2 and CVD, and worsened anxiety, depression, and quality of life [57,58]. Insulin resistance underpins PCOS and is present even in lean women with PCOS in a form mechanistically distinct from obesity-associated insulin resistance [59]. Weight gain, and particularly elevated central adiposity, then further worsens insulin resistance and associated clinical features [60,61]. Insulin resistance and hyperinsulinemia contribute to hyperandrogenism in PCOS through augmenting ovarian androgen production and reducing hepatic sex-hormone binding globulin production [62]. Overweight and obesity consequently worsen the reproductive features of PCOS including ovulatory or menstrual disorders and poorer assisted reproductive technologies (ART) outcomes such as less successful ovulation induction, lower pregnancy rate, and adverse pregnancy outcomes including gestational diabetes and stillbirth [63–65].

Overweight and obesity are associated with female infertility independent of PCOS or anovulatory infertility. Overweight and obese women are less likely to conceive after 12 months of unprotected intercourse [66] and to have either anovulatory or ovulatory infertility [4,5,67,68]. Specifically, it is estimated that 25% of ovulatory infertility in the US can be attributed to being overweight [5], and the relative risk of anovulation increases from 1.3 times in women with a BMI of 24–31 kg/m^2 to 2.7 times in women with a BMI >32 kg/m^2 compared with women of normal weight [54]. An elevated risk of primary ovulatory infertility has been observed for women with an elevated body mass index (BMI) at 18 years of age either with or without PCOS [54]. Obesity is also associated with reduced conception rates and fecundability in women with regular cycles or normal ovulation [69–71]. Additional effects of obesity may include impaired oocyte development, embryo development,

and endometrial receptivity [72,73]. Overweight or obese women also have an increased risk of miscarriage [74–78]. Excess weight may also contribute to psychological features including sexual dysfunction and reduced sexual frequency [79], further impacting on reproductive success. Within ART, maternal overweight and obesity are also associated with increased duration of ovulation induction, higher clomiphene citrate and gonadotropin doses to achieve ovarian response, impaired response to ovarian stimulation, decreased follicle number and oocytes retrieved, increased cycle cancellation rate, and reduced pregnancy and live birth rates [75–77,80,81]. The adverse effects of excess adiposity also relate to the site of deposition of the adipose tissue. Women with elevated upper body fat (assessed by methods including the waist-to-hip ratio or waist circumference) have an elevated risk of menstrual irregularity, a history of infertility, and a decrease in the probability of conception with ART [82–85].

Overweight and Obesity, Pregnancy Complications and Offspring Health

Fertility relates to the ability to achieve and sustain a healthy pregnancy. There is also an association between overweight or obesity in women and obstetric complications that can affect the chance of sustaining a pregnancy and having a live birth. Overweight and obesity in the mother are associated with adverse maternal and infant outcomes including infertility, preeclampsia, gestational diabetes, induction of labor, cesarean birth, excess infant birth weight, and resuscitation at birth [86]. As a subset of infertile women, PCOS is also associated with an elevated risk of pregnancy complications including miscarriage, gestational diabetes, pregnancy-related hypertension, preeclampsia and preterm birth, and cesarean section [87], and it is of additional interest that these risks are independent of the elevated obesity commonly present in PCOS. Fertility finally relates to the ability to deliver a healthy baby. Children of women who are overweight or obese during pregnancy have an elevated prevalence of childhood and adulthood obesity [88]. There is also increasing evidence that children born to mothers who were overweight or obese prior to or during pregnancy have a worsened cardiometabolic profile ranging from elevated surrogate markers of insulin resistance, inflammation, cytokines or lipids in cord blood or early or later adulthood [89–92], elevated blood pressure in childhood, adolescence or adulthood [92–94], worsened glucose tolerance in early adulthood [91], higher metabolic syndrome prevalence in adolescence [95], or a higher risk of all-cause mortality or cardiovascular mortality and morbidity in adulthood [96,97]. This suggests there may be a programming influence of maternal obesity on offspring obesity and cardiometabolic health including altered carbohydrate and lipid metabolism.

Weight Management and Fertility

Given the association between overweight and obesity and infertility, maternal and infant pregnancy outcomes, and child health, either in PCOS, anovulatory women, or ovulatory women, weight management (prevention of excess weight gain, achieving a modest weight loss, and sustaining a reduced weight long term) is a logical treatment strategy. A multidisciplinary lifestyle management program in obese anovulatory infertile women, either with or without PCOS, achieved a modest weight loss (6.3–10.2 kg) in association with ovulation resumption, menstrual improvements, spontaneous or medical treatment associated pregnancies, and live birth in over two thirds of the women [98–100], and in association with improvements in insulin resistance and hyperandrogenism [100]. Similarly, a lifestyle program during ART in overweight women without PCOS resulted in a significant reduction in weight (3.8 ± 3.0 kg) and an increased pregnancy rate by 53% overall (20/38 participants) with the reduction in waist circumference being most strongly associated with increased odds of pregnancy (OR 1.286; $p = 0.042$) [101].

While weight management is a key treatment strategy for improving the fertility of overweight and obese women, maintaining a reduced weight long term is very difficult and weight regain is common [102]. Alternative obesity treatments include surgical treatments such as bariatric gastrointestinal surgical procedures (gastric banding, gastric bypass including Roux-en-Y, biliopancreatic diversion, and sleeve gastrectomy). These are increasingly being used particularly in combination with lifestyle strategies or after lifestyle strategies have proved unsuccessful. Bariatric surgery is associated with improvements in menstrual regularity, ovulation, and fertility [103,104], and the resolution of infertility and PCOS [105,106], which may be mechanistically related to improvements in the hormonal profile, specifically decreased insulin resistance, hyperandrogenism, and estradiol or increased LH and FSH in women with or without PCOS [106–109], luteal function [110], and sexual function [109]. Obstetric outcomes such as excess gestational weight gain, gestational diabetes, macrosomia, and hypertensive disorders of pregnancy are improved following bariatric surgery in comparison to weight-matched controls although there may be an increase in deleterious outcomes such as intrauterine growth restriction [103,104], preterm births, and small for gestational age births [111]. A number of guidelines for bariatric surgery recommend eligibility with a BMI $\geq$40 kg/m^2 or $\geq$35 kg/m^2 if present with other comorbid conditions in conjunction with prior weight management programs [112,113]. However, the 2009 American Congress of Obstetrician and Gynecologists clinical practice guidelines do not specifically recommend bariatric surgery with respect to fertility [114].

OBESITY IMPACTS ON MALE FERTILITY

Basic Science Research

Obesity Reduced Male Fecundity

Animal models have been useful in gaining information about the effects that paternal obesity has on reproductive function in a controlled environment, especially considering the many potential confounders that hinder human studies. The first evidence that associated male obesity with reproductive dysfunction is a genetic model of leptin deficiency ($Ob^{-/-}$) that results in obese male mice that are infertile [115], which can be rescued by leptin treatment [116]. Rodent studies that measure traditional sperm measures demonstrate that HFD-induced obesity reduces sperm motility, decreases sperm count, reduces the proportion of sperm with normal morphology, and increases epididymal transit time [117–123]. However, it should be noted that these sperm parameters are often reported concomitant with reduced testosterone concentrations, increased estradiol concentrations, and altered glucose homeostasis in the HFD fed groups, which independent of obesity are known to impair male reproductive health. Furthermore, sperm function, as measured by oocyte binding, is also impaired in a rodent model of HFD feeding that was directly related to reduced sperm capacitation and reduced fertilization [118].

Obesity Alters the Molecular Makeup of Sperm

Besides alterations to traditional sperm measures, animal models also demonstrate that the molecular makeup of sperm is altered by the consumption of an HFD. For example, increased sperm DNA damage has been reported in rodent models of obesity [118,122,123]. Furthermore, rodent studies that have modeled male obesity have linked increased oxidative stress (as measured by reactive oxygen species, ROS) in sperm with the obese state, collectively concluding that adiposity positively associated with increased sperm oxidative stress [118,122]. Male obesity has also been associated with alterations to the epigenetic state of mature sperm, specifically microRNA content and global methylation status. A shift in sperm microRNA content and hypomethylation are associated with a HFD-feeding model of obesity [124]. Interestingly, increased DNA damage, increased ROS, altered sperm microRNA content, and sperm DNA global hypomethylation are also associated with subfertility/infertility independent of obesity [125–128]. Animal models indicate that male obesity is associated with alterations to the molecular/epigenetic composition of sperm. This not only holds implications for sperm function but there is also a growing body of evidence from animal models that there is a significant increase to the burden of noncommunicable disease in offspring [124,129,130], which likely results from paternal programming of embryo/fetal development [131,132]. Understanding the interrelationship between ROS, DNA damage, and altered epigenetic state will be important to know how male obesity not only impairs fertility but also increases the risk of disease in offspring.

Individual HFD Components Can Impair Sperm

The effects on male reproductive function of a number of components that are elevated by the consumption of a typical HFD have been investigated individually. For example, fried/baked carbohydrate-rich food (e.g., potato chips, bread) contains elevated concentrations of acrylamide [133], which is known to cross the blood–testis barrier [134] and impair male fertility [135,136]. This reproductive dysfunction was evident by a reduction in copulation combined with a cessation of spermatogenesis [136] and reduced litter sizes combined with increased embryo resorptions and reduced offspring survival [135]. Furthermore, increased acrylamide exposure exacerbated male reproductive dysfunction caused by HFD feeding [117]. It has been reported that acrylamide is metabolized to glycidamide, which damages sperm DNA by the formation of glycidamide adducts [137]. Elevated serum cholesterol concentration due to an HFD has been correlated with reduced sperm motility, reduced oocyte binding by sperm, and increased DNA damage [24]. When elevated dietary cholesterol intake was investigated in isolation to an HFD increased sperm membrane cholesterol content impaired sperm osmotic resistance, acrosomal reaction, and capacitation, which was associated with reduced motility and increased morphological defects [138]. Additionally, increased serum triglycerides have been correlated with increased sperm head defects [24]. Presumably the lipid makeup of sperm membranes is altered by increased serum cholesterol/triglyceride concentrations, impairing sperm function.

It has been demonstrated that diet/exercise interventions can somewhat ameliorate male obesity-induced reproductive dysfunction in rodents [24], which suggests that further investigations surrounding the efficacy of interventions to restore obesity-induced reproductive dysfunction in humans are warranted.

Human Research

Impaired Reproductive Health is Associated with Male Obesity

Studies in human cohorts that aim to assess the effect that male obesity has on fertility are easily confounded by numerous external factors that may also modulate fertility. Confounding factors can include diet (calorie content, composition, micronutrient content), environmental exposure (toxicant exposure, radiation exposure, endocrine disruptors, scrotal heat, vocational chemical exposure), genetic background, illicit/prescribed drug use, smoking, age, activity levels, serum testosterone/estrogen concentration, and idiopathic subfertility. This may in part explain the lack of consensus in literature regarding the impact of male obesity on traditional World Health Organization sperm parameters (sperm concentration, motility, and morphology) [139]. Nonetheless, there is an increasing body of evidence that male obesity is associated with reproductive dysfunction, specifically altering the physical structure, molecular structure, and function of spermatogenic cells in the testes and mature sperm [140–142].

Since the 1970s the rates of obese men of typical reproductive age has more than tripled [1,143] and obese/overweight men are overrepresented in couples seeking IVF treatment [10]. The increase in male obesity has occurred concomitantly with an increased reliance on ART, especially intracytoplasmic sperm injection (ICSI) [144,145]. Couples that include an obese male partner typically experience increased time to conception [11,12], decreased pregnancy rates, and increased pregnancy loss in those undergoing ART [10,13–17]. This is in part due to reduced sperm–egg binding and impaired embryo development (to blastocyst), resulting in reductions to fertilization, implantations, and live births from ART [10,15,16,18].

Sperm oxidative stress has been correlated to male obesity in humans with a positive association reported between increasing adiposity and increased sperm oxidative stress [19]. Sperm oxidative stress independently correlates with decreased motility, increased DNA damage, decreased proportions of sperm undergoing acrosomal reaction, and reduced embryo implantation from IVF [146–148]. Furthermore, a consensus of human studies in an ART setting has determined a relationship between obesity and reduced sperm DNA integrity despite the various technologies used to measure sperm DNA integrity (i.e., TUNEL, COMET, and SCSA) [19,21,149–155]. Reduced sperm DNA integrity is also associated with male reproductive dysfunction, independent of obesity [156,157], despite insufficient evidence to date that it can be used as a diagnostic indicator of IVF outcome [158].

Potential Modulators of Male Obesity-Induced Reproductive Dysfunction

The precise mechanism that drives reproductive dysfunction that can occur as a result of male obesity remains unknown. There is evidence in human cohorts that mirror animal studies that serum cholesterol and phospholipids impair semen quality, whereby elevated serum cholesterol and phospholipid concentrations associated with an abnormal sperm head morphology and decreased proportion of sperm with intact acrosomes [159]. It must be noted that elevated serum cholesterol and phospholipid concentrations occur in the majority of obese men [159]. The potential modes of pathology include, but are not limited to, altered hormonal profiles (reduced SHBG, testosterone, FSH/LH ratios, inhibin B, and increased estrogen [14,19,150,151,153,160–165] – possibly resulting from similar insults to the male HPG axis as per Fig. 11.1), which potentially arise from increased adipose depots [166,167] or aberrant glucose homeostasis [168–172]; increased testicular temperature due to increases in scrotal adipose depots [173–179]; altered seminal plasma composition (increased fructose and adipocytokines) [155,180]; altered sperm proteomics [181,182], and erectile dysfunction [163,183]. The mechanism that underlies male obesity-induced reproductive dysfunction is likely a combination of the mentioned modes.

Weight Loss Strategies

Excitingly, diet/exercise interventions appear to be able to somewhat ameliorate male obesity-induced reproductive dysfunction in humans [23]. But it must be noted that the drastic weight loss achieved by surgical intervention (i.e., gastric banding), which does not necessarily occur in conjunction with improved eating/exercise habits, although reported to improve hormone profiles (testosterone, inhibin B, SHBG, and estrogens) and erectile dysfunction, are not yet proven to improve sperm function with limited data available [107,184–186]. Although bariatric surgical interventions are effective for weight loss and restoring the health and well-being of an individual, they are as yet under-researched for outcomes that measure male fertility. Perhaps interventions that achieve weight loss by relatively simple and noninvasive lifestyle interventions that improve eating/exercise habits might be preferable.

PRACTICAL ADVICE AND IMPLICATIONS

With the rising prevalence of obesity in the developed and developing world, the reproductive impacts of an adverse lifestyle are clearly emerging as a major public health issue. With earlier onset of excess weight and advancing maternal age, excess weight is now overlapping considerably with the reproductive years. This is impacting on reproductive health across the lifespan for women and in men appears to impact on fertility and on the health of the next generation. As outlined, increased weight exacerbates PCOS prevalence and severity, increased infertility, adversely impacts on assisted reproduction outcomes, has adverse effects on the mother and baby during pregnancy, and has long-term impacts for the children. Despite this increased burden of disease and health and economic costs, evidence from human trials on how best to manage excess weight is not yet

clear. Exercise and modest weight loss in mothers improve PCOS features including fertility, likewise in obese women. Lifestyle changes and weight loss have benefits in men. Yet in women, extreme weight loss through very low calorie diets preconception or through bariatric surgery may not have health benefits, and from animal studies and recent epidemiological data, may have adverse impacts on the health of the baby.

The evidence is clearer during pregnancy where a comprehensive systematic review of 7278 pregnant women showed prevention of excess gestational weight gain, reduced preeclampsia, and shoulder dystocia [187], with no impact on birth weight nor any safety concerns [187]. Success factors included focusing on diet or combined interventions rather than exercise alone. Behavioral strategies, such as self-monitoring and interventions that include motivational interviewing and modern technology, are also useful [187–190].

However, ultimately the optimal solution would be to take a population-wide approach to prevention of weight gain. Prevention is feasible, requiring minor energy balance adjustments (~220 kJ/day) [191], yet established obesity requires intensive, multidisciplinary, costly treatment with poor efficacy and sustainability [192]. Prevention is also preferable as health impacts may not necessarily be entirely reversible with weight loss [193]. Effective simple interventions are available to prevent weight gain, and community and population health initiatives are also effective including those with a strong focus on the prevention of childhood obesity.

Overall, current evidence would suggest that all young women especially, but also young men, should be aware of the adverse impact of excess weight on reproduction across infertility, pregnancy, and outcomes for the baby and parents longer term. A strong focus on prevention of weight gain is needed; nonextreme lifestyle interventions should be recommended for those overweight seeking optimal fertility and pregnancy health, and extreme rapid weight loss should probably be avoided pending further evidence.

CONCLUSIONS

Overall there is an emerging body of evidence that both male and female obesity can cause reproductive dysfunction and that the concentration of particular serum metabolites/hormones may indicate how severe this dysfunction might be. Furthermore, being overweight or obese is inextricably linked to PCOS, the most common cause of anovulatory infertility in women. Given the ever-increasing prevalence of obesity, this predicts a somber future for fertility worldwide and potentially threatens the health of future generations. Low to moderate intensity lifestyle interventions that improve dietary intake and physical activity in both men and women represent a potential circuit breaker that promises to not only improve overall health but also restore fertility. Unfortunately, this type of intervention has failed to mitigate the increasing prevalence of overweight or obesity in any population worldwide to date. Given the increasing contribution of overweight and obesity to the burden of disease worldwide, it is evident that interventions that reduce body weight and adiposity are needed that are effective on a population scale and that would ideally be readily accessible to developing nations.

References

[1] Ng M, Fleming T, Robinson M, Thomson B, Graetz N, Margono C, et al. Global, regional, and national prevalence of overweight and obesity in children and adults during 1980–2013: a systematic analysis for the Global Burden of Disease Study 2013. Lancet 2014;384(9945):766–81.

[2] CDCP. Prevalence of overweight and obesity among adults: United States, 2003–2004. US Department of Health and Human Services 2003; Accessed March 12, 2014 http://www.cdc.gov/nchs/data/hestat/overweight/overweight_adult_03.htm.

[3] World Health Organization. Obesity: preventing and managing the global epidemic. Geneva: World Health Organization; 1997.

[4] Grodstein F, Goldman MB, Cramer DW. Body mass index and ovulatory infertility. Epidemiology 1994;5(2):247–50.

[5] Rich-Edwards JW, Spiegelman D, Garland M, Hertzmark E, Hunter DJ, Colditz GA, et al. Physical activity, body mass index, and ovulatory disorder infertility. Epidemiology 2002;13(2):184–90.

[6] Cameron AJ, Welborn TA, Zimmet PZ, Dunstan DW, Owen N, Salmon J, et al. Overweight and obesity in Australia: the 1999–2000 Australian Diabetes. Obesity and Lifestyle Study (AusDiab). Med J Aust 2003;178(9):427–32.

[7] O'Brien K, Webbie K. Australian Institute of Health and Welfare. Health, well-being and bodyweight: characteristics of overweight and obesity in Australia, 2001. AIHW Bulletin No. 13 Canberra AIHW 2001;. Viewed December 4, 2014 http://www.aihw.gov.au/publication-detail/?id=6442467582.

[8] St-Onge MP, Roberts AL, Chen J, Kelleman M, O'Keeffe M, RoyChoudhury A, et al. Short sleep duration increases energy intakes but does not change energy expenditure in normal-weight individuals. Am J Clin Nutr 2011;94(2):410–6.

[9] Sermondade N, Dupont C, Faure C, Boubaya M, Cedrin-Durnerin I, Chavatte-Palmer P, et al. Body mass index is not associated with sperm–zona pellucida binding ability in subfertile males. Asian J Androl 2013;15(5):626–9.

[10] Bakos HW, Henshaw RC, Mitchell M, Lane M. Paternal body mass index is associated with decreased blastocyst development and reduced live birth rates following assisted reproductive technology. Fertil Steril 2011;95(5):1700–4.

[11] Nguyen RH, Wilcox AJ, Skjaerven R, Baird DD. Men's body mass index and infertility. Hum Reprod 2007;22(9):2488–93.

[12] Ramlau-Hansen CH, Thulstrup AM, Nohr EA, Bonde JP, Sorensen TI, Olsen J. Subfecundity in overweight and obese couples. Hum Reprod 2007;22(6):1634–7.

[13] Keltz J, Zapantis A, Jindal SK, Lieman HJ, Santoro N, Polotsky AJ. Overweight men: clinical pregnancy after ART is decreased in IVF but not in ICSI cycles. J Assist Reprod Genet 2010;27(9–10):539–44.

[14] Hinz S, Rais-Bahrami S, Kempkensteffen C, Weiske WH, Miller K, Magheli A. Effect of obesity on sex hormone levels, antisperm antibodies, and fertility after vasectomy reversal. Urology 2010;76(4):851–6.

[15] Colaci DS, Afeiche M, Gaskins AJ, Wright DL, Toth TL, Tanrikut C, et al. Men's body mass index in relation to embryo quality and clinical outcomes in couples undergoing in vitro fertilization. Fertil Steril 2012;98(5):1193–9. e1.

[16] Merhi ZO, Keltz J, Zapantis A, Younger J, Berger D, Lieman HJ, et al. Male adiposity impairs clinical pregnancy rate by in vitro fertilization without affecting day 3 embryo quality. Obesity (Silver Spring) 2013;21(8):1608–12.

[17] Ramasamy R, Bryson C, Reifsnyder JE, Neri Q, Palermo GD, Schlegel PN. Overweight men with nonobstructive azoospermia have worse pregnancy outcomes after microdissection testicular sperm extraction. Fertil Steril 2013;99(2):372–6.

[18] Hwang K, Walters RC, Lipshultz LI. Contemporary concepts in the evaluation and management of male infertility. Nat Rev Urol 2011;8(2):86–94.

[19] Tunc O, Bakos HW, Tremellen K. Impact of body mass index on seminal oxidative stress. Andrologia 2011;43(2):121–8.

[20] Aitken RJ, Gibb Z, Mitchell LA, Lambourne SR, Connaughton HS, De Iuliis GN. Sperm motility is lost in vitro as a consequence of mitochondrial free radical production and the generation of electrophilic aldehydes but can be significantly rescued by the presence of nucleophilic thiols. Biol Reprod 2012;87(5):110.

[21] Fariello RM, Pariz JR, Spaine DM, Cedenho AP, Bertolla RP, Fraietta R. Association between obesity and alteration of sperm DNA integrity and mitochondrial activity. BJU Int 2012;.

[22] Ribas-Maynou J, Garcia-Peiro A, Fernandez-Encinas A, Amengual MJ, Prada E, Cortes P, et al. Double stranded sperm DNA breaks, measured by Comet assay, are associated with unexplained recurrent miscarriage in couples without a female factor. PLoS One 2012;7(9):e44679.

[23] Hakonsen LB, Thulstrup AM, Aggerholm AS, Olsen J, Bonde JP, Andersen CY, et al. Does weight loss improve semen quality and reproductive hormones? Results from a cohort of severely obese men. Reprod Health 2011;8:24.

[24] Palmer NO, Bakos HW, Owens JA, Setchell BP, Lane M. Diet and exercise in an obese mouse fed a high-fat diet improve metabolic health and reverse perturbed sperm function. Am J Physiol Endocrinol Metab 2012;302(7):E768–80.

[25] Kanasaki K, Koya D. Biology of obesity: lessons from animal models of obesity. J Biomed Biotechnol 2011;. doi: 10.1155/2011/197636.

[26] de Luca C, Kowalski TJ, Zhang Y, Elmquist JK, Lee C, Kilimann MW, et al. Complete rescue of obesity, diabetes, and infertility in db/db mice by neuron-specific LEPR-B transgenes. J Clin Invest 2005;115(12):3484–93.

[27] Castracane VD, Henson MC. When did leptin become a reproductive hormone? Semin Reprod Med 2002;20(2):89–92.

[28] Wu LL, Dunning KR, Yang X, Russell DL, Lane M, Norman RJ, et al. High-fat diet causes lipotoxicity responses in cumulus–oocyte complexes and decreased fertilization rates. Endocrinology 2010;151(11):5438–45.

[29] Igosheva N, Abramov AY, Poston L, Eckert JJ, Fleming TP, Duchen MR, et al. Maternal diet-induced obesity alters mitochondrial activity and redox status in mouse oocytes and zygotes. PLoS One 2010;5(4):e10074.

[30] Tortoriello DV, McMinn JE, Chua SC. Increased expression of hypothalamic leptin receptor and adiponectin accompany resistance to dietary-induced obesity and infertility in female C57BL/6J mice. Int J Obes 2007;31(3):395–402.

[31] Minge CE, Bennett BD, Norman RJ, Robker RL. Peroxisome proliferator-activated receptor-gamma agonist rosiglitazone reverses the adverse effects of diet-induced obesity on oocyte quality. Endocrinology 2008;149(5):2646–56.

[32] Michalakis K, Mintziori G, Kaprara A, Tarlatzis BC, Goulis DG. The complex interaction between obesity, metabolic syndrome and reproductive axis: a narrative review. Metabolism 2013;62(4):457–78.

[33] van Herpen NA, Schrauwen-Hinderling VB. Lipid accumulation in non-adipose tissue and lipotoxicity. Physiol Behav 2008;94(2):231–41.

[34] Eizirik DL, Cardozo AK, Cnop M. The role for endoplasmic reticulum stress in diabetes mellitus. Endocr Rev 2008;29(1):42–61.

[35] Ozcan L, Ergin AS, Lu A, Chung J, Sarkar S, Nie D, et al. Endoplasmic reticulum stress plays a central role in development of leptin resistance. Cell Metab 2009;9(1):35–51.

[36] Kleinridders A, Lauritzen HP, Ussar S, Christensen JH, Mori MA, Bross P, et al. Leptin regulation of Hsp60 impacts hypothalamic insulin signaling. J Clin Invest 2013;123(1):4667–80.

[37] Sharma S, Morinaga H, Hwang V, Fan W, Fernandez MO, Varki N, et al. Free fatty acids induce Lhb mRNA but suppress Fshb mRNA in pituitary LbetaT2 gonadotropes and diet-induced obesity reduces FSH levels in male mice and disrupts the proestrous LH/FSH surge in female mice. Endocrinology 2013;154(6):2188–99.

[38] Quennell JH, Howell CS, Roa J, Augustine RA, Grattan DR, Anderson GM. Leptin deficiency and diet-induced obesity reduce hypothalamic kisspeptin expression in mice. Endocrinology 2011;152(4):1541–50.

[39] Sagae SC, Menezes EF, Bonfleur ML, Vanzela EC, Zacharias P, Lubaczeuski C, et al. Early onset of obesity induces reproductive deficits in female rats. Physiol Behav 2012;105(5):1104–11.

[40] Wu LL, Russell DL, Norman RJ, Robker RL. Endoplasmic reticulum (ER) stress in cumulus-oocyte complexes impairs pentraxin-3 secretion, mitochondrial membrane potential (DeltaPsi m), and embryo development. Mol Endocrinol 2012;26(4):562–73.

[41] Luzzo KM, Wang Q, Purcell SH, Chi M, Jimenez PT, Grindler N, et al. High fat diet induced developmental defects in the mouse: oocyte meiotic aneuploidy and fetal growth retardation/brain defects. PLoS One 2012;7(11):e49217.

[42] Shankar K, Zhong Y, Kang P, Lau F, Blackburn ML, Chen JR, et al. Maternal obesity promotes a proinflammatory signature in rat uterus and blastocyst. Endocrinology 2011;152(11):4158–70.

[43] Wutz A, Theussl HC, Dausman J, Jaenisch R, Barlow DP, Wagner EF. Non-imprinted *Igf2r* expression decreases growth and rescues the Tme mutation in mice. Development 2001;128(10):1881–7.

[44] Jungheim ES, Schoeller EL, Marquard KL, Louden ED, Schaffer JE, Moley KH. Diet-induced obesity model: abnormal oocytes and persistent growth abnormalities in the offspring. Endocrinology 2010;151(8):4039–46.

[45] Rattanatray L, MacLaughlin SM, Kleemann DO, Walker SK, Muhlhausler BS, McMillen IC. Impact of maternal periconceptional overnutrition on fat mass and expression of adipogenic and lipogenic genes in visceral and subcutaneous fat depots in the postnatal lamb. Endocrinology 2010;151(11):5195–205.

[46] Nicholas LM, Morrison JL, Rattanatray L, Ozanne SE, Kleemann DO, Walker SK, et al. Differential effects of exposure to maternal obesity or maternal weight loss during the periconceptional period in the sheep on insulin signalling molecules in skeletal muscle of the offspring at 4 months of age. PloS One 2013;8(12):e84594.

[47] Teede H, Zoungas S, Deeks A, Farrell E, Moran L, Vollenhoven B. Check: independent learning program for GPs. Polycystic ovary syndrome. In: Practitioners RACoG, editor. South Melbourne, Victoria: Royal Australian College of General Practitioners; 2008. comp: Please set this ref as book ref.

[48] Kaplowitz PB. Link between body fat and the timing of puberty. Pediatrics 2008;121(Suppl 3):S208–17.

[49] Freedman DS, Khan LK, Serdula MK, Dietz WH, Srinivasan SR, Berenson GS. Relation of age at menarche to race, time period, and anthropometric dimensions: the Bogalusa Heart Study. Pediatrics 2002;110(4):e43.

[50] Davison KK, Susman EJ, Birch LL. Percent body fat at age 5 predicts earlier pubertal development among girls at age 9. Pediatrics 2003;111(4 Pt 1):815–21.

[51] Ong KK, Northstone K, Wells JC, Rubin C, Ness AR, Golding J, et al. Earlier mother's age at menarche predicts rapid infancy growth and childhood obesity. PLoS Medicine 2007;4(4):e132.

[52] Lakshman R, Forouhi N, Luben R, Bingham S, Khaw K, Wareham N, et al. Association between age at menarche and risk of diabetes in adults: results from the EPIC-Norfolk cohort study. Diabetologia 2008;51(5):781–6.

[53] Lakshman R, Forouhi NG, Sharp SJ, Luben R, Bingham SA, Khaw KT, et al. Early age at menarche associated with cardiovascular disease and mortality. J Clin Endocrinol Metab 2009;94(12):4953–60.

[54] Rich-Edwards JW, Goldman MB, Willett WC, Hunter DJ, Stampfer MJ, Colditz GA, et al. Adolescent body mass index and infertility caused by ovulatory disorder. Am J Obstet Gynecol 1994;171(1):171–7.

[55] March WA, Moore VM, Willson KJ, Phillips DI, Norman RJ, Davies MJ. The prevalence of polycystic ovary syndrome in a community sample assessed under contrasting diagnostic criteria. Hum Reprod 2010;25(2):544–51.

[56] Norman RJ, Dewailly D, Legro RS, Hickey TE. Polycystic ovary syndrome. Lancet 2007;370(9588):685–97.

[57] Moran LJ, Misso ML, Wild RA, Norman RJ. Impaired glucose tolerance, type 2 diabetes and metabolic syndrome in polycystic ovary syndrome: a systematic review and meta-analysis. Hum Reprod Update 2010;16(4):347–63.

[58] Barry JA, Kuczmierczyk AR, Hardiman PJ. Anxiety and depression in polycystic ovary syndrome: a systematic review and meta-analysis. Hum Reprod 2011;26(9):2442–51.

[59] Dunaif A, Finegood DT. Beta-cell dysfunction independent of obesity and glucose intolerance in the polycystic ovary syndrome. J Clin Endocrinol Metabol 1996;81(3):942–7.

[60] Lim SS, Norman RJ, Davies MJ, Moran LJ. The effect of obesity on polycystic ovary syndrome: a systematic review and meta-analysis. Obes Rev 2013;14(2):95–109.

[61] Goverde AJ, van Koert AJ, Eijkemans MJ, Knauff EA, Westerveld HE, Fauser BC, et al. Indicators for metabolic disturbances in anovulatory women with polycystic ovary syndrome diagnosed according to the Rotterdam consensus criteria. Hum Reprod 2009;24(3):710–7.

[62] Teede H, Hutchison SK, Zoungas S. The management of insulin resistance in polycystic ovary syndrome. Trends Endocrinol Metab 2007;18(7):273–9.

[63] Balen AH, Conway GS, Kaltsas G, Techatrasak K, Manning PJ, West C, et al. Polycystic ovary syndrome: the spectrum of the disorder in 1741 patients. Hum Reprod 1995;10(8):2107–11.

[64] Kiddy DS, Sharp PS, White DM, Scanlon MF, Mason HD, Bray CS, et al. Differences in clinical and endocrine features between obese and non-obese subjects with polycystic ovary syndrome: an analysis of 263 consecutive cases. Clin Endocrinol 1990;32(2):213–20.

[65] Al-Azemi M, Omu FE, Omu AE. The effect of obesity on the outcome of infertility management in women with polycystic ovary syndrome. Arch Gynecol Obstet 2004;270(4):205–10.

[66] Lake JK, Power C, Cole TJ. Women's reproductive health: the role of body mass index in early and adult life. Int J Obes Relat Metab Disord 1997;21(6):432–8.

[67] Green BB, Weiss NS, Daling JR. Risk of ovulatory infertility in relation to body weight. Fertil Steril 1988;50(5):721–6.

[68] Hartz AJ, Barboriak PN, Wong A, Katayama KP, Rimm AA. The association of obesity with infertility and related menstrual abnormalities in women. Int J Obes 1979;3(1):57–73.

[69] Jensen TK, Scheike T, Keiding N, Schaumburg I, Grandjean P. Fecundability in relation to body mass and menstrual cycle patterns. Epidemiology 1999;10(4):422–8.

[70] van der Steeg JW, Steures P, Eijkemans MJ, Habbema JD, Hompes PG, Burggraaff JM, et al. Obesity affects spontaneous pregnancy chances in subfertile, ovulatory women. Hum Reprod 2008;23(2):324–8.

[71] Gesink Law DC, Maclehose RF, Longnecker MP. Obesity and time to pregnancy. Hum Reprod 2007;22(2):414–20.

[72] Brewer CJ, Balen AH. The adverse effects of obesity on conception and implantation. Reproduction 2010;140(3):347–64.

[73] Robker RL. Evidence that obesity alters the quality of oocytes and embryos. Pathophysiology 2008;15(2):115–21.

[74] DeZee KJ, Jackson JL, Hatzigeorgiou C, Kristo D. The Epworth sleepiness scale: relationship to sleep and mental disorders in a sleep clinic. Sleep Med 2006;7(4):327–32.

[75] Wang JX, Davies M, Norman RJ. Body mass and probability of pregnancy during assisted reproduction treatment: retrospective study. BMJ 2000;321(7272):1320–1.

[76] Fedorcsak P, Dale PO, Storeng R, Ertzeid G, Bjercke S, Oldereid N, et al. Impact of overweight and underweight on assisted reproduction treatment. Hum Reprod 2004;19(11):2523–8.

[77] Asghari A, Mohammadi F, Kamrava SK, Tavakoli S, Farhadi M. Severity of depression and anxiety in obstructive sleep apnea syndrome. Eur Arch Otorhinolaryngol 2012;269(12):2549–53.

[78] Boots C, Stephenson MD. Does obesity increase the risk of miscarriage in spontaneous conception: a systematic review. Semin Reprod Med 2011;29(6):507–13.

[79] Sarwer DB, Lavery M, Spitzer JC. A review of the relationships between extreme obesity, quality of life, and sexual function. Obes Surg 2012;22:668–76.

[80] Tamer Erel C, Senturk LM. The impact of body mass index on assisted reproduction. Curr Opin Obstet Gynecol 2009;21(3):228–35.

[81] Dodson WC, Kunselman AR, Legro RS. Association of obesity with treatment outcomes in ovulatory infertile women undergoing superovulation and intrauterine insemination. Fertil Steril 2006;86(3):642–6.

[82] Hartz AJ, Rupley DC, Rimm AA. The association of girth measurements with disease in 32,856 women. Am J Epidemiol 1984;119(1):71–80.

[83] Kaye SA, Folsom AR, Prineas RJ, Potter JD, Gapstur SM. The association of body fat distribution with lifestyle and reproductive factors in a population study of postmenopausal women. Int J Obes 1990;14(7):583–91.

[84] Zaadstra BM, Seidell JC, Van Noord PA, te Velde ER, Habbema JD, Vrieswijk B, et al. Fat and female fecundity: prospective study of effect of body fat distribution on conception rates. BMJ 1993;306(6876):484–7.

[85] Wass P, Waldenstrom U, Rossner S, Hellberg D. An android body fat distribution in females impairs the pregnancy rate of *in-vitro* fertilization-embryo transfer. Hum Reprod 1997;12(9):2057–60.

[86] Dodd JM, Grivell RM, Ngyuen MA, Chan A, Robinson JS. Maternal and perinatal health outcomes by maternal body mass index category. ANZJOG 2011;51(2):136–40.

[87] Qin JZ, Pang LH, Li MJ, Fan XJ, Huang RD, Chen HY. Obstetric complications in women with polycystic ovary syndrome: a systematic review and meta-analysis. Reprod Biol Endocrinol 2013;11:56.

[88] Schack-Nielsen L, Michaelsen KF, Gamborg M, Mortensen EL, Sorensen TI. Gestational weight gain in relation to offspring body mass index and obesity from infancy through adulthood. Int J Obes 2010;34(1):67–74.

[89] Catalano PM, Farrell K, Thomas A, Huston-Presley L, Mencin P, de Mouzon SH, et al. Perinatal risk factors for childhood obesity and metabolic dysregulation. Am J Clin Nutr 2009;90(5):1303–13.

[90] Group HSCR. Hyperglycaemia and Adverse Pregnancy Outcome (HAPO) Study: associations with maternal body mass index. BJOG 2010;117(5):575–84.

[91] Mingrone G, Manco M, Mora ME, Guidone C, Iaconelli A, Gniuli D, et al. Influence of maternal obesity on insulin sensitivity and secretion in offspring. Diabetes Care 2008;31(9):1872–6.

[92] Hochner H, Friedlander Y, Calderon-Margalit R, Meiner V, Sagy Y, Avgil-Tsadok M, et al. Associations of maternal prepregnancy body mass index and gestational weight gain with adult offspring cardiometabolic risk factors: the Jerusalem Perinatal Family Follow-up Study. Circulation 2012;125(11):1381–9.

[93] Laor A, Stevenson DK, Shemer J, Gale R, Seidman DS. Size at birth, maternal nutritional status in pregnancy, and blood pressure at age 17: population based analysis. BMJ 1997;315(7106):449–53.

[94] Lawlor DA, Najman JM, Sterne J, Williams GM, Ebrahim S, Davey Smith G. Associations of parental, birth, and early life characteristics with systolic blood pressure at 5 years of age: findings from the Mater-University study of pregnancy and its outcomes. Circulation 2004;110(16):2417–23.

[95] Boney CM, Verma A, Tucker R, Vohr BR. Metabolic syndrome in childhood: association with birth weight, maternal obesity, and gestational diabetes mellitus. Pediatrics 2005;115(3):e290–6.

[96] Forsen T, Eriksson JG, Tuomilehto J, Teramo K, Osmond C, Barker DJ. Mother's weight in pregnancy and coronary heart disease in a cohort of Finnish men: follow up study. BMJ 1997;315(7112):837–40.

[97] Reynolds RM, Allan KM, Raja EA, Bhattacharya S, McNeill G, Hannaford PC, et al. Maternal obesity during pregnancy and premature mortality from cardiovascular event in adult offspring: follow-up of 1 323 275 person years. BMJ 2013;347:f4539.

[98] Clark AM, Ledger W, Galletly C, Tomlinson L, Blaney F, Wang X, et al. Weight loss results in significant improvement in pregnancy and ovulation rates in anovulatory obese women. Hum Reprod 1995;10(10):2705–12.

[99] Clark AM, Thornley B, Tomlinson L, Galletley C, Norman RJ. Weight loss in obese infertile women results in improvement in reproductive outcome for all forms of fertility treatment. Hum Reprod 1998;13(6):1502–5.

[100] Hollmann M, Runnebaum B, Gerhard I. Effects of weight loss on the hormonal profile in obese, infertile women. Hum Reprod 1996;11(9):1884–91.

[101] Moran L, Tsagareli V, Norman R, Noakes M. Diet and IVF pilot study: short-term weight loss improves pregnancy rates in overweight/obese women undertaking IVF. Aust N Z J Obstet Gynaecol 2011;51(5):455–9.

[102] Ayyad C, Andersen T. Long-term efficacy of dietary treatment of obesity: a systematic review of studies published between 1931 and 1999. Obes Rev 2000;1(2):113–9.

[103] Maggard MA, Yermilov I, Li Z, Maglione M, Newberry S, Suttorp M, et al. Pregnancy and fertility following bariatric surgery: a systematic review. JAMA 2008;300(19):2286–96.

[104] Guelinckx I, Devlieger R, Vansant G. Reproductive outcome after bariatric surgery: a critical review. Hum Reprod Update 2009;15(2):189–201.

[105] Neto RM, Herbella FA, Tauil RM, Silva FS, de Lima SE Jr. Comorbidities remission after Roux-en-Y gastric bypass for morbid obesity is sustained in a long-term follow-up and correlates with weight regain. Obes Surg 2012;22(10):1580–5.

[106] Escobar-Morreale HF, Botella-Carretero JI, Alvarez-Blasco F, Sancho J, San Millan JL. The polycystic ovary syndrome associated with morbid obesity may resolve after weight loss induced by bariatric surgery. J Clin Endocrinol Metabol 2005;90(12):6364–9.

[107] Bastounis EA, Karayiannakis AJ, Syrigos K, Zbar A, Makri GG, Alexiou D. Sex hormone changes in morbidly obese patients after vertical banded gastroplasty. Eur Surg Res 1998;30(1):43–7.

[108] Gerrits EG, Ceulemans R, van Hee R, Hendrickx L, Totte E. Contraceptive treatment after biliopancreatic diversion needs consensus. Obes Surg 2003;13(3):378–82.

[109] Lopez-Garcia E, Faubel R, Leon-Munoz L, Zuluaga MC, Banegas JR, Rodriguez-Artalejo F. Sleep duration, general and abdominal obesity, and weight change among the older adult population of Spain. Am J Clin Nutr 2008;87(2):310–6.

[110] Rochester D, Jain A, Polotsky AJ, Polotsky H, Gibbs K, Isaac B, et al. Partial recovery of luteal function after bariatric surgery in obese women. Fertil Steril 2009;92(4):1410–5.

[111] Roos N, Neovius M, Cnattingius S, Trolle Lagerros Y, Saaf M, Granath F, et al. Perinatal outcomes after bariatric surgery: nationwide population based matched cohort study. BMJ 2013;347:f6460.

[112] Dixon JB, le Roux CW, Rubino F, Zimmet P. Bariatric surgery for type 2 diabetes. Lancet 2012;379(9833):2300–11.

[113] SIGN. Management of obesity: a national clinical guideline. Edinburgh, Scotland: Scottish Intercollegiate Guidelines Network (SIGN); 2010.

[114] ACOG. ACOG Practice Bulletin No. 105: bariatric surgery and pregnancy. Obstet Gynecol 2009;113:1405–13.

[115] Chehab FF, Mounzih K, Lu R, Lim ME. Early onset of reproductive function in normal female mice treated with leptin. Science 1997;275(5296):88–90.

[116] Mounzih K, Lu R, Chehab FF. Leptin treatment rescues the sterility of genetically obese *ob/ob* males. Endocrinology 1997;138(3):1190–3.

[117] Ghanayem BI, Bai R, Kissling GE, Travlos G, Hoffler U. Diet-induced obesity in male mice is associated with reduced fertility and potentiation of acrylamide-induced reproductive toxicity. Biol Reprod 2010;82(1):96–104.

[118] Bakos HW, Mitchell M, Setchell BP, Lane M. The effect of paternal diet-induced obesity on sperm function and fertilization in a mouse model. Int J Androl 2011;34(5 Pt 1):402–10.

[119] Fernandez CD, Bellentani FF, Fernandes GS, Perobelli JE, Favareto AP, Nascimento AF, et al. Diet-induced obesity in rats leads to a decrease in sperm motility. Reprod Biol Endocrinol 2011;9(1):32.

[120] Palmer NO, Fullston T, Mitchell M, Setchell BP, Lane M. SIRT6 in mouse spermatogenesis is modulated by diet-induced obesity. Reprod Fertil Dev 2011;23(7):929–39.

[121] Fernandes GS, Arena AC, Campos KE, Volpato GT, Anselmo-Franci JA, Damasceno DC, et al. Glutamate-induced obesity leads to decreased sperm reserves and acceleration of transit time in the epididymis of adult male rats. Reprod Biol Endocrinol 2012;10:105.

[122] Chen XL, Gong LZ, Xu JX. Antioxidative activity and protective effect of probiotics against high-fat diet-induced sperm damage in rats. Animal 2013;7(2):287–92.

[123] Vendramini V, Cedenho AP, Miraglia SM, Spaine DM. Reproductive function of the male obese Zucker rats: alteration in sperm production and sperm DNA damage. Reprod Sci 2013;21(2):221–9.

[124] Fullston T, Ohlsson Teague EM, Palmer NO, DeBlasio MJ, Mitchell M, Corbett M, et al. Paternal obesity initiates metabolic disturbances in two generations of mice with incomplete penetrance to the F2 generation and alters the transcriptional profile of testis and sperm microRNA content. FASEB J 2013;27(10):4226–43.

[125] Koppers AJ, De Iuliis GN, Finnie JM, McLaughlin EA, Aitken RJ. Significance of mitochondrial reactive oxygen species in the generation of oxidative stress in spermatozoa. J Clin Endocrinol Metab 2008;93(8):3199–207.

[126] Tunc O, Tremellen K, Oxidative DNA. damage impairs global sperm DNA methylation in infertile men. J Assist Reprod Genet 2009;26(9-10):537–44.

[127] Abu-Halima M, Hammadeh M, Schmitt J, Leidinger P, Keller A, Meese E, et al. Altered microRNA expression profiles of human spermatozoa in patients with different spermatogenic impairments. Fertil Steril 2013;99(5):1249–55. e16.

[128] Aitken RJ, Smith TB, Jobling MS, Baker MA, De Iuliis GN. Oxidative stress and male reproductive health. Asian J Androl 2014;16(1):31–8.

[129] Ng SF, Lin RC, Laybutt DR, Barres R, Owens JA, Morris MJ. Chronic high-fat diet in fathers programs beta-cell dysfunction in female rat offspring. Nature 2010;467(7318):963–6.

[130] Fullston T, Palmer NO, Owens JA, Mitchell M, Bakos HW, Lane M. Diet-induced paternal obesity in the absence of diabetes diminishes the reproductive health of two subsequent generations of mice. Hum Reprod 2012;27(5):1391–400.

[131] Mitchell M, Bakos HW, Lane M. Paternal diet-induced obesity impairs embryo development and implantation in the mouse. Fertil Steril 2011;95(4):1349–53.

[132] McPherson NOB, H W, Setchell BP, Owens JA, Lane M. Improving metabolic health in obese male mice via diet and exercise restores embryo development and fetal growth. PLoS One 2013;8(8):e71459.

[133] Tareke E, Rydberg P, Karlsson P, Eriksson S, Tornqvist M. Analysis of acrylamide, a carcinogen formed in heated foodstuffs. J Agric Food Chem 2002;50(17):4998–5006.

[134] Marlowe C, Clark MJ, Mast RW, Friedman MA, Waddell WJ. The distribution of [14C]acrylamide in male and pregnant Swiss–Webster mice studied by whole-body autoradiography. Toxicol Appl Pharmacol 1986;86(3):457–65.

[135] Sakamoto J, Hashimoto K. Reproductive toxicity of acrylamide and related compounds in mice – effects on fertility and sperm morphology. Arch Toxicol 1986;59(4):201–5.

[136] Tyl RW, Friedman MA. Effects of acrylamide on rodent reproductive performance. Reprod Toxicol 2003;17(1):1–13.

[137] Nixon BJ, Katen AL, Stanger SJ, Schjenken JE, Nixon B, Roman SD. Mouse spermatocytes express *CYP2E1* and respond to acrylamide exposure. PLoS One 2014;9(5):e94904.

[138] Saez Lancellotti TE, Boarelli PV, Romero AA, Funes AK, Cid-Barria M, Cabrillana ME, et al. Semen quality and sperm function loss by hypercholesterolemic diet was recovered by addition of olive oil to diet in rabbit. PLoS One 2013;8(1):e52386.

[139] Palmer NO, Bakos HW, Fullston T, Lane M. Impact of obesity on male fertility, sperm function and molecular composition. Spermatogenesis 2012;2(4):253–63.

[140] Du Plessis SS, Cabler S, McAlister DA, Sabanegh E, Agarwal A. The effect of obesity on sperm disorders and male infertility. Nat Rev Urol 2010;7(3):153–61.

[141] MacDonald AA, Herbison GP, Showell M, Farquhar CM. The impact of body mass index on semen parameters and reproductive hormones in human males: a systematic review with meta-analysis. Hum Reprod Update 2010;16(3):293–311.

[142] Teerds KJ, de Rooij DG, Keijer J. Functional relationship between obesity and male reproduction: from humans to animal models. Hum Reprod Update 2011;17(5):667–83.

[143] Dixon T, Waters AM. A growing problem. Trends and patterns in overweight and obesity among adults in Australia 1980 to 2001; 2003. AIHW bulletin no. 8, AUS 36; 19 pp.

[144] Wang YA, Dean J, Badgery-Parker T, Sullivan EA. Assisted reproduction technology in Australia and New Zealand 2006. Assisted Reproduction Technology series 2008;12(Cat. No. PER 43). Canbera AIHW.

[145] Sunderam S, Chang J, Flowers L, Kulkarni A, Sentelle G, Jeng G, et al. Assisted reproductive technology surveillance – United States, 2006. MMWR Surveill Summ 2009;58(5):1–25.

[146] Zorn B, Vidmar G, Meden-Vrtovec H. Seminal reactive oxygen species as predictors of fertilization, embryo quality and pregnancy rates after conventional *in vitro* fertilization and intracytoplasmic sperm injection. Int J Androl 2003;26(5):279–85.

[147] Aziz N, Saleh RA, Sharma RK, Lewis-Jones I, Esfandiari N, Thomas AJ Jr, et al. Novel association between sperm reactive oxygen species production, sperm morphological defects, and the sperm deformity index. Fertil Steril 2004;81(2):349–54.

[148] Aitken RJ, Baker MA. Oxidative stress, sperm survival and fertility control. Mol Cell Endocrinol 2006;250(1-2):66–9.

[149] Kort HI, Massey JB, Elsner CW, Mitchell-Leef D, Shapiro DB, Witt MA, et al. Impact of body mass index values on sperm quantity and quality. J Androl 2006;27(3):450–2.

[150] Chavarro JE, Toth TL, Wright DL, Meeker JD, Hauser R. Body mass index in relation to semen quality, sperm DNA integrity, and serum reproductive hormone levels among men attending an infertility clinic. Fertil Steril 2010;93(7):2222–31.

[151] Paasch U, Grunewald S, Kratzsch J, Glander HJ. Obesity and age affect male fertility potential. Fertil Steril 2010;94(7):2898–901.

[152] Rybar R, Kopecka V, Prinosilova P, Markova P, Rubes J. Male obesity and age in relationship to semen parameters and sperm chromatin integrity. Andrologia 2011;43(4):286–91.

[153] La Vignera S, Condorelli RA, Vicari E, Calogero AE. Negative effect of increased body weight on sperm conventional and nonconventional flow cytometric sperm parameters. J Androl 2012;33(1):53–8.

[154] Dupont C, Faure C, Sermondade N, Boubaya M, Eustache F, Clement P, et al. Obesity leads to higher risk of sperm DNA damage in infertile patients. Asian J Androl 2013;15(5):622–5.

[155] Thomas S, Kratzsch D, Schaab M, Scholz M, Grunewald S, Thiery J, et al. Seminal plasma adipokine levels are correlated with functional characteristics of spermatozoa. Fertil Steril 2013;99(5):1256–63. e3.

[156] Aitken RJ, De Iuliis GN. On the possible origins of DNA damage in human spermatozoa. Mol Hum Reprod 2010;16(1):3–13.

[157] Wright C, Milne S, Leeson H, Sperm DNA. damage caused by oxidative stress: modifiable clinical, lifestyle and nutritional factors in male infertility. Reprod BioMed Online 2014;28(6):684–703.

[158] Pfeifer S, Goldberg J, McClure RD, R L, Thomas M, Widra E, et al. Diagnostic evaluation of the infertile male: a committee opinion. Fertil Steril 2012;98(2):294–301.

[159] Schisterman EF, Mumford SL, Chen Z, Browne RW, Boyd Barr D, Kim S, et al. Lipid concentrations and semen quality: the LIFE study. Andrology 2014;2(3):408–15.

[160] Jensen TK, Andersson AM, Jorgensen N, Andersen AG, Carlsen E, Petersen JH, et al. Body mass index in relation to semen quality and reproductive hormones among 1,558 Danish men. Fertil Steril 2004;82(4):863–70.

[161] Fejes I, Koloszar S, Zavaczki Z, Daru J, Szollosi J, Pal A. Effect of body weight on testosterone/estradiol ratio in oligozoospermic patients. Arch Androl 2006;52(2):97–102.

[162] Aggerholm AS, Thulstrup AM, Toft G, Ramlau-Hansen CH, Bonde JP. Is overweight a risk factor for reduced semen quality and altered serum sex hormone profile? Fertil Steril 2008;90(3):619–26.

[163] Pauli EM, Legro RS, Demers LM, Kunselman AR, Dodson WC, Lee PA. Diminished paternity and gonadal function with increasing obesity in men. Fertil Steril 2008;90(2):346–51.

[164] Hofny ER, Ali ME, Abdel-Hafez HZ, Kamal Eel D, Mohamed EE, Abd El-Azeem HG, et al. Semen parameters and hormonal profile in obese fertile and infertile males. Fertil Steril 2010;94(2):581–4.

[165] Ramlau-Hansen CH, Hansen M, Jensen CR, Olsen J, Bonde JP, Thulstrup AM. Semen quality and reproductive hormones according to birthweight and body mass index in childhood and adult life: two decades of follow-up. Fertil Steril 2010;94(2):610–8.

[166] Cohen PG. The hypogonadal-obesity cycle: role of aromatase in modulating the testosterone-estradiol shunt –a major factor in the genesis of morbid obesity. Med Hypotheses 1999;52(1):49–51.

[167] Meinhardt U, Mullis PE. The essential role of the aromatase/p450arom. Semin Reprod Med 2002;20(3):277–84.

[168] Clarke IJ, Horton RJ, Doughton BW. Investigation of the mechanism by which insulin-induced hypoglycemia decreases luteinizing hormone secretion in ovariectomized ewes. Endocrinology 1990;127(3):1470–6.

[169] Laing I, Olukoga AO, Gordon C, Boulton AJ. Serum sex-hormone-binding globulin is related to hepatic and peripheral insulin sensitivity but not to beta-cell function in men and women with Type 2 diabetes mellitus. Diabet Med 1998;15(6):473–9.

[170] Medina CL, Nagatani S, Darling TA, Bucholtz DC, Tsukamura H, Maeda K, et al. Glucose availability modulates the timing of the luteinizing hormone surge in the ewe. J Neuroendocrinol 1998;10(10):785–92.

[171] Kasturi SS, Tannir J, Brannigan RE. The metabolic syndrome and male infertility. J Androl 2008;29(3):251–9.

[172] La Vignera S, Condorelli R, Vicari E, D'Agata R, Calogero AE. Diabetes mellitus and sperm parameters. J Androl 2012;33(2):145–53.

[173] Shafik A, Olfat S. Lipectomy in the treatment of scrotal lipomatosis. Br J Urol 1981;53(1):55–61.

[174] Setchell BP. The Parkes Lecture. Heat and the testis. J Reprod Fertil 1998;114(2):179–94.

[175] Southorn T. Great balls of fire and the vicious cycle: a study of the effects of cycling on male fertility. J Fam Plann Reprod Health Care 2002;28(4):211–3.

[176] Yaeram J, Setchell BP, Maddocks S. Effect of heat stress on the fertility of male mice *in vivo* and *in vitro*. Reprod Fertil Dev 2006;18(6):647–53.

[177] Ivell R. Lifestyle impact and the biology of the human scrotum. Reprod Biol Endocrinol 2007;5:15.

[178] Mariotti A, Di Carlo L, Orlando G, Corradini ML, Di Donato L, Pompa P, et al. Scrotal thermoregulatory model and assessment of the impairment of scrotal temperature control in varicocele. Ann Biomed Eng 2011;39(2):664–73.

[179] Wise LA, Cramer DW, Hornstein MD, Ashby RK, Missmer SA. Physical activity and semen quality among men attending an infertility clinic. Fertil Steril 2011;95(3):1025–30.

[180] Martini AC, Tissera A, Estofan D, Molina RI, Mangeaud A, de Cuneo MF, et al. Overweight and seminal quality: a study of 794 patients. Fertil Steril 2010;94(5):1739–43.

[181] Kriegel TM, Heidenreich F, Kettner K, Pursche T, Hoflack B, Grunewald S, et al. Identification of diabetes- and obesity-associated proteomic changes in human spermatozoa by difference gel electrophoresis. Reprod BioMed Online 2009;19(5):660–70.

[182] Pyttel S, Zschornig K, Nimptsch A, Paasch U, Schiller J. Enhanced lysophosphatidylcholine and sphingomyelin contents are characteristic of spermatozoa from obese men-A MALDI mass spectrometric study. Chem Phys Lipids 2012;165(8):861–5.

[183] Christensen BS, Gronbaek M, Pedersen BV, Graugaard C, Frisch M. Associations of unhealthy lifestyle factors with sexual inactivity and sexual dysfunctions in Denmark. J Sex Med 2011;.

[184] Reis LO, Favaro WJ, Barreiro GC, de Oliveira LC, Chaim EA, Fregonesi A, et al. Erectile dysfunction and hormonal imbalance in morbidly obese male is reversed after gastric bypass surgery: a prospective randomized controlled trial. Int J Androl 2010;33(5):736–44.

[185] Reis LO, Zani EL, Saad RD, Chaim EA, de Oliveira LC, Fregonesi A. Bariatric surgery does not interfere with sperm quality – a preliminary long-term study. Reprod Sci 2012;19(10):1057–62.

[186] Savastano S, Di Somma C, Pivonello R, Tarantino G, Orio F, Nedi V, et al. Endocrine changes (beyond diabetes) after bariatric surgery in adult life. J Endocrinol Invest 2013;36(4):267–79.

[187] Thangaratinam S, Rogozińska E, Jolly K, Glinkowski S, Roseboom T, Tomlinson JW, et al. Effects of interventions in pregnancy on maternal weight and obstetric outcomes: meta-analysis of randomised evidence. BMJ 2012;344:e2088. doi: 10.1136/bmj.e2088.

[188] Muktabhant B, Lumbiganon P, Ngamjarus C, Dowswell T. Interventions for preventing excessive weight gain during pregnancy. Cochrane Database Syst Rev 2012; 4: CD007145.

[189] Hill B, Skouteris H, Fuller-Tyszkiewicz M. Interventions designed to limit gestational weight gain: a systematic review of theory and meta-analysis of intervention components. Obes Rev 2013;14(6):435–50.

[190] van der Pligt P, Wilcox J, Hesketh KD, Ball K, Wilkinson S, Crawford D, et al. Systematic review of lifestyle interventions to limit postpartum weight retention: implications for future opportunities to prevent maternal overweight and obesity following childbirth. Obes Rev 2013;14(10):792–805.

[191] Hill J. Understanding and addressing the epidemic of obesity: an energy balance perspective. Endocr Rev 2006;27(7):750–61.

[192] Wadden TA, Butryn ML, Wilson C. Lifestyle modification for the management of obesity. Gastroenterology 2007;132(6):2226–38.

[193] Australian Institute of Health and Welfare. Indicators for chronic disease and their determinants. In: Government, A., editor. Canberra: AIHW; 2008. Cat. no. PHE 75.

12

Obesity and Gestational Outcomes

Aoife M. Egan, MB, BCh, BAO,*
Michael C. Dennedy, MD, PhD,***
*Galway Diabetes Research Centre, Galway, Ireland
**National University of Ireland, Galway, Ireland

INTRODUCTION: OBESITY IN WOMEN OF CHILDBEARING AGE

Obesity can be defined as excessive fat accumulation that may impair health. Body mass index (BMI) is a simple index of weight-for-height that is commonly used to classify obesity in adults and is defined as a person's weight in kilograms divided by the square of her height in meters (kg/m^2). A BMI of greater than or equal to 30 kg/m^2 defines obesity. Using this definition, worldwide obesity has nearly doubled since 1980 and in 2008 the World Health Organization (WHO) reported that nearly 300 million women were obese [1]. The incidence of maternal obesity at the start of pregnancy is increasing and accelerating [2]. In the United States, obesity has a prevalence of 21% in prepregnant females and European studies have identified a prevalence of approximately 19% among pregnant women [3,4]. Many factors can contribute to obesity; however, the rapidity with which obesity is increasing suggests that behavioral and environmental influences, rather than biological changes, have fueled this problem [5]. We will explore the effects of this global pandemic in relation to maternal and fetal pregnancy outcomes, examine the mechanisms leading to adverse gestational outcomes, and evaluate methods to reduce the impact of obesity on women and their offspring.

OBESITY AND ADVERSE MATERNAL OUTCOMES

Even prior to pregnancy, obesity has a negative effect on women's reproductive health and in particular is linked to polycystic ovarian syndrome of which one of the hallmarks is oligomenorrhea [6]. Weight loss results in significant improvement in pregnancy and ovulation rates in anovulatory obese women [7], while obesity increases the risk of spontaneous abortion during infertility treatment [8].

Obese mothers have an increased risk of many medical disorders during pregnancy. Preeclampsia is a multisystem disease that usually arises after 32 weeks of gestation and classically presents with hypertension and proteinuria [9]. A systematic review of 13 cohort studies noted that the risk of preeclampsia typically doubled with each 5–7 mg/m^2 increase in prepregnancy BMI [10]. This evaluation included nearly 1.4 million women, and the adverse association has been replicated in subsequent independent studies [4,10,11]. Unfortunately, preeclampsia is associated with a large increase in maternal and fetal morbidity and mortality during the pregnancy [12] and women with a history of preeclampsia are at risk of long-term cardiovascular complications [9].

Gestational diabetes mellitus is defined as any degree of glucose intolerance with onset or first recognition during pregnancy [13]. It is well established that obesity in pregnancy is a major risk factor for the development of this condition and that the noted increase in its frequency can be attributed to the increase in obesity in women of reproductive age [14,15]. The treatment of gestational diabetes during pregnancy typically involves diet and exercise, daily monitoring of glycemic control, and regular outpatient reviews. If glycemic targets are not achieved, pharmacological therapy including insulin is recommended. Along with the burden of day-to-day management, gestational diabetes mellitus is associated with multiple adverse maternofetal outcomes, including greater odds of birth weight, newborn percent body fat and cord C-peptide >90th centile, primary cesarean delivery, and preeclampsia [15,16]. Moreover, the combination of gestational diabetes mellitus and obesity are associated with substantially higher odds ratios for adverse outcomes compared with those for gestational diabetes mellitus or obesity alone [15].

Handbook of Fertility. http://dx.doi.org/10.1016/B978-0-12-800872-0.00012-3

Long-term consequences of gestational diabetes mellitus include an increased risk of developing type 2 diabetes mellitus. Bellamy et al., in a systematic review and meta-analysis, demonstrated this increased risk compared to those who had a normoglycemic pregnancy (RR 7.43, 95% CI 4.79–11.51) [17]. Postpartum glucose testing in an Irish cohort revealed that 19% of women with gestational diabetes mellitus had postpartum dysglycemia compared with 2.7% of those women with normal glucose tolerance in pregnancy. A recent study demonstrated that women with previous gestational diabetes mellitus demonstrate a greater than threefold prevalence of metabolic syndrome compared to women with normal glucose tolerance in pregnancy. Metabolic syndrome was defined using adult treatment panel-III criteria and insulin resistance using HOMA2-IR [18].

Further maternal complications associated with obesity during pregnancy include an increased risk of venous thrombosis. A large case control study described an adjusted odds ratio of 62.3 for antenatal venous thromboembolism and 40.1 for postnatal venous thromboembolism in women with BMI >25 kg/m^2 who were immobilized during pregnancy when compared to women of normal BMI who remained mobile throughout [19]. Understandably, there is also an increased incidence of pain in the back and pelvis in obese women due to the mechanical burden of obesity compounded by the weight of the expanding uterus [20]

In relation to cesarean delivery, increasing pregravid BMI independently increases the risk for cesarean delivery and reduces vaginal birth after cesarean delivery success [21,22]. One evaluation of a European population noted that elective cesarean delivery was undertaken for 18.5% of obese mothers, compared with 7.1% of mothers with a normal BMI, and 17.4% of obese mothers had an emergency cesarean delivery, compared with 10.6% mothers with normal BMI [4]. A recent systematic review of 11 cohort studies reported that the risk of cesarean delivery increased by 50% in women with a BMI of 30–35 kg/m^2 and more than doubled in women with a BMI >35 kg/m^2 compared to women with a normal BMI [23]. Although maternal obesity is not an actual indication for labor induction, obese women are more likely to have induction of labor, require higher doses of oxytocin and prostaglandin doses, and longer duration of labor when compared to their normal BMI counterparts [20,24]. These findings combined with fewer successful instrumental deliveries in women with a raised BMI likely explain this increased cesarean rate [4].

The increased cesarean rate in obese women is also associated with difficulty in anesthesia management and an increase in postoperative complications such as wound infection and excessive blood loss [25]. While the use of spinal or epidural anesthesia is recommended, it may be technically difficult to administer due to obscured landmarks and may result in a need for general anesthesia, which is also associated with difficulties including intraoperative respiratory events [26]. A retrospective analysis of 287,213 completed singleton pregnancies in the United Kingdom established that compared to women with normal BMI, a multitude of adverse outcomes were significantly more common in obese pregnant women. These included postpartum hemorrhage (odds ratio 1.16), genital tract infection (odds ratio 1.24), urinary tract infection (odds ratio 1.17), and wound infection (odds ratio 1.27) [27]. The increased rate of wound infections in the obese population has even led experts to challenge the established practice of a transverse incision, which is aesthetically appealing but may be associated with poor wound healing due to the incision being covered by a large panniculus. Although a randomized, controlled trial is required for further evaluation, a recent study of 3200 women demonstrated that in morbidly obese women undergoing a primary cesarean delivery, vertical skin incision was associated with a lower wound complication rate [28].

Due to its association with the multiple additional comorbidities described earlier, healthcare professionals endorse that maternal obesity has a major impact on services and resources [29]. This has significant implications for hospital budgets and resource allocation. Finally, the stark reality of maternal obesity is highlighted in the most recent report of maternal deaths in the United Kingdom that revealed that approximately half of maternal deaths were in obese or overweight women [30].

OBESITY AND ADVERSE NEONATAL OUTCOMES

The risk that maternal obesity poses to the developing fetus and neonate is significant. It is suggested that while the relationship between maternal BMI and adverse pregnancy outcomes is exponential, there is a threshold for unacceptable risk at 28 kg/m^2 [4]. There is an increased risk of early pregnancy loss and recurrent miscarriage [31]. One cohort study demonstrated that the odds ratio of a history of miscarriage was 1.4 in obese women compared to their normal BMI counterparts [11].

Later in pregnancy, it has been consistently demonstrated that maternal obesity is associated with fetal macrosomia and large for gestational age birth weights [4,11,32]. Macrosomia is typically defined as a birth weight of greater than 4–4.5 kg; however, it must be appreciated that its true clinical significance lies on a continuum [33]. Both macrosomia and large for gestational age birth weight are associated with a two- to threefold higher risk of intrauterine death, shoulder dystocia, and consequent brachial plexus injuries in addition to higher rates of cesarean section [34,35]. Longitudinal observational studies have also demonstrated poorer long-term outcomes for these infants, including childhood obesity and asthma, and in later life the metabolic syndrome, type 2 diabetes, and cancer [20,35].

A case-control study in 2007 noted that mothers of offspring with spina bifida, heart defects, anorectal atresia, hypospadias, limb reduction defects, diaphragmatic hernia, and omphalocele were significantly more likely to be obese than mothers of controls, with odds ratios ranging between 1.33 and 2.10 [36]. Stothard et al., in a systematic review and meta-analysis, noted an association between maternal obesity and an increased risk of structural abnormalities including neural tube defects, spina bifida, cardiovascular anomalies, septal anomalies, cleft palate, anorectal atresia, hydrocephaly, and limb reduction anomalies [37]. Worryingly, increasing maternal weight is associated with increases in risks of stillbirth and very preterm (less than 32 weeks) birth [38]. Due to the increase in these complications, infants born to obese mothers require admission to the neonatal intensive care unit more frequently than offspring of normal weight mothers [6, 32]. The immediate impact of maternal obesity on short-term obstetric resources was highlighted in a meta-analysis by Heslehurst et al. in 2008 with the authors calling for urgent national clinical guidelines for the management of obese pregnant women and public health interventions to help safeguard the health of mothers and their babies [6].

Unfortunately, the adverse consequences of maternal obesity are not restricted to the pregnancy or the time period immediately postpartum. The extent of childhood obesity is concerning with 9.5% of infants and toddlers at or above the 95th percentile of the weight-for-recumbent-length growth charts in a representative sample of the US population in 2007–2008. This percentage rose to 11.9% when evaluating children and adolescents [39]. In the adolescent population in the United States, obesity appears to affect ethnic groups differently with one study revealing a significantly greater prevalence of high BMI in non-Hispanic black girls than in non-Hispanic white girls [40]. There is an increasing body of evidence that associates maternal obesity during pregnancy with long-term raised BMI and adiposity in the offspring. A longitudinal study of the northern Finland birth cohort for 1966 evaluated the associations between BMI at 31 years of age and multiple factors including maternal BMI before pregnancy. The authors concluded that the heavier the mother, the heavier the offspring from birth to 31 years [41]. These findings were replicated in a prospective cohort study based in the United Kingdom, which identified eight factors in early life to be associated with an increased risk of obesity in childhood including parental obesity (both parents: adjusted odds ratio, 10.44). Other identified factors included more than 8 h spent watching television per week at age 3 years (1.55) and very early BMI or adiposity rebound

(15.00) [42]. It is unclear if this long-term effect is due to priming of the fetus during the obese pregnancy or is due to the indirect environmental influence of an "over-fed" home [20].

MECHANISMS OF ADVERSE OUTCOMES

The exact process by which maternal obesity results in the plethora of adverse outcomes as described earlier is poorly understood. It is likely multifactorial involving insulin resistance, altered glucose and lipid metabolism, and a proinflammatory state. Vitamin D and folic acid deficiency are also felt to play a significant role in the development of complications related to pregnancy in the obese mother.

While maternal obesity has been noted to be associated with poor outcomes independent of diabetes mellitus in pregnancy, insulin and glucose homeostasis are undoubtedly key elements in developing pregnancy-related complications for many obese women. It is established that during pregnancy in the nonobese woman there are significant changes in insulin release and insulin sensitivity. Throughout gestation, there is a significant increase in the insulin/glucose ratio during oral glucose tolerance testing and also a significant 3.0–3.5-fold increase in first-phase and second-phase insulin release. Through to 36 weeks gestation, there is a 56% decrease in insulin sensitivity [43]. Obesity itself, independent of pregnancy, also causes insulin resistance and therefore in the obese, pregnant woman there is an exaggerated response with marked maternal hyperinsulinemia and hyperglycemia, which may or may not reach the cutoff for gestational diabetes mellitus [44]. The Pedersen hypothesis described in 1952 forms the basis for our understanding of the pathophysiological consequences of hyperglycemia during pregnancy. Jorgen Pedersen postulated that maternal hyperglycemia results in fetal hyperglycemia with subsequent hypertrophy of fetal islet tissue with insulin hypersecretion [45]. Fetal hyperinsulinemia subsequently causes macrosomia as the metabolic effects of insulin in its putative receptor increases fat and glycogen deposition and the excess insulin acts on the insulin-like growth factor receptors to further stimulate fetal growth [20]. The result is fetal macrosomia and on delivery an increase in the risk of neonatal hyperglycemia. The Pedersen hypothesis was based on observations in women with frank hyperglycemia often in the setting of type 1 diabetes and so the hyperglycemia and adverse pregnancy outcomes (HAPO) study was conducted to clarify the risk of adverse outcomes associated with various degrees of maternal glucose intolerance less severe than that in overt diabetes mellitus. The results indicated strong, continuous associations of maternal glucose levels below those diagnostic of diabetes with increased birth weight and increased cord-blood serum C-peptide levels [46]. This study provided the basis for the creation of new diagnostic criteria for gestational diabetes mellitus with overall lower cut-off values and highlights the significance of even mild glucose derangements during pregnancy [47].

Despite adequate glycemic control fetal macrosomia may still occur. Langer et al. noted that obese women with gestational diabetes mellitus had a two- to three-fold higher risk for adverse pregnancy outcome when compared with overweight and normal weight patients with well-controlled gestational diabetes mellitus [48]. This highlights the point that insulin resistance affects more than glucose metabolism. In the obese individual increased fatty acids and triglycerides are hallmarks of insulin resistance and may have a role in modulating fetal growth and adiposity in this setting [49]. Clinical studies support this assumption including work by Di Cianni et al., which demonstrated that prepregnancy BMI and fasting maternal serum triglycerides were independently associated with neonatal birth weight in women with normal glucose tolerance but a positive screening test [50], and a study by Nolan et al. that demonstrated that triglycerides were also predictive of neonatal birth weight ratio in subjects with gestational diabetes mellitus [51]. Catalano and Hauguel-de Mouzon highlighted the extent of the metabolic derangement in the obese pregnancy in a review paper in 2011 and cited work by Freinkel that suggested that fetal overgrowth is a function of multiple nutritional factors in addition to glucose [49,52].

Pregnancy is a proinflammatory state and it is also evident that obesity is associated with a state of chronic low-grade inflammation [53,54]. In human pregnancy, maternal obesity is associated with metabolic inflammation, characterized by changes in cytokines, protein hormones, and acute phase proteins with an increase in leptin and high sensitive C-reactive protein with increasing BMI category [55]. It is also demonstrated that in the obese pregnancy, there is an accumulation of a heterogeneous macrophage population and proinflammatory mediators in the placenta [56]. While these proinflammatory mediators do not necessarily transfer directly across the placenta into the fetal circulation, this metabolic inflammation may play an indirect role at the maternal fetal interface to modify the availability of energy substrates for fetal needs [49,57]. Evidence that there is a strong positive correlation with maternal pregravid BMI and fetal insulin resistance is concerning and demonstrates that metabolic compromise is already present at birth continuing the vicious cycle of obesity [57]. Our knowledge on the contribution of chronic inflammation to the pathogenesis of obesity-related adverse pregnancy outcome is rapidly expanding. One area of research interest is that of microbiota, with data accumulating in animal models and humans suggesting that obesity and type 2 diabetes are associated with a profound dysbiosis that directly contributes to chronic inflammation [58]. Basu et al. obtained blood and subcutaneous abdominal adipose tissue from 120 lean and obese pregnant women at delivery and found doubling of plasma endotoxin concentrations in obese compared with lean women suggesting that the increased endotoxemia relates to the obesity rather than to the pregnancy. The authors propose that the recognition and sensing of bacterial derived pathogens initiates the transduction of immunomodulatory signals within the adipose cells and that the increased circulating endotoxemia is an indication of maternal environmental changes in pregnancy with obesity [59]. Alteration of microbiota may offer a therapeutic option in future management of obesity in pregnancy.

Obesity is associated with vitamin D deficiency due to sequestration of vitamin D in body fat leading to reduced bioavailability [60]. Vitamin D is a fat-soluble vitamin that preferentially deposits in adipose tissue but once deposited becomes functionally useless. Additionally, the normal conversion of the precursor to active vitamin D in the skin is also reduced in obese individuals due to decreased skin exposure to the sun [20]. The latter may be related to reduced physical function associated with obesity, which has a significant relationship with vitamin D status [61]. Pregnant women must sustain their own vitamin D stores as well as those of their fetuses and therefore the obese woman is at an immediate disadvantage with pregravid obese women having lower adjusted mean vitamin D concentrations and a higher prevalence of vitamin D deficiency compared with lean women (61 versus 36%). Vitamin D status of neonates of obese mothers is also poorer than neonates of lean mothers (50.1 versus 56.3 nmol/l) [62]. Maternal vitamin D deficiency is directly associated with multiple adverse pregnancy outcomes including preeclampsia. As the active form of vitamin D has been shown to regulate the transcription and function of genes associated with placental invasion, normal implantation, and angiogenesis, deficiency may have an adverse effect on this process. Vitamin D deficiency may therefore facilitate the pathogenesis of preeclampsia, including reduced placental perfusion secondary to abnormal implantation, and encourage production of materials from the poorly perfused placenta that initiate the ensuing sequelae of the condition [63]. Low maternal vitamin D in late pregnancy is also associated with reduced intrauterine long bone growth and slightly shorter gestation (0.7 weeks) [64]. Finally, in pregnant women an inverse correlation was found between vitamin D and insulin resistance, measured using homeostasis assessment suggesting that deficiency may contribute to insulin resistance [65].

In relation to the fetus, it is felt that sufficiency of vitamin D is critical for development, particularly for the fetal brain and immunological functions along with adequate skeletal formation [66]. Future research will fully define the true extent of the role of vitamin D in adverse gestational outcomes associated with maternal obesity.

Folic acid is a crucial element in fetal development and periconceptual folate supplementation has a strong protective effect against neural tube defects, which arise during the development of the brain and spinal cord [67]. BMI is inversely associated with serum folate among women of childbearing age and therefore deficiency of folic acid undoubtedly accounts for the increased prevalence of neonatal neural tube defects in obesity [68].

OBESITY MANAGEMENT: PREPREGNANCY

While prevention of obesity in the first instance is the ideal, there is an urgency to treat those already affected. Management of obese women of reproductive age should commence prior to pregnancy. The authors recommend that women are made aware of the pregnancy-related complications associated with obesity and public health campaigns have a potential role in this regard. Based on current best evidence, we recommend that obese women are provided with access to specialist weight management programs in a multidisciplinary setting to support weight loss preconceptually. Over the last several years, our understanding of the mechanisms that regulate body weight has progressed significantly. Many of our previous presumptions regarding obesity have been questioned and a recent review clarifies many false and scientifically unsupported beliefs about obesity that are pervasive in both scientific literature and the popular press [69]. Despite this increased understanding, treatment of obesity in order to realize sustained weight loss in all patient groups remains a challenge.

Dietary modifications that reduce energy intake remain central to any weight loss strategy. Although many dietary strategies are effective in weight reduction, there remains the question of how we can help people adhere to a dietary program for the long term [70]. It is disillusioning to note that an analysis of structured weight-loss programs indicated that weight-loss maintenance 4 or 5 years after the intervention averages just 3.0 kg or 23% of initial weight loss [71]. Increased focus is on physical activity that produces not just measurable improvements in weight but also improves overall cardiovascular health [70]. Pedometers have become popular as a tool for motivating physical activity with one systematic review suggesting that the use of a pedometer is associated with significant increases in physical activity and significant decreases in BMI and blood pressure [72]. Health care providers should be aware that patients often encounter personal obstacles to exercise, including medical and psychological. One evaluation of obese patients with type 2 diabetes highlights the varied nature of perceived barriers to exercise and makes the point that a "catch all" approach to improving exercise habits is unlikely to succeed and that each patient should be addressed uniquely [73].

The development of pharmacological therapy for obesity has been especially disappointing with side effects frequently outweighing the benefits. Two weight loss medications, rimonabant and sibutramine, have been withdrawn from the market in recent years due to safety concerns in relation to psychiatric illness and cardiovascular disease, respectively. Lorcaserin, a serotonin type 2c receptor agonist, was officially launched in the United States in June 2013. Lorcaserin is well tolerated and its use is associated with a modest weight loss of approximately 3.2 kg at 1 year [74]. Another available agent in the United States is a combination of phentermine, a nonselective stimulator of synaptic norepinephrine (noradrenaline), dopamine, and serotonin release, and topiramate, an anticonvulsant [75]. The CONQUER study demonstrated a significant weight loss with 37% patients achieving a >10% weight loss on the lower dose and 48% on the higher dose. Again the medication was well tolerated in the trial setting but it has not received a European license due to long-term safety concerns, relating in particular to cardiovascular, central nervous system, and teratogenic effects [75,76]. It must be noted that none of these medications are suitable in pregnancy and should be avoided in those women actively planning pregnancy. Appropriate contraception should be utilized when weight loss medications are used in women of childbearing years. Other promising agents for obesity treatment include glucagon-like peptide 1 receptor agonists, which are already available for treatment of type 2 diabetes mellitus; a combination of bupropion and naltrexone, which is in clinical trial stages, and developing compounds including RM-493, a melanocortin type 4 receptor agonist that targets central hypothalamic pathways [75].

Finally, bariatric surgery can effectively reduce body weight and is a viable option for certain obese women of reproductive age. Currently, bariatric surgery is generally reserved for patients with severe or complex obesity. It is the most effective weight loss treatment in this population, both in terms of amount of weight loss achieved and durability of weight loss or "weight maintenance" as well as amelioration of obesity-related comorbidities [75,77]. Surgical options vary and include gastric bypass procedures, sleeve gastrectomy, and banding. Prior to such interventions, careful patient selection should take place, including a thorough medical assessment, as well as psychological and dietetic assessments [78]. Because nutritional deficiencies are common postbariatric surgery, women of childbearing age should be counseled to avoid getting pregnant while actively losing weight and until the serum nutrient levels have stabilized [79]. Post-bariatric surgery dietary guidelines recommend supplementation with multivitamins and iron. The importance of adherence to these treatments must be emphasized to the pregnant woman as adverse outcomes including neural tube defects that have been associated with noncompliance [80]. Based on best available evidence, the authors recommend increased monitoring of iron and folate levels throughout pregnancy. Of note, low red cell folate, thought to indicate tissue deficiency, may be more important than a low serum folate in the diagnosis of folate deficiency. However, as measurement of red cell folate is more expensive; it is not recommended routinely but should be completed if there is still a high clinical suspicion of folate deficiency in the setting of a normal serum folate [81]. During pregnancy, patients who have gastric bands should be monitored by their surgeon as adjustment may be required and it should be noted that a history of bariatric surgery is not classified as an independent indication for cesarean delivery [80].

MANAGEMENT OF OBESITY DURING PREGNANCY

Postconception, pregnancy in the obese woman should be treated as high risk. The authors recommend that all women have their weight and height measured and BMI calculated at the booking visit with a weight check at each encounter thereafter. Close attention to weight gain during pregnancy is of utmost importance and collection of simple bedside measurables may be one of the most effective ways of directing management of obesity during pregnancy. In the context of a busy clinical environment, this old-fashioned process of collecting and monitoring may be facilitated through the use of modern database systems. These statements are supported by recent data showing that excessive gestational weight gain (GWG) represents a potential risk factor for adverse outcomes [82–84]. This risk is independent of prepregnancy BMI and diabetes in pregnancy [85,86], and in 2009, the Institute of Medicine in the United States published guidelines describing BMI appropriate thresholds for GWG [87]. These guidelines recommend a total weight gain of 11.5–16.0 kg in normal weight women but suggest a weight gain of just 5.0–9.0 kg in the obese woman. Unfortunately, more than half of women exceed these recommendations and it is now evident that this is associated with an increased risk of a postpartum weight retention of greater than 2 kg with obvious negative implications for future pregnancy [85,88]. Management programs to prevent excessive GWG have been suggested to improve obstetric outcomes; however, these studies include selected groups of women and many may not be feasible to provide as part of routine care

[89,90]. Additionally, other studies would suggest that in the obese population, while lifestyle intervention may lead to restricted GWG, there is no clear evidence that there are improved obstetric outcomes as a result [91]. We do not advocate severe weight restriction however, as it was demonstrated in overweight and obese women that weight loss or gain less than or equal to 5 kg is associated with increased risk of small for gestational age infants [92]. The authors do recommend that attention should be paid to GWG during antenatal care to ensure that it is within the limits set by the institute of medicine to ensure optimal outcomes for mother and infant. Postpartum, further counseling and support for weight loss should be encouraged. Even modest interpregnancy weight reduction among women with obesity has been shown to significantly reduce the risk of developing gestational diabetes mellitus in the subsequent pregnancy. Glazer et al. demonstrated in a population-based cohort study that women who lost at least 10 lbs between pregnancies had a decreased risk of gestational diabetes relative to women whose weight changed by less than 10 lbs (RR: 0.63) [93].

In relation to gestational diabetes mellitus for which obesity is a major risk factor, screening should take place in all women with a BMI of greater than or equal to 30 kg/m^2 at 24–28 weeks. Criteria for diagnosis of gestational diabetes mellitus have changed frequently over the past number of years but the authors recommend a one-step approach using a 75 g oral glucose tolerance test as per the International Association of Diabetes in Pregnancy Study Group [47]. Appropriate treatment for gestational diabetes mellitus appears to lower the risk for some perinatal complications including preeclampsia and shoulder dystocia as revealed in a meta-analysis of five randomized controlled trials [94]. Post-diagnosis, dietary and lifestyle changes along with blood glucose monitoring advice are necessary. Insulin therapy is required if tight glycemic goals are not achieved. The role of metformin in gestational diabetes mellitus is not fully clear but recent data are convincing from a safety and efficacy viewpoint with or without supplemental insulin [95]. Some studies have also suggested that use of metformin in the mother can increase insulin sensitivity and improve longer-term outcomes in offspring [96,97]. Other work has revealed that women treated with metformin gained less weight in pregnancy but also lost less weight in the first year postpartum and their offspring weighed more at 1 year than those not exposed [98]. We await with interest the results of an ongoing trial to evaluate metformin in pregnant women to reduce the risk of obesity of metabolic syndrome in their babies [99]. As per recommendations from the American Diabetes Association, women should have a glucose tolerance test at 6–12 weeks postpartum to ensure that persistent diabetes is not present and women with a history of gestational diabetes mellitus should have lifelong screening for the development of diabetes or prediabetes at least every 3 years [100].

Supplementation with higher doses of folate (5 mg) and vitamin D (10 mcg) should ideally be commenced 1 month prepregnancy but at the latest in early pregnancy in women with a BMI of greater than or equal to 27 kg/m^2 [101]. As previously described, obese women are categorized as high risk for fetal neural tube defects and therefore should be prescribed this higher dose of folate. In high-risk women due to a previous pregnancy with a neural tube defect, a randomized controlled trial has shown that higher dose supplementation reduces the risk of a subsequent neural tube defect-affected pregnancy by 72% [102]. Vitamin D should continue while breastfeeding [101].

During the pregnancy, regular blood pressure monitoring is necessary as part of preeclampsia surveillance. The authors recommended use of a size-appropriate arm cuff to ensure the most accurate blood pressure measurement. The National Institute for Health and Clinical Excellence (NICE) in the United Kingdom recommend aspirin 75 mg once daily from 12 weeks until the birth of the baby in the presence of more than one moderate risk factor for preeclampsia. Identified moderate risk factors include first pregnancy, age 40 years or older, pregnancy interval of more than 10 years, family history of preeclampsia, multiple pregnancy, and a BMI of 35 kg/m^2 or more at first visit [103].

The American College of Obstetricians and Gynecologists (ACOG) have published clear opinions on the management of obesity in pregnancy and it is advised that the patient have an anesthesiology consultation antepartum or if not, early in labor to allow adequate time for construction of an appropriate anesthetic plan. For those obese women who require cesarean delivery, consideration should be given to using a higher dose of preoperative antibiotics for surgical prophylaxis [26]. This latter recommendation is based on an ACOG practice bulletin on the use of prophylactic antibiotics in labor and delivery [104]. The use of suture closure of the subcutaneous layer after cesarean delivery in obese patients may lead to a significant reduction in the incidence of postoperative wound disruption [26]. Regarding management of the increased risk of thromboembolism, all women in the obese category should have placement of pneumatic compression devices and the ACOG advise that individual risk assessment may also lead to the use of heparin thromboprophylaxis [104]. A joint guideline by the Centre for Maternal and Child Enquiries and the Royal College of Obstetricians and Gynecologists in the United Kingdom advises that women with a booking BMI of greater than or equal to 30 kg/m^2 who also have two or more additional risk factors for thromboembolism should be considered for prophylactic low molecular weight heparin (LMWH) antenatally. This should generally continue until 6 weeks postpartum. Women with a BMI of greater than 35–40 kg/m^2 should be offered postnatal thromboprophylaxis [101]. All women with a BMI of greater or equal to 30 kg/m^2 should be recommended to have active management in the third stage of labor [101]. Home birth is not an option for this group of women due to the unacceptable risk of emergency cesarean section and the increased likelihood that neonatal special care or intensive care admission is required [20].

Finally, maternal obesity is associated with reduced breastfeeding rates. It is noted that obese mothers often feel uncomfortable breastfeeding in public and perceive that their milk supply is insufficient. Worryingly, despite greater breastfeeding difficulties, obese mothers were less likely to seek support for breastfeeding in the first 3 months [105]. The authors recommend that these women receive extra support to overcome barriers to breastfeeding given the clear benefits to the mother and child, which include a reduction in postpartum dysglycemia in women with a history of gestational diabetes mellitus [106].

CONCLUSIONS

Obesity is one of the most common health problems in the United States and worldwide. There is an overwhelming body of evidence that links obesity to many of the complications associated with pregnancy and treatment remains a challenge. We recommend appropriate prepregnancy support for obese women to facilitate preconceptual weight loss. The obese pregnancy must be treated as high risk and managed in a structured setting by an expert multidisciplinary team. Special consideration should be given to GWG, nutritional supplementation, blood pressure monitoring, thromboprophylaxis, and a planned delivery with anesthesiology consultation antenatally or early in labor. Interpregnancy weight gain should be avoided and consultation with a weight-reduction specialist before the next pregnancy should be an available option to all women. It is hoped that future research will better inform clinicians and patients on ways to improve outcomes for this high-risk group of women.

References

[1] WHO. Fact Sheet N0 311: Obesity and Overweight. 2014. Available from: http://www.who.int/mediacentre/factsheets/fs311/en/; [accessed 08.08.14].

[2] Heslehurst N, Ells LJ, Simpson H, Batterham A, Wilkinson J, Summerbell CD. Trends in maternal obesity incidence rates, demographic predictors, and health inequalities in 36,821 women over a 15-year period. BJOG 2007;114:187–94.

[3] Huda SS, Brodie LE, Sattar N. Obesity in pregnancy: prevalence and metabolic consequences. Semin Fetal Neonatal Med 2010;15:70–6.

[4] Dennedy MC, Avalos G, O'Reilly MW, O'Sullivan EP, Gaffney G, Dunne F. ATLANTIC-DIP: raised maternal body mass index (BMI) adversely affects maternal and fetal outcomes in glucose-tolerant women according to International Association of Diabetes and Pregnancy Study Groups (IADPSG) criteria. J Clin Endocrinol Metab 2012;97:E608–12.

[5] Stein CJ, Colditz GA. The epidemic of obesity. J Clin Endocrinol Metab 2004;89:2522–5.

[6] Heslehurst N, Simpson H, Ells LJ, Rankin J, Wilkinson J, Lang R, et al. The impact of maternal BMI status on pregnancy outcomes with immediate short-term obstetric resource implications: a meta-analysis. Obes Rev 2008;9:635–83.

[7] Clark AM, Ledger W, Galletly C, Tomlinson L, Blaney F, Wang X, et al. Weight loss results in significant improvement in pregnancy and ovulation rates in anovulatory obese women. Hum Reprod 1995;10:2705–12.

[8] Wang JX, Davies MJ, Norman RJ. Obesity increases the risk of spontaneous abortion during infertility treatment. Obes Res 2002;10:551–4.

[9] Ahmed R, Dunford J, Mehran R, Robson S, Kunadian V. Pre-eclampsia and future cardiovascular risk among women: a review. J Am Coll Cardiol 2014;63:1815–22.

[10] O'Brien TE, Ray JG, Chan WS. Maternal body mass index and the risk of preeclampsia: a systematic overview. Epidemiology 2003;14:368–74.

[11] Owens LA, O'Sullivan EP, Kirwan B, Avalos G, Gaffney G, Dunne F. ATLANTIC DIP: the impact of obesity on pregnancy outcome in glucose-tolerant women. Diabetes Care 2010;33:577–9.

[12] Abalos E, Cuesta C, Carroli G, Qureshi Z, Widmer M, Vogel JP, Souza JP. Network WMSoMaNHR. Pre-eclampsia, eclampsia and adverse maternal and perinatal outcomes: a secondary analysis of the World Health Organization Multicountry Survey on Maternal and Newborn Health. BJOG 2014;121(Suppl 1):14–24.

[13] American Diabetes, Association. Diagnosis and classification of diabetes mellitus. Diabetes Care 2004;27(Suppl 1):S5–S10.

[14] Weiss JL, Malone FD, Emig D, Ball RH, Nyberg DA, Comstock CH, et al. Obesity, obstetric complications and cesarean delivery rate – a population-based screening study. Am J Obstet Gynecol 2004;190:1091–7.

[15] Catalano PM, McIntyre HD, Cruickshank JK, McCance DR, Dyer AR, Metzger BE, et al. The hyperglycemia and adverse pregnancy outcome study: associations of GDM and obesity with pregnancy outcomes. Diabetes Care 2012;35:780–6.

[16] O'Sullivan EP, Avalos G, O'Reilly M, Dennedy MC, Gaffney G, Dunne F. Atlantic Diabetes in Pregnancy (DIP): the prevalence and outcomes of gestational diabetes mellitus using new diagnostic criteria. Diabetologia 2011;54:1670–5.

[17] Bellamy L, Casas JP, Hingorani AD, Williams D. Type 2 diabetes mellitus after gestational diabetes: a systematic review and meta-analysis. Lancet 2009;373:1773–9.

[18] Noctor E, Crowe C, Carmody LA, Kirwan B, O'Dea A, Glynn LG, et al. ATLANTIC-DIP: prevalence of metabolic syndrome and insulin resistance in women with previous gestational diabetes mellitus by International Association of Diabetes in Pregnancy Study Groups criteria. Acta Diabetol 2014;. [epub ahead of print].

[19] Larsen TB, Sørensen HT, Gislum M, Johnsen SP. Maternal smoking, obesity, and risk of venous thromboembolism during pregnancy and the puerperium: a population-based nested case-control study. Thromb Res 2007;120:505–9.

[20] Dennedy MC, Dunne F. The maternal and fetal impacts of obesity and gestational diabetes on pregnancy outcome. Best Pract Res Clin Endocrinol Metab 2010;24:573–89.

[21] Durnwald CP, Ehrenberg HM, Mercer BM. The impact of maternal obesity and weight gain on vaginal birth after cesarean section success. Am J Obstet Gynecol 2004;191:954–7.

[22] Ehrenberg HM, Durnwald CP, Catalano P, Mercer BM. The influence of obesity and diabetes on the risk of cesarean delivery. Am J Obstet Gynecol 2004;191:969–74.

[23] Poobalan AS, Aucott LS, Gurung T, Smith WC, Bhattacharya S. Obesity as an independent risk factor for elective and emergency caesarean delivery in nulliparous women – systematic review and meta-analysis of cohort studies. Obes Rev 2009;10:28–35.

[24] Pevzner L, Powers BL, Rayburn WF, Rumney P, Wing DA. Effects of maternal obesity on duration and outcomes of prostaglandin cervical ripening and labor induction. Obstet Gynecol 2009;114:1315–21.

[25] Catalano PM, Ehrenberg HM. The short- and long-term implications of maternal obesity on the mother and her offspring. BJOG 2006;113:1126–33.

[26] The American College of Obstetricians and Gynaecologists. Committee Opinion. Number 549; 2013. Available from: http://www.acog.org/~/media/Committee%20Opinions/Committee%20on%20Obstetric%20Practice/co549.pdf?dmc(1&ts(20140808T1246379340 [accessed 08.08.14].

[27] Sebire NJ, Jolly M, Harris JP, Wadsworth J, Joffe M, Beard RW, et al. Maternal obesity and pregnancy outcome: a study of 287,213 pregnancies in London. Int J Obes Relat Metab Disord 2001;25:1175–82.

[28] Marrs CC, Moussa HN, Sibai BM, Blackwell SC. The relationship between primary cesarean delivery skin incision type and wound complications in women with morbid obesity. Am J Obstet Gynecol 2014;210:319. e1-4.

[29] Heslehurst N, Lang R, Rankin J, Wilkinson JR, Summerbell CD. Obesity in pregnancy: a study of the impact of maternal obesity on NHS maternity services. BJOG 2007;114:334–42.

[30] Cantwell R, Clutton-Brock T, Cooper G, Dawson A, Drife J, Garrod D, et al. Saving Mothers' Lives: reviewing maternal deaths to make motherhood safer: 2006–2008. The Eighth Report of the Confidential Enquiries into Maternal Deaths in the United Kingdom. BJOG 2011;118(Suppl 1):1–203.

[31] Lashen H, Fear K, Sturdee DW. Obesity is associated with increased risk of first trimester and recurrent miscarriage: matched case-control study. Hum Reprod 2004;19:1644–6.

[32] Jensen DM, Damm P, Sørensen B, Mølsted-Pedersen L, Westergaard JG, Ovesen P, et al. Pregnancy outcome and prepregnancy body mass index in 2459 glucose-tolerant Danish women. Am J Obstet Gynecol 2003;189:239–44.

[33] Dennedy MC, Dunne F. Macrosomia: defining the problem worldwide. Lancet 2013;381:435–6.

[34] Henriksen T. The macrosomic fetus: a challenge in current obstetrics. Acta Obstet Gynecol Scand 2008;87:134–45.

[35] Koyanagi A, Zhang J, Dagvadorj A, Hirayama F, Shibuya K, Souza JP, et al. Macrosomia in 23 developing countries: an analysis of a multicountry, facility-based, cross-sectional survey. Lancet 2013;381:476–83.

[36] Waller DK, Shaw GM, Rasmussen SA, Hobbs CA, Canfield MA, Siega-Riz AM, Study NBDP. et al. Prepregnancy obesity as a risk factor for structural birth defects. Arch Pediatr Adolesc Med 2007;161:745–50.

[37] Stothard KJ, Tennant PW, Bell R, Rankin J. Maternal overweight and obesity and the risk of congenital anomalies: a systematic review and meta-analysis. JAMA 2009;301:636–50.

[38] Cnattingius S, Lambe M. Trends in smoking and overweight during pregnancy: prevalence, risks of pregnancy complications, and adverse pregnancy outcomes. Semin Perinatol 2002;26:286–95.

[39] Ogden CL, Carroll MD, Curtin LR, Lamb MM, Flegal KM. Prevalence of high body mass index in US children and adolescents, 2007–2008. JAMA 2010;303:242–9.

[40] Flegal KM, Ogden CL, Yanovski JA, Freedman DS, Shepherd JA, Graubard BI, et al. High adiposity and high body mass index-for-age in US children and adolescents overall and by race-ethnic group. Am J Clin Nutr 2010;91:1020–6.

[41] Laitinen J, Power C, Järvelin MR. Family social class, maternal body mass index, childhood body mass index, and age at menarche as predictors of adult obesity. Am J Clin Nutr 2001;74:287–94.

[42] Reilly JJ, Armstrong J, Dorosty AR, Emmett PM, Ness A, Rogers I, Team ALSoPaCS. et al. Early life risk factors for obesity in childhood: cohort study. BMJ 2005;330:1357.

[43] Catalano PM, Tyzbir ED, Roman NM, Amini SB, Sims EA. Longitudinal changes in insulin release and insulin resistance in nonobese pregnant women. Am J Obstet Gynecol 1991;165:1667–72.

[44] Kahn SE, Hull RL, Utzschneider KM. Mechanisms linking obesity to insulin resistance and type 2 diabetes. Nature 2006;444:840–6.

[45] Pederson J. Diabetes and pregnancy: blood sugar of newborn infants. Copenhagen: Danish Science Press; 1952. 230.

[46] Metzger BE, Lowe LP, Dyer AR, Trimble ER, Chaovarindr U, Coustan DR, et al. Hyperglycemia and adverse pregnancy outcomes. N Engl J Med 2008;358:1991–2002.

[47] Metzger BE, Gabbe SG, Persson B, Buchanan TA, Catalano PA, Damm P, et al. International association of diabetes and pregnancy study groups recommendations on the diagnosis and classification of hyperglycemia in pregnancy. Diabetes Care 2010;33:676–82.

[48] Langer O, Yogev Y, Xenakis EM, Brustman L. Overweight and obese in gestational diabetes: the impact on pregnancy outcome. Am J Obstet Gynecol 2005;192:1768–76.

[49] Catalano PM, Hauguel-De Mouzon S. Is it time to revisit the Pedersen hypothesis in the face of the obesity epidemic? Am J Obstet Gynecol 2011;204:479–87.

[50] Di Cianni G, Miccoli R, Volpe L, Lencioni C, Ghio A, Giovannitti MG, et al. Maternal triglyceride levels and newborn weight in pregnant women with normal glucose tolerance. Diabet Med 2005;22:21–5.

[51] Nolan CJ, Riley SF, Sheedy MT, Walstab JE, Beischer NA. Maternal serum triglyceride, glucose tolerance, and neonatal birth weight ratio in pregnancy. Diabetes Care 1995;18:1550–6.

[52] Freinkel N. Banting Lecture 1980. Of pregnancy and progeny. Diabetes 1980;29:1023–35.

[53] Hotamisligil GS. Inflammation and metabolic disorders. Nature 2006;444:860–7.

[54] Segovia SA, Vickers MH, Gray C, Reynolds CM. Maternal obesity, inflammation, and developmental programming. Biomed Res Int 2014;2014:418975.

[55] Madan JC, Davis JM, Craig WY, Collins M, Allan W, Quinn R, et al. Maternal obesity and markers of inflammation in pregnancy. Cytokine 2009;47:61–4.

[56] Challier JC, Basu S, Bintein T, Minium J, Hotmire K, Catalano PM, et al. Obesity in pregnancy stimulates macrophage accumulation and inflammation in the placenta. Placenta 2008;29:274–81.

[57] Catalano PM, Presley L, Minium J, Hauguel-de Mouzon S. Fetuses of obese mothers develop insulin resistance *in utero*. Diabetes Care 2009;32:1076–80.

[58] Tilg H, Moschen AR. Microbiota and diabetes: an evolving relationship. Gut 2014;63:1513–21. doi: 10.1136/gutjnl-2014-306928.

[59] Basu S, Haghiac M, Surace P, Challier JC, Guerre-Millo M, Singh K, et al. Pregravid obesity associates with increased maternal endotoxemia and metabolic inflammation. Obesity (Silver Spring) 2011;19:476–82.

[60] Holick MF. Vitamin D deficiency. N Engl J Med 2007;357:266–81.

[61] Ahern T, Khattak A, O'Malley E, Dunlevy C, Kilbane M, Woods C, et al. Association between vitamin D status and physical function in the severely obese. J Clin Endocrinol Metab 2014;99:E1327–31.

[62] Bodnar LM, Catov JM, Roberts JM, Simhan HN. Prepregnancy obesity predicts poor vitamin D status in mothers and their neonates. J Nutr 2007;137:2437–42.

[63] Bodnar LM, Catov JM, Simhan HN, Holick MF, Powers RW, Roberts JM. Maternal vitamin D deficiency increases the risk of preeclampsia. J Clin Endocrinol Metab 2007;92:3517–22.

[64] Morley R, Carlin JB, Pasco JA, Wark JD. Maternal 25-hydroxyvitamin D and parathyroid hormone concentrations and offspring birth size. J Clin Endocrinol Metab 2006;91:906–12.

[65] Maghbooli Z, Hossein-Nezhad A, Karimi F, Shafaei AR, Larijani B. Correlation between vitamin D3 deficiency and insulin resistance in pregnancy. Diabetes Metab Res Rev 2008;24:27–32.

[66] Lapillonne A. Vitamin D deficiency during pregnancy may impair maternal and fetal outcomes. Med Hypotheses 2010;74:71–5.

[67] Lumley J, Watson L, Watson M, Bower C. Periconceptional supplementation with folate and/or multivitamins for preventing neural tube defects. Cochrane Database Syst Rev 2001; CD001056.

[68] Tinker SC, Hamner HC, Berry RJ, Bailey LB, Pfeiffer CM. Does obesity modify the association of supplemental folic acid with folate status among nonpregnant women of childbearing age in the United States? Birth Defects Res A Clin Mol Teratol 2012;94:749–55.

[69] Casazza K, Pate R, Allison DB. Myths, presumptions, and facts about obesity. N Engl J Med 2013;368:2236–7.

[70] Bessesen DH. Update on obesity. J Clin Endocrinol Metab 2008;93:2027–34.

[71] Anderson JW, Konz EC, Frederich RC, Wood CL. Long-term weight-loss maintenance: a meta-analysis of US studies. Am J Clin Nutr 2001;74:579–84.

[72] Bravata DM, Smith-Spangler C, Sundaram V, Gienger AL, Lin N, Lewis R, et al. Using pedometers to increase physical activity and improve health: a systematic review. JAMA 2007;298:2296–304.

[73] Egan AM, Mahmood WA, Fenton R, Redziniak N, Kyaw Tun T, Sreenan S, et al. Barriers to exercise in obese patients with type 2 diabetes. QJM 2013;106:635–8.

[74] Chan EW, He Y, Chui CS, Wong AY, Lau WC, Wong IC. Efficacy and safety of lorcaserin in obese adults: a meta-analysis of 1-year randomized controlled trials (RCTs) and narrative review on short-term RCTs. Obes Rev 2013;14:383–92.

[75] Manning S, Pucci A, Finer N. Pharmacotherapy for obesity: novel agents and paradigms. Ther Adv Chronic Dis 2014;5:135–48.

[76] Gadde KM, Allison DB, Ryan DH, Peterson CA, Troupin B, Schwiers ML, et al. Effects of low-dose, controlled-release, phentermine plus topiramate combination on weight and associated comorbidities in overweight and obese adults (CONQUER): a randomised, placebo-controlled, phase 3 trial. Lancet 2011;377:1341–52.

[77] Sjöström L, Narbro K, Sjöström CD, Karason K, Larsson B, Wedel H, Study SOS. et al. Effects of bariatric surgery on mortality in Swedish obese subjects. N Engl J Med 2007;357:741–52.

[78] Neff KJ, le Roux CW. Bariatric surgery: a best practice article. J Clin Pathol 2013;66:90–8.

[79] Shah M, Simha V, Garg A. Review: long-term impact of bariatric surgery on body weight, comorbidities, and nutritional status. J Clin Endocrinol Metab 2006;91:4223–31.

[80] Maggard MA, Yermilov I, Li Z, Maglione M, Newberry S, Suttorp M, et al. Pregnancy and fertility following bariatric surgery: a systematic review. JAMA 2008;300:2286–96.

[81] Galloway M, Rushworth L. Red cell or serum folate? Results from the National Pathology Alliance benchmarking review. J Clin Pathol 2003;56:924–6.

[82] Chu SY, Callaghan WM, Bish CL, D'Angelo D. Gestational weight gain by body mass index among US women delivering live births, 2004–2005: fueling future obesity. Am J Obstet Gynecol 2009;200:e1–7. 271.

[83] Mamun AA, O'Callaghan M, Callaway L, Williams G, Najman J, Lawlor DA. Associations of gestational weight gain with offspring body mass index and blood pressure at 21 years of age: evidence from a birth cohort study. Circulation 2009;119:1720–7.

[84] Nohr EA, Vaeth M, Baker JL, Sorensen T, Olsen J, Rasmussen KM. Combined associations of prepregnancy body mass index and gestational weight gain with the outcome of pregnancy. Am J Clin Nutr 2008;87:1750–9.

B. BARIATRIC SURGERY, OBESITY, AND FERTILITY

[85] Egan AM, Dennedy MC, Al-Ramli W, Heerey A, Avalos G, Dunne F. ATLANTIC-DIP: excessive gestational weight gain and pregnancy outcomes in women with gestational or pregestational diabetes mellitus. J Clin Endocrinol Metab 2014;99:212–9.

[86] Black MH, Sacks DA, Xiang AH, Lawrence JM. The relative contribution of prepregnancy overweight and obesity, gestational weight gain, and IADPSG-defined gestational diabetes mellitus to fetal overgrowth. Diabetes Care 2013;36:56–62.

[87] Institute of Medicine (US) and National Research Council (US) Committee to Reexamine IOM Pregnancy Weight Guidelines. In: Rasmussen KM, Yaktine AL, editors. Weight gain during pregnancy: reexamining the guidelines. National Academies Press, Washington, DC; 2009.

[88] Haugen M, Brantsæter AL, Winkvist A, Lissner L, Alexander J, Oftedal B, et al. Associations of pre-pregnancy body mass index and gestational weight gain with pregnancy outcome and postpartum weight retention: a prospective observational cohort study. BMC Pregnancy Childbirth 2014;14:201.

[89] Kinnunen TI, Raitanen J, Aittasalo M, Luoto R. Preventing excessive gestational weight gain – a secondary analysis of a cluster-randomised controlled trial. Eur J Clin Nutr 2012;66: 1344–50.

[90] Thangaratinam S, Rogozinska E, Jolly K, Glinkowski S, Roseboom T, Tomlinson JW, et al. Effects of interventions in pregnancy on maternal weight and obstetric outcomes: meta-analysis of randomised evidence. BMJ 2012;344:e2088.

[91] Oteng-Ntim E, Varma R, Croker H, Poston L, Doyle P. Lifestyle interventions for overweight and obese pregnant women to improve pregnancy outcome: systematic review and meta-analysis. BMC Med 2012;10:47.

[92] Catalano PM, Mele L, Landon MB, Ramin SM, Reddy UM, Casey B, Network EKSNIoCHaHDM-FMU. et al. Inadequate weight gain in overweight and obese pregnant women: what is the effect on fetal growth? Am J Obstet Gynecol 2014;.

[93] Glazer NL, Hendrickson AF, Schellenbaum GD, Mueller BA. Weight change and the risk of gestational diabetes in obese women. Epidemiology 2004;15:733–7.

[94] Horvath K, Koch K, Jeitler K, Matyas E, Bender R, Bastian H, et al. Effects of treatment in women with gestational diabetes mellitus: systematic review and meta-analysis. BMJ 2010;340: c1395.

[95] Rowan JA, Hague WM, Gao W, Battin MR, Moore MP, Investigators MT. Metformin versus insulin for the treatment of gestational diabetes. N Engl J Med 2008;358:2003–15.

[96] Rowan JA, Rush EC, Obolonkin V, Battin M, Wouldes T, Hague WM. Metformin in gestational diabetes: the offspring follow-up (MiG TOFU): body composition at 2 years of age. Diabetes Care 2011;34:2279–84.

[97] Barrett HL, Nitert MD, Jones L, O'Rourke P, Lust K, Gatford KL, et al. Determinants of maternal triglycerides in women with gestational diabetes mellitus in the metformin in gestational diabetes (MiG) study. Diabetes Care 2013;36:1941.

[98] Carlsen SM, Martinussen MP, Vanky E. Metformin's effect on first-year weight gain: a follow-up study. Pediatrics 2012;130:e1222–6.

[99] Norman J. A multicentre randomised placebo controlled clinical trial of metformin versus placebo in pregnant women to reduce the risk of obesity and metabolic syndrome in their babies. Current Controlled Trials. http://www.controlled-trials.com/ISRCTN51279843; [accessed 08.08.14].

[100] American Diabetes Association. Standards of medical care in diabetes – 2013. Diabetes Care 2013;36(Suppl 1):S11–66.

[101] CMACE/RCOG Joint Guideline. Management of Women with Obesity in Pregnancy; 2010. http://www.rcog.org.uk/files/rcog-corp/CMACERCOGJointGuidelineManagementWomenObesityPregnancya.pdf; [accessed 08.08.2014].

[102] MRC Vitamin Study Research Group. Prevention of neural tube defects: results of the Medical Research Council Vitamin Study. Lancet 1991;338:131–7.

[103] National Institute for Health and Care Excellence. Hypertension in pregnancy: NICE clinical guideline 107; 2010. http://www.nice.org.uk/guidance/CG107; [accessed 08.08.2014].

[104] American College of Obstetricians and Gynecologists (ACOG). ACOG Practice Bulletin No. 120: use of prophylactic antibiotics in labor and delivery. Obstet Gynecol 2011;117:1472–83.

[105] Mok E, Multon C, Piguel L, Barroso E, Goua V, Christin P, et al. Decreased full breastfeeding, altered practices, perceptions, and infant weight change of prepregnant obese women: a need for extra support. Pediatrics 2008;121:e1319–24.

[106] O'Reilly MW, Avalos G, Dennedy MC, O'Sullivan EP, Dunne F. Atlantic DIP: high prevalence of abnormal glucose tolerance postpartum is reduced by breast-feeding in women with prior gestational diabetes mellitus. Eur J Endocrinol 2011;165:953–9.

EXERCISE AND LIFESTYLE IN FERTILITY

13

Lifestyle Factors and Reproductive Health

Ashok Agarwal, PhD, Damayanthi Durairajanayagam, PhD***

*Center for Reproductive Medicine, Cleveland Clinic, Cleveland, Ohio, USA
**Department of Physiology, Faculty of Medicine, MARA University of Technology,
Selangor, Malaysia

INTRODUCTION

Infertility is a worldwide problem that affects about 10–15% of couples within their reproductive ages [1]. One third of these couples could attribute their infertility issues to the male partner, another one third to the female partner, and the remaining one third due to a combined effect. A further 10% of infertile couples face idiopathic infertility, when they are unable to conceive due to no apparent reason. Whatever the reason for infertility, there are an increasing number of couples who are unable to conceive despite having regular, unprotected intercourse for over a year.

In fact, human fertility rates are arguably declining with time [2,3] and the exact causes for the decline has not yet been determined [4]. However, there are many factors that may be attributed to this gradual decrease in fertility, including lifestyle habits, environmental factors, occupational exposure, and medical causes. Lifestyle habits refer to those habits that are mostly within our control and those that are resulting from our daily habits and rituals, behavior, and personal choices that invariably affect reproductive parameters in men and women. In this review, we will discuss the effects of common lifestyle factors on the reproductive health of couples within the reproductive age group.

EFFECTS OF LIFESTYLE FACTORS ON FERTILITY

The ability to conceive, either spontaneously or through assisted reproduction, ensure a healthy pregnancy, and carry a fetus to live birth may be influenced by multiple factors, including modifiable factors and habitual practices that could influence the fertility potential of an individual. Lifestyle factors may affect the male partner's fertility, which could manifest as poor semen quality, or it may affect the female partner, who may present with longer time to achieve pregnancy or be susceptible to miscarriages.

In this section, individual lifestyle factors, such as age, smoking tobacco, extremes of body weight, physical activity or a lack of it, stress, antioxidant supplementation and dietary practices, alcohol and caffeine consumption, use of illicit or prescription drugs, and sexually transmitted infections, are examined. Other factors, such as lubricant use and lack of sleep and the evidence that each of these factors may pose a threat to reproductive health in men and women, are also discussed.

Age

The age at which a couple decides to pursue conception is an important factor as fertility reduces with time. In men, testosterone levels decrease with age and result in hypogonadism [5]. Demographic data from a study suggests that male fertility begins a steady decline from 35 to 39 years onward, with a decline of about 22% per annum [6]. Semen quality deteriorates with aging, as sperm displays age-related reduction in motility and viability [7]. Advanced male age is associated with lower semen volume, sperm motility, and morphologically normal sperm but not with sperm concentration [8]. Infertile men aged 40 and above were found to have increased levels of DNA damage [9]. As men grow older, their time to achieve pregnancy also increases: men >45 years of age take a longer time to conceive and it could be more than 1 to 2 years compared to men <25 years of age [10]. Advanced paternal age seems

Handbook of Fertility. http://dx.doi.org/10.1016/B978-0-12-800872-0.00013-5

to be associated with an increased risk of spontaneous abortion and autosomal dominant conditions, autism spectrum disorders, and schizophrenia [11]. In assisted reproduction, advanced paternal age was found to be related with declining implantation rates and had a poorer assisted reproductive technology (ART) outcome compared to younger men [12].

Women nowadays tend to postpone childbearing into their 30s and 40s. This trend could be due to multiple reasons – placing greater emphasis on advancing their education or building a career, financial instability, or late marriages [13]. In women, oocyte quantity and quality declines with time [14]. Furthermore, as women age, their time to pregnancy (TTP) increases [15] while their ability to maintain pregnancy declines [6]. Women below 30 years of age have a higher likelihood of conceiving (71%) compared to women over 36 years of age (41%) [15].

Child-bearing at a later age is associated with increased obstetrical and perinatal complications. Pregnant women, 35 years and older, had a higher incidence rate of pregnancy complications (hypertensive disorders of pregnancy and breech presentation), cesarean delivery, and perinatal death compared to women aged between 20 and 34 years [16]. Advanced maternal age (>35 years of age) also predisposes women to adverse pregnancy outcomes such as severe maternal outcome, early neonatal or perinatal mortality, low birth weight, and stillbirth [17].

Cigarette Smoking

According to the Centers for Disease Control and Prevention, tobacco use remains the leading preventable cause of disease and mortality in the United States, and is well documented to cause cardiovascular and respiratory diseases as well as cancer. About 22% of the world's population aged 15 years and older are cigarette smokers [18]. Among the reproductive age group in the United States, about 35% of males and 30% of women are smokers [19]. Exposure to tobacco may come in the active form of smoking cigarettes, cigars, and pipes or the use of smokeless tobacco, or by passive exposure to second-hand smoke. Passive smoking by either partner exerts a negative effect on sperm parameters that almost equals the adverse effects of active smoking [20]. Chewing tobacco or the use of smokeless tobacco also caused a reduction in sperm count, motility, viability, and morphology [21].

In fact, smoking worsens sperm quality in both fertile and infertile men. Regardless of fertility status, men who smoke demonstrated lower total sperm count and sperm motility, and a higher percentage of sperm with abnormal morphology [22]. In addition, male smokers have lower antioxidant activity [23], sperm density [24], and

sperm penetration [25] compared to that of nonsmokers. Among couples expecting a baby, the study of Hull et al. found that if the male partner smoked actively and the female partner smoked either actively or passively, then these couples had a longer TTP compared to women who were free from tobacco smoke exposure [20]. Regardless of the smoking status of the female, men who smoke heavily experienced delayed conception [20]. In assisted reproduction, male smokers have significantly lower success rates with intracytoplasmic sperm injection and *in vitro* fertilization cycles [26].

Several meta-analyses have been carried out to determine how smoking can influence fertility and the clinical outcome of assisted reproduction in women. Female smokers were found to have higher odds of infertility compared to nonsmokers (OR 1.6, 95% CI, 1.34–1.91) [27]. Women smokers attempting ART cycles have higher odds of spontaneous miscarriage (OR 0.65, 95% CI, 1.33–3.50) and lower live birth per cycle (OR 0.54, 95% CI, 0.3–0.99) [28]. These women also had higher odds of an ectopic pregnancy (OR 15.69, 95% CI, 2.87–85.76) and lower odds of clinical pregnancy (OR 0.56, 95% CI, 0.43–0.73) [28].

Smoking can potentially affect female fertility in several ways: by increasing FSH levels (leading to a shorter follicular phase) and decreasing progesterone levels during the luteal phase of a cycle [29], by reducing the ovarian reserve [30], by targeting the uterine tube (poor oocyte pick up and carriage of fertilized embryo along the oviduct) [31], and by disrupting the uterine environment [32].

Despite strong evidence pointing to a dose-dependent effect on poorer sperm parameters in smokers compared to nonsmokers, currently available data are still insufficient to elucidate the exact mechanisms on how smoking decreases male fertility [19]. However, given the multiple health risks associated with cigarette smoking in men and women, smoking cessation is warranted to maintain or restore the fertility potential in current smokers [33]. To compound matters further, smoking is associated with alcohol intake among men [34]. The effects of alcohol intake on reproductive health are discussed further on in the article (section on "Alcohol and Consumption").

Extremes of Body Weight and Exercise

Overweight and Obesity

There is an increasing trend of being overweight and obese in populations across the globe. This trend could be attributed to changing dietary composition, more energy intake, and less exercise [35]. Extremes of body weight (being obese or being underweight) could disrupt the hormonal balance and cause ovulatory dysfunction.

Besides affecting reproductive health, obesity also increases the risks of cardiovascular diseases, diabetes mellitus, metabolic syndrome, and certain cancers [36]. Diabetes mellitus and metabolic syndrome are systemic diseases known to have a negative impact on fertility [37].

In a meta-analysis of 21 studies ($n = 13,077$), compared to men with normal body weight, the association of total sperm count being oligozoospermic or azoospermic in underweight men is 1.15, overweight men is 1.11, obese men is 1.28, and morbidly obese men is 2.04. The study concluded that being overweight and obese were associated with an increased prevalence of oligozoospermia or azoospermia [38]. However, a previous meta-analysis of five studies investigating total sperm count and sperm concentration did not show any evidence of association of semen parameters with body mass index (BMI). Instead, researchers found that an increase in male BMI was associated with a reduction in testosterone [39]. Another study reviewed that increase in male BMI also causes a reduction in sperm concentration, increases sperm with poor motility and abnormal morphology, and compromises chromatin integrity [40]. In men, a high BMI is associated with impaired semen parameters. Similarly, a large waist circumference (WC) is associated with poor semen parameters in infertile men [41].

BMI is a measure of body fat in height and weight: a BMI of <18.5 denotes an underweight individual, BMI between 18.5 and 24.9 represents an individual with normal body weight, BMI >25 denotes an overweight individual, while a BMI >30 represents an obese individual [42]. In a study among married infertile women ($n = 2527$), the relative risk for ovulatory infertility was 1.3 in women with BMI between 24 and 31 kg/m^2 compared to 2.7 in women with BMI >32 kg/m^2, suggesting that excess body weight is a major cause of ovulatory infertility [43] along with a sedentary lifestyle [44].

Obese women were also associated with increasing levels of insulin, lactate, triglycerides, and C-reactive protein, and decreasing levels of sex-hormone-binding globulin (SHBG) in their follicular fluid, which could impair oocyte development [45]. Overweight and obese women have higher total testosterone levels compared to women with normal weight [46,47]. Excess body weight is also associated with polycystic ovulatory syndrome (PCOS), which is linked to hyperandrogenism, anovulation, and infertility [47]. In addition, excess body weight exacerbates problems related to PCOS, such as hyperandrogenism, menstrual disturbances, infertility [48], insulin resistance, and dyslipidemia [46].

Studies of obese women have reported that high BMI levels are associated with reduced fecundity [49–53]. Obesity in men is associated with lower testosterone and inhibin B levels [54,55] and reduced fecundity [53,56]. Semen quality was found to be poorer in some studies [55,57] but not others [54,58].

Underweight

Women with low body fat percentage (BMI <18.5) are more likely to have ovarian disorder and infertility [59]. Women who were severely underweight (BMI <17) had an increased risk of ovulatory infertility (RR 1.6) [60]. Fertility in underweight women could be hampered by increased FSH levels [61], secondary amenorrhea, and shortened luteal phase [62]. Women who are underweight (BMI <18.5–20) were found to have poorer clinical pregnancy rates compared to women with normal weight [63]. In addition, underweight women were found to have higher pregnancy loss (11%) compared to women with normal BMI (0.5%), suggesting that underweight women were at a higher risk of miscarriage [64]. A meta-analysis of 78 studies ($n = 1,025,794$) found that underweight women (BMI <20.0) had an increased risk of preterm birth and delivery of a low-birth-weight infant compared to women with normal weight [65].

Physical Activity and Weight Loss

Physical activity has a protective effect on fertility along with weight loss. In infertile obese women who experienced weight loss (about 10 kg/m^2) through diet and exercise intervention, resumed spontaneous ovulation, achieved spontaneous pregnancy, had lower miscarriage rates, and increased live birth [66]. Obese, infertile women (BMI >30) who previously could not achieve conception and were engaged in a diet and exercise regime for at least 6 months (reduction in BMI of about 9.6) had improved spontaneous conception rate and with ovulation induction therapy had improved pregnancy outcome [67]. Overweight infertile patients (BMI ≥25 mg/m^2) who achieved a 10% weight loss through diet and exercise regime had higher conception and live birth rates compared to the women who did not [68]. In women with PCOS, engaging in aerobic exercise helped improve ovarian volume and the number of follicles [69].

In a study on young, healthy men ($n = 189$) evaluating the relationship between physical activity (in the form of moderate/vigorous exercise) versus sedentary behavior (watching TV) on semen quality, men who had longer hours of physical activity (≥15 h/week vs. <5 h/week) had 73% more sperm concentration and total count, while men who were sedentary for longer periods (>20 h/week vs. 0 h/week of watching TV) had 44% less sperm concentration and total count. However, there was no significant relationship obtained for sperm motility or morphology [70]. Physically active men who exercised in moderation (at least 1 h × 3 sessions/week) had better sperm quality (mostly in morphology) compared to men who exercised more vigorously and frequently [71]. Men from couples attending ART clinics ($n = 2261$) who engaged in bicycling (≥5 h/week) had low total motile sperm counts (OR 2.05) and sperm concentration (OR 1.92) compared to those who did not exercise regularly [72]. Regardless of BMI, physical activity in moderate amounts is weakly correlated to better fecundity in women [73].

Psychological Stress

Stress can occur in multiple forms, be it psychological or mental, social, physical or biological, and trauma. Stressors may range from the routine pressures of daily life (work responsibilities and family life), life events or sudden negative lifestyle changes (moving homes, contracting an illness, employment issues, loss of a loved one, marital troubles) or trauma (involvement in serious accidents or natural disasters). It is safe to assume that in this day and age, most people experience multiple forms of stress either acutely, intermittently or chronically, during their everyday living.

In fact, infertility in itself is the most stressful occurrence among couples who desire parenthood but are unable to conceive naturally [74]. Couples are also faced with highly stressful situations during the various steps involved in assisted reproduction [75], especially when undergoing oocyte retrieval, embryo transfer, and failure to achieve pregnancy [76,77]. Unsuccessful treatment of a male partner's infertility is also a cause for his prolonged sadness [78].

In general, stress impacts the endocrine and immune systems and the autonomic nervous system [79]. In the male, stress has been shown to reduce testosterone levels [80] and spermatogenesis [81]. Furthermore, mental stress has a negative effect on sperm quality, and causes a reduction in sperm concentration and motility (or progressive sperm motility), and increases the number of sperm with abnormal morphology [82–87].

Physical forms of stress have been found to disrupt the female menstrual cycle [88]. In response to stress, the hypothalamus–pituitary–adrenal axis is activated, leading to an increase in plasma cortisol levels and a corresponding increase in salivary cortisol. On the other hand, the body's sympathetic response to stressors causes an increase in circulatory norepinephrine, which leads to a corresponding increase in salivary alpha-amylase. Therefore, salivary cortisol and alpha-amylase could serve as biomarkers of stress [89]. Lynch et al. reported that women with higher amounts of salivary alpha-amylase had a nearly 30% decrease in fecundity (longer TTP) compared to women with lower amounts of alpha-amylase. Thus, women who were psychologically stressed had a greater than twofold increased risk of infertility (RR 2.07) [89].

Nutritional Status and Dietary Practices

Antioxidants

In the female, reactive oxygen species play a physiological role in oocyte maturation, steroidogenesis in the ovaries, ovulation and implantation, formation of fluid-filled cavity, blastocyst, luteolysis, and maintenance of the corpus luteum in pregnancy. In the male, physiological levels of reactive oxygen species play a role in sperm maturation, capacitation and hyperactivation, acrosome reaction, and sperm–oocyte fusion [90]. However, an imbalance in reactive oxygen species and antioxidant levels lead to a state of oxidative stress, either due to an increase in reactive oxygen species or a reduction in antioxidant capacity. Oxidative stress causes protein and lipid peroxidation, disruption of sperm membrane fluidity, and poor DNA integrity, which leads to sperm with poor motility and DNA damage [90].

Antioxidants are scavengers of reactive oxygen species and reactive nitrogen species, which are derivatives of oxygen and nitrogen, respectively. Antioxidants are naturally available *in vivo* in the seminal plasma [91] in males and in the female reproductive tract. In males, antioxidant intake lessens oxidative stress and DNA damage in infertile men, especially in asthenospermic men [92]. Fertility problems in the female could be caused by ovulatory problems, eggs of poor quality, damaged fallopian tube, and endometriosis. The state of oxidative stress brought on by these conditions in the female may be alleviated by antioxidant supplementation [93]. By quenching the reactive oxygen species to maintain it at homeostatic levels, antioxidants are able to prevent oxidative stress and its deleterious effects, thereby protecting the gamete.

In males, a Cochrane review of 34 studies ($n = 2876$) showed that compared to control, oral antioxidant intake in male partners of ART couples is associated with a significant increase in pregnancy rates (pooled OR 4.18, 95% CI 2.65–6.59, $P < 0.00001$, $n = 964$) and live birth rates (pooled OR 4.85, 95% CI 1.92–12.24, $P = 0.0008$, $n = 214$) [94]. An earlier systematic review of 17 randomized studies found that oral antioxidant supplementation (vitamins C and E, zinc, selenium, folate, carnitine, and carotenoids) in infertile men increased either sperm quality or pregnancy rates [95]. A meta-analysis of male infertility patients showed that oral intake of coenzyme Q_{10} ($n = 149$) enhanced sperm parameters by increasing coenzyme Q_{10} concentration in seminal plasma (RR 49.55), sperm concentration (RR 5.33), and sperm motility (RR 4.50), but not pregnancy or live birth rates [96]. Other studies showed varying results; for example, vitamin C improved semen quality, but not vitamin E or selenium [97]. However, vitamin E and selenium were found to decrease MDA levels (a marker of oxidative stress damage) more than vitamin B [98].

In females, a Cochrane review of 28 studies ($n = 3548$) showed that compared to control, standard treatment, or no treatment, supplementation with oral antioxidants (pentoxifylline, *N*-acetyl-cysteine, melatonin,

L-arginine, myoinositol, vitamins C and E, vitamin D + calcium, and omega-3-polyunsaturated fatty acids) in subfertile women were not associated with increased live birth rate (OR 1.25, 95% CI, $P = 0.82$, 2 RCTs, $n = 97$) nor was it associated with increased clinical pregnancy rate (OR 1.30, 95% CI, $P = 0.92$, 13 RCTs, $n = 2441$). However, a few studies found that oral intake of pentoxifylline was associated with increased clinical pregnancy rate versus control or no treatment (OR 2.03, 95% CI $P = 0.009$, 3 RCTs, $n = 276$). In the same meta-analysis, 14 studies reported of women on oral antioxidant therapy having adverse effects such as miscarriage, ectopic or multiple pregnancy, and gastrointestinal disturbances, but these did not differ from women who were not on antioxidant supplementation [93]. In women with unexplained infertility, increased oral intake of antioxidants from dietary supplements was associated with shorter TTP. The following groups of women had a shorter TTP: women (BMI <25 kg/m^2) with a higher intake of vitamin C (HR 1.09); women (BMI ≥ 25 kg/m^2) with a higher intake of β-carotene (HR 1.29); women aged <35 years with a higher intake of β-carotene (HR 1.19) and vitamin C (HR 1.10); and women aged $\geq$ 35 years with a higher intake of vitamin E (HR 1.07) had shorter TTP [99].

In summary, in females, there was low quality evidence in associating oral antioxidant therapy with no antioxidants/different types of antioxidants [93]. However, in infertile males, some of the effective antioxidants supplemented are carnitine, selenium, coenzyme Q$_{10}$, vitamin C, and glutathione [100].

Diet

In males, a diet high in fiber, carbohydrates, lycopene and folate [97], fruits, and vegetables [101], and low in proteins and fats [97], are correlated to higher semen quality. Men who consumed more processed red meat had lower total sperm count and fewer morphologically normal sperm compared to men who consumed less processed meat. However, men who ate more fish had better sperm counts and more morphologically normal sperm compared to men who ate less fish [102,103]. Men who had higher total fat intake had poorer total sperm count and sperm concentration, while men who had higher intake of omega-3 polyunsaturated fats had more morphologically normal sperm [104]. Higher intake of trans fatty acids was found to lower total sperm count in young healthy men [105] and in infertile men [106]. Independent of overall dietary patterns, increased intake of total full fat dairy food (mainly cheese) reduced sperm progressive motility and increased the number of sperm with abnormal morphology [107]. Smokers (past or current) who had high intake of cheese had poorer sperm concentration compared to those who consumed less cheese. Instead, men who have higher intake of low-fat dairy, especially low-fat milk, have better sperm concentration and motility compared to those who consumed less low-fat dairy [108]. A study proposed that young healthy men who consumed more sugar-sweetened beverages were associated with poorer sperm motility, especially in lean men. However, this association was not seen in overweight/obese men [109].

In women, substituting carbohydrates with animal protein (namely chicken or turkey meat) lowered ovulatory fertility (OR 1.18), but substituting carbohydrates with vegetable-based protein protected ovulatory fertility (OR 0.5) [110]. The risk of ovulatory infertility increased in women who consumed trans-fats instead of carbohydrates (RR 1.73) or monounsaturated fats (RR 1.31) [111]. Women who took thiamine (vitamin B1) (RR 0.84), vitamin C (RR 0.81), folate (vitamin B9) (RR 0.79), and vitamin E (RR 0.70) sourced from only their food had lower risk of endometriosis, but those who took these multivitamins as oral supplements showed no documented link to the risk of endometriosis. The authors suggested that the protective mechanism from endometriosis was conferred as a whole from foods rich not only in these multivitamins (vitamins B, C, and E) but from the other components also present in those foods [112]. Women who had a higher intake of total and low-fat dairy foods and those who had higher predicted vitamin D (25(OH)D) levels were associated with a lower risk of endometriosis (18 and 24%, respectively) [113].

In summary, couples who are planning for conception should pay attention to their diet as it could affect ovulation and sperm quality. On the whole, a diet high in fruits and vegetables, legumes and whole grains, fish, and chicken were associated with sperm with better progressive motility [114]. Women, who chose vegetable over animal protein and high-fat over low-fat dairy, who ate more monounsaturated fats compared to trans-fats and more iron and multivitamins, plus who had an overall lower glycemic load, had lower infertility rates due to ovulatory disorders [115].

Alcohol Consumption

The effect of alcohol consumption on human fertility is dependent on the quantity consumed. There is no current threshold dose to indicate how much alcohol consumed will increase the risk of infertility. However, not all types of alcoholic drinks may have a similar negative effect on sperm parameters. For example, beer and spirits were found to have a negative association with sperm concentration and total sperm count, but not wines [116]. In general, most studies find that alcohol intake affects male infertility by reducing semen volume, increasing the number of sperm with abnormal morphology, and by causing leukocytospermia. However, cessation of alcohol intake may cause a partial reversal of these parameters [117]. Heavy alcohol consumption has a borderline association with sperm motility, but not with sperm count, concentration, or morphology [34]. Smokers who drink have abnormal sperm morphology compared to smokers who do not [34]. Men who consume high amounts of alcohol (>8 drinks/week or >40 g/day) are reported to have disrupted spermatogenesis [118]. Men whose alcohol intake was >80 g/day ($n = 44$) displayed partial or complete spermatogenic arrest ($n = 23$) or Sertoli-only cell syndrome ($n = 5$) [119].

In women, alcohol consumption modulates estrogen and progesterone levels causing menstrual disorders and ovulatory irregularities [120]. Alcohol consumption may also lengthen the time to achieve pregnancy [121]. In a study looking at alcohol intake in fertile ($n = 3833$) and women seeking infertility treatment ($n = 1050$), researchers found an increase in ovulatory disorders with heavy alcohol use (OR 1.6) compared to nondrinkers. Endometriosis risk was about 50% higher in infertile women who consumed either moderate (OR 1.6) or high (OR 1.5) amounts of alcohol compared to those who did not. Thus, alcohol consumption of about 7–8 drinks per day (moderate levels) in women is associated with lower fecundity and a higher risk of spontaneous abortion [122]. Consumption of moderate amounts of alcohol is also associated with decreased concentration of plasma antioxidants and increased concentration of isoprostanes (a marker of oxidative stress *in vivo*) [123]. Young women (25–35 years) trying to achieve pregnancy for the first time displayed lower conception rates with an alcohol intake of five or less drinks a week [124]. In fact, even one alcoholic drink per week could lower conception rates [125].

Alcohol ingestion during pregnancy increases the risk of spontaneous abortion, premature birth, low birth weight, fetal death, and fetal alcohol spectrum disorder [126,127]. In fact, the American College of Obstetricians and Gynecologists and the American Academy of Pediatrics have concluded that alcohol abstinence during pregnancy should be recommended to women planning for pregnancy and to pregnant women [128].

Caffeine Use

Caffeine (1,3,7-trimethylxanthine), an alkaloid, is present naturally in the seeds, fruit, and leaves of coffee, tea, and cacao trees and many other plants [129]. The commonly consumed foods and beverages that contain caffeine are coffee, tea, cola/soft drinks, energy drinks, chocolate, and cocoa products [130]. Intake of dietary caffeine (2–3 cups of brewed coffee/3.3 mg/kg) stimulates the central nervous system, elevates stroke volume and

cardiac output in women, increases vascular resistance in men, and increases blood pressure in both [131].

In men, there is no conclusive evidence as yet on the effect of caffeine consumption relating to sperm parameters [34,132]. Caffeine and alcohol intake are also unrelated to semen parameters [34]. However, in a study of 2554 men, high caffeine intake (>800 mg/day) or high consumption of cola (>14 0.5 L bottle/week) were negatively associated with sperm concentration and total sperm count, but this relationship was significant only in cola drinkers [133]. In an earlier study on 343 sons born of women who consumed coffee during pregnancy, sons whose current caffeine consumption (from coffee and cola) was between 175 and 1075 mg ($n = 62$) had higher testosterone levels compared to sons who drank moderate or low amounts of caffeine. However, sons whose mother's daily coffee intake was high during pregnancy (825–1225 mg, $n = 47$), had lower testosterone levels than sons whose mothers had lower daily coffee intake during pregnancy [134].

In women, high caffeine intake (>500 mg/day) has been associated with prolonged TTP (OR 1.45), with a higher risk in smokers (OR 1.56) than in nonsmokers [135]. Among nonsmoking women, the risk for spontaneous abortion increased in relation to the amount of caffeine consumed: moderate 100–299 mg/day (OR 1.3) versus high >500 mg/day (OR 2.2). However, the risk for spontaneous abortion was not higher in smokers who consumed the same amount of caffeine compared to nonsmokers [136]. Greater consumption of coffee is also linked to a higher risk of fetal death; pregnant women who drank eight cups or more of coffee per day had higher risk for fetal death (OR 1.59) compared to mothers who drank less between 1 or 2–3 cups/day (OR 1.03). Fetal death in these women who consumed large amounts of coffee during pregnancy occurred mainly after the completion of 20 weeks of gestation [137].

In general, while the relationship between caffeine intake and sperm quality is not evident, high caffeine intake may likely be detrimental to women's reproductive health by increasing time to achieve pregnancy and causing miscarriages, spontaneous abortion, or fetal deaths.

Drugs

Illicit Drugs

Among the illicit drugs that have a negative impact on male fertility are marijuana, opioid narcotics, cocaine, methamphetamines and Ecstasy, and anabolic-androgenic steroids (testosterone derivatives) [138].

Marijuana is the most common illicit drug used among men in the reproductive age group in the United States [138] and around the world [139]. Smoking marijuana releases psychoactive cannabinoid compounds, which can act centrally and peripherally to impair the reproductive process. The cannabinoids contained in marijuana bind to endogenous cannabinoid receptors on sperm causing lower sperm motility by inhibition of sperm capacitation and acrosome reaction [138]. Besides sperm cells, endogenous cannabinoid receptors are found in the vas deference and testis in males and in the ovary and uterus in females [36]. In males, these cannabinoids block the hypothalamic-pituitary-gonadal (HPG) axis causing a reduction of LH and subsequently lower testosterone production in the Leydig cells. Furthermore, it also influences apoptosis of Sertoli cells causing a disruption in the spermatogenesis process [139]. The cannabinoid-induced lowering of testosterone levels occur in a dose-dependent manner and could reverse after 3 months of discontinuation [40].

While acute marijuana use reduces LH levels in women, chronic users tend to maintain their LH levels having developed tolerance to it [140]. Regular female users of marijuana have a higher risk of primary infertility compared to nonusers (RR 1.7) [141]. In women, cannabinoids can affect fertilization, transport in the oviduct, and the development of the fetus and placenta. However, it does not affect the quality of the egg [139]. Exposure to marijuana *in utero* is correlated with congenital abnormalities, low birth rates, and stillbirth [140].

Anabolic Steroids

Anabolic-androgenic steroids are taken by men for enhancing muscle strength and performance [36]. Anabolic steroids suppress the HPG axis via negative feedback [142], leading to hypogonadotropic hypogonadism and inhibition of endogenous testosterone production, which in turn results in disrupted spermatogenesis and erectile dysfunction (ED) [143]. High intake of anabolic steroids reduces sperm count and motility, and increases the number of sperm with abnormal morphology [143]. The negative impact on sperm parameters are reversible upon discontinuation, but may take anywhere between 4 months and over a year to recover [142].

Prescription Drugs

Medications, such as antibiotics, antidepressants, and psychotropics, have a negative impact on semen parameters [144]. Nitrofurantoin is an antibiotic used to treat bacterial infections of the urinary tract. While short-term, low doses of nitrofurantoin do not affect spermatogenesis, high doses of nitrofurantoin causes maturation arrest in the testis [145]. Minocyclin (a broad-spectrum tetracycline) is sperm-toxic at any concentration [145], while erythromycin could negatively affect sperm concentration and motility [145,146]. Aminoglycoside antibiotics, gentamicin, and neomycin could directly inhibit spermatogenesis [143]. Antidepressants could impair sexual interest and disrupt erectile function [147,148]. The use of selective serotonin reuptake inhibitors (SSRIs) or tricyclic antidepressants could result in ED and reduced libido [149], impaired ejaculation, or cause hyperprolactinemia, which could reversibly suppress spermatogenesis [143].

Sexually Transmitted Infections

Sexually transmitted infections (STIs) are caused by bacteria, viruses, and parasites. STIs spread mainly through sexual contact, but can also infect through blood products, tissue transfer, and from mother to child during pregnancy and parturition. Most STIs are present without symptoms and can lead to chronic diseases; certain STIs can increase the risk of acquiring HIV by more than threefold [150]. In general, STIs in women are associated with tubal infertility and results in pregnancy complications, which include miscarriage, preterm labor and still birth [151].

Chlamydia (caused by the bacterium *Chlamydia trachomatis*), gonorrhea (caused by the bacterium *Neisseria gonorrhea*), and syphilis (caused by the spirochete *Treponema pallidum*) are common types of STI that can occur in men and women. In women, Chlamydia infections can damage the fallopian tubes, resulting in infertility. Chlamydia could also cause an ectopic pregnancy [152]. Chlamydial and gonorrheal infections (caused by the bacterium *N. gonorrhea*) can lead to pelvic inflammatory disease (PID), which is associated with female tubal infertility [151]. In males, chronic gonorrhea can cause urethral strictures and inflammation of the epididymis and/or testis [153]. Syphilis can be passed on from an infected mother to her unborn child. Chlamydia, gonorrhea, and syphilis are STIs treatable by antibiotics. However, if left untreated, syphilis can bring about serious, prolonged complications in women [152].

Human papillomavirus (HPV) infection is prevalent in sperm from sexually active men with and without risk factors for HPV [154] and may also be present in the semen of asymptomatic men [155]. The incidence of HPV infection is higher in infertile men and causes a decrease in sperm motility and sperm with normal morphology [156]. In women, HPV infections increase her risk of cancer of the cervix.

The human immunodeficiency virus (HIV) affects the T cells of the immune system and can lead to acquired immunodeficiency syndrome (AIDS). HIV can be passed on from an infected mother to her unborn child. A meta-analysis of 31 studies reported that the risks of adverse perinatal outcomes in mothers infected with HIV are spontaneous abortion (OR 4.05), stillbirth (OR 3.91), infant mortality (OR 3.69), low birth weight (OR 2.09), preterm delivery (OR 1.83), perinatal mortality (OR 1.79), intrauterine growth retardation (OR 1.7), neonatal mortality (OR 1.10), and fetal abnormality (OR 1.08) [157]. The incidence of an STI escalates the likelihood of HIV transmission [158].

Infections of the male accessory organs reduces sperm count [159] while prostatitis causes sexual dysfunction and poor sperm parameters [160]. Infections of the testis and epididymis are particularly detrimental to sperm cells as they have a longer contact time to the adverse effects of reactive oxygen species without adequate antioxidant protection [158].

OTHER FACTORS

Lubricants

The choice of lubricants used during intercourse may affect chances of conception. Certain types of commonly used vaginal, coital, or surgical lubricants are toxic to sperm and disrupt sperm motility and viability [161,162].

Sleep

A cross-sectional study on 953 men showed that sleep disturbances (self-reported) during the past month could lead to poorer sperm quality compared to men who had lesser sleep disturbances [163]. Although further studies are required to investigate the possibility of disturbed sleep reducing male fecundity, it does add another potential factor to the pool of risk factors of male fertility [164].

CONCLUSIONS AND RECOMMENDATIONS

Daily life comprises of multiple negative lifestyle factors that have been shown to contribute toward the decline in the fertility potential of men and women. Although the evidence available from studies performed on these factors vary in strength, any combination of negative lifestyle choices could have significant and cumulative impact on fecundity and warrants due attention. In an observational study in pregnant women ($n = 2112$) looking at the effects of different lifestyles on TTP, couples who had more than four negative lifestyle habits (e.g., smoking >15 cigarettes a day + >20 alcohol units/day + BMI >25 mg/kg^2 + >6 cups of coffee/tea, had a progressively longer time to achieve pregnancy (sevenfold longer), lower conception rates (less 60%), and greater relative risk to subfecundity (7.3-fold more likely) than those without negative variables [51]. To further add to the problem, the effect of other contributing factors that could influence reproductive health besides lifestyle practices, namely, occupational and environmental, clinical, and genetic factors, should also be taken into account in the overall success of a couple's quest to parenthood (Fig. 13.1).

In this sense, awareness about factors that contribute to subfertility and counseling on how to avoid or at least minimize its impact in each partner could increase the chances of a more favorable rate of conception and pregnancy, and eventually live birth. Knowledge about these potential causes of infertility is essential in men and women alike, in order to understand their own fertility potential. With knowledge comes the key ability to recognize the need for and to facilitate clinical help, if required, in the face of issues impacting fertility. It is equally important to know about factors that could boost optimal conditions for conceiving (the do's), as well as appreciate the factors that could contribute toward a decline in fertility potential (the don'ts). Not indulging in these negative lifestyle factors do not "increase" fertility per se; rather, it maintains baseline fertility [165], and avoids its premature deterioration in order to give natural conception a fighting chance.

Self-awareness of one's lifestyle choices and habits will allow for the identification of a particular set of risk factors that is most likely to pose a threat to one's fertility. Despite the mixed level of evidences associating each factor with reproductive potential, given that these factors may act in concert whatever the individual lifestyle of a couple is, its effect on fertility should be given due importance. Furthermore, most of the factors could be modified with a little conscious effort to change poor habits into healthier options. While these may not guarantee a successful pregnancy, adopting a healthier lifestyle would have an overall significant impact on the health status of the individual.

Physicians could also contribute by helping couples take the important initial step to identify the major modifiable lifestyle factors in couples having trouble conceiving and to educate patients on the various factors that could potentially impact their fertility and how interventional methods could help optimize their natural fertility. The goal would be to boost natural fertility for an optimal chance at natural conception, and only resort to clinical intervention or assisted reproduction when natural fertility ceases to be an option. Preconception lifestyle advice is essential in educating couples and the physician could recommend interventions to improve the overall health of the couple, for example, by incorporating moderate levels of physical activity, having a well-balanced and nutritious diet plan, to engage in various stress-relieving practices and get enough sleep, to reduce coffee and alcohol consumption, cessation of smoking, and so on. However, as important as it is to know the risk factors that could threaten fertility, it seems even more challenging to go

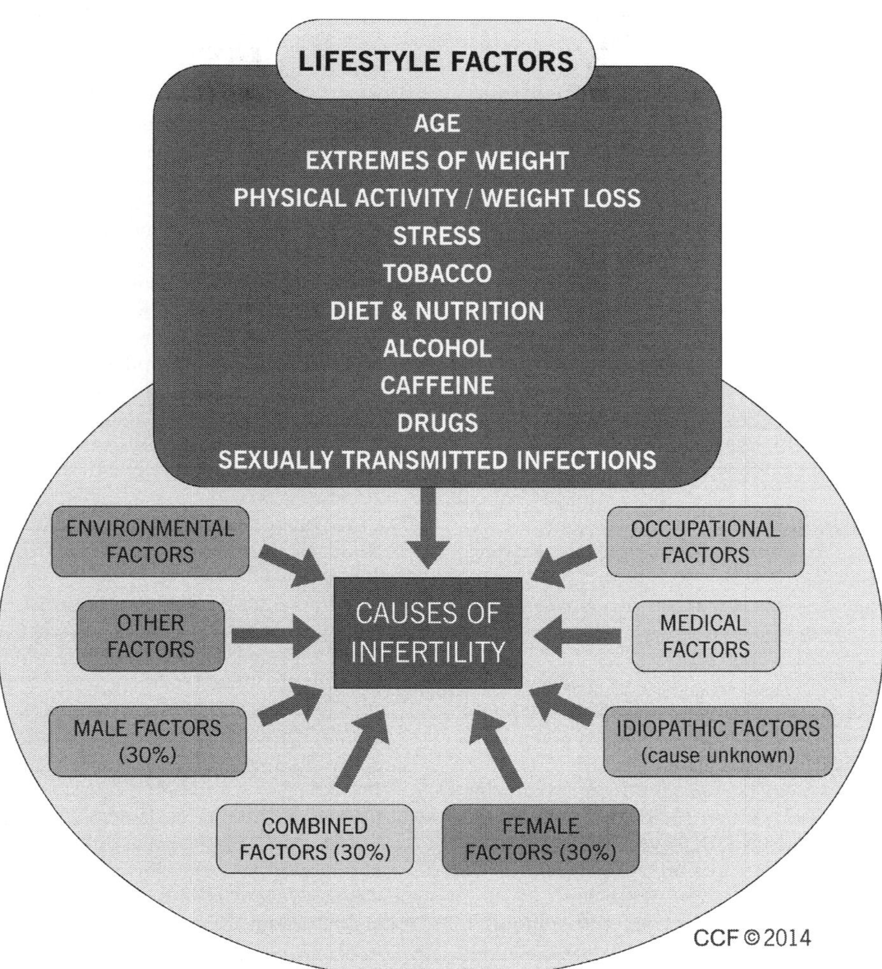

FIGURE 13.1 Factors that could contribute to infertility and hence negatively influence the reproductive health of a couple. Reprinted with permission, Cleveland Clinic Center for Medical Art & Artwork © 2014. All Rights Reserved.

beyond a short-term phase of change to successfully maintain the positive lifestyle changes in both partners permanently.

References

[1] Homan G, Litt J, Norman RJ. The FAST study: fertility assessment and advice targeting lifestyle choices and behaviours: a pilot study. Hum Reprod 2012;27(8):2396–404.

[2] Carlsen E, Giwercman A, Keiding N, Skakkebaek NE. Evidence for decreasing quality of semen during past 50 years. BMJ 1992;305(6854):609–13.

[3] Rostad B, Schmidt L, Sundby J, Schei B. Has fertility declined from mid-1990s to mid-2000s? Acta Obstet Gynecol Scand 2013;92(11):1284–9.

[4] Jurewicz J, Radwan M, Sobala W, Ligocka D, Radwan P, Bochenek M, et al. Lifestyle and semen quality: role of modifiable risk factors. Syst Biol Reprod Med 2014;60(1):43–51.

[5] Stewart AF, Kim ED. Fertility concerns for the aging male. Urology 2011;78(3):496–9.

[6] Matorras R, Matorras F, Exposito A, Martinez L, Crisol L. Decline in human fertility rates with male age: a consequence of a decrease in male fecundity with aging? Gynecol Obstet Invest 2011;71(4): 229–35.

[7] Tubman A, Check JH, Bollendorf A, Wilson C. Low hypo-osmotic swelling tests correlate with low percent motility and age of the male. Clin Exp Obstet Gynecol 2013;40(1):35–6.

[8] Kidd SA, Eskenazi B, Wyrobek AJ. Effects of male age on semen quality and fertility: a review of the literature. Fertil Steril 2001;75(2):237–48.

[9] Varshini J, Srinag BS, Kalthur G, Krishnamurthy H, Kumar P, Rao SB, et al. Poor sperm quality and advancing age are associated with increased sperm DNA damage in infertile men. Andrologia 2012;44(Suppl. 1):642–9.

[10] Hassan MA, Killick SR. Effect of male age on fertility: evidence for the decline in male fertility with increasing age. Fertil Steril 2003;79(Suppl. 3):1520–7.

[11] Liu K, Case A, Reproductive Endocrinology and Infertility Committee Family Physicians Advisory Committee Maternal-Fetal Medicine Committee Executive and Council of the Society of Obstetricians. et al. Advanced reproductive age and fertility. J Obstet Gynecol Can 2011;33(11):1165–75.

[12] Robertshaw I, Khoury J, Abdallah ME, Warikoo P, Hofmann GE. The effect of paternal age on outcome in assisted reproductive technology using the ovum donation model. Reprod Sci 2014;21(5):590–3.

[13] Gindoff PR, Jewelewicz R. Reproductive potential in the older woman. Fertil Steril 1986;46(6):989–1001.

[14] Nelson SM, Telfer EE, Anderson RA. The aging ovary and uterus: new biological insights. Hum Reprod Update 2013;19(1):67–83.

[15] Mutsaerts MA, Groen H, Huiting HG, Kuchenbecker WK, Sauer PJ, Land JA, et al. The influence of maternal and paternal factors on time to pregnancy – a Dutch population-based birth-cohort study: the GECKO Drenthe study. Hum Reprod 2012;27(2):583–93.

[16] Giri A, Srivastav VR, Suwal A, Tuladhar AS. Advanced maternal age and obstetric outcome. Nepal Med Coll J 2013;15(2):87–90.

[17] Laopaiboon M, Lumbiganon P, Intarut N, Mori R, Ganchimeg T, Vogel JP, et al. Advanced maternal age and pregnancy outcomes: a multicountry assessment. BJOG 2014;121(Suppl. 1):49–56.

[18] WHO. Prevalence of smoking. http://www.who.int/gho/tobacco/en/; 2014 [accessed 31.05.14].

[19] Pfeifer S, Fritz M, Goldberg J, McClure RD, Thomas M, Practice Committee of the American Society for Reproductive Medicine. et al. Smoking and infertility: a committee opinion. Fertil Steril 2012;98(6):1400–6.

[20] Hull MG, North K, Taylor H, Farrow A, Ford WC. Delayed conception and active and passive smoking. The Avon Longitudinal Study of Pregnancy and Childhood Study Team. Fertil Steril 2000;74(4):725–33.

[21] Said TM, Ranga G, Agarwal A. Relationship between semen quality and tobacco chewing in men undergoing infertility evaluation. Fertil Steril 2005;84(3):649–53.

[22] Li Y, Lin H, Li Y, Cao J. Association between socio-psycho-behavioral factors and male semen quality: systematic review and meta-analyses. Fertil Steril 2011;95(1):116–23.

[23] Pasqualotto FF, Umezu FM, Salvador M, Borges E Jr, Sobreiro BP, Pasqualotto EB. Effect of cigarette smoking on antioxidant levels and presence of leukocytospermia in infertile men: a prospective study. Fertil Steril 2008;90(2):278–83.

[24] Vine MF, Margolin BH, Morrison HI, Hulka BS. Cigarette smoking and sperm density: a meta-analysis. Fertil Steril 1994;61(1):35–43.

[25] Sofikitis N, Miyagawa I, Dimitriadis D, Zavos P, Sikka S, Hellstrom W. Effects of smoking on testicular function, semen quality and sperm fertilizing capacity. J Urol 1995;154(3):1030–4.

[26] Zitzmann M, Rolf C, Nordhoff V, Schräder G, Rickert-Föhring M, Gassner P, et al. Male smokers have a decreased success rate for *in vitro* fertilization and intracytoplasmic sperm injection. Fertil Steril 2003;79(Suppl. 3):1550–4.

[27] Augood C, Duckitt K, Templeton AA. Smoking and female infertility: a systematic review and meta-analysis. Hum Reprod 1998;13(6):1532–9.

[28] Waylen AL, Metwally M, Jones GL, Wilkinson AJ, Ledger WL. Effects of cigarette smoking upon clinical outcomes of assisted reproduction: a meta-analysis. Hum Reprod Update 2009;15(1):31–44.

[29] Windham GC, Mitchell P, Anderson M, Lasley BL. Cigarette smoking and effects on hormone function in premenopausal women. Environ Health Perspect 2005;113(10):1285–90.

[30] Sharara FI, Beatse SN, Leonardi MR, Navot D, Scott RT Jr. Cigarette smoking accelerates the development of diminished ovarian reserve as evidenced by the clomiphene citrate challenge test. Fertil Steril 1994;62(2):257–62.

[31] Talbot P, Riveles K. Smoking and reproduction: the oviduct as a target of cigarette smoke. Reprod Biol Endocrinol 2005;3:52.

[32] Soares SR, Simon C, Remohi J, Pellicer A. Cigarette smoking affects uterine receptiveness. Hum Reprod 2007;22(2):543–7.

[33] Barazani Y, Katz BF, Nagler HM, Stember DS. Lifestyle, environment, and male reproductive health. Urol Clin North Am 2014;41(1):55–66.

[34] Karmon AE, Toth TL, Afeiche M, Tanrikut C, Hauser R, Chavarro JE. Alcohol and caffeine intake in relation to semen parameters among fertility patients. ASRM 2013;100(Suppl. 3). S12.

[35] Norman RJ, Noakes M, Wu R, Davies MJ, Moran L, Wang JX. Improving reproductive performance in overweight/obese women with effective weight management. Hum Reprod Update 2004;10(3):267–80.

[36] Anderson K, Nisenblat V, Norman R. Lifestyle factors in people seeking infertility treatment – a review. Aust N Z J Obstet Gynecol 2010;50(1):8–20.

[37] Verit A, Verit FF, Oncel H, Ciftci H. Is there any effect of insulin resistance on male reproductive system? Arch Ital Urol Androl 2014;86(1):5–8.

[38] Sermondade N, Dupont C, Faure C, Boubaya M, Cédrin-Durnerin I, Chavatte-Palmer P, et al. Body mass index is not associated with sperm-zona pellucida binding ability in subfertile males. Asian J Androl 2013;15(5):626–9.

[39] MacDonald AA, Herbison GP, Showell M, Farquhar CM. The impact of body mass index on semen parameters and reproductive hormones in human males: a systematic review with meta-analysis. Hum Reprod Update 2010;16(3):293–311.

[40] Cabler S, Agarwal A, Flint M, du Plessis SS. Obesity: modern man's fertility nemesis. Asian J Androl 2010;12(4):480–9.

[41] Eisenberg ML, Li S, Behr B, Cullen MR, Galusha D, Lamb DJ, et al. Semen quality, infertility and mortality in the USA. Hum Reprod 2014;29(7):1567–74.

[42] CDC. Interpretation of BMI for Adults. http://www.cdc.gov/healthyweight/assessing/bmi/adult_bmi/index.html#Used; 2014 [accessed 31.05.14].

[43] Rich-Edwards JW, Goldman MB, Willett WC, Hunter DJ, Stampfer MJ, Colditz GA, et al. Adolescent body mass index and infertility caused by ovulatory disorder. Am J Obstet Gynecol 1994;171(1):171–7.

[44] Rich-Edwards JW, Spiegelman D, Garland M, Hertzmark E, Hunter DJ, Colditz GA, et al. Physical activity, body mass index, and ovulatory disorder infertility. Epidemiology 2002;13(2):184–90.

[45] Robker RL, Akison LK, Bennett BD, Thrupp PN, Chura LR, Russell DL, et al. Obese women exhibit differences in ovarian metabolites, hormones, and gene expression compared with moderate-weight women. J Clin Endocrinol Metab 2009;94(5):1533–40.

[46] Holte J, Bergh T, Berne C, Lithell H. Serum lipoprotein lipid profile in women with the polycystic ovary syndrome: relation to anthropometric, endocrine and metabolic variables. Clin Endocrinol 1994;41(4):463–71.

[47] Lim SS, Norman RJ, Davies MJ, Moran LJ. The effect of obesity on polycystic ovary syndrome: a systematic review and meta-analysis. Obes Rev 2013;14(2):95–109.

[48] Balen AH, Conway GS, Kaltsas G, Techatrasak K, Manning PJ, West C, et al. Polycystic ovary syndrome: the spectrum of the disorder in 1741 patients. Hum Reprod 1995;10(8):2107–11.

[49] Zaadstra BM, Seidell JC, Van Noord PA, te Velde ER, Habbema JD, Vrieswijk B, et al. Fat and female fecundity: prospective study of effect of body fat distribution on conception rates. BMJ 1993;306(6876):484–7.

[50] Bolumar F, Olsen J, Rebagliato M, Saez-Lloret I, Bisanti L. Body mass index and delayed conception: a European Multicenter Study on Infertility and Subfecundity. Am J Epidemiol 2000;151(11):1072–9.

[51] Hassan MA, Killick SR. Negative lifestyle is associated with a significant reduction in fecundity. Fertil Steril 2004;81(2):384–92.

[52] Gesink Law DC, Maclehose RF, Longnecker MP. Obesity and time to pregnancy. Hum Reprod 2007;22(2):414–20.

[53] Jensen TK, Scheike T, Keiding N, Schaumburg I, Grandjean P. Fecundability in relation to body mass and menstrual cycle patterns. Epidemiology 1999;10(4):422–8.

[54] Aggerholm AS, Thulstrup AM, Toft G, Ramlau-Hansen CH, Bonde JP. Is overweight a risk factor for reduced semen quality and altered serum sex hormone profile? Fertil Steril 2008;90(3):619–26.

[55] Jensen TK, Andersson AM, Jorgensen N, Andersen AG, Carlsen E, Petersen JH, et al. Body mass index in relation to semen quality and reproductive hormones among 1,558 Danish men. Fertil Steril 2004;82(4):863–70.

[56] Ramlau-Hansen CH, Thulstrup AM, Nohr EA, Bonde JP, Sorensen TI, Olsen J. Subfecundity in overweight and obese couples. Hum Reprod 2007;22(6):1634–7.

[57] Chavarro JE, Toth TL, Wright DL, Meeker JD, Hauser R. Body mass index in relation to semen quality, sperm DNA integrity, and serum reproductive hormone levels among men attending an infertility clinic. Fertil Steril 2010;93(7):2222–31.

[58] Magnusdottir EV, Thorsteinsson T, Thorsteinsdottir S, Heimisdottir M, Olafsdottir K. Persistent organochlorines, sedentary occupation, obesity and human male subfertility. Hum Reprod 2005;20(1):208–15.

[59] Kirchengast S, Huber J. Body composition characteristics and fat distribution patterns in young infertile women. Fertil Steril 2004;81(3):539–44.

[60] Grodstein F, Goldman MB, Cramer DW. Body mass index and ovulatory infertility. Epidemiology 1994;5(2):247–50.

[61] Cramer DW, Barbieri RL, Xu H, Reichardt JK. Determinants of basal follicle-stimulating hormone levels in premenopausal women. J Clin Endocrinol Metab 1994;79(4):1105–9.

[62] Frisch RE. Body fat, menarche, fitness, and fertility. Hum Reprod 1987;2(6):521–33.

[63] Li S, Zhao JH, Luan J, Ekelund U, Luben RN, Khaw KT, et al. Physical activity attenuates the genetic predisposition to obesity in 20,000 men and women from EPIC-Norfolk prospective population study. PLoS Med 2010;7(8):1–9. pii: e1000332. doi: 10.1371/journal.pmed.1000332.

[64] Davies MJ. Evidence for effects of weight on reproduction in women. Reprod Biomed Online 2006;12(5):552–61.

[65] Han Z, Mulla S, Beyene J, Liao G, McDonald SD. Knowledge Synthesis Group. Maternal underweight and the risk of preterm birth and low birth weight: a systematic review and meta-analyses. Int J Epidemiol 2011;40(1):65–101.

[66] Clark AM, Thornley B, Tomlinson L, Galletley C, Norman RJ. Weight loss in obese infertile women results in improvement in reproductive outcome for all forms of fertility treatment. Hum Reprod 1998;13(6):1502–5.

[67] Khaskheli MN, Baloch S, Baloch AS. Infertility and weight reduction: influence and outcome. J Coll Physicians Surg Pak 2013;23(10):798–801.

[68] Kort JD, Winget C, Kim SH, Lathi RB. A retrospective cohort study to evaluate the impact of meaningful weight loss on fertility outcomes in an overweight population with infertility. Fertil Steril 2014;101(5):1400–3.

[69] Costa EC, Azevedo GD. Defining exercise prescription in lifestyle modification programs for overweight/obese polycystic ovary syndrome women. Fertil Steril 2012;97(2). e5; author reply e6.

[70] Zini A, de Lamirande E, Gagnon C. Reactive oxygen species in semen of infertile patients: levels of superoxide dismutase- and catalase-like activities in seminal plasma and spermatozoa. Int J Androl 1993;16(3):183–8.

[71] Vaamonde D, Da Silva-Grigoletto ME, Garcia-Manso JM, Vaamonde-Lemos R, Swanson RJ, Oehninger SC. Response of semen parameters to three training modalities. Fertil Steril 2009;92(6):1941–6.

[72] Wise LA, Cramer DW, Hornstein MD, Ashby RK, Missmer SA. Physical activity and semen quality among men attending an infertility clinic. Fertil Steril 2011;95(3):1025–30.

[73] Sharma R, Biedenharn KR, Fedor JM, Agarwal A. Lifestyle factors and reproductive health: taking control of your fertility. Reprod Biol Endocrinol 2013;11:66.

[74] Freeman EW, Boxer AS, Rickels K, Tureck R, Mastroianni, Jr. L. Psychological evaluation and support in a program of in vitro fertilization and embryo transfer. Fertil Steril 1985;43(1):48–53.

[75] Klonoff-Cohen H. Female and male lifestyle habits and IVF: what is known and unknown. Hum Reprod Update 2005;11(2):179–203.

[76] Demyttenaere K, Nijs P, Evers-Kiebooms G, Koninckx PR. Coping, ineffectiveness of coping and the psychoendocrinological stress responses during in-vitro fertilization. J Psychosom Res 1991;35 (2-3):231–43.

[77] Connolly KJ, Edelmann RJ, Bartlett H, Cooke ID, Lenton E, Pike S. An evaluation of counselling for couples undergoing treatment for in-vitro fertilization. Hum Reprod 1993;8(8):1332–8.

[78] Fisher JR, Hammarberg K. Psychological and social aspects of infertility in men: an overview of the evidence and implications for psychologically informed clinical care and future research. Asian J Androl 2012;14(1):121–9.

[79] Hjollund NH, Jensen TK, Bonde JP, Henriksen TB, Andersson AM, Kolstad HA, et al. Distress and reduced fertility: a follow-up study of first-pregnancy planners. Fertil Steril 1999;72(1):47–53.

[80] Kreuz LE, Rose RM, Jennings JR. Suppression of plasma testosterone levels and psychological stress. A longitudinal study of young men in Officer Candidate School. Arch Gen Psychiatry 1972;26(5):479–82.

[81] McGrady AV. Effects of psychological stress on male reproduction: a review. Arch Androl 1984;13(1):1–7.

[82] Hall E, Burt VK. Male fertility: psychiatric considerations. Fertil Steril 2012;97(2):434–9.

[83] Pook M, Krause W, Rohrle B. Coping with infertility: distress and changes in sperm quality. Hum Reprod 1999;14(6):1487–92.

[84] Clarke RN, Klock SC, Geoghegan A, Travassos DE. Relationship between psychological stress and semen quality among in-vitro fertilization patients. Hum Reprod 1999;14(3):753–8.

[85] Morelli G, De Gennaro L, Ferrara M, Dondero F, Lenzi A, Lombardo F, et al. Psychosocial factors and male seminal parameters. Biol Psychol 2000;53(1):1–11.

[86] Eskiocak S, Gozen AS, Kilic AS, Molla S. Association between mental stress & some antioxidant enzymes of seminal plasma. Indian J Med Res 2005;122(6):491–6.

[87] Gollenberg AL, Liu F, Brazil C, Drobnis EZ, Guzick D, Overstreet JW, et al. Semen quality in fertile men in relation to psychosocial stress. Fertil Steril 2010;93(4):1104–11.

[88] Chrousos GP, Torpy DJ, Gold PW. Interactions between the hypothalamic-pituitary-adrenal axis and the female reproductive system: clinical implications. Ann Intern Med 1998;129(3):229–40.

[89] Lynch CD, Sundaram R, Maisog JM, Sweeney AM, Buck Louis GM. Preconception stress increases the risk of infertility: results from a couple-based prospective cohort study – the LIFE study. Hum Reprod 2014;29(5):1067–75.

[90] Kothari S, Thompson A, Agarwal A, du Plessis SS. Free radicals: their beneficial and detrimental effects on sperm function. Indian J Exp Biol 2010;48(5):425–35.

[91] Zini A, de Lamirande E, Gagnon C. Reactive oxygen species in semen of infertile patients: levels of superoxide dismutase- and catalase-like activities in seminal plasma and spermatozoa. Int J Androl 1993;16(3):183–8.

[92] Gharagozloo P, Aitken RJ. The role of sperm oxidative stress in male infertility and the significance of oral antioxidant therapy. Hum Reprod 2011;26(7):1628–40.

[93] Showell MG, Brown J, Clarke J, Hart RJ. Antioxidants for female subfertility. Cochrane Database Syst Rev 2013;8. CD007807.

[94] Showell MG, Brown J, Yazdani A, Stankiewicz MT, Hart RJ. Antioxidants for male subfertility. Cochrane Database Syst Rev 2011;(1). CD007411.

[95] Ross C, Morriss A, Khairy M, Khalaf Y, Braude P, Coomarasamy A, et al. A systematic review of the effect of oral antioxidants on male infertility. Reprod Biomed Online 2010;20(6):711–23. doi: 10.1016/j.rbmo.2010.03.008. Epub 2010, Mar 10.

[96] Lafuente R, Gonzalez-Comadran M, Sola I, López G, Brassesco M, Carreras R, et al. Coenzyme Q10 and male infertility: a meta-analysis. J Assist Reprod Genet 2013;30(9):1147–56.

[97] Mendiola J, Torres-Cantero AM, Vioque J, Moreno-Grau JM, Ten J, Roca M, et al. A low intake of antioxidant nutrients is associated

with poor semen quality in patients attending fertility clinics. Fertil Steril 2010;93(4):1128–33.

[98] Keskes-Ammar L, Feki-Chakroun N, Rebai T, Sahnoun Z, Ghozzi H, Hammami S, et al. Sperm oxidative stress and the effect of an oral vitamin E and selenium supplement on semen quality in infertile men. Arch Androl 2003;49(2):83–94.

[99] Ruder EH, Hartman TJ, Reindollar RH, Goldman MB. Female dietary antioxidant intake and time to pregnancy among couples treated for unexplained infertility. Fertil Steril 2014;101(3): 759–66.

[100] Agarwal A, Sekhon LH. The role of antioxidant therapy in the treatment of male infertility. Hum Fertil 2010;13(4):217–25.

[101] Wong WY, Zielhuis GA, Thomas CM, Merkus HM, Steegers-Theunissen RP. New evidence of the influence of exogenous and endogenous factors on sperm count in man. Eur J Obstet Gynecol Reprod Biol 2003;110(1):49–54.

[102] Afeiche MC, Gaskins AJ, Williams PL, Toth TL, Wright DL, Tanrikut C, et al. Processed meat intake is unfavorably and fish intake favorably associated with semen quality indicators among men attending a fertility clinic. J Nutr 2014;144(7):1091–8.

[103] Afeiche MC, Williams PL, Gaskins AJ, Mendiola J, Jørgensen N, Swan SH, et al. Meat intake and reproductive parameters among young men. Epidemiology 2014;25(3):323–30.

[104] Attaman JA, Toth TL, Furtado J, Campos H, Hauser R, Chavarro JE. Dietary fat and semen quality among men attending a fertility clinic. Hum Reprod 2012;27(5):1466–74.

[105] Chavarro JE, Minguez-Alarcon L, Mendiola J, Cutillas-Tolin A, Lopez-Espin JJ, Torres-Cantero AM. Trans fatty acid intake is inversely related to total sperm count in young healthy men. Hum Reprod 2014;29(3):429–40.

[106] Chavarro JE, Furtado J, Toth TL, Ford J, Keller M, Campos H, et al. Trans-fatty acid levels in sperm are associated with sperm concentration among men from an infertility clinic. Fertil Steril 2011;95(5):1794–7.

[107] Afeiche M, Williams PL, Mendiola J, Gaskins AJ, Jørgensen N, Swan SH, et al. Dairy food intake in relation to semen quality and reproductive hormone levels among physically active young men. Hum Reprod 2013;28(8):2265–75.

[108] Afeiche MC, Bridges ND, Williams PL, Gaskins AJ, Tanrikut C, Petrozza JC, et al. Dairy intake and semen quality among men attending a fertility clinic. Fertil Steril 2014;101(5):1280–7. e2.

[109] Chiu YH, Afeiche MC, Gaskins AJ, Williams PL, Mendiola J, Jørgensen N, et al. Sugar-sweetened beverage intake in relation to semen quality and reproductive hormone levels in young men. Hum Reprod 2014;29(7):1575–84.

[110] Chavarro JE, Rich-Edwards JW, Rosner BA, Willett WC. Protein intake and ovulatory infertility. Am J Obstet Gynecol 2008;198(2):210e1–7.

[111] Chavarro JE, Rich-Edwards JW, Rosner BA, Willett WC. Dietary fatty acid intakes and the risk of ovulatory infertility. Am J Clin Nutr 2007;85(1):231–7.

[112] Darling AM, Chavarro JE, Malspeis S, Harris HR, Missmer SA. A prospective cohort study of vitamins B, C, E, and multivitamin intake and endometriosis. J Endometr 2013;5(1):17–26.

[113] Harris HR, Chavarro JE, Malspeis S, Willett WC, Missmer SA. Dairy-food, calcium, magnesium, and vitamin D intake and endometriosis: a prospective cohort study. Am J Epidemiol 2013;177(5):420–30.

[114] Gaskins AJ, Colaci DS, Mendiola J, Swan SH, Chavarro JE. Dietary patterns and semen quality in young men. Hum Reprod 2012;27(10):2899–907.

[115] Chavarro JE, Rich-Edwards JW, Rosner BA, Willett WC. Diet and lifestyle in the prevention of ovulatory disorder infertility. Obstet Gynecol 2007;110(5):1050–8.

[116] Ramlau-Hansen CH, Toft G, Jensen MS, Strandberg-Larsen K, Hansen ML, Olsen J. Maternal alcohol consumption during pregnancy and semen quality in the male offspring: two decades of follow-up. Hum Reprod 2010;25(9):2340–5.

[117] La Vignera S, Condorelli RA, Balercia G, Vicari E, Calogero AE. Does alcohol have any effect on male reproductive function? A review of literature. Asian J Androl 2013;15(2):221–5.

[118] Pajarinen J, Karhunen PJ, Savolainen V, Lalu K, Penttila A, Laippala P. Moderate alcohol consumption and disorders of human spermatogenesis. Alcohol Clin Exp Res 1996;20(2): 332–7.

[119] Pajarinen JT, Karhunen PJ. Spermatogenic arrest and 'Sertoli cell-only' syndrome – common alcohol-induced disorders of the human testis. Int J Androl 1994;17(6):292–9.

[120] Becker U, Tonnesen H, Kaas-Claesson N, Gluud C. Menstrual disturbances and fertility in chronic alcoholic women. Drug Alcohol Depend 1989;24(1):75–82.

[121] Gude D. Alcohol and fertility. J Hum Reprod Sci 2012;5(2):226–8.

[122] Grodstein F, Goldman MB, Cramer DW. Infertility in women and moderate alcohol use. Am J Public Health 1994;84(9):1429–32.

[123] Hartman TJ, Baer DJ, Graham LB, Stone WL, Gunter EW, Parker CE, et al. Moderate alcohol consumption and levels of antioxidant vitamins and isoprostanes in postmenopausal women. Eur J Clin Nutr 2005;59(2):161–8.

[124] Jensen TK, Hjollund NH, Henriksen TB, Scheike T, Kolstad H, Giwercman A, et al. Does moderate alcohol consumption affect fertility? Follow up study among couples planning first pregnancy. BMJ 1998;317(7157):505–10.

[125] Hakim RB, Gray RH, Zacur H. Alcohol and caffeine consumption and decreased fertility. Fertil Steril 1998;70(4):632–7.

[126] Mukherjee RA, Hollins S, Abou-Saleh MT, Turk J. Low level alcohol consumption and the fetus. BMJ 2005;330(7488):375–6.

[127] Windham GC, Fenster L, Swan SH. Moderate maternal and paternal alcohol consumption and the risk of spontaneous abortion. Epidemiology 1992;3(4):364–70.

[128] Sokol RJ, Delaney-Black V, Nordstrom B. Fetal alcohol spectrum disorder. JAMA 2003;290(22):2996–9.

[129] Reid TR. Caffeine. National Geographic 2005; p. 15–32.

[130] Nawrot P, Jordan S, Eastwood J, Rotstein J, Hugenholtz A, Feeley M. Effects of caffeine on human health. Food Addit Contam 2003;20(1):1–30.

[131] Hartley TR, Lovallo WR, Whitsett TL. Cardiovascular effects of caffeine in men and women. Am J Cardiol 2004;93(8):1022–6.

[132] Povey AC, Clyma JA, McNamee R, Moore HD, Baillie H, Pacey AA, et al. Modifiable and non-modifiable risk factors for poor semen quality: a case-referent study. Hum Reprod 2012;27(9): 2799–806.

[133] Jensen TK, Swan SH, Skakkebaek NE, Rasmussen S, Jorgensen N. Caffeine intake and semen quality in a population of 2,554 young Danish men. Am J Epidemiol 2010;171(8):883–91.

[134] Ramlau-Hansen CH, Thulstrup AM, Bonde JP, Olsen J, Bech BH. Semen quality according to prenatal coffee and present caffeine exposure: two decades of follow-up of a pregnancy cohort. Hum Reprod 2008;23(12):2799–805.

[135] Bolumar F, Olsen J, Rebagliato M, Bisanti L. Caffeine intake and delayed conception: a European multicenter study on infertility and subfecundity. European Study Group on Infertility Subfecundity. Am J Epidemiol 1997;145(4):324–34.

[136] Cnattingius S, Signorello LB, Anneren G, Clausson B, Ekbom A, Ljunger E, et al. Caffeine intake and the risk of first-trimester spontaneous abortion. N Engl J Med 2000;343(25):1839–45.

[137] Bech BH, Nohr EA, Vaeth M, Henriksen TB, Olsen J. Coffee and fetal death: a cohort study with prospective data. Am J Epidemiol 2005;162(10):983–90.

[138] Fronczak CM, Kim ED, Barqawi AB. The insults of illicit drug use on male fertility. J Androl 2012;33(4):515–28.

[139] Battista N, Pasquariello N, Di Tommaso M, Maccarrone M. Interplay between endocannabinoids, steroids, and cytokines

in the control of human reproduction. J Neuroendocrinol 2008;20(Suppl. 1):82–9.

[140] Park B, McPartland JM, Glass M. Cannabis, cannabinoids and reproduction. Prostaglandins Leukot Essent Fatty acids 2004;70(2):189–97.

[141] Mueller BA, Daling JR, Weiss NS, Moore DE. Recreational drug use and the risk of primary infertility. Epidemiology 1990;1(3):195–200.

[142] Thompson ST. Prevention of male infertility: an update. Urol Clin North Am 1994;21(3):365–76.

[143] Pasqualotto FF, Lucon AM, Sobreiro BP, Pasqualotto EB, Arap S. Effects of medical therapy, alcohol, smoking, and endocrine disruptors on male infertility. Rev Hosp Clin Fac Med Sao Paulo 2004;59(6):375–82.

[144] Hendrick V, Gitlin M, Altshuler L, Korenman S. Antidepressant medications, mood and male fertility. Psychoneuroendocrinology 2000;25(1):37–51.

[145] Schlegel PN, Chang TS, Marshall FF. Antibiotics: potential hazards to male fertility. Fertil Steril 1991;55(2):235–42.

[146] King K, Chan PJ, Patton WC, King A. Antibiotics: effect on cryopreserved-thawed human sperm motility *in vitro*. Fertil Steril 1997;67(6):1146–51.

[147] Rosen RC, Lane RM, Menza M. Effects of SSRIs on sexual function: a critical review. J Clin Psychopharmacol 1999;19(1):67–85.

[148] Gitlin MJ. Psychotropic medications and their effects on sexual function: diagnosis, biology, and treatment approaches. J Clin Psychiatry 1994;55(9):406–13.

[149] Buffum J. Pharmacosexology update: prescription drugs and sexual function. J Psychoact drugs 1986;18(2):97–106.

[150] WHO. Sexually transmitted infections (STIs). 2014. www.who.int/mediacentre/factsheets/fs110/en/ (accessed 31 May 2014).

[151] Brookings C, Goldmeier D, Sadeghi-Nejad H. Sexually transmitted infections and sexual function in relation to male fertility. Korean J Urol 2013;54(3):149–56.

[152] CDC. Sexually Transmitted Diseases (STDs). 2014. www.cdc.gov/std/default.htm (accessed 31 May 2014).

[153] Ochsendorf FR. Sexually transmitted infections: impact on male fertility. Andrologia 2008;40(2):72–5.

[154] Garolla A, Pizzol D, Foresta C. The role of human papillomavirus on sperm function. Curr Opin Obstet Gynecol 2011;23(4):232–7.

[155] Laprise C, Trottier H, Monnier P, Coutlee F, Mayrand MH. Prevalence of human papillomaviruses in semen: a systematic review and meta-analysis. Hum Reprod 2014;29(4):640–51.

[156] Yang Y, Jia CW, Ma YM, Zhou LY, Wang SY. Correlation between HPV sperm infection and male infertility. Asian J Androl 2013;15(4):529–32.

[157] Brocklehurst P, French R. The association between maternal HIV infection and perinatal outcome: a systematic review of the literature and meta-analysis. Br J Obstet Gynecol 1998;105(8):836–48.

[158] Ochsendorf FR. Infections in the male genital tract and reactive oxygen species. Hum Reprod Update 1999;5(5):399–420.

[159] Rolf C, Kenkel S, Nieschlag E. Age-related disease pattern in infertile men: increasing incidence of infections in older patients. Andrologia 2002;34(4):209–17.

[160] De Sanctis V, Soliman A, Mohamed Y. Reproductive health in young male adults with chronic diseases in childhood. Pediatr Endocrinol Rev 2013;10(3):284–96.

[161] Vargas J, Crausaz M, Senn A, Germond M. Sperm toxicity of "nonspermicidal" lubricant and ultrasound gels used in reproductive medicine. Fertil Steril 2011;95(2):835–6.

[162] Agarwal A, Deepinder F, Cocuzza M, Short RA, Evenson DP. Effect of vaginal lubricants on sperm motility and chromatin integrity: a prospective comparative study. Fertil Steril 2008;89(2):375–9.

[163] Jensen TK, Andersson AM, Skakkebaek NE, Joensen UN, Jensen MB, Lassen TH, et al. Association of sleep disturbances with reduced semen quality: a cross-sectional study among 953 healthy young Danish men. Am J Epidemiol 2013;177(10):1027–37.

[164] Slama R. Invited commentary: sleep disturbances – another threat to male fecundity? Am J Epidemiol 2013;177(10):1038–41.

[165] Bunting L, Boivin J. Knowledge about infertility risk factors, fertility myths, and illusory benefits of healthy habits in young people. Hum Reprod 2008;23(8):1858–64.

14

Effect of Diet and Physical Activity of Farm Animals on their Health and Reproductive Performance

Anna Wilkanowska, MSc, Dariusz Kokoszyński, PhD***

*Department of Agricultural, Environmental and Food Sciences,
University of Molise, Campobasso, Italy
**Department of Poultry Breeding and Animal Products Evaluation,
University of Technology and Life Sciences, Bydgoszcz, Poland

INTRODUCTION

Humans have been using animals, primarily for food, transport, and companionship, for a very long time. Nowadays, a pivotal role is played by animals reared for the production of food or other agricultural products. Globally, the farm animal sector is highly dynamic. In developing countries, it is evolving in response to rapidly increasing demand for livestock products. Animal products can serve as an important source of nutritional energy, protein, and micronutrients, and are important factors for physical and cognitive development and health. Strong demands for the quality and safety of food for reasons of human health necessitate a search for new approaches to the enhancement of health and reproductive potential in farm animals.

FARM ANIMAL VIGOR AND HEALTH STATUS

Vigor is a term encompassing those properties that determine the potential level of activity, performance, and longevity of animals. Vigor largely depends on health. For many veterinarians focusing on animals' physical health and their role in helping farmers to provide affordable food, the term "animal health" appears to be predominantly a question of the presence or absence of disease [1]. For farmers, good animal health is an explicit goal of livestock farming and is fundamental to the maximization of animal productivity and well-being [2,3]. Healthy animals produce profitable amounts of milk, meat, and eggs, are capable of reproducing, and are free of disease – both infectious and metabolic.

Farm animals are subject to the same diseases and parasitic problems encountered in animals not used for food production. The main difference is that during periods of reduced production, these animals may become stressed and are more susceptible to disease, especially if undernourished [4]. For this reason, an important step in maintaining animal health is the development of a health plan for each farm. The main goal of such a plan is to promote and optimize the health and well-being of individual animals and groups by minimizing mortality, illness, and injury over time. This involves planning, in order to encourage disease resistance through prevention, detection, and management.

There is a close relationship between animal health and animal welfare. The welfare of agricultural research animals depends on good health, on the presence of positive emotional states, and the absence of adverse or unpleasant subjective states [5]. This is a contentious issue in which judgments are based not solely on scientific evidence but also on philosophical values and economic considerations [6,7]. "Welfare" describes the state of an animal at a specific time and can be "poor" or "good" regardless of what people think about the morality of using animals in a particular way [8]. Among the main issues involved in the concept of welfare are the concepts of "suffering" and "need" as well as the following "freedoms": (1) freedom from thirst and hunger; (2) freedom from discomfort; (3) freedom from pain, injury, and disease; (4) freedom to express normal behavior; and (5) freedom from fear and distress [9,10].

Handbook of Fertility. http://dx.doi.org/10.1016/B978-0-12-800872-0.00014-7

During the past few decades, interest in the welfare of farm animals has grown. Welfare has been the subject of many experiments carried out on cattle, pigs, sheep, poultry, and so on [10–13]. These studies, besides making it possible to obtain a deeper knowledge of animals' minds, also give a clear picture of how animals perceive the world and how environmental stimuli may affect their welfare [14]. Moreover, methods are being developed that can be used to assess the level of welfare in farm animals. As of now, there are several different methods to estimate the well-being of animals, generally based on a range of welfare parameters. Welfare is multidimensional and cannot be measured directly, but rather is inferred from external parameters. In principle, these parameters can be divided into two categories. One, environmental parameters, describes features of the environment and management. Measurements in the other category, animal-based parameters, record animals' reactions to specific environments [15]. The assessment of welfare at the farm level can be used as an advisory tool by farmers, as a source of information for legislation, and as a component of quality assurance schemes for consumers [16].

Welfare and health can also be assessed by means of reproductive performance [1], which is one of the most important components of breeding strategy and genetic gain in farm animal production systems. However, reproductive traits have received increased attention not only due to biological reasons. Reproduction is also a critically important aspect of economic efficiency in the livestock industry, as failures in this area cause extensive economic losses [17].

REPRODUCTIVE PERFORMANCE OF LIVESTOCK SPECIES

Reproduction is a complex process with many components, several of which have been used as measures of reproductive performance. However, related traits cannot be readily measured directly from the animal itself, other than through the success or failure of conception or pregnancy [18]. Estimated breeding values for reproductive traits in farm animals are difficult to evaluate because the expression of reproductive potential is often constrained by the management system employed [19,20]. Most animals will reproduce when managerial and nutritional conditions are optimal; however, in less favorable conditions, only those with the highest genetic traits for reproductive fitness will reproduce [21]. Additionally, the low heritability generally associated with conventional fertility measurements makes this quality difficult to detect, thus limiting the development of selection aimed at obtaining a significant genetic gain in reproductive efficiency. An alternative approach may be the identification of physiological parameters, such as endocrine factors that are both related to fertility and heritable [22].

The relevance of different reproductive traits differs among species. However, for all farm animals the most important reproductive trait is fertility, which they express in different ways. Another common feature is that all reproductive characteristics are correlated in some way with reproductive efficiency. Measurements of the reproductive performance of selected farm animals are presented next.

Cattle

Reproductive success is of paramount importance for the economic efficiency of cattle production [20]. High reproductive efficiency is necessary for efficient milk production and therefore has an important influence on herd profitability [23]. Bovine reproduction is less efficient than that of other farm species, for example, pigs and sheep. This means that the rate of genetic progress is likely to be relatively slow [24]. In beef cattle, observed fertility is an inherent trait that cannot be fully defined by phenotypic measurements made directly on the animals. Accordingly, many studies have been conducted to find reliable indicators of fertility [18]. Until now, female fertility in beef cattle has been recorded and measured in many ways, including age at first calving (the period needed for a cow to reach maturity and to reproduce for the first time), calving date (the day within the calving season in which a heifer or cow calves), first-service conception rate (the percentage of cows that conceive as a result of their *first* service), pregnancy rate (the percentage of cows in a herd that become pregnant every 21-day period after the voluntary waiting period), calving interval (the period that elapses between the time a cow gives birth and the time it reproduces again), longevity (the period between first calving and culling), and stayability (the length of the productive life of females). However, from a biological point of view, the calving rate ($h^2 \leq 0.20$) is perhaps the most appropriate measure of fertility [20,24]. The calving rate is defined as the number of calves actually produced by a cow divided by the number of potential calves [20].

Measures of reproductive performance need to be considered not only in females, but also in males [25]. Bulls play a key role in calf production and represent an important source of bioeconomic capital in this activity [26]. Thus, impairment of bull fertility results in great economic losses, particularly in extensive cattle production systems.

Bull fertility is a complex characteristic made up of several physiological processes such as development of the reproductive system from birth to puberty, spermatogenesis, ejaculation, and mating behavior (e.g., libido and copulation). Semen volume, concentration of spermatozoa, proportion of dead and abnormal spermatozoa, and motility of spermatozoa are recognized as important indices of bull semen quality and significantly correlated with freezability [27]. Additionally, one of the parameters estimated during the clinical examination, which is part of the reproductive evaluation of a bull, is scrotal circumference. Measurement of scrotal circumference is significant because it assesses testicular volume and is highly correlated with sperm output [26]. Among all measures of fertility in beef cattle, scrotal circumference presents several advantages: it is inexpensive and easy to measure, is characterized by moderate heritability, and is reported to be favorably associated with female fertility [28–30]. Estimates of heritability for scrotal circumference in bulls ranges from 0.00 to 0.88 [31–35].

Sheep

Unlike most domestic livestock species, sheep are widely known as animals with marked reproductive seasonality. However, similarly to other farm animals, reproductive traits in sheep are the most important features affecting profitability [36]. These characteristics are of a categorical nature, but in practice the continuous distribution of traits is analyzed [37]. Measures of reproduction commonly used in sheep include estrus rate (ewes estrus/ewes mated × 100), lambing rate (ewes lambed/ewes mated × 100), infertility rate (infertile ewes/ewes exposed × 100), fecundity (lambs born/ewes mated), litter size (lambs born/number of lambing ewes), single lambing (single-born lambs/lambs born × 100), twin lambing (twins born/lambs born × 100), and survival rate (weaning lambs/lambs born × 100) [38]. These characteristics are the most important factors in determining sheep productivity and the economic efficiency of a lamb production enterprise. Estimates of heritability for these reproductive traits in the literature range from 0.01 to 0.42 [39–41].

Moreover, the reproductive potentials of sheep are determined to a large extent by ram fertility. The reproductive capacity of rams, as in the case of ewes, is influenced by seasonal factors such as temperature, relative humidity, and the number of sunny hours [42]. However, in contrast to females, which become anovulatory outside the breeding season, rams are not azoospermic during the nonbreeding season despite a significant reduction in sperm production [43,44]. Also, physiological and behavioral variations in rams are less pronounced than in ewes [45].

Pigs

Compared to other farm animals, the pig is considered to be a highly prolific species with a high ovulation rate and, if mated at the right moment, the sow also has a very high probability of becoming pregnant [46]. Reproductive performance of the sow herd is the pivotal factor in controlling the efficiency of swine production and is considered economically important for the swine industry. Low levels of reproductive performance may result in a low per-sow profit and may limit attempts to improve the herd genetically [47]. The most frequently evaluated trait of pigs is the number of live piglets born in the litter. Other parameters include the number of stillborn piglets and the number of weaned/reared piglets per litter [48]. The latter characteristic is a trait of major economic importance in pig production. The main components of this trait are the ovulation rate of sows, early-embryonic survival, fetal survival until and around farrowing, and survival of piglets from birth until weaning. The number of piglets weaned can be increased by improving any of these components [49].

The success and efficiency of pig farming largely depends on the reproductive performance of males as well. The main criterion for keeping a boar at an insemination station is the production of ejaculate containing a high quantity of spermatozoa with high fertilization capacity [50]. The number and quality of spermatozoa inseminated determine the boar's impact on litter size. The relationship between these characteristics and litter size tends to be unique for boars and is best described as a fertility pattern or curve [51].

Poultry

For birds the major parameters of reproductive performance are fertility and hatchability, considered the major determinants of profitability in the hatchery industry. Fertility refers to the percentage of incubated eggs that are fertile [52], whereas hatchability can be expressed either relative to the total number of eggs laid, in which case it encompasses both fertility and embryo survival, or relative to the number of fertile eggs (determined by candling) [53]. Hatchability is a typical fitness trait with low heritability, which suggests that optimization of breeder-farm and hatchery management is the most promising route for improvement [54]. Genetic variation in the hatchability of a fertile egg is derived mostly from the female laying the egg. This significant effect of the dam is attributable to the quality of the laid egg, which affects the successful development of the embryo into a chick during incubation as well as the emergence of the chick from the egg at hatching. Hatch results are most considerably affected by egg weight, shape index, shell thickness and porosity, and the share of yolk, albumen, and shell in eggs. However, of all the morphological traits of an egg, albumen is of greatest importance, as it is an important reservoir for water, essential ions, and protein, the latter forming 99% of the dry matter of albumen and also possessing useful antimicrobial properties [55].

A marked impact on the reproductive efficiency of the poultry industry is exerted by semen quality. Assessment of the semen quality characteristics of birds gives an excellent indicator of their reproductive potential and has been reported to be a major determinant of the fertility and subsequent hatchability of eggs [56,57]. It is also noteworthy that poultry spermatozoa are characterized by a unique structure and chemical composition. The most important feature of the lipid composition of avian semen is the presence of an extremely high proportion of polyunsaturated fatty acids, such as arachidonic and docosatetraenoic acids, in the phospholipid fraction of spermatozoa [58–60]. Additionally, researchers have observed a remarkable variation in semen parameters among avian species, even species of similar size. The differences in volumes and sperm concentration of domestic fowl semen depend largely on the relative contribution of the various reproductive glands, the number of spermatozoa, and the extent to which their genetic potentials can be exploited [61]. Moreover, in all poultry species, semen quality parameters change with age in males, leading to a progressive decline in fertility [62,63].

INFLUENCE OF DIET ON THE HEALTH AND REPRODUCTION PERFORMANCE OF LIVESTOCK

Several factors are known to affect the health and reproductive performance of farm animals, among them biological type, physical environment, and nutrition [64]. However, evidence from the literature and practical experience suggests that nutritional factors are perhaps the most crucial, in terms of their direct effects on health status and reproductive phenomena. Additionally, nutritional factors, more than any others, readily lend themselves to manipulation in order to ensure positive outcomes [65].

Diet plays a pivotal role in controlling animal health status through influencing the immune system. Nutrients can influence several, if not all, aspects of immune response [66]. Nutrient deficiencies increase susceptibility to most infectious diseases, including bacterial, viral, and parasitic diseases. Once disease has developed, nutritional deficiencies exacerbate the severity of the disease and increase the probability of secondary infections [67]. Nutrition also has a great impact on health through its effect on oxidative stress which, due to the oxidation of important biological molecules, leads to damage to and dysfunction of tissues and organ systems and, obviously, to decreased productivity. Many disease states (e.g., coccidiosis and mastitis) and nonoptimal environments (e.g., heat stress) are also associated with oxidative stress. Thus, effective antioxidant protection plays a key role in ensuring the health of farm animals [68].

The relationship between nutrition and reproduction is also well known. It is generally accepted that nutritional management is the main controlling factor for reproduction in many types of domestic livestock, both ruminant and monogastric species. Nutrient intake (i.e., feed, energy, and protein) may affect reproduction indirectly through metabolic processes that include changes in metabolites (e.g., glucose and triglycerides) and metabolic hormones (e.g., insulin, insulin-like growth factor, and glucagon) [69]. Nutrition affects all aspects of the chain of reproductive events from gametogenesis to puberty in both males and females. Insufficient intake of energy, protein, vitamins, and micro- and macrominerals have all been associated with suboptimal reproductive performance. It is noteworthy that the reproductive organs are a lower priority in terms of essential diet components than other tissues and organs. In times of nutrient deficiency, growth and development of the reproductive organs are retarded [70]. Thus, undernutrition delays the onset of puberty. Malnutrition also prolongs the postpartum interval to conception, interferes with normal ovarian cyclicity by decreasing gonadotropin secretion, and increases infertility [71,72].

Feeding strategy is an important tool in livestock production. A more complete understanding of the "how," "when," and "which" factors in animal diets may significantly improve health status, vigor, reproductive performance, and production efficiency. It has been observed that in reproductive flocks of poultry, especially broiler chickens, duck, and guinea fowl, there is a disturbing trend involving birds consuming excessive amounts of feed in the rearing period (before laying). Therefore, after a period of intensive growth, the amount of feed should be limited, which will prevent excessive weight gain and excessive obesity in birds. Studies conducted on a reproductive herd of Peking ducks [73] have shown that a reduction of 10% in the quantity of feed (between the 7th and 20th weeks of life) increased the production of eggs to 164 (1.2%), egg fertilization to 80.8%, and survival of birds during reproduction (no deaths) compared with ducks fed *ad libitum*. Also, studies with pigs indicated that the intake of gilts and sows must be restricted more severely before breeding. In an experiment comparing the effects of level of feed intake at the time of breeding and during the last third of gestation on reproductive performance, it was observed that sows receiving a low level of feed (2.67 kg/day) at breeding farrowed significantly more live pigs than those receiving a higher level (3.18 kg/day) [74]. However, the restricted feeding strategy normally used during gestation does not always permit the recovery of body reserves in the subsequent parity cycle. This is particularly harmful in gilts, since they have lower body reserves and a 20% lower voluntary feed intake during lactation compared to multiparous sows [75]. Therefore, it seems necessary to review feeding strategies and to adapt them to meet the feeding requirements of leaner sows, particularly in gilts [76].

Similar to feeding strategy, an important role affecting the health status and reproductive function in animals is played by the composition of feed. The most important nutritional factor is energy balance, which can usually be defined as net energy intake minus net energy expenditure for maintenance, activity, and production (e.g., milk) [77,78]. When their net nutrient requirement exceeds the net nutrient intake, the animals will use their energy stores to meet the deficit. An animal in this state is in "negative energy balance." Similarly, when the net nutrient requirement is less than the net nutrient intake, the animal will store the excess nutrients and/or disperse them as metabolic heat. An animal in this state is in a so-called "positive energy balance" [79].

One way of viewing the relationship between nutrition and animal health status is through energy balance. Researchers carrying out experiments on cattle have found that excessive energy intakes during late lactation and dry periods can lead to "fat cow" problems. Cows that are overconditioned when calving have a higher incidence of retained placenta, more uterine infections, and more cystic ovaries [80]. Meanwhile, epidemiological studies have related negative energy balance directly or indirectly via milk yield to laminitis, leg problems, mastitis, and metabolic disorders such as ketosis, ruminal acidosis, and displaced abomasa [81–83].

Energy status is also considered to be the main nutritional factor influencing reproductive processes, with prolonged low energy intake impairing fertility. The influence of negative energy balance on reproductive performance is seen primarily at the hypothalamo-pituitary level of reproductive control [84] and is distinguished by hypoinsulinemia, hypoglycemia, elevated plasma growth hormone (GH) and suppressed plasma insulin-like growth factor I (IGF-I), changes related to the inhibition of gonadotropin-releasing hormone (GnRH) pulsatility, and anovulation and anestrus in females. There is also evidence that negative energy balance has certain direct ovarian effects in ewes that are independent of its effects on the hypothalamo-pituitary axis [85,86]. Positive energy balance is associated with alterations in the hepatic metabolism of steroids, which can lead to disturbances in negative feedback between the ovary and the hypothalamo-pituitary system and, theoretically, to increased folliculogenesis. Positive energy balance also leads to increased glucose uptake and increased insulin and leptin concentrations in the blood. These changes appear to affect the ovary directly and are connected with increased folliculogenesis and increased ovulation rates in sheep [79,87,88].

The primary dietary sources of energy are carbohydrates and fats. Carbohydrates are usually a cheaper source of energy than fats, which, however, are often used to increase the caloric concentration of the diet. Fats are important constituents of a diet and, in addition to their functions as sources of energy and carriers for fat-soluble vitamins, they supply essential fatty acids that are vital components of cell membranes and precursors for potent hormone-like compounds such as leukotrienes and prostaglandins [89]. Fatty acids act as mediators in a series of processes in several reproductive tissues, including fluidity of cell membranes, intracellular signaling, and susceptibility to oxidative damage [90]. Changes in chain length, degree of unsaturation, and position of the double bonds in the acyl chain of fatty acids may have a major impact on reproductive function and play a role in farm animals' reproduction [91].

Improving fatty acid status, especially that of unsaturated fatty acids, may have positive effects on cattle reproduction by enhancing ovary function and follicular development [91], improving the quality of oocytes [92], and increasing first-service conception rates [93]. Fatty acids play an important role in embryonic development as well. Through desaturation and chain elongation reactions, extensive conversion of linoleic and alpha-linolenic acids to longer chain polyunsaturated fatty acids, which are essential for the embryo, takes place [94]. In poultry, the alteration of dietary fatty acid composition can also have undesirable effects on fertility, hatchability, embryonic survival, and posthatch growth [95]. It has been observed that cyclopropene fatty acids in quails' diet drastically increased levels of 18:0, decreased 18:1 $(n - 9)$, and decreased embryonic mortality (by 54%, one week after treatment) [96]. Moreover, experiments carried out on pigs showed that piglets born and reared by sows fed a diet supplemented with ω-3 PUFAs (from day 60 of gestation to day 21 of lactation) weigh more at weaning compared to controls (6.01 and 5.49 kg, respectively) [97].

The effect of dietary fatty acids on semen quality was investigated as well. Supplementation of boars' diets with 30 g/kg of tuna oil containing 25% DHA may improve sperm characteristics, including concentration (increase of 2.15×10^8 cells/mL), vitality (9.1% increase), and the proportion of spermatozoa with progressive motility (4.6% increase) and a normal acrosome score (9.1% increase) [98]. Also, studies with some poultry species [99,100] have indicated the beneficial influence of ω-3 fatty acids on male reproductive capacity. However, other experiments on rabbits [101], boars [102], and turkeys [103] showed no effects of ω-3-PUFA supplementation on semen quality.

Moreover, there is evidence that the immune system is supported by dietary fatty acids in different livestock species. The mechanism by which fatty acids modulate the immune response was investigated in numerous studies conducted on farm animals. It was suggested that some of the immunomodulatory effects of ω-3 PUFAs may result from their effects on intracellular signaling pathways and transcription factor activity on lipid rafts, and that the immunomodulatory effect of these fatty acids may be mediated by their effect on lipid raft structure and composition [104]. Researchers observed that feeding laying chickens diets rich in ω-3 PUFAs promoted the growth of the thymus, spleen, and bursa of Fabricius (organs of the avian immune system) up to the fourth week of age [105]. A study conducted on pigs demonstrated that plasma, colostrum, milk, and immune components, including immunoglobulin and cytokines, were clearly affected by altering the ratio of ω-6:ω-3 PUFAs in lactating sow diets; the litter's average daily gain between d 0 and d 14 tended to increase when the dietary ratio of ω-6:ω-3 was 9:1 [106].

The second limiting nutrient in most diets is protein, which is also the most important structural element of cells even though animal organisms do not synthesize the numerous amino acids that compose proteins. For this reason the requirement for protein is also a requirement for exogenous amino acids contained in dietary protein. The proteins of live cells are continuously subjected to degradation, creating a pool of free amino acids that circulate in the organism and are used for the synthesis of new proteins, depending on the current requirements of the organism [107]. It is noteworthy that if dietary energy is inadequate to meet demands, it can be supplemented by the breakdown of body fat and muscle. However, there is no way for the body to compensate for prolonged low levels of dietary protein.

Prolonged inadequate protein intake has been reported to reduce reproductive performance. It has also been found that reproductive performance may be impaired if protein is fed in amounts that greatly exceed the animals' requirements. In the case of dairy cows, high dietary protein intake stimulates milk production [108] but has also been related to decreased fertility [109]. Cows fed excess dietary protein experienced increased blood urea, altered uterine fluid composition, decreased uterine pH, and reduced conception rates [80,110]. Urea nitrogen concentrations greater than 19 mg/dL in plasma and milk have been associated with decreased pregnancy rates in dairy cattle [111]. Moreover, in cows fed high dietary protein, plasma progesterone concentrations were reportedly lower [112]. Levels of protein in ewes' diets also have effects on reproductive performance. Research results suggest that the supply of a high level of bypass protein (rumen undegradable protein) is essential to increase colostrum and milk production in ewes [113]. The amount of colostrum available at lambing and the milk production of ewes after lambing are important factors influencing lamb survival and growth rates. Furthermore, it has been observed that lamb survival at weaning is greater for lambs from ewes with a high protein intake during late gestation [114].

In birds, protein level in the diet is also one of the necessary tools involved in the regulation of reproduction. It has been demonstrated that in laying hens, reduction of the dietary protein level in the rearing period (after the active growth period) makes it possible to shorten the time of production of low-weight eggs (not useful in clutches and/or for consumption), as well as reducing the frequency of fallopian tube prolapse. Moreover, at the start of egg production by laying hens (typically slightly earlier), it is necessary to increase the levels of protein (to about 16%) and calcium (3.5%) necessary for the production of eggs. Dietary protein has a significant impact on the reproductive performance of bird males as well. It is recommended to use a diet with a relatively low protein content (for meat-type birds: 11–12%), which ensures proper development of testes and production of good-quality semen. An excess of protein in the diet of males harms their reproductive ability, shortening the period of fertility and reducing semen quality [115].

Additionally, it has been indicated that in regulating avian reproductive performance, a pivotal role is played by the amino acid composition of feed. Lysine, methionine, methionine plus cystine, and tryptophan are the major amino acids that can be limiting in practical feeds for laying hens [116]. Dietary digestible amino acids affect egg weight, production, albumen weight and height, and Haugh unit (a measurement of an egg's protein quality based on the height of its egg white). Scientists also noticed that formulating a broiler breeder diet based on digestible amino acids in feed better predicts dietary protein quality and bird performance than one based on total amino acids [117].

In cattle, protein is one of the nutritional factors affecting retained placenta, a failure of the fetal membranes. Extreme deficiency of dietary energy, protein, or both can result in this metabolic disorder, which is indirectly associated with higher occurrences of cystic ovaries, lower milk yield, and higher culling rates. Cows fed diets low in dietary crude protein (8%) for the entire dry period had a higher incidence (50%) of retained placenta compared with cows fed 15% crude protein (20%) [118]. Deficiencies or excesses of dietary protein are also among the factors that may predispose animals toward ketosis, a lactation disorder in dairy cows usually associated with intense milk production and negative energy balance [119].

An important role in the reproductive performance of animals is played by vitamins. Dietary supplementation with vitamins is required during gestation and lactation to prevent reproductive failure [120]. Moreover, a number of vitamins are involved in the immune system. Impaired immunity can increase the susceptibility of animals to infectious diseases. Inadequate levels of certain vitamins in the diets of livestock species can result also in metabolic disorders that impair animal health. Vitamins vary widely, with some soluble in water, others in fat; moreover, some influence animals' vigor, health status, and reproduction more than others.

One of the most important vitamins for the maintenance of reproductive function is vitamin A. It is also highly essential for maintaining the health and integrity of epithelial tissues in the body. Vitamin A and β-carotene, which acts as a precursor to vitamin A, play a protective role against periparturient diseases by significantly decreasing lymphocyte proliferation during parturition. A lack of vitamin A reduces an animal's resistance to invading pathogens and makes it more susceptible to infections. With special reference to reproduction, this microelement affects ovarian steroidogenesis and directly or indirectly influences the uterine environment and early embryo and fetal development [121]. In cattle, β-carotene plays a specific role in progesterone synthesis and is necessary for the maintenance of estrous cycles [122]. Studies on the possible role of β-carotene in sows have shown the positive effect of a precursor to vitamin A on the number of piglets born alive (10.30 for the control group vs. 10.72 for the experimental group) and litter weight at birth (17.5 kg control vs. 18.3 kg experimental) [120].

There is also important evidence on the particular effects of vitamin E on the health and reproductive performance of livestock species. However, opinions differ among researchers as well as practical livestock producers regarding the conditions under which vitamin E supplementation is required and the levels at which it should be administered [123]. This vitamin is important for various hormone biosynthesis associated with fertility and sexual drive. Vitamin E deficiency exerts a suppressive effect directly on the gonadal function to decrease hormone synthesis in Leydig cells, causes increased secretion of pituitary LH due to the feedback mechanism [124], and affects the structural differentiation of epithelial cells of the epididymis [125].

The role of vitamin E in animal reproduction continues to be studied. There is no documented evidence that vitamin E deficiency is a significant cause of reproductive failure in dairy cows [126]. In a study conducted on cows fed low vitamin E rations for four generations, no significant influence of this vitamin on reproductive performance was observed [127]. However, in experiments carried out on pigs, increased litter size and reduced preweaning piglet mortality resulted from increasing sows' dietary vitamin E intake during gestation [128]. Moreover, α-tocopheryl acetate also improves spermatogenesis and semen quality in boars [129,130] and possibly the fertilization of oocytes in females [131]. It has also been suggested that supplementing vitamin E in boar diets appears to be most beneficial in optimizing libido [132]. In fowls, vitamin E levels reduced infertility [133], and

played a role in reversing the sterility of male chickens fed a diet with high linoleic acid [134]. Vitamin E is supposed to play an important role in antioxidant protection of seminal plasma in birds [135]. Additionally, vitamin E supplements improved egg production in hens confined under high temperatures, and markedly improved egg fertility and hatchability [136,137].

Vitamin E plays a pivotal role in controlling animal health status through influencing the immune system. It is noteworthy that this vitamin is also an important lipid-soluble antioxidant in the body. In young calves, a deficiency of vitamin E results in white muscle disease. Moreover, low vitamin E levels in cows increase the incidence of certain diseases, especially mastitis [138].

Vitamin C (ascorbic acid) is another vitamin promoting reproduction and counteracting infections by pathogenic bacteria and viruses [139]. Vitamin C is characterized by three biological functions of particular relevance to reproduction, each dependent on its role as a reducing agent: it is required for the biosynthesis of collagen, the biosynthesis of steroid and peptide hormones, and the prevention or reduction of the oxidation of biomolecules [140,141]. Early experiments showed direct effects of ascorbate deficiency on male fertility in laboratory and farm species. Inadequate ascorbic acid levels have been associated with low sperm count, increased numbers of abnormal sperm, reduced motility, and agglutination [141].

Vitamin C is especially useful as a dietary supplement for animals under heat stress (high air temperature) causing oxidative stress, which decreases plasma levels of ascorbic acid. Beneficial effects of vitamin C added to feed on birds' reproductive performance have been observed. Researchers have reported the influence of vitamin C supplements on the improvement of egg production and hatchability in poultry [142–145]. Dietary supplementation with vitamin C may be beneficial for lactating cows in hot weather as well [146]. Moreover, vitamin C, as an antioxidant, plays a pivotal role in immune cell function. Ascorbic acid protects cells from lipid peroxidation and increases the fluidity of cell membranes, thus enhancing immune response [147]. In macrophages, vitamin C also increases chemotaxis, phagocytosis, and cell adherence, all of which are important in immune cell function [148].

For optimum reproductive performance and maintenance of good health status in livestock species, microminerals (selenium, zinc, manganese, copper, and molybdenum, among others) are necessary. They are also important for the functioning of various components of the immune system [149]. Their deficiency reduces disease resistance and increases susceptibility to disease. Microminerals are involved in several biological processes and are components of metalloenzymes and enzyme cofactors. Due to their involvement in carbohydrate, protein, and nucleic acid metabolism, any change in their level may alter the production of reproductive and other hormones [150].

One of the most controversial trace elements is selenium (Se). On one hand, it is toxic at high doses and there is a great deal of information related to issues of environmental contamination with Se. On the other hand, selenium deficiency is a global problem related to increased susceptibility to various animal diseases and decreased productive and reproductive performance in farm animals [151]. Selenium's beneficial effects, which alleviate reproductive problems in dairy animals, have been associated with increased glutathione peroxidise activity in blood and tissues [150]. Frequently noted selenium-responsive reproductive disorders of cattle include retained placenta, abortion, irregular estrous cycle, early embryonic mortality, cystic ovaries, mastitis, and metritis, all of which can be reduced by supplementation of selenium [152]. Moreover, vitamin E plays an important role in the reproductive processes of males as well. High concentrations of selenium in testes and epididymides indicate its importance for the processes of production and maturation of spermatozoa [129,153]. Studies performed on boars [129,154,155] have showed that the addition of 0.5 ppm selenium to the feed of young boars markedly improved and positively influenced both semen quality (sperm motility, concentration, and morphology) and conception rates.

An important role in the reproductive performance of males is also played by zinc (Zn), a micromineral essential in the production of many sex hormones (e.g., testosterone and gonadotrophin-releasing hormone) [156]. Supplementation with zinc increases daily sperm production and reduces the proportion of abnormal spermatozoa. It is noteworthy that zinc requirements for testicular growth and development and for spermatogenesis are greater than those for body growth and appetite [157]. Moreover, feeding with organic zinc may enhance resistance to mastitis-causing pathogens due to improvements in the skin integrity and keratin lining of the teat canal [158].

Like zinc, manganese (Mn) also plays a significant role in animal reproduction. The principal disorders associated with manganese deficiency are infertility, congenital limb deformity, and poor growth rates in calves. In males, inadequate levels of manganese lead to decreased libido, decreased motility of spermatozoa, and reduced counts of sperm in the ejaculate [159]. Manganese supplementation has proven effective in shortening postpartum anestrus and increasing conception rates in dairy cows [150].

Another particularly important trace element is copper (Cu). There are marked variations in the tolerance of domestic animals to increased levels of dietary copper [160]. Pigs are highly tolerant, whereas sheep are the species most susceptible to chronic copper toxicity. Reported reproductive disorders associated with copper deficiency in grazing ruminants include low fertility associated with delayed or depressed estrus, delayed postpartum return to estrus, infertility associated with anestrus, abortion, and fetal resorption [161,162]. Moreover, copper is interdependent with molybdenum (Mo) with reference to the organisms of ruminants. Generally, a toxic level of one occurs in the presence of a lower level of the other. Molybdenum deficiency decreases libido, reduces spermatogenesis, and causes sterility in males, and is responsible for delayed puberty, reduced conception rate, and anestrus in females [159].

THE EFFECT OF PHYSICAL ACTIVITY ON HEALTH AND REPRODUCTION OF LIVESTOCK SPECIES

Farm animals have been undergoing human-managed selection since their original domestication. Initially, selection was limited to docility and manageability, but in the last 60 years, breeding programs have focused on the genetic improvement of production traits, such as growth rate, milk yield, and number of eggs [163]. However, maximizing productivity has also resulted in serious threats to the welfare of the animals involved [164]. The reduction or abandonment of pasturing, and thus locomotor activity, has resulted in frequent cases of health and reproductive problems [165].

In farm animals, locomotion has been identified as a prerequisite for several behaviors. Moreover, locomotion is influenced by physical abilities, as determined by morphology, lameness, and pain. It is also affected by environmental conditions such as temperature, lighting schedule and light intensity, flock size, and space per individual. A small amount of space provided to animals is one easily recognizable aspect of husbandry systems perceived by the public as a sign that welfare is poor. It has been suggested that there are both qualitative and quantitative space requirements. Especially important is qualitative space, that is, the space required for the performance of normal activities such as eating, exploring, carrying out social behavior, or withdrawal from visual contact with others [166]. By providing more floor space per animal, it is possible to influence the animals' level of physical activity and development, and firmness of the skeleton, especially legs. For example, in broilers, physical activity influences morphometric bone parameters, that is, cross-section of the cortex, and in this way improves their mechanical characteristics through an enhanced supply of blood to the epiphysis of long bones and adequate mineralization [167]. Meanwhile, in dairy cows, exercise improves blood circulation, develops the muscular system, and promotes health [168,169]. Increased walking also reduces blood levels of NEFA, potentially reducing the risk of metabolic and digestive disorders [170]. Some researchers have speculated that locomotor activity in cattle may increase at higher stocking densities, because increased competition may force cows to move in order to access resources or escape competitors [171].

Numerous studies on the consequences of "crowding" showed that the most significant problem related to high stocking density is aggression, which occurs predominantly due to competition for access to a limited resource, or to establish social relationships between unfamiliar animals. Aggression has a negative effect on the physical condition of animals. Most wounds caused by aggression are scratches or cuts on the skin. Aggression may also result in leg problems, in particular when the design of the pen is such that animals cannot easily avoid an incident or when the floor provides insufficient support to the feet and legs during interaction; however, the incidence of leg problems as a consequence of fighting is substantially lower than that of skin lesions [172]. Apart from aggression, at higher stocking densities the likelihood of disease outbreak rapidly rises.

For poultry producers the most marked problem is feather pecking, a behavior that often leads to extensive damage to plumage. The actual pulling of feathers is painful and can result in damaged feathers and wounded birds [173]. Therefore, feather pecking is regarded as a major welfare problem for birds who are attacked. Higher stocking densities have been associated with higher levels of feather pecking in laying hens in different systems [174]. It has been observed that the use of an outdoor range in flocks of laying hens reduced the risk of feather pecking [175]. However, levels of aggression are generally low in large flocks of laying hens [174], probably due to the formation of subgroups within a large flock [176].

Moreover, some studies have shown the effects of stocking density, rearing system, and group size on the reproductive performance in animals. Noted adverse effects include decreased gonadal activity in males and declines in reproductive parameters in general [166]. Experiments conducted on domesticated birds demonstrated that egg weight, fertility, hatchability, and late embryonic mortality vary significantly between modern and traditional breeding management systems [177]. Low fertility and high embryonic mortality values have been observed in traditional chicken-rearing systems permitting a high rate of locomotor activity [178]. Also, in a study carried out on turkeys, a higher reproductive performance was obtained through an intensive system of management followed by a semi-intensive and free-range system [179].

CONCLUSIONS

The past century of livestock research has focused on the discovery of factors affecting animals' health and the identification of their effects on reproductive performance. It has been observed that the reproductive performance and vigor of farm animals is largely dependent on genetic merit, physical environment, nutrition, and management. Evidence from practical experience and the literature suggests that nutritional factors are perhaps the most crucial, in terms of their direct effects on the reproductive phenomenon and their potential to moderate the effects of other factors. Moreover, so far studies have also confirmed the effects of physical activity on the health of farm animals. Therefore, focusing on improving these factors influencing the health and reproduction of farm animals holds great promise for future livestock management strategies.

References

[1] Fall N. Health and reproduction and in organic and conventional Swedish dairy cows. PhD thesis. Uppsala, Swedish University of Agricultural Sciences; 2009.

[2] Starkey P, Jaiyesimi-Njobe F, Hanekom D. Animal traction in South Africa: overview of the key issues. In: Starkey P, editor. Animal Traction in South Africa – Empowering Rural Communities. Africa: Development Bank of Southern Africa. A DBSA-SANAT; 1995. p. 17–33.

[3] Pearson RA, Nengomasha E, Krecek RC. The challenges in using donkeys for work in Africa. In: Meeting the challenges of animal traction. Proceedings of an ATNESA workshop. Debre Zeit, Ethiopia, May 5–9, 1997. p. 190–198.

[4] James M, Krecek RC. Management of draught animals a welfare and health perspective in South Africa. Proceedings of an ATNESA workshop. South Africa, Sep 1999. p. 153–161.

[5] King LA. Behavioral evaluation of the psychological welfare and environmental requirements of agricultural research animals: theory, measurement, ethics, and practical implications. ILAR J 2003;44(3):211–21.

[6] Fraser D. Science in a value-laden world: keeping our thinking straight. Appl Anim Behav Sci 1997;54:29–32.

[7] Stafleu FR, Grommers FJ, Vorstenbosch J. Animal welfare: evolution and erosion of a moral concept. Anim Welfare 1996;225–34.

[8] Engebretson M. The welfare and suitability of parrots as companion animals: a review. Anim Welfare 2006;15:263–76.

[9] Brambell FWR, Barbour DS, Barnett MB, Ewer TK, Hobson A, Pitchforth H, et al. Report of the Technical Committee to Enquire Into the Welfare of Animals Kept Under Intensive Livestock Husbandry Systems. London: Her Majesty's Stationery Office; 1965.

[10] FAWC. Second Report on Priorities for Research and Development in Farm Animal welfare. London: MAFF Publ; 1993.

[11] Sørensen JT, Hindhede J, Rousing T, Fossing C. Assessing animal welfare in a dairy cattle herd with an automatic milking system. Proceeding of the first North American conference on robotic milking. Mar 20–22, 2002. p. 54–60.

[12] Carlsson F, Frykblom P, Lagerkvist CJ. Farm animal welfare – testing for market failure. J Agric Appl Econ 2007;39(1):61–73.

[13] Popescu S, Borda C, Lazar EA, Hegedüs CI. Assessment of dairy cow welfare in farms from Transylvania. Proceeding of 4th

Croatian & 4th international symposium on agriculture. 2009. p. 752–6.

[14] Carenzi C, Verga M. Animal welfare: review of the scientific concept and definition. Ital J Anim Sci 2009;8(Suppl.1):21–30.

[15] Johnsen PF, Johannesson T, Sandøe P. Assessment of farm animal welfare at herd level: many goals, many methods. Acta Agric Scand 2001;Sect. A(Suppl. 30):26–33.

[16] Napolitano F, Grasso F, Bordi A, Tripaldi C, Saltalamacchia F, Pacelli C, et al. On-farm welfare assessment in dairy cattle and buffaloes: evaluation of some animal-based parameters. Ital J Anim Sci 2005;4:223–31.

[17] Inchaisri C, Jorritsma R, Vos PLAM, van der Weijden GC, Hogeveen H. Economic consequences of reproductive performance in dairy cattle. Theriogenology 2010;74:835–46.

[18] Eler JP, Silva JA, Ferraz JB, Dias F, Oliveira HN, Evans JL, et al. Genetic evaluation of the probability of pregnancy at 14 months for Nellore heifers. J Anim Sci 2002;80:951–4.

[19] Notter DR, Johnson MH. Simulation of genetic control of reproduction in beef cows. IV. Within-herd breeding value estimation with pasture mating. J Anim Sci 1988;66:280–6.

[20] Meyer K, Hammond K, Parnell PF, Mackinnon MJ, Sivarajasingam S. Estimates of heritability and repeatability for reproductive traits in Australian beef cattle. Livest Prod Sci 1990;25:15–30.

[21] Morris CA. A review of relationships between aspects of reproduction in beef heifers and their lifetime production. 2. Associations with relative calving date and with dystocia. Anim Breed Abstracts 1980;48:753–67.

[22] Darwash AO, Lamming GE, Woolliams JA. The potential for identifying heritable endocrine parameters associated with fertility in post-partum dairy cows. Anim Sci 1999;68:333–47.

[23] Pryce JE, Royal MD, Garnsorthy PC, Mao IL. Fertility in the high-producing dairy cow. Livest Prod Sci 2004;86:125–35.

[24] Ball PJH, Peters AR. Reproduction in cattle. London: Blackwell Publishing; 2004. p. 5–6.

[25] Cammack KM, Thomas MG, Enns RM. Reproductive traits and their heritabilities in beef cattle. Professional Anim Sci 2009;25:517–28.

[26] Menegassi SRO, Barcellos JOJ, Peripolli V, Pereira PRRX, Borges JBS, Lampert VN. Measurement of scrotal circumference in beef bulls in Rio Grande do Sul. Arquivo Brasileiro de Medicina Veterinária e Zootecnia 2011;63(1):87–93.

[27] Fiaz M, Usmani RH, Abdullah M, Ahmad T. Evaluation of semen quality maintained under subtropical environment. Pak Vet J 2010;30(2):75–8.

[28] Brinks JS, McInerney MJ, Chenoweth PJ. Relationship of age at puberty in heifers to reproductive traits in young bulls. JT Proc West Sec Am Soc Anim Sci 1978;29:28–30.

[29] Morris CA, Baker RL, Cullen NG. Genetic correlations between pubertal traits in bulls and heifers. Livest Prod Sci 1992;31:221–33.

[30] Vargas CA, Elzo MA, Chase CC Jr, Chenoweth PJ, Olson TA. Estimation of genetic parameters for scrotal circumference, age at puberty in heifers, and hip height in Brahman cattle. J Anim Sci 1998;76:2536–40.

[31] Coulter GH, Rounsaville TR, Foote RH. Heritability of testicular size and consistency in Holstein bulls. J Anim Sci 1976;43:9–12.

[32] Latimer FG, Wilson LL, Cain MF. Scrotal measurements in beef bulls: heritability estimates, breed and test station effects. J Anim Sci 1982;54:473–9.

[33] King RG, Kress DD, Anderson DC, Doornbos DE, Burfening PJ. Genetic parameters in Herefords for puberty in heifers and scrotal circumference in bulls. Proc West Sec Am Soc Anim Sci 1983;34:11–3.

[34] Knights SA, Baker RL, Gianola D, Gibb JB. Estimates of heritabilities and of genetic and phenotypic correlations among growth and reproductive traits in yearling Angus bulls. J Anim Sci 1984;58:887–93.

[35] Coulter GH, Mapletoft RJ, Kozub GC, Bailey DRC, Cates WF. Heritability of scrotal circumference in one- and two-year-old bulls of different beef breeds. Can J Anim Sci 1987;67:645–51.

[36] Matos CAP, Thomas DL, Gianola D, Pérez-Enciso M, Young LD. Genetic analysis of discrete reproductive traits in sheep using linear and nonlinear models. II. Goodness of fit and predictive ability. J Anim Sci 1997;75:88–94.

[37] Mohammadi H, Moradi Shahrebabak M, Moradi Sharebabak H, Vatankhah M. Estimation of genetic parameters of reproductive traits in Zandi sheep using linear and threshold models. Czech J Anim Sci 2012;57(7):382–8.

[38] Ceyhan A, Sezenler T, Yidirir M, Erdogan I. Reproductive performance and lamb growth characteristics of Ramlıç sheep. J Kafkas Universitesi Veteriner Fakultesi Dergisi 2010;16(2):213–6.

[39] Brash LD, Fogarty NM, Barwick SA, Gilmour AR. Genetic parameters for Australian maternal and dual-purpose meat sheep breeds. I. Live weight, wool production and reproduction in Border Leicester and related types. Aust J Agric Res 1994;45:459–68.

[40] Snyman MA, Erasmus GJ, van Wyk JB. The possible genetic improvement of reproduction and survival rate in Afrino sheep using a threshold model. South Afr J Anim Sci 1998;28:120–4.

[41] Vatankhah M, Talebi MA. Heritability estimates and correlations between production and reproductive traits in Lori-Bakhtiari sheep in Iran. South Afr J Anim Sci 2008;38(2):110–8.

[42] Oláh J, Kusza S, Harangi S, Posta J, Kovács A, Pécsi A, et al. Seasonal changes in scrotal circumference, the quantity and quality of ram semen in Hungary. Archiv Tierzucht 2013;56(10):102–8.

[43] Dacheux JL, Pisselet C, Blanc MR, Hocherau De Reviers MT, Courot M. Seasonal variation in rat testis fluid secretion and sperm production in different breeds of rams. J Reprod Fertil 1981;61:363–71.

[44] Aurich C, Burgmann F, Hoppe H. Opioid regulation of LH and prolactin release in the horse identical or independent endocrine pathways? Anim Reprod Sci 1996;44:127–34.

[45] Rosa HJD, Bryant MJ. Seasonality of reproduction in sheep. Small Ruminant Res 2003;48:155–71.

[46] Peltoniemi OA, Oliviero C, Hälli O, Heinonen M. Feeding affects reproductive performance and reproductive endocrinology in the gilt and sow. Acta Veterinaria Scandinavica 2007;49 (Suppl. 1):S6.

[47] Rekwot PI, Jegede JO, Ehoche OW, Tegbe TSB. Reproductive performance in smallholder piggeries in northern Nigeria. Trop Agric 2001;78(2):130–3.

[48] Popovac M, Radojkovic′ D, Petrovic′ M, Mijatovic′ M, Gogic′ M, Stanojevic′ D, et al. Heritability and connections of sow fertility traits. Biotech Anim Husbandry 2012;28(3):469–75.

[49] Lund MS, Puonti M, Rydhmer L, Jensen J. Relationship between litter size and perinatal and pre-weaning survival in pigs. Anim Sci 2002;74:217–22.

[50] Wysokińska A, Kondracki S, Banaszewska D. Morphometrical characteristics of spermatozoa in Polish Landrace boars with regard to the number of spermatozoa in an ejaculate. Reprod Biol 2009;9(3):271–82.

[51] Flowers WL. Increasing fertilization rate of boars: influence of number and quality of spermatozoa inseminated. J Anim Sci 2002;80:47–53.

[52] King'ori AM. Review of the factors that influence egg fertility and hatchability in poultry. Int J Poult Sci 2011;10(6):483–92.

[53] Wolc A, White IMS, Hill WG, Olori VE. Inheritance of hatchability in broiler chickens and its relationship to egg quality traits. Poult Sci 2010;89:2334–40.

[54] Förster A, Kalm E, Flock DK. Schlupffähigkeit aus züchterischer Sicht (Hatchability from breeding perspective). (in German). Information 1992;(September/October):13–6.

[55] Deeming DC. Embryonic development and utilization of egg components. In: Deeming DC, editor. Avian Incubation, Behavior,

Environment, and Evolution. Oxford: Oxford University Press; 2002. p. 43–53.

[56] Peters SO, Omidiji EA, Ikeobi CON, Ozoje MO, Adebambo OA. Effect of naked neck and hatch ability in local chicken. In: Self sufficiency of animal protein in Nigeria. Proceedings of the 9th annual conference of Animal Science Association, Ebonyi State University, Abakaliki, Nigeria, Sep 13–16, 2004. p. 262–64.

[57] Orunmuyi M, Livinus AC, Ifeanyi NB. Semen quality characteristics and effect of mating ratio on reproductive performance of Hubbard broiler breeders. J Agric Sci 2013;5(1):154–9.

[58] Surai PF. Effect of selenium and vitamin E content of the maternal diet on the antioxidant system of the yolk and the developing chick. Br Poult Sci 2000;41(2):235–43.

[59] Surai PF. Selenium in poultry nutrition. 1. Antioxidant properties, deficiency and toxicity. World's Poult Sci J 2002;58(3):333–47.

[60] Cerolini S, Pizzi F, Gliozzi T, Maldjian A, Zaniboni L, Parodi L. Lipid manipulation of chicken semen by dietary means and its relation to fertility: a review. World's Poult Sci J 2003;59(1):65–75.

[61] Peters SO, Shoyebo OD, Ilori BM, Ozoje MO, Ikeobi CON, Adebambo OA. Semen quality traits of seven strain of chickens raised in the humid tropics. Int J Poult Sci 2008;7(10):949–53.

[62] Bakst MR, Cecil HC. Effect of modifications of semen diluent with cell culture serum replacements on fresh and stored turkey semen quality and hen fertility. Poult Science 1992;71:754–64.

[63] Kelso KA, Cerolini RC, Noble NH, Sparks C, Speake BK. Lipid and antioxidant changes in the semen of broiler fowl from 25 to 60 wk of age. J Reprod Fertil 1996;106:201–6.

[64] Smith OB, Somade B. Nutrition reproduction interactions in farm animals. Proceedings of the international foundation seminar on animal reproduction. Niamey, Niger, Jan 17–21, 1994. p. 1–31.

[65] Smith OB, Akinbamijo OO. Micronutrients and reproduction in farm animals. Anim Reprod Sci 2000;60–61:549–60.

[66] Ingvartsen KL, Moyes K. Nutrition, immune function and health of dairy cattle. Animal 2013;7(Suppl.1):112–22.

[67] Dubeski P. Alberta Feedlot Management Guide. 2nd ed Alberta: Feeder Associations of Alberta Ltd; 2000.

[68] Salobir J, Frankič T, Rezar V. Animal nutrition for health of animals, human and environment. Proceedings of the 20th international symposium animal science days. Kranjska Gora, Slovenia, Sep 19–21, 2012. p. 41–49.

[69] Koketsu Y, Dial GD, Pettigrew JE, King VL. The influence of nutrient intake on biological measures of breeding herd productivity. Swine Health Prod 1996;4(2):85–93.

[70] Aherne FX, Williams IH, Head RH. Nutrition-Reproduction Interactions in Swine. Recent Advances in Animal Nutrition in Australia. University of New England Publishing Unit; 1991. p. 185–202.

[71] Capuco AV, Wood DL, Bright SA, Miller RH, Britman J. Regeneration of teat canal keratin in lactating dairy cows. J Dairy Sci 1990;73:1051–7.

[72] Boland MP, Lonergan P. Effects of nutrition on fertility in dairy cows. Adv Dairy Technol 2005;15:19–33.

[73] Mazanowski A, Kokoszyński D. Effect of restricted feeding of ducks from 7 to 20 weeks of age on their reproductive traits. Roczniki Nauk Zootechnicznych 1998;25(4):147–57.

[74] Mayrose VB, Speer VC, Hays VW. Effect of feeding levels on the reproductive performance of swine. J Anim Sci 1966;25:701–5.

[75] Young MG, Tokach MD, Aherne FX, Main RG, Dritz SS, Goodband RD, et al. Effect of sow parity and weight at service on target maternal weight and energy for gain in gestation. J Anim Sci 2005;83:255–61.

[76] Cerisuelo A, Sala R, Gasa J, Chapinal N, Carrión D, Coma J, et al. Effects of extra feeding during mid-pregnancy on gilts productive and reproductive performance. Span J Agric Res 2008;6(2):219–29.

[77] Van Knegsel ATM, Van Den Brand H, Dijkstra J, Tamminga S, Kemp B. Effect of dietary energy source on energy balance, production, metabolic disorders and reproduction in lactating dairy cattle. Reprod Nutr Dev 2005;45:665–88.

[78] Garcia-Garcia RM. Integrative control of energy balance and reproduction in females. ISRN Vet Sci 2012;1–13.

[79] Scaramuzzi RJ, Campbell BK, Downing JA, Kendell NR, Khalid M, Munoz-Gutierrez M. A review of the effects of supplementary nutrition in the ewe on the concentrations of reproductive and metabolic hormones and the mechanisms that regulate folliculogenesis and ovulation rate. Reprod Nutr Dev 2006;46(4):339–54.

[80] Elrod CC, Butler WR. Reduction of fertility and alteration of uterine pH in heifers fed excess ruminally degradable protein. J Anim Sci 1993;71:694–701.

[81] Grohn YT, Erb HN, McCulloch CE, Saloniemi HS. Epidemiology of metabolic disorders in dairy cattle: association among host characteristics, disease, and production. J Dairy Sci 1989;72: 1876–85.

[82] Heuer C, Schukken Y, Dobbelaar P. Postpartum body condition score and results from the first test day milk as predictors of disease, fertility, yield, and culling in commercial dairy herds. J Dairy Sci 1999;82:295–304.

[83] Collard BL, Boettcher PJ, Dekkers JCM, Petitclerc D, Schaeffer LR. Relationships between energy balance and health traits of dairy cattle in early lactation. J Dairy Sci 2000;83:2683–90.

[84] Wade GN, Jones JE. Neuroendocrinology of nutritional infertility. Am J Physiol Regul Integr Comp Physiol 2005;287:1277–96.

[85] Lozano JM, Lonergran P, Boland MP, O'Callaghan D. Influence of nutrition on the effectiveness of superovulation programmes in ewes: effect on oocyte quality and postfertilization development. Reproduction 2003;125:543–53.

[86] Kiyma Z, Alexander BM, Van Kirk EA, Murdoch WJ, Hallford DM, Moss GE. Effects of feed restriction on reproductive and metabolic hormones in ewes. J Anim Sci 2004;82:2548–57.

[87] Parr RA, Davis IF, Fairclough RJ, Miles MA. Overfeeding during early pregnancy reduces peripheral progesterone concentration and pregnancy rate in sheep. J Reprod Fertil 1987;80:317–20.

[88] Parr RA, Davis IF, Miles MA, Squires TJ. Feed intake affects metabolic clearance rate of progesterone in sheep. Res Vet Sci 1993;55:306–10.

[89] Hanis T, Zidek V, Sachova J, Klir P, Deyl Z. Effects of dietary trans-fatty acids on reproductive performance of Wistar rats. Br J Nutr 1989;61:519–29.

[90] Wathes DC, Abayasekara DR, Aitken RJ. Polyunsaturated fatty acids in male and female reproduction. Biol Reprod 2007;77(2):190–201.

[91] Mattos R, Staples CR, Thatcher WW. Effects of dietary fatty acids on reproduction in ruminants. Rev Reprod 2000;5:38–45.

[92] Fouladi-Nashta AA, Gutierrez CG, Gong JG, Garnsworthy P, Webb R. Impact of dietary fatty acids on oocyte quality and development in lactating dairy cows. Biol Reprod 2007; 77:9–17.

[93] Lopes CN, Scaropa AB, Cappellozza BI, Cooke RF, Vasconcelos JLM. Effects of rumen-protected polyunsaturated fatty acid supplementation on reproductive performance of Bos indicus beef cows. J Anim Sci 2009;87:3935–43.

[94] Bordoni A, Cocchi M, Lodi R, Turchetto E, Ruggeri F. Metabolism and distribution of linoleic and alpha-linolenic acid derivatives in chick embryo development. Progr Lipid Res 1986;25:407–11.

[95] Aydin R, Pariza MW, Cook ME. Olive oil prevents the adverse effects of dietary conjugated linoleic acid on chick hatchability and egg quality. J Nutr 2001;131:800–6.

[96] Donaldson WE, Fites BL. Embryo mortality in quail induced by cyclopropene fatty acids: reduction by maternal diets high in unsaturated fatty acids. J Nutr 1970;100:605–10.

[97] Mateo RD, Carroll JA, Hyun Y, Smith S, Kim SW. Effect of dietary supplementation of omega-3 fatty acids and high levels of dietary protein on performance of sows. J Anim Sci 2009;87:948–59.

[98] Rooke JA, Shao CC, Speake BK. Effects of feeding tuna oil on the lipid composition of pig spermatozoa and *in vitro* characteristics of semen. Reproduction 2001;121:315–22.

[99] Surai PF, Noble RC, Sparks NHC, Speake BK. Effect of long-term supplementation with arachidonic or docosahexaenoic acids on sperm production in the broiler chicken. J Reprod Fertil 2000;120:257–64.

[100] Blesbois E, Douard V, Germain M, Boniface P. Effect of n-3 polyunsaturated dietary supplementation on the reproductive capacity of male turkeys. Theriogenology 2004;61:537–49.

[101] Gliozzi TM, Zaniboni L, Maldjian A, Luzi F, Maertens L, Cerolini S. Quality and lipid composition of spermatozoa in rabbits fed DHA and vitamin E rich diets. Theriogenology 2009;71:910–9.

[102] Paulenz H, Taugbol O, Hofmo P, Saarem K. A preliminary study on the effect of dietary supplementation with cod liver oil on the polyunsaturated fatty acid composition of boar semen. Vet Res Commun 1995;19:273–84.

[103] Zaniboni L, Rizzi R, Cerolini S. Combined effect of DHA and α-tocopherol enrichment on sperm quality and fertility in the turkey. Theriogenology 2006;65:1813–27.

[104] Al-Khalifa H, Givens DI, Rymer C, Yaqoob P. Effect of n-3 fatty acids on immune function in broiler chickens. Poult Sci 2012;91:74–88.

[105] Wang YW, Field CJ, Sim JS. Dietary polyunsaturated fatty acids alter lymphocyte subset proportion and proliferation, serum immunoglobulin G concentration, and immune tissue development in chicks. Poult Sci 2000;79:1741–8.

[106] Yao W, Li J, Wang JJ, Zhou W, Wang Q, Zhu R, et al. Effects of dietary ratio of n-6 to n-3 polyunsaturated fatty acids on immunoglobulins, cytokines, fatty acid composition, and performance of lactating sows and suckling piglets. J Anim Sci Biotechnol 2012;3(1):43.

[107] Timmerman KL, Volpi E. Amino acid metabolism and regulatory effects in aging. Curr Opin Clin Nutr Metab Care 2008;11:45–9.

[108] Grings EE, Roffler RE, Deitelhoff DP. Response of dairy cows in early lactation to additions of cottonseed meal in alfalfa-based diets. J Dairy Sci 1991;74:2580–7.

[109] Canfield RW, Sniffen CJ, Butler WR. Effects of excess degradable protein on postpartum reproduction and energy balance in dairy cattle. J Dairy Sci 1990;73:2342–9.

[110] Jordan ER, Chapman TE, Holtan DW, Swanson LV. Relationship of dietary crude protein to composition of uterine secretions and blood in high-producing dairy cows. J Dairy Sci 1983;66:1854–62.

[111] Butler WR, Calaman JJ, Beam SW. Plasma and milk urea nitrogen in relation to pregnancy rate in lactating dairy cattle. J Anim Sci 1996;74:858–65.

[112] Sonderman JP, Larson LL. Effect of dietary protein and exogenous gonadotropin-releasing hormone on circulating progesterone concentrations and performance of Holstein cows. J Dairy Sci 1989;72:2179–83.

[113] Hinch GN, Lynch JJ, Nolan JV, Leng RA, Bindon BM, Piper LR. Supplementation of high fecundity Border Leicester × Merino ewes with a high protein feed: its effect on lamb survival. Aust J Exp Agric 1996;36:129–36.

[114] Christenson RK, Prior RL. Influence of dietary protein and energy on reproductive performance and nitrogen metabolism in Finn-cross ewes. J Anim Sci 1976;43:1104–13.

[115] Larbier M, Leclercq B. Żywienie Drobiu (Poultry Nutrition). Warsaw: PWN Publisher; 1995. p. 227–228 (in Polish).

[116] Olomu JM. Monogastric Animal Nutrition: Principles and Practice. Nigeria: Jachem Publishers; 1995. p. 68–69.

[117] Nasr J, Yaghobfar A, Nezhad YE, Adl KN. Effects of amino acids and metabolizable energy on egg characteristics and broiler breeder performance. Afr J Biotechnol 2011;10(49):10066–71.

[118] Bačić G, Karadjole T, Mačešić N, Karadjole M. A brief review of etiology and nutritional prevention of metabolic disorders in dairy cattle. Vet Arch 2007;77(6):567–77.

[119] Littledike ET, Young JW, Beitz DC. Common metabolic diseases of cattle: ketosis, milk fever, grass tetany, and downer cow complex. J Dairy Sci 1981;64:1465–82.

[120] Coffey MT, Britt JH. Enhancement of sow reproductive performance by beta carotene or vitamin A. J Anim Sci 1993;71:1198–202.

[121] Kumar S, Pandey AK, Mutha Rao M, Razzaque WAA. Role of β carotene/vitamin A in animal reproduction. Vet World 2010;3(5):236–7.

[122] Ball GFM. Chemical and biological nature of the fat-soluble vitamins. Fat-soluble Vitamin Assays in Food Analysis. A Comprehensive Review, vol. 1988. London: Elsevier, Applied Science Publisher; 1988. p. 7–24.

[123] Pour HA, Sis NM, Razlighi SN, Azar MS, Babazadeh MH, Maddah MT, et al. Effects of vitamin E on ruminant animal. Ann Biol Res 2011;2(4):244–51.

[124] Akazawa N, Mikami S, Kimura S. Effects of vitamin E deficiency on the hormone secretion of the pituitary-gonadal axis of the rat. Tohoku J Exp Med 1987;152:221–9.

[125] Bensoussan KC, Morales R, Hermo L. Vitamin E deficiency causes incomplete spermatogenesis and affects the structural differentiation of epithelial cells of the epididymis in the rat. J Androl 1998;19:266–88.

[126] Williams GL. Modulation of luteal activity in postpartum beef cows through changes in dietary lipid. J Anim Sci 1989;67:785–93.

[127] Schweigert FJ, Zucker H. Concentration of vitamin A, beta-carotene and vitamin E in individual bovine follicles of different quality. J Reprod Fertil 1988;82:575–9.

[128] Mahan DC. Assessment of the influence of dietary vitamin E on sows and offspring in three parities: reproductive performance, tissue tocopherol, and effects on progeny. J Anim Sci 1991;69:2904–17.

[129] Marin-Guzman J, Mahan DC, Pate JL. Effect of dietary selenium and vitamin E on spermatogenic development in boars. J Anim Sci 2000;78:1537–43.

[130] Wallock LM, Tamura T, Mayr CA, Johnston KE, Ames BN. Low seminal plasma folate concentrations are associated with low sperm density and count in male smokers and nonsmokers. Fertil Steril 2001;75:252–9.

[131] Umesiobi DO. Vitamin E supplementation and effects on fertility rate and subsequent body development of their weanling piglets. J Agric Trop Subtrop 2009;110:155–68.

[132] Umesiobi DO. The effect of vitamin E supplementation on the libido and reproductive capacity of Large White boars. South Afr J Anim Sci 2012;42(Suppl. 1):559–63.

[133] Arscott GH, Parke JE, Dickinson EM. Effect of dietary linoleic acid, vitamin E and ethoxyguin on fertility of male chickens. J Nutr 1965;87:63–8.

[134] Arscott GH, Parker JE. Effectiveness of vitamin E in reversing sterility of male chickens fed a diet high in linoleic acid. J Nutr 1967;91:219–22.

[135] Surai PF, Cerolini S, Wishart GJ, Speake BK, Noble RC, Sparks NHC. Lipid and antioxidant composition of chicken semen and its susceptibility to peroxidation. Poult Avian Biol Rev 1998;9:11–23.

[136] Muduuli DS, Marquardt RR, Guenter W. Effect of dietary vicine and vitamin E supplementation on the productive performance of growing and laying chickens. Br J Nutr 1982;47:53–60.

[137] Kirunda DF, Scheideler SE, McKee SR. The efficacy of vitamin E (DL-alpha-tocopheryl acetate) supplementation in hen diets to alleviate egg quality deterioration associated with high temperature exposure. Poult Sci 2001;80:1378–83.

[138] Spears JW. Role of mineral and vitamin status on health of cows and calves. WCDS Adv Dairy Technol 2011;23:287–97.

[139] Hassan RA, Morsy WA, Abd El-Lateif AI. 2012. Effect of dietary ascorbic acid and betaine supplementation on semen characteristics bucks under high ambient temperature. Proceedings 10th world rabbit congress. Sharm El- Sheikh, Egypt, Sep 3–6, 2012. p. 273–277.

[140] Zreik TG, Kodaman PH, Jones EE, Olive DL, Behrman H. Identification and characterization of an ascorbic acid transporter in human granulosa lutein cells. Mol Hum Reprod 1999;5(4):299–302.

[141] Luck MR, Jeyaseelan I, Scholes RA. Ascorbic acid and fertility. Biol Reprod 1995;52:262–6.

[142] Peebles ED, Brake J. Relationship of dietary ascorbic acid to broiler breeder performance. Poult Sci 1985;64:2041–8.

[143] Kontecka K, Witkiewicz K, Sobczak J. Wpływ dodatku witaminy C do paszy na wskaźniki produkcyjne i hematologiczne kur nieśnych (Effect of dietary vitamin C supplement on productive and hematological indicators of laying hens). (in Polish). Przegląd Hodowlany Zeszyty Naukowe PTZ 1997;32:139–45.

[144] Kontecka H, Książkiewicz J, Nowaczewski S, Kisiel T, Krystianiak S. Effect of feed supplementation with vitamin C on reproduction results in Pekin ducks. Proceeding of 9th International Symposium on Current Problems of Breeding, Health and Production of Poultry. České Budějovice, Czech Republic, Feb 6–7, 2001. p. 42.

[145] Nowaczewski S, Kontecka H. Effect of dietary vitamin C supplement on reproductive performance of aviary pheasants. Czech J Anim Sci 2005;50(5):208–12.

[146] Padilla L, Matsui T, Kamiya Y, Kamiya M, Tanaka M, Yano H. Heat stress decreases plasma vitamin C concentration in lactating cows. Livest Sci 2006;101(1):300–4.

[147] Bendich A. Physiological role of antioxidants in the immune system. J Dairy Sci 1993;76:2789–94.

[148] Del Rio M, Ruedas G, Medina S, Victor VM, De la Fuente M. Improvements by several antioxidants of macrophage function *in vitro*. Life Sci 1998;63:871–81.

[149] Andrieu S. Is there a role for organic trace element supplements in transition cow health? Vet J 2008;176:77–83.

[150] Kumar S, Pandey AK, Razzaque WAA, Dwivedi DK. Importance of micro minerals in reproductive performance of livestock. Vet World 2011;4(5):230–3.

[151] Lyons MP, Papazyan TT, Surai PF. Selenium in food chain and animal nutrition: lessons from nature-review. Asian Austral J Anim Sci 2007;20(7):1135–55.

[152] Randhawa SS, Randhawa CS. Trace element imbalances as a cause of infertility in farm animals. Recent advances in animal reproduction and Gynaecology. Proceedings of the summer institute, held at PAU. Ludhiana, July 25–Aug 13, 1994. p. 103–121.

[153] Kotowska E, Kotowski B. Znaczenie selenu i witaminy E w reprodukcji świń (The importance of selenium and vitamin E in the reproduction of pigs). (in Polish). Przegląd Hodowlany 2001; 7:9–11.

[154] Marin-Guzman J, Mahan DC, Chung YK, Pate JL, Pope WF. Effects of dietary selenium and vitamin E on boar performance and tissue responses, semen quality, and subsequent fertilization rates in mature gilts. J Anim Sci 1997;75:2994–3003.

[155] Marin-Guzman J, Mahan DC, Whitmoyer R. Effect of dietary selenium and vitamin E on the ultrastructure and ATP concentration of boar spermatozoa, and the efficacy of added sodium selenite in extended semen on sperm motility. J Anim Sci 2000;78: 1544–50.

[156] Hambidge KM, Casey CE, Krebs NF. Zinc. In: Mertz WD, editor. Trace Elements in Human and Animal Nutrition, vol. 2. Orlando, USA: Academic Press; 1986. p. 1–138.

[157] Underwood EJ, Somers M. Studies of zinc nutrition in sheep: 1. The relation of zinc to growth, testicular development, and spermatogenesis in young rams. Aust J Agric Res 1969;20: 889–97.

[158] Tomlinson DJ, Mulling CH, Fakler TM. Invited review: formation of keratins in the bovine claw: roles of hormones, minerals, and vitamins in functional claw integrity. J Dairy Sci 2004;87: 797–809.

[159] Kumar S. Management of infertility due to mineral deficiency in dairy animals. Proceedings of ICAR summer school on "Advance diagnostic techniques and therapeutic approaches to metabolic and deficiency diseases"; 2003.

[160] Howell JM, Gooneratne RS. The pathology of trace minerals toxicity in animals. In: Howell JM, Gawthorne JM, editors. Trace Minerals in Animals and Man, vol. 2. Florida, USA: CRC Press; 1987. p. 53–78.

[161] Annenkov BN. Mineral feeding of cattle. In: Geogievski VI, Annekov BN, Samokhin VI, editors. Mineral Nutrition of Animals. London, England: Butterworth; 1981. p. 285–305.

[162] Corah LR, Ives S. The effects of essential trace minerals on reproduction in beef cattle. Vet Clin North Am Food Anim Pract 1991;7:40–57.

[163] Oltenacu PA, Broom DM. The impact of genetic selection for increased milk yield on the welfare of dairy cows. Anim Welfare 2010;19:39–49.

[164] D'Silva J. Adverse impact of industrial animal agriculture on the health and welfare of farmed animals. Integr Zool 2006;1:53–8.

[165] Kołacz R. Dobrostan zwierząt a postęp genetyczny (Animal welfare and the genetic progress). (in Polish). Przegląd Hodowlany 2006;9:8–11.

[166] Whittaker AL. Space requirements to optimize welfare and performance in group housed pigs – a review. Am J Anim Vet Sci 2012;7(2):48–54.

[167] Vitorović D. Anatomske karakteristike kostiju i mišića pilića lakog i teškog tipa gajenih na podu i u kavezima. PhD thesis. Belgrade, Faculty of Veterinary Medicine; 1992.

[168] Gustafson GM. Effects of daily exercise on the health of tied dairy cows. Prev Vet Med 1993;17:209–23.

[169] Davidson JA, Beede DK. Exercise training of late-pregnant and nonpregnant dairy cows affects physical fitness and acid-base homeostasis. J Dairy Sci 2009;92:548–62.

[170] Adewuyi AA, Roelofs JB, Gruys E, Toussaint MJM, van Eerdenburg FJCM. Relationship of plasma nonesterified fatty acids and walking activity in postpartum dairy cows. J Dairy Sci 2006;89:2977–9.

[171] Bøe KE, Færevik G. Grouping and social preferences in calves, heifers and cows. Appl Anim Behav Sci 2003;80:175–90.

[172] Spoolder HAM, Geudeke MJ, Van der Peet-Schwering CMC, Soede NM. Group housing of sows in early pregnancy: a review of success and risk factors. Livest Sci 2009;125:1–14.

[173] Gentle MJ, Hunter LN. Physiological and behavioural responses associated with feather removal in *Gallus gallus* var *domesticus*. Res Vet Sci 1991;50:95–101.

[174] Nicol CJ, Gregory NG, Knowles TG, Parkman ID, Wilkins LJ. Differential effects of increased stocking density, mediated by increased flock size, on feather pecking and aggression in laying hens. Appl Anim Behav Sci 1999;65:137–52.

[175] Nicol CJ, Potzsch C, Lewis K, Green LE. Matched concurrent case–control study of risk factors for feather pecking in hens on free-range commercial farms in the UK. Br Poult Sci 2003;44: 515–23.

[176] Grigor PN, Hughes BO, Appleby MC. Social inhibition of movement in domestic hens. Appl Anim Behav Sci 1995;49:1381–8.

[177] Lariviere J-M, Michaux C, Famir F, Detilleux J, Verleyen V, Leroy P. Reproductive performance of the Ardennaise chicken breed under traditional and Morden breeding management system. Int J Poult Sci 2009;8:446–51.

[178] Hocking PM, Robertson GW, Teverson D. Fertility and early embrionic mortality in traditional breeds of chicken in the UK. Proceeding 5th European poultry genetic symposium. Braedstrup, Denmark. 2007. p. 108.

[179] Anandh MA, Jagatheesan PNR, Kumar PS, Paramasivam A, Rajarajan G. Effect of rearing systems on reproductive performance of turkey. Vet World 2012;5(4):226–9.

15

Prenatal Physical Activity and Gestational Weight Gain

Jennifer L. Kraschnewski, MD, MPH, Cynthia H. Chuang, MD, MSc

Penn State College of Medicine, Hershey, Pennsylvania, USA

INTRODUCTION

Excessive gestational weight gain (GWG) has significant public health implications due to the increased risk of birth complications and cesarean section, and ultimately, increased rates of long-term obesity for the mother [1–6]. Furthermore, exceeding GWG recommendations is harmful to the offspring as well, conferring increased risk of childhood and adult obesity. This suggests that pregnancy is a critical period of time to affect weight outcomes across generations [1,7–11]. Unfortunately, the majority of US women exceed the Institute of Medicine (IOM) guidelines for recommended GWG [12,13]. In this chapter, we discuss prenatal physical activity, including current recommendations, pregnant women's levels of activity, and interventions to attempt to improve engagement. In addition, we discuss other predictors of GWG, evidence for GWG interventions, and recommendations for future approaches.

GESTATIONAL WEIGHT GAIN GUIDELINES

Understanding the optimal amount of weight that should be gained during pregnancy has been shaped by research and epidemiology over the last century. While women in the early 1900s would have been encouraged to gain 15–20 pounds, guidelines for weight gain increased to 20–25 pounds in the 1970s, given a growing recognition that lower GWG was associated with premature birth and small-for-gestational-age infants [14]. Concern for low birth weight infants remained the focus of the IOM 1990 pregnancy weight guidelines, which were additionally stratified by prepregnancy maternal body mass index (BMI) [15].

The rise of the obesity epidemic has affected reproductive-aged women as well, leading to a growing concern for excessive GWG. The majority of women of childbearing age are already overweight or obese, which elevates the risk of gestational diabetes, preeclampsia, eclampsia, cesarean delivery, macrosomic infants, and failure to initiate and sustain breastfeeding [1,16]. Risk of adverse obstetrical outcomes (i.e., preeclampsia, gestational diabetes mellitus, cesarean delivery, and macrosomic infants) is further increased if women have excessive GWG [17]. Research has demonstrated that excessive GWG also leads to increased postpartum weight retention and long-term overweight and obesity after pregnancy. Recognizing higher prepregnancy BMI as a predictor of poor birth outcomes, the IOM revisited the GWG guidelines in 2009 [1]. The new guidelines provide a range for recommended GWG, tailored for each prepregnancy BMI category (see Table 15.1). These guidelines recommend that women with normal prepregnancy BMI gain 25–35 pounds during pregnancy, whereas overweight and obese women are advised to gain 15–25 pounds and 11–20 pounds, respectively [1]. The updated IOM guidelines differ from the 1990 version by providing a specific upper GWG limit for obese women.

Unfortunately, similar to increasing rates of overweight and obesity, rates of excessive GWG have been increasing over time. Nearly half of normal-weight women and 2/3 of overweight and obese women exceed the IOM guidelines for weight gain during pregnancy [1,12,13,18]. This trend has continued since the introduction of the new IOM guidelines, with prospective cohort studies ($n = 1388$–3006) demonstrating between 51 and 78% of pregnant women exceeding the recommended weight gain [12,19,20]. Particularly concerning is the finding that overweight and obese women are at significantly increased risk for excessive GWG, with studies

Handbook of Fertility. http://dx.doi.org/10.1016/B978-0-12-800872-0.00015-9

TABLE 15.1   2009 IOM Recommendations for Total Weight Gain During Pregnancy, by Prepregnancy BMI [1]

Prepregnancy BMI	Total weight gain	
	Range (kg)	Range (lbs)
Underweight (< 18.5 kg/m^2)	12.5–18	28–40
Normal weight (18.5–24.9 kg/m^2)	11.5–16	25–35
Overweight (25.0–29.9 kg/m^2)	7–11.5	15–25
Obese (≥ 30.0 kg/m^2)	5–9	11–20

demonstrating they are two to six times more likely to exceed weight gain guidelines than women of normal weight [21,22]. In addition to a higher prepregnancy BMI, women who are younger, pregnant with their first child, and have a lower income are all at increased risk of excessive GWG [13,18,23]. Behavioral predictors that have been associated with excessive GWG include watching >2 h of television viewing per day [24] and sedentary behavior (see Section "Physical Activity in Pregnancy").

Excessive GWG is an important contributor to postpartum weight retention and long-term overweight and obesity. Although on average, postpartum retention is modest (only 0.5 kg), it may be an important cause of weight gain for women on an individual basis [25,26]. For example, 1/4 of women with a normal prepregnancy BMI who gain more than 20 kg during pregnancy will move up one BMI category (i.e., from normal weight to overweight) at 6 months postpartum [27,28]. More than 60% of women who exceed the GWG guidelines retain greater than 10 pounds at 6 months postpartum [1]. This is especially concerning when nearly 2/3 of women of childbearing age are already overweight.

Gestational weight gain is the single strongest predictor of postpartum weight retention at 1 year. This finding has led to the suggestion that women who are overweight and obese prior to pregnancy do not need to comply with any minimum weight gain recommendations during pregnancy [29]. Greater postpartum weight retention is also associated, but to a lesser degree, with smoking cessation during pregnancy, decreased physical activity frequency, change in caloric intake, and not breastfeeding [29–35]. In addition to the negative health effects, higher postpartum weight is associated with a significant increase in mother's body dissatisfaction [36].

At least as concerning as the effect of excessive GWG on maternal obesity risk is that excessive GWG additionally confers a 40% increased risk of obesity throughout the life course of the offspring [7–10,37–39]. As a result, excessive GWG threatens the health of two generations. As the second most preventable cause of death [40], obesity is a risk factor for cardiovascular disease, stroke, several types of cancer, and diabetes mellitus [41].

Achieving the recommended GWG requires surprisingly few additional calories in the pregnant woman's diet. In fact, pregnant women are not recommended to increase their caloric intake at all during the first trimester, with only 300 and 450 additional calories per day, respectively, in the second and third trimesters [42]. Unfortunately, this is in stark contrast to social norms, which encourage pregnant women to "eat for two" [43]. Women report their peer support structures as strongly influential on their dietary behaviors [44]. As a result, women have come to embrace pregnancy as a time to gain weight freely. For some women, pregnancy is seen as an opportunity to eat without limitations, with excessive dietary intake viewed as healthy for their baby [44,45]. When social norms and the medical evidence differ, we rely on health care providers to tackle these important issues and provide evidence-based recommendations to individual patients. Unfortunately, providers have fallen largely silent on weight counseling in general [46], and perhaps even more so during pregnancy [45,47].

HEALTH CARE PROVIDER ADVICE ON GWG

The Centers for Disease Control and Prevention's 2006 Recommendations to Improve Preconception Health and Health Care identify obesity as an important risk factor for adverse pregnancy outcomes, and recommends appropriate weight loss before pregnancy to reduce these risks [48]. Although women would ideally enter pregnancy at a healthy preconception weight, the high prevalence of overweight and obesity makes this increasingly unlikely. While we need to continue to encourage obese women to achieve weight loss prior to pregnancy as appropriate, opportunities for prepregnancy weight loss counseling do not often occur since preconception visits and management is not commonplace. As a result, GWG may be a more practical target for health care provider advice than preconception weight. Furthermore, all women, regardless of prepregnancy weight, benefit from healthy GWG [1,49]. Pregnancy may also serve as an ideal time for prenatal care providers to target weight control and healthful behaviors, such as meeting physical activity guidelines, given its characterization as a "teachable moment" [50]. In fact, increased patient knowledge and practitioner advice about guideline-recommended weight gain are two factors associated with appropriate GWG, suggesting this is a targetable intervention during pregnancy [47,51].

Although the rationale for GWG guidelines are well-delineated in the IOM report, how health care providers advise pregnant women about GWG goals, and whether this advice is effective in helping women to gain an appropriate amount of weight, remain underexplored. Studies have estimated that between 33 and 81% of women receive no advice from practitioners on appropriate GWG [47,52,53]. Qualitative studies have revealed that prenatal care providers perceive GWG counseling to be useless, and generally only approach the topic of GWG when asked owing to fear of offending or causing stress to the patient [54]. Even if obstetricians are providing GWG advice, only 64% individualize their advice based on a woman's prepregnancy weight [55]. Also concerning is that nearly one in four overweight or obese women who received GWG advice report being told to gain more than the IOM guidelines [45,47,56].

These findings that pregnant women are not receiving appropriate advice about GWG are further supported by similar results in qualitative studies of patients. Women report receiving inappropriate and insufficient GWG advice, and that information was often contradictory, which is unlikely to positively influence how women shape their goals and expectations for weight gain in pregnancy [44,45,57]. Most pregnant women also reported a general lack of concern about GWG and suggested that if it were important, their health care provider would have discussed it with them (which they did not) [57]. In contrast to what women report receiving, obstetricians more frequently report discussing appropriate weight gain, nutrition, and physical activity [58]. Unfortunately, physicians consistently self-report higher levels of adherence to evidence-based guidelines when compared to objective measures, resulting in an overestimation of their performance [59].

Providers may be giving inaccurate or no advice for a number of reasons, including feeling awkward acknowledging women's weight, particularly those who are overweight or obese, in fear of embarrassing them [54]. In addition, health care providers may fail to calculate their patient's prepregnancy BMI, and thus fail to recognize overweight or obesity and not adjust GWG recommendations appropriately [44,54]. Some providers avoid offering GWG counseling for fear of causing anxiety in the patient or "victimizing" them [44,54]. As a result, providers may employ a "reactive" approach to counseling patients on weight gain, that is, waiting for cues from the patient to address the issue [54,58]. Waiting for patients to bring up the issue is problematic as

overweight women may not recognize themselves to be overweight or fail to ask their provider appropriate questions to result in health information receipt [45,58]. In addition, time limitations during clinical encounters may restrict providers' ability to counsel patients on appropriate weight gain, a barrier to counseling frequently identified in primary care clinical settings [57,60,61]. Furthermore, providers may feel inadequately trained to appropriately address weight and physical activity or believe that such counseling is ineffective [54,57,62,63]. As a result, providers are unlikely to change practice in counseling women about weight gain without further tools to help women achieve their recommended GWG.

A lack of counseling on appropriate GWG leaves women to establish their own expectations for the course of their weight gain and on their perception of what is acceptable [45]. Most women report having goals for their GWG, which are closely tied to actual weight gain during pregnancy [47,51,64]. Despite not being informed by their health care provider about recommended weight gain, most women (77%) have ideal weight gain goals within the IOM guidelines, suggesting they receive information from other sources [47,65]. Unfortunately, women who are overweight or obese prior to pregnancy have been shown to be least likely to identify guideline-concordant GWG goals [47,52]. Studies have found that overweight and obese women have up to seven times the odds of reporting their ideal GWG above the IOM guidelines as compared to their normal-weight counterparts [47,52]. As overweight and obese women are also more likely to gain weight above the IOM guidelines, modifying goals for weight gain in this population may be an effective means of limiting excessive GWG [56].

The updated IOM guidelines recommend that health care providers tailor women's GWG goals and chart their weight gain, sharing this information with them to inform and support their progress to their weight goal [1]. Additionally, the American College of Obstetricians and Gynecologists (ACOG) recommend pregnant women be provided with individualized advice about diet and physical activity [66]. Others have argued that recommendations for women, particularly those who are overweight and obese, should not focus on weight gain but rather dietary quality and physical activity [29]. Although individualized treatment is likely required to help women achieve guideline-concordant GWG, it is unlikely to be sufficient. The IOM guidelines additionally call for further research on multiple levels, including the family and community, to help women succeed in achieving appropriate weight gain [1]. Particular attention is recommended for low-income and minority women, who have an increased risk of being overweight and obese prepregnancy, as well as increased rates of poorer nutrition and sedentary behaviors [1].

PHYSICAL ACTIVITY IN PREGNANCY

Physical activity, specifically 30 min or more on most (if not all) days during pregnancy is recommended as part of a healthy pregnancy by the ACOG [66]. Furthermore, key recommendations from the United States Federal Guidelines encourage healthy women to attain at least 150 min of moderate-intensity aerobic activity each week [67]. In addition, pregnant women who have previously been engaged in vigorous intensity aerobic activity may continue this activity with their health care provider's approval [67]. The current guidelines are a far cry from the first prenatal guidelines, issued by ACOG in 1985, which restricted activity by duration (<15 min for strenuous activity), heart rate (≤140 beats/min), and core body temperature (<100.4°F), although they endorsed most aerobic physical activity as safe within these limits [68].

Exercise is known to have tremendous health benefits for adults, including reducing the risk of premature death, and diseases such as coronary heart disease, stroke, some cancers, type 2 diabetes mellitus, osteoporosis, depression, high blood pressure, and high cholesterol [67]. Evidence continues to accumulate in support of physical activity benefits for both mother and baby during pregnancy. Women who engage in at least 30 min of moderate physical activity per day in their last trimester have better cardiovascular fitness than less active women [69]. In addition, physical activity reduces the risk of certain medical complications associated with pregnancy, including hypertensive complications, such as preeclampsia [70–72]. Epidemiologic data also suggest that there may be an additional benefit in the primary prevention of gestational diabetes mellitus, particularly for women who are obese [73]. Furthermore, exercise has been endorsed by the American Diabetes Association as an adjunctive therapy to dietary intervention for treating gestational diabetes mellitus [74,75].

In addition to these benefits, regular vigorous exercise in pregnancy has been shown to improve women's energy levels, appearance, and satisfaction with their physical stamina [76]. A Cochrane systematic review of randomized controlled trials to promote aerobic exercise during pregnancy confirmed that regular exercise improves women's physical fitness, but identified the need for future studies given that there was insufficient evidence to conclude the overall risks and benefits for the woman and the baby [77]. Contrary to common misconceptions, moderate-intensity activity in healthy pregnant women does not result in an increased risk of low birth weight, preterm labor, or early pregnancy loss [67,78–80].

Whether prenatal physical activity has an important role in preventing excessive GWG remains unclear. In fact, the 2009 IOM report specifically calls for studies to be conducted on the effect of physical activity on weight gain during pregnancy [1]. Subsequent studies have suggested that physical activity, both preconception and prenatal, is inversely associated with excessive GWG [12,19,50,53,81,82]. A meta-analysis of GWG interventions evaluating physical activity found a statistically significant mean difference of −0.61 kg (95% CI: −1.17 to −0.06 kg) in the exercise intervention groups as compared to the control groups, suggesting that physical activity "might be a successful strategy in restricting GWG" [83].

The relationship between physical activity and mode of delivery in a modern day population of women with significant rates of overweight and obesity is not entirely clear. Historically, studies have found that women who participated in high levels of exercise throughout pregnancy had a lower frequency of cesarean delivery and vaginal operative delivery [84,85]. More recent investigations, which include a greater proportion of overweight and obese women, have not shown statistically significant reduced risks of cesarean delivery [80,86].

Despite convincing evidence of the health benefits of prenatal physical activity and supporting guidelines, pregnancy is a time of decreased physical activity for most women [44,80,87,88]. In fact, nearly twice as many women are sedentary during pregnancy compared to the already low average rates of physical activity among US adults [89]. More than 1/3 of pregnant women report not engaging in any moderate or vigorous-intensity leisure activities in the past month [90]. A recent national study found that only 14% of pregnant women met the minimum national recommendations for physical activity for moderate-intensity activities [91]. Other population-based studies have found that only 6–13% of women engage in moderate or vigorous physical activity during pregnancy [92,93]. The most common type of leisure physical activity reported by women is walking and swimming [88,91]. The odds of meeting physical activity guidelines were significantly higher in the first trimester as compared to the third [91].

Pregnancy may serve as an ideal time for reproductive-age women to receive counseling on physical activity, as healthful exercise behaviors during pregnancy may lead to persistent healthy behaviors beyond pregnancy. However, it remains unclear whether pregnant women are advised by their prenatal care providers on healthy goals for physical activity during pregnancy. Qualitative work with postpartum women demonstrated that most report receiving insufficient and inappropriate advice from their prenatal providers on exercise during pregnancy [45]. If they received any advice regarding physical activity, most women were advised not to exercise more intensely than before pregnancy. Because most women were not exercising before pregnancy, this advice was interpreted to mean that they should not exercise at all [45]. Our current understanding on health care provider advice suggests that even if pregnant women receive counseling, the advice may not be consistent with the federal guidelines for physical activity; a recent survey found that more than 1/4 of obstetricians did not believe moderate intensity exercise during pregnancy was beneficial [94]. This is supported by findings in qualitative interviews with women, in which the most common forms of exercise recommended where stretching and walking, which may not be sufficiently intense to constitute "moderate intensity" physical activity as recommended by the federal guidelines [45,67,89].

Identifying which women are physically active during pregnancy has been examined in a limited number of studies [95]. Having higher socioeconomic status, health insurance, higher levels of education, and being older have all been associated with increased physical activity during pregnancy [89,91,93,95,96]. Additionally, women who were physically active prior to their pregnancy are more likely to be active during their pregnancy [87]. There is also a positive association between women's physical activity levels during pregnancy and their partner's physical activity [88]. Predictors of less physical activity during pregnancy include having other children at home, a higher prepregnancy BMI, or smoking [89,93,97]. Unfortunately, there is a paucity of data regarding the unique determinants of prenatal physical activity among overweight and obese women, a population of specific interest based on increased potential for benefit [98]. Further research is therefore warranted to determine how best to engage women, and specifically those with overweight and obesity, in physical activity during pregnancy. Understanding which women are likely to engage in physical activity, in addition to the typical barriers for participation, can help tailor such future interventions.

A handful of studies have investigated the barriers pregnant women face in trying to be physically active. Typically, the most commonly reported barriers to prenatal physical activity are intrapersonal, including lacking time, lacking energy, and having general physical discomfort [44,99,100]. Another commonly cited reason in up to 1/3 of women is the unfounded concern for potential harm to the baby, suggesting awareness of the federal guidelines for physical activity and the supporting evidence are poorly disseminated [44,45,99,101]. Other women report not being physically active due to not knowing what to do, a barrier that could potentially be overcome with appropriate health care provider advice [45,102].

In addition to intrapersonal barriers, physical activity is further challenged by a woman's social and built environment. For example, neighborhood environments, specifically those with physical incivilities and decreased access to exercise facilities and built environments, have been associated with decreased physical activity both before and during pregnancy [92]. Women's behaviors also appear to be strongly influenced by the views of their social support network, which frequently encourage rest instead of activity and suggest physical activity may, in fact, be unsafe [44,103]. The influences on physical activity outside the individual woman (i.e., social support network, neighborhood environment) suggests interventions need to address more than one-on-one counseling to successfully encourage pregnant women to be physically active.

Identifying successful ways to engage women in physical activity during pregnancy is an opportunity to establish healthful habits for beyond pregnancy, similar to helping women quit smoking. Physical activity rates in the postpartum period have been inconsistently reported in research studies [104]. One cohort study ($n = 471$) found that, although physical activity levels decreased from the second to third trimester, they rebounded at 3 months postpartum and remained stable at 12 months postpartum [104]. However, most of this physical activity was low intensity and may not be enough to confer health benefits or postpartum weight loss efforts [104]. Health care providers may have an additional opportunity to re-engage women postpartum by providing tailored physical activity advice. Consideration is also necessary for the creation of new physical activity guidelines specifically targeting otherwise sedentary overweight and obese woman. Tailored guidelines for this at-risk population have the opportunity to provide safe and supportive recommendations while easing health care provider counseling [105].

GWG INTERVENTIONS

Given that the majority of women gain too much weight during pregnancy, effective interventions are needed to prevent excessive GWG. Furthermore, effective interventions have the potential to prevent long-term weight retention and obesity in both reproductive-age women and their children. In a time when no effective obesity prevention strategies exist, efforts to reduce GWG as a strategy for obesity prevention is a promising and logical approach. Unfortunately, many interventions have been conducted with the attempt to help women achieve healthy GWG but have had limited success. A number of systematic reviews have been published on this topic, including a 2012 Cochrane review, and suggest that interventions to date have had limited effectiveness [83,106,107].

The 2012 Cochrane Review of 28 interventions for preventing excessive GWG found interventions for normal weight women with promising but modest results. Four of the included trials ($n = 444$) reported excessive weight gain as an outcome. Two studies comparing regular weight monitoring and an intervention focused on diet and exercise, as compared to usual care, showed no significant difference in GWG. However, two studies ($n = 247$) investigating behavioral counseling as compared with standard of care found a positive treatment effect (RR 0.72, 95% CI 0.54–0.95) [81,108–110]. Unfortunately, none of the included interventions were effective in preventing excessive GWG in high-risk groups (i.e., overweight and obese women). Specifically, this review found that four randomized controlled trials investigating behavioral counseling, regular weight measurement, and nutritional advice from a brochure, were ineffective at preventing excessive GWG in women with overweight and obesity [81,108,109,111].

The largest and most notable randomized trial was the Fit for Delivery Study, a behavioral intervention to prevent excessive GWG in pregnant women ($n = 401$) that enrolled participants during their first prenatal visit (mean gestational age: 13.5 ±1.8 weeks) [81]. Women were randomized to receive either standard of care or the behavioral intervention, which consisted of a detailed face-to-face visit with an interventionist, three phone calls from a dietician, scales, food records, pedometers, automated weekly postcards promoting healthy eating and exercise, and personalized mailings of weight graphs with feedback. There was no difference in total GWG in the intervention group as compared to the control group. In subgroup analysis of normal weight women, participants in the intervention group were less likely to exceed IOM recommendations for GWG compared with usual care (40 vs. 52%, $p = 0.003$); however, the absolute percent of women with excessive GWG was still high, and there was no effect in overweight and obese women. Since the 2012 Cochrane Review, an intervention by Bogaerts and colleagues investigated the effects of a lifestyle intervention in obese pregnant women and demonstrated a 9.6% reduction in excessive GWG in the lifestyle intervention group compared with usual care [112]. However, a greater reduction of 18% compared with usual care was seen in the third arm, which received an information-only brochure, adding further confusion to whether behavioral interventions are truly effective.

Given the lack of effectiveness of most GWG interventions, Hill et al. conducted a review of the theory and intervention components utilized in interventions as a next step in addressing excessive GWG [113]. In this review, eight studies were identified as theory-based, with six of the eight reporting positive effects on the outcome of GWG [113]. A wide variety of theoretical underpinnings and health behavior change strategies were used, however, making it challenging to compare across studies. This review concluded that theory-based studies were as effective as non-theory-based studies, with key strategies for success including education, motivational interviewing, behavioral self-monitoring, and rewards for successful behavior [113]. Similarly to GWG interventions, most prenatal physical activity intervention studies fail to report theoretical underpinnings to strategies for promoting physical activity as well [98].

The general lack of effectiveness of most GWG interventions likely results from an inability of previous studies to overcome the significant barriers to achieving healthy GWG for pregnant women. Specifically missing from randomized clinical trials to date are interventions to educate and inform the social network surrounding pregnant women, including their partners, families, and peers. Interventions are also needed to engage health care providers to effectively counsel women to achieve healthy GWG [44].

ACHIEVING HEALTHY GWG

Unfortunately, research has not yet determined the best way to help pregnant women, particularly those who are overweight or obese, achieve healthy GWG. Lifestyle interventions trialed have employed multiple behavioral change techniques, including goal setting, action planning, and self-monitoring, largely modeled after approaches used in weight loss interventions [113]. The theoretical underpinnings of these types of behavioral change techniques assume that the participant believes that guideline-adherent GWG is a desirable goal. Smokers enter smoking cessation trials because they want to quit smoking, overweight persons enter weight loss studies because they want to lose weight; however, we cannot similarly assume this is the case for pregnant women and research suggests the opposite may be true. Therefore, innovative approaches are necessary to be successful in aiding women in achieving appropriate GWG.

Changing social norms around pregnancy weight gain and the culture of "eating for two" are likely necessary to successfully help pregnant women understand the importance of, and ultimately achieve, guideline-adherent weight gain. Even intensive lifestyle interventions (i.e., weekly meetings) will be overpowered when women spend the vast majority of their pregnancy receiving opposing messages from the real world. Unlike obesity or tobacco use, excessive GWG is not recognized to be harmful by society at large, including at the individual, health care provider, and community levels. Despite this reality, nearly all GWG interventions to date have engaged women on an individual basis, ignoring the social network within which they operate. These findings have led the IOM to call for approaches to address family- and community-level factors to allow women to succeed gaining weight within guidelines [1].

Public health messaging is a power tool that has been utilized with significant success in the "Back to Sleep" campaign. Despite existing social norms that placing an infant on its stomach was the safest approach, the "Back to Sleep" campaign placed the American Academy of Pediatrics 1992 recommendations on nearly every mother's radar, resulting in a sharp decline in sudden infant death syndrome [114]. A similar campaign may be useful to call attention to the importance of achieving GWG within guidelines. By targeting the entire population, there is an opportunity to shape the current social norms to support healthy lifestyles in pregnant women.

Providers will also need to take a more active role in advising women on healthy GWG and the importance of physical activity during pregnancy. Further research is necessary to assist providers in how to effectively counsel pregnant women on GWG and integrate this counseling into clinical practice. Simply being advised to lose weight by a health care provider is associated with successful weight loss attempts in overweight individuals, and similar success may occur from providing GWG counseling [115]. However, determining and providing the appropriate availability of resources and tools to allow providers to counsel effectively on GWG will be necessary [43].

It will also be necessary to improve our understanding of how women who achieve guideline-concordant weight are successful [64]. Most behavioral interventions for GWG to date have been guided by a deductive "top–down" approach, in that they incorporate interventions selected empirically by investigators and derived from the weight loss literature. However, there may be different behaviors successfully used by women in achieving guideline-concordant GWG. Developing future interventions through an inductive "bottom–up" approach – that is, through identifying behaviors and habits of women who have successfully achieved healthy GWG on their own – may introduce a more effective intervention [64,116,117]. Although the basic principles of healthy nutrition and physical activity apply to both pregnant and nonpregnant persons, pregnancy symptoms and opposing social norms surrounding these behaviors may strongly influence lifestyle choices and will need to be addressed to help women successfully achieve guideline-concordant GWG.

References

[1] IOM (Institute of Medicine), NRC (National Research Council). Weight Gain During Pregnancy: Reexamining the Guidelines. Washington, DC, 2009.

[2] Rooney BL, Schauberger CW. Excess pregnancy weight gain and long-term obesity: one decade later. Obstet Gynecol 2002;100(2):245–52.

[3] Rooney BL, Schauberger CW, Mathiason MA. Impact of perinatal weight change on long-term obesity and obesity-related illnesses. Obstet Gynecol 2005;106(6):1349–56.

[4] Linne Y, Dye L, Barkeling B, Rossner S. Weight development over time in parous women – the SPAWN study – 15 years follow-up. Int J Obes Relat Metab Disord 2003;27(12):1516–22.

[5] Amorim AR, Rossner S, Neovius M, Lourenco PM, Linne Y. Does excess pregnancy weight gain constitute a major risk for increasing long-term BMI? Obesity (Silver Spring) 2007;15(5):1278–86.

[6] Mamun AA, Kinarivala M, O'Callaghan MJ, Williams GM, Najman JM, Callaway LK. Associations of excess weight gain during pregnancy with long-term maternal overweight and obesity: evidence from 21 y postpartum follow-up. Am J Clin Nutr 2010;91(5):1336–41.

[7] Oken E, Rifas-Shiman SL, Field AE, Frazier AL, Gillman MW. Maternal gestational weight gain and offspring weight in adolescence. Obstet Gynecol 2008;112(5):999–1006.

[8] Oken E, Taveras EM, Kleinman KP, Rich-Edwards JW, Gillman MW. Gestational weight gain and child adiposity at age 3 years. Am J Obstet Gynecol 2007;196(4). 322 e1–322 e8.

[9] Mamun AA, O'Callaghan M, Callaway L, Williams G, Najman J, Lawlor DA. Associations of gestational weight gain with offspring body mass index and blood pressure at 21 years of age: evidence from a birth cohort study. Circulation 2009;119(13):1720–7.

[10] Mamun AA, Mannan M, Doi SA. Gestational weight gain in relation to offspring obesity over the life course: a systematic review and bias-adjusted meta-analysis. Obes Rev 2014;15(4):338–47.

[11] Margerison Zilko CE, Rehkopf D, Abrams B. Association of maternal gestational weight gain with short- and long-term maternal and child health outcomes. Am J Obstet Gynecol 2010;202(6):574e1–8.

[12] Kraschnewski JL, Chuang CH, Downs DS, Weisman CS, McCamant EL, Baptiste-Roberts K, et al. Association of prenatal physical activity and gestational weight gain: results from the first baby study. Women's Health Issues 2013;23(4):e233–8.

[13] Chu SY, Callaghan WM, Bish CL, D'Angelo D. Gestational weight gain by body mass index among US women delivering live births, 2004–2005: fueling future obesity. Am J Obstet Gynecol 2009;200(3):271e1–7.

[14] National Research Council. Maternal nutrition and the course of pregnancy. Washington, DC: National Academy Press; 1970.

[15] Institute of Medicine. Nutrition during pregnancy. Washington, DC: National Academy Press; 1990.

[16] Baeten JM, Bukusi EA, Lambe M. Pregnancy complications and outcomes among overweight and obese nulliparous women. Am J Public Health 2001;91(3):436–40.

[17] Flick AA, Brookfield KF, de la Torre L, Tudela CM, Duthely L, Gonzalez-Quintero VH. Excessive weight gain among obese women and pregnancy outcomes. Am J Perinatol 2010;27(4):333–8.

[18] Olson CM, Strawderman MS. Modifiable behavioral factors in a biopsychosocial model predict inadequate and excessive gestational weight gain. J Am Diet Assoc 2003;103(1):48–54.

[19] Stuebe AM, Oken E, Gillman MW. Associations of diet and physical activity during pregnancy with risk for excessive gestational weight gain. Am J Obstet Gynecol 2009;201(1):58e1–8.

[20] Ferrari RM, Siega-Riz AM. Provider advice about pregnancy weight gain and adequacy of weight gain. Matern Child Health J 2013;17(2):256–64.

[21] Chasan-Taber L, Schmidt MD, Pekow P, Sternfeld B, Solomon CG, Markenson G. Predictors of excessive and inadequate gestational weight gain in Hispanic women. Obesity (Silver Spring) 2008;16(7):1657–66.

[22] Wells CS, Schwalberg R, Noonan G, Gabor V. Factors influencing inadequate and excessive weight gain in pregnancy: Colorado, 2000–2002. Matern Child Health J 2006;10(1):55–62.

[23] Brawarsky P, Stotland NE, Jackson RA, Fuentes-Afflick E, Escobar GJ, Rubashkin N, et al. Pre-pregnancy and pregnancy-related factors and the risk of excessive or inadequate gestational weight gain. Int J Gynaecol Obstet 2005;91(2):125–31.

[24] Skouteris H, McCabe M, Milgrom J, Kent B, Bruce LJ, Mihalopoulos C, et al. Protocol for a randomized controlled trial of a specialized health coaching intervention to prevent excessive gestational weight gain and postpartum weight retention in women: the HIPP study. BMC Public Health 2012;12:78.

[25] Sichieri R, Field AE, Rich-Edwards J, Willett WC. Prospective assessment of exclusive breastfeeding in relation to weight change in women. Int J Obes Relat Metab Disord 2003;27(7):815–20.

[26] Williamson DF, Madans J, Pamuk E, Flegal KM, Kendrick JS, Serdula MK. A prospective study of childbearing and 10-year weight gain in US white women 25 to 45 years of age. Int J Obes Relat Metab Disord 1994;18(8):561–9.

[27] Nohr EA, Vaeth M, Baker JL, Sorensen T, Olsen J, Rasmussen KM. Combined associations of prepregnancy body mass index and gestational weight gain with the outcome of pregnancy. Am J Clin Nutr 2008;87(6):1750–9.

[28] Viswanathan M, Siega-Riz AM, Moos MK, Deierlein A, Mumford S, Knaack J, et al. Outcomes of maternal weight gain. Evid Rep Technol Assess (Full Rep) 2008;(168):1–223.

[29] Melzer K, Schutz Y. Pre-pregnancy and pregnancy predictors of obesity. Int J Obes (Lond) 2010;34(Suppl 2):S44–52.

[30] Olson CM, Strawderman MS, Hinton PS, Pearson TA. Gestational weight gain and postpartum behaviors associated with weight change from early pregnancy to 1 y postpartum. Int J Obes Relat Metab Disord 2003;27(1):117–27.

[31] Baker JL, Gamborg M, Heitmann BL, Lissner L, Sorensen TI, Rasmussen KM. Breastfeeding reduces postpartum weight retention. Am J Clin Nutr 2008;88(6):1543–51.

[32] Krause KM, Lovelady CA, Peterson BL, Chowdhury N, Ostbye T. Effect of breast-feeding on weight retention at 3 and 6 months postpartum: data from the North Carolina WIC Programme. Public Health Nutr 2010;13(12):2019–26.

[33] Ostbye T, Krause KM, Swamy GK, Lovelady CA. Effect of breastfeeding on weight retention from one pregnancy to the next: results from the North Carolina WIC program. Prev Med 2010;51(5):368–72.

[34] Ostbye T, Peterson BL, Krause KM, Swamy GK, Lovelady CA. Predictors of postpartum weight change among overweight and obese women: results from the Active Mothers Postpartum Study. J Women's Health (Larchmt) 2012;21(2):215–22.

[35] Wiklund P, Xu L, Lyytikainen A, Saltevo J, Wang Q, Völgyi E, et al. Prolonged breast-feeding protects mothers from later-life obesity and related cardio-metabolic disorders. Public Health Nutr 2012;15(1):67–74.

[36] Gjerdingen D, Fontaine P, Crow S, McGovern P, Center B, Miner M. Predictors of mothers' postpartum body dissatisfaction. Women Health 2009;49(6):491–504.

[37] Fraser A, Tilling K, Macdonald-Wallis C, Sattar N, Brion MJ, Benfield L, et al. Association of maternal weight gain in pregnancy with offspring obesity and metabolic and vascular traits in childhood. Circulation 2010;121(23):2557–64.

[38] Schack-Nielsen L, Michaelsen KF, Gamborg M, Mortensen EL, Sorensen TI. Gestational weight gain in relation to offspring body mass index and obesity from infancy through adulthood. Int J Obes (Lond) 2010;34(1):67–74.

[39] Wrotniak BH, Shults J, Butts S, Stettler N. Gestational weight gain and risk of overweight in the offspring at age 7 y in a multicenter, multiethnic cohort study. Am J Clin Nutr 2008;87(6):1818–24.

[40] Mokdad AH, Ford ES, Bowman BA, Dietz WH, Vinicor F, Bales VS, et al. Prevalence of obesity, diabetes, and obesity-related health risk factors. JAMA 2001;289(1):76–9.

[41] Thompson D, Edelsberg J, Colditz GA, Bird AP, Oster G. Lifetime health and economic consequences of obesity. Arch Intern Med 1999;159(18):2177–83.

[42] Butte NF, King JC. Energy requirements during pregnancy and lactation. Public Health Nutr 2005;8(7A):1010–27.

[43] Kraschnewski JL, Chuang CH. Eating for two: excessive gestational weight gain and the need to change social norms. Women's Health Issues 2014;24(3):e257–9.

[44] Campbell F, Johnson M, Messina J, Guillaume L, Goyder E. Behavioural interventions for weight management in pregnancy: a systematic review of quantitative and qualitative data. BMC Public Health 2011;11:491.

[45] Stengel MR, Kraschnewski JL, Hwang SW, Kjerulff KH, Chuang CH. What my doctor didn't tell me: examining health care provider advice to overweight and obese pregnant women on gestational weight gain and physical activity. Women's Health Issues 2012;22(6):e535–40.

[46] Kraschnewski JL, Sciamanna CN, Stuckey HL, Chuang CH, Lehman EB, Hwang KO, et al. A silent response to the obesity epidemic: decline in US physician weight counseling. Med Care 2013;51(2):186–92.

[47] Phelan S, Phipps MG, Abrams B, Darroch F, Schaffner A, Wing RR. Practitioner advice and gestational weight gain. J Women's Health (Larchmt) 2011;20(4):585–91.

[48] Johnson K, Posner SF, Biermann J, Cordero JF, Atrash HK, Parker CS, et al. Recommendations to improve preconception health and health care – United States. A report of the CDC/ATSDR Preconception Care Work Group and the Select Panel on Preconception Care. MMWR Recomm Rep 2006;55(RR–6):1–23.

[49] Olander EK, Atkinson L, Edmunds JK, French DP. The views of pre- and post-natal women and health professionals regarding gestational weight gain: an exploratory study. Sex Reprod Healthc 2011;2(1):43–8.

[50] Phelan S. Pregnancy: a "teachable moment" for weight control and obesity prevention. Am J Obstet Gynecol 2010;202(2). 135e1–135e8.

[51] Strychar IM, Chabot C, Champagne F, Ghadirian P, Leduc L, Lemonnier MC, et al. Psychosocial and lifestyle factors associated with insufficient and excessive maternal weight gain during pregnancy. J Am Diet Assoc 2000;100(3):353–6.

[52] Stotland NE, Haas JS, Brawarsky P, Jackson RA, Fuentes-Afflick E, Escobar GJ. Body mass index, provider advice, and target gestational weight gain. Obstet Gynecol 2005;105(3):633–8.

[53] Herring SJ, Nelson DB, Davey A, Klotz AA, Dibble LV, Oken E, et al. Determinants of excessive gestational weight gain in urban, low-income women. Women's Health Issues 2012;22(5):e439–46.

[54] Stotland NE, Gilbert P, Bogetz A, Harper CC, Abrams B, Gerbert B. Preventing excessive weight gain in pregnancy: how do prenatal care providers approach counseling? J Women's Health (Larchmt) 2010;19(4):807–14.

[55] Power ML, Cogswell ME, Schulkin J. Obesity prevention and treatment practices of U.S. obstetrician-gynecologists. Obstet Gynecol 2006;108(4):961–8.

[56] Cogswell ME, Scanlon KS, Fein SB, Schieve LA. Medically advised, mother's personal target, and actual weight gain during pregnancy. Obstet Gynecol 1999;94(4):616–22.

[57] Olander EK, Atkinson L, Edmunds JK, French DP. The views of pre- and post-natal women and health professionals regarding gestational weight gain: an exploratory study. Sex Reprod Healthc 2011;2(1):43–8.

[58] Duthie EA, Drew EM, Flynn KE. Patient-provider communication about gestational weight gain among nulliparous women: a qualitative study of the views of obstetricians and first-time pregnant women. BMC Pregnancy Childbirth 2013;13:231.

[59] Adams AS, Soumerai SB, Lomas J, Ross-Degnan D. Evidence of self-report bias in assessing adherence to guidelines. Int J Qual Health Care 1999;11(3):187–92.

[60] Orleans CT, George LK, Houpt JL, Brodie KH. Health promotion in primary care: a survey of U.S. family practitioners. Prev Med 1985;14(5):636–47.

[61] Yarnall KS, Pollak KI, Ostbye T, Krause KM, Michener JL. Primary care: is there enough time for prevention? Am J Public Health 2003;93(4):635–41.

[62] Foster GD, Wadden TA, Makris AP, Davidson D, Sanderson RS, Allison DB, et al. Primary care physicians' attitudes about obesity and its treatment. Obes Res 2003;11(10):1168–77.

[63] Sussman AL, Williams RL, Leverence R, Gloyd PW Jr, Crabtree BF. Self determination theory and preventive care delivery: a Research Involving Outpatient Settings Network (RIOS Net) study. J Am Board Fam Med 2008;21(4):282–92.

[64] Chuang CH, Stengel M, Hwang SW, Velott D, Kjerulff KH, Kraschnewski JL. Dietary and physical activity habits in overweight and obese pregnant women who achieve and exceed recommended gestational weight gain. Obes Res Clin Pract 2014;8(6):e577–83.

[65] Kraschnewski JL, Chuang CH, Poole ES, Peyton T, Blubaugh I, Pauli J, et al. Paging "dr. Google": does technology fill the gap created by the prenatal care visit structure? Qualitative focus group study with pregnant women. J Med Internet Res 2014;16(6):e147.

[66] ACOG committee opinion.. Exercise during pregnancy and the postpartum period. Number 267, January 2002. American College of Obstetricians and Gynecologists. Int J Gynaecol Obstet 2002;77(1):79–81.

[67] Physical Activity Guidelines Advisory Committee. Physical activity guidelines advisory committee report. Washington, DC: U.S. Department of Health and Human Services; 2008.

[68] ACOG. Technical Bulletin: Exercise during pregnancy and the postnatal period. Washington DC: American College of Obstetricians and Gynecologists; 1985.

[69] Melzer K, Schutz Y, Soehnchen N, Othenin-Girard V, Martinez de Tejada B, Irion O, et al. Effects of recommended levels of physical activity on pregnancy outcomes. Am J Obstet Gynecol 2010;202(3):266e1–6.

[70] Saftlas AF, Logsden-Sackett N, Wang W, Woolson R, Bracken MB. Work, leisure-time physical activity, and risk of preeclampsia and gestational hypertension. Am J Epidemiol 2004;160(8):758–65.

[71] Sorensen TK, Williams MA, Lee IM, Dashow EE, Thompson ML, Luthy DA. Recreational physical activity during pregnancy and risk of preeclampsia. Hypertension 2003;41(6):1273–80.

[72] Martin CL, Brunner Huber LR. Physical activity and hypertensive complications during pregnancy: findings from 2004 to 2006 North Carolina Pregnancy Risk Assessment Monitoring System. Birth 2010;37(3):202–10.

[73] Dye TD, Knox KL, Artal R, Aubry RH, Wojtowycz MA. Physical activity, obesity, and diabetes in pregnancy. Am J Epidemiol 1997;146(11):961–5.

[74] Jovanovic-Peterson L, Peterson CM. Exercise and the nutritional management of diabetes during pregnancy. Obstet Gynecol Clin North Am 1996;23(1):75–86.

[75] Bung P, Artal R. Gestational diabetes and exercise: a survey. Semin Perinatol 1996;20(4):328–33.

[76] Marquez-Sterling S, Perry AC, Kaplan TA, Halberstein RA, Signorile JF. Physical and psychological changes with vigorous exercise in sedentary primigravidae. Med Sci Sports Exerc 2000;32(1):58–62.

[77] Kramer MS, McDonald SW. Aerobic exercise for women during pregnancy. Cochrane Database Syst Rev 2006;(3). CD000180.

[78] Barakat R, Stirling JR, Lucia A. Does exercise training during pregnancy affect gestational age? A randomised controlled trial. Br J Sports Med 2008;42(8):674–8.

[79] Gavard JA, Artal R. Effect of exercise on pregnancy outcome. Clin Obstet Gynecol 2008;51(2):467–80.

[80] Tinloy J, Chuang CH, Zhu J, Pauli J, Kraschnewski JL, Kjerulff KH. Exercise during pregnancy and risk of late preterm birth, cesarean delivery, and hospitalizations. Women's Health Issues 2014;24(1):e99–e104.

[81] Phelan S, Phipps MG, Abrams B, Darroch F, Schaffner A, Wing RR. Randomized trial of a behavioral intervention to prevent excessive gestational weight gain: the Fit for Delivery Study. Am J Clin Nutr 2011;93(4):772–9.

[82] Weisman CS, Hillemeier MM, Downs DS, Chuang CH, Dyer AM. Preconception predictors of weight gain during pregnancy: prospective findings from the Central Pennsylvania Women's Health Study. Women's Health Issues 2010;20(2):126–32.

[83] Streuling I, Beyerlein A, Rosenfeld E, Hofmann H, Schulz T, von Kries R. Physical activity and gestational weight gain: a meta-analysis of intervention trials. BJOG 2011;118(3):278–84.

[84] Clapp JF 3rd. The course of labor after endurance exercise during pregnancy. Am J Obstet Gynecol 1990;163(6 Pt 1):1799–805.

[85] Hall DC, Kaufmann DA. Effects of aerobic and strength conditioning on pregnancy outcomes. Am J Obstet Gynecol 1987;157(5):1199–203.

[86] Bovbjerg ML, Siega-Riz AM. Exercise during pregnancy and cesarean delivery: North Carolina PRAMS, 2004–2005. Birth 2009;36(3):200–7.

[87] Pereira MA, Rifas-Shiman SL, Kleinman KP, Rich-Edwards JW, Peterson KE, Gillman MW. Predictors of change in physical activity during and after pregnancy: Project Viva. Am J Prev Med 2007;32(4):312–9.

[88] Liu J, Blair SN, Teng Y, Ness AR, Lawlor DA, Riddoch C. Physical activity during pregnancy in a prospective cohort of British women: results from the Avon longitudinal study of parents and children. Eur J Epidemiol 2011;26(3):237–47.

[89] Zhang J, Savitz DA. Exercise during pregnancy among US women. Ann Epidemiol 1996;6(1):53–9.

[90] Evenson KR, Savitz DA, Huston SL. Leisure-time physical activity among pregnant women in the US. Paediatr Perinat Epidemiol 2004;18(6):400–7.

[91] Evenson KR, Wen F. National trends in self-reported physical activity and sedentary behaviors among pregnant women: NHANES 1999–2006. Prev Med 2010;50(3):123–8.

[92] Laraia B, Messer L, Evenson K, Kaufman JS. Neighborhood factors associated with physical activity and adequacy of weight gain during pregnancy. J Urban Health 2007;84(6):793–806.

[93] Petersen AM, Leet TL, Brownson RC. Correlates of physical activity among pregnant women in the United States. Med Sci Sports Exerc 2005;37(10):1748–53.

[94] Evenson KR, Pompeii LA. Obstetrician practice patterns and recommendations for physical activity during pregnancy. J Women's Health (Larchmt) 2010;19(9):1733–40.

[95] Foxcroft KF, Rowlands IJ, Byrne NM, McIntyre HD, Callaway LK. Bambino group. Exercise in obese pregnant women: the role of social factors, lifestyle and pregnancy symptoms. BMC Pregnancy Childbirth 2011;11:4.

[96] Ning Y, Williams MA, Dempsey JC, Sorensen TK, Frederick IO, Luthy DA. Correlates of recreational physical activity in early pregnancy. J Matern Fetal Neonatal Med 2003;13(6):385–93.

[97] Hinton PS, Olson CM. Predictors of pregnancy-associated change in physical activity in a rural white population. Matern Child Health J 2001;5(1):7–14.

[98] Downs DS, Chasan-Taber L, Evenson KR, Leiferman J, Yeo S. Physical activity and pregnancy: past and present evidence and future recommendations. Res Q Exerc Sport 2012;83(4):485–502.

[99] Evenson KR, Moos MK, Carrier K, Siega-Riz AM. Perceived barriers to physical activity among pregnant women. Matern Child Health J 2009;13(3):364–75.

[100] Symons Downs D, Ulbrecht JS. Understanding exercise beliefs and behaviors in women with gestational diabetes mellitus. Diabetes Care 2006;29(2):236–40.

[101] Clarke PE, Gross H. Women's behaviour, beliefs and information sources about physical exercise in pregnancy. Midwifery 2004;20(2):133–41.

[102] Rutkowska E, Lepecka-Klusek C. The role of physical activity in preparing women for pregnancy and delivery in Poland. Health Care Women Int 2002;23(8):919–23.

[103] Leiferman J, Swibas T, Koiness K, Marshall JA, Dunn AL. My baby, my move: examination of perceived barriers and motivating factors related to antenatal physical activity. J Midwifery Women's Health 2011;56(1):33–40.

[104] Borodulin K, Evenson KR, Herring AH. Physical activity patterns during pregnancy through postpartum. BMC Women's Health 2009;9:32.

[105] Mottola MF. Exercise prescription for overweight and obese women: pregnancy and postpartum. Obst Gynecol Clin North Am 2009;36(2):301–16. viii.

[106] Skouteris H, Hartley-Clark L, McCabe M, Milgrom J, Kent B, Herring SJ, et al. Preventing excessive gestational weight gain: a systematic review of interventions. Obes Rev 2010;11(11):757–68.

[107] Muktabhant B, Lumbiganon P, Ngamjarus C, Dowswell T. Interventions for preventing excessive weight gain during pregnancy. Cochrane Database Syst Rev 2012;4. CD007145.

[108] Jeffries K, Shub A, Walker SP, Hiscock R, Permezel M. Reducing excessive weight gain in pregnancy: a randomised controlled trial. Med J Aust 2009;191(8):429–33.

[109] Polley BA, Wing RR, Sims CJ. Randomized controlled trial to prevent excessive weight gain in pregnant women. Int J Obes Relat Metab Disord 2002;26(11):1494–502.

[110] Hui AL, Ludwig SM, Gardiner P, Sevenhuysen G, Murray R, Morris M. Community-based exercise and dietary intervention during pregnancy: a pilot study. Can J Diabetes 2006;30(2):169–75.

[111] Guelinckx I, Devlieger R, Mullie P, Vansant G. Effect of lifestyle intervention on dietary habits, physical activity, and gestational weight gain in obese pregnant women: a randomized controlled trial. Am J Clin Nutr 2010;91(2):373–80.

[112] Bogaerts AF, Devlieger R, Nuyts E, Witters I, Gyselaers W, Van den Bergh BR. Effects of lifestyle intervention in obese pregnant women on gestational weight gain and mental health: a randomized controlled trial. Int J Obes (Lond) 2013;37(6):814–21.

[113] Hill B, Skouteris H, Fuller-Tyszkiewicz M. Interventions designed to limit gestational weight gain: a systematic review of theory and meta-analysis of intervention components. Obes Rev 2013;14(6):435–50.

[114] Pollack HA, Frohna JG. Infant sleep placement after the back to sleep campaign. Pediatrics 2002;109(4):608–14.

[115] Pool AC, Kraschnewski JL, Cover LA, Lehman EB, Stuckey HL, Hwang KO, et al. The impact of physician weight discussion on weight loss in US adults. Obes Res Clin Pract 2014;8(2):e131–9.

[116] Kraschnewski JL, Stuckey HL, Rovniak LS, Lehman EB, Reddy M, Poger JM, et al. Efficacy of a weight-loss website based on positive deviance. A randomized trial. Am J Prev Med 2011;41(6):610–4.

[117] Stuckey HL, Boan J, Kraschnewski JL, Miller-Day M, Lehman EB, Sciamanna CN. Using positive deviance for determining successful weight-control practices. Qual Health Res 2011;21(4):563–79.

16

Lifestyle Intervention for Resumption of Ovulation in Anovulatory Women with Obesity and Infertility

Walter K. Kuchenbecker, MBChB, PhD, Annemieke Hoek, MD, PhD***

*Isala Fertility Center, Isala Clinics, Zwolle, The Netherlands
**Department of Obstetrics and Gynecology, University Medical Center Groningen,
Groningen, The Netherlands

INTRODUCTION

The prevalence of obesity (body mass index, BMI ≥ 30 kg/m^2) is reaching epidemic proportions in many developed countries but is also on the rise in the developing world [1] (Fig. 16.1). Obesity significantly contributes to the overall burden of disease worldwide, and the effect of obesity on female reproduction is an additional growing concern [2].

Obesity-related pregnancy complications are of great concern because of an increase in perinatal mortality and even maternal mortality [3,4]. Maternal obesity may also contribute to long-term disease of the offspring, like childhood obesity, increased risk of developing cardiovascular disease, and diabetes [5,6]. Obesity in women is an important contributor to anovulation, with an exponential increase in anovulation with increasing body weight [7,8] (Fig. 16.2). Polycystic ovary syndrome (PCOS), a very common cause of infertility due to anovulation, affects 4–7% of women. In 40–60% of women with PCOS, overweight, or obesity aggravates the mechanisms of anovulation [9].

The increase in the prevalence of obesity worldwide is attributable to an increase in energy consumption, change in dietary composition, and a more sedentary lifestyle resulting in reduced physical activity (PA). Lifestyle intervention in anovulatory women with obesity and infertility may contribute to resumption of ovulation and improvement in pregnancy outcomes. In this chapter we will critically evaluate the role and limitations of lifestyle intervention in anovulatory women with obesity and infertility.

DEFINING OBESITY AND ITS CONTRIBUTION TO ANOVULATION

Obesity has traditionally been defined as a BMI ≥ 30 kg/m^2, based on the risk of developing chronic conditions like diabetes mellitus and cardiovascular disease [10]. Defining obesity by BMI categories alone is, however, insufficient to assess and predict the female reproductive consequences of obesity. Obesity-related risk assessment based on BMI alone does not account for differences in bone and muscle mass and ethnic differences. In South Asian women, for example, a BMI >25 kg/m^2 already confers increased metabolic risks [11].

Accumulation of fat in the abdominal area (abdominal or central obesity), expressed by increased waist–hip ratio or waist circumference, has a detrimental effect on female reproduction, independent of BMI [12,13]. Obesity and especially central obesity is associated with insulin resistance (IR) and the resulting hyperinsulinemia contributes to anovulation by increased ovarian androgen secretion [14,15] leading to lower sex hormone binding globulin (SHBG) concentrations [16,17] and consequently higher free androgen levels. Insulin and high androgen levels directly influence intraovarian steroidogenesis, which lead to arrest of follicle growth, as has been shown in women with PCOS [18,19].

Increased androgens in turn contribute to central obesity [20], thus a vicious circle evolves where androgens favor abdominal fat accumulation, which in turn facilitates IR and androgen production [21].

IR and the resulting hyperinsulinemia due to obesity can be considered to play a pivotal role in anovulation,

Handbook of Fertility. http://dx.doi.org/10.1016/B978-0-12-800872-0.00016-0

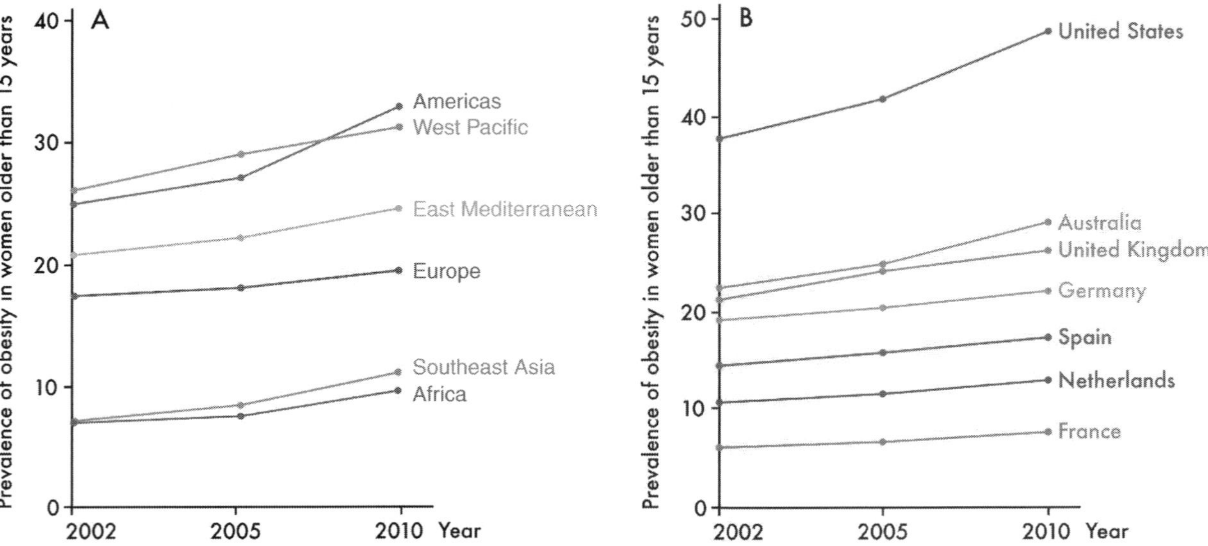

FIGURE 16.1   Prevalence of obesity in women older than 15 years of age in different regions (A) and in different developed countries (B). Derived from the global WHO database [1].

but adipokines secreted by adipose tissue may also contribute to anovulation. Adipose tissue is a complex and highly active endocrine organ with the expression and secretion of various proteins and peptides, collectively called adipokines [22]. Leptin, an adipokine mainly produced by subcutaneous fat, induces anovulation by its effect on the hypothalamic function and direct effects on the ovary [23]. With increasing BMI, adipose tissue becomes dysfunctional with a change in the secreted adipokine profile, deposition of fat in ectopic organs like the liver, pancreas, heart and skeletal muscle, and hypothalamic resistance to leptin [24,25].

IR and the consequences of adipose tissue dysfunction may be considered the focus of lifestyle intervention for the resumption of ovulation in anovulatory women with obesity and infertility.

THE ROLE OF LIFESTYLE INTERVENTION FOR RESUMPTION OF OVULATION

Weight Reduction

The studies of weight reduction in anovulatory women with obesity are mainly small observational studies often limited to women with PCOS. Data of a randomized controlled trial on lifestyle intervention in women with obesity and infertility are forthcoming [26]. Weight reduction programs in women with anovulation have shown that 5–10% loss of initial body weight leads to improvement in menstrual pattern (from 40 to 80%) and resumption of regular ovulatory cycles (from 25 to 70%) by reducing circulating insulin levels and reducing hyperandrogenemia [27–36] (Table 16.1).

Loss of Abdominal Fat

In 18 overweight/obese women with PCOS undergoing a 6 month lifestyle intervention program, the nine women who resumed ovulation lost 11% abdominal fat with 71% improvement in insulin sensitivity in spite of a modest weight loss of 2–5% of the initial body weight [37]. In 32 anovulatory women with obesity and PCOS undergoing a 6 month lifestyle intervention program, loss of intra-abdominal fat was 18.5 and 8.6%, respectively, in women who resumed ovulation (RO+) compared to those who remained anovulatory (RO−). In this study, loss of intra-abdominal fat and not subcutaneous abdominal fat was required for the resumption of ovulation in women with PCOS [33] (Fig. 16.2). Previous studies in different patient populations revealed that preferential loss of intra-abdominal fat contributes to improvement in insulin sensitivity [38]. Loss of IAF and hence improvement of insulin sensitivity seems to be important for the resumption of ovulation in anovulatory women with obesity and infertility (Fig. 16.3).

Increased Physical Activity (PA)

Lack of PA contributes to obesity by decreasing energy expenditure. Increased PA improves not only cardiopulmonary function, but also insulin sensitivity and markers of adipose tissue dysfunction [39,40]. Furthermore, exercise has a favorable effect on the general well-being and health-related quality of life [41]. In combination with dietary intervention, increased PA adds little additional weight loss than diet alone, but it prevents loss of muscle mass [42]. Skeletal muscle is an important site of insulin-mediated glucose disposal, and preserving and building muscle mass by increased PA

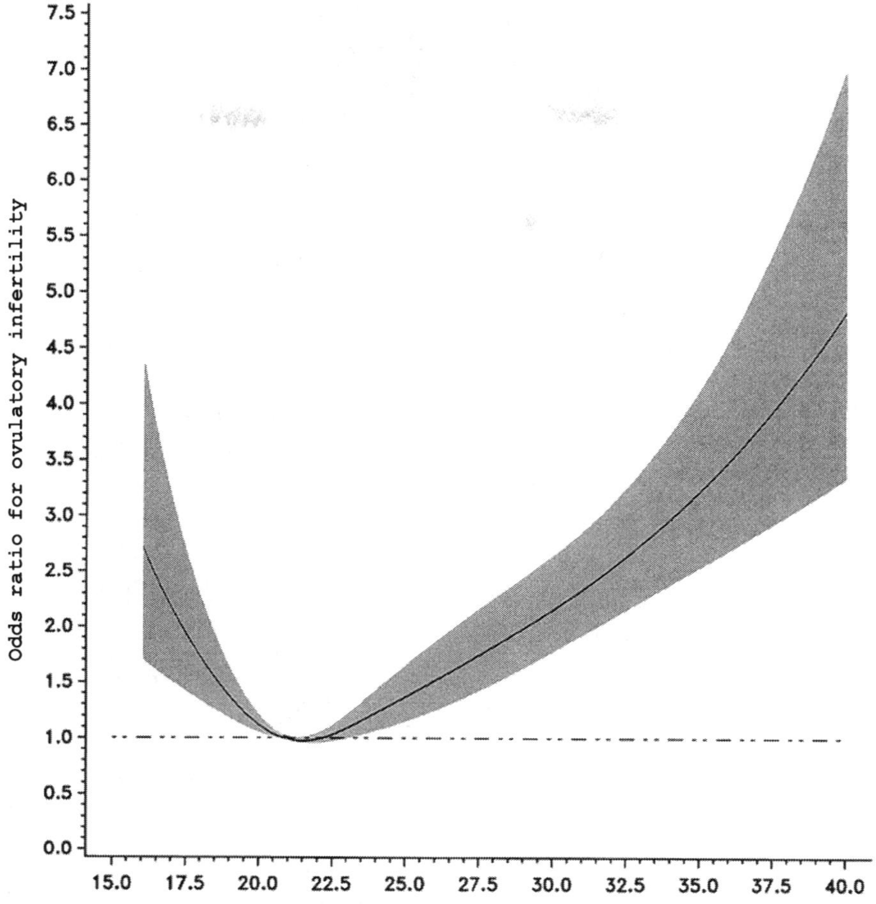
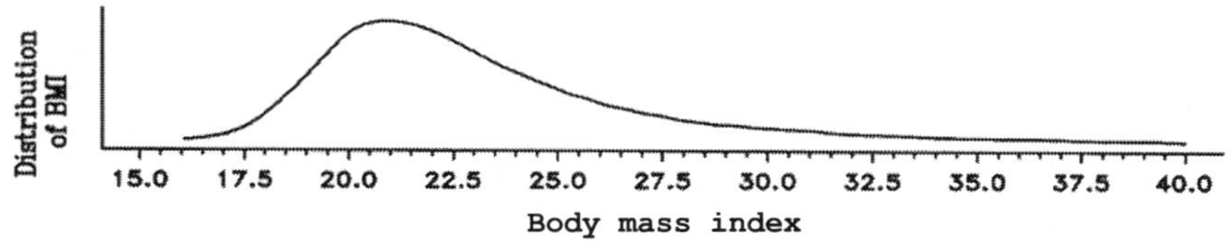

FIGURE 16.2 Multivariate odds ratios and 95% confidence intervals of ovulatory disorder infertility by body mass index (BMI, kg/m^2) [8].

improves insulin sensitivity [40]. Moderate-intensity PA contributes to weight loss and has beneficial effects on body composition and loss of intra-abdominal fat [43].

Recent publications indicate that adipose tissue in adult humans does not only contain white adipocytes but also brown adipocytes, which can proliferate under certain conditions like during PA. This browning effect of adipose tissue has a protective effect against metabolic disorders and can reverse adipose tissue dysfunction [44]. A recently discovered hormone, irisin, produced by murine and human skeletal muscles during PA, has been shown to be a potent inducer of the browning effect of adipose tissue [45].

In view of the mentioned effects of PA, it can be postulated that PA can contribute to the resumption of ovulation by improvement of insulin sensitivity and by decreasing adipose tissue dysfunction.

Vigorous PA performed by female athletes or by women in subsistence economies triggers menstrual irregularities by hypothalamic suppression, but it is not clear whether moderate- to vigorous-intensity PA also contributes to menstrual irregularities in women of developed countries [46,47]. PA activity may have a positive effect on the resumption of ovulation up to a certain level, and above a certain threshold it may be deleterious [48]. Based on the Nurses' Health Study II prospective cohort, vigorous PA was associated with a reduction of ovulatory infertility partially based on its contribution to weight loss. After adjusting for BMI, there was still a 5% reduction in ovulatory infertility for every hour of vigorous

TABLE 16.1   Studies Showing the Effect of Weight Reduction on Resumption of Ovulation in Anovulatory Women with Obesity

Study	N	BMI (kg/m^2), mean, SD	Intervention duration	Weight loss (kg) or % weight loss	Resumption of ovulation and resumption of regular menstrual cycles
Pasquali et al. [36]	20	32 ± 4.7	1,000–1,500 kcal/day, 8 ± 2.4 months	4.8–15.2 kg	14/20 – RO
Kiddy et al. [32]	24	34.1 ± 4.9	1,000 kcal/day, 6–7 months	13 women lost, >5% weight loss	>5% weight loss – RO
Clark et al. [27]	18	38.7 ± 6.8	Lifestyle program, 6 months	6.3 kg	12/18 – RO
Hollmann et al. [31]	35	34.6, (30.8–37.3), min/max	5,000–10,000 calorie reduction/week, 32 ± 14 weeks	10.3 ± 6.3 kg	80% – RM
Palomba et al. [34]	20	33.2 ± 1.4	800 kcal deficit/day, 24 weeks	10.5 ± 4.1 kg ovulatory, 2.3 ± 3.1 kg anovulatory	5/20 – RO
Crosignani et al. [29]	27	32.1 ± 4.2	1200 kcal/day, exercise recommended, 6 months	75% of women achieved, 5% weight loss	15/27 – RO (all-in women with ≥10% weight loss)
Tang et al. [35]	74	38.9 ± 9.5	500 kcal reduction/day, increased physical activity, 24 weeks	1.5 kg (−0.3 to 2.6), (mean, 95% CI)	58.1% – RM
Kuchenbecker et al. [33]	32	37.8 ± 5.2	≥500 kcal deficit/day, lifestyle program, 24 weeks	6.3% weight loss in women who resumed ovulation	17/32 – RO

PA activity per week [8]. An Internet-based prospective cohort study of women planning a pregnancy revealed that moderate intensity PA increased fecundability but vigorous PA decreased fecundability and cycle regularity in the total group of women [49]. More than 4 h of moderate intensity PA per week without vigorous PA improved fecundability in all women. In women who were overweight or obese, 2 h of vigorous PA increased fecundability but a further increase in duration of vigorous PA had no further effect on fecundability.

Diet and Dietary Composition

According to original recommendations, the dietary composition of diets during lifestyle intervention for weight reduction and weight maintenance in the general population should entail a low fat (approximately 30% of energy containing 10% saturated fat), high carbohydrate (approximately 55%), and moderate protein (approximately 15%) content and high fiber diet [10]. Existing literature does not clearly solve the question on which dietary composition incorporated in lifestyle intervention achieves maximal weight loss but indicates that a diet with low carbohydrate intake and of low glycemic index may lead to greater weight loss [50–55]. This effect may be attributable to reduced appetite by increased protein intake [56]. A recent systematic review indicates that on the long term, a diet of lower total fat intake leads to small but statistically significant sustained weight loss [57].

Studies on the benefits of energy-restricted diets in overweight and obese women of reproductive age have consistently shown an improvement in menstrual irregularities, ovulation, and spontaneous conception rates. In these studies, the effect of different dietary compositions on reproductive outcomes could not be evaluated due to the heterogeneity of the study populations, small sample size, and differences in lifestyle interventions. The role of dietary composition during lifestyle intervention in women with PCOS has however received much attention and has recently been evaluated in a systematic review of six publications (a total of 137 participants) [58]. In the majority of the studies, weight loss was the determining factor for the improvement of PCOS presentation independent of dietary composition. This systematic review revealed subtle differences between diets with greater weight loss for a monounsaturated fat – enriched diet, improved menstrual irregularity for a low glycemic index diet, and improvement of insulin sensitivity for a low glycemic index diet or a low carbohydrate diet. Data from the prospective cohort Nurses' Health Study II reveal that a diet with greater intake of monounsaturated fat, supplementing animal proteins with vegetable proteins, and a higher intake of low glycemic carbohydrates decreases the risk of ovulatory infertility [59].

Other Lifestyle Factors

Smoking has an adverse effect on fertility but no association with anovulation has been shown. The evidence on the impact of increased alcohol and caffeine consumption on female fertility is inconclusive [60]. The adverse effect of smoking, alcohol, and caffeine consumption is

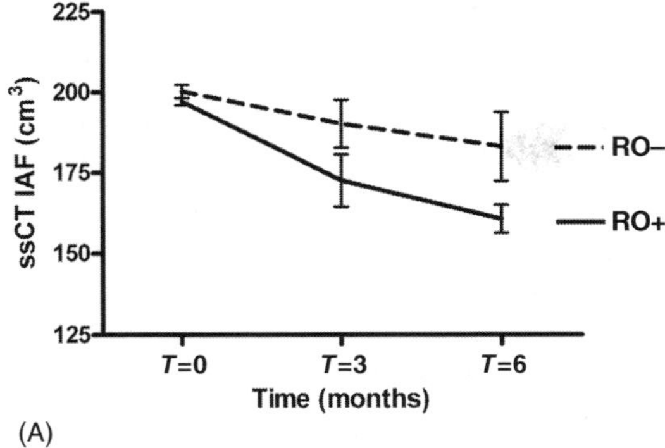

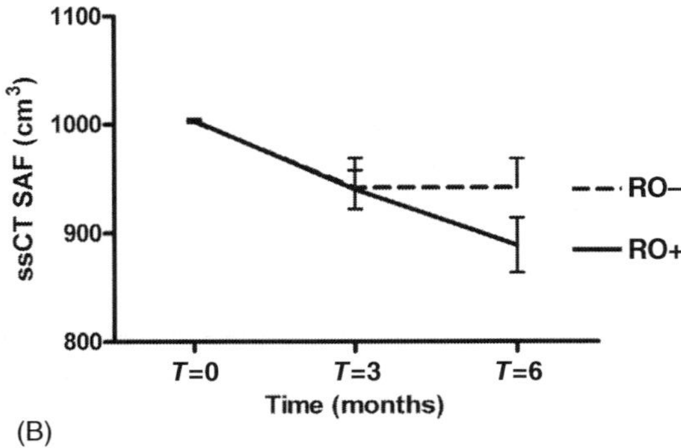

FIGURE 16.3 Differences in loss of (A) intra-abdominal fat (IAF) and (B) subcutaneous abdominal fat (SAF) in 32 anovulatory women with PCOS and obesity undergoing a lifestyle program between those who resumed ovulation (RO+) and those that remained anovulatory (RO−) during a 6 month lifestyle program. RO+ (solid line) and RO− (dashed line) participants. Error bars indicate standard errors [33].

dose-dependent with a cumulative effect with each additional lifestyle factor [61]. Considering the detrimental effect of smoking, alcohol consumption, and increased caffeine intake on fecundity and pregnancy outcome, avoiding and limiting these substances during lifestyle intervention is commonly advised.

IMPLEMENTATION OF LIFESTYLE INTERVENTION

Most individuals who are overweight or obese experience great difficulty to achieve and maintain weight loss, and therefore a multifactorial approach based on diet, increased PA, and behavior modification is advised to help patients lose weight [10]. The combination of cognitive-behavior interventions with diet and increased PA results in greater weight loss than the combination of diet and increased PA alone [10,62].

Obese individuals have an intrinsic resistance to changing behavior, and low self-esteem and low self-efficacy are limiting factor's in achieving weight loss [63]. Individual guidance and cognitive-behavioral strategies are advised and will contribute to more and sustained weight loss [62,64]. Professionals, like nurses, dieticians, and psychologists trained in cognitive-behavior therapy, can be employed as lifestyle counselors. Counseling sessions should be scheduled more than once per month during the first three months (high intensity counseling) and thereafter the frequency of counseling sessions may be reduced [65] and combined with telephone-delivered counseling [66]. In-person counseling may be combined with interactive computer-based interventions to improve self-monitoring and adherence to diet and increased PA [67].

Lifestyle counselors are encouraged to enhance social support of the patient by including spouses, family members, or friends (buddies) to support and motivate the patient during the difficult process of changing their lifestyle.

The aim of cognitive-behavioral therapy is to help patients change their lifestyle and eating behavior by modifying and monitoring their food consumption and controlling stimuli that trigger eating and by monitoring increase in PA. Changing the routine of meal planning and food consumption will help to slow down the rate of eating, which triggers a sense of satiety. Behavioral contracting by defining specific behavior change and realistic weight loss goals in combination with rewards for achieving these goals can be applied as a tool during the counseling sessions. Defining realistic weight loss goals at the onset of the lifestyle intervention program is essential to reinforce the motivation and perseverance to change adverse behavior. Unrealistic weight loss goals and too little weight loss during the initial phase of the lifestyle program increases the risk of drop-out [63,68,69]. Drop-out in women with obesity and infertility undergoing a lifestyle intervention program is a major limiting factor in achieving the maximal benefit of weight loss on female reproductive outcome. According to a systematic review, 24% (range 0–31%) of women with obesity and infertility undergoing a lifestyle intervention program experience drop-out [70]. This study could not identify intervention- or patient-related risk factors associated with drop-out.

Considering the difficulties of behavior change during lifestyle intervention programs, the lifestyle counselors are advised to apply interviewing and counseling techniques to achieve positive behavior change. Motivational interviewing is a counseling technique described as a directive patient-centered counseling style for eliciting behavioral change focused on exploring and resolving ambivalence. A meta-analysis of 11 RCTs revealed that applying motivational interviewing in lifestyle intervention programs contributes to greater weight loss [71]. During motivational interviewing, lifestyle counselors are advised to express empathy, enhance the awareness of the inconsistencies between the patient's unhealthy behavior and the treatment goals, and support self-efficacy. The lifestyle counselor should avoid opposing any resistance from the patient and attempt to direct the patient in identifying barriers to change and exploring their own solutions. Patients who try to change health behavior are known to follow the following stages of change: precontemplation, contemplation, preparation, action, maintenance, relapse, and termination. During the counseling sessions the stages of change model can be applied allowing the counselor to use stage-specific strategies in order to achieve behavioral change [72].

Before starting to increase the levels of PA, it is prudent to counsel patients about the risk of musculoskeletal strain and joint injury. Gradual increase in the levels of PA is advised to avoid injury. Medical evaluation prior to starting an exercise program is only necessary in the context of preexisting musculoskeletal complaints. Increased PA that includes aerobic exercise and resistance training is preferred [66]. The schedule of increasing levels of PA should be individualized according to the personal preferences of the participants.

Many anovulatory women with obesity may not be able to access or attend a structured lifestyle program as mentioned earlier due to travel distances or occupational and social commitments. These women may benefit from a commercially available weight loss program. These programs achieve more weight loss and are more cost effective than other primary care weight loss initiatives [73,74].

CONCLUSIONS

Lifestyle intervention in anovulatory women with obesity contributes to resumption of ovulation mediated by improving insulin sensitivity and decreasing adipose tissue dysfunction. Lifestyle intervention based on the combination of diet, increased PA, and behavior modification achieves maximal weight loss and decreases the risk of drop-out. Weight loss can be achieved by calorie reduction aiming at an energy deficit of 600 kcal/day while ensuring adequate and balanced nutritional intake. Dietary composition with low carbohydrate intake of a low glycemic index, intake of monounsaturated fats, and vegetable proteins may offer some benefit for resumption of ovulation. Anovulatory women with obesity are to be encouraged to engage in moderate intensity PA of >4 h/week for resumption of ovulation. Two hours per week of vigorous PA may contribute to further increase in resumption of ovulation. Behavior modification during a lifestyle intervention program is essential to achieve and maintain change in lifestyle behavior. Behavior modification may be implemented by trained

counselors during a period of high intensity counseling. Trained lifestyle counselors should aim to apply motivational interviewing techniques as part of cognitive-behavioral therapy. During lifestyle intervention in anovulatory women with obesity, the lifestyle counselors are also advised to address other adverse lifestyle factors, like smoking, alcohol, and caffeine consumption in order to improve fecundity and pregnancy outcome.

References

[1] WHO Global Infobase. 2012. https://apps.who.int/infobase/

[2] Haslam DW, James WP. Obesity. Lancet 2005;366:1197–209. (1474-547; 9492).

[3] Nelson SM, Fleming R. Obesity and reproduction: impact and interventions. Curr Opin Obstet Gynecol 2007;19:384–9. (1040-872; 4).

[4] Balen AH, Anderson RA. Policy, Practice Committee of the BFS. Impact of obesity on female reproductive health: British Fertility Society, Policy and Practice Guidelines. Hum Fertil (Camb) 2007;10(4):195–206.

[5] Gaillard R, Steegers EA, Duijts L, Felix JF, Hofman A, Franco OH, et al. Childhood cardiometabolic outcomes of maternal obesity during pregnancy: the Generation R Study. Hypertension 2014;63(4):683–91.

[6] Wadhwa PD, Buss C, Entringer S, Swanson JM. Developmental origins of health and disease: brief history of the approach and current focus on epigenetic mechanisms. Semin Reprod Med 2009;27(5):358–68.

[7] Rich-Edwards JW, Goldman MB, Willett WC, Hunter DJ, Stampfer MJ, Colditz GA, et al. Adolescent body mass index and infertility caused by ovulatory disorder. Am J Obstet Gynecol 1994;171:171–7. (0002-9378; 1).

[8] Rich-Edwards JW, Spiegelman D, Garland M, Hertzmark E, Hunter DJ, Colditz GA, et al. Physical activity, body mass index, and ovulatory disorder infertility. Epidemiology 2002;13(2):184–90.

[9] Pasquali R, Gambineri A, Pagotto U. The impact of obesity on reproduction in women with polycystic ovary syndrome. BJOG 2006;113(10):1148–59.

[10] National Institutes of, Health. Clinical guidelines on the identification, evaluation, and treatment of overweight and obesity in adults – the evidence report. Obes Res 1998;6(Suppl. 2):51S–209S. (1071-7323).

[11] Wijeyaratne CN, Balen AH, Barth JH, Belchetz PE. Clinical manifestations and insulin resistance (IR) in polycystic ovary syndrome (PCOS) among South Asians and Caucasians: is there a difference? Clin Endocrinol (Oxf) 2002;57(3):343–50.

[12] Wass P, Waldenstrom U, Rossner S, Hellberg D. An android body fat distribution in females impairs the pregnancy rate of *in-vitro* fertilization-embryo transfer. Hum Reprod 1997;12:2057–60. (0268-1161; 9).

[13] Zaadstra BM, Seidell JC, Van Noord PA, te Velde ER, Habbema JD, Vrieswijk B, et al. Fat and female fecundity: prospective study of effect of body fat distribution on conception rates. BMJ 1993;306:484–7. (0959-8138; 6876).

[14] Poretsky L. On the paradox of insulin-induced hyperandrogenism in insulin-resistant states. Endocr Rev 1991;12:3–13. (0163-769; 1).

[15] Dunaif A. Insulin resistance and the polycystic ovary syndrome: mechanism and implications for pathogenesis. Endocr Rev 1997;18:774–800. (0163-769; 6).

[16] Pasquali R, Casimirri F, Venturoli S, Antonio M, Morselli L, Reho S, et al. Body fat distribution has weight-independent effects on clinical, hormonal, and metabolic features of women with polycystic ovary syndrome. Metabolism 1994;43:706–13. (0026-0495; 6).

[17] Pasquali R, Pelusi C, Genghini S, Cacciari M, Gambineri A. Obesity and reproductive disorders in women. Hum Reprod Update 2003;9:359–72. (1355-4786; 4).

[18] Franks S, Stark J, Hardy K. Follicle dynamics and anovulation in polycystic ovary syndrome. Hum Reprod Update 2008;14(4):367–78.

[19] Willis D, Mason H, Gilling-Smith C, Franks S. Modulation by insulin of follicle-stimulating hormone and luteinizing hormone actions in human granulosa cells of normal and polycystic ovaries. J Clin Endocrinol Metab 1996;81:302–9. (0021-972; 1).

[20] Escobar-Morreale HF, San Millan JL. Abdominal adiposity and the polycystic ovary syndrome. Trends Endocrinol Metab 2007;18:266–72. (1043-2760; 7).

[21] Garg A. Regional adiposity and insulin resistance. J Clin Endocrinol Metab 2004;89:4206–10. (0021-972; 9).

[22] Ahima RS. Adipose tissue as an endocrine organ. Obesity (Silver Spring) 2006;14(Suppl. 5):242S–9S.

[23] Mitchell M, Armstrong DT, Robker RL, Norman RJ. Adipokines: implications for female fertility and obesity. Reproduction 2005;130:583–97. (1470-1626; 5).

[24] Villa J, Pratley RE. Adipose tissue dysfunction in polycystic ovary syndrome. Curr Diab Rep 2011;11(3):179–84.

[25] Weiss R. Fat distribution and storage: how much, where, and how? Eur J Endocrinol 2007;157(Suppl. 1):S39–45. (0804-4643).

[26] Mutsaerts MA, Groen H, ter Bogt NC, Bolster JH, Land JA, Bemelmans WJ, et al. The LIFESTYLE study: costs and effects of a structured lifestyle program in overweight and obese subfertile women to reduce the need for fertility treatment and improve reproductive outcome. A randomised controlled trial. BMC Women's Health 2010;10:22.

[27] Clark AM, Ledger W, Galletly C, Tomlinson L, Blaney F, Wang X, et al. Weight loss results in significant improvement in pregnancy and ovulation rates in anovulatory obese women. Hum Reprod 1995;10:2705–12. (0268-1161; 10).

[28] Falsetti L, Pasinetti E, Mazzani MD, Gastaldi A. Weight loss and menstrual cycle: clinical and endocrinological evaluation. Gynecol Endocrinol 1992;6(1):49–56.

[29] Crosignani PG, Colombo M, Vegetti W, Somigliana E, Gessati A, Ragni G. Overweight and obese anovulatory patients with polycystic ovaries: parallel improvements in anthropometric indices, ovarian physiology and fertility rate induced by diet. Hum Reprod 2003;18:1928–32. (0268-1161; 9).

[30] Guzick DS, Wing R, Smith D, Berga SL, Winters SJ. Endocrine consequences of weight loss in obese, hyperandrogenic, anovulatory women. Fertil Steril 1994;61:598–604. (0015-0282; 4).

[31] Hollmann M, Runnebaum B, Gerhard I. Effects of weight loss on the hormonal profile in obese, infertile women. Hum Reprod 1996;11:1884–91. (0268-1161; 9).

[32] Kiddy DS, Hamilton-Fairley D, Bush A, Short F, Anyaoku V, Reed MJ, et al. Improvement in endocrine and ovarian function during dietary treatment of obese women with polycystic ovary syndrome. Clin Endocrinol (Oxf) 1992;36:105–11. (0300-0664; 1).

[33] Kuchenbecker WK, Groen H, van Asselt SJ, Bolster JH, Zwerver J, Slart RH, et al. In women with polycystic ovary syndrome and obesity, loss of intra-abdominal fat is associated with resumption of ovulation. Hum Reprod 2011;26(9):2505–12.

[34] Palomba S, Giallauria F, Falbo A, Russo T, Oppedisano R, Tolino A, et al. Structured exercise training programme versus hypocaloric hyperproteic diet in obese polycystic ovary syndrome patients with anovulatory infertility: a 24-week pilot study. Hum Reprod 2008;23:642–50. (1460-2350; 0268-1161; 3).

[35] Tang T, Glanville J, Hayden CJ, White D, Barth JH, Balen AH. Combined lifestyle modification and metformin in obese patients with polycystic ovary syndrome. A randomized, placebo-controlled, double-blind multicentre study. Hum Reprod 2006;21:80–9. (0268-1161; 1).

[36] Pasquali R, Antenucci D, Casimirri F, Venturoli S, Paradisi R, Fabbri R, et al. Clinical and hormonal characteristics of obese amenorrheic hyperandrogenic women before and after weight loss. J Clin Endocrinol Metab 1989;68:173–9. (0021-972; 1).

[37] Huber-Buchholz MM, Carey DG, Norman RJ. Restoration of reproductive potential by lifestyle modification in obese polycystic ovary syndrome: role of insulin sensitivity and luteinizing hormone. J Clin Endocrinol Metab 1999;84:1470–4. (0021-972; 4).

[38] Park HS, Lee K. Greater beneficial effects of visceral fat reduction compared with subcutaneous fat reduction on parameters of the metabolic syndrome: a study of weight reduction programmes in subjects with visceral and subcutaneous obesity. Diabet Med 2005;22:266–72. (0742-3071; 3).

[39] Tremblay A, Despres JP, Maheux J, Pouliot MC, Nadeau A, Moorjani S, et al. Normalization of the metabolic profile in obese women by exercise and a low fat diet. Med Sci Sports Exerc 1991;23(12):1326–31.

[40] Vigorito C, Giallauria F, Palomba S, Cascella T, Manguso F, Lucci R, et al. Beneficial effects of a three-month structured exercise training program on cardiopulmonary functional capacity in young women with polycystic ovary syndrome. J Clin Endocrinol Metab 2007;92(4):1379–84.

[41] Lemoine S, Rossell N, Drapeau V, Poulain M, Garnier S, Sanguignol F, et al. Effect of weight reduction on quality of life and eating behaviors in obese women. Menopause 2007;14(3 Pt 1):432–40.

[42] Wu T, Gao X, Chen M, van Dam RM. Long-term effectiveness of diet-plus-exercise interventions vs. diet-only interventions for weight loss: a meta-analysis. Obes Rev 2009;10(3):313–23.

[43] Irwin ML, Yasui Y, Ulrich CM, Bowen D, Rudolph RE, Schwartz RS, et al. Effect of exercise on total and intra-abdominal body fat in postmenopausal women: a randomized controlled trial. JAMA 2003;289(3):323–30.

[44] Smorlesi A, Frontini A, Giordano A, Cinti S. The adipose organ: white-brown adipocyte plasticity and metabolic inflammation. Obes Rev 2012;13(Suppl. 2):83–96.

[45] Bostrom P, Wu J, Jedrychowski MP, Korde A, Ye L, Lo JC, et al. A PGC1-alpha-dependent myokine that drives brown-fat-like development of white fat and thermogenesis. Nature 2012;481(7382):463–8.

[46] Bullen BA, Skrinar GS, Beitins IZ, von Mering G, Turnbull BA, McArthur JW. Induction of menstrual disorders by strenuous exercise in untrained women. N Engl J Med 1985;312(21):1349–53.

[47] Green BB, Daling JR, Weiss NS, Liff JM, Koepsell T. Exercise as a risk factor for infertility with ovulatory dysfunction. Am J Public Health 1986;76(12):1432–6.

[48] Gudmundsdottir SL, Flanders WD, Augestad LB. Physical activity and fertility in women: the North-Trondelag Health Study. Hum Reprod 2009;24(12):3196–204.

[49] Wise LA, Rothman KJ, Mikkelsen EM, Sorensen HT, Riis AH, Hatch EE. A prospective cohort study of physical activity and time to pregnancy. Fertil Steril 2012;97(5). 11362-1142.e1-4.

[50] Yu-Poth S, Zhao G, Etherton T, Naglak M, Jonnalagadda S, Kris-Etherton PM. Effects of the National Cholesterol Education Program's Step I and Step II dietary intervention programs on cardiovascular disease risk factors: a meta-analysis. Am J Clin Nutr 1999;69(4):632–46.

[51] Thomas DE, Elliott EJ, Baur L. Low glycaemic index or low glycaemic load diets for overweight and obesity. Cochrane Database Syst Rev 2007;3(3). CD005105.

[52] Sacks FM, Bray GA, Carey VJ, Smith SR, Ryan DH, Anton SD, et al. Comparison of weight-loss diets with different compositions of fat, protein, and carbohydrates. N Engl J Med 2009;360(9):859–73.

[53] Larsen TM, Dalskov SM, van Baak M, Jebb SA, Papadaki A, Pfeiffer AF, et al. Diets with high or low protein content and glycemic index for weight-loss maintenance. N Engl J Med 2010;363(22):2102–13.

[54] Gardner CD, Kiazand A, Alhassan S, Kim S, Stafford RS, Balise RR, et al. Comparison of the Atkins, Zone, Ornish, and LEARN diets for

change in weight and related risk factors among overweight premenopausal women: the A TO Z Weight Loss Study: a randomized trial. JAMA 2007;297(9):969–77.

[55] Dansinger ML, Gleason JA, Griffith JL, Selker HP, Schaefer EJ. Comparison of the Atkins, Ornish, Weight Watchers, and Zone diets for weight loss and heart disease risk reduction: a randomized trial. JAMA 2005;293(1):43–53.

[56] Bowen J, Noakes M, Clifton PM. Appetite regulatory hormone responses to various dietary proteins differ by body mass index status despite similar reductions in ad libitum energy intake. J Clin Endocrinol Metab 2006;91(8):2913–9.

[57] Hooper L, Abdelhamid A, Moore HJ, Douthwaite W, Skeaff CM, Summerbell CD. Effect of reducing total fat intake on body weight: systematic review and meta-analysis of randomised controlled trials and cohort studies. BMJ 2012;345:e7666.

[58] Moran LJ, Ko H, Misso M, Marsh K, Noakes M, Talbot M, et al. Dietary composition in the treatment of polycystic ovary syndrome: a systematic review to inform evidence-based guidelines. J Acad Nutr Diet 2013;113(4):520–45.

[59] Chavarro JE, Rich-Edwards JW, Rosner BA, Willett WC. Diet and lifestyle in the prevention of ovulatory disorder infertility. Obstet Gynecol 2007;110(5):1050–8.

[60] Homan GF, Davies M, Norman R. The impact of lifestyle factors on reproductive performance in the general population and those undergoing infertility treatment: a review. Hum Reprod Update 2007;13(3):209–23.

[61] Hassan MA, Killick SR. Negative lifestyle is associated with a significant reduction in fecundity. Fertil Steril 2004;81(2):384–92.

[62] Shaw K, O'Rourke P, Del Mar C, Kenardy J. Psychological interventions for overweight or obesity. Cochrane Database Syst Rev 2005;2(2). CD003818.

[63] Fabricatore AN, Wadden TA, Moore RH, Butryn ML, Heymsfield SB, Nguyen AM. Predictors of attrition and weight loss success: results from a randomized controlled trial. Behav Res Ther 2009;47(8):685–91.

[64] Wadden TA, Butryn ML. Behavioral treatment of obesity. Endocrinol Metab Clin North Am 2003;32(4):981. 1003, x.

[65] Whitlock EP, Orleans CT, Pender N, Allan J. Evaluating primary care behavioral counseling interventions: an evidence-based approach. Am J Prev Med 2002;22(4):267–84.

[66] Wadden TA, Webb VL, Moran CH, Bailer BA. Lifestyle modification for obesity: new developments in diet, physical activity, and behavior therapy. Circulation 2012;125(9):1157–70.

[67] Wieland LS, Falzon L, Sciamanna CN, Trudeau KJ, Brodney S, Schwartz JE, et al. Interactive computer-based interventions for weight loss or weight maintenance in overweight or obese people. Cochrane Database Syst Rev 2012;8. CD007675.

[68] Messier V, Hayek J, Karelis AD, Messier L, Doucet E, Prud'homme D, et al. Anthropometric, metabolic, psychosocial, and dietary factors associated with dropout in overweight and obese postmenopausal women engaged in a 6-month weight loss programme: a MONET study. Br J Nutr 2010;103(8):1230–5.

[69] Colombo O, Ferretti VV, Ferraris C, Trentani C, Vinai P, Villani S, et al. Is drop-out from obesity treatment a predictable and preventable event? Nutr J 2014;13. 13,2891-13-13.

[70] Mutsaerts MA, Kuchenbecker WK, Mol BW, Land JA, Hoek A. Dropout is a problem in lifestyle intervention programs for overweight and obese infertile women: a systematic review. Hum Reprod 2013;28(4):979–86.

[71] Armstrong MJ, Mottershead TA, Ronksley PE, Sigal RJ, Campbell TS, Hemmelgarn BR. Motivational interviewing to improve weight loss in overweight and/or obese patients: a systematic review and meta-analysis of randomized controlled trials. Obes Rev 2011;12(9):709–23.

[72] Prochaska JO, Velicer WF. The transtheoretical model of health behavior change. Am J Health Promot 1997;12(1):38–48.

[73] Finkelstein EA, Kruger E. Meta- and cost-effectiveness analysis of commercial weight loss strategies. Obesity (Silver Spring) 2014;1942–51.

[74] Jolly K, Lewis A, Beach J, Denley J, Adab P, Deeks JJ, et al. Comparison of range of commercial or primary care led weight reduction programmes with minimal intervention control for weight loss in obesity: lighten up randomised controlled trial. BMJ 2011;343. d6500.

17

Effects of Lifestyle on Female Reproductive Features and Success: Lessons from Animal Models

Antonio Gonzalez-Bulnes, MVD, PhD,
Susana Astiz, MVD, PhD, Dip ECBHM

INIA, Madrid, Spain

INTRODUCTION

Nonoptimal nutrition, either in excess or deficiency, affects reproduction; it is well-known that diet and lifestyle have a strong impact on reproductive function. Traditionally, food scarcity has been the main determinant of reproductive success in human societies. In the last century, food shortage continued to be the main limiting factor in nondeveloped countries; conversely, due to the high rate of individuals affected by overnutrition and obesity, food abundance started to appear as the main limiting factor for reproductive success in developed countries. In this way, in 1952, Rogers and Mitchell [1] reported obesity as the common and most noteworthy feature in around 43% of the women with reproductive disorders. Some years later, the incidence of menstrual disorders, anovulatory cycles, and infertility was confirmed to be significantly higher in obese than in normal-weight women; these studies also pointed to negative effects on pre- and postnatal development of the offspring if pregnancy was achieved [2]. These early evidences induced the development of numerous observational and interventional studies and nowadays the link between obesity and reproductive dysfunction is well-documented.

At present, reproductive disorders associated to obesity are increasing worldwide due to profound changes in the diet and lifestyle of individuals living in rapidly developing areas; these changes are not only related to food abundance but also to rapid urbanization and therefore a strong trend to lack of exercise and sedentary habits [3]. In fact, the highest absolute number of people affected by obesity and associated diseases are found in developing countries like China and India. Furthermore, the highest incidence in relative percentages are found in countries of the Middle East; more than 20% of adults from Qatar, Saudi Arabia, and the United Arab Emirates have developed diabetes (World Health Organization, WHO). The WHO predicts that by 2015, approximately 2.3 billion adults will be overweight and more than 700 million will be obese; on the other hand, diabetes deaths will double in 2030 the incidence of 2005.

Progress and modern lifestyle are also linked to other risks for reproductive function. First, modern living is associated with postponement of childbearing [4]. The most important causes for childbearing delay are the economic uncertainty; changes in values, housing conditions, and partnership dynamics; the absence of family-supportive policies; women's education level and women's participation in the labor market; the search for gender equity; and the increase in the effectiveness of contraception methods [5]. In any case, childbearing postponement is associated with decreased fertility in males and females and adverse pregnancy outcomes in women [6].

However, reproductive failures are increasing even when malnutrition and intention to delay childbearing are absent. This fact suggests other causal factors and evidences are pointing to the increasingly polluted environment and to the contaminants known as "endocrine disruptors (EDs)" or "ED compounds or chemicals (EDCs)." The EDCs (mainly phytoestrogens, pesticides, heavy metals, plasticizers, and lacquers) are ubiquitous in the environment, food, and consumer products. These contaminants, hallmarks of modern life, may antagonize or mimic endogenous hormones and therefore may affect

Handbook of Fertility. http://dx.doi.org/10.1016/B978-0-12-800872-0.00017-2

reproductive processes. Moreover, nutritional imbalances and EDCs exposures may share common pathways [7] and thus, there may be mutually reinforcing effects.

Thus, having in mind the impact on worldwide public health, the research in this area is imperative. Obviously, interventional research in humans is limited by ethical issues; hence, investigations need to be conducted in animal models. The present review aims to outline the main animal models for translational studies on metabolic and endocrine issues and to summarize the highlights, main results, and future lines of translational research on the effects of diet and lifestyle on female reproduction.

ANIMAL MODELS

Most of the scientific research is carried out on laboratory rodents, mainly rats and mice but also on hamsters and guinea pigs. The rodents need little space, are relatively inexpensive to maintain, have a sequenced genome, and are easily modified by genetic engineering. Overall, rodents are good models for studies on fundamental processes. However, there are also numerous disadvantages when using rodents in systemic studies on nutrition and lifestyle, on reproductive features, or combining both areas.

In studies on the effects of diet and lifestyle, the main weaknesses are related to the marked differences in behavior (nocturnal habits, physical activity, and intake patterns), metabolism, and adipose tissue biology between rodents and humans. The comparative research on further disorders like metabolic syndrome or diabetes indicates cumulative differences in glucose homeostasis between rodents and humans; from gene expression to cellular signaling to phenotypic expression [8]. After that, endocrine processes largely vary from one species to another. In research on the determinants of reproductive processes, either diet and lifestyle or EDCs, the differences among species are even more pronounced [9]. First, in studies regarding the environmental effects on fertility, we need to take into account that follicular differentiation occurs prenatally in most of the species, except rodents. In rodents, the primordial follicular pool is established at postnatal stages; thus, the effects of prenatal and postnatal exposures are quite different from humans and other mammals [10–13].

Another crucial issue is the small body size of rodents, which flaws the translational application (or increase the cost) of imaging techniques and impedes the serial sampling of large amounts of blood and tissues. Hence, rodents may be the elective model for studies on a concrete mechanism, but systemic translational studies are more and more preferred in larger animals. In this sense, different species of large animals fulfill research necessities and offer numerous profitable characteristics [14–17].

Briefly, housing and management are easy and not too expensive, behavioral patterns are diurnal, and body size allows application of imaging techniques and serial sampling of large amounts of blood and tissues. Afterward, pathways regulating appetite, energy balance, and adipogenesis are more similar to humans in large animals than in rodents [18,19]. Finally, in recent years, the genomic analysis is well-advanced and it is possible to obtain targeted gene mutations for specific models [20]. Having in mind body-size, handling-easiness, and cost-efficiency, the main large species used as models are rabbits, sheep, and swine.

Rodents

Currently, most of the studies on obesity and metabolic disorders are performed in mice and to a lesser extent in rats. This is because the most abundant genetic variants producing monogenic obesity in humans have been replicated in mice (which include leptin, LEP; leptin receptor, LEPR; melanocortin 4 receptor, MC4R; and proopiomelanocortin, POMC); mutations for orexigenic neuropeptide Y (NPY) and agouti and agouti-related protein (AGRP) (agouti/yellow obese and AGRP mice) have been also obtained [21]. Among monogenic rat models, we should also make reference to the Zucker Fatty and Zucker Diabetic Fatty strains [22]. Most of these species demonstrate hyperphagia, marked obesity, and varying levels of hyperglycemia and hyperinsulinemia [23].

The most-used monogenic models for severe obese phenotypes are based on mice with mutations for genes encoding LEP (Lep$^{ob/ob}$ mouse) and LEPR (Lepr$^{db/db}$ mouse) [24–27]. In fact, leptin was identified after the obese mouse hereditable syndrome (*ob/ob* mouse; currently Lep$^{ob/ob}$) was discovered in 1949 [28–29]. Obese mice are characterized by being grossly overweight due to high food consumption and scarce physical activity, hyperglycemic, hyperlipidemic, and hyperinsulinemic. Thus, these mice have been extensively used as a model for diabetes and obesity. The obese syndrome, as a consequence of leptin deficiency, also affects immune function, cardiovascular system, and reproductive function [30,31]; male and female *ob/ob* mice are infertile. The administration of leptin restores weight, metabolic function, and fertility [32]. Hence, this model has been extensively used for experimentation in the links among obesity and reproductive function.

However, monogenic disorders, although having a clear and drastic effect on the phenotype, are very infrequent (it only accounts for 5% of morbid human obesity) [33]. Moreover, monogenic disorders only lie in a small number of genes; mainly, *LEP*, *LEPR*, *MC4R*, and *POMC* [34–38]. Thus, research is increasingly more focused on polygenic and diet-induced models of obesity and diabetes [39].

Rabbit

Rabbits are larger than mice and rats, allowing nonlethal studies on physiology by serial sampling of blood and tissues and, mainly, by application of imaging techniques. Furthermore, the necessity of space and the economic costs are lower than other species of large mammals. In the case of studies on obesity and associated metabolic and cardiovascular diseases, the rabbit is a well-recognized model due to their similarities in the lipoprotein metabolism with humans [40–41]. Obesity, diabetes, dyslipidemia, and cardiovascular diseases may be induced by diets with high content of fat and energy, by pharmacological treatments and, currently, by application of targeted gene mutation techniques [42]. The rabbit is also a well-considered model in reproductive physiology, and mainly in embryology and prenatal programming [43,44] due to its closer resemblance in embryogenesis, placental structure, and nutrient transport to humans [45,46].

Sheep

The sheep have a convenient size and temperament, which facilitate housing, handling, and sampling. Sheep have been traditionally used in biomedical research as an excellent model for the study of reproductive endocrinology, ovarian biology, and fertility therapies [47–50]. Screening of follicular and luteal dynamics by noninvasive sequential imaging (ultrasonography) and hormone assessment are well-established techniques in this species [51] and protocols for either contraception or cycle management or controlled ovarian stimulation are well-defined and commonly applied.

Besides the mentioned distinctive qualities, the sheep is especially interesting for current studies on developmental programming of adult obesity and associated disorders, due to singleton pregnancies and similar developmental trajectory to human beings [52]. Hence, the sheep is commonly used for studies on the effect of exogenous steroids on fetal programming [53–55] and mainly on programming of reproductive dysfunctions like ovarian polycystic syndrome [56]. Sheep are also extensively used for studies on nutritional programming, either by deficiency or excess, since the timing of development of adipose tissue and of the hypothalamic neural network regulating appetite and energy balance occurs at a comparable period to humans [57,58]. Furthermore, prenatal malnutrition in sheep has long-term effects on homeostasis and endocrine and cardiovascular systems [59–64], like in humans.

Pig

The pig has been considered an outstanding animal model for studies linking nutrition, metabolism, reproduction, and development for a long time [65]. It is becoming increasingly apparent that pigs are the translational model of choice to bridge the gap between preclinical rodent models research and clinical research in humans in most of the biomedical disciplines [66].

First, pigs share many more anatomical and physiological similarities with humans than any other animal species [42,67–69], including proportional organ sizes. Additional advantages are a convenient size and nature for housing, early puberty, short weaning to mating interval, and a high prolificacy for breeding and maintenance of strains. There are also other important resemblances with humans like omnivorous habits, propensity for a sedentary behavior and obesity, and characteristics of metabolism and cardiovascular system [70–72]. Pigs have been also used in research on reproductive physiology; mainly during pregnancy in studies on intrauterine growth retardation (IUGR) and developmental programming. The first advantage is the prolificacy of the species, which allows the sampling of different littermates [73], either at pre- and postnatal stages, and their exposure to different treatments after birth. Prolificacy is also related to the appearance of IUGR in some of the components of the litters [74–76]; mainly in high-prolific breeds, since placental growth is endangered in case of limited space *in utero* with a high number of developing embryos [77,78]. Incidence of IUGR may be increased by maternal nutritional deficiencies [79,80], prejudicing nutrient supply of the fetuses.

Pigs used for biomedical research can be either random- or purpose-bred. Pigs from a random-bred source are typically farm pigs, where genetic selection is focused on production parameters. However, some random-bred pigs may develop cardiometabolic disorders without other interferences than dietary changes. Purpose-bred pigs are originated from a closed colony with a selection for a specific phenotype. The pigs mostly used for research on obesity and associated diseases are Göttingen minipigs [81,82] and Yucatan and Ossabaw Island pigs [83,84].

In fact, Yucatan and Ossabaw pigs are directly derived from one primary gene pool, a random-bred pig, the Iberian pig. The Iberian pig was introduced in America by Spanish and Portuguese colonizers, since there were no pigs in the New World before their arrival in the fifteenth century, and it is supposed to have remained isolated without introduction of further genetic material from other breeds thereafter. The Göttingen breed was created for biomedical research produced by breeding three different races: Minnesota minipigs, Vietnamese, and Landrace pigs. In turns, Minnesota minipigs were produced using Guinea hogs from Alabama (descendant from pigs that were imported from Canary Island, which are descendants from Iberian pigs) and Piney Woods pigs from Louisiana (which are also descendants from Iberian pigs introduced by the first colonizers).

Thus, purpose-bred pigs used as models of obesity have Iberian ancestors. The Iberian breed is the most abundant and representative type of the Mediterranean swine (*Sus scrofa meridionalis*), which is genetically different from modern commercial breeds (*Sus scrofa ferus*) [85,86]. Mediterranean pigs have been reared in semiferal conditions for centuries and have developed a *thrifty genotype* for the accommodation to seasonal cycles of feasting and famine to which they are exposed in their environment [87]. Hence, these animals show much higher voluntary food intake in comparison to other breeds [88] and store excess fat during food abundance that enables survival during periods of scarcity [89]. The abundance of fat causes a high secretion of leptin [90], which should diminish food intake. However, the Iberian pig has a gene polymorphism for the leptin receptor (*LEPR*) with effects on food intake, body weight, and fat deposition [91,92]. As a consequence, Iberian *LEPR* alleles increase insatiability and pigs become more and more obese when food is in excess (i.e., Iberian swine shows leptin resistance).

Hence, Mediterranean pigs and descendants are optimal models for studies on obesity, leptin resistance, and metabolic syndrome [93,94]. Functional alterations by DNA sequence variations on key genes as *LEPR*, as found in the pig, are more common than monogenic disorders [95,96]. Another important advantage of the Iberian model is that the inbreeding of Ossabaw and Göttingen is not found in the Iberian breed, since Iberian pigs are numerous (the estimated census of breeding sows is around 238,000). Moreover, Iberian pigs are currently still reared, and have been reared for centuries, in semiferal conditions; thus, influences of domestication and "life-comfort" may be not so apparent like in animals reared at biomedical animal facilities. Hence, findings in the Iberian breed would be more translational to the human population, which have also been exposed to noncontrolled environments for generations. In fact, the Iberian pig faces similar conditions to humans living in developing countries. Specifically, the Iberian pig has an adaptive *thrifty phenotype* owing to its gene polymorphism for the leptin receptor, has a background of exposure to harsh environment and food scarcity, and a current availability of nutrients in excess (when exposed to modern animal production systems).

EFFECTS OF DIET AND LIFESTYLE ON PUBERTY

The menarche is preceded by changes in body form and composition of the girl due to a significant increase of absolute and relative amounts of body fat [97]; the lean-body-weight-to-fat ratio changes from 1:3 in the period preceding puberty to 1:5 at puberty [98]. The endocrine signal linking age, body weight, fat reserves, and puberty onset is the hormone leptin. The hormone leptin is a protein produced in the adipocytes [29], which communicates the levels of peripheral energy stores to the central nervous system (CNS) and hence, regulates satiety, energy homeostasis, and fat metabolism but also other functions like reproduction [99]. Specifically, the onset of puberty is related to increases in body fat and thus in leptin secretion by the adipocytes. Obese adolescent girls, due to the increased amount of body fat and the increased secretion of leptin, may display a precocious onset of ovulatory cyclic activity (menarche) when compared to normal-weight adolescents [100].

The relationship between leptin and the onset of puberty was first demonstrated in *ob/ob* mouse, which remain prepubertal [32,101] in a hypogonadotrophic-hypogonadal state [102]. After that, studies in large animals (heifers and gilts) confirmed that leptin concentrations and expression of its receptors increase with age and adiposity, until puberty is reached [103,104]. The comparison between gilts of lean (Large White × Landrace) and obese genotype (Iberian) reared under similar conditions has shown a significant earlier onset of puberty in the Iberian breed [105], related to higher and earlier increase in leptin concentrations [106]. In the same way, the onset of puberty in gilts prone to obesity was significantly determined by diet, with a high fat intake level during the juvenile period inducing early puberty with lower body weight and size but with a higher body fat content [94].

Based on these results, it is possible to hypothesize that exogenous leptin supply would advance puberty attainment. This hypothesis was earlier confirmed in mice; experiments performed in female mice demonstrated a significant advancement of puberty onset after leptin administration supply [107,108]. However, these results were not translatable to large animals and humans, which therefore reinforce the differences among small and large mammals [109]. In large mammals, treatment with leptin does not trigger puberty onset, in spite of the determinant changes in the relationship among adiposity–leptin–puberty, which supports the hypothesis that leptin is just one of several factors initiating puberty.

Thus, the most current studies in animal models are focused on the identification of the essential role of the interaction between leptin and kisspeptin and the recognition of the potential roles of epigenetics and miRNA-related pathways in the central control of puberty [110,111].

EFFECTS OF DIET AND LIFESTYLE ON FERTILITY

The epidemiological information on the high incidence of subfertility and anovulatory infertility in obese women is abundant, although most overweight and obese women can achieve pregnancy. Women of childbearing age may be affected by menstrual alterations (from ovulatory irregularities to chronic anovulation), polycystic ovarian syndrome (PCOS), and hence, reduced pregnancy rates and even infertility [112–125].

Epidemiological studies and interventional experiments in animal models have linked obesity with three main factors that may impair reproductive function; leptin resistance, insulin resistance, and pro-inflammatory state.

The hormone leptin is necessary for reproduction; obese mice with monogenic deficiencies in leptin ($Lep^{ob/ob}$) or LEPR ($Lepr^{db/db}$) are infertile. Such infertility is related to a hypogonadotrophic-hypogonadal state [102]. In the $Lep^{ob/ob}$ mice, the administration of a gonadotrophin-replacement therapy succeeds in inducing ovulations; however, fertile ovulations are only obtained after applying a leptin-replacement therapy, which supports the existence of direct, local effects on the ovary [126]. However, the influence of leptin on ovarian function may be also detrimental, as hyperleptinemia and leptin resistance in women are related to menses irregularities, chronic oligo-anovulation, and infertility [127]. Mechanistic experiments in animal models have demonstrated adverse effects on granulosa and theca cells, ultimately affecting follicular development and ovulation [128–130].

The role of insulin resistance in fertility disorders is increasingly evident; studies in Zucker Fatty rats (a monogenic rodent model) [131] have shown that a state of continuous insulin resistance is associated with changes in ovarian morphology and hormone profile, and thus, abnormal estrous cyclicity and infertility [132]. Insulin resistance is highly related with one of the most prominent causes of anovulation and infertility in women – the PCOS. In fact, this disorder is estimated to affect around 10% of women of childbearing age worldwide [133]. The syndrome is defined by the presence of hyperandrogenism (clinical and/or biochemical; although there is a minority of PCOS without overt hyperandrogenism), ovarian dysfunction (oligo-anovulation and/or polycystic ovaries), and the exclusion of related disorders [134]. The occurrence of PCOS is often linked to obesity, metabolic syndrome, and type 2 diabetes [135,136]. Traditionally, rodent models (mice and rats), either genetically modified or hormonally induced, have been used for the study on the causes and biological mechanisms linked to the development of the disease [137]; however, the most intensive and translatable research has been performed by prenatal exposure to androgens in nonhuman primates [138,139] and sheep [140].

Finally, obesity is also associated with a pro-inflammatory state. Obesity is therefore related to high levels of pro-inflammatory cytokines in follicular fluid [141], which may induce deficiencies in the activity of the hypothalamus-pituitary-gonadal axis, and may impair the occurrence of regular ovulatory processes [142]. Concomitantly with this pro-inflammatory state, a higher oxidative stress in obese males has been found to be related to infertility [143].

Most of the women affected by reproductive problems are submitted to assisted reproduction treatments (ART); however, some authors have reported that the success of the induced cycles is compromised [144–148]. The main cause for unsuccessful assisted cycles is a low ovarian response to the exogenous hormones, which may lead to cycle cancellation or collection of fewer oocytes [149,150]. Some studies have pointed to the pro-inflammatory state and the higher oxidative stress of the patients, which clearly affect ART yields [151]. Then, if pregnancy is achieved, ART is associated to higher incidence of induced labor, cesarean section, premature birth, small-for-gestational-age babies, pediatric cancer, imprinting disorders, and congenital abnormalities [152–155]. Several experimental and clinical studies [156] identify placental dysfunction as the determinant factor, with studies in mice pointing to deficiencies in hormone and metabolite signaling, and excessive placental inflammation and oxidative stress [156,157].

Concomitantly, some studies on ART yields in humans also indicate that lower reproductive outputs in obese females, with a possible lower oocyte/embryo developmental competence, may be coincident with alterations in oviduct/uterine environment that lead to deficiencies in early pregnancy and implantation [158,159]. The same has been found in the Iberian swine model of leptin resistance; the lower reproductive efficiency of the breed has been related to increased embryo losses in the first third of pregnancy rather than to deficiencies in follicle development and ovulatory efficiency [160–162].

EFFECTS OF DIET AND LIFESTYLE ON PREGNANCY AND OFFSPRING DEVELOPMENT

Obese pregnant women may be affected by implantation failures and miscarriages in the first trimester of pregnancy; after that, the risks of abortion, miscarriage, and preterm delivery is increased and the offspring may be finally more prone to different diseases at postnatal life [163–166]. The mechanisms responsible for these events have also been related to the maternal states of leptin resistance, insulin-resistance, inflammation, and oxidative stress like in subfertility. The combination of these factors modifies the feto-maternal environment and may impede adequate development of the offspring in this case.

The women affected by leptin resistance and insulin resistance develop hyperglycemia and triglyceridemia [167]. Hence, their fetuses are exposed, through placental transport, to high concentrations of glucose and lipids and, in consequence, the development of such fetuses is accelerated, and the newborns are large-for-gestational-age (LGA) and obese. However, in some obese or overweight women, changes in maternal metabolic status induce changes in placental structure; mainly in the form of vascular alterations affecting the placental development and subsequent function. The main function of the placenta is to supply nutrients and oxygen to the fetus and the removal of waste substances. Deficiencies in placental development and function affect conceptus development by deficiencies in the supply of nutrients and oxygen, causing fetal losses or compromising the growth of conceptus in a process named as IUGR; in consequence, offspring become small-for-gestational-age (SGA) [168–170].

Any alteration in the feto-maternal environment turns on adaptive changes in the placental structure and function for assuring offspring development and viability [171]; failure or success in these adaptive changes will determine pregnancy failure or success. Hence, the study of placental function is currently a highly important area for research. There are different animal models for studies on placental efficiency, as elegantly revised by Desoye and Shafrir [172], with diverse similarities to humans. When compared to humans, the placenta of rodents has a discoideal shape with a hemomonochorial barrier in the guinea pig (as found in humans); however, the barrier is hemotrichorial in the mouse and the rat. These three species have a labyrinthine feto-maternal interdigitation and a countercurrent blood flow. Conversely, humans have villous interdigitation and multivillous blood flow, features found in sheep, but in which the shape is cotiledonary and the barrier is epitheliochorial. Finally, the placenta of rabbits is discoideal and hemodichorial with labyrinthine interdigitations. Thus, the election of the adequate model will be based on the target of the study.

The state of insulin resistance affects placental development by disturbing the mechanisms of vascular dilation and neoangiogenesis and therefore the uteroplacental blood flow [173], which is modulated by the placental growth factor (PlGF), the vascular endothelial growth factor (VEGF), and endothelial nitric oxide [174]. These changes cause hypoxia and, at the same time, are aggravated by the increase in oxidative stress and inflammatory state caused by hypoxia [175]. The effects of deficiencies in endothelial nitric oxide have been widely studied in the Nos3-knockout mice. Inhibition of NOS3 in Nos3-knockout mice is related to reduced growth and a higher mortality of the fetuses [176, 177], causing reduced prolificacy and deficient prenatal development [178, 179], with a lower weight of the neonates at delivery [180]. The Nos3-knockout placentas have increased hypoxia and elevated superoxide dismutase activity than wild-type placentas [181]. Thus, these mice

are a good model for altered vascular adaptation to pregnancy, preeclampsia, uteroplacental hypoxia, placental insufficiency, and IUGR.

However, the most frequently used models for IUGR studies are pig and sheep. In pig, IUGR is commonly found in some littermates of highly prolific lines [182–184], due to restrictions in uterine space availability prejudicing adequate placental development. Incidence of IUGR may be also increased by maternal nutritional deficiencies [185], distorting nutrient supply of the fetuses. Specifically, obese Iberian pigs are characterized by a lower prolificacy and a lower uterine capacity than modern commercial breeds, and there are emerging evidences of a high IUGR incidence in larger litters or in case of nutritional limitations [186]. The piglets affected by IUGR are predisposed to high neonatal morbidity and mortality rates (as found in humans) and to the development of obesity and metabolic and endocrine disorders [187,188]; all alterations that can be modulated by adequate diet and exercise [189]. Differences in fetal development among lean and obese genotypes seem to be related to differences in fetal availability of triglycerides, cholesterol, and IL-6 [190]. In this sense, currently, a high number of the studies on adverse clinical outcomes during obese pregnancies points to alterations in the inflammatory response [191]. Maternal obesity induces inflammation in the placenta [192]; these results, arising from studies on human pregnancies at term, have been confirmed to be initiated early in pregnancy using sheep models [193].

In any case, surviving offspring, either LGA or SGA, will become, by prenatal programming, more prone to obesity and associated diseases than those that were born from healthy women. The effects of developmental programming turn on long-lasting perturbations in endocrine and neuronal signals regulating the hypothalamus–adipose tissue axis of the offspring [194]; mainly increased circulating leptin levels, hypothalamic leptin resistance, and enhanced orexigenic pathways (especially elevated circulating NPY levels and activated adipogenic NPY system in adipose tissue) leading to hyperphagia. A higher energy intake and altered sympathetic activity, concomitantly with increased adipogenesis and lipogenesis capacities, promote obesity.

The same is found in animals and thus, the use of animal models for studies on prenatal programming is abundant [195]. In our laboratory, the use of the Iberian swine model for obesity has shown narrow similarities in developmental and metabolic profiles with people from developing countries. The intake of obesogenic diets during juvenile periods induces the prodrome of metabolic syndrome (obesity with impairments of glucose regulation) and an earlier puberty [94]; however, young animals are still able to develop adaptive responses for regulating homeostasis. On the contrary, during adulthood, the animals display the five components of the metabolic syndrome (central obesity, dyslipidemia, insulin resistance, impaired glucose tolerance, and elevated blood pressure) and even the prodrome of type 2 diabetes [93]; there is no other animal model that develops more than three components at the same time. In case of pregnancy, the dyslipidemic state of the females induces excess of triglycerides and cholesterol at the fetoplacental unit, affecting the developmental trajectory of the fetuses. The offspring exposed to maternal overnutrition during pregnancy are prone to obesity at so early postnatal stages as during lactation; in case of obesogenic diets at juvenile stages, metabolic changes are even more severe than in nonchallenged piglets, with increased adiposity, evidences of metabolic syndrome and even prodrome of type 2 diabetes, and a precocious puberty [187,188].

At this time, there is increasing evidence noting that prenatal programming of the offspring developmental is clearly driven by sex (revised by Aiken and Ozanne [196]); in this scenario, the possibility of comparing littermates is highly interesting and therefore animal models, like rodents, rabbits, and pigs, as well as sheep with twin pregnancies, are widely used. An interesting output of the research on this area is the substantiation of maternal influences driving sex-related differences in developmental programming. At the beginning, a different adaptive response of males and females to inadequate prenatal and early postnatal environments was assumed; currently, the question is whether the responses of male and female offspring are identical and, on the contrary, the mother modulates environmental conditions depending on the sex that she is carrying. However, this is a novel research topic that has only just got off the ground.

CONCLUSIONS

The deleterious effects of nonoptimal diet and lifestyle on reproductive features and success, not only of current but also of future generations, are a highly concerning issue in countries in which the obesity epidemic is present; the number of which is increasing daily. Two main actions need to be undertaken in parallel in the near future. The first is the development of adequate strategies for the improvement of individualized health care and population-wide diagnosis, prevention, and treatment. The second, which is essential for undertaking the first one, is a profound research of the mechanisms causing reproductive failures and disorders, and the strategies for prevention and treatment. Such action needs to rely on interventional studies in adequate animal models for a rapid translation to human medicine. Currently, animal research is characterized by two main pillars: a high knowledge of the physiology and the translational value of the different species, and a strong respect for adequate conditions of animal welfare and health.

ACKNOWLEDGMENTS

AGB is a member of the EU COST-Action BM1308 "Sharing Advances on Large Animal Models (SALAAM)".

References

[1] Rogers J, Mitchell GW. The relation of obesity to menstrual disturbances. New Engl J Med 1952;247:53–6.

[2] Cochrane WA. Overnutrition in prenatal and neonatal life: a problem? Can Med Assoc J 1965;93:893–9.

[3] Hu FB. Globalization of diabetes: the role of diet, lifestyle, and genes. Diab Care 2011;34:1249–57.

[4] Huang L, Sauve R, Birkett N, Fergusson D, van Walraven C. Maternal age and risk of stillbirth: a systematic review. CMAJ 2008;178:165–72.

[5] Mills M, Rindfuss RR, McDonald P, Velde E. ESHRE reproduction and society task force. Why do people postpone parenthood? Reasons and social policy incentives. Hum Reprod Update 2011;17:848–60.

[6] Wong-Taylor LA, Lawrence A, Cowen S, Jones H, Nauta M, Ramsay-Marcelle Z, et al. Maternal and neonatal outcomes of spontaneously conceived pregnancies in mothers over 45 years: a review of the literature. Arch Gynecol Obstet 2012;285:1161–6.

[7] Barouki R, Gluckman PD, Grandjean P, Hanson M, Heindel JJ. Developmental origins of non-communicable disease: implications for research and public health. Environ Health 2012;11:42.

[8] Chandrasekera PC, Pippin JJ. Of rodents and men: species-specific glucose regulation and type 2 diabetes research. ALTEX 2014;31:157–76.

[9] Habert R, Muczynski V, Grisin T, Moison D, Messiaen S, Frydman R, et al. Concerns about the widespread use of rodent models for human risk assessments of endocrine disruptors. Reproduction 2014;147:119–29.

[10] McLaren A. Germ and somatic cell lineages in the developing gonad. Mol Cell Endocrinol 2000;163:3–9.

[11] Pepling ME, Spradling AC. Mouse ovarian germ cell cysts undergo programmed breakdown to form primordial follicles. Dev Biol 2001;234:339–51.

[12] Mazaud-Guittot S, Guigon CJ, Coudouel N, Magre S. Consequences of fetal irradiation on follicle histogenesis and early follicle development in rat ovaries. Biol Reprod 2006;75:749–59.

[13] Guigon CJ, Magre S. Contribution of germ cells to the differentiation and maturation of the ovary: insights from models of germ cell depletion. Biol Reprod 2006;74:450–8.

[14] Clarke IJ. Models of "obesity" in large animals and birds. Front Horm Res 2008;36:107–17.

[15] Lunney JK. Advances in swine biomedical model genomics. Intern J Biol Sci 2007;3:179–84.

[16] Ireland JJ, Roberts RM, Palmer GH, Bauman DE, Bazer FW. A commentary on domestic animals as dual-purpose models that benefit agricultural and biomedical research. J Anim Sci 2008;86:2797–805.

[17] Jawerbaum A, White V. Animal models in diabetes and pregnancy. Endocr Rev 2010;31:680–701.

[18] Speakman J, Hambly C, Mitchell S, Król E. The contribution of animal models to the study of obesity. Lab Anim 2008;42:413–32.

[19] Gonzalez-Bulnes A, Pallares P, Ovilo C. Ovulation, implantation and placentation in females with obesity and metabolic disorders: life in the balance. Endocr Metab Immune Disord Drug Targets 2011;11:285–301.

[20] Aigner B, Renner S, Kessler B, Klymiuk N, Kurome M, Wunsch A, et al. Transgenic pigs as models for translational biomedical research. J Mol Med (Berl) 2010;88:653–64.

[21] Augustine KA, Rossi RM. Rodent mutant models of obesity and their correlations to human obesity. Anat Rec 1999;257:64–72.

[22] Zucker LM, Antoniades HN. Insulin and obesity in the Zucker genetically obese "fatty" rat. Endocrinology 1972;90:1320–30.

[23] Shafrir E. Diabetes in animals. In: Porte JR, Sherwin RS, editors. Diabetes Mellitus D. Stamford: Appleton & Lange; 1995. p. 301–48.

[24] Tartaglia LA, Dembski M, Weng X, Deng N, Culpepper J, Devos R, et al. Identification and expression cloning of a leptin receptor, OB-R. Cell 1995;83:1263–71.

[25] Chen H, Charlat O, Tartaglia LA, Woolf EA, Weng X, Ellis SJ, et al. Evidence that the diabetes gene encodes the leptin receptor: identification of a mutation in the leptin receptor gene in *db/db* mice. Cell 1996;84:491–5.

[26] Chua SC, Chung WK, Wu-Peng XS, Zhang Y, Liu SM, Tartaglia L, et al. Phenotypes of mouse diabetes and rat fatty due to mutations in the OB (leptin) receptor. Science 1996;271:994–6.

[27] Lee GH, Proenca R, Montez JM, Carroll KM, Darvishzadeh JG, Lee JI, et al. Abnormal splicing of the leptin receptor in diabetic mice. Nature 1996;379:632–5.

[28] Ingalls AM, Dickie MM, Snell GD. Obese, a new mutation in the house mouse. J Hered 1950;41:317–8.

[29] Zhang Y, Proenca R, Maffei M, Barone M, Leopold L, Friedman JM. Positional cloning of the mouse obese gene and its human homologue. Nature 1994;372:425–32.

[30] Jones N, Harrison GA. Genetically determined obesity and sterility in the mouse. Proc Soc Study Fertil 1957;51–6.

[31] Lindström P. The physiology of obese-hyperglycaemic mice [*ob/ob* mice]. Sci World J 2007;7:666–85.

[32] Barash IA, Cheung CC, Weigle DS, Ren H, Kabigting EB, Kuijper JL, et al. Leptin is a metabolic signal to the reproductive system. Endocrinol 1996;137:3144–7.

[33] Ranadive SA, Vaisse C. Lessons from extreme human obesity: monogenic disorders. Endocrinol Metabol Clin North Am 2008;37:733–51.

[34] Montague CT, Farooqi IS, Whitehead JP, Soos MA, Rau H, Wareham NJ, et al. Congenital leptin deficiency is associated with severe early-onset obesity in humans. Nature 1997;387:903–8.

[35] Clement K, Vaisse C, Lahlou N, Cabrol S, Pelloux V, Cassuto D, et al. A mutation in the human leptin receptor gene causes obesity and pituitary dysfunction. Nature 1998;392:398–401.

[36] Vaisse C, Clement K, Guy-Grand B, Froguel P. A frameshift mutation in human MC4R is associated with a dominant form of obesity. Nat Genet 1998;20:113–4.

[37] Yeo GS, Farooqi IS, Aminian S, Halsall DJ, Stanhope RG, O'Rahilly S. A frameshift mutation in MC4R associated with dominantly inherited human obesity. Nat Genet 1998;20:111–2.

[38] Krude H, Biebermann H, Luck W, Horn R, Brabant G, Gruters A. Severe early-onset obesity; adrenal insufficiency and red hair pigmentation caused by POMC mutations in humans. Nat Genet 1998;19:155–7.

[39] King AJF. The use of animal models in diabetes research. Br J Pharmacol 2012;166:877–94.

[40] Kobayashi T, Ito T, Shiomi M. Roles of the WHHL rabbit in translational research on hypercholesterolemia and cardiovascular diseases. J Biomed Biotechnol 2011;2011. 406473.

[41] Ozaki MR, de Almeida EA. Evolution and involution of atherosclerosis and its relationship with vascular reactivity in hypercholesterolemic rabbits. Exp Toxicol Pathol 2013;65:297–304.

[42] Bahr A, Wolf E. Domestic animal models for biomedical research. Reprod Domest Anim 2012;47(Suppl. 4):59–71.

[43] Palinski W, Napoli C. The fetal origins of atherosclerosis: maternal hypercholesterolemia, and cholesterol-lowering or antioxidant treatment during pregnancy influence *in utero* programming and postnatal susceptibility to atherogenesis. FASEB J 2002;16:1348–60.

[44] Picone O, Laigre P, Fortun-Lamothe L, Archilla C, Peynot N, Ponter AA, et al. Hyperlipidic hypercholesterolemic diet in prepubertal rabbits affects gene expression in the embryo, restricts fetal growth and increases offspring susceptibility to obesity. Theriogenology 2011;75:287–99.

[45] Püschel B, Daniel N, Bitzer E, Blum M, Renard JP, Viebahn C. The rabbit (*Oryctolagus cuniculus*): a model for mammalian reproduction and early embryology. Cold Spring Harb Protoc 2010;. 2010: pdb.emo139.

[46] Fischer B, Chavatte-Palmer P, Viebahn C, Navarrete Santos A, Duranthon V. Rabbit as a reproductive model for human health. Reproduction 2012;144:1–10.

[47] Campbell BK, Souza C, Gong J, Webb R, Kendall N, Marsters P, et al. Domestic ruminants as models for the elucidation of the mechanisms controlling ovarian follicle development in humans. Reproduction 2003;(Suppl. 61):429–43.

[48] Bloch M, Rubinow DR, Schmidt PJ, Lotsikas A, Chrousos GP, Cizza G. Cortisol response to ovine corticotropin-releasing hormone in a model of pregnancy and parturition in euthymic women with and without a history of postpartum depression. J Clin Endocrinol Metab 2005;90:695–9.

[49] Bordes A, Lornage J, Demirci B, Franck M, Courbiere B, Guerin JF, et al. Normal gestations and live births after orthotopic autograft of vitrified warmed hemi-ovaries into ewes. Hum Reprod 2005;20:2745–8.

[50] Denschlag D, Knobloch C, Kockrow A, Baessler A, Goebel H, Wellens E, et al. Autologous heterotopic transplantation of ovarian tissue in sheep. Fertil Steril 2005;83:501–3.

[51] Gonzalez-Bulnes A, Baird DT, Campbell BK, Cocero MJ, García-García RM, Inskeep EK, et al. Multiple factors affecting the efficiency of multiple ovulation and embryo transfer in sheep and goats. Reprod Fertil Dev 2004;16:421–35.

[52] Poore KR, Boullin JP, Cleal JK, Newman JP, Noakes DE, Hanson MA, et al. Sex- and age-specific effects of nutrition in early gestation and early postnatal life on hypothalamo-pituitary-adrenal axis and sympathoadrenal function in adult sheep. J Physiol 2010;588:2219–37.

[53] Dodic M, Abouantoun T, O'Connor A, Wintour EM, Moritz KM. Programming effects of short prenatal exposure to dexamethasone in sheep. Hypertension 2002;40:729–34.

[54] Manikkam M, Crespi EJ, Doop DD, Herkimer C, Lee JS, Yu S, et al. Fetal programming: prenatal testosterone excess leads to fetal growth retardation and postnatal catch-up growth in sheep. Endocrinology 2004;145:790–8.

[55] Veiga-Lopez A, Steckler TL, Abbott DH, Welch KB, MohanKumar PS, Phillips DJ, et al. Developmental programming: impact of excess prenatal testosterone on intrauterine fetal endocrine milieu and growth in sheep. Biol Reprod 2011;84:87–96.

[56] Padmanabhan V, Veiga-Lopez A. Sheep models of polycystic ovary syndrome phenotype. Mol Cell Endocrinol 2013;373:8–20.

[57] McMillen IC, Adam CL, Mühlhäusler BS. Early origins of obesity: programming the appetite regulatory system. J Physiol 2005;565:9–17.

[58] Gardner DS, Rhodes P. Developmental origins of obesity: programming of food intake or physical activity? Adv Exp Med Biol 2009;646:83–93.

[59] Bloomfield FH, Oliver MH, Giannoulias CD, Gluckman PD, Harding JE, Challis JR. Brief undernutrition in late-gestation sheep programs the hypothalamic-pituitary-adrenal axis in adult offspring. Endocrinology 2003;144:2933–40.

[60] Cleal JK, Poore KR, Newman JP, Noakes DE, Hanson MA, Green LR. The effect of maternal undernutrition in early gestation on gestation length and fetal and postnatal growth in sheep. Pediatr Res 2007;62:422–7.

[61] Gilbert JS, Ford SP, Lang AL, Pahl LR, Drumhiller MC, Babcock SA, et al. Nutrient restriction impairs nephrogenesis in a gender-specific manner in the ovine fetus. Pediatr Res 2007;61:42–7.

[62] Ojeda NB, Grigore D, Alexander BT. Developmental programming of hypertension: insight from animal models of nutritional manipulation. Hypertension 2008;52:44–50.

[63] Metges CC. Early nutrition and later obesity: animal models provide insights into mechanisms. Adv Exp Med Biol 2009;646:105–12.

[64] Zhang S, Rattanatray L, Morrison JL, Nicholas LM, Lie S, McMillen IC. Maternal obesity and the early origins of childhood obesity: weighing up the benefits and costs of maternal weight loss in the periconceptional period for the offspring. Exp Diabetes Res 2011;2011. 585749.

[65] Houpt KA, Houpt TR, Pond WG. The Pig as a model for the study of obesity and of control of food intake: a review. Yale J Biol Med 1979;52:307–29.

[66] Litten-Brown JC, Corson AM, Clarke L. Porcine models for the metabolic syndrome, digestive and bone disorders: a general overview. Animal 2010;4:899–920.

[67] Spurlock ME, Gabler NK. The development of porcine models of obesity and the metabolic syndrome. J Nutr 2008;138:397–402.

[68] Aigner T, Cook JL, Gerwin N, Glasson SS, Laverty S, Little CB, et al. Histopathology atlas of animal model systems - overview of guiding principles. Osteoarthritis Cartilage 2010;18:S2–6.

[69] Aigner B, Renner S, Kessler B, Klymiuk N, Kurome M, Wünsch A, et al. Transgenic pigs as models for translational biomedical research. J Mol Med (Berl) 2010;88:653–64.

[70] Douglas WR. Of pigs and men and research: a review of applications and analogies of the pig (*Sus scrofa*) in human medical research. Space Life Sci 1972;3:226–34.

[71] Mahley RW, Weisgraber KH, Inneraty T, Brewer HB Jr, Assmann G. Swine lipoproteins and atherosclerosis: changes in the plasma lipoproteins and apoproteins induced by cholesterol feeding. Biochemistry 1975;14:2817–23.

[72] Bell FP, Gerrity RG. Evidence for an altered lipid metabolic state in circulating blood monocytes under conditions of hyperlipemia in swine and implications in arterial lipid metabolism. Arterioscler Thromb 1992;12:155–62.

[73] Wessels JM, Linton NF, Croy A, Tayade C. A review of molecular contrasts between arresting and viable porcine attachment sites. Am J Reprod Immunol 2007;58:470–80.

[74] Ashworth CJ, Finch AM, Page KR, Nwagwu MO, McArdle HJ. Causes and consequences of fetal growth retardation in pigs. Reprod Suppl 2001;58:233–46.

[75] Foxcroft GR, Dixon WT, Novak S, Putman CT, Town SC, Vinsky MD. The biological basis for prenatal programming of postnatal performance in pigs. J Anim Sci 2006;84:E105–12.

[76] Wu G, Bazer FW, Wallace JM, Spencer TE. Board-invited review: intrauterine growth retardation: implications for the animal sciences. J Anim Sci 2006;84:2316–37.

[77] Town SC, Putman CT, Turchinsky NJ, Dixon WT, Foxcroft GR. Number of conceptuses *in utero* affects porcine fetal muscle development. Reproduction 2004;128:443–54.

[78] Van der Waaij EH, Hazeleger W, Soede NM, Laurenssen BFA, Kemp B. Effect of excessive, hormonally induced intrauterine crowding in the gilt on fetal development on day 40 of pregnancy. J Anim Sci 2010;88:2611–9.

[79] Metges CC, Lang IS, Hennig U, Brüssow KP, Kanitz E, Tuchscherer M, et al. Intrauterine growth retarded progeny of pregnant sows fed high protein:low carbohydrate diet is related to metabolic energy deficit. PLoS One 2012;7:e31390.

[80] Metges CC. Early nutrition and later obesity: animal models provide insights into mechanisms. Adv Exp Med Biol 2009;646:105–12.

[81] Johansen T, Hansen HS, Richelsen B, Malmlöf R. The obese Göttingen minipig as a model of the metabolic syndrome. Dietary effects on obesity; insulin sensitivity; and growth hormone profile. Comp Med 2001;51:150–5.

[82] Christoffersen BO, Grand N, Golozoubova V, Svendsen O, Raun K. Gender-associated differences in metabolic syndrome-related parameters in Göttingen minipigs. Comp Med 2007;57:493–504.

[83] Guillerm-Regost C, Louveau I, Sébert SP, Damon M, Champ MM, Gondret F. Cellular and biochemical features of skeletal muscle in obese Yucatan minipigs. Obesity 2006;14:1700–7.

[84] Dyson MC, Alloosh M, Vuchetich JP, Mokelke EA, Sturek M. Components of metabolic syndrome and coronary artery disease in female Ossabaw swine fed excess atherogenic diet. Comp Med 2006;56:35–45.

[85] SanCristobal M, Chevalet C, Haley CS, Joosten R, Rattink AP, Harlizius B, et al. Genetic diversity within and between European pig breeds using microsatellite markers. Anim Genet 2006;37:187–98.

[86] Ollivier L. European pig genetic diversity: a minireview. Animal 2009;3:915–24.

[87] Morales J, Pérez JF, Baucells MD, Mourot J, Gasa J. Comparative digestibility and lipogenic activity in Landrace and Iberian finishing pigs fed *ad libitum* corn- and corn–sorghum–acorn-based diets. Livest Prod Sci 2002;77:195–205.

[88] Nieto R, Miranda A, García MA, Aguilera JF. The effect of dietary protein content and feeding level on the rate of protein deposition and energy utilization in growing Iberian pigs from 15 to 50kg body weight. Br J Nutr 2002;88:39–49.

[89] López-Bote C. Sustained utilization of Iberian pig breed. Meat Sci 1998;49:S17–27.

[90] Fernandez-Figares I, Lachica M, Nieto R, Rivera-Ferre MG, Aguilera JF. Serum profile of metabolites and hormones in obese (Iberian) and lean (Landrace) growing gilts fed balanced or lysine deficient diets. Livest Sci 2007;110:73–81.

[91] Ovilo C, Fernández A, Noguera JL, Barragán C, Letón R, Rodríguez C, et al. Fine mapping of porcine chromosome 6 QTL and LEPR effects on body composition in multiple generations of an Iberian by Landrace intercross. Genet Res 2005;85:57–67.

[92] Muñoz G, Óvilo C, Silió L, Tomás A, Noguera JL, Rodríguez MC. Single and joint population analyses of two experimental pig crosses to confirm QTL on SSC6 and LEPR effects on fatness and growth traits. J Anim Sci 2009;87:459–68.

[93] Torres-Rovira L, Astiz S, Caro A, Lopez-Bote C, Ovilo C, Pallares P, et al. Diet-induced swine model with obesity/leptin resistance for the study of metabolic syndrome and type 2 diabetes. Sci World J 2012;2012. 510149.

[94] Torres-Rovira L, Gonzalez-Añover P, Astiz S, Caro A, Lopez-Bote C, Ovilo C, et al. Effect of an obesogenic diet during the juvenile period on growth pattern, fatness and metabolic, cardiovascular and reproductive features of swine with obesity/leptin resistance. Endocr Metab Immune Disord Drug Targets 2013;13:143–51.

[95] Ukkola O, Bouchard C. Role of candidate genes in the responses to long-term overfeeding: review of findings. Obes Rev 2004;5:3–12.

[96] Zabeau L, Defeau D, Van der Heyden J, Iserentant H, Vandekerckhove J, Tavernier J. Functional analysis of leptin receptor activation using a Janus kinase/signal transducer and activator of transcription complementation assay. Mol Endocrinol 2004;18:150–61.

[97] Frisch RE. Body fat; puberty and fertility. Biol Rev 1984;59:161–88.

[98] Frisch RE, Revelle R, Cook S. Components of weight at menarche and the initiation of the adolescent growth spurt in girls: estimated total water; lean body weight and fat. Human Biol 1973;45:469–83.

[99] Lindstrom P. The physiology of obese-hyperglycaemic mice [*ob/ob* mice]. Sci World J 2007;7:666–85.

[100] Pelusi C, Pasquali R. Polycystic ovary syndrome in adolescents. Pathophysiology and treatment implication. Treat Endocrinol 2003;2:215–30.

[101] O'Rahilly S. Life without leptin. Nature 1998;392:330–1.

[102] Hamm ML, Bhat GK, Thompson WE, Mann DR. Folliculogenesis is impaired and granulosa cell apoptosis is increased in leptin-deficient mice. Biol Reprod 2004;71:66–72.

[103] Barb CR, Kraeling RR, Rampacek GB. Metabolic regulation of the neuroendocrine axis in pigs. Reprod Suppl 2002;59:203–17.

[104] Barb CR, Kraeling RR. Role of leptin in the regulation of gonadotropin secretion in farm animals. Anim Reprod Sci 2004;82–83:155–67.

[105] Gonzalez-Añover P, Encinas T, Gomez-Izquierdo E, Sanz E, Letelier CA, Torres-Rovira L, et al. Advanced onset of puberty in gilts of thrifty genotype (Iberian pig). Reprod Domest Anim 2010;45:1003–7.

[106] Gonzalez-Añover P, Vigo E, Encinas T, Torres-Rovira L, Pallares P, Gomez-Izquierdo E, et al. Prepuberal evolution of plasma leptin levels in gilts of thrifty genotype (Iberian pig) and lean commercial crosses (Large White × Landrace). Res Vet Sci 2012;93:100–2.

[107] Ahima RS, Dushay J, Flier SN, Prabakaran D, Flier JS. Leptin accelerates the onset of puberty in normal female mice. J Clin Invest 1997;99:391–5.

[108] Chehab FF, Mounzih K, Ronghua L, Lim ME. Early onset of reproductive function in normal female mice treated with leptin. Science 1997;275:88–90.

[109] Zieba DA, Amstalden M, Williams GL. Regulatory roles of leptin in reproduction and metabolism: a comparative review. Domest Anim Endocrinol 2005;29:166–85.

[110] Sanchez-Garrido MA, Tena-Sempere M. Metabolic control of puberty: roles of leptin and kisspeptins. Horm Behav 2013;64:187–94.

[111] Tena-Sempere M. Keeping puberty on time: novel signals and mechanisms involved. Curr Top Dev Biol 2013;105:299–329.

[112] Metwally M, Li TC, Ledger WL. The impact of obesity on female reproductive function. Obes Rev 2007;8:515–23.

[113] Green BB, Weiss NS, Daling JR. Risk of ovulatory infertility in relation to body weight. Fertil Steril 1988;50:721–6.

[114] Zaadstra BM, Seidell JC, Van Noord PA. Fat and female fecundity: prospective study of effect of body fat distribution on conception rates. BMJ 1993;306:484–7.

[115] Grodstein F, Goldman MB, Cramer DW. Body mass index and ovulatory infertility. Epidemiology 1994;5:247–50.

[116] Rich-Edwards JW, Goldman MB, Willett WC. Adolescent body mass index and infertility caused by ovulatory dysfunction. Am J Obstet Gynecol 1994;71:171–7.

[117] Jensen TK, Scheike T, Keiding N, Schaumburg I, Grandjean P. Fecundability in relation to body mass and menstrual cycle patterns. Epidemiology 1999;10:422–8.

[118] Bolumar F, Olsen J, Rebagliato M, Saez-Lloret I, Bisanti L. Body mass index and delayed conception: a European Multicenter Study on Infertility and Subfecundity. Am J Epidemiol 2000;151:1072–9.

[119] Hassan MA, Killick SR. Negative lifestyle is associated with a significant reduction in fecundity. Fertil Steril 2004;81:384–92.

[120] Linne Y. Effects of obesity on women's reproduction and complications during pregnancy. Obes Rev 2004;5:137–43.

[121] Pasquali R. Obesity; fat distribution and infertility. Maturitas 2006;54:363–71.

[122] Pasquali R, Gambineri A. Metabolic effects of obesity on reproduction. Reprod Biomed Online 2006;12:542–51.

[123] Pasquali R, Gambineri A, Pagotto U. The impact of obesity on reproduction in women with polycystic ovary syndrome. Br J Obstet Gynaecol 2006;113:1148–59.

[124] Metwally M, Ledger WL, Li TC. Reproductive endocrinology and clinical aspects of obesity in women. Ann N Y Acad Sci 2008;1127:140–6.

[125] Brewer CJ, Balen AH. The adverse effects of obesity on conception and implantation. Reproduction 2010;140:347–64.

[126] Pallares P, Garcia-Fernandez RA, Criado LM, Letelier CA, Fernandez-Toro JM, Esteban D, et al. Substantiation of ovarian effects of leptin by challenging a mouse model of obesity/type 2 diabetes. Theriogenology 2010;73:1088–95.

[127] Pasquali R. Obesity; fat distribution and infertility. Maturitas 2006;54:363–71.

[128] Duggal PS, Van Der Hoek KH, Milner CR, Ryan NK, Armstrong DT, Magoffin DA, et al. The *in vivo* and *in vitro* effects of exogenous leptin on ovulation in the rat. Endocrinology 2000;141:1971–6.

[129] Spicer LJ. Leptin: a possible metabolic signal affecting reproduction. Domest Anim Endocrinol 2001;21:251–70.

[130] Swain JE, Dunn RL, McConnell D, Gonzalez-Martinez J, Smith GD. Direct effects of leptin on mouse reproductive function: regulation of follicular; oocyte; and embryo development. Biol Reprod 2004;71:1446–52.

[131] Marin Bivens CL, Olster DH. Abnormal estrous cyclicity and behavioral hyporesponsiveness to ovarian hormones in genetically obese Zucker female rats. Endocrinology 1997;138:143–8.

[132] Honnma H, Endo T, Kiya T, Nagasawa K, Baba T, Fujimoto T, et al. Remarkable features of ovarian morphology and reproductive hormones in insulin-resistant Zucker fatty (*fa/fa*) rats. Reprod Biol Endocrinol 2010;8:73.

[133] Franks S. Polycystic ovary syndrome. N Engl J Med 1995;333:853–61.

[134] Azziz R, Carmina E, Dewailly D, Diamanti-Kandarakis E, Escobar-Morreale HF, Futterweit W, et al. The Androgen Excess and PCOS Society criteria for the polycystic ovary syndrome: the complete task force report. Fertil Steril 2009;91:456–88.

[135] Goodarzi MO, Dumesic DA, Chazenbalk G, Azziz R. Polycystic ovary syndrome: etiology, pathogenesis and diagnosis. Nat Rev Endocrinol 2011;7:219–31.

[136] Pasquali R, Stener-Victorin E, Yildiz BO, Duleba AJ, Hoeger K, Mason H, et al. PCOS Forum: research in polycystic ovary syndrome today and tomorrow. Clin Endocrinol (Oxf) 2011;74:424–33.

[137] Walters KA, Allan CM, Handelsman DJ. Rodent models for human polycystic ovary syndrome. Biol Reprod 2012;86:149. 1–12.

[138] Abbott DH, Dumesic DA, Eisner JR, Colman RJ, Kemnitz JW. Insights into the development of polycystic ovary syndrome (PCOS) from studies in prenatally androgenised female rhesus monkeys. Trends Endocrinol Metab 1998;9:62–7.

[139] Eisner JR, Barnett MA, Dumesic DA, Abbott DH. Ovarian hyperandrogenism in adult female rhesus monkeys exposed to prenatal androgen excess. Fertil Steril 2002;77:167–72.

[140] Recabarren SE, Padmanabhan V, Codner E, Lobos A, Durán C, Vidal M, et al. Postnatal developmental consequences of altered insulin sensitivity in female sheep treated prenatally with testosterone. Am J Physiol Endocrinol Metab 2005;289:E801–6.

[141] La Vignera S, Condorelli R, Bellanca S, La Rosa B, Mousaví A, Busà B, et al. Obesity is associated with a higher level of proinflammatory cytokines in follicular fluid of women undergoing medically assisted procreation (PMA) programs. Eur Rev Med Pharmacol Sci 2011;15:267–73.

[142] Tersigni C, Nicuolo FD, D'Ippolito S, Veglia M, Castellucci M, Simone ND. Adipokines: new emerging roles in fertility and reproduction. Obstet Gynecol 2011;66:47–63.

[143] Michalakis K, Mintziori G, Kaprara A, Tarlatzis BC, Goulis DG. The complex interaction between obesity, metabolic syndrome and reproductive axis: a narrative review. Metabolism 2013;62:457–78.

[144] Wang JX, Davies M, Norman RJ. Body mass and probability of pregnancy during assisted reproduction treatment: retrospective study. BMJ 2000;321:1320–1.

[145] Loveland JB, McClamrock HD, Malinow AM, Sharara FI. Increased body mass index has a deleterious effect on *in vitro* fertilization outcome. J Assist Reprod Genet 2001;18:382–6.

[146] Spandorfer SD, Kump L, Goldschlag D, Brodkin T, Davis OK, Rosenwaks Z. Obesity and *in vitro* fertilization: negative influences on outcome. J Reprod Med 2004;49:973–7.

[147] Lintsen AM, Pasker-de Jong PC, de Boer EJ, et al. Effects of subfertility cause; smoking and body weight on the success rate of IVF. Hum Reprod 2005;20:1867–75.

[148] Reverchon M, Maillard V, Froment P, Ramé C, Dupont J. Adiponectin and resistin: a role in the reproductive functions? Med Sci (Paris) 2013;29:417–24.

[149] Fedorcsak P, Storeng R, Dale PO, Tanbo T, Abyholm T. Obesity is a risk factor for early pregnancy loss after IVF or ICSI. Acta Obstet Gynecol Scand 2000;79:43–8.

[150] Fedorcsak P, Dale PO, Storeng R, Ertzeid G, Bjercke S, Oldereid N, et al. Impact of overweight and underweight on assisted reproduction treatment. Hum Reprod 2004;19:2523–8.

[151] Gupta S, Sekhon L, Kim Y, Agarwal A. The role of oxidative stress and antioxidants in assisted reproduction. Curr Womens Health Rev 2010;6:227–38.

[152] Kallen B, Finnstrom O, Nygren KG, Olausson PO. In vitro fertilization in Sweden: child morbidity including cancer risk. Fertil Steril 2005;84:605e10.

[153] Shevell T, Malone FD, Vidaver J, Porter TF, Luthy DA, Comstock CH, et al. Assisted reproductive technology and pregnancy outcome. Obstet Gynecol 2005;106. 1039e45.

[154] Allen VM, Wilson RD, Cheung A. Pregnancy outcomes after assisted reproductive technology. J Obstet Gynaecol Can 2006;28:220–50.

[155] Buckett WM, Chian RC, Holzer H, Dean N, Usher R, Tan SL. Obstetric outcomes and congenital abnormalities after in vitro maturation, in vitro fertilization, and intracytoplasmic sperm injection. Obstet Gynecol 2007;110:885–91.

[156] Raunig JM, Yamauchi Y, Ward MA, Collier AC. Placental inflammation and oxidative stress in the mouse model of assisted reproduction. Placenta 2011;32:852–8.

[157] Raunig JM, Yamauchi Y, Ward MA, Collier AC. Assisted reproduction technologies alter steroid delivery to the mouse fetus during pregnancy. J Steroid Biochem Mol Biol 2011;126: 26–34.

[158] Bellver J, Melo MA, Bosch E, Serra V, Remohí J, Pellicer A. Obesity and poor reproductive outcome: the potential role of the endometrium. Fertil Steril 2007;88:446–51.

[159] Bellver J, Ayllón Y, Ferrando M, Melo M, Goyri E, Pellicer A, et al. Female obesity impairs in vitro fertilization outcome without affecting embryo quality. Fertil Steril 2010;93: 447–54.

[160] Gonzalez-Añover P, Encinas T, Sanz E, Letelier CA, Torres-Rovira L, de Mercado E, et al. Preovulatory follicle dynamics and ovulatory efficiency in sows with thrifty genotype and leptin resistance due to leptin receptor gene polymorphisms (Iberian pig). Gen Comp Endocrinol 2011;170:200–6.

[161] Gonzalez-Añover P, Encinas T, Torres-Rovira L, Sanz E, Pallares P, Ros JM, et al. Patterns of corpora lutea growth and progesterone secretion in sows with thrifty genotype and leptin resistance due to leptin receptor gene polymorphisms (Iberian pig). Reprod Domest Anim 2011;46:1011–6.

[162] Gonzalez-Añover P, Encinas T, Torres-Rovira L, Pallares P, Muñoz-Frutos J, Gomez-Izquierdo E, et al. Ovulation rate, embryo mortality and intrauterine growth retardation in obese swine with gene polymorphisms for leptin and melanocortin receptors. Theriogenology 2011;75:34–41.

[163] Hamilton-Fairley D, Kiddy D, Watson H, Paterson C, Franks S. Association of moderate obesity with a poor pregnancy outcome in women with polycystic ovary syndrome treated with low dose gonadotrophin. Br J Obstet Gynaecol 1992;99:128–31.

[164] Galtier-Dereure F, Mantepeyreoux F, Boulat P, Bringer J, Jaffol C. Weight excess before pregnancy: complications and costs. Int J Obes Rel Metab Dis 1995;19:4432–8.

[165] Wang JX, Davies MJ, Norman RJ. Obesity increases the risk of spontaneous abortion during infertility treatment. Obes Res 2002;10:551–4.

[166] Lashen H, Fear K, Sturdee DW. Obesity is associated with increased risk of first trimester and recurrent miscarriage: matched case-control study. Hum Reprod 2004;19:1644–6.

[167] Herrera E, Ortega-Senovilla H. Disturbances in lipid metabolism in diabetic pregnancy – are these the cause of the problem? Best Pract Res Clin Endocrinol Metab 2010;24:515–25.

[168] Nohr EA, Bech BH, Davies MJ, Frydenberg M, Henriksen TB, Olsen J. Pre-pregnancy obesity and fetal death: a study within the Danish National Birth Cohort. Obstet Gynecol 2005;106:250–9.

[169] Nohr EA, Bech BH, Vaeth M, Rasmussen KM, Henriksen TB, Olsen J. Obesity, gestational weight gain and preterm birth: a study within the Danish National Birth Cohort. Paediatr Perinat Epidemiol 2007;21:5–14.

[170] Nohr EA, Vaeth M, Bech BH, Henriksen TB, Cnattingius S, Olsen J. Maternal obesity and neonatal mortality according to subtypes of preterm birth. Obstet Gynecol 2007;110:1083–90.

[171] Sandovici I, Hoelle K, Angiolini E, Constancia M. Placental adaptations to the maternal–fetal environment: implications for fetal growth and developmental programming. Reprod Biomed Online 2012;25:68–89.

[172] Desoye G, Shafrir E. Placental metabolism and its regulation in health and diabetes. Mol Aspects Med 1994;15:505–682.

[173] Bird IM, Zhang LB, Magness RR. Possible mechanisms underlying pregnancy-induced changes in uterine artery endothelial function. Am J Physiol 2003;284:245–58.

[174] Marliss EB, Chevalier S, Gougeon R, et al. Elevations of plasma methylarginines in obesity and ageing are related to insulin sensitivity and rates of protein turnover. Diabetologia 2006;49:351–9.

[175] Li HP, Chen X, Li MQ. Gestational diabetes induces chronic hypoxia stress and excessive inflammatory response in murine placenta. Int J Clin Exp Pathol 2013;6:650–9.

[176] Hefler LA, Reyes CA, O'Brien WE, Gregg AR. Perinatal development of endothelial nitric oxide synthase-deficient mice. Biol Reprod 2001;64:666–73.

[177] Van der Heijden OW, Essers YP, Fazzi G, Peeters LL, De Mey JG, van Eys GJ. Uterine artery remodeling and reproductive performance are impaired in endothelial nitric oxide synthase-deficient mice. Biol Reprod 2005;72:1161–8.

[178] Pallares P, Gonzalez-Bulnes A. Intrauterine growth retardation in endothelial nitric oxide synthase-deficient mice is established from early pregnancy stages. Biol Reprod 2008;78:1002–6.

[179] Pallares P, Garcia-Fernandez RA, Criado LM, Letelier CA, Esteban D, Fernandez-Toro JM, et al. Disruption of the endothelial nitric oxide synthase gene affects ovulation, fertilization and early embryo survival in a knockout mouse model. Reproduction 2008;136:573–9.

[180] Longo M, Jain V, Langenveld J, Vedernikov YP, Garfield RE, Hankins GD, et al. Enhanced growth and improved vascular function in offspring from successive pregnancies in endothelial nitric oxide synthase knockout mice. Am J Obstet Gynecol 2004;191:1470–6.

[181] Kusinski LC, Stanley JL, Dilworth MR, Hirt CJ, Andersson IJ, Renshall LJ, et al. eNOS knockout mouse as a model of fetal growth restriction with an impaired uterine artery function and placental transport phenotype. Am J Physiol Regul Integr Comp Physiol 2012;303:R86–93.

[182] Foxcroft GR, Dixon WT, Novak S, Putman CT, Town SC, Vinsky MD. The biological basis for prenatal programming of postnatal performance in pigs. J Anim Sci 2006;84:E105–12.

[183] Ashworth CJ, Finch AM, Page KR, Nwagwu MO, McArdle HJ. Causes and consequences of fetal growth retardation in pigs. Reprod Suppl 2001;58:233–46.

[184] Wu G, Bazer FW, Wallace JM, Spencer TE. Board-invited review: intrauterine growth retardation: implications for the animal sciences. J Anim Sci 2006;84:2316–37.

[185] Metges CC, Lang IS, Hennig U, Brüssow KP, Kanitz E, Tuchscherer M, et al. Intrauterine growth retarded progeny of pregnant sows fed high protein:low carbohydrate diet is related to metabolic energy deficit. PLoS One 2012;7:e31390.

[186] Gonzalez-Bulnes A, Ovilo C, Lopez-Bote CJ, Astiz S, Ayuso M, Perez-Solana ML, et al. Gender-specific early postnatal catch-up growth after intrauterine growth retardation by food restriction in swine with obesity/leptin resistance. Reproduction 2012;144:269–78.

[187] Barbero A, Astiz S, Lopez-Bote C, Perez-Solana ML, Ayuso M, Garcia-Real I, et al. Maternal malnutrition and offspring sex determine juvenile obesity and metabolic disorders in a swine model of leptin resistance. PLoS One 2013;8:e78424.

[188] Gonzalez-Bulnes A, Astiz S, Sanchez-Sanchez R, Perez-Solana ML, Gomez-Fidalgo E. Maternal diet-induced obesity in swine with leptin resistance modifies puberty and pregnancy outputs of the adult offspring. J Dev Orig Health Dis 2013;4:290–5.

[189] Barbero A, Astiz S, Ovilo C, Lopez-Bote CJ, Perez-Solana ML, Ayuso M, et al. Prenatal programming of obesity in a swine model of leptin resistance: modulatory effects of controlled postnatal nutrition and exercise. J Dev Orig Health Dis 2014;5(3):248–58.

[190] Torres-Rovira L, Tarrade A, Astiz S, Mourier E, Perez-Solana M, de la Cruz P, et al. Sex and breed-dependent organ development and metabolic responses in fetuses from lean and obese/leptin resistant swine. PLoS One 2013;8:e66728.

[191] Denison FC, Roberts KA, Barr SM, Norman JE. Obesity, pregnancy, inflammation, and vascular function. Reproduction 2010;140:373–85.

[192] Challier JC, Basu S, Bintein T, Minium J, Hotmire K, Catalano PM, et al. Obesity in pregnancy stimulates macrophage accumulation and inflammation in the placenta. Placenta 2008;29:274–81.

[193] Zhu MJ, Dua M, Nathanielsz PW, Ford SP. Maternal obesity up-regulates inflammatory signaling pathways and enhances cytokine expression in the mid-gestation sheep placenta. Placenta 2010;31:387–91.

[194] Breton C. The hypothalamus-adipose axis is a key target of developmental programming by maternal nutritional manipulation. J Endocrinol 2013;216:R19–31.

[195] Gonzalez-Bulnes A, Ovilo C. Genetic basis, nutritional challenges and adaptive responses in the prenatal origin of obesity and type-2 diabetes. Curr Diabetes Rev 2012;8:144–54.

[196] Aiken CE, Ozanne SE. Sex differences in developmental programming models. Reproduction 2013;145:R1–R13.

NUTRITION AND REPRODUCTION

18

Green Leafy Vegetables: A Health Promoting Source

Muhammad Atif Randhawa, BSc, MSc, PhD,
Ammar Ahmad Khan, BSc, MSc, Muhammad Sameem Javed, BSc, MSc,
Muhammad Wasim Sajid, BSc, MSc

University of Agriculture, Faisalabad, Pakistan

INTRODUCTION

Man has had a tremendous pressure on green leafy vegetables (GLVs) since the start of civilization and these vegetables can survive in adverse environmental conditions such as droughts. Traditional vegetables are valuable sources of nutrition in rural areas where exotic species (sp.) are not available. They provide good nutrition at low cost, in contrast to the costly exotic sp. Different parts of the world have a valuable heritage of various indigenous leafy vegetables and some of them are also used as medicinal plants traditionally (Table 18.1). Forest dwellers are still using wild edible plants for sustainable edible values and also for culinary purposes. Therefore, they are enjoying a healthier life [1].

Nutrition is a basic human need and a prerequisite for a healthy life. It is a discipline that gives information about the different aspects of food composition and presents many approaches to get the appropriate nourishment from food. The agricultural sector and public organizations extensively use nutritional information to encourage fresh food consumption by the masses. A proper diet is very much essential for all the stages of human life from the fetus to old age. Nowadays, people are becoming well aware of the nutritional benefits of fresh vegetables and fruits, and thereby searching for more diversity in their food to get the utmost nutrition. The main focus of consumers is on those foods that are rich in vitamins (A, C, and E), antioxidants, and minerals like potassium (K), calcium (Ca), and magnesium (Mg). It is believed that GLVs and fresh fruits are gaining a significant place in the food pyramid, being a good source of trace elements and other bioactive compounds [2]. A diverse meal, comprising five vegetable servings a day, is essential for attaining and retaining a decent health. It is that part of food that mainly provides the noncaloric portion and gives health benefits. Vitamins, fiber, antioxidants, minerals, and other bioactive compounds play a major role in the nutritive significance of vegetables.

Vegetables are the fresh and edible parts of herbaceous plants. They are an essential foodstuff and extremely valuable for health maintenance and disease prevention. They comprise valuable foodstuff that can be effectively used for the development and repair of different parts of the body. Vegetables are a very valuable material for maintaining the alkaline reserve of the human body. Vegetables are mainly appreciated due to their high vitamins, carbohydrates, and especially their mineral content. Vegetables of different kinds, including leaves, edible roots, fruits, stems, or seeds, are available and every group contributes in its own way to the diet.

Vitamins known as organic compounds are present in naturally available foods particularly in vegetables or present as "precursors." Vitamins are required for the maintenance and proper functioning of vision, mucous membranes, bones, hairs, teeth, and the skin. They support the body for proper absorption of phosphorous and calcium required for the growth and maintenance of bones. Vitamins play a very important role for the normal working of endocrine glands, the nervous system, and clotting of blood. They are also required for macromolecule metabolism. Half of the recommended daily allowance (RDA) for some vitamins (C, E, B9) and β-carotene, Ca, and Fe can be achieved by 60 g of spinach, 50 g of lettuce, and 100 g of broccoli. As a whole, vegetables are considered as a natural reserve of nutrients

Handbook of Fertility. http://dx.doi.org/10.1016/B978-0-12-800872-0.00018-4

TABLE 18.1   Popular Green Leafy Vegetables Grown in Different Continents

	Common name	Botanical name
Americas	Turnip greens	*Brassica rapa*
	Swiss chard	*Beta vulgaris*
	Spinach	*Spinacia oleracea*
	Mustard greens	*Brassica juncea*
	Broccoli	*Brassica oleracea*
	Romaine lettuce	*L. sativa*
Australia	Broccoli	*B. oleracea*
	Asparagus	*Asparagus densiflorus*
	Bok Choy	*Brassica chinensis*
	Spinach	*S. oleracea*
	Silver beet	*Be. vulgaris*
	Brussels sprouts	*B. oleracea*
	Cauliflower	*B. oleracea*
	snow peas	*B. chinensis*
	Lettuce	*L. sativa*
Africa	Taro	*Colocasia esculenta*
	Eru	*G. africanum*
	Cow pea	*Vigna unguiculata*
	Horseradish	*Moringa oleifera*
	Groundcherry	*Physalis viscose*
	Gherkin	*Cucumis angora*
	Hell's curse	*Amaranthus cruentus*
	African cabbage	*Cleome gynandra*
	Chinese cabbage	*B. rapa*
	Nightshade	*Solanum nigrum*
	Jew's mallow plant	*Cor. olitorius*
	bitter water melon	*Citrullus lanatus*
Asia	Spinach	*S. oleracea*
	Fenugreek Leaves	*Trigonella foenum*
	Water spinach	*Aranthus tricolor*
	Amaranth	*Amaranthus spinosus*
	Matar saag	*T. foenum-graecum*
	Mustard greens	*B. juncea*
	Sorel leaves	*Hibiscus cannabinus*
	Colocacia or taro leaves	*Col. esculenta*
	Cabbages	*B. oleracea*
	Bok choy	*B. chinensis*
	Choy sum	*B. rapa*
	Tatsoi	*B. narinosa*

TABLE 18.1 Popular Green Leafy Vegetables Grown in Different Continents (*cont.*)

	Common name	Botanical name
Asia	Mizuna	*B. japonica*
	Kang kong	*Ipomoea aquatica*
Europe	Italian rucola	(*Eruca sativa*)
	Lettuce	*L. sativa*
	Spinach	*S. oleracea*
	Chicory	*Cichorus intybus*
	Arugula	*E. sativa*
	Mache	*Valerianella locusta*
	Glasswort	*Salicornia europaea*
	Common purslane	*P. oleracea*
	Pepper leaf	*Capsicum annumm*

gifted by God to human beings, such as carrots, a good vitamin A source, which is required for normal vision function. Similarly, spinach and other leafy vegetables have sufficient amounts of vitamin C to prevent and cure scurvy. Some vegetables have a high dietary fiber amount, such as spinach, cabbage, and lettuce, to prevent constipation.

THERAPEUTIC VALUE AND HEALTH BENEFITS OF GREEN LEAFY VEGETABLES

GLVs play a very significant role in our nutrition and diet and have also been used as medicine since ancient times; they are also important sources of protective foods [3]. They are the best instant accessible sources of essential amino acids, minerals, vitamins, and fiber [4,5]. Their bioactive elements have a variety of biological usage, such as antimicrobial and antioxidant activities [6–9], and very supportive for managing age-related ailments and oxidative strains [10]. Vegetables contain huge amounts of ascorbic acid, folic acids, carotenes, riboflavin, and minerals like iron, phosphorous, and calcium [11]. Due to their photosynthetic material, higher amounts of vitamin K are present in GLVs as compared to other vegetables and fruits due to the direct involvement of vitamin K in the photosynthetic process. Usually, less toxic effects are involved when vegetables are used as medicine [12,13]. Vegetables also have the capability to produce many other secondary metabolites of comparatively complex structure having the ability to work as antimicrobial agents [14–16]. GLVs are also rich in compounds having antidiabetic [17], antihistaminic [18], anticarcinogenic [19], and hypolipidemic [20] features, and having the ability to work as curative and preventive agents against aging, insomnia, obesity, hypertension,

and cardiovascular diseases (CVDs) [21–23]. GLVs are a rich source of phytochemicals and have an enormous amount of antioxidants (Table 18.2) [24].

In all plant cells, secondary metabolites of bioactive compounds are present, and their concentration varies based on the different parts of plants, climates, particular growth phase, and seasons. The accumulation of these compounds is very high in leaves and is very beneficial too [25,26]. In test samples, the majority of secondary metabolites identified (steroids, flavonoids, tannins, alkaloids, and saponins) are phyto-protectants and are very necessary for body building and cell growth and repair [27]. The role of vegetables as medicine is due to the availability of chemical compounds that have the ability to produce a physiological action in the human body with the help of antibacterial, antiviral, antioxidant, antiinflammatory, and detoxification activities, and immune system repair (Table 18.3) [28].

There are many studies that strongly prove that diets rich in vegetables and fruits have a good effect on human health, providing protection against age-related degenerative diseases such as CVD, Alzheimer's disease, several forms of cancers, and cataracts [29–34]. Besides the major constituents of food (carbohydrate, protein, and fat) and micronutrients (minerals, vitamins, and trace elements), there are many compounds in fruits and vegetables that are suggested to be a part of health-promoting effects. These compounds include groups of phytochemicals (Fig. 18.1) like flavonoids, carotenoids and other polyphenols, glucosinolates, isothiocyanates, allylic sulfides, phytosterols, monoterpenes dietary fibers, and phenolic acids [35,36].

Gupta and coworkers [37] reported that several GLVs are rich sources of antioxidant vitamins. Vegetables are a great source of phytochemicals, and some antinutritional content, like saponins and so on, have potential in reducing some diseases in man [38]. Some of these

TABLE 18.2   Phenolics, Antioxidant Activity, Ascorbic Acid, and Carotenoids Contents of Some Green Leafy Vegetables

Vegetable	Total phenolic content	Antioxidant activity	Ascorbic acid content	Carotenoids	References
L. sativa	75.88 ± 0.54 mg GA/g	69.50 ± 1.00 µg AA/g	7.3 mg/100 g	2.22 mg/100 g	Zdravkovic et al. [174]
T. foenum-graecum	158.33 ± 20.41 mg of tannic acid/100 g	1,292.28 ± 92.86 (%) µmol of ascorbic acid/g	101.36 ± 0.00 mg/ 100 g	34.78 ± 0.01 mg/100 g	Gupta and Prakash [175]
Murayya koenigii	387.50 ± 30.62 mg of tannic acid/100 g	2,691.78 ± 91.73 (%) µmol of ascorbic acid/g	29.31 ± 0.00 mg/ 100 g	64.51 ± 0.00 mg/100 g	Gupta and Prakash [175]
Centella asiatica	150.00 ± 0.00 mg of tannic acid/100 g	623.78 ± 34.46 (%) µmol of ascorbic acid/g	15.18 ± 0.80 mg/ 100 g	36.40 ± 0.01 mg/100 g	Gupta and Prakash [175]
M. oleifera	208.5 ± 42.5 mg TAE/100 g(1)	4.12 ± 0.16 µmol TE g^{-1}(1)	162 ± 63 mg/ 100 g (2)	1286 ± 689 µg RAE (2)	(1) Mathiventhan and Sivakanesan [176], (2) Yang et al. [177]
S. oleracea	397.00 ± 1.00 µg GAE/g (1)	21.83 ± 0.76 µ mol trolax /g (1)	14.22 ± 1.04 mg/ 100 g (2)	36.96 ± 1.74 µg/g (2)	(1)Yadav et al. [178], (2) Melo et al. [179]
C. sativum	1.12 mg GAE/100 mL (1)	26.82% (DPPH) (1)	361 mg/ 100 g (2)	51.4 µg/g (2)	(1) Al-Juhaimi and Ghafoor [180], (2) Clyde et al. [181]
P. crispum	1.22 mg GAE/100 mL (1)	30.35% (DPPH) (1)	133 mg/ 100 g (2)	56.7 mg/100 g (3)	(1) Al-Juhaimi and Ghafoor [180], (2) Caunii et al. [182], (3) Yahia et al. [183]

TABLE 18.3   Medicinal Uses of Some Popular Green Leafy Vegetables

Sr no.	Leafy vegetable	Medicinal use	References
1.	*Basella rubra* (spinach)	Fertility enhancement in women	Mensah et al. [26]
2.	*B. oleracea* Acephala (kale)	Help to manage type 2 diabetes and is a terrific addition to any weight-loss plan	Kimiywe et al. [184]
3.	Lettuce	Prevent arthritis, cataracts, and macular degeneration	Cai et al. [185]
4.	*Ba. rubra* L. (sag)	Reduce the risk of heart disease, enhance memory, and improve mood,	Sonkar et al. [186]
5.	*B. oleracea* (Collard Greens)	Prevent bone fractures	Mensah et al. [26]
6.	*C. sativum* L.	Antianxiety activity used for Vitamin deficiency and disorders	Mahendra and Bisht [187]
7.	*Mu. koenigii* L. Spreng (currry leaves)	Antiulcer, antimicrobial, cytotoxic activity, phagocytic activity. Good for bones and eyes in children. Reduce depression and supply calcium to brain	Handral et al. [188]
8.	*T. foenum-graecum* L. (fenugreek)	Antidiabetic, anticarcinogenic, anti-inflammatory, antioxidant	Toppo et al. [189]
9.	*Allium sativum* (garlic green)	Antioxidant, antimicrobial, cholesterol-lowering and blood-thinning	Blumenthal [190]
10.	*B. oleracea* L. var.botrytis L. (broccoli)	Anticancer	Reddy et al. [191]
11.	*Mentha spicata* L. (mint)	Prevent hysteria	Raju et al. [192]

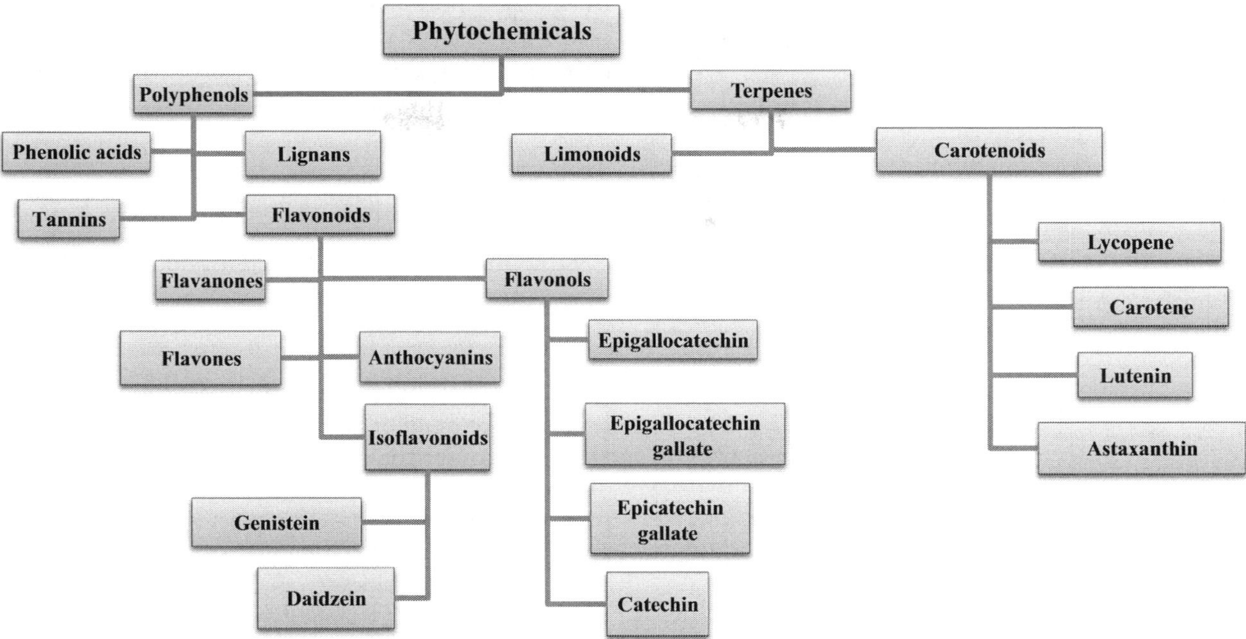

FIGURE 18.1 Classification of dietary phytochemicals.

diseases include high blood pressure, heart disease, stroke, and other CVDs [39]. Despite the health benefits of GLVs, they can also absorb pesticide residues during growing in the field or green house, and so on. Fortunately, the fast growing nature of most of the leafy vegetables dissipates the pesticide residues quickly and residues at harvest are mostly within maximum allowable limits [40,41]. Furthermore, these pesticide residues incurred from the fields can be removed from vegetables by thorough washing with potable tap water [42,43]. The pesticide residues can also be very effectively removed from vegetables by dipping the vegetables in organic acid solutions [44,45].

The bulk in diet is provided by the fiber in vegetables and helps to decrease the starchy food intake, prevent constipation, enhance gastrointestinal function, and decrease the chance of metabolic diseases like hypercholesterolemia and diabetes mellitus. Many vegetables have the ability to act as antibiotics, blood-building agents, and antihypertensive agents. Vegetables also increase fertility in females when used as a soup [26].

Vegetables also contain many non-nutritive compounds besides their major nutrients like carbohydrate, proteins, and fat, which play a significant role against chronic diseases. Those compounds have very low amount of fat contents like all other plants or contain no cholesterol. In vegetable crops, many phytochemicals are present in very low amounts. However, when a sufficient amount of vegetables is consumed, a reasonable amount of phytochemicals is obtained and it

significantly contributes against chronic diseases to protect living cells [46]. Protection against age-related disorders can be achieved by consuming more amounts of vegetables and fruits [47]. Most of the antioxidants, like β-carotene and vitamins C and E, play a positive role against age-related disorders.

Antidiabetic Properties

Diabetes mellitus is a noncommunicable disease and is found in most countries of the world. In 1995, about 135 million people were affected by this disease globally and the number is estimated to rise to about 300 million people by 2025 [53]. There are two main types of diabetes mellitus: type 1 and 2. Type 1 diabetes mellitus is also known as juvenile onset disease or insulin-dependent diabetes. This type is recognized by an autoimmune-mediated destruction of beta cells of the pancreas. The rate of destruction is fast in some individuals and slow in others [54]. Type 2 diabetes is also known as adult-onset diabetes or non-insulin-dependent diabetes. In this type patients have a relative deficiency of insulin. Individuals with type 2 diabetes are frequently resistant to the action of insulin [55,56]. Type 2 diabetes mellitus is considered as the most familiar disease and in 2010 its presence among adults was about 6.4% in the world. The prevalence of type 2 diabetes mellitus will increase up to 7.7% by 2030 [57].

The most common feature of this disorder is known as hyperglycemia. Different drugs are used in controlling diabetes, including alpha-amylase and beta-amylase inhibitors. These drugs are used orally and help to prevent the digestion of complex carbohydrates [58,59]. Different studies were conducted in several areas of India, and results showed that there was an increasing trend in the prevalence of diabetes mellitus disease. The Diabetes Research Centre in Chennai carried out different studies and findings revealed that the prevalence of this disease had steadily enhanced in adults of urban areas from 5.2% in 1984 to 13.9% in 2000 [60].

Reactive oxygen species and oxygen free radicals are commonly produced in the human body. Oxidative damage to all biochemicals in the body is usually accompanied in type 2 diabetes patients [61]. Due to excessive oxidative stress, lipid peroxidation is increased under diabetic conditions. Therefore, prevention from this oxidative damage was considered to play an important role in diabetes or its complications that originate from peroxidation of lipids [62]. Different GLVs, such as cabbage, spinach, kale, lettuce, and broccoli, have different contents of antioxidants, vitamins, carbohydrates, phytochemicals, and minerals. These compounds play a vital role to protect against CVDs and diabetes. Different biological mechanisms are responsible for controlling the risk of diabetes.

GLVs contain high amounts of magnesium, potassium, antioxidant vitamins, phytochemicals, and plant proteins, which are also responsible for lowering the risk of type 2 diabetes in addition to the high fiber content and low glycemic loads and low energy intake. Among overweight women, a high intake of green and dark yellow leafy vegetables may be useful for preventing type 2 diabetes disease [63]. GLVs are a cheap and rich source of phytochemicals and micronutrients with antioxidant properties. Lettuce leaves, reddish leaves, and omum leaves particularly have high values of these important compounds [64].

Phenolic compounds possess therapeutic properties and high antioxidant activity including antihypertensive and antidiabetic activity [65]. Polyphenolic antioxidants are mostly present in fruits and GLVs. These antioxidant vitamins including alpha-tocopherol, β-carotene, phenolics, and ascorbic acids are most commonly present in vegetables [66–68]. These compounds are present in fresh GLVs like wild basil, fluted pumpkin, waterleaf, bitter leaf, jute, and chaya. Among these vegetables jute and wild basil have the highest amount of phenolic content while waterleaf has the lowest amount of this compound. Ferric-reducing properties are also present in these GLVs. Ferric ions accumulate in the islets of Langerhans and in acinar cells, which result in the destruction of beta-cells associated with diabetes mellitus [69]. Lipid peroxidation and alpha-amylase activity is also reduced by the compounds present in GLVs. Inhibition of enzymes associated with hypertension and diabetes

by the free polyphenol-rich extract of leafy vegetables have been observed. So these vegetables are a safer and cheaper alternative of synthetic drugs that can be used for the prevention of hypertension and diabetes mellitus disease [70].

Antimicrobial and Anti-Inflammatory Activity

GLVs are widely consumed around the world in various forms such as fresh and processed. For centuries these vegetables have been consumed by humans and are declared as generally recognized as safe (GRAS). GLVs are well-known for their esthetic features that involve flavor, color, and so on, and therapeutic values [71]. Moreover, their bioactive substances and phytonutrients are putative to perform several biological processes, which include antimicrobial and antioxidant activities [72]. Leafy vegetables are being used for their antimicrobial activities in food processing industries. Their use as antibacterial agents is becoming popular as they have GRAS status. There are several leafy vegetables that are believed to perform certain pharmacological effects in the body. For example, a few of them have a very strong tendency to act as antibiotics and some are beneficial for inducing low blood pressure, while a few of them have elucidated their impacts in enhancing fertility in females [26].

Leafy vegetables, namely, *Coriandrum sativum*, *Lactuca sativa*, *Mentha piperita*, *Portulaca oleracea*, and *Raphanus sativus* have high antibacterial activity in alcoholic extract [73]. Antibacterial activities of freeze-dried and irradiated parsley (*Petroselinum crispum*) and cilantro (*C. sativum*) leaves were determined on methanolic and aqueous extracts against *Bacillus subtilis* and *Escherichia coli*. Prooxidant activity of the GLV extracts was significant ($p < 0.05$) on the bacterial cell damage and growth inhibition of the studied bacterial species. Leafy vegetables can be a good source for antimicrobial activity in the future. However, there is a need to explore the antimicrobial potential of leafy vegetables against different pathogens in various food systems.

Since the last few decades, it has been elucidated that the biological effects of phenolics and flavonoids in green vegetables are attracting increasing demands for research. They act like plant hormone regulators, an integral part of the natural defense system, which provides protection against certain microbial agents. In addition, phenolic compounds have been widely studied to explore their health-promoting benefits, which mainly involve anti-inflammatory response. Spinach and the mustard plant have good anti-inflammatory potential [74]. *Euphorbia hirta* and *Ipomoea involucrata* contain saponins, which have anti-inflammatory, antifungal/antiyeast, antibacterial, antiparasitic, cytotoxicity, and antitumor, antiviral, and some other biological functions

[75]. According to Guil-Guerrero and coworkers [76], the leaf extracts of *Urtica dioica* could induce an anti-inflammatory response as it dissipates the activation of NF-6B, which is involved in many inflammatory processes.

Antioxidant Activity

Most of the cellular organelles are sensitive to oxidative damage. All substances, like lipids, proteins, nucleic acids, and carbohydrates, can undergo oxidative modifications. The lipid peroxidation process is initiated when the hydrogen atom is removed from the unsaturated fatty acyl chain and initiates a chain reaction. This is why lipid peroxidation inhibition is very essential in diseases that involve free radicals [77]. The antioxidants have the tendency to play a crucial role in the protection against oxidative damage disorders. The term "antioxidant" refers to compounds that can prolong or diminish the oxidation of lipids by inhibiting the promotion of oxidative chain reaction [78].

It has been suggested in a number of studies that natural antioxidants from plant materials can replace synthetic ones. It is assumed that natural antioxidants are safe to use and they are widely distributed in all parts of the plants such as the stem, leaves, flowers, and seeds. Certain typical compounds that could involve antioxidant activity are vitamins, carotenoids, and phenolic compounds. Therefore, the regular daily intake of vegetables and fruits that are naturally enriched with these nutrients should be recommended in the diet to avoid several pathological complications and to improve the overall health standard [79–82]. Vitamin C has been designated as a powerful biological antioxidant and it was explored that it can act as a chain-breaking scavenger for peroxy radicals as well as act in synergy with vitamin E. Vitamin E can also act as a chain-breaking oxidant and is very important in quenching the singlet oxygen.

Metabolic carcinogen stimulation is free radical dependent and DNA damage by mediated free radicals plays a key role in carcinogenesis [83,84]. Regular consumption of dietary oxidants, which are found in leafy vegetables and fruits, could be very helpful in reducing cancer-related oxidative damage. Some studies have revealed that there are some phytochemicals present in common vegetables and fruits and they can perform many functions, which include stimulation of the immune system, scavenging of oxidative agents, gene expression regulation in cell proliferation, and apoptosis and antimicrobial effects [85].

GLVs are a good source of antioxidants because they are a rich source of ascorbic acids, carotenoids, and phenolic contents and these all contribute to the antioxidant activity of leafy vegetables. Mustard greens, coriander, and spinach are abundant in vitamin E, C, and A. These nutrients team up to feed on the free radicals, which usually damage the cell membranes. Vitamin E and β-carotene put in protecting actions against harmful free radicals within the lipid-soluble parts of the body, while vitamin C balances it out within the water-soluble parts of the body. Together they provide wonderful benefits to people struggling with asthma, heart diseases, and menopausal symptoms.

The amount of polyphenol content in cabbage and spinach varies from 5 mg/100 g to 69.5 mg/100 g, respectively. The order of leafy vegetables that terminates the free radical chain reactions is spinach < cabbage < coriander leaves < hongone leaves. Leafy vegetables also have a capability to provide a protective effect on peroxide formation while storing heated oils. Furthermore, leafy vegetables are good antioxidant agents that remain stable at high temperatures and could act as substitutes for synthetic antioxidants [86].

Cardiovascular Disease and Leafy Vegetables

In the United States, adolescent and childhood obesity has reached an endemic extent, and the rate of this condition continues to increase in an immense manner. In South Africa and other African countries, obesity among adolescents and children is increasing due to Western influence and vegetable consumption and decline in physical activity with age [87]. The commonness of chronic diseases, such as obesity, can be decreased with regular use of leafy vegetables in the sense of weight loss and it is confirmed through different research studies [88]. Fiber intake from GLVs and the chances of the development of coronary heart disease (CHD)show an inverse relationship, so CHD risk decreases with the consumption of fiber from these vegetables. Research also indicates that there is an inverse relationship between the occurrence of stroke and vegetable consumption and this concept supports the consumption of vegetables and fruits for preventing CVDs [89]. Daucher and coworkers [90] conducted meta-analysis studies and revealed that by the intake of an additional portion of fruits and vegetables per day the risk of developing CVD reduces by about 4%.

Carotenoids

A well-known class of pigments known as carotenoids is responsible for imparting orange, red, and yellow color to plants. β-Carotene, lycopene, zeaxanthin, and lutein are some examples of this class and these compounds are a very important part of the human diet. For best absorption of carotenoids in the body, cooking with little fat makes these compounds paramount for consumption, after they have been pureed or chopped. Vitamin A is best obtained from carotenoids. Research studies have been done for determining the role of carotenoids in cancer, eye disease, and heart disease. In some fruits and tomatoes, lycopene imparts the

TABLE 18.4 Nutritive Constituents of Green Leafy Vegetables that have a Positive Impact on Human Health and their Sources

	Constituent	Source	Established or proposed effects on human wellness
1.	Vitamin C (ascorbic acid)	Broccoli, cabbage, spinach, leafy greens, chard, turnip greens	Protect scurvy, aids wound healing, healthy immune-system, CVDs
2.	Vitamin A (carotenoids)	Dark-green vegetables such as collards, broccoli, arugula, spinach, turnip greens	Night blindness prevention, chronic fatigue, psoriasis, heart disease, stroke, cataracts
3.	Vitamin K	Lettuce, spinach, lentils, green onions, crucifers (cabbage, broccoli, Brussels sprouts), leafy greens	Synthesis of procoagulant factors, osteoporosis
4.	Vitamin E (tocopherols)	Kale, spinach, salad greens, dark-green leafy vegetables	Heart disease, LDL oxidation, immune system, diabetes, cancer
5.	Fiber	Most fresh vegetables such as Brussels sprouts, cabbage, broccoli, cooked dry beans, and peas	Diabetes, heart disease
6.	Folate (folicin or folic acid)	Dark-green leafy vegetables (such as spinach, mustard greens, butterhead lettuce, broccoli, Brussels sprouts,), asparagus	Birth defects, cancer, heart disease, nervous system
7.	Calcium	Vegetables (such as beans greens, okra, and tomatoes) peas, cauliflower	Osteoporosis, muscular/skeletal, teeth, blood pressure
8.	Magnesium	Spinach, collard, cabbage	Osteoporosis, nervous system, teeth, immune system

red color. Lycopene is most commonly obtained from processed and cooked tomato products. This compound also plays a vital role in preventing heart disease and cancer. Lutein and zeaxanthin in the human diet are obtained from dark green and leafy vegetables. These two compounds are also very important for preventing age-related macular degeneration, oxidative damage to eyes, and CVD (Table 18.4).

Flavonoids

Flavonoids are produced by plants, belonging to the family of phytochemical compounds. The risk of CVDs decreases with the high intake of these compounds [91]. Different clinical studies have been conducted and results obtained from these studies highlighted that the risk factor of CVD, such as dyslipidemia, hypertension, and diabetes, can be reduced by the micro- and macronutrients present in fruits and vegetables [92].

According to different population studies, a healthy life style and ample consumption of vegetables and fruits are correlated well to lowering the incidence of CVD. It is generally assumed that people who consume vegetables and fruits exercise more, smoke less, and are usually better educated than those people not consuming fruits and vegetables [93].

A healthy diet pattern that includes high fruits and vegetables intake is inversely associated with consumption of fatty foods and also lowers the risk of CVD [94]. Ness and Powles [95] also discovered evidence about fruits and vegetables intake and the risk of CVD and they showed that there is a considerable inverse relationship between vegetable consumption and incidence of CVD. A similar relationship was also reported by

Alonso et al. [96]. wherein the consumption of vegetables and fruits decreases blood pressure. He et al. [89] reported that the risk of stroke development decreases with the higher consumption of fruits and vegetables. Utilization of individual fruits and vegetables by 600 g/day can decrease the threat of CVD by 31% and the risk of stroke by 19%.

Hypertension and Leafy Vegetables

The main cause of CVD is hypertension, and it affects about 1 billion individuals in the world [97]. About 72 million people in the Americas (one in three adults) are expected to have hypertension and only about 34% will maintain a normal blood pressure (BP) [98–100]. The risk of prehypertension is developing in nearly 70 million adults, and blood pressure is expected to be between 120/80 mmHg and 140/90 mmHg. The possibility of hypertension development, especially systolic elevation, will happen in about 90% of adults in the United States by the age of 65 [99]. There is also a relationship between the high risk of mortality and morbidity from CHD, stroke cerebrovascular accident (CVA), congestive heart failure, end-stage renal disease, and myocardial infarction. Low blood pressure control is more dangerous for patients with chronic kidney disease and diabetes [101]. The most common reason for patient visits to clinics is hypertension and it is the main cause of antihypertensive medicine use with the cost reaching about $20 billion per annum. According to different epidemiological studies diet has been recognized as playing a pivotal role in blood pressure [102–105]. Blood pressure is reduced by dietary therapies like increased potassium and magnesium intake and reduced sodium intake. Various antioxidants, designer foods, and nutritional supplements present in fruit and vegetable diets are known for reducing BP [106–110]. Potassium intake is high in industrialized countries and this intake is inversely related to CVD and BP [111,112]. Dietary potassium is associated with lower risk of stroke and hypertension, while dietary sodium is related with elevation of blood pressure. The incidence of CVA mortality in young women is inversely related to the increased dietary intake of potassium of about 60–80 mmol/day [113].

A study was conducted on 43,738 US men aged 40–75 years, which revealed a considerable inverse relationship between the risk of CVA and the intake of potassium [114]. Independent of the effect of potassium intake on BP, it has a direct effect on preventing the chance of CVA and different mechanisms used for this [113,114]. The risk of hypertension is only 1% of the population in an isolated society, consuming a diet high in fruits and vegetables. While this risk is about one third in populations of industrialized countries consuming a diet high in trans fats, processed foods, refined carbohydrates, low fiber, saturated fats, large amounts of dietary sodium, and reduced dietary magnesium and potassium [115]. It is demonstrated through epidemiological studies that the populations consuming a primeval diet have lower BP and this reduction in BP decreases CVD risk among vegetarians in industrialized countries [102,103,116].

The homeostasis of potassium and sodium plays a vital role in endothelium-dependent vasodilatation. Synthesis of nitric oxide decreases with sodium retention. Nitric oxide is an arteriolar vasodilator elaborated by epithelial cells and this retention of sodium also increases the plasma level of asymmetric dimethyl L-arginine, which is an endogenous inhibitor of nitric oxide production [117]. Opposite effects are induced by sodium restriction. Endothelium-dependent vasodilatation by hyperpolarizing the endothelial cell through opening of potassium channels and stimulation of sodium pumps occurs due to the diet rich in serum sodium and potassium [118,119]. Decreased cytosolic calcium, which in turn promotes vasodilatation, happen due to endothelial hyperpolarization, which is transmitted to the vascular smooth muscle cells [118]. Moreover, to increase vasodilation, some other mechanism by which BP is influenced by potassium include alteration in intracellular sodium and tonicity, reduced vasoconstrictive sensitivity to norepinephrine and angiotensin II, natriuresis, increased sodium/potassium ATPase activity, increased serum and urinary kallikrein, proliferation in vascular smooth muscle, alteration in DNA synthesis, improved insulin sensitivity, sympathetic nervous system cells, reduction in cardiac diastolic dysfunction, reduction in transforming growth factor (TGF)-beta, decrease in vascular neointimal formation, and decrease in NADPH oxidase, inflammation, and oxidative stress [120–126].

Dietary Fiber

Dietary fibers (DFs) are present naturally in fruits, vegetables, nuts, and cereals, and the composition and amount of these fibers vary from one food to another [127]. Different foods without starch give about 20–35 g fiber/100 g of their dry mass while food that contain starchy material provides 10 g/100 g of their dry weight. The content of dietary fiber in vegetables and fruits is 1.5–2.5 g of their dry mass [128]. Cereals are one of the most important sources of dietary fiber and in Western countries they contribute about 50% of the fiber intake [129]. About 3% from minor sources, 16% DF may come from fruits and 30–40% from vegetables [130,131].

DF content in vegetables in the range of 1.5 to 2.5 g/100g of dry matter, and the amount of DF is much higher in some products like red and white beans. Similarly, fruits also contain a large amount of DF [132]. Curry leaves have soluble fiber (4.4%) and insoluble fiber (55.6%), on the basis of fresh weight of leaves [133]. *Chorus olitorius* (jute), containing crude fiber (0.33%), is a local plant from Australia, South America, some parts of Europe, and tropical Asia and Africa, and also commonly used in medicines to treat different diseases like tumors, cystitis, fever, and cold [134]. High levels of vitamin C and protein are present in young shoot tips of this vegetable and it is eaten raw or cooked [135] and also in soup preparation. Another leafy vegetable known as waterleaf (*Talinum triangulare*) grown in different parts of Nigeria. The leaves of *Telifaria accidentlis*, along with edible shoots, have crude protein, moisture, carbohydrate, iron, oil, and ash. The crude fiber contents of this vegetable are 0.32%. *Amaranthus hybridusis* is known to people of southeastern Nigeria as "inene" and it belongs to the family Amaranthacae. *A. hybridusis* is widely grown in West Africa, Indonesia, and Malaysia [136]. *A. hybridusis* is an annual plant, it is spineless and up to 80 m high with grooves. The leaves are green and variable in shape and size. Its fiber content is 19.60%. *Gnetum africana* belongs to the family of plants called Gnetaceae. In Nigeria it is known as "Ukazi" or "Afang." In Southern Nigeria *G. africana* is used for the preparation of soup and salad. The salad is often eaten as a mild laxative and its high fiber content (27.25%) helps to prevent constipation [137]. As a potent medicinal plant it is used in the treatment of piles and high blood pressure [138]. Bako and workers [137] reported that the leaves cure stomach disorder and prevents constant urination in nursing mothers.

Raw vegetables, when compared with those eaten cooked, have lower total dietary fiber in general. The value of soluble dietary fiber ranges from 0.10 to 0.77% and insoluble dietary fiber value ranges from 0.88 to 3.06% for raw vegetables. TDF varies between 3.50% for broccoli and 0.98% for iceberg lettuce [139].

EFFECT OF GREEN LEAFY VEGETABLES ON MOTHER AND FETUS HEALTH

During pregnancy women are highly conscious about their diet because a healthy diet is beneficial for the mother and the developing fetus. Foods of plant origin are abundant in micronutrients such as antioxidants, phytochemicals, minerals, vitamins, and dietary fiber. The risk of metabolic syndrome, CVD, and inflammation is reduced by a high intake of vegetable in the diet [140].

Roman and coworkers [141] studied the impacts of consumption of vegetables on pregnancy and anthropometric measures at birth in a general population mother–infant cohort. A diet composed of large portions of vegetables had considerable impact on birth weight and length. Literature has elaborated the fact that vegetable consumption during pregnancy may have a beneficial effect on the growth of fetus. More fetus growth and weight gain was observed in mothers consuming leafy vegetables in comparison with mothers consuming a nonvegetable diet [142]. In another study strong association was found for fruit and vegetable intake of mothers in which infant birth weight increased significantly [143].

Recently, a study was conducted in India about the health of pregnant women and special concentration was paid on underweight women. A healthy pregnancy period and optimum birth weight of the fetus are strongly linked with consumption of such foods having high contents of micronutrients (vitamins A and C, folacin, calcium, and iron). Food rich in energy and protein content has no considerable impact on the birth size of the infant. Vegetable should be at a necessary proportion of the diet of pregnant women as it is beneficial for the proper growth size of the fetus. Women who consumed leafy vegetables, fruits, and milk products 3–4 times per week gave birth to healthy infants. A daily portion of 200 g (2.6 mg of β-carotene) of African eggplant leaves were given to mothers in the intervention groups for three months while those in the control group did not receive any additional vegetables. Mothers receiving leafy vegetables showed a good profile for vitamin A. Dietary modification and nutrition education for women of childbearing age to include natural food sources rich in provitamin A may provide a long-term solution to prevent vitamin A deficiency in developing countries [144]. Organic food consumption with high sources of vegetables and fruits is considered healthy during pregnancy [145].

Fiber has a vital importance in human nutrition. The human gastrointestinal tract cannot digest fiber. It helps in bowel movement by absorbing water. During pregnancy, the daily recommended intake of fiber is 25–35 g. However, in some countries of the world the consumed fiber is less than 14 g/day during pregnancy. Sufficient water intake and hydration is also necessary along with proper intake of fiber. It is an important part of pregnancy diet for mother and fetus health. Increase of fiber content in the diet during pregnancy should be gradual and should start with soluble fiber to avoid bloating and flatulence. Proper fiber intake prevents constipation and hemorrhoids during pregnancy [146].

Antioxidants balance related to fiber consumption has a provital role in the development of the fetus during pregnancy. It has been shown that after consumption of a fiber-rich diet, the antioxidant capacity of superoxide dismutase (SOD) increases significantly but placental MDA (malondialdehyde) level decreases significantly. Moreover, the antioxidant's scavenging potential of free radicals get increased with a fiber-rich diet. All these have an effect on normal fetal development and health. Oxidative stress is known to effect the pathogenesis of hypertension and preeclampsia during pregnancy so it is postulated that the consumption of a high amount of fiber and vegetables rich in antioxidants could be beneficial in ameliorating hypertension. Antioxidants could pass through the placenta and selenium antioxidant has been shown to be effective in the structure of glutathione peroxidase, so it is related to the lower rate of miscarriage and preeclampsia [147–149].

Allergies developing in early childhood are a direct consequence of low fruit and vegetable consumption of the mother during pregnancy. Pregnant women who had a higher consumption of fiber were shown to have babies with low incidence of developing asthma [150]. Eating whole grains during pregnancy would reduce the risk of tube defects, spina bifida, and anencephaly during fetal development.

Folic Acid

Folic acid is the term used for naturally occurring folates referred to as "folate" and also known as pteroylmonoglutamic acid. These naturally occurring folates have pteroylmonoglutamic acids with 2–8 glutamic acid groups attached to the primary structure [151]. Folic acid and closely related compounds are found abundantly in GLVs. Folic acid characterized as vitamin B9 was first isolated from spinach in 1941. It is water-soluble and an essential vitamin member of the vitamin B complex [152]. Folic acid is found in abundance in various GLVS like asparagus, lettuce, broccoli, lentils, and spinach. Folic acid is essential for promoting the transport of amino acids in the protein synthesis process to their specific location [153]. It is also required for the methylation of amino acids, DNA, and RNA [154,155]. Folic acid is requisite in the regeneration of methionine from homocysteine, which in turn is essential to maintain cardiovascular well-being [156]. It also contributes in cell division, regulation, and differentiation as well as helps to regulate mood, sleep, appetite, and the central nervous system [157].

Folic acid plays a vital role in the synthesis of nucleic acids and amino acids and is also required for fetus growth [158]. The risk of having babies with neural tube birth defects, including anencephaly, encephalocele, and spina bifida, can be addressed with sufficient maternal dietary intake of folic acid before conception and during the early stages of pregnancy [159]. Folic acid is very important for cell proliferation and DNA synthesis among the nutrients required during gestation.

During gestation folate requirements are 5- to 10-fold higher for nonpregnant women. For pregnant women the RDA for folate is 600 µg/day for the development of maternal tissues, placenta, and the fetus [160]. These requirements for folate can only be met by adequate maternal dietary intake of folate. Clinical observations [161,162] indicate that the fetus can consume the folate stores to a point of reduction and as a result maternal folate deficiency occurs, which leads to cell death, particularly of high proliferative somatic cells, and megaloblastosis arises. This will result in undesirable consequences to the fetus during gestation. In addition, the time it takes for neural crest cells to migrate to the embryonic neural tube is as short as 5 h; deficiency of folate will result in the most intense effects on these neural crest cells particularly during the critical stages of fetus development, which will result in neurocritopathies and neural tube defects such as spina bifida and anencephaly. In order to avoid birth defects in babies, the current RDA of folate for nonpregnant women is (400 µg/day). An RDA of 4 mg/day is suggested for such women with a history of giving birth to an infant with neural tube defects.

In one study, maternal folate was limited by restricting dietary folate before conception and during pregnancy by using a murine model [163], and results were poor reproductive outcomes with decreased fetal weight, delay in heart and palate development, and increased fetal deaths [164]. Additionally, a number of earlier studies [165–167] showed that a remarkably increased percentage of the population in South India is suffering from folate deficiency with the evidence of megaloblastic anemia in about 60% of pregnant women. According to a national pilot program on the control of micronutrient malnutrition [168], the estimated daily intake of folic acid in women belonging to low economic status [169] in urban and rural areas of different Indian states ranged from 75 to 167 µg, which is much less than the RDA of 400 µg/day. The recommended daily intake is even higher than 400 µg in most countries. Although the risk of neural tube birth defect is lowered by folic acid up to 50–75%, these defects are not solely due to the deficiency in folic acid. Toxicity

due to excessive intake of folate from fortified foods or food supplements is very uncommon [170] because the excess folic acid present in the body is lost in the urine due to its solubility in water. Elevating the concentration of these beneficial components to human health in fruits and GLVs is the current concern of plant scientists through improved practices of plant breeding and genetics. Moreover, it has been revealed that the use of specific commercial cultivars by the growers can significantly increase the concentration of these health-promoting components as much as fivefold just by selection of good quality soil type for growing of fruits and vegetables [171,172].

Fertility and Green Leafy Vegetables

Dark leafy green vegetables are rich in vital minerals, vitamins, antioxidants, fiber, and chlorophyll. They can be eaten raw or cooked; however, eating raw or half cooked is better as cooking depletes the vitamins and antioxidants, especially the ones important for fertility such as zinc, vitamin C, and folate. Greens, such as broccoli, Brussels sprouts, and cabbage, contain diindolylmethane, which helps the body to get rid of excess "bad" estrogen. These types of greens should be eaten at least three times a week. Folic acid is very important for the proper development of the fetus. Eating folic acid while a woman is preparing for conception is better, because the baby will need this vital nutrient before the woman is even able to get a positive pregnancy test. Women who do not get sufficient amounts of iron may suffer anovulation and possibly poor egg health, which can inhibit pregnancy at 60% higher than those with sufficient iron stores in their blood. GlVs (such as spinach and asparagus) are rich sources for increasing iron in the body. Vitamin C is important for the absorption of iron and helps to protect the cells from free-radical damage among its many other important functions in the body [173].

Leafy greens and vegetables supply the body with important minerals and nutrients. The body needs to have a healthy acid/alkaline balance to function at its best. Most diets today are very acidic (full of meat, sugar, and white processed foods) and need to have more fruits and vegetables incorporated in them. Spinach is known as a super food that provides the body with an enormous amount of healthy vitamins and nutrients that are important for fertility and pregnancy. Spinach is rich in iron and folic acid, two nutrients that are important for reproduction and fetal health. Folic acid is important in preventing neural tube defects in the fetus, and iron helps to promote oxygen levels in cells, organs, and the fetus.

Having a healthy alkaline environment is important for allowing the sperm to survive and make it to the egg. A mineral in the vaginal fluids activates the whipping of a sperm's tail so it can begin to travel and make its long journey possible. If this mineral is not present the snapping action may be weak or not activated. Creating a healthy body begins with eating mineral-rich foods like GLVs.

CONCLUSIONS

GLVs are considered as natural caches of nutrients for human beings as they are a rich source of vitamins, such as ascorbic acid, folic acid, tocopherols, β-carotene, and riboflavin, as well as minerals such as iron, calcium, and phosphorous. They also contain an enormous range of bioactive health-promoting compounds such as antioxidants and antimicrobial agents, which provide health benefits beyond basic nutrition. Over the years, efforts have been made to domesticate these GLVs as they are a rich source of these vital and important nutrients. The leafy vegetables contain a high amount of dietary fiber; thus, they help in alleviating obesity and diabetes. Micronutrients are essential for growth, and maternal micronutrient deficiency in developing countries may be an important cause of abortion. Supplementation of a mother's diet with GLVs can cover the deficiency of micronutrients and also affect the health and weight gain of the fetus. During the last few decades, cancer and CVDs have become prevalent due to the change of dietary patterns from fiber-rich food to fast food. However, if we add GLVs, we can decrease the prevalence of such diseases. GLVs and fruits are rich in antioxidants and GRAS and economically feasible. Therefore, the application of such ingredients from plant origin, which possesses antioxidant and antimicrobial properties, may prove to be advantageous in reducing economic losses and adding to the shelf life of food products.

References

[1] Kamble VS, Jadhav VD. Traditional leafy vegetables: a future herbal medicine. Int J Agric Food Sci 2013;3(2):56–8.

[2] Gibson RS, Hotz C. Dietary diversification/modification strategies to enhance micronutrient content and bioavailability of diets in developing countries. Bri J Nutr 2001;85:159–66.

[3] Nnamani CV, Oselebe HO, Agbatutu A. Assessment of nutritional values of three underutilized indigenous leafy vegetables of Ebonyi State Nigeria. Afr J Biotechnol 2009;8(9):2321–4.

[4] Sharma HP, Kumar RA. Health security in ethnic communities through nutraceutical leafy vegetables. J Environ Res Dev 2013;7(4):1423–9.

[5] Adenipenkun CO, Oyetunji OJ. Nutritional values of some tropical vegetables. J Appl Biosci 2010;35:2294–300.

[6] Burt S. Essential oils: their antibacterial properties and potential applications in foods. A review. Int J Food Microbiol 2004;94(3):223–53.

[7] Chanda S, Baravalia Y, Kaneria M, Rakholiya K. Fruit and vegetable peels – strong natural source of antimicrobics. In: Mendez-Vilas A, editor. Current Research, Technology and Education Topics in Applied Microbiology and Microbial Biotechnology. Spain: Formatex Research Center; 2010.

[8] Gutierrez J, Barry-Ryan C, Bourke P. The antimicrobial efficacy of plant essential oil combinations and interactions with food ingredients. Int J Food Microbiol 2008;124(1):91–7.

[9] Kim SJ, Cho AR, Han J. Antioxidant and antimicrobial activities of leafy green vegetable extracts and their applications to meat product preservation. Food Control 2013;29:112–20.

[10] Gacch RN, Kabaliye VN, Dhole NA, Jadhav AD. Antioxidant potential of selected vegetables commonly used in diet in Asian subcontinent. Indian J Nat Prod Resour 2010;1(3):306–13.

[11] Fasuyi AO. Nutritional potentials of some tropical vegetable leaf meals: chemical characterization and functional properties. Afr J Biotechnol 2006;5(1):49–53.

[12] Matasyoh JC, Maiyo ZC, Ngure RM, Chepkorir R. Chemical composition and antimicrobial activity of the essential oil of *Coriandrum sativum*. Food Chem 2009;113:526–9.

[13] Evarando LS, Oliveira LE, Freire LKR, Sousa PC. Inhibitory action of some essential oils and phytochemicals on growth of various moulds isolated from foods. Braz Arch Biol Technol 2005;48:234–41.

[14] Kubo I, Fujita K, Kubo A, Nihei K, Ogura T. Antibacterial activity of coriander volatile compounds against *Salmonella choleraesuis*. J Agric Food Chem 2004;52:3329–32.

[15] Hedges LJ, Lister CE. Nutritional Attributes of Some Exotic and Lesser Known Vegetables. Plant & Food Research Confidential Report No. 2325. New Zealand: New Zealand Institute for Plant & Food Research Limited; 2009.

[16] Dhiman K, Gupta A, Sharma DK, Gill NS, Goyal A. A review on the medicinal important plants of the family of Cucurbitaceae. Asian J Clin Nutr 2012;4:16–26.

[17] Kesari AN, Gupta RK, Watal G. Hypoglycemic effects of *Murraya koengii* on normal and alloxan-diabetic rabbits. J Ethnopharmacol 2005;97:247–51.

[18] Yamamura S, Ozawa K, Ohtani K, Kasai R, Yamasaki K. Antihistaminic flavones and aliphatic glycosides from *Mentha spicata*. Phytochemistry 1998;48:131–6.

[19] Rajeshkumar NV, Joy KL, Kuttan G, Ramsewak RS, Nair MG, Kuttan R. Antitumour and anticarcinogenic activity of *Phyllanthus amarus* extract. J Ethnopharmacol 2002;81:17–22.

[20] Khanna AK, Rizvi F, Chander R. Lipid lowering activity of *Phyllanthus niruri* in hyperlipemic rats. J Ethnopharmacol 2002;82:19–22.

[21] Iyer SR, Sethi R, Abroham AA. Analysis of nitrogen and phosphate in enriched and non enriched vermicompost. J Environ Res Dev 2012;7(2):899–904.

[22] Vishwakarma KL, Dubey V. Nutritional analysis of indigenous wild edible herbs used in eastern Chhattisgarh. India Emir J Food Agric 2011;23(6):554–60.

[23] Patro HK, Kumar A, Shukla DK, Mahapatra BS. Total productivity nutrient uptake and economics of rice wheat cropping system as influenced by *Crotalaria junea* green manuring. J Environ Res Dev 2011;5(3):532–41.

[24] Elias KM, Nelson KO, Simon M, Johnson K. Phytochemical and antioxidant analysis of methanolic extracts of four African indigenous leafy vegetables. Ann Food Sci Technol 2012;13(1):37–42.

[25] Jain AK, Tiwari P. Nutritional value of some traditional edible plants used by tribal communities during emergency with reference to Central India. Indian J Tradit Know 2012;11(1):51–7.

[26] Mensah JK, Okoli RI, Ohaju-Obodo JO, Eifediyi K. Phytochemical, nutritional and medical properties of some leafy vegetables consumed by Edo people of Nigeria. Afr J Biotech 2008;7(14):2304–9.

[27] Kubmarawa D, Khan ME, Punah AM, Hassan M. Phytochemical screening and antibacterial activity of extracts from *Parkia clappertoniana* Keay against human pathogenic bacteria. J Med Plant Res 2008;2(12):352–5.

[28] Johanna WL. Spicing up a vegetarian diet: chemopreventive effects of phytochemicals. Am J Clin Nutr 2003;78:579–83.

[29] Williamson G. Protective effects of fruit and vegetables in the diet. J Nutr Food Sci 1996;96:6–10.

[30] Liu S, Manson JE, Lee IM, Cole SR, Hennekens CH, Willett WC, et al. Fruit and vegetable intake and risk of cardiovascular disease: the women's health study. Amr J Clin Nutr 2000;72:922–8.

[31] Gandini S, Merzenich H, Robertson C, Boyle P. Meta-analysis of studies on breast cancer risk and diet: the role of fruit and vegetable consumption and the intake of associated micronutrients. Eur J Cancer 2000;36:636–46.

[32] Liu S, Lee IM, Ajani U, Cole SR, Buring JE, Manson JE. Intake of vegetables rich in carotenoids and risk of coronary heart disease in men: the physicians' health study. Int J Epidemiol 2001;30:130–5.

[33] Joshipura KJ, Hu FB, Manson JE, Stampfer MJ, Rimm EB, Speizer FE, et al. The effect of fruit and vegetable intake on risk for coronary heart disease. Ann Intern Med 2001;134:1106–14.

[34] Kang JH, Ascherio A, Grodstein F. Fruit and vegetable consumption and cognitive decline in aging women. Ann Neurol 2005;57:713–20.

[35] Lister CE. The medicine of life. Vitamins and other bioactive food components. Food Technol 1999;18:20–35.

[36] Kris-Etherton PM, Hecker KD, Bonanome A, Coval SM, Binkoski AE, Hilpert KF, et al. Bioactive compounds in foods: their role in the prevention of cardiovascular disease and cancer. Am J Med 2002;113:71–88.

[37] Gupta S, Lakshmi JA, Manjunath MN, Prakash J. Analysis of nutrient and antinutrient content of underutilized green leafy vegetables. Food Sci Technol (LWT) 2005;38:339–45.

[38] Alta MVA, Adeogun AO. Nutrient components of some tropical leafy vegetables. Food Chem 1995;53:375–9.

[39] Williamson G, Dupont MS, Heaney RK, Roger G, Rhodes MJ. Induction of glutathione S transferase activity in hepG2 cells by extracts of fruits and vegetables. Food Chem 1997;2:157–60.

[40] Randhawa MA, Anjum FM, Asi MR, Butt MS, Ahmed A, Randhawa MS. Removal of endosulfan residues from vegetables by household processing. J Sci Ind Res 2007;66:849–52.

[41] Randhawa MA, Anjum FM, Ahmed A, Randhawa MS. Field incurred chlorpyrifos and 3,5,6-trichloro-2-pyridinol residues in fresh and processed vegetables. Food Chem 2007;103:1016–23.

[42] Randhawa MA, Anjum FM, Randhawa MS, Ahmed A, Farooq U, Abrar M, et al. Dissipation of deltamethrin on supervised vegetables and removal of its residue by household processing. J Chem Soc Pak 2008;30:227–31.

[43] Randhawa MA, Anjum FM, Asi MR, Ahmed A, Nawaz H. Field incurred endosulfan residues in fresh and processed vegetables and dietary intake assessment. Int J Food Prop 2014;17:1109–15.

[44] Randhawa MA, Anjum MN, Butt MS, Yasin M, Imran M. Minimization of imidacloprid residues in cucumber and bell pepper through washing with citric acid and acetic acid solutions and their dietary intake assessment. Int J Food Prop 2014;17:978–86.

[45] Zafar S, Ahmed A, Ahmad R, Randhawa MA, Gulfraz M, Ahmad A, et al. Chemical residues of some pyrethroid insecticides in egg plant and okra fruits: effect of processing and chemical solutions. J Chem Soc Pak 2012;34:1169–75.

[46] Pradeep KS, Mallikarjuna KR. Phytochemicals in vegetables and their health benefits. Asian J Agric Rural Dev 2012;2:177–83.

[47] Ames BM, Shigena MK, Hagen TM. Oxidants, antioxidants and the degenerative diseases of aging. Proc Natl Acad Sci USA 1993;90:7915–22.

[48] Buring JE, Hennekens CH. Antioxidant vitamins and cardiovascular disease. Nutr Rev 1997;55:53–60.

[49] Gey KF, Puska P, Jordan P, Moser UK. Inverse correlation between plasma vitamin E and mortality from ischemic heart disease in cross-cultural epidemiology. Am J Clin Nutr 1999;53:326–34.

[50] Stahelin HB, Gey KF, Eichholzer M, Ludin E. β-Carotene and cancer prevention: The Basel Study. Am J Clin Nutr 1991;53:265–9.

[51] Steinberg D. Antioxidants and atherosclerosis: a current assessment. Circulation 1991;84:1420–5.

[52] Willett WC. Micronutrients and cancer risk. Am J Clin Nutr 1994;59:162–5.

[53] King H, Aubert RE, Herman WH. Global burden of diabetes 1995–2025: prevalence, numerical estimates, and projection. Diabetes Care 1998;21:1414–31.

[54] Zimmet PZ, Tuomi T, Mackay R, Rowley MJ, Knowles W, Cohen M, et al. Latent autoimmune diabetes mellitus in adults (LADA): the role of antibodies to glutamic acid decarboxylase in diagnosis and prediction of insulin dependency. Diabetic Med 1994;11:299–303.

[55] DeFronzo RA, Bonadonna RC, Ferrannini E. Pathogenesis of NIDDM. In: Alberti KGMM, Zimmet P, DeFronzo RA, editors. International Textbook of Diabetes Mellitus. 2nd ed. Chichester: John Wiley & Sons; 1997. p. 635–712.

[56] Lillioja S, Mott DM, Spraul M, Ferraro R, Foley JE, Ravussin E, et al. Insulin resistance and insulin secretory dysfunction as precursors of non-insulin-dependent diabetes. Prospective Study of Pima Indians. N Engl J Med 1993;329:1988–92.

[57] Shaw JE, Sicree RA, Zimmet PZ. Global estimates of the prevalence of diabetes for 2010 and 2030. Diabetes Res Clin Pract 2010;87:4–14.

[58] Harris MI, Zimmer P. International Textbook of Diabetes Mellitus. 1st ed. John Wiley & Sons: London; 1992. p. 3–18.

[59] Nishikawa T, Edelstein D, Du XL, Yamagishi S, Matsumura T, Kaneda Y, et al. Normalizing mitochondrial superoxide production blocks three pathways of hyperglycaemic damage. Nat 2000;404:787–90.

[60] Ramachandran A, Snehalatha C, Vijay V. Temporal changes in prevalence of type 2 diabetes and impaired glucose tolerance in urban southern India. Diabetes Res Clin Pract 2002;58:55–60.

[61] Giugliano D, Ceriello A, Paolisso G. Oxidative stress and diabetic vascular complications. Diabetes Care 1996;19:257–67.

[62] Stanely MP, Menon VP. Antioxidant action of *Tinospora cordifolia* root extract in alloxan diabetic rats. Phytother Res 2001;15:213–8.

[63] Liu S, Serdula M, Janket S, Cook NR, Sesso HD, Willett WC, et al. A prospective study of fruit and vegetable intake and the risk of type 2 diabetes in women. Diabetes Care 2004;27:2993–6.

[64] Tarwadi K, Agte V. Potential of commonly consumed green leafy vegetables for their antioxidant capacity and its linkage with the micronutrient profile. Int J Food Sci Nutr 2003;54:417–25.

[65] Kwon YI, Jang HD, Shetty K. Evaluation of *Rhodiola crenulata* and *Rhodiola rosea* for management of type II diabetes and hypertension. Asia Pac J Clin Nutr 2006;15:425–32.

[66] Chu YF, Sun J, Wu X, Liu RH. Antioxidant and antiproliferative activities of common vegetables. J Agric Food Chem 2002;50:6910–6.

[67] Oboh G, Rocha JBT. Antioxidant in Foods: A New Challenge for Food Processors: Leading Edge Antioxidants Research. New York: Nova Science Publishers Inc; 2007. 35–64.

[68] Oboh G, Raddatz H, Henle T. Antioxidant properties of polar and non-polar extracts of some tropical green leafy vegetables. J Sci Food Agric 2008;88:2486–92.

[69] Pulido R, Bravo L, Saura-Calixto F. Antioxidant activity of dietary polyphenols as determined by a modified ferric reducing/antioxidant power assay. J Agric Food Chem 2000;48:3396–402.

[70] Saliu JA, Oboh G. *In vitro* antioxidative and inhibitory actions of phenolic extract of some tropical green leafy vegetables on key enzymes linked to type 2 diabetes and hypertension. J Chem Pharm Res 2013;5:148–57.

[71] Faller ALK, Fialho E. The antioxidant capacity and polyphenol content of organic and conventional retail vegetables after domestic cooking. Food Res Int 2009;42:210–5.

D. NUTRITION AND REPRODUCTION

[72] Gutierrez j, Barry-Ryan C, Bourke P. The antimicrobial efficacy of plant essential oil combinations and interactions with food ingredients. Int J Food Microbiol 2008;124:91–7.

[73] Bhat RS, Al-Daihan S. Phytochemical constituents and antibacterial activity of some green leafy vegetables. Asian Pac J Trop Biomed 2014;4:189–93.

[74] Garcia-Lafuente A, Guillamon E, Villares A, Rostagno MA, Martinez JA. Flavonoids as anti-inflammatory agents: implications in cancer and cardiovascular disease. Inflamm Res 2009;58:537–52.

[75] Sparg SG, Light ME, Staden JV. Biological activities and distribution of plant saponins. J Ethnopharmacol 2004;94:219–43.

[76] Guil-Guerrero JL, Rebolloso-Fuentes MM, TorijaIsasa ME. Fatty acids and carotenoids from Stinging Nettle (*Urtica dioica* L.). J Food Compos Anal 2003;16:111–9.

[77] Vayalil PK. Antioxidant and antimutagenic properties of aqueous extract of date fruit (*Phoenix dactylifera* L. Arecaceae). J Agric Food Chem 2002;50:610–7.

[78] Velioglu YS, Mazza G, Gao L, Oomah BD. Antioxidant activity and total phenolics in selected fruits, vegetables, and grain products. J Agric Food Chem 1998;46:4113–7.

[79] Klipstein-Grobusch K, Launer LJ, Geleijnse JM, Boeing H, Hofman A, Witterman JC. Serum carotenoids and atherosclerosis: the Rotterdam study. Atherosclerosis 2000;148:49–56.

[80] Moeller SM, Jacques PF, Blumberg JB. The potential role of dietary xanthophylls in cataract and age-related macular degeneration. J Am Coll Nutr 2000;19:522–7.

[81] Morris MC, Bekett LA, Scherr PA, Hebert LE, Bennett DA, Field TS, et al. Vitamin E and vitamin C supplement use and risk of incident Alzheimer disease. Alzheimer Dis Assoc Disord 1998;12: 121–6.

[82] Slattery ML, Benson J, Curtin K, Ma KN, Schaeffer D, Potter JD. Carotenoids and colon cancer. Am J Clin Nutr 2000;71:575–82.

[83] Feig DI, Sowers LC, Loeb LA. Reverse chemical mutagenesis: identification of the mutagenic lesions resulting from reactive oxygen species-mediated damage to DNA. Proc Natl Acad Sci USA 1994;91:6609–13.

[84] Guyton KZ, Kensler TW. Oxidative mechanism in carcinogenesis. Br Med Bull 1993;49:523–44.

[85] Waladkhani AR, Clemens MR. Effect of dietary phytochemicals on cancer development. Int J Mol Med 1998;1:747–53.

[86] Shyamala BN, Gupta S, Lakshmi AJ, Prakash J. Leafy vegetable extracts antioxidant activity and effect on storage stability of heated oils. Innov Food Sci Emerg Technol 2005;6:239–45.

[87] Mauriello LM, Driskell MH, Sherman KJ, JohnsonSS, Prochaska JM, Prochaska JO. Acceptability of a school-based intervention for prevention of adolescent obesity. J Sch Nurs 2006;22:269–77.

[88] Tohill BC, Seymour J, Serdula M, Kettel-Khan L, Rolls BJ. What epidemiological studies tell us about the relationship between fruit and vegetable consumption and body weight. Nutr Rev 2004;62:4365–74.

[89] He FJ, Nowson CA, Macgregor GA. Fruits and vegetables consumption and stoke: meta-analysis of cohort studies. Lancet 2006;367:320–6.

[90] Daucher L, Amouye P, Hercberg S, Dallongeville J. Fruit and vegetable consumption and risk of coronary heart disease: a meta-analysis of cohort studies. J Nutr 2006;136:2588–92.

[91] Ali M, Al-Qattan KK, Al-Enezi F, Khanafer RMA, Mustafa T. Effect of allicin from garlic powder on serum lipids and blood pressure in rats fed with a high cholesterol diet. Prostaglandins Leukot Essent Fatty Acids 2000;62(4):253–9.

[92] Bazzano LA, Serdula MK, Liu S. Dietary intake of fruits and vegetables and risks of cardiovascular disease. Curr Atheroscler Rep 2003;5:492–9.

[93] Joshipura KJ, Ascherio A, Manson JE, Stampter MJ, Rim EB. Fruit and vegetable intake in relation to the risk of ischaemic stroke. JAMA 1999;282:1233–9.

[94] Fung TT, Willett WC, Stampter MJ, Manson JE, Hu FB. Dietary patterns and the risk of coronary heart disease in women. Arch Intern Med 2001;161:1857–62.

[95] Ness AR, Powles JW. Fruit and vegetables and cardiovascular disease: a review. Int J Epidemiol 1997;26(1):1–13.

[96] Alonso A, Fuente DC, Martin-Arnau AM, de Irala J, Martinez JA, Gonzalez MA. Fruit and vegetable consumption is inversely associated with blood pressure in a Mediterranean population with a high vegetable-fat intake. Br J Nutr 2004;92:311–9.

[97] Israili ZH, Hernandez-Hernandez R, Valasco M. The future of antihypertensive treatment. Am J Ther 2007;14:121–34.

[98] Hajja RI, Kotchne TA. Trends in prevalence, awareness, treatment and control of hypertension in the United States. 1988–2000. The National Health and Nutrition Examination Survey (NHANES) 1999–2000. JAMA 2003;16:407–43.

[99] Svetke YLP, Simons-Morton DG, Proschan MA. Effect of the dietary approaches to stop hypertension diet and reduced sodium intake on blood pressure control. J Clin Hypertens 2004;6: 373–81.

[100] Hajja RI, Kotchen JM, Kotchen TA. Hypertension: trends in prevalence, incidence, and control. Annu Rev Public Health 2006;27:465–90.

[101] Carte RBL. Implementing the new guidelines for hypertension: JNC7, ADA, WHO-ISH. J Manag Care Pharm 2004;10(5):S18–25.

[102] Young DB, Lin H, McCabe RD. Potassium's cardiovascular protective mechanisms. Am J Physiol 1995;268(4):825–37.

[103] Swales JD. Intersalt: an international study of electrolyte excretion and blood pressure. Results for 24 h urinary sodium and potassium excretion. Intersalt Cooprative Research Group. Br Med J 1988;297(6644):319–28.

[104] Cassel J. Studies of hypertension in migrants. In: Paul O, editor. Epidemiology and Control of Hypertension. Miami: Symposium Specialists; 1975. p. 41–58.

[105] Elfor J, Phillips A, Thomson AG, Shaper AG. Migration and geographic variations in blood pressure in Britain. Br Med J 1990;300:291–4.

[106] Appel LJ, Moore TJ, Obarzanek E. A clinical trial of the effects of dietary patterns on blood pressure. DASH collaborative research group. N Engl J Med 1997;336:1117–24.

[107] Svetkey LP, Simons-Morton D, Vollmer WM. Effects of dietary patterns on blood pressure – subgroup analysis of the Dietary Approaches to Stop Hypertension (DASH) randomized clinical trial. Arch Intern Med 1999;159:285–93.

[108] Sacks FM, Svetkey LP, Vollmer WM, Appel LJ. For the DASH-sodium collaborative research group. Effects on blood pressure or reduced dietary sodium and the Dietary Approaches to Stop Hypertension (DASH) diet. N Engl J Med 2001;344:3–10.

[109] Sacks FM, Campos H. Dietary therapy in hypertension. N Engl J Med 2010;362:2102–12.

[110] Houston MC. Nutrition and nutraceutical supplements in the treatment of hypertension. Expert Rev Cardiovasc Ther 2010;8(6):821–33.

[111] Kesteloot H, Joossens JV. Relationship of dietary sodium, potassium, calcium, and magnesium with blood pressure. Belgian Interuniversity Research on Nutrition and Health. Hypertension 1988;12(6):594–9.

[112] Yamori Y, Kihara M, Nara Y, et al. Hypertension and diet: multiple regression analysis in a Japanese farming community [letter]. Lancet 1981;1(8231):1204–5.

[113] Khaw KT, Barrett-Connor E. Dietary potassium and stroke associated mortality. N Engl J Med 1987;316:235–40.

[114] Ascherio A, Rimm EB, Hernan MA, et al. Intake of potassium, magnesium, calcium, and fiber and risk of stroke among US men. Circulation 1998;98:1198–204.

[115] Adrogue HJ, Madias NE. Sodium and potassium in the pathogenesis of hypertension. N Engl J Med 2007;356:1966–78.

[116] He J, Tell GS, Tang YC, Mo PS, He GQ. Relation of electrolytes to blood pressure in men. The Yi people study. Hypertension 1991;17(3):378–85.

[117] Fujiwara N, Osanai T, Kamada T, Katoh T, Takahashi K, Okumura K. Study on the relationship between plasma nitrite and nitrate level and salt sensitivity in human hypertension: modulation of nitric oxide synthesis by salt intake. Circulation 2000;101(8):856–61.

[118] Haddy FJ, Vanhoutte PM, Feletou M. Role of potassium in regulating blood flow and blood pressure. Am J Physiol Regul Integr Comp Physiol 2006;290(3):546–52.

[119] Amberg GC, Bonev AD, Rossow CF, Nelson MT, Santana LF. Modulation of the molecular composition of large conductance, Ca(2+) activated K(+) channels in vascular smooth muscle during hypertension. J Clin Invest 2003;112(5):717–24.

[120] Undurti DN. Nutritional factors in the pathobiology of human essential hypertension. Nutrition 2001;17:337–46.

[121] Pruess HG. Diet, genetics and hypertension. J Am Coll Nutr 1997;16:296–305.

[122] Barri YM, Wingo CS. The effects of potassium depletion and supplementation on blood pressure: a clinical review. Am J Med Sci 1997;3:37–40.

[123] Ma G, Mamaril JLC, Young DB. Increased potassium concentration inhibits stimulation of vascular smooth muscle proliferation by PDGF-BB and BFGF. Am J Hypertens 2000;13:1055–60.

[124] Ando K, Matsui H, Fujita M, Fujita T. Protective effect of dietary potassium against cardiovascular damage in salt-sensitive hypertension: possible role of its antioxidant actions. Curr Vasc Pharmacol 2010;8(1):59–63.

[125] Kido M, Katsuyuki A, Onozato ML, Tojo A, Yoshikawa M, Ogita T, et al. Protective effect of dietary potassium against vascular injury in salt-sensitive hypertension. Hypertension 2008;51:225–31.

[126] Ying WZ, Aaron K, Want PX, Sanders PW. Potassium inhibits dietary salt-induced transforming growth factor-beta production. Hypertension 2009;54(5):115–63.

[127] Desmedt A, Jacobs H. Soluble fiber in guide to functional food ingredients. England: Food RA Leatherhead Publishing Surrey; 2001.

[128] Selvendran RR, Robertson JA. Dietary fiber in foods: amount and type. In: Metabolic, Physiological Aspects of Dietary Fiber in Food. Luxembourg, Commission of the European Communities, Brussels; 1994.

[129] Lambo AM, Oste R, Nyman ME. Dietary fiber in fermented oat and barley β-glucan rich concentrates. Food Chem 2005;89:283–93.

[130] Gregory J, Foster K, Tyler H, Wiseman M. The Dietary and Nutritional Study of British Adults. London: HMSO; 1990.

[131] Cummings JH. Metabolic and Physiological Aspects of Dietary Fiber. Brussels: Commission of the European Communities; 1996.

[132] Johnson IT, Southgate DAT. Dietary fiber and related substances. Food Safety Series No. 3. London Chapman & Hall; 1994.

[133] Gopalan C, Ramasastri BV, Balasubramaniyam SC, Rao BSN, Deosthale YG, Pant KC. Nutritive Value of Indian Foods. Hyderabad, India: Indian Council of Medical Research; 1996.

[134] Oboh G, Radatz H, Henle T. Characterization of the antioxidant properties of hydrophilic and lipophilic extracts of Jute (*Corchorus olitorius*) leaf. Int J Food Sci Nutrition 2009;60:124–34.

[135] Shittu TA, Ogunmoyela OA. Water blanching treatment and nutrient retention in some Nigerian green leafy vegetables. Proceedings 25th Annual Conference of the Nigerian Institute of Food Science and Technology, Lagos. 2001; p. 64–65.

[136] Hugues D, Philip DL. Afarican Garden and Orchard. London: Macmillan Publisher; 1989. p. 30–35.

[137] Bako SP, Luka SA, Bedo EB, Aula J. Ethanobotany and Nutrient Content of *Gnetum africana* in Nigeria. USA: SCITECH Publisher; 2002. p. 79–84.

[138] Okafor JO. Horticultural promising indigenous wild plant species of the Nigeria forest zone. Acta Horticulturae 1983;123:165–76.

[139] Li BW, Andrews KW, Pehrsson PR. Individual sugars, soluble, and insoluble dietary fiber contents of 70 high consumption foods. J Food Compos Anal 2002;15:715–23.

[140] Meltzer HM, Anne LB, Roy MN, Per M, Jan A, Margareta H. Effect of dietary factors in pregnancy on risk of pregnancy complications: results from the Norwegian Mother and Child Cohort Study. Am J Clin Nutr 2011;94:1970–4.

[141] Ramón R, Ballester F, Iñiguez C, Rebagliato M, Murcia M, Esplugues A, et al. Vegetable but not fruit intake during pregnancy is associated with newborn anthropometric measures. J Nutr 2009;139:561–7.

[142] Knudsen VK, Orozova-Bekkevold IM, Mikkelsen TB, Wolff S, Olsen SF. Major dietary patterns in pregnancy and fetal growth. Eur J Clin Nutr 2008;62:463–70.

[143] Mikkelsen TB, Merete O, Orozova-Bekkevold I, Knudsen VK, Olsen SF. Association between fruit and vegetable consumption and birth weight: a prospective study among 43,585 Danish women. Scand J Public Health 2006;34(6):616–22.

[144] Tchum SK, Newton S, Tanumihardjo SA, Fareed KNA, Tetteh A, Agyei SO. Evaluation of a green leafy vegetable intervention in Ghanaian postpartum mothers. Afr J Food Agric Nutr Dev 2009;9(6):1294–308.

[145] Torjusen H, Lieblein G, Næs T, Haugen M, Meltzer HM, Brantsater AL. Food patterns and dietary quality associated with organic food consumption during pregnancy; data from a large cohort of pregnant women in Norway. BMC Public Health 2012;12:612.

[146] Bradley CS, Kennedy CM, Turcea AM, Rao SS, Nygaard IE. Constipation in pregnancy: prevalence, symptoms, and risk factors. Obstet Gynecol 2007;110(6):1351–7.

[147] Lin Y, Han XF, Fang ZF, Che LQ, Wu D, Wu XQ, et al. The beneficial effect of fiber supplementation in high- or low-fat diets on fetal development and antioxidant defense capacity in the rat. Eur J Nutr 2012;51(1):19–27.

[148] Watanabe K, Mori T, Iwasaki A, Kimura C, Matsushita H, Shinohara K, et al. Increased oxygen free radical production during pregnancy may impair vascular reactivity in preeclamptic women. Hypertens Res 2013;36(4):356–60.

[149] Qiu C, Coughlin KB, Frederick IO, Sorensen TK, Williams MA. Dietary fiber intake in early pregnancy and risk of subsequent preeclampsia. Am J Hypertens 2008;21(8):903–9.

[150] Peroni DG, Bonomo B, Casarotto S, Boner AL, Piacentini GL. How changes in nutrition have influenced the development of allergic diseases in childhood. Ital J Pediatr 2012;38:22.

[151] Herbert V. Folic acid. In: Shils ME, Olson JA, Shike M, Ross AC, editors. Modern nutrition in health and disease. 9th ed Philadelphia: Williams & Wilkins; 1999. p. 433–46.

[152] Cossins E. The fascinating world of folate and one-carbon metabolism. Can J Bot 2000;78:691–708.

[153] Kelly GS. Folates: supplemental forms and therapeutic applications. Alt Med Rev 1998;3:208–20.

[154] Lucock MD, Daskalakis I, Schorah CJ, Levene MI, Hartly J. Analysis and biochemistry of blood folate. Biochem Mol Med 1996;58:93–112.

[155] Ma J, Stampfer MJ, Giovannucci E, Artigas C, Hunter DJ, Fuchs C, et al. Methylenetetrahydrofolate reductase polymorphism, dietary interactions, and risk of colorectal cancer. Cancer Res 1997;57:1098–102.

[156] Laclerc D, Wilson A, Dumas R, Gafuik C, Song D, Watkins D, et al. Genetic cloning and mapping of cDNA for methionine synthase reductase, a flavoprotein defective in patients with homocystinuria. Proc Natl Acad Sci USA 1998;95:3059–64.

[157] Bottiglieri T, Laundry M, Crellin R, Toone BK, Carney MW, Reynolds EH. Homocystine, folate, methylation, and mono-amine metabolism in depression. J Neurol Neurosurg Psychol 2000;69:228–32.

[158] Antony AC. *In utero* physiology: role of folic acid in nutrient delivery and fetal development. Am J Clin Nutr 2007;85:598–603.

[159] Food and Drug Administration. Food labeling: health claims and label statements; folate and neural tube defects; proposed rules. Federal Register 1993;58:53254–88.

[160] Chitambar CR, Antony AC. Nutritional aspects of hematologic diseases. In: Shils ME, Shike M, Ross AC, Caballero C, Cousins RJ, editors. Modern Nutrition in Health and Disease. 10th ed. Philadelphia, PA: Lippincott Williams & Wilkins; 2005. p. 1436–61.

[161] Antony AC. Megaloblastic anemias. In: Hoffman R, Benz EJ Jr, Shattil SJ, editors. Hematology. Basic Principles and Practice. 4th ed. Philadelphia, PA: Elsevier Churchill Livingstone; 2005. p. 519–56.

[162] Antony AC. Folate receptors: reflections on a personal odyssey and a perspective on unfolding truth. Theme issue: folate receptor-targeted drugs for cancer and inflammatory diseases. In: Low PS, Antony AC, editors. Advanced Drug Delivery Reviews: Folate Receptor Targeted Drugs for Cancer and Inflammatory Diseases. New York, NY: Elsevier; 2004. p. 1059–66.

[163] Heid MK, Bills ND, Hinrichs SH, Clifford AJ. Folate deficiency alone does not produce neural tube defects in mice. J Nutr 1992;122:888–94.

[164] Burgoon JM, Selhub J, Nadeau M, Sadler TW. Investigation of the effects of folate deficiency on embryonic development through the establishment of a folate deficient mouse model. Teratology 2002;65:219–27.

[165] Tamura T, Picciano M. Folate and human reproduction. Am J Clin Nutr 2006;83:993–1016.

[166] Karthigaini S, Gnanasundaram D, Baker SJ. Megaloblastic erythropoiesis and serum vitamin B12 and folic acid levels in pregnancy in South Indian women. J Obstet Gynaecol Br Commonw 1964;71:115–22.

[167] Mathan VI, Baker JS. Epidemic tropical sprue and other epidemics of diarrhea in South Indian villages. Am J Clin Nutr 1968;21:1077–87.

[168] Chakravarty I, Sinha R. Prevalence of micronutrient deficiency based on results obtained from the national pilot program on control of micronutrient malnutrition. Nutr Rev 2002;60:53–8.

[169] Misra A, Vikram NK, Pandey RM, Dwivedi M, Ahmad FU, et al. Hyperhomocysteinemia and low intakes of folic acid and vitamin B 12 in urban North India. Eur J Nutr 2002;41:68–77.

[170] Hathcock JN. Vitamins and minerals: efficacy and safety. Am J Clin Nutr 1997;66:427–37.

[171] Lester GE, Eischen F. Beta-carotene content of postharvest orange-fleshed muskmelon fruit: effect of cultivar, growing location and fruit size. Plant Foods Human Nutr 1996;49:191–7.

[172] Lester GE, Crosby KM. Ascorbic acid, folic acid, and potassium content in post harvest green-flesh honeydew muskmelons: Influence of cultivar, fruit size, soil type and year. J Am Soc Hort Sci 2002;127:843–7.

[173] Rodriguez H. The 21 Day Fertility Diet Challenge. California: KMF Publishing, Inc; 2011. p. 34–36.

[174] Zdravkovic JM, Acamovic-Djokovic GS, Mladenovic JD, Pavlovic RM, Zdravkovic MS. Antioxidant capacity and contents of phenols, ascorbic acid, β-carotene and lycopene in lettuce. Hem Ind 2014;68(2):193–8.

[175] Gupta S, Prakash J. Studies on Indian green leafy vegetables for their antioxidant activity. Plant Foods Hum Nutr 2009;64:39–45.

[176] Mathiventhan U, Sivakanesan R. Total phenolic content and total antioxidant activity of sixteen commonly consumed green leafy vegetables stored under different conditions. Eur Int J Sci Technol 2013;2:123–32.

[177] Yang RU, Samson T, Lee T, Chang L, Kuo G, Lai P. Moringa, a novel plant rich in antioxidants, bioavailable iron, and nutrients. In: Ho CT, ed. Challenges in chemistry and biology of herbs. Am Chem Soc 2006;224–39.

[178] Yadav RK, Kalia P, Kumar R, Jain V. Antioxidant and nutritional activity studies of green leafy vegetables. Int J Agric Food Sci Technol 2013;4:707–12.

[179] Melo EA, Lima VL, Maciel MIS, Caetano ACS, Leal FLL. Polyphenol ascorbic acid and total carotenoid contents in common fruits and vegetables. Braz J Food Technol 2006;9:89–94.

[180] Al-Juhaimi F, Ghafoor K. Total phenols and antioxidant activities of leaf and stem extracts from coriander, mint and parsley grown in Saudi Arabia. Pak J Bot 2011;43(4):2235–7.

[181] Clyde D, Bertini DJ, Dmochowski R, Koop H. The vitamin A and C content of *Coriandrum sativum* and the variations in the loss of the latter with various methods of food preparation and preservation. Qual Plant 1979;28:317–22.

[182] Caunii A, Cuciureanu R, Zakar AM, Tonea E, Giuchici C. Chemical composition of common leafy vegetables. Studia Universitatis "Vasile Goldiş", Seria Ştiinţele Vieţii 2010;20:45–8.

[183] Yahia EM, Ramirez-Padilla GK, Carrillo-Lopez A. Carotenoid content of five fruits and vegetables and their bioconversion to vitamin a measured by retinol accumulation in rat livers. Acta Horticulturae 2009;841:619–24.

[184] Kimiywe J, Waudo J, Mbithe D, Maundu P. Utilization and medicinal value of indigenous leafy vegetables consumed in urban and peri-urban Nairobi. Afr J Food Agric Nutr Dev 2007;7.(4).

[185] Cai Y, Luo Q, Sun M, Corke H. Antioxidant activity and phenolic compounds of 112 traditional Chinese medicinal plants associated with anticancer. Life Sci 2004;74(17):2157–84.

[186] Sonkar SD, Gupta R, Saraf SA. Effect of *Basella rubra* L. leaf extract on haematological parameters and amylase activity. Pharmacogn Commun 2012;2(3):10–3.

[187] Mahendra P, Bisht S, et al. Antianxiety activity of *Coriandrum sativum* assessed using different experimental anxiety models. Indian J Pharmacol 2011;43(5):574–7.

[188] Handral HK, Pandith A, Shruthi SD. A review on *Murraya koenigii*: multipotential medicinal plant. Asian J Pharm Clin Res 2012;5(4):5–14.

[189] Toppo FA, Akhand R, Pathak AK. Pharmacological actions and potential uses of *Trigonella foenum graecum*: a review. Asian J Pharm Clin Res 2009;2(4):29–38.

[190] Blumenthal M. Herbal Medicine: Expanded Commission E Monographs. Newton, MA: Integrative Medicine Communications; 2000.

[191] Reddy BS, Kawamori T, Lubet R, Steele V, Kelloff G, Rao CV. Chemopreventive effect of S-methylmethnethiosulfonate and sulindac administered together during the promotion/progression stages of colon carcinogenesis. Carcinogenesis 1999;20:1645–8.

[192] Raju M, Varakumar S, Lakshminarayana R, Krishnakantha TP, Baskaran V. Carotenoid composition and vitamin A activity of medicinally important green leafy vegetables. Food Chem 2007;101(4):1598–605.

Nutrition, Lifestyle, and Obesity in Urology

Leonardo Reis, MD, MSc, PhD,**, Ricardo Miyaoka, MD, PhD*,†,††*

*University of Campinas, Campinas SP, Brazil
**Pontifical Catholic University of Campinas, Campinas SP, Brazil
†ANDROFERT, Referral Center for Male Reproduction, Campinas SP, Brazil
††Sumaré State Hospital, Sumaré SP, Brazil

OBESITY AND BENIGN UROLOGICAL DISEASE

Lower Urinary Tract Symptoms (LUTS)

Obesity has been correlated with a higher incidence of benign prostatic enlargement (BPE) and therefore lower urinary tract symptoms (LUTS) [1–4]. The exact mechanism through which obesity leads to prostate hyperplasia remains to be fully elucidated but it seems that an increase in the proportion of estrogen/testosterone, increase in fasting glycemia, and hyperinsulinemia occur in obese men compared with nonobese subjects [4,5].

Our group has recently showed in a cross-sectional study including 490 men originating from a general urologic clinic that the odds ratio for moderate or severe LUTS among patients diagnosed with metabolic syndrome (MS) was 2.1. MS was more common in those patients presenting older age, higher body mass index (BMI), and larger prostate size. Also, older age and BMI had significant relative risk for lower urinary tract symptoms. However, only age remained as an independent factor for LUTS on multivariate analysis, suggesting that the association of male LUTS, prostate volume, and MS might be coincidental and related to older age [6]. Larger and prospective studies are warranted for definitive conclusions.

Stress Urinary Incontinence

Obesity is related to an increase in urinary incontinence in women [4,7]. It also correlates with increased intra-abdominal pressure, which enhances bladder pressure and may ultimately lead to stress urinary incontinence [7]. The long-term effect of weight loss over urinary comorbidities has not been deeply investigated although preliminary data suggest that both urinary incontinence and BPE may be reversed [4,8,9]. Several authors recommend weight loss as part of the therapeutic approach before either medical or surgical intervention in obese incontinent women [8,9].

Sexual Dysfunction

Both incidence and severity of sexual dysfunction in men increase along with obesity [4]. A morbidly obese man has an estimated level of sexual dysfunction of a nonobese man 20 years older [10]. Although data are still scarce, there is some evidence that female sexual dysfunction may also be related to obesity [4]. Obesity-related sexual dysfunction seems to be derived from multiple factors. In morbidly obese patients, high levels of diabetes, MS, and hypertension have been clearly associated with endothelial dysfunction and erectile dysfunction (ED) [11]. Obesity-related ED is also a consequence of psychological factors. Obesity and pronounced abdominal fat often make normal-sized penis to look smaller or shorter than they really are and may negatively impact the subject's self-esteem [2].

Boyanov et al. showed that testosterone replacement improves ED in aged men with type II diabetes and visceral obesity, decreasing visceral obesity, and improving glucose homeostasis [12].

In a randomized study involving 20 morbidly obese patients, Reis et al. showed that weight reduction following bariatric surgery improved sexual function assessed by the IIEF-5 questionnaire (International Index

Handbook of Fertility. http://dx.doi.org/10.1016/B978-0-12-800872-0.00019-6

of Erectile Function), improved levels of total and free testosterone, and follicle-stimulating hormone and reduced levels of prolactin. On the other hand, behavioral and dietary changes in lifestyle were able to reduce the BMI but not improve hormones or sexual function in morbidly obese patients [13].

Another study showed significant improvement in sexual hormone levels (luteinizing hormone, FSH, leptin, TT, and estradiol) following biliopancreatic bypass [14]. Dallal et al. reported on a significant improvement in sexual function following gastric bypass reaching the expected levels for age-stratified groups; and 60% of diabetic patients and 40% of hypertensive patients no longer needed medication [15].

In a randomized controlled trial involving 110 obese men with a complaint of ED without associated diabetes or hypertension, the group of patients treated on a restricted hypocaloric diet improved their IIEF-5 scores from 13.9 to 17 [16].

Fertility

In young women, overweight and obesity are among the major causes of subfertility. Weight loss improves fertility in these patients and should be considered a first-line therapeutic option. A 10–15% weight loss increases spontaneous pregnancy rates up to 30%, with an additional 40% chance when pharmacologically induced ovulation is used [2,17]. Moreover, there is evidence that extreme BMI in women (very obese or low weight) lead to reduced implantation and pregnancy rates following *in vitro* fertilization (IVF). As such, it is recommended that these patients normalize their weight before being started on a fertility treatment [18].

Obesity might also negatively impact male fertility directly and indirectly. It can induce sleep disorders and sexual dysfunction, change hormonal profiles and scrotal temperature, and ultimately compromise seminal parameters [19,20]. Additionally, varicocele is more common in obese compared to nonobese men and may also play a role in obese-related infertility [2].

Inhibin B, an important marker of spermatogenesis, is 25–32% lower in obese men than their normal-weight counterparts [17]. The testosterone/estradiol rate is consistently reduced in infertile obese men, with a 6% increase in estradiol levels and 25–32% decrease in testosterone levels when compared with nonobese men [19].

Weight loss may be achieved by behavioral changes (diet and/or exercise) or surgical intervention, and is associated with an increase in androgen and inhibin B levels and seminal parameters improvement [18]. SHBG, testosterone, and FSH levels have been shown to increase while a reduction in insulin and leptin levels could be seen [21].

Some authors have shown that sudden substantial weight loss either by diet or surgery might lead to infertility, as absorption of nutrients is intentionally decreased but not always properly compensated with diet or vitamins reposition, which may impair spermatogenesis [19].

In a recent work, Reis et al. found that morbidly obese patients show baseline discrete alterations in seminal parameters and massive weight loss after bariatric surgery does not alter the quality of sperm in a significant way, but an improvement in sexual function and hormonal levels was detected. Lifestyle changes and diet adjustments positively affected BMI only. They concluded that, while bariatric surgery might not impair fertility, future studies should include infertile men and adopt childbirth rate as a complementary endpoint [21].

In a review article the same authors argued that despite weight loss being the basis of obesity treatment and associated comorbidities, bariatric surgery impact in male reproductive potential is yet to be defined. There is limited evidence to support preoperative counseling currently and a judicious analysis must be done in a case-by-case basis with participation of the patient himself until the literature can offer more solid data on this issue [22].

Calculi Formation

It has long been suggested that eating habits and lifestyle changes have significantly contributed to the rise in both prevalence and incidence of urinary calculi in the last decades. Lack of adequate liquid volume intake and animal protein-rich diets are considered important risk factors for stone disease development.

Several studies have suggested that obesity enhances the risk of developing kidney stone disease, especially in women [4]. Obese patients are known to present excessive urinary stone elements such as sodium, calcium, magnesium, sulfate, phosphate, oxalate, uric acid, cysteine, and low citrate, which is stone protective [23]. Obesity is also associated with high urinary osmolality when compared with nonobese subjects. Even overweight children present a higher risk of calculi formation in both boys and girls and this has been associated with alterations in urine composition (hyperoxaluria and hypercalciuria) [24].

Long-term effects of weight loss in urinary calculi disease have not yet been completely clarified. It is still not clear if reduction in food ingestion following bariatric procedures may compensate malabsorption and reduce the risk of urolithiasis in long-term follow-up [4]. Jejunoileal intestinal bypass surgeries, which lead to intense malabsorption, may cause hyperoxaluria and calculi formation, and surgical techniques that have been proposed to substitute jejunoileal bypass (gastric band, Roux-in-Y and biliopancreatic derivation) might have the same impact, though, gastric band seems to be less associated with calculi formation since it does not induce malabsorption [25]. Extreme fasting and weight loss diets that are based on high protein and low carbohydrate ingestion, such as the Atkin's diet, can also favor the risk of calculi formation.

Besides, morbid obesity imposes several different physiological and technical challenges for urinary calculi treatment [26]. Extracorporeal shockwave lithotripsy (ESWL) may not be feasible as the machine bed cannot support superobese patients and long (>10 cm) skin-to-stone distance can make treatment unsuccessful. In many cases, percutaneous nephrolithotomy is the only possible choice but depends on the availability of specific extra-long tools [27]. For patients who harbor a not so large calculi mass, a transureteroscopic approach is preferable [28].

Renal Disease

Obese or overweight individuals are more likely to develop chronic renal failure and end-stage renal failure, albuminuria, diabetes, and hypertension [29]. Preliminary data suggest that several clinical and nephrologic changes secondary to obesity can be either reversed or attenuated with reduction of corporal fat achieved through diet and surgical intervention [30].

OBESITY AND UROLOGICAL MALIGNANCY: POSSIBLE MECHANISMS

Fat tissue has a role in immunological response and as such may indirectly affect carcinogenesis. Fat tissue is a lipid deposit and may interfere with macrophage function and be a source of mutagenic agents, lipid peroxidation, and gonadal steroid precursors [2].

Furthermore, mitochondrial function in cells of elderly and obese patients is less efficient in lipid metabolism and therefore more prone to accumulate nocive oxidative agents [31]. A rise in circulating lipids enhances the stress over mitochondrial aging. As such, the combination of aging and obesity represents a suitable environment for neoplasia development [31].

Prostate Cancer

A meta-analysis with 22 prospective series concluded that obesity is associated with advanced (RR 1.12 for each 5 kg/m^2) and not with localized disease (RR 0.96) [32]. Three large prospective studies showed that obesity is positively associated with high risk of high-grade high-stage disease and inversely correlated with low-grade low-stage disease [33]. In the Prostate Cancer Prevention Trial men with BMI <25 kg/m^2 presented an 18% lower

risk of low-grade prostate cancer (PCa) compared with obese men. However, they presented a higher risk (78%) of high-grade (Gleason 8–10) disease [33].

Currently, there is a wide range of options for PCa treatment depending on cancer stage and patient profile. As obese patients seem to suffer from more aggressive diseases as shown by radical prostatectomy specimens, evaluation of cure rates and functional results in this specific population is fundamental [1,33]. In addition, obese men have a higher risk of positive surgical margins after radical prostatectomy [34] and higher biochemical recurrence rates regardless of adverse clinicopathological features when compared with normal-weight homolog counterparts [34]. These results might be related to inherent tumor biology in obese patients. Regarding surgical aspects, obese men present longer operative time and higher estimated blood loss during radical prostatectomy when compared with normal-weight men [1].

A large cohort study involving 5041 men revealed an alarming conclusion. Even after adjusting subjects according to comorbidities, the likelihood of men to receive, instead of radical prostatectomy, a nonsurgical treatment, such as brachytherapy, external radiotherapy, primary androgen deprivation, and watchful waiting, increases in parallel with BMI increase. Very obese men in particular (BMI >35 kg/m^2) were more likely to undergo brachytherapy and androgen deprivation-only therapy [35]. Other influencing factors must be ruled out since current tendencies herein used may not be ideal for obese men, considering that they tend to present a more aggressive disease.

Similarly to what happens after surgical treatment, obese men present a higher risk for biochemical recurrence [36] and lower cancer-specific survival time [37], even after controlling for pathological features. A possible mechanism for radiotherapy increased failure rate in obese men relies on the naturally occurring daily variation of prostate localization, which is wider in obese men [1].

Some studies have reported on the obesity-related increased risk of death for PCa (2.5 times in a study involving 6,763 men [38], increased risk of 20–34% in more than 900,000 men in another report [39]) and this risk progressively increases along with BMI [39].

Renal Cancer

Several studies have demonstrated an increase in renal clear cell carcinoma in people who are fat-rich diet consumers [2]. Other reports have shown that obesity enhances the risk of having a renal carcinoma (RC) [1,2,40]. Increased relative risk has been estimated in 1.07 (95% CI 1.05–1.09) for each additional unit in BMI, regardless of gender. This means that about 3 kg of corporal weight gain correlates with an increase in risk of developing RC of about 7% [41]. Interestingly, obesity may offer a longer cancer-specific survival rate after nephrectomy for treated localized tumors [42].

Currently, it is a consensus that a laparoscopic approach is viable for morbidly obese patients offering less morbidity against open nephrectomy [1].

Bladder Cancer

The influence of BMI over bladder cancer is still poorly understood. Although there seems to be a relationship, definitive conclusions cannot be drawn from currently existing data [1]. Perioperative risks are much higher following open radical cystectomy in obese patients as surgery becomes technically more challenging in regard to the increase in operative time and blood loss as well as stoma-related complications for patients undergoing ileal conduit urinary derivation [1,43].

Despite the scarce data available in the literature, current knowledge suggests no significant association between BMI and overall survival or cancer-specific survival following either radical or partial cystectomy [43].

Testicular Cancer

Data on the correlation between testicular cancer and obesity are also limited. Obesity has been linked to increased testicular cancer risk in a study performed in the Pacific Basin. One of the possible causes is a hormonal imbalance secondary to obesity with estrogen in excess, and a failure in thermal control of the testicle [44]. Other authors, however, have claimed the opposite [45].

PROTECTING NUTRITIONAL FACTORS

Nutritional factors, such as fat intake reduction and increase in vitamins A, E, C, and D, lycopene, and selenium intake, may have a protective role against PCa [46].

Just as high-fat-rich diets with high caloric value seem to stimulate some types of cancer development, low-fat diets seem to play a protective role as shown in *in vitro* studies [2]. Giovannucci et al. [47] have reported on a reduced risk of PCa associated with greater consumption of fruit and vegetables, and lycopene was associated with a lower risk of PCa in an epidemiological study. Isoflavonoid extracts from soy foods, such as genistein, have the ability to inhibit PCa cellular growth *in vitro*. This is in agreement with the fact that Asian men present a lower incidence of PCa and are regular consumers of high-level soy diets. No study, however, has shown that soy supplementation can lower the incidence of PCa in humans [48]. Active ingredients of garlic and allicin have also been able to reduce tumoral cell growth *in vitro,* but had no effect in clinical trials whatsoever [49].

Daviglus et al. reported on longer overall survival in PCa patients associated with beta-carotene and vitamin C intake [50]. Selenium has been proposed as an effective agent to reduce PCa progression. Clark et al. [51] reported on a reduction of 63% in PCa incidence using selenium supplements in the diet. Beta-carotene and vitamin E have also been analyzed in regard to their potential to inhibit cancer growth.

The Selenium and Vitamin E Cancer Prevention Trial, a randomized prospective double-blind trial designed to determine if selenium and vitamin E could reduce PCa risk, showed that both selenium and vitamin E should not be taken at doses that exceed recommended dietary intakes under the risk of increasing high-grade PCa incidence [52].

PHYSICAL EXERCISE AND WEIGHT LOSS

Prospective studies have shown that overweight patients who exercise have lower morbidity and mortality than their sedentary counterparts [53]. Although there is a lack of well-controlled studies comparing the incidence of cancer in sedentary versus physically active overweight subjects, physical activities seem to promote a degree of protection against breast and prostate cancers [54].

Few studies found no impact of weight loss in the risk of cancer in obese patients. Despite the aforementioned benefits, weight loss may have unwanted effects such as biliary calculi formation and nutrient deficiency.

Weight loss, however, benefits the cardiovascular system, psychological self-esteem, and overall well-being [2]. Besides, benign urological conditions, including urinary incontinence, benign prostate hyperplasia, ED, and fertility issues, can be minimized following weight loss [13–16,18–21,55,56].

Patients who are under weight loss programs should keep a physical activity and controlled diet program that they can cope with. The "yo-yo" effect where weight loss and gain alternate sequentially may present a higher risk to cardiovascular system health than maintaining a constant slightly overweight [2].

KEY POINTS

- Obesity represents a rising epidemic worldwide; about 300,000 deaths occur annually in the United States associated with overweight and obesity, as well as 10 of the most common overall causes of death.
- Obesity seems to be associated with an increasing incidence of benign prostatic enlargement and its secondary lower urinary tract symptoms as well as a

higher incidence of urinary incontinence in women; weight loss may revert these problems.

- Sexual dysfunction severity and incidence increase with obesity in men; surgery-induced weight loss improves sexual function quality as assessed by the IIEF-5 questionnaire and sexual hormone levels.
- Obesity affects both male and female fertility; weight loss might improve fertility rates in both genders and must be a first-line therapeutic option, mainly in women.
- Obesity increases the risk of developing urinary calculi, especially in women, and also makes more complex the stone treatment. Bariatric surgery, notably the jejunoileal bypass that causes significant malabsorption, may lead to severe oxaluria and therefore urinary calculi formation. Weight loss achieved through extreme fasting or high-protein low-carbohydrate diets can also increase renal calculi formation.
- Obesity is a risk factor for prostate and kidney cancers. Obese men may harbor more aggressive PCa, have a higher risk of presenting positive surgical margins following radical prostatectomy, as well as biochemical recurrence after surgery regardless of pathological features. Obesity also correlates with a higher risk of death for PCa.
- Nutritional factors including reduced fat and increased vitamins A, C, D, and lycopene intake may be protective against PCa. Vitamin E and selenium supplementations are not recommended previously "normal" dietary intake as they have been linked to a higher risk of presenting high-grade disease.

References

[1] Stewart SB, Freedland SJ. Influence of obesity on the incidence and treatment of genitourinary malignancies. Urol Oncol 2011;29(5):476–86.

[2] Mydlo JH. The impact of obesity in urology. Urol Clin North Am 2004;31(2):275–87.

[3] Kiess W, Galler A, Reich A, Muller G, Kapellen T, Deutscher J, et al. Clinical aspects of obesity in childhood and adolescence. Obes Rev 2001;2(1):29–36.

[4] Natarajan V, Master V, Ogan K. Effects of obesity and weight loss in patients with nononcological urological disease. J Urol 2009;181(6):2424–9.

[5] Dahle SE, Chokkalingam AP, Gao YT, Deng J, Stanczyk FZ, Hsing AW. Body size and serum levels of insulin and leptin in relation to the risk of benign prostatic hyperplasia. J Urol 2002;168(2):599–604.

[6] Zamuner M, Laranja WW, Alonso JC, Simões FA, Rejowski RF, Reis LO. Is metabolic syndrome truly a risk factor for male lower urinary tract symptoms or just an epiphenomenon? Adv Urol 2014;2014:203854.

[7] Sugerman H, Windsor A, Bessos M, Wolfe L. Intra-abdominal pressure, sagittal abdominal diameter and obesity comorbidity. J Intern Med 1997;241(1):71–9.

[8] Subak LL, Whitcomb E, Shen H, Saxton J, Vittinghoff E, Brown JS. Weight loss: a novel and effective treatment for urinary incontinence. J Urol 2005;174(1):190–5.

[9] Cummings JM, Rodning CB. Urinary stress incontinence among obese women: review of pathophysiology therapy. Int Urogynecol J Pelvic Floor Dysfunct 2000;11(1):41–4.

[10] Kratzik CW, Schatzl G, Lunglmayr G, Rucklinger E, Huber J. The impact of age, body mass index, and testosterone on erectile dysfunction. J Urol 2005;174(1):240–3.

[11] Corona G, Mannucci E, Forti G, Maggi M. Hypogonadism, ED, metabolic syndrome and obesity: a pathological link supporting cardiovascular diseases. Int J Androl 2009;32(6):587–98.

[12] Boyanov MA, Boneva Z, Christov VG. Testosterone supplementation in men with type 2 diabetes, visceral obesity and partial androgen deficiency. Aging Male 2003;6(1):1–7.

[13] Reis LO, Favaro WJ, Barreiro GC, de Oliveira LC, Chaim EA, Fregonesi A, et al. Erectile dysfunction and hormonal imbalance in morbidly obese male is reversed after gastric bypass surgery: a prospective randomized controlled trial. Int J Androl 2010;33(5):736–44.

[14] Alagna S, Cossu ML, Gallo P, Tilocca PL, Pileri P, Alagna G, et al. Biliopancreatic diversion: long-term effects on gonadal function in severely obese men. Surg Obes Relat Dis 2006;2(2):82–6.

[15] Dallal RM, Chernoff A, O'Leary MP, Smith JA, Braverman JD, Quebbemann BB. Sexual dysfunction is common in the morbidly obese male and improves after gastric bypass surgery. J Am Coll Surg 2008;207(6):859–64.

[16] Esposito K, Giugliano F, Di Palo C, Giugliano G, Marfella R, D'Andrea F, et al. Effect of lifestyle changes on erectile dysfunction in obese men: a randomized controlled trial. JAMA 2004;291(24):2978–84.

[17] Crosignani PG, Vegetti W, Colombo M, Ragni G. Resumption of fertility with diet in overweight women. Reprod BioMed Online 2002;5(1):60–4.

[18] Nichols JE, Crane MM, Higdon HL, Miller PB, Boone WR. Extremes of body mass index reduce *in vitro* fertilization pregnancy rates. Fertil Steril 2003;79(3):645–7.

[19] Pasqualotto FF, Medeiros GS, Carolina C, Pasqualotto EB. Efeito da obesidade na função reprodutiva masculina. Urol Essen 2011;1(4):27–40.

[20] Hakonsen LB, Thulstrup AM, Aggerholm AS, Olsen J, Bonde JP, Andersen CY, et al. Does weight loss improve semen quality and reproductive hormones? Results from a cohort of severely obese men. Reprod Health 2011;8:24.

[21] Reis LO, Zani EL, Saade RD, Chaim EA, de Oliveira LC, Fregonesi A. Bariatric surgery does not interfere with sperm quality – a preliminary long-term study. Reprod Sci 2012;19(10):1057–62.

[22] Reis LO, Dias FG. Male fertility, obesity, and bariatric surgery. Reprod Sci 2012;19(8):778–85.

[23] Taylor EN, Curhan GC. Body size and 24-hour urine composition. Am J Kidney Dis 2006;48(6):905–15.

[24] Sarica K, Eryildirim B, Yencilek F, Kuyumcuoglu U. Role of overweight status on stone-forming risk factors in children: a prospective study. Urology 2009;73(5):1003–7.

[25] Asplin JR, Coe FL. Hyperoxaluria in kidney stone formers treated with modern bariatric surgery. J Urol 2007;177(2):565–9.

[26] Semins MJ, Matlaga BR, Shore AD, Steele K, Magnuson T, Johns R, et al. The effect of gastric banding on kidney stone disease. Urology 2009;74(4):746–9.

[27] Pearle MS, Nakada SY, Womack JS, Kryger JV. Outcomes of contemporary percutaneous nephrostolithotomy in morbidly obese patients. J Urol 1998;160(3 Pt 1):669–73.

[28] Fabrizio MD, Behari A, Bagley DH. Ureteroscopic management of intrarenal calculi. J Urol 1998;159(4):1139–43.

[29] Kopple JD, Feroze U. The effect of obesity on chronic kidney disease. J Ren Nutr 2011;21(1):66–71.

[30] Navaneethan SD, Yehnert H, Moustarah F, Schreiber MJ, Schauer PR, Beddhu S. Weight loss interventions in chronic kidney disease: a systematic review and meta-analysis. Clin J Am Soc Nephrol 2009;4(10):1565–74.

[31] Hietanen E, Bartsch H, Bereziat JC, Camus AM, McClinton S, Eremin O, et al. Diet and oxidative stress in breast, colon and prostate cancer patients: a case-control study. Eur J Clin Nutr 1994;48(8):575–86.

[32] MacInnis RJ, English DR. Body size and composition and prostate cancer risk: systematic review and meta-regression analysis. Cancer Causes Control 2006;17(8):989–1003.

[33] Gong Z, Neuhouser ML, Goodman PJ, Albanes D, Chi C, Hsing AW, et al. Obesity, diabetes, and risk of prostate cancer: results from the prostate cancer prevention trial. Cancer Epidemiol Biomarkers Prev 2006;15(10):1977–83.

[34] Freedland SJ, Aronson WJ, Kane CJ, Presti JC Jr, Amling CL, Elashoff D, et al. Impact of obesity on biochemical control after radical prostatectomy for clinically localized prostate cancer: a report by the Shared Equal Access Regional Cancer Hospital database study group. J Clin Oncol 2004;22(3):446–53.

[35] Davies BJ, Walsh TJ, Ross PL, Knight SJ, Sadetsky N, Carroll PR, et al. Effect of BMI on primary treatment of prostate cancer. Urology 2008;72(2):406–11.

[36] Stroup SP, Cullen J, Auge BK, L'Esperance JO, Kang SK. Effect of obesity on prostate-specific antigen recurrence after radiation therapy for localized prostate cancer as measured by the 2006 Radiation Therapy Oncology Group-American Society for Therapeutic Radiation and Oncology (RTOG-ASTRO) Phoenix consensus definition. Cancer 2007;110(5):1003–9.

[37] Palma D, Pickles T, Tyldesley S. Obesity as a predictor of biochemical recurrence and survival after radiation therapy for prostate cancer. BJU Int 2007;100(2):315–9.

[38] Snowdon DA, Phillips RL, Choi W. Diet, obesity, and risk of fatal prostate cancer. Am J Epidemiol 1984;120(2):244–50.

[39] Calle EE, Rodriguez C, Walker-Thurmond K, Thun MJ. Overweight, obesity, and mortality from cancer in a prospectively studied cohort of U.S. adults. N Engl J Med 2003;348(17):1625–38.

[40] Chow WH, Gridley G, Fraumeni JF Jr, Jarvholm B. Obesity, hypertension, and the risk of kidney cancer in men. N Engl J Med 2000;343(18):1305–11.

[41] Bergstrom A, Hsieh CC, Lindblad P, Lu CM, Cook NR, Wolk A. Obesity and renal cell cancer – a quantitative review. Br J Cancer 2001;85(7):984–90.

[42] Parker AS, Lohse CM, Cheville JC, Thiel DD, Leibovich BC, Blute ML. Greater body mass index is associated with better pathologic features and improved outcome among patients treated surgically for clear cell renal cell carcinoma. Urology 2006;68(4):741–6.

[43] Maurer T, Maurer J, Retz M, Paul R, Zantl N, Gschwend JE, et al. Influence of body mass index on operability, morbidity, and disease outcome following radical cystectomy. Urol Int 2009;82(4):432–9.

[44] Kolonel LN, Ross RK, Thomas DB, Thompson DJ. Epidemiology of testicular cancer in the Pacific Basin. Natl Cancer Inst Monogr 1982;62:157–60.

[45] Davies TW, Prener A, Engholm G. Body size and cancer of the testis. Acta Oncol 1990;29(3):287–90.

[46] Ansari MS, Gupta NP, Hemal AK. Chemoprevention of carcinoma prostate: a review. Int Urol Nephrol 2002;34(2):207–14.

[47] Giovannucci E, Ascherio A, Rimm EB, Stampfer MJ, Colditz GA, Willett WC. Intake of carotenoids and retinol in relation to risk of prostate cancer. J Natl Cancer Inst 1995;87(23):1767–76.

[48] Perabo FG, Von Low EC, Ellinger J, von Rucker A, Muller SC, Bastian PJ. Soy isoflavone genistein in prevention and treatment of prostate cancer. Prostate Cancer Prostatic Dis 2008;11(1):6–12.

[49] Lee ES, Steiner M, Lin R. Thioallyl compounds: potent inhibitors of cell proliferation. Biochim Biophys Acta 1994;1221(1):73–7.

[50] Daviglus ML, Dyer AR, Persky V, Chavez N, Drum M, Goldberg J, et al. Dietary beta-carotene, vitamin C, and risk of prostate cancer: results from the Western Electric Study. Epidemiology 1996;7(5):472–7.

[51] Clark LC, Combs GF Jr, Turnbull BW, Slate EH, Chalker DK, Chow J, et al. Effects of selenium supplementation for cancer prevention in patients with carcinoma of the skin. A randomized controlled trial. Nutritional Prevention of Cancer Study Group. JAMA 1996;276(24):1957–63.

[52] Klein EA, Thompson IM, Lippman SM, Goodman PJ, Albanes D, Taylor PR, et al. SELECT: the selenium and vitamin E cancer prevention trial. Urol Oncol 2003;21(1):59–65.

[53] Shephard RJ. Exercise and cancer: linkages with obesity? Int J Obes Relat Metab Disord 1995;Suppl 4:S62–8.

[54] Whittemore AS, Kolonel LN, Wu AH, John EM, Gallagher RP, Howe GR, et al. Prostate cancer in relation to diet, physical activity, and body size in blacks, whites, and Asians in the United States and Canada. J Natl Cancer Inst 1995;87(9):652–61.

[55] Kinzl JF, Trefalt E, Fiala M, Hotter A, Biebl W, Aigner F. Partnership, sexuality, and sexual disorders in morbidly obese women: consequences of weight loss after gastric banding. Obes Surg 2001;11(4):455–8.

[56] Norman RJ, Clark AM. Obesity and reproductive disorders: a review. Reprod Fertil Dev 1998;10(1):55–63.

20

Pregnancy with Multiple Micronutrients: Perinatal Mortality Reduction

Batool A. Haider, MD, MSc, ScD,*
*Zulfiqar A. Bhutta, PhD, MBBS, FRCPCH, FAAP***

*Harvard School of Public Health, Boston, Massachusetts, USA
**Robert Harding Chair in Global Child Health and Policy, Centre for Global Child Health, Hospital for Sick Children, Toronto, Canada

INTRODUCTION

Micronutrients are vitamins and minerals that are required in minute amounts for normal functioning, growth, and development of the human body [1]. These vitamins and minerals are found in various foods and an appropriate consumption of a variety of foods may prevent deficiencies of these micronutrients. Women in low- and middle-income countries develop multiple micronutrient (MMN) deficiencies as their consumption of micronutrient-rich and fortified foods is low [2].

Micronutrient deficiencies lead to a host of clinical and subclinical conditions including iron deficiency anemia, megaloblastic anemia, congenital abnormalities in neonates, mental retardation, cretinism, night blindness, and skin conditions [1]. The requirement of vitamins and minerals is increased in women during pregnancy due to the increased demands of the growing fetus. Women in low- and middle-income countries who are already deficient in many of these micronutrients are at an increased risk of exacerbation of the condition with resultant impact on the health and well-being of the mother and the baby.

According to the 2011 estimates, the worldwide prevalence of anemia in pregnant women was 38% (95% confidence interval 33–43%), translating into 32 (28–36) million pregnant women globally [3]. In 2011, more than 50% of anemia in pregnant women was due to iron deficiency. Iron deficiency anemia during pregnancy has been associated with low birth weight and preterm birth (PTB) [4]. Deficiencies of vitamin A, zinc, iodine, and folic acid are also prevalent among pregnant women. The role of folic acid deficiency in congenital defects is well-established. But the role of folic acid deficiency in the risk of spontaneous abortions and other birth outcomes has been inconclusive [5]. A recent review of the global burden of maternal malnutrition concluded that 10–19% of women in the reproductive age were seriously malnourished [6]. A low body mass index (BMI), which is BMI <18.5, in women at the time of conception has been associated with an increased risk of birth of a baby with low birth weight.

In low- and middle-income countries, the micronutrient deficiencies are exacerbated by high rates of fertility, low interpregnancy intervals, low maternal age, poverty, and poor diet of the women. Because of the persistently high burden of iron deficiency anemia in pregnant women, the World Health Organization (WHO) has long recommended the prenatal use of iron supplements in low- and middle-income countries, and this is also recommended in many high-income countries [7–9]. In 1998, the UNICEF/UNU/WHO proposed a MMN supplement tablet providing one recommended daily allowance of vitamin A, vitamin B1, vitamin B2, niacin, vitamin B6, vitamin B12, vitamin C, vitamin D, vitamin E, copper, selenium, iodine, zinc, with iron, and folic acid [10]. Given the high burden of MMN deficiencies globally, supplementation with MMN could prove beneficial. There are some concerns regarding an increase in the risk of perinatal mortality in women who consume these MMN supplements [11]. Hence, in this chapter we evaluated the evidence available from the trials of MMN supplementation during pregnancy and conducted a meta-analysis to estimate the effect of these supplements on the risk of perinatal and neonatal mortality and stillbirth outcomes.

Handbook of Fertility. http://dx.doi.org/10.1016/B978-0-12-800872-0.00020-2

METHODS

We used the Cochrane Collaborations methodology to conduct this review. The findings have also been published as part of a larger Cochrane review [12].

Criteria for Considering Studies for this Review

All prospective randomized controlled trials evaluating MMN supplementation in women during pregnancy and its effects on maternal and pregnancy outcomes were included. There was no limit on the length of gestation at the time of enrollment in the study. Trials including HIV-positive women were excluded. However, if trial included both HIV-positive and HIV-negative women, then data for HIV-negative women were included in the review if possible.

Studies comparing supplementation with MMN supplements containing three or more micronutrients compared with placebo, no supplementation, or supplementation with two or less micronutrients were included. We evaluated the effects of micronutrients that were different in the two groups and no cointerventions. We compared MMN supplements containing at least three micronutrients with supplements with two or less, or none, as iron with folic acid or folic acid alone are standard recommendations for pregnant women in many countries. Trials that used fewer than three supplements in the intervention group were excluded regardless of their outcome. There were no limits on the duration of supplementation. Since WHO recommends the use of iron folic acid supplementation in women during pregnancy as a part of routine antenatal care, we also evaluated the effect of MMN supplementation versus supplementation with iron and folic acid. Outcomes included in the review were perinatal mortality, stillbirths, and neonatal mortality. Perinatal mortality was defined as the death of a fetus after 28 weeks of gestation till the first 7 days of life. Stillbirth was defined as the death of a fetus after 28 weeks of gestation but before birth, and neonatal mortality was defined as the death of a neonate within the first 7 days of life.

We included all trials irrespective of language or publication status. Trials that were not published in the English language were translated into English before inclusion in the review. Both individual and cluster-randomized trials were included; however, quasi-randomized trials were excluded.

Search Methods for Identification of Studies

We searched the Cochrane Pregnancy and Childbirth Group's Trials Register by contacting the Trials Search Coordinator (February 17, 2012). The Cochrane Pregnancy and Childbirth Group's Trials Register contains trials identified from the quarterly searches of the Cochrane Central Register of Controlled Trials (CENTRAL); weekly searches of MEDLINE and EMBASE; hand searches of 30 journals and the proceedings of major conferences; weekly current awareness alerts for a further 44 journals; plus monthly BioMed Central email alerts. Details of the search strategies can be found within the editorial information about the Cochrane Pregnancy and Childbirth Group.

We also searched reference lists of retrieved articles and key reviews. We also contacted experts in the field for additional and ongoing trials. We did not apply any language restrictions.

Data Collection and Analysis

Study Selection

The review authors screened the titles and the abstracts of the studies. The full texts of the studies were reviewed to assess eligibility. Any disagreement between the reviewers was resolved through discussion.

Data Extraction and Management

The authors extracted the data from studies that were found to satisfy the inclusion criteria using a data extraction form. Any discrepancy between the reviewers was resolved through discussion. Data were analyzed using the Review Manager software [13]. The authors of the original studies were contacted in case of any query or to obtain further details about the study.

Unit of Analysis Issues

There were five cluster-randomized trials included in this review [14,15,22,23,25]. The generic inverse variance method was used to include the data from these trials. The generic inverse variance method utilizes the cluster-adjusted estimates from the trials. We synthesized the relevant information from both cluster-randomized and individually randomized trials. As there was little heterogeneity between the study designs and the interaction between the effect of the intervention and the choice of randomization unit was considered to be unlikely, we considered it reasonable to combine the results from both designs.

RESULTS

Literature Search

We identified a total of 23 trials (involving 76,532 women) eligible for inclusion in this review but only 12 trials measured perinatal mortality. 13 trials measured stillbirths and 10 presented data for the outcome of neonatal mortality. We excluded 64 trials that did not satisfy the inclusion criteria.

Study Characteristics

A brief description of the included trials is presented next.

Bhutta 2009 [14]

This cluster-randomized trial was conducted in urban and rural areas in Pakistan. Maternal malnutrition, vitamin A deficiency, anemia, and iron deficiency were common in the study setting. Pregnant women with gestational age <16 weeks were eligible for enrollment and were randomized to either multiple-micronutrient supplements ($n = 1148$) or iron with folic acid supplement group ($n = 1230$). Multiple-micronutrient supplements included vitamins A 800 μg, D 200 IU, E 10 mg, C 70 mg, B1 1.4 mg, B2 1.4 mg, niacin 18 mg, B6 1.9 mg, B12 2.6 mg, folic acid 400 μg, iron 30 mg, zinc 15 mg, copper 2 mg, selenium 65 μg, and iodine 150 μg. Iron with folic acid group received 60 mg iron and 400 μg folic acid. Iron and folic acid were given to all participants; however, the dose of iron was 30 mg in the intervention group and 60 mg in the control group. These intervention groups were also randomized to nutritional education. In this review, we compared MMN and MMN + nutritional education groups with iron folic acid and iron folic acid + nutritional education group. The outcomes measured in this trial included size at birth, gestational age at birth, perinatal mortality, and maternal anemia (Hb < 11 g/dL).

There was no significant difference in baseline characteristics between the two groups.

Christian 2003 [15]

This was a double blind cluster-randomized trial, carried out in rural Nepal from December 1998 to April 2001. A total of 4926 pregnant women were enrolled in the study. Women who were pregnant, breastfeeding an infant less than 9 months old, were menopausal, sterilized, or widowed were excluded from the study. Eligible women were randomized into five groups: group 1 included 941 women who received folic acid 400 μg and vitamin A; group 2 ($n = 957$) received folic acid 400 μg, iron 60 mg as ferrous fumerate, and vitamin A; group 3 ($n = 999$) contained the same minerals as group 2 in addition to 30 mg of zinc as zinc sulfate; group 4 ($n = 1050$) received similar micronutrients as group 3 in addition to vitamin D 10 μg, vitamin E 10 mg, vitamin B1 1.6 mg, vitamin B2 1.8 mg, niacin 20 mg, vitamin B6 2.2 mg, vitamin B12 2.6 μg, vitamin C 100 mg, vitamin K 65 μg, copper 2 mg, and magnesium 100 mg. The control group ($n = 1051$) received 1000 μg of vitamin A only. All supplements were given orally from the time of pregnancy detection until 12 weeks after a live birth or 5 weeks after a still birth or a miscarriage. All women were offered two 400 mg single-dose albendazole in the second and third trimester of pregnancy because of the high prevalence of hookworm infestation in this population. Hookworm infestation and vitamin A deficiency were one of the major causes of anemia in this population. Due to this reason, vitamin A was given to all the participants including the control group. The outcomes measured included PTBs, SGA (weight <10 percentile of gestational age), LBW (<2500 g), side-effects, fetal loss, perinatal mortality, neonatal mortality, and 3 month infant mortality.

For the purpose of the review, we included groups 2, 3, and 4 in the MMN group, whereas groups 1 and 5 were in the control group. Baseline characteristics did not differ significantly among the various randomization groups except for ethnicity and land holding.

Dieckmann 1943 [16]

The study was carried out in Chicago, USA. A total of 554 women were selected at random and assigned to four groups: group 1 control ($n = 175$); group 2 received a cereal containing calcium, phosphorus, and iron ($n = 179$); group 3 received vitamin A and D ($n = 98$); group 4 received cereal along with vitamin A and D ($n = 102$). Supplements were given throughout pregnancy. The intervention group (groups 2 and 4) received 100 g of cereal containing calcium 0.78 g, phosphorus 0.62 g, and iron 30 mg, but, on an average, 30–50 g of cereal was consumed each day. The women were also given vitamin A 39,900 IU and vitamin D 5,500 IU daily. The other group (groups 1 and 3) was the control. We included groups 2 and 4 in the MMN group and groups 1 and 3 in the control group. The study measured hemoglobin, serum calcium, phosphorus and protein, PTB, toxemia in pregnancy, pregnancy loss, perinatal mortality, anemia, and placental abruption. The groups were comparable at baseline.

Fawzi 2007 [17]

This study was a double-blind trial in Dar es Salam, Tanzania. Pregnant women who attended antenatal clinics between August 2001 and July 2004, had a negative test for HIV infection, and planned to stay in the city until delivery and for 1 year thereafter with gestational age between 12 and 27 weeks were included in the trial. The study groups were similar with respect to baseline characteristics. The MMN group ($n = 4214$) received vitamin B1 20 mg, B2 20 mg, B6 25 mg, B12 50 μg, C 500 mg, E 30 mg, niacin 100 mg, and folic acid 0.8 mg. The control group ($n = 4214$) received iron and folic acid. Women were randomly assigned to receive either MM or control from the time of enrollment until 6 weeks after delivery. All women irrespective of group received iron 60 mg and folic acid 0.25 mg. Malaria prophylaxis (sulfadoxine-pyrimethamine tablets) at 20 and 30 weeks of gestation was given to all. The study groups were similar with respect to baseline characteristics: low birth weight (<2500 g), preterm delivery (<37 weeks of gestation), very low birth weight (<2000 g), extremely preterm delivery (<34 weeks of gestation), small for gestational age (<10 percentile for gestational age), fetal death, death in first 6 weeks, length, head circumference, placental weight, risk of cesarean section, maternal mortality, hematologic status (Hb < 11 and < 8.5 g/dL), and immune status (CD4 count <775 mm^{-3}, CD8 count <480 mm^{-3}, and CD3 count <1350 mm^{-3}).

Kaestel 2005 [18]

This study was conducted in Guinea-Bissau. Pregnant women with less than 37 weeks of gestation were eligible for enrollment. A total of 2100 women were randomized into three groups: the MMN RDA group, MMN 2 RDA group, and 60 mg iron 400 µg folic acid group. Fifteen micronutrients were included in the supplement at RDA level, except for folic acid, which was included at 400 µg level. The supplement consisted of vitamin A 800 µg, D 200 IU, E 10 mg, C 70 mg, B1 1.4 mg, B2 1.4 mg, niacin 18 mg, B6 1.9 mg, B12 2.6 µg, folic acid 400 µg, iron 30 mg, zinc 15 mg, copper 2 mg, selenium 65 µg, and iodine 150 µg. The intervention group ($n = 1392$) received MMN supplements (supplement RDA $n = 695$, supplement 2 RDA $n = 697$) while the other group received folic acid 400 µg and iron 60 mg $n = 708$. Malaria was endemic but HIV prevalence was relatively low in the study setting. Iron folic acid was given to all participants in this study. There was no significant difference in baseline characteristics between randomization groups. We used the 1 RDA and control groups in this review. Size at birth, gestational age at birth, PTB (<37 weeks of gestation), low birth weight (<2500 g), miscarriage (fetal loss before 28 completed weeks of gestation), perinatal mortality (fetal loss between 28 weeks of gestation and first 7 days of life), neonatal mortality (deaths within the first 28 days of life), maternal hemoglobin, anemia (Hb < 100 g/L) and maternal death (death during pregnancy or within 42 days after termination of pregnancy), and childhood mortality.

Osrin 2005 [19]

This study was undertaken in Janakpur, Nepal. All women attending a designated antenatal clinic at the zonal hospital were considered for enrollment. Women were eligible for enrollment if an ultrasound examination confirmed a singleton pregnancy, a gestational age between 12 and 20 completed weeks, no notable fetal abnormality, no existing maternal illness of a severity that could compromise the outcome of pregnancy, and the participant lived in an area of Dhanusha or the adjoining district of Mohattari accessible for home visits. Participants received supplements throughout pregnancy until delivery. The MMN group ($n = 600$) received tablets containing vitamin A 800 µg, vitamin E 10 mg, vitamin D 5 µg, B1 1.4 mg, B2 1.4 mg, niacin 18 mg, B6 1.9 mg, B12 2.6 µg, folic acid 400 µg, vitamin C 70 mg, iron 30 mg, zinc 15 mg, copper 2 mg, selenium 65 µg, and iodine 150 µg. The control group ($n = 600$) received tablets containing iron 60 mg and folic acid 400 µg. There were two prespecified deviations from the protocol: if a participant's enrollment blood hemoglobin concentration was less than 70 g/L, she was given an extra 60 mg of iron daily, anthelmintic treatment, and her hemoglobin was rechecked after 1 month; and if a participant described night blindness at any time, she was given 2000 µg of vitamin A daily and referred for medical follow-up. Infants were followed up to 3 months. Both groups of participants were comparable at baseline. Birth weight, low birth weight (<2500 g), gestational duration, preterm delivery (<37 weeks of gestation), miscarriage, stillbirth, early and late neonatal death, infant length, and head circumference.

Ramakrishnan 2003 [20]

This randomized controlled trial was carried out during 1997–2000 in Mexico. Pregnant women who were less than 13 weeks pregnant, were not receiving MMN supplementation, and who agreed to participate were included in the study. A total of 873 women were randomized into the MMN group ($n = 435$) and the iron only group ($n = 438$). MMN tablets included the following vitamins and minerals: iron 60 mg as ferrous sulfate, folic acid 215 µg, vitamin A 2150 IU, vitamin D3 309 IU, vitamin E 5.73 IU, thiamin 0.93 mg, riboflavin 1.87 mg, niacin 15.5 mg, vitamin B6 1.94 mg, vitamin B12 2.04 µg, vitamin C 66.5 mg, zinc 12.9 mg, and magnesium 252 mg. The controls were given iron only tablets with 60 mg of iron as iron sulfate. All were given orally, from recruitment 6 days a week until delivery. The two groups did not differ significantly in most of the characteristics at recruitment, except for marital status (more single mothers in MMN supplementation group) and mean BMI (significantly lower in the MMN supplementation group). PTBs (<37 weeks of gestation), small-for-gestational age (below the 10th percentile for birth-weight-for-gestational age), low birth weight (<2500 g), perinatal mortality, mean hemoglobin concentration, and mean serum ferritin were the outcomes measured.

Roberfroid 2008 [21]

This was a factorial, double-blind, randomized controlled trial from March 2004 to October 2006 in the Hounde health district of Burkina Faso. Pregnant women irrespective of gestational age were included. Women who planned to leave the area within 2 years were excluded. The intervention group ($n = 714$) received vitamin A 800 µg, D 200 IU, E 10 mg, B1 1.4 mg, B2 1.4 mg, niacin 18 mg, folic acid 400 mg, B6 1.9 mg, B12 2.6 µg, C 70 mg, zinc 15 mg, iron 30 mg, copper 2 mg, selenium 65 µg, and iodine 150 µg. The control group ($n = 712$)

received folic acid 400 µg and iron 60 mg. Supplement intake was observed directly and were given till 3 months after delivery. Participants were also randomly assigned to receive either malaria chemoprophylaxis (300 mg cholorquine/week) or intermittent preventive treatment (1500 mg sulfadoxine and 75 mg pyrimethamine once in the second and third trimester). All participants received albendazole 400 mg during the second and third trimester. Severely anemic women received ferrous sulfate 200 mg and folic acid 0.25 mg twice daily for 3 months regardless of their allocation groups. The study groups were similar with respect to baseline characteristics except for a small difference in hemoglobin (lower in intervention group) and BMI (lower in control group). Stillbirths (fetal death between 28 weeks of gestation till birth), neonatal deaths, perinatal death, gestation age, PTBs (<37 weeks of gestation), birth weight, low birth weight (<2500 g), small for gestational age (birth weight less than 10 percentile of a reference population), large for gestational age, birth length, Rohrer index, arm circumference, chest circumference, head circumference, hemoglobin in cord blood, and soluble serum transferrin receptor were the outcomes measured.

Summit 2008 [22]

SUMMIT was a double-blind cluster randomized trial in Lombok island of Indonesia. Pregnant women of any gestational age assessed by physical exam and reported LMP between July 1, 2001 and April 1, 2004 were enrolled. The MMN group (n = 15,804) received iron 30 mg, folic acid 400 µg, vitamin A 800 µg, D 200 IU, E 10 mg, C 70 mg, B1 1.4 mg, B6 1.9 mg, B12 2.6 µg, zinc 15 mg, copper 2 mg, selenium 65 µg, iodine 150 µg, and niacin 18 mg. The placebo group (n = 15,486) received iron 30 mg and folic acid 400 µg. Women in both groups received supplements throughout pregnancy until 90 days postpartum. Both groups were comparable in terms of baseline characteristics. Outcomes measured included early infant mortality (death within 12 weeks of birth), neonatal mortality (death within 28 days of birth), early neonatal mortality (death within 7 days of birth), late neonatal mortality (death between 7 and 28 days of birth), postneonatal mortality (death between 28 days and 12 weeks of birth), fetal loss, abortions (fetal loss before 28 weeks of gestation), still births (death between 28 weeks and before delivery), perinatal mortality (still birth or death within 7 days of birth), and maternal mortality related to pregnancy up to 12 weeks postpartum. This study was stopped early due to insufficient funds.

Sunawang 2009 [23]

This was a cluster-randomized trial conducted in two subdistricts of Indramayu district of the west Java province of Indonesia. Pregnant women irrespective of gestational age from May 2000 till August 2003 were included.

Women suffering from diabetes mellitus, coronary heart disease, and tuberculosis were excluded. The intervention group (n = 432) received UNICEF/UNU/WHO recommended formula including 15 micronutrients. The control group (n = 411) received ferrous sulfate 60 mg and folic acid 0.25 mg. Birth weight, birth length, head and chest circumference, hemoglobin, serum ferritin, serum zinc, serum retinol and urinary iodine, miscarriage, stillbirths, and neonatal mortality were the outcomes measured in this trial. Study groups were similar with respect to baseline characteristics. Supplements were given from the time of enrollment at 12–20 weeks gestation and continued up to 30 days postpartum.

Tofail 2008 [24]

The study was conducted in Matlab, a rural subdistrict in the east central plain of Bangladesh from May 2002 till December 2003. Pregnant women with gestational age 6–8 weeks, hemoglobin greater than or equal to 80 g/L, and no serious disease were eligible for enrollment. The MMN group (n = 1224) received vitamin A 800 µg, D 200 IU, E 10 mg, C 70 mg, B1 1.4 mg, B2 1.4 mg, niacin 18 mg, B6 1.9 mg, B12 2.6 mg, folic acid 400 µg, iron 30 mg, zinc 15 mg, copper 2 mg, selenium 65 µg, and iodine 150 µg, while the other group received folic acid and iron (60 mg iron, 400 µg folic acid n = 1265 and 30 mg iron, 400 µg folic acid n = 1248). The control group with 30 mg iron is included in this review. Size at birth, gestational age at birth, perinatal mortality, maternal hemoglobin, motor development and behavioral development, and infant micronutrient status were the outcomes measured. Women were divided into two groups, that is, early food group and usual food group. Each food group was divided into three subgroups of MMN and iron-folic acid groups. Iron folic acid was given to all participants. There was no significant difference in baseline characteristics between randomization groups. Maternal malnutrition was prevalent in this study setting.

Zeng 2008 [25]

This was a community-based cluster randomized trial in two poor rural counties in Shaanxi province of northwest China between August 2002 and January 2006. Pregnant women of less than 28 weeks gestation between August 2002 and January 2006 were included. Group A (n = 2017) received folic acid 0.4 mg, group B (n = 1912) received iron 60 mg and folic acid 0.4 mg, and group C (n = 1899) received iron 30 mg, folic acid 0.4 mg, zinc 15 mg, copper 2 mg, selenium 0.65 mg, iodine 0.15 mg, vitamin A 0.8 mg, B1 1.4 mg, B2 1.4 mg, B6 1.9 mg, B12 0.026 mg, D 0.05 mg, C 70 mg, E 10 mg, and niacin 18 mg. Outcomes measured included birth weight, low birth weight (<2500 g), small for gestational age (weight <10 percentile for gestational age), PTB (<37 weeks of gestation), very PTB (<34 weeks of gestation), gestational age, birth length, head circumference, hemoglobin, anemia (Hb <110 g/L in third trimester), neonatal deaths (death within 28 days of delivery), early neonatal deaths (death within 7 days of delivery), perinatal deaths (fetal death after 28 weeks of gestation plus early neonatal deaths), and mental and psychomotor development outcomes until 1 year of age by using the Bayley Scales of Infant Development (BSID). For review purposes, MMN and iron folate groups are used. Intervention was administered until 6 weeks postpartum. Baseline characteristics at enrollment and both cluster and individual level baseline characteristics were balanced by treatment groups. We used the comparisons of MMN versus iron folate in this review.

Evidence from Trials

Perinatal Mortality

Twelve studies were included in the analysis of perinatal mortality. MMN supplementation did not show any significant impact on the risk of perinatal mortality as compared to supplementation with two or less micronutrients, no supplementation, or a placebo (RR 0.96; 95% CI 0.84–1.10; 12 studies; Fig. 20.1). Eleven trials compared MMN supplementation with iron and folic acid. The analysis demonstrated no significant association of MMN supplementation during pregnancy with perinatal mortality as compared to iron and folic acid supplementation (RR 0.99; 95% CI 0.84–1.16; 11 studies; Fig. 20.2).

Stillbirth

Thirteen trials were included in the analysis of stillbirths. MMN supplementation was not found to be significantly associated with the risk of stillbirths (RR 0.95; 95% CI 0.85, 1.06; 13 studies; Fig. 20.3). Thirteen trials also compared MMN supplementation with iron folic acid. We did not find any significant association between MMN supplementation and stillbirth as compared to iron and folic acid supplementation (RR 0.96; 95% CI 0.86, 1.07; 13 studies; Fig. 20.4).

Neonatal Mortality

Ten trials evaluated the effect of MMN on the risk of neonatal mortality. MMN supplementation was not found to be significantly associated with an increased risk of neonatal mortality as compared to supplementation with two or less micronutrients, no supplementation, or a placebo (RR 1.01; 95% CI 0.89, 1.16; 10 trials; Fig. 20.5). We also did not find any significant association of MMN supplementation as compared to iron and folic acid supplementation (RR 1.01; 95% CI 0.89, 1.15, nine trials; Fig. 20.6).

Study or subgroup	log [Risk ratio] (SE)	Risk ratio IV, Random, 95% CI	Weight (%)	Risk ratio IV, Random, 95% CI
I All trials				
Bhutta, 2009a	−0.5749 (0.22696)		6.7	0.56 [0.36, 0.88]
Christian, 2003	−0.01 (0.1348)		12.9	0.99 [0.76, 1.29]
Dieckmann, 1943	0.5877 (0.552)		1.5	1.80 [0.61, 5.31]
Fawzi, 2007	−0.1508 (0.0836)		18.5	0.86 [0.73, 1.01]
Kaestel, 2005	−0.1863 (0.2099)		7.6	0.83 [0.55, 1.25]
Osrin, 2005	0.1906 (0.2719)		5.1	1.21 [0.71, 2.06]
Ramakrishnan, 2003	0.2151 (0.6601)		1.0	1.24 [0.34, 4.52]
Roberfroid, 2008	0.5766 (0.318)		3.9	1.78 [0.95, 3.32]
SUMMIT, 2008	−0.10536 (0.06651)		20.6	0.90 [0.79, 1.03]
Sunawang, 2009	−0.2485 (0.3125)		4.1	0.78 [0.42, 1.44]
Tofail, 2008	−0.04082 (0.19895)		8.1	0.96 [0.65, 1.42]
Zeng, 2008	0.3221 (0.1711)		9.9	1.38 [0.99, 1.93]
Subtotal (95% CI)			**100.0**	**0.96 [0.84, 1.10]**

Heterogeneity: Tau^2 = 0.02; Chi^2 = 18.91, df = 11 (P = 0.06); I^2 = 42%

Test for overall effect: Z = 0.61 (P = 0.54)

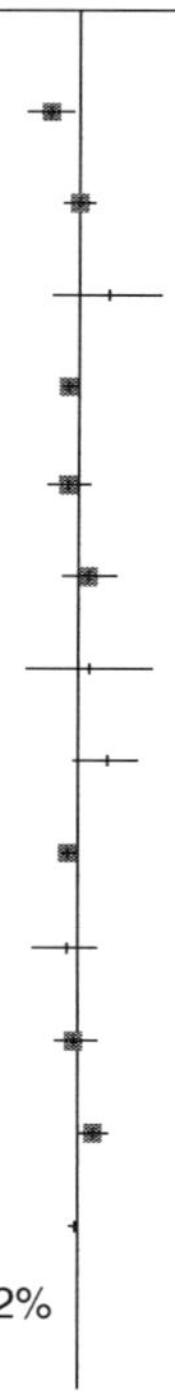

FIGURE 20.1   Randomized controlled trials evaluating the effect of MMN versus control on the risk of perinatal mortality.

Study or subgroup	log [Risk ratio] (SE)	Risk ratio IV, Random, 95% CI	Weight (%)	Risk ratio IV, Random, 95% CI
I All trials				
Bhutta, 2009a	−0.5749 (0.22696)		7.9	0.56 [0.36, 0.88]
Christian, 2003	0.3293 (0.1753)		10.5	1.39 [0.99, 1.96]
Fawzi, 2007	−0.1508 (0.08361)		17.1	0.86 [0.73, 1.01]
Kaestel, 2005	−0.1863 (0.2099)		8.7	0.83 [0.55, 1.25]
Osrin, 2005	0.1906 (0.2719)		6.2	1.21 [0.71, 2.06]
Ramakrishnan, 2003	0.2151 (0.6601)		1.4	1.24 [0.34, 4.52]
Roberfroid, 2008	0.5766 (0.318)		4.9	1.78 [0.95, 3.32]
SUMMIT, 2008	−0.10536 (0.06651)		18.3	0.90 [0.79, 1.03]
Sunawang, 2009	−0.2485 (0.3125)		5.1	0.78 [0.42, 1.44]
Tofail, 2008	−0.04082 (0.19895)		9.2	0.96 [0.65, 1.42]
Zeng, 2008	0.3221 (0.1711)		10.8	1.38 [0.99, 1.93]
Subtotal (95% CI)			**100.0**	**0.99 [0.84, 1.16]**

Heterogeneity: $Tau^2 = 0.03$; $Chi^2 = 22.47$, df = 10 ($P = 0.01$); $I^2 = 56\%$
Test for overall effect: $Z = 0.13$ ($P = 0.90$)

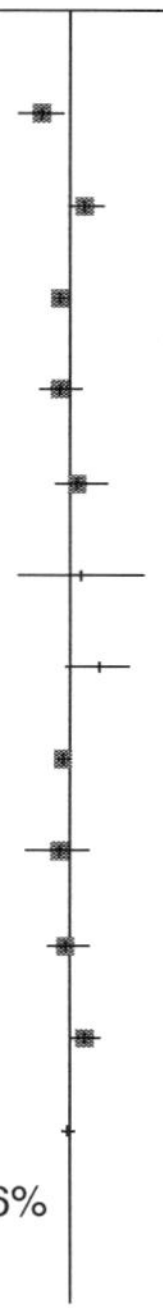

FIGURE 20.2 Randomized controlled trials evaluating the effect of MMN versus iron folic acid on the risk of perinatal mortality.

Study or subgroup	log [Risk ratio] (SE)	Risk ratio IV, Fixed, 95% CI	Weight (%)	Risk ratio IV, Fixed, 95% CI
I All trials				
Kaestel, 2005	−0.7133 (0.304)		3.3	0.49 [0.27, 0.89]
Friis, 2004	−0.5978 (0.6299)		0.8	0.55 [0.16, 1.89]
Osrin, 2005	−0.1863 (0.3475)		2.5	0.83 [0.42, 1.64]
Bhutta, 2009a	−0.17585 (0.2115)		6.7	0.84 [0.55, 1.27]
Fawzi, 2007	−0.1392 (0.1183)		21.6	0.87 [0.69, 1.10]
Sunawang, 2009	−0.1053 (0.4818)		1.3	0.90 [0.35, 2.31]
SUMMIT, 2008	−0.1053 (0.09302)		34.9	0.90 [0.75, 1.08]
Tofail, 2008	0 (0.2867)		3.7	1.00 [0.57, 1.75]
Christian, 2003	0.157 (0.2141)		6.6	1.17 [0.77, 1.78]
Zegre, 2007	0.1655 (0.1982)		7.7	1.18 [0.80, 1.74]
Ramakrishnan, 2003	0.2151 (0.6601)		0.7	1.24 [0.34, 4.52]
Zeng, 2008	0.3001 (0.1902)		8.3	1.35 [0.93, 1.96]
Roberfroid, 2008	0.5539 (0.3835)		2.1	1.74 [0.82, 3.69]
Subtotal (95% CI)			**100.0**	**0.95 [0.85, 1.06]**

Heterogeneity: $Chi^2 = 15.14$, df = 12 ($P = 0.23$); $I^2 = 2.1\%$
Test for overall effect: $Z = 0.94$ ($P = 0.35$)

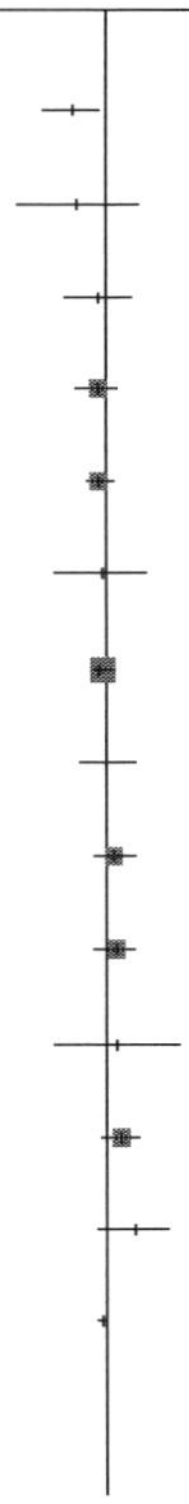

FIGURE 20.3 Randomized controlled trials evaluating the effect of MMN versus control on the risk of stillbirth.

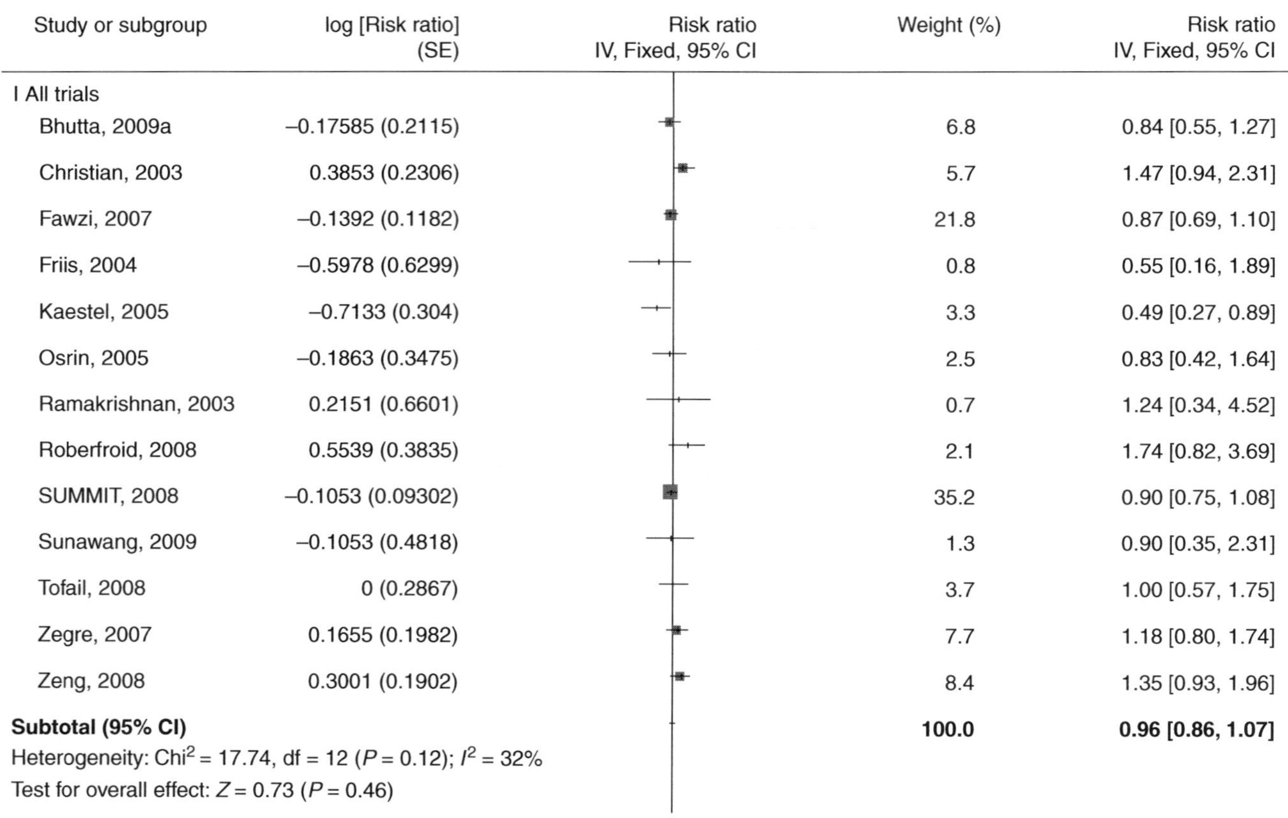

Study or subgroup	log [Risk ratio] (SE)	Risk ratio IV, Fixed, 95% CI	Weight (%)	Risk ratio IV, Fixed, 95% CI
I All trials				
Bhutta, 2009a	−0.17585 (0.2115)		6.8	0.84 [0.55, 1.27]
Christian, 2003	0.3853 (0.2306)		5.7	1.47 [0.94, 2.31]
Fawzi, 2007	−0.1392 (0.1182)		21.8	0.87 [0.69, 1.10]
Friis, 2004	−0.5978 (0.6299)		0.8	0.55 [0.16, 1.89]
Kaestel, 2005	−0.7133 (0.304)		3.3	0.49 [0.27, 0.89]
Osrin, 2005	−0.1863 (0.3475)		2.5	0.83 [0.42, 1.64]
Ramakrishnan, 2003	0.2151 (0.6601)		0.7	1.24 [0.34, 4.52]
Roberfroid, 2008	0.5539 (0.3835)		2.1	1.74 [0.82, 3.69]
SUMMIT, 2008	−0.1053 (0.09302)		35.2	0.90 [0.75, 1.08]
Sunawang, 2009	−0.1053 (0.4818)		1.3	0.90 [0.35, 2.31]
Tofail, 2008	0 (0.2867)		3.7	1.00 [0.57, 1.75]
Zegre, 2007	0.1655 (0.1982)		7.7	1.18 [0.80, 1.74]
Zeng, 2008	0.3001 (0.1902)		8.4	1.35 [0.93, 1.96]
Subtotal (95% CI)			**100.0**	**0.96 [0.86, 1.07]**

Heterogeneity: Chi^2 = 17.74, df = 12 (P = 0.12); I^2 = 32%
Test for overall effect: Z = 0.73 (P = 0.46)

FIGURE 20.4 Randomized controlled trials evaluating the effect of MMN versus iron folic acid on the risk of stillbirth.

Study or subgroup	log [Risk ratio] (SE)	Risk ratio IV, Fixed, 95% CI	Weight (%)	Risk ratio IV, Fixed, 95% CI
Bhutta, 2009a	−0.0266 (0.2023)		11.1	0.97 [0.66, 1.45]
Christian, 2003	0.4055 (0.2334)		8.3	1.50 [0.95, 2.37]
Kaestel, 2005	0.2231 (0.3996)		2.8	1.25 [0.57, 2.74]
Osrin, 2005	0.4253 (0.3829)		3.1	1.53 [0.72, 3.24]
Roberfroid, 2008	0.7419 (0.5041)		1.8	2.10 [0.78, 5.64]
SUMMIT, 2008	−0.1053 (0.08626)		60.9	0.90 [0.76, 1.07]
Sunawang, 2009	0.61618 (0.5067)		1.8	0.54 [0.20, 1.46]
Tofail, 2008	0.2927 (0.2814)		5.7	1.34 [0.77, 2.33]
Vadillo-Ortega, 2011	0.4055 (0.9079)		0.5	1.50 [0.25, 8.89]
Zeng, 2008	0.1398 (0.3401)		3.9	1.15 [0.59, 2.24]
Total (95% CI)			**100.0**	**1.01 [0.89, 1.16]**

Heterogeneity: Chi^2 = 11.13, df = 9 (P = 0.27); I^2 = 19%
Test for overall effect: Z = 0.19 (P = 0.85)

FIGURE 20.5 Randomized controlled trials evaluating the effect of MMN versus control on the risk of neonatal mortality.

Study or subgroup	log [Risk ratio] (SE)	Risk ratio IV, Fixed, 95% CI	Weight (%)	Risk ratio IV, Fixed, 95% CI
Bhutta, 2009a	−0.0266 (0.2023)		11.1	0.97 [0.66, 1.45]
Christian, 2003	0.3988 (0.2324)		8.4	1.49 [0.94, 2.35]
Kaestel, 2005	0.2231 (0.3996)		2.9	1.25 [0.57, 2.74]
Osrin, 2005	0.4253 (0.3829)		3.1	1.53 [0.72, 3.24]
Roberfroid, 2008	0.7419 (0.5041)		1.8	2.10 [0.78, 5.64]
SUMMIT, 2008	−0.1053 (0.08626)		61.2	0.90 [0.76, 1.07]
Sunawang, 2009	−0.61618 (0.5067)		1.8	0.54 [0.20, 1.46]
Tofail, 2008	0.2927 (0.2814)		5.8	1.34 [0.77, 2.33]
Zeng, 2008	0.1398 (0.3401)		3.9	1.15 [0.59, 2.24]
Total (95% CI)			100.0	**1.01 [0.89, 1.15]**

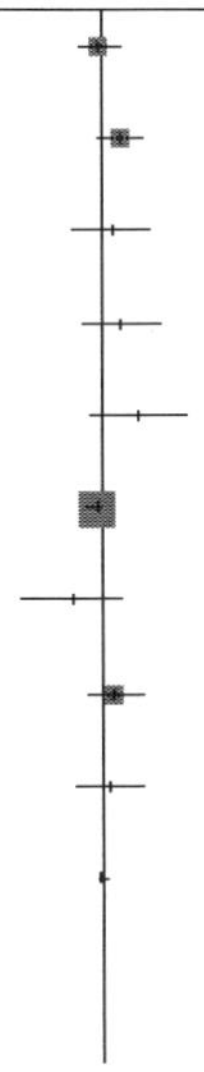

Heterogeneity: Chi^2 = 10.87, df = 8 (P = 0.21); I^2 = 26%
Test for overall effect: Z = 0.16 (P = 0.88)

FIGURE 20.6 Randomized controlled trials evaluating the effect of MMN versus iron folic acid on the risk of neonatal mortality.

DISCUSSION

This systematic review and meta-analysis did not find any significant effect of MMN supplementation on the risk of perinatal mortality, stillbirth, or neonatal mortality as compared to iron folic acid supplementation, supplementation with two or less micronutrients, no supplementation, or a placebo.

These findings corroborate those from other reviews [26–28]. An earlier version of the Cochrane review, which included data from nine trials evaluating the benefits of MMN supplementation, also did not report any significant association between MMN and the risk of perinatal mortality, stillbirth, or neonatal mortality as compared to iron folic acid supplementation, supplementation with two or less micronutrients, no supplementation, or a placebo [26]. Similar findings were observed in other reviews evaluating this association [27,28].

There were concerns regarding the use of MMN supplements during pregnancy [11]. A pooled analysis of two trials of MMN supplements conducted in Nepal showed an increase in the risk of perinatal and neonatal mortality in the MMN group. This increase in the risk was postulated to be due to the shift in the birth weight distribution and potentially higher rates of birth asphyxia in these neonates [11]. These two trials individually had found a nonsignificant increase in the risk of neonatal and perinatal mortality; however, the pooled effect estimate indicated a significant association. These concerns have not been supported by other reviews or studies. [29,30] One of the largest trials of MMN supplementation during pregnancy was conducted in Indonesia and administered through routine health services [22]. This trial provided evidence of a reduction in the risk of neonatal mortality in the MMN supplemented group as compared to the control. As in the SUMMIT trial supplementation was administered through routine health services; it included the presence of skilled birth attendants at the time of delivery, which was however not the case in trials conducted in Nepal.

Another assessment of the MMN supplements using the methodology of the Child Health Epidemiology Research Group (CHERG) also did not support the findings of an increase in neonatal mortality with MMN supplements [26]. As indicated earlier, an increase in the risk could be attributed to the absence of skilled care at delivery. This hypothesis is supported by the findings of an increase in the risk of neonatal mortality in subgroups of populations where majority of the births occurred at home in the absence of skilled care, while no effect was observed where majority of the births occurred in a health facility. A large trial was recently conducted in Bangladesh including around 44,000 pregnant women. The findings from this trial will further consolidate the evidence of the impact of MMN supplements during pregnancy on these outcomes.

CONCLUSIONS

A review of trials evaluating MMN supplementation during pregnancy does not provide evidence of an increase in the risk of perinatal mortality, neonatal mortality, or stillbirths in low- or middle-income countries. Future trials should include careful surveillance of the study population to measure these outcomes.

This Cochrane Review is published in the Cochrane Database of Systematic Reviews 2012, Issue 11. Cochrane

Reviews are regularly updated as new evidence emerges and in response to feedback, and the Cochrane Database of Systematic Reviews should be consulted for the most recent version of the Review [12].

References

[1] Bowman BA, Russell RM. Present Knowledge in Nutrition. 8th ed. Washington, DC: ILSI Press; 2001.

[2] Huffman SL, Baker J, Shumann J, Zehner ER. The case for promoting multiple vitamin/mineral supplements for women of reproductive age in developing countries. The LINKAGES Project (www.linkagesproject.org); 1998.

[3] Stevens G, Finucane M, De-Regil L, Paciorek C, Flaxman S, Branca F, et al. Global, regional, and national trends in total and severe anaemia prevalence in children and pregnant and non-pregnant women. Lancet Glob Health 2013;1(1):e16–25.

[4] Haider BA, Olofin I, Wang M, Spiegelman D, Ezzati M, Fawzi WW, et al. Anaemia, prenatal iron use, and risk of adverse pregnancy outcomes: systematic review and meta-analysis. BMJ 2013; 346(f3443) doi: 10.1136/bmj.f3443.

[5] Lassi ZS, Salam RA, Haider BA, Bhutta ZA. Folic acid supplementation during pregnancy for maternal health and pregnancy outcomes. Cochrane Database Syst Rev 2013;3. CD006896, DOI 10.1002/14651858.CD006896.pub2.

[6] Black RE, Allen LH, Bhutta ZA, Caulfield LE, de Onis M, Ezzati M, et al. Maternal and child undernutrition: global and regional exposures and health consequences. Lancet 2008;371:243–60.

[7] Stoltzfus R, Dreyfuss M. Guidelines for the Use of Iron Supplements to Prevent and Treat Iron Deficiency Anaemia. Washington, DC: ILSI Press; 1998.

[8] Centers for Disease Control and Prevention. Recommendations to prevent and control iron deficiency in the United States. MMWR Recomm Rep 1998;47(RR-3):1–29.

[9] Institute of Medicine. Iron Deficiency Anemia: Guidelines for Prevention, Detection, and Management Among U.S. Children and Women of Childbearing Age. Washington, DC: National Academy Press; 1993.

[10] UNICEF. Composition of a multi-micronutrient supplement to be used in pilot programmes among pregnant women in developing countries 1999. UNICEF (www.UNICEF.org); 1999.

[11] Christian P, Osrin D, Manandhar DS, Khatry SK, de L Costello AM, West KP Jr. Antenatal micronutrient supplements in Nepal. Lancet 2005;366:711–2.

[12] Haider BA, Bhutta ZA. Multiple-micronutrient supplementation for women during pregnancy. Cochrane Database Syst Rev 2012;11. CD004905, DOI 10.1002/14651858.CD004905.pub3.

[13] Review Manager (RevMan) [Computer program]. Version 5.0. Copenhagen, The Nordic Cochrane Centre: The Cochrane Collaboration; 2008.

[14] Bhutta ZA, Rizvi A, Raza F, Hotwani S, Zaidi S, Soofi S, et al. A comparative evaluation of multiple micronutrient and iron-folate supplementation during pregnancy in Pakistan: impact on pregnancy outcomes. Food Nutr Bull 2009;30(4):S496–505.

[15] Christian P, West KP Jr, Khatry SK, LeClerq SC, Pradhan EK, Katz J, et al. Effects of maternal micronutrient supplementation on fetal loss and infant mortality: a cluster-randomised trial in Nepal. Am J Clin Nutr 2003;78:1194–202.

[16] Dieckmann WJ, Adair FL, Michel H, Kramer S, Dunkle F, Arthur B, et al. Calcium, phosphorus, iron, and nitrogen balances in pregnant women. Am J Obstet Gynecol 1943;47:357–68.

[17] Fawzi WW, Msamanga GI, Urassa W, Hertzmark E, Petraro P, Willett WC, et al. Vitamins and perinatal outcomes among HIV-negative women in Tanzania. N Engl J Med 2007;356(14):1423–31.

[18] Kaestel P, Michaelsen KF, Aaby P, Friis H. Effects of prenatal multimicronutrient supplements on birth weight and perinatal mortality: a randomised, controlled trial in Guinea-Bissau. Eur J Clin Nutr 2005;59(9):1081–9.

[19] Osrin D, Vaidya A, Shrestha Y, Baniya RB, Manandhar DS, Adhikari RK, et al. Effects of antenatal micronutrient supplementation on birthweight and gestational duration in Nepal: double-blind, randomised controlled trial. Lancet 2005;365:955–62.

[20] Ramakrishnan U, Gonzalez-Cossio T, Neufeld LM, Rivera J, Martorell R. Multiple micronutrient supplementation during pregnancy does not lead to greater infant birth size than does iron-only supplementation: a randomized controlled trial in a semirural community in Mexico. Am J Clin Nutr 2003;77:720–5.

[21] Roberfroid D, Huybregts L, Lanou H, Henry MC, Meda N, Menten J, et al. Effects of maternal multiple micronutrient supplementation on fetal growth: a double-blind randomized controlled trial in rural Burkina Faso. Am J Clin Nutr 2008;88:1330–40.

[22] Shankar AH, Jahari AB, Sebayang SK, Aditiawarman, Apriatni M, et al. The Supplementation with Multiple Micronutrients Intervention Trial (SUMMIT) Study Group. Effect of maternal multiple micronutrient supplementation on fetal loss and infant death in Indonesia: a double-blind cluster-randomised trial. Lancet 2008;371(9608):215–27.

[23] Sunawang, Utomo B, Hidayat A, Kusharisupeni, Subarkah. Preventing low birth weight through maternal multiple micronutrient supplementation: a cluster-randomized controlled trial in Indramayu, West Java. Food Nutr Bull 2009;30(4):S488–95.

[24] Persson LA, Eneroth H, Ekstrom EC. Multiple micronutrient supplementation during pregnancy: a review of effects on birth size, maternal haemoglobin, and perinatal mortality demonstrated in trials in Bangladesh, Guinea-Bissau, and Pakistan. Report for UNICEF/UNU/WHO; 2004.

[25] Zeng L, Cheng Y, Dang S, Yan H, Dibley MJ, Chang S, et al. Impact of micronutrient supplementation during pregnancy on birth weight, duration of gestation and perinatal mortality in rural western China: double blind cluster randomised controlled trial. BMJ 2008;337:2001.

[26] Haider BA, Bhutta ZA. Multiple-micronutrient supplementation for women during pregnancy. Cochrane Database Syst Rev 2006;4. CD004905.

[27] Haider BA, Yakoob MY, Bhutta ZA. Effect of multiple micronutrient supplementation during pregnancy on maternal and birth outcomes. BMC Public Health 2011;11(Suppl. 3):S19.

[28] Ronsmans C, Fisher DJ, Osmond C, Margetts BM, Adou P, Aguayo VM, et al. Effect of multiple micronutrient supplementation during pregnancy on stillbirths and early and late neonatal mortality: a meta-analysis. Food Nutr Bull 2009;30(4):S547–55.

[29] Bhutta ZA, Haider BA. Maternal micronutrient deficiencies in developing countries. Lancet 2008;371:186–7.

[30] Bhutta ZA, Haider BA. Prenatal micronutrient supplementation: are we there yet? Can Med Assoc J 2009;180(12):1188–9.

21

Nutrition and Hormones: Role in Male Infertility

Lakshana Sreenivasan, BSc
The University of Arizona, Arizona, USA

INTRODUCTION

Spermatogenesis is a reproductive process that occurs in men who attain puberty. Spermatozoa are produced from primordial germ cells. Male germ cells – the spermatogonia – are regulated by BMP8B. Optimum concentration of BMP8B leads to the differentiation of the male germ cells. Failure to produce spermatozoa could lead to infertility [1]. Male infertility accounts for approximately half of the infertility cases in the world. About 7% of the general male population is infertile. Male infertility can be caused by hormonal imbalances, genetic factors, congenital disorders, and environmental factors that affect the genes involved in metabolism or hormone production. The etiology of male infertility is not understood in approximately 50% of the cases and this is called "idiopathic infertility" [2]. Apart from the genetic factors, environment, diet, and lifestyle play an enormous role in the causation of male infertility. Drugs and alcohol abuse, lack of exercise, vitamin deficiency, exposure to environmental toxins like tobacco smoke, radiation, and pesticide, or stress are some of the factors that influence spermatogenesis.

Amino acids in the diet are significant for entire body homeostasis and necessary for reproduction and growth. Arginine-deficiency in adult males decreases the sperm count by 90% and also increases the number of nonmotile sperm by 10 times. Arginine is an important dietary constituent for adult males. Previous studies showed 18% increase in sperm and about 7.6% increase in sperm motility. Arginine causes increased concentrations of nitric oxide and polyamines, which are vital for spermatogenesis and sperm motility. Arginine supplements could improve male fertility [3]. Modifying testicular physiology could lead to male infertility by decreasing the sperm count. Seminiferous tubules and Sertoli cells offer structural support for the nascent germ cells with the help of N-cadherin, connexin43, or E-cadherin. Improper cell–cell interactions between Sertoli and germ cells cause apoptosis of the germ cells preventing spermatogenesis. Recent studies have depicted bile acids as endocrine factors that are influenced by the nuclear receptor FXRα and the G protein-coupled receptor TGR5. These two receptors are located in the testes and have significant function in metabolism making them a good therapeutic target. Adult male mice were used to demonstrate the influence of bile acids on testicular physiology and fertility. Dietary cholic acid, a derivative of bile acids was administered to adult male mice. About 24% of the mice were found to be infertile post-treatment and there was a reduction in the number of spermatozoa. Mice treated with cholic acid also showed increased apoptosis of germ cells. In the absence of the bile acid receptor TGR5, there was no change in fertility of the mice. Testicular steroidogenesis was repressed after exposure to FXRα agonists. Another finding showed that decrease in connexin43 and N-cadherin showed sloughing of germ cells in response to bile acid exposure [4].

Twenty-one studies were conducted with 13,077 men in relation to obesity and infertility. The experimental findings suggest that there was an association between obesity and azoospermia or oligozoospermia. A higher body mass index (BMI) does lower the sperm count but scientists are not certain if a lower BMI could reverse the effect. Another result in this study suggested that the alteration of the hypothalamic-pituitary-gonadal axis decreases the level of free testosterone and increases estradiol levels [5]. Sugar-sweetened beverage has been previously studied in relation to obesity. Obesity is speculated to cause a decrease in sperm count. A study was conducted with 305 men. Semen analyses, reproductive hormone measurement, and diet were assessed before the men were approved for the study. The results suggest that a higher intake of sugar-sweetened beverages decreased sperm motility and viability. The result was only significant in lean men and was not observed in

Handbook of Fertility. http://dx.doi.org/10.1016/B978-0-12-800872-0.00021-4

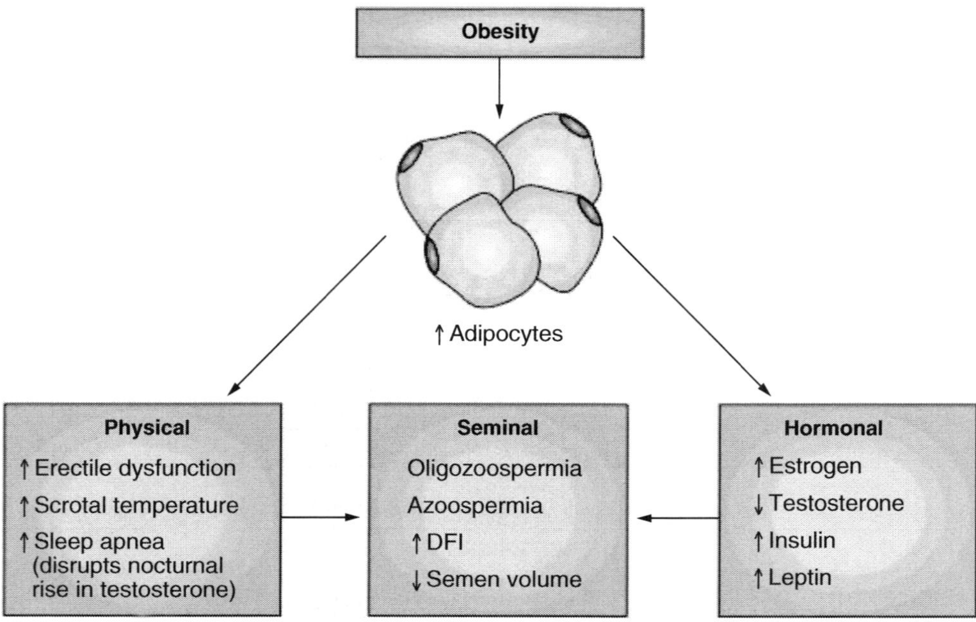

FIGURE 21.1   Factors affected by obesity and its role in male infertility.

obese men. The follicle-stimulating hormone levels were inversely related to the intake of sugar-sweetened beverages. In a previous study conducted, it was found that the intake of sugar-sweetened beverage causes insulin resistance, which in turn leads to oxidative stress. Oxidative stress is also known to cause decreased sperm motility and influence male infertility (Fig. 21.1) [6].

Obesity is hypothesized to play a significant role in the quality of the sperm and male infertility. The underlying mechanism of this relationship is not very well understood. Obesity is due to the high number and size of adipocytes. As mentioned earlier in this review, an increase in total body fat was associated with lower levels of testosterone. Aromatase cytochrome P459 enzyme plays an important role in the biosynthesis of estrogen. Obese men show an increase in aromatase enzyme activity leading to increased production of estrogen. Higher concentrations of estrogen affect spermatogenesis and decrease the free testosterone levels by negative feedback [7]. In another study, mice models were used to check the association of male infertility with P450 aromatase enzyme expression and compared them to male infertility caused by testicular fibrosis. AROM+ mice are infertile due to overexpression of P450 aromatase enzyme. The increase in enzyme levels caused testicular inflammation and is also characterized by fibrosis. The results showed that hormonal imbalance plays a key role in chronic inflammation and testicular fibrosis [8]. Leptin is a protein hormone secreted by adipose tissues. Previous studies have shown an increase in leptin concentration in obese men. Leptin deficiency in mice was found to be associated with poor spermatogenesis. Another study conducted showed that decreasing the leptin concentrations could improve the

spermatogenesis and sperm quality. Improper orientation of the hypothalamus-pituitary-gonadal axis could affect the feedback regulation of testosterone levels. The FSH levels are also affected by negative regulation of the testosterone levels, thereby contributing to male infertility [9].

Zinc (Zn) is a very important component for proper testicular function. Earlier studies showed that zinc deficiency leads to poor Leydig cell function and lower testosterone levels. Dietary zinc supplementation could reverse the effect and promote high testosterone levels. Zn is primarily found in the Leydig cells and influence the testosterone levels in association with the follicle-stimulating hormone. Previous studies have shown that Zn deficiency is associated with decreased sperm count, sperm viability, sperm motility, improper spermatogenesis, and reduced angiotensin converting enzyme activity and oxidative damage (Fig. 21.2) [10].

Nigella sativa L. seed, often known as the black seed, is used as a preservative in Asia, Africa, and the Middle East. *N. sativa* was administered to diabetic patients in a previous study. It was not toxic and had no fatal effects. In the study conducted by Kolahdooz et al., optimum intake of *N. sativa* by men with abnormal semen characteristics led to enhanced sperm count, sperm viability, and sperm motility. The black seed also has antioxidant properties. It reduces the reactive oxygen species. Inactivation of free radicals reduces oxidative stress and improves sperm morphology and other factors [11]. Oxidative stress and damage are known to cause several major diseases like cancer and neurodegenerative and cardiovascular diseases. Oxidative damage is mostly caused by reactive oxygen species [12]. Previous studies showed that the motility of the sperm decreases in the presence of reactive oxygen species. Reactive

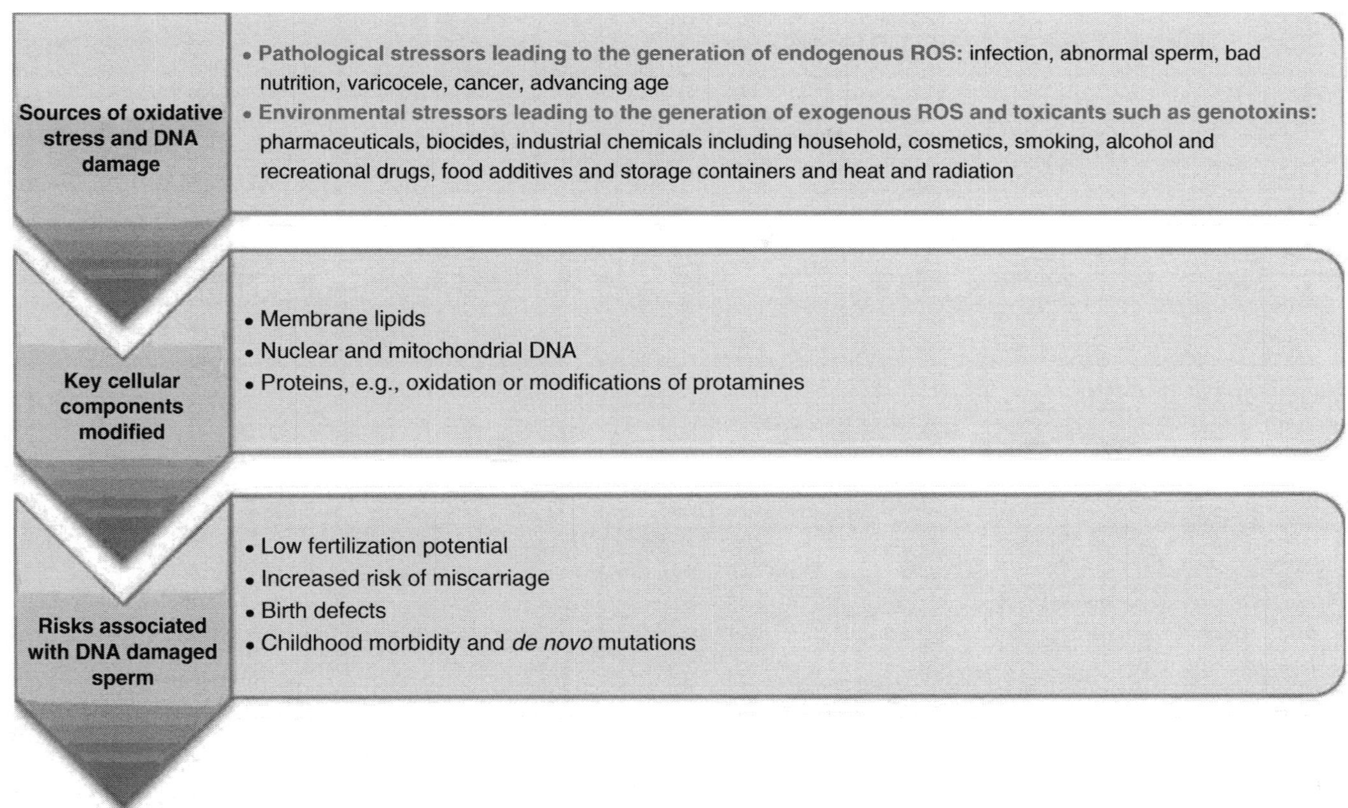

FIGURE 21.2 Oxidative stress and DNA damage potentially lead to male infertility [12].

oxygen species causes DNA damage and fragmentation. There have been strong associations between antioxidant compounds and oxidative DNA damage. Increased levels of antioxidants like vitamin C have shown to improve semen quality [13]. Overall this compendium reflects on the challenges faced with male infertility. The underlying mechanisms of how obesity affects male infertility are not very well understood. Vitamin deficiency and hormonal imbalance also influence sperm production and motility. Decrease in testosterone levels, sperm count, sperm viability, and poor sperm motility are the most common factors leading to male infertility. Further experimentation and promising therapeutic targets need to be ventured into for better male infertility treatment options. Based on current research knowledge, we can deduce that diet and hormones play a significant role in maintaining homeostasis and proper reproductive function.

References

[1] Gilbert SF. Developmental Biology. 6th edn. Sunderland, MA: Sinauer Associates; 2000.

[2] Krausz C. Male infertility: pathogenesis and clinical diagnosis. Best Pract Res Clin Endocrinol Metab 2011;25:271–85.

[3] Wu G. Amino acids: metabolism, functions, and nutrition. Amino Acids 2009;37:1–17.

[4] Baptissart M, Vega A, Martinot E, Pommier AJ, Houten SM, Marceau G, et al. Bile acids alter male fertility through TGR5 signaling pathways. Hepatology 2014;60(3):1054–65.

[5] Sermondade N, Faure C, Fezeu L, Shayeb AG, Bonde JP, Jensen TK. BMI in relation to sperm count: an updated systematic review and collaborative meta-analysis. Hum Reprod Update 2013;19(3): 221–31.

[6] Chiu YH, Afeiche MC, Gaskins AJ, Williams PL, Mendiola J, Jorgensen N, et al. Sugar-sweetened beverage intake in relation to semen quality and reproductive hormone levels in young men. Hum Reprod 2014;29(7):1575–84.

[7] Hammoud AO, Gibson M, Peterson CM, Hamilton BD, Carrell DT. Obesity and male reproductive potential. J Androl 2006;27(5): 619–26.

[8] Strauss L, Adam M, Mayerhofer A., Poutanen M P450 aromatase over-expressing mice (AROM +) as a model for infertility in men. In: Endocrine Abstracts. Proceedings of European Society of Endocrinology, The Netherlands, Rotterdam. N.p.: Bioscientifica, n.d. p. 142.

[9] Du Plessis S, Cabler S, McAlister DA, Sabanegh E, Agarwal A. The effect of obesity on sperm disorders and male infertility. Nat Rev Urol 2010;7(3):153–61.

[10] Croxford TP, McCormick NH, Kelleher SL. Moderate zinc deficiency reduces testicular Zip6 and Zip10 abundance and impairs spermatogenesis in mice. J Nutr 2011;141(3):359–65.

[11] Kolahdooz M, Nasri S, Modarres SZ, Kianbakht S, Huseini HF. Effects of *Nigella sativa* L. seed oil on abnormal semen quality in infertile men: a randomized, double-blind, placebo-controlled clinical trial. Phytomedicine 2014;21(6):901–5.

[12] Gharagozloo P, Aitken RJ. The role of sperm oxidative stress in male infertility and the significance of oral antioxidant therapy. Hum Reprod 2011;26(7):1628–40.

[13] Catak Z, Aydin S, Sahin I, Kuloglu T, Aksoy A, Dagli AF. Regulatory neuropeptides (ghrelin, obestatin and nesfatin-1) levels in serum and reproductive tissues of female and male rats with fructose-induced metabolic syndrome. Neuropeptides 2014;48(3):167–77.

22

Nutrition and Developmental Programming of Central Nervous System (CNS)

Sayali Chintamani Ranade, PhD

National Brain Research Centre, Manesar, Haryana, India

INTRODUCTION

The development of all body systems in an individual is an outcome of the interaction between genetic and environmental factors. The adverse *in utero* environment can therefore alter the trajectory of this development. Recent advances in research have identified early life as a critical developmental window in which an organism is particularly vulnerable to perturbations. The factors affecting gestational and early postnatal development include maternal diet, lifestyle, maternal disease, their treatment, and exposure to environmental pollutants.

The last two decades have seen a considerable increase in the incidence of metabolic syndrome worldwide. The metabolic abnormalities characteristic of this pathological entity include abdominal obesity, glucose intolerance, dyslipidemia, hypertension, and atherosclerosis [1]. Evidence from animal studies and human population data has offered substantial evidence that most of these pathological conditions can be traced back to their *in utero* origins.

DEVELOPMENTAL ORIGINS OF ADULT DISEASE

During the late 1980s, Barker and colleagues proposed the hypothesis called the "developmental origins of adult disease hypothesis" (DoAD). The hypothesis suggested that an adverse intrauterine environment can induce permanent changes in the metabolism and physiology of the offspring. These alterations in turn lead to an increased risk of disease in adulthood. The *in utero* environment not only controls the growth and development of the fetus, but also programs its subsequent health. It is now evident that the origins of many adult diseases lie in their early life environment. Since then, substantial experimental evidence from different mammalian species has reliably supported the hypothesis that an adverse *in utero* environment has a programming role in postnatal physiology and pathophysiology [2].

According to Barker's hypothesis, the period of pregnancy and the intrauterine environment are crucial to the tendency to develop diseases like hypertension, diabetes, coronary heart disease, metabolic disorders, pulmonary, renal, and mental illnesses [3]. The spectrum of such postnatal consequences is increasing and now includes cancer, immune system disorders, and pathologies in brain functions.

ROLE OF MATERNAL NUTRITION IN EARLY LIFE PROGRAMMING

One of the important parameters that affect gestational and early postnatal development is nutrition. Maternal nutrition largely impacts the *in utero* environment available to the offspring, thereby becoming one of the most important factors sculpting various developing systems. Nutrition, in addition to providing energy during development, also modulates the developing systems of the offspring. Optimal supply of essential nutrients during development ensures proper maternal health and also normal, healthy development of the fetus. Unlike other environmental factors, nutrition can exert its effects directly by modifying the gene structure and expression of genetic factors. The clinical data and results from animal experimentation clearly suggest the

Handbook of Fertility. http://dx.doi.org/10.1016/B978-0-12-800872-0.00022-6

decisive role of early life malnutrition in the outcome of health during adulthood.

Initial evidence correlating nutritional factors and developmental programming came from "pseudo-experiments" of maternal nutritional manipulation in human pregnancy such as the epidemiological data from the Dutch Hunger Winter or Great Chinese famine condition of 1959–1962. These results from such pseudo-experiments or epidemiological studies are complex to interpret since those were not controlled studies and there were multiple factors involved. The defects reported cannot be associated to any particular nutrient or deficiency. However, they do suggest possible involvement of multiple nutritional factors. These results have been used to carry out well-planned animal studies. Animal experimentation offers a reasonable solution, wherein even a single nutrient deficiency or a specific developmental window can be studied in great detail by modifying experimental parameters.

The effects of nutrition become more pronounced if encountered during development versus in adulthood. This early life vulnerability of the developing organism to the nutritional insults can be attributed to the limited availability of essential nutrients stores, developing metabolic pathways, and immature physiological functions. All these factors make the developing system more susceptible to the developmental insults. Adverse nutritional environment during intrauterine life can result in long lasting and sometimes permanent changes in the physiology and metabolism of the offspring. These altered parameters in turn leads to an increased risk of diseases in adulthood. Perinatal maternal malnutrition sensitizes the offspring to the development of several chronic diseases including metabolic syndrome disorders, breathing disorders, anemia, and psychiatric-neurological disorders. The mechanisms controlling nutritional programming are very complex and will be discussed elsewhere in this chapter.

NUTRITIONAL DEFICIENCIES

The term "nutrition" refers to the optimum, balanced intake of nutrients, essential for proper functioning of all body systems. Any deviation from this optimum intake is termed as "malnutrition." This could be either qualitative or quantitative in nature. Quantitative nutritional deviation can be of two types: undernutrition refers to insufficient caloric intake (chronic energy deficiency, CED, with BMI <18.5) and overnutrition is increased food intake accompanied by reduced energy expenditure leading to the storage of excess energy in the form of fat.

The term qualitative malnutrition includes imbalance in intake of one or more nutritional constituents. These nutritional constituents could be either macronutrient/s or micronutrient/s. The intake, excess, or lack of specific nutrient/s during pregnancy influence pregnancy outcome and future health of an infant. These nutritional deviations could further be classified into macronutrient deficiencies, like protein deficiency, and micronutrient deficiencies such as iron, folic acid, vitamin D, zinc, and calcium deficiency.

Although the last two decades have seen a small decline in the cases of undernutrition, both under- and malnutrition continues to be prevalent in low-income and middle-income countries encompassing major portions of Asia and Africa [4]. In economically developed countries malnutrition and undernutrition continue to be a problem in certain sections of society [5]. The diet of people who live below the poverty line is not adequate in terms of caloric value. In addition it is highly imbalanced and contains very little protein or unbalanced proportions of one or more micronutrient content. The incidence of anemia among pregnant women can be as high as 80% in certain developing countries [6]. In addition to socioeconomic strata, other reasons of maternal malnutrition are maternal disease conditions such as diabetes mellitus, hypertension, and eating disorders. Maternal and child malnutrition in low-income and middle-income countries encompasses both undernutrition and a growing problem of excess body weight and obesity [4]. Health risks associated with maternal obesity will be discussed later in this chapter.

THE DEVELOPING NERVOUS SYSTEM

Brain development, like any other body system, also occurs in the privileged compartment of the mother's body – the uterus and any unfavorable *in utero* change is perceived as a potential risk factor affecting normal brain development. The developing nervous system in general or brain in particular is highly vulnerable to the developmental insults due to several reasons. Major events in the brain development occur around the time of birth, both prenatally and postnatally. Since the brain has a critical developmental window centered around the time of birth, it is logical to think that any perturbation in the *in utero* environment or postnatal nutrition would significantly affect the brain development. Many nutritional insults exert their effects through stress pathways. The brain is also the central organ responsible for stress responses, determining the adaptive or maladaptive responsiveness to stress thereby altering its structure and/or function [7].

Brain development is a highly organized process in which the pattern of events is very crucial in terms of their sequence and timing. These events include neurogenesis, neuronal migration, synaptogenesis, gliogenesis, and neuronal wiring. Each of these events either sequential or overlapping has to occur in a proper temporal window. Any deviation from this pattern has the potential to lead to an aberrant brain development and significantly compromised brain function.

Perinatal brain development is associated with different developmental processes and a high growth rate. Each of these developmental processes exhibits specific requirements of nutrients such as amino acids (protein synthesis) or fatty acids (e.g., brain growth). Depending on the specific requirement of nutrient/s during a particular developmental window and brain region involved therein, certain nutritional factors have greater impact on the particular brain region development compared to others. The effect of any nutrient deficiency or overabundance on brain development will be governed by the principle of timing, dose, and duration [8]. Each of these will be discussed in detail in the course of this chapter.

Type of Deficiency

The type of deficiency or deficiencies encountered during development also decides the future of specific brain region or brain function. Prenatal and early postnatal protein calorie malnutrition (PCM) is associated with neurotransmitter, cellular, electrophysiological, and behavioral disruptions that have been demonstrated in schizophrenia [9]. Involvement of folate deficiency in neural tube defects during prenatal periods is one of the most strongly established findings in the literature. Many experimental evidences show the association between periconceptional folate supplementation and diminished neural tube defects [10,11]. Maternal supplementation with cod liver oil, which contains very long-chain n-3 fatty acids, during pregnancy has been associated with higher IQ at the age 4 years [12]. Different types of malnutritional deficiencies have been associated with dysfunctions of different types of memories associated with the hippocampus [13]. Each of these factors can affect brain development either in isolation or in conjunction with other factors.

Timing of Deficiency

Different developmental processes occur in different brain regions, in a very tightly regulated temporal window. Therefore, the timing of the exposure to the deficiency becomes one of the important parameters affecting brain development. These effects could be species-specific, brain-region-specific or even process-specific. Maternal malnutrition is thought to have its most severe impact on the offspring during the period of rapid brain growth, which largely differs among species. In rodents, the period of rapid brain growth is marked by the last week of gestation, and peaks thereafter during lactation [14]. In humans 24–42 weeks of gestation is particularly

vulnerable to nutritional insults due to the rapid trajectory of several neurologic processes, including synapse formation and myelination [15,16].

The effect of timing of deficiency exposure also depends upon the developmental trajectory of a particular brain region coinciding with the deficiency timing. The hippocampus is the brain region exhibiting corticocortical connectivity and functionality during the late gestational period [17]. So any nutritional deficiency encountered during late gestation will significantly affect the hippocampus compared to other brain regions. On the other hand the large proportion of cerebellar maturation occurs postnatally [18]. So any postnatal nutritional insult, like protein deficiency, will have a larger impact on cerebellum maturation versus other brain regions [19].

The timing of deficiency exposure also shows specificity toward a particular developmental processes. During early periods of development the predominant developmental process is cell proliferation, so gestational nutritional insults will significantly affect cell number [20,21]. Nutritional deficiencies encountered during late gestation or early postnatal period will have an effect on processes like differentiation, synaptogenesis, and neuronal connectivity.

In some cases maternal nutritional status during the preconception period also affects the growth and development of the young one [22]. When sows enter pregnancy in poor nutritional condition with decreased body reserves, this can negatively affect the growth and development of early embryos [23]. Maternal protein restriction in the rat [24] during periconceptional or preimplantation periods resulted in modifications of embryo development leading to abnormalities of the offspring in late gestation and postnatal periods.

Dose or Severity

The dose or severity of a particular nutritional insult is another factor that decides the outcome of future brain development. The rule of thumb is, the stronger the deficiency, the higher the impact. Every dietary factor or nutrient has its optimum range in which it can favor normal development. Anything significantly above or below that level could be detrimental to the normal development of an offspring. Severe quantitative reduction in food intake during pregnancy has long been associated with permanent brain dysfunction, particularly at cognitive and behavior levels [25]. The effect of moderate reduction in maternal nutrition on the fetus have been poorly examined, since they are less detrimental with lesser severity because there is prioritization of nutrients at the maternal level, wherein the fetus gets the available nutrition at the expense of the mother's requirements.

Some nutrients, like iron, are regulated within a relatively narrow range, wherein a slight deviation from the normal range induces abnormal brain development. Others, like zinc, iodine, and choline, have wider ranges of tolerance [8]. A nutrient at one concentration may promote normal brain development or at another concentration may be harmful.

Optimum energy intake of the mother is essential for the proper growth and development of the offspring. Conditions of CED or protein energy malnutrition (PEM) have shown varying degrees of defects depending on severity of malnutrition. In our own laboratory, pups born to dams maintained on 70% calorie restriction showed lower brain-to-body ratio compared to their 50% calorie restricted counterparts (unpublished data).

A study showing the effects of maternal protein deficiency of varying degrees on brain oxidative status has reported that severe protein calorie restriction (4%) shows the highest oxidative damage in the cortex and cerebellum compared to moderate protein restriction (12%) and control (24% of dietary protein) [26].

Duration

Duration is another important parameter deciding the outcome of brain development. Exposure to deficiency for longer periods will have a stronger effect with many regions involved. Since a longer period of exposure covers the critical developmental window of many different regions and also many developmental processes, this will have a larger impact with more global distribution. Our own data involving maternal protein deficiency and its effect on cerebellar development show that extended periods of maternal protein malnutrition, including both prenatal and postnatal developmental window, have the cerebellar defects more universally distributed across the cerebellum. These defects are not cerebellar lobe specific or cell type specific [27]. The defects induced by intrauterine or prenatal protein malnutrition, however, show very specific defects, which include delayed cerebellar neurogenesis [28], loss of Granule cell layer [29], and aberrant Purkinje cell dendritic arborization [30].

The hippocampus, a brain region associated with spatial memory function, also exhibit varying degrees of defects on this function depending on the duration for which it is exposed to maternal iron deficiency [31]. The pups exposed to maternal iron deficiency showed the structural and molecular defects in the hippocampus, which were correlated with radial arm maze performance – a measure of hippocampus-associated memory. The reference memory function in particular was affected. The pups exposed to iron deficiency throughout pregnancy and lactation displayed the complete spectrum of defects, while pups from dams that were iron deficient only during pregnancy or during lactation displayed subsets of defects [31].

Thus, it is evident from the above mentioned data that longer exposures to nutritional deficiencies show stronger impact on brain development compared to the shorter duration. In the case of shorter duration of deficiencies, the immediate replenishment of the deficient factor may be responsible for the rescued phenotypes since during that period the brain still remains amenable to external influences like replenishment or fortification.

In addition to deficiency, qualitative or quantitative, excess of any nutritional factor or overnutrition also plays a significant role in metabolic programming [32]. About 68% of the US population is obese [33] and 60% of the UK population is predicted to be obese by 2050 [34]. This obesity epidemic becomes a major health challenge, attributable to obesity-associated health risks. The increased prevalence of metabolic disorders in recent years is a reflection of an adverse lifestyle favoring high fat intake and low physical activity. This imbalance between energy intake and expenditure becomes one of the major detrimental factors of a hazardous lifestyle. Maternal obesity becomes one important risk factor for many metabolic diseases for the mother during gestation and postpartum, and for the child later in life. With this surge in maternal obesity and its increasingly deleterious effects, recent studies have focused on the potential detrimental effects of fetal overnutrition. These studies have established a strong cause and effect relationship between maternal obesity and the onset of metabolic syndrome in the offspring [32].

Recent evidences show a significant role of maternal obesity in brain programming as well. The hypothalamus–obesity axis is highly vulnerable to maternal nutritional status. Maternal obesity during the pre- and postnatal period has been shown to sensitize the hypothalamus–obesity axis leading to altered appetite and reprogrammed fat cell metabolism [35]. Another important region affected by maternal obesity is the hippocampus. A high fat diet during pregnancy and postweaning has been shown to alter the expression of genes mediating the hippocampal synaptic efficacy and impairs spatial learning and memory in adulthood [36]. Maternal obesity during pregnancy has been shown to alter specific brain cell networks. Diminished proliferation and neuronal maturation has been observed in certain brain regions of the infants born to obese mothers. The brain regions involved are the third ventricle, hypothalamus, and cerebral cortex [37].

PROBABLE MECHANISM OF ACTION

It is now evident from the discussion so far that the suboptimal nutritional status of *in utero* environment remodels many developing systems of the fetus, thereby predisposing them to health risks manifested in the later part of life. The programming potential of the adverse *in utero* environment proposed originally by Barker and colleagues [38,39] was based on epidemiological studies. The associative nature of the *in utero* environment and fetal programming was further supported by animal experimentation. Although the physiological mechanisms involved in such fetal programming remain largely unknown, some hypotheses were proposed to explain the association between suboptimal nutrition and fetal remodeling.

The "thrifty gene hypothesis" was proposed to explain the influence of perinatal perturbations on fetal programming and susceptibility of the offspring to adult diseases [40,41]. When the fetus is exposed to an altered intrauterine environment, it makes certain changes or adaptations, which then ensures its survival. These adaptations are in the form of alerted set points of various developing systems. These adaptive changes are beneficial during the intrauterine period because they adapt to the fetus's current needs. These adaptations are carried by the fetus into its postnatal environment. Now if the postnatal environment is different than that of the *in utero* environment, then the altered set point does not match with it. This mismatch leads to the diseased condition. The larger this mismatch is, the higher is the risk involved.

An extension of the thrifty genes hypothesis came in the form of the predictive adaptive response (PAR). The PAR is a developmental trajectory taken by an organism during a period of developmental plasticity in response to perceived environmental cues [42]. This hypothesis suggests that based on its intrauterine environment a fetus predicts a certain postnatal environment. Based on this prediction, changes or adaptations are made. Since these adaptations happen during the periods of developmental plasticity, they remain permanent and are carried into the offspring's adult life. Now if there is a mismatch between the predicted postnatal environment and the actual one, the offspring is predisposed to a diseased condition. Although both these hypotheses, which are extensions of each other, give a conceptual insight into the developmental programming of adult disease, they do not offer much information about the actual mechanisms involved.

The mechanisms controlling nutritional programming are very complex and none of them occur in isolation or exclusion of the other. Many or all of these influences act synergistically to get the adapted phenotype of the fetus. The influences fetal programming exerts can act at the genetic, epigenetic, and endocrine level.

Genetic and environmental factors together shape up the phenotype of a new-born and are involved in the intrauterine programming process. Nutritional intervention greatly influences this programming at the genetic level. The best example of genetic regulation comes from perinatal programming of obesity. Human epidemiological studies demonstrate a strong influence of the mother's body weight on infant obesity [43,44]. The obesity programming acts through critical pathways of increased appetite and adipogenesis. It has been demonstrated that in response to maternal undernutrition, the offspring may express an increased ratio of appetite/satiety gene expression [45–47]. In addition, IUGR male offspring (who have higher propensity to develop obesity in the later part of life) show upregulated adipogenic signaling cascade as evidenced by an increased expression of enzymes promoting adipocyte [48]. These changes occur at the level of the hypothalamus.

In endocrine regulation of fetal programming, many hormones and neuropeptides have been implicated. Two types of hormones play a significant role: stress hormones and metabolism controlling hormones.

Glucocorticoids (GCs) act at multiple sites and show a spectrum of effects during fetal programming. Elevated plasma cortisol levels *in utero* predisposes the offspring to the risk of glucose intolerance, hypertension, and dyslipidemia (metabolic syndrome) [49–51], and can be linked to the development of cardiovascular disease, insulin resistance, and diabetes in later life [52,53]. Many brain regions, including the hypothalamus [54] and hippocampus [55], express a significant number of receptors for these glucocorticoids. Glucocorticoids play a major role in altering the hippocampal-hypothalamus-pituitary-adrenal (HPA) axis sensitivity. The fetal levels of these hormones are highly sensitive to nutritional perturbation. Elevated levels of glucocorticoids *in utero* can bring about changes in the levels of receptors, and associated channels, enzymes, and transporters, thereby reprogramming the HPA axis. Glucocorticoids regulate behavior and neuroendocrine activities by acting through specific intracellular receptors in the brain and periphery. In addition, GCs are necessary for hippocampal neuronal survival, memory acquisition and consolidation [56]. Thus, stress hormones both regulate and are regulated by developmental programming in response to nutritional manipulation. Early life stress has also been known to affect cognitive development of the offspring and may also predispose it to psychiatric disorders. Thus, prenatal stress or exposure to excess glucocorticoids is one of the pathways by which nutritional fetal programming occurs [31] and thus studying the role of stress in fetal programming might also provide the insight into nutrition-induced fetal programming and adult pathophysiology.

Metabolic hormones such as leptin, insulin, IGF-1 are closely associated with and programmed during metabolic remodeling. Several nutritional manipulations, like maternal food restriction [57], have been shown to affect the hypothalamic circuitry. This effect has been shown to be associated with a drastic reduction or even absence of postnatal leptin surge in the offspring [58]. The prenatal or neonatal level of leptin can modify neuronal plasticity, thus enabling the animal to adapt to a lower nutrient supply. However, if postnatal conditions are with a higher nutrient supply, the organism can develop the metabolic syndrome since it has adapted to poor nutritional conditions [59]. The action of leptin is mainly through hypothalamic circuitry programming [60]. Fetus exposure to over- [37] or undernutrition [61] also resulted in an increased level of circulating leptin in later life.

Prenatal undernutrition was found to be associated with unexpected endocrine responses of leptin, IGF1, and cortisol during fasting. The fetal programming is suggested to interfere with the hypothalamic integration of important endocrine axis [62].

Epigenetic mechanisms have been implicated in mediating the long-lasting effects seen in perinatal programming in response to nutritional modification. The epigenome is more dynamic and amenable to changes compared to the genome. The epigenetic aberrations represent long-term consequences of exposure to inadequate or inappropriate nutritional status during critical periods of development. These epigenetic modifications change the binding of transcription factors to specific gene promoters, and/or alter the conformation and function of chromatin, which modulates gene expression. The predominant epigenetic modifications include methylation, acetylation, phosphorylation, and ubiquitination [63].

The epigenome could be affected globally or sometimes specifically with respect to the gene involved, region, or organ. Individuals who were exposed to the Dutch Hunger Winter prenatally exhibited less DNA methylation of IGF gene clusters, compared to their unexposed counterparts [64].

The remodeled HPA axis, which underlies many late onset disease conditions, is one very good example of epigenetic modifications. Programming of the HPA axis in response to nutritional manipulations involves alterations in the expression of genes attributable to epigenetic remodeling of chromatin in the hippocampus and many peripheral tissues [65]. The genes include glucocorticoid receptor genes and 11-β-HSD2 [66].

Maternal calorie restriction models have demonstrated hypermethylation of the oncogene H Ras in adult offspring [67]. Changes in histone modifications include losses in acetylation and increases in dimethylation of H3 at the glucose transporter type 4 (GLUT4) locus. Periconceptional undernutrition in sheep have been shown to result in epigenetic changes in the hypothalamic proopiomelanocortin and GR genes in the fetus [68].

Thus, many of the mechanisms discussed earlier play part in remodeling the fetal system exposed to early life nutritional manipulation. These mechanisms mostly act synergistically to adapt to the adverse *in utero* situation. These are those adaptations that predispose a fetus to many late onset diseases. Proper understanding of these mechanisms may help us prevent or at least delay the onset of these diseases with developmental origin.

NUTRITIONAL INADEQUACIES AND THEIR NEUROLOGICAL CONSEQUENCES

Brain development is primarily regulated by nutrients and growth factors. So any deviation from the optimum for any of these two factors is detrimental to brain development. The developing brain is highly vulnerable to nutritional insults and can get affected at various levels including neuroanatomical, cellular, molecular, behavioral, neurochemical, and oxidative status.

Adequate nutrition is necessary from the beginning even before conception; however, the importance of nutrients like folic acid [69], copper [70], and vitamin A [71], B12 [69] becomes apparent at the early stage of brain development – the neural tube formation – which is about 3 weeks from the day of conception. The essential amino acid tryptophan is required in optimum amounts in embryonic development as early as E11. The reduction or absence of this amino acid causes a less efficient serotonergic system [72].

Polyunsaturated fatty acids, like docosahexaenoic acid DHA-EPA and acideicosa-pentanoic acid, are essential nutrients not produced in the body. They play a protective role in neurological development including development of the brain and retina. Fish is an important source of these fatty acids so a diet having sufficient intake of fish ensures proper neurological development. Low dietary intake of n-3 PUFAs during gestation causes increased proinflammatory cytokines and modification of neuronal plasticity-related gene expression. These effects may contribute to neurodevelopmental alterations leading to neurodevelopmental disorders [73]. Excess dietary intake of omega 3 fatty acid on the other hand has also been known to adversely affect sensory/neurological function possibly through neurotoxicity mechanism [74].

Gestational vitamin D insufficiency has been linked to altered brain development and adult mental health [75]. Deficiency of vitamin B12 can cause delayed myelination, demyelination of nerves, imbalance of neurotropic and neurotoxic cytokines, and/or accumulation of lactate in brain cells [76].

Early life deficiencies of trace elements, like iron, zinc, copper, and selenium, have been implicated in several brain pathologies and disorders [77]. Limbic system pathologies associated with early life iron deficiency include hippocampal alterations, myelination defects, and altered neurotransmitter synthesis. The brain disorders like Friedreich's ataxia and neurodegenerative diseases like Alzheimer's and Parkinson's on the other hand have been associated with excess iron [77]. Rodent data show that zinc deficiency during the critical period of development is involved in delayed CNS maturation and reduced brain size thereby affecting behavior and memory [78]. Copper deficiency manifests itself in the form of neurological alterations such as spastic gait and sensorial ataxia. CNS demyelination, optical neuritis, and isolated peripheral neuropathy have also been reported [79]. Selenium deficiency show altered spatial memory defects, long-term potentiation deficits in the hippocampus [80], and motor function deficits as evidenced by spastic pathology [81].

However, managing the absolute amount of these nutrients is not sufficient since they interact among themselves. It is therefore essential to maintain their relative amounts or ratio.

Reduction in total calorie intake due to poverty, disease, or lifestyle has significant effects on the development of the nervous system. The consequences of severe quantitative reduction in food intake during pregnancy has long been associated with permanent brain dysfunction, particularly at cognitive and behavior levels. Some structural changes at brain level and excitability level modification have also been reported in some rodent studies [82,83].

Even moderate malnutrition has now been shown to be involved in causing major cerebral developmental disturbances. The several mechanisms suggested include neurotrophic factor suppression, cell proliferation and cell death imbalance, impaired glial maturation and neuronal process formation, downregulation of gene ontological pathways and related gene products, and upregulated transcription of cerebral catabolism [84].

Rats malnourished for prolonged periods, during gestation, lactation, and adulthood showed alterations in the power spectra of the EEG of areas of the neocortex and limbic system [85].

Protein malnutrition during the perinatal period is among the leading nongenetic factors affecting brain development [86]. Human studies and animal experimentation show that prenatal protein malnutrition predisposes the offspring to neurocognitive deficits [87]. Perinatal protein malnutrition is another nutritional deficiency affecting various brain regions including cerebrum, brainstem, spinal cord, hippocampus, and cerebellum. The effects of protein deficiency on the brain are both global as well as local. The defects range from decreased brain size and weight [88] to decreased velocity of impulse conduction [89]. Depending on the timing encountered protein deficiency affects processes like neurogenesis [90], cellular migration, differentiation [91], synaptogenesis, plasticity [92,93], and nerve myelination [94]. Prenatal protein malnutrition leads to a decrease in neuronal and glial density in the cerebral cortex and cerebellum [95]. Early life protein deficiency also reduces spine density in brain areas like the hippocampus [96]

and cerebellum [97]. At the biochemical level it affects various neurotransmitter systems [89,98]. At the behavioral level performance at different behavioral tasks is affected. These tasks include spatial memory task [99], social interaction, reactivity to aversive stimuli [100], and rotarod [27]. In addition, protein deficiency also causes delayed velocity of impulse conduction [89] and motor coordination [27]. Overnutrition has recently been implicated in metabolic syndrome defects and even in some neuroinflammatory and neurodegenerative disorders [101]. Obesity is not only an imbalance of calorie intake and expenditure but it also involves dysregulation of neurohormones and neurotransmitter. In addition, recent evidence also suggests the role of neuroinflammation of CNS (hypothalamus in particular) on the etiology of metabolic syndrome. Perturbations in neuroendocrine signaling [102,103] and neuroinflammation [104] lie central to pathologies induced by overnutrition to the CNS. A mechanistic insight suggests a close association between overnutrition-induced metabolic diseases and neurodegenerative diseases like Alzheimer's and Parkinson's. This is a recent advancement in nutritional research and needs to be further looked into.

NUTRITIONAL DEFICIENCIES AND PSYCHIATRIC DISORDERS

The evidence for the neurodevelopmental origin of psychiatric illnesses, like schizophrenia [105,106], depression [107], and bipolar disorder [108], has come from the postmortem brain studies of patients suffering from these disorders. The prenatal and postnatal developmental windows therefore become the critical stages of development wherein the brain is highly susceptible to environmental perturbations. Nutritional insult during critical stages of development has a potential to lead to an inappropriately developed nervous system, which ultimately shows a compromised brain function.

Malnutrition early in life has been implicated in the subsequent development of cognitive and behavioral impairments in childhood and adolescence; decreased attention, conduct problems, and decreased IQ have all been documented [107]. The limbic system, which controls emotions, cognition, and motor functions, is highly susceptible to early life nutritional perturbation. These limbic system pathologies and defects in many other brain structures, like the hippocampus, hypothalamus, and cerebellum, may underlie malnutrition-induced cognitive alterations such as depression, anxiety, attention, and memory deficits.

Converging evidence suggests that *in utero* exposure to nutritional deficiency is a determinant of schizophrenia [109]. The first evidence for it came from early gestational exposure to the Dutch Hunger Winter of

1944–1945 and to a severe famine in China and their association with increased risk of schizophrenia in offspring. Nutrient candidates folate, essential fatty acids, retinoids, vitamin D, and iron may be implicated in the etiology of schizophrenia [109].

ADHD etiopathology also shows the involvement of nutritional deprivation. IUGR, which is a common consequence of *in utero* nutritional deficiency, elevates the risk for negative neurobehavioral developmental outcomes of ADHD by 10% [110,111]. Mouse models of ADHD when fed a protein-deficient diet throughout pregnancy and lactation show eightfold elevated expression levels of dopamine-related genes in PFA and the hypothalamus and a behavior similar to ADHD [112]. Prenatal dietary-restricted animals showed endophenotypes similar to ADHD symptoms [113].

Low birth weight or IUGR has also been implicated in child and adolescent depression in epidemiologic studies in the United States [114] as well as in adults from the Dutch Famine cohort [109]. Individuals with histories of early childhood malnutrition endorse a higher level of depressive symptoms than peers who did not experience such an episode [115].

In addition to other nutritional disturbances imbalances in the levels of these trace elements have also been shown to be linked with psychiatric disorders such as depression, anxiety, and attention and memory deficits [77].

EARLY LIFE MALNUTRITION AND AGING BRAIN

There is a close relationship between early life malnutrition and predisposition to late onset diseases. Early life malnutrition is known to durably modify the developing nervous system and its effects run not only into adulthood but also contribute significantly to the aging process. Early life nutritional insults act mainly by programming the HPA axis through neural and hormonal inputs. The nervous system thus gets a stronger impact of early life malnutrition in terms of altered threshold or sensitivity. Since the aging brain is generally debilitating in terms of its sensory, motor responses, and many other aspects such as cognition and motivation, the important question that remains to be answered is whether this age-dependent decline is different in the brain malnourished during developmental periods versus the well-nourished one.

One of the many theories of Alzheimer's disease includes early life malnutrition [116]. Vitamin D deficiency in early life is known to affect neuronal differentiation, axonal connectivity, dopamine ontogeny, and brain structure and functions. It has been implicated in many late onset neurological disorders including depression and Alzheimer's disease [117]. Early life undernutrition has been shown to alter the levels of reduced glutathione in the mouse forebrain [118]. The sensitivity of the aging HPA axis to exogenous stimulation is found to be altered in sheep exposed to early life malnutrition [119]. Except for these isolated studies not much is known about the effect of early life malnutrition on the aging brain. Since much of the metabolic programming also happens through the neural route, more work should focus on studying the effects of gestational and lactational nutritional insults on the aging brain. The mechanisms involved in these effects also should be studied in great detail.

NUTRITIONAL INTERVENTION

Increasing incidences of metabolic syndrome, neurodegenerative disorders, and psychiatric illnesses impose significant health and economic burden in developing and developed countries. Human epidemiological studies and data from animal experimentation has strongly established the role of lifestyle on the etiology of these pathological conditions.

Diet or nutrition is emerging as one robust factor underlying these pathologies. Nutrition becomes more important during pregnancy and infancy, which are critical periods for the formation of the brain wherein the foundation for the cognitive, motor, and socio-emotional skills is laid. The nervous tissue is a metabolically highly demanding tissue, so nutrition has a significant effect on its development. Diet, however, is a multidimensional exposure and thus the prevalence or symptomology of these pathologies cannot be attributed to any single nutrient or food group. In addition, none of these nutrients act in isolation; they interact with each other in a very precise manner to shape up developing systems. Similarly, many brain regions develop simultaneously and also form functional connections for optimum brain function. Since these deficiencies are encountered during the formative period, when the brain is exhibiting its highest level of plasticity, most of these adaptations remain long lasting or in many cases irreversible.

The two major challenges nutritional neuroscience is facing in the current scenario therefore are as follows:

1. Whether the irreversible nature of these *in vitro* maladaptations can be fully or partially restored by using nutritional intervention therapy or by any other therapeutic measure.
2. Whether predisposition to many late onset neurological disorders induced by *in utero* nutritional insults can be reversed or prevented or at least delayed by nutritional modification during the perinatal period or during adulthood.

Although reports from animal studies and a few randomized trials on a human population have yielded some data on nutritional interventions and reversibility of these defects to a certain extent [120–122], the most effective timing of this nutritional intervention is not very clearly defined. More studies are required, which will enable us to define the precise critical window for the interventions. Energy restriction has been known to reverse or recover some defects of early life nutritional insults by acting through modulation of the HPA axis [123]. Physical training for longer periods has also been found to be beneficial to reverse these effects of early life nutritional deficiencies [124]. The energy restriction and physical activity can thus be used as potential therapies. More research should focus on delineating the detailed mechanisms involved in these alternative therapies.

Since the mechanism of action of nutritional insult on the developing brain involves an intricate interplay of hormones, neurotropic factors, and their receptors, more attention should be diverted to establish the role of each of these factors independently and of their interactions. A clear and detailed insight into the mechanism will give us a better understanding of the levels at which the nutritional factors are acting, which will help us design better intervention therapies.

In conclusion, early life nutrition is an important factor modulating adult phenotype. The developing nervous system is highly susceptible to the nutritional perturbations during the critical periods of development. The significant effect of early life nutritional insults manifests itself in later stages of life as neurodevelopmental, neurodegenerative, or psychiatric disorders. Detailed studies using appropriate animal models are required to delineate the mechanisms involved in these malnutrition-induced defects. These mechanistic details may provide the knowledge required for optimizing preventative and rehabilitative strategies to reverse, prevent, or delay these disorders.

References

[1] Orozco-Solis R, Matos RJB, Guzman-Quevaedo O, Lopes de Souza S, Bihouee A, Houlgatte R, et al. Nutritional programming in the rat is linked to long-lasting changes in nutrient sensing and energy homeostasis in the hypothalamus. PLoS One 2010;5(10):e13537–42.

[2] Chen M, Zang L. Epigenetic mechanisms in developmental programming of adult disease. Drug Discov Today 2011;16(23–24): 1007–18.

[3] Capra L, Tezza G, Mazzai F, Boner AL. The origins of health and disease: the influence of maternal diseases and lifestyle during gestation. Ital J Pediatr 2013;39:7.

[4] Black RE, Victora GC, Walker SP, Bhutta ZA, Mercedes PC, de Onis ME, et al. The Maternal and Child Nutrition Study Group Maternal and child undernutrition and overweight in low-income and middle-income countries. Lancet 2013;382:427–51.

[5] International Dietary Energy Consultancy Group (IDECG) Annual Report 1997.

[6] Sharma JB, Soni D, Murthy NS, Malhotra M. Effect of dietary habits on prevalence of anemia in pregnant women of Delhi. J Obstet Gynaecol Res 2003;29(2):73–8.

[7] McEwen BS. Central effects of stress hormones in health and disease: understanding the protective and damaging effects of stress and stress mediators. Eur J Pharmacol 2008;583:174–85.

[8] Georgieff MK. Nutrition and the developing brain: nutrient priorities and measurement. Am J Clin Nutr 2008;85(Suppl.): 614S–20S.

[9] Brown AS, Susser ES. Prenatal nutritional deficiency and risk of adult schizophrenia. Schizophr Bull 2008;34(6):1054–63.

[10] Laurence KM, James N, Miller MH, Tennant GB, Campbell H. Double-blind randomised controlled trial of folate treatment before conception to prevent recurrence of neural-tube defects. BMJ (Clin Res Ed) 1981;282:1509–11.

[11] MRC Vitamin Study Research Group. Prevention of neural tube defects: results of the Medical Research Council Vitamin Study. Lancet 1991;338:131–7.

[12] Helland IB, Smith L, Saarem K, Saugstad OD, Drevon CA. Maternal supplementation with very-long-chain n-3 fatty acids during pregnancy and lactation augments children's IQ at 4 years of age. Pediatrics 2003;111:e39–44.

[13] Ranade S, Rose AJ, Rao M, Galego J, Gressens P, Mani S. Different types of nutritional deficiencies affect different domains of spatial memory function checked in radial arm maze. Neuroscience 2008;152:859–66.

[14] Galler JR, Shumsky JS, Morgane PJ. Malnutrition and brain development. In: Walker WA, editor. Nutrition in pediatrics. 2nd ed. London: BC Decker Inc; 1997. p. 196–212.

[15] Rao R, Georgieff MK. Early nutrition and brain development. In: Nelson CA, editor. The effects of early adversity on neurobehavioral development. Minnesota Symposium on Child Psychology, 31. Hillsdale, NJ: Erlbaum Associates; 2000. p. 1–30.

[16] Kretchmer N, Beard JL, Carlson S. The role of nutrition in the development of normal cognition. Am J Clin Nutr 1996;63(6):997S–1001S.

[17] Thompson RA, Nelson CA. Developmental science and the media: early brain development. Am Psychol 2001;56:5–15.

[18] Hatten ME, Alder J, Zimmerman K. Genes involved in cerebellar cell specification and differentiation. Curr Opin Neurobiol 1997;7:40–7.

[19] Dobbing J. Vulnerable periods of brain development. In: Lipids, malnutrition and the developing brain. Ciba Foundation Symposium; 1971. p. 9–29.

[20] Winick M, Rosso P. The effect of severe early malnutrition on cellular growth of the human brain. Pediatr Res 1969;3:181–4.

[21] Stead JD, Neal C, Meng F. Transcriptional profiling of the developing rat brain reveals that the most dramatic regional differentiation in gene expression occurs postpartum. J Neurosci 2006;26:345–53.

[22] Guilloteau P, Zabielski R, Hammon HM, Metges CC. Adverse effects of nutritional programming during prenatal and early postnatal life, some aspects of regulation and potential. J Physiol Pharmacol 2009;60(3):17–35.

[23] Vinsky MD, Novak S, Dixon WT, Dyck MK, Foxcroft GR. Nutritional restriction in lactating primiparous sows selectively affects female embryo survival and overall later development. Reprod Fertil Dev 2006;18:347–55.

[24] Kwong WY, Wild AE, Roberts P, Willis AC, Fleming TP. Maternal undernutrition during the preimplantation period of rat development causes blastocyst abnormalities and programming of postnatal hypertension. Development 2000;127(19):4195–202.

[25] Olness K. Effects on brain development leading to cognitive impairment: a worldwide epidemic. J Dev Behav Pediatr 2003;24:120–30.

[26] Tatli M, Guzel A, Kizil G, Kavak V, Yavuz M, Kizil M. Comparison of the effects of maternal protein malnutrition and intrauterine growth restriction on redox state of central nervous system in offspring rats. Brain Res 2007;1156:21–30.

[27] Ranade SC, Nawaz S, Rambtla PK, Rose AJ, Gressens P, Mani S. Early protein malnutrition disrupts cerebellar development and impairs motor coordination. British J Nutr 2012;107:1167–75.

[28] Wallingford JC, Shreder RE, Zeman FJ. Effect of maternal protein-calorie malnutrition on fetal rat cerebellar neurogenesis. J Nutr 1980;110:543–51.

[29] Clos J, Favre C, Selme-Matrat M. Effects of undernutrition on cell formation in the rat brain and specifically on cellular composition of the cerebellum. Brain Res 1977;123:13–26.

[30] Chen S, Hillman D. Giant spines and enlarged synapses induced in Purkinje cells by malnutrition. Brain Res 1980;187:487–93.

[31] Ranade SC, Nawaz S, Chakrabarti A, Gressens P, Mani S. Spatial memory deficits in maternal iron deficiency paradigms are associated with altered glucocorticoid levels. Horm Behav 2013;64:26–36.

[32] Alfaradhi MZ, Ozanne SE. Developmental programming in response to maternal overnutrition. Front Genet 2011;2(27):1–13.

[33] Flegal KM, Carroll MD, Ogden CL, Curtin LR. Prevalence and trends in obesity among US adults: 1999–2008. JAMA 2010;303:235–41.

[34] Musingarimi P. Obesity in the UK: are view and comparative analysis of policies within the devolved administrations. Health Policy 2009;91:10–6.

[35] Brenton C. The hypothalamus-adipose axis is a key target of developmental programming by maternal nutritional manipulation. J Endocrinol 2013;216(2):R19–31.

[36] Page KC, Jones EK, Anday EK. Maternal and postweaning high-fat diets disturb hippocampal gene expression, learning, and memory function. Am J Physiol Regul Integr Comp Physiol 2014;306(8):R527–537.

[37] Stachowiak EK, Oommen S, Vasu VT, et al. Maternal obesity affects gene expression and cellular development in fetal brains. Nutr Neurosci 2013;16(3):96–103.

[38] Barker DJ, Osmond C. Infant mortality, childhood nutrition, and ischaemic heart disease in England and Wales. Lancet 1986;1:1077–81.

[39] Barker DJ. Fetal nutrition and cardiovascular disease in adult life. Lancet 1993;341:938–41.

[40] Hales CN, Barker DJ. The thrifty phenotype hypothesis. Br Med Bull 2001;60:5–20.

[41] Gluckman PD. Predictive adaptive responses and human evolution. Trends Ecol Evol 2005;20:527–33.

[42] Low FM, Gluckman PD, Hanson MA. Developmental plasticity, epigenetics, and human health. Evol Biol 2012;39:650–65.

[43] Emanuel I, Filakti H, Alberman E, Evans SJ. Intergenerational studies of human birthweight from the 1958 birth cohort. 1. Evidence for a multigenerational effect. Br J Obstetr Gynaecol 1992;99:67–74.

[44] Selling KE, Carstensen J, Finnstrom O, Sydsio G. Intergenerational effects of preterm birth and reduced intrauterine growth: a population-based study of Swedish mother–offspring pairs. Br J Obstet Gynaecol 2006;113:430–40.

[45] Remmers F, Verhagen LA, Adan RA. Hypothalamic neuropeptide expression of juvenile and middle-aged rats after early postnatal food restriction. Endocrinology 2008;149:3617–25.

[46] Garcia AP, Palou M, Priego T. Moderate caloric restriction during gestation results in lower arcuate nucleus NPY- and alphaMSH-neurons and impairs hypothalamic response to fed/fasting conditions in weaned rats. Diabetes Obes Metab 2010;12:403–13.

[47] Plagemann A, Harder T, Rake. A hypothalamic nuclei are malformed in weanling offspring of low protein malnourished rat dams. J Nutr 2000;130:2582–9.

[48] Desai M, Guang H, Ferelli M. Programmed upregulation of adipogenic transcription factors in intrauterine growth-restricted offspring. Reprod Sci 2008;15:785–96.

[49] Levitt NS, Lambert EV, Woods D. Impaired glucose tolerance and elevated blood pressure in low birth weight, nonobese, young South African adults: early programming of cortisol axis. J Clin Endocrinol Metab 2000;85:4611–8.

[50] Barker DJ. Fetal programming of coronary heart disease. Trends Endocrinol Metab 2002;13:364–8.

[51] Ward AM, Syddall HE, Wood PJ. Fetal programming of the hypothalamic-pituitaryadrenal (HPA) axis: low birth weight and central HPA regulation. J Clin Endocrinol Metab 2004;89:1227–33.

[52] Ward AM, Moore V, Steptoe A, Cockington RA, Robinson JS, Phillips DI. Size at birth and cortisol responses to psychological stress: evidence for fetal programming. J Hypertens 2004;I22:2295–301.

[53] Phillips DI, Bennett FI, Wilks R, Thame M, Boyne M, Osmond C, et al. Maternal body composition, offspring blood pressure and the hypothalamic-pituitary adrenal axis. Paediatr Perinat Epidemiol 2005;19:294–302.

[54] Whitnall MH. Regulation of the corticotropin releasing hormone neurosecretory system. Prog Neurobiol 1993;40:573–629.

[55] Jacobson L, Sapolsky R. The role of the hippocampus in feedback regulation of the hypothalamic-pituitary-adrenocortical axis. Endocr Rev 1991;12:118–34.

[56] Uchoa ET, Aguilera G, Herman JP, Fiedler JL, Deak T, de Sousa MB. Novel aspects of glucocorticoid actions. J Neuroendocrinol 2014;26(9):557–72. (ahead of print).

[57] Delahaye F, Breton C, Risold PY, Enache M, Dutriez-Casteloot I, Laborie C, et al. Maternal perinatal undernutrition drastically reduces postnatal leptin surge and affects the development of arcuate nucleus proopiomelanocortin neurons in neonatal male rat pups. Endocrinology 2008;149:470–5.

[58] Palou M, Konieczna J, Torrens JM, Sánchez J, Priego T, Fernandes ML, et al. Impaired insulin and leptin sensitivity in the offspring of moderate caloric-restricted dams during gestation is early programmed. J Nutr Biochem 2012;23:1627–39.

[59] de Moura EG, Lisboa PC, Passos MC. Neonatal programming of neuroimmunomodulation – role of adipocytokines and neuropeptides. Neuroimmunomodulation 2008;15:176–88.

[60] Bouret SG, Simerly RB. Minireview: leptin and development of hypothalamic feeding circuits. Endocrinology 2004;145:2621–6.

[61] Lee S, Lee KA, Choi GY, Desai M, Lee SH, Pang MG, et al. Feed restriction during pregnancy/lactation induces programmed changes in lipid, adiponectin and leptin levels with gender differences in rat offspring. J Matern Fetal Neonatal Med 2013;26(9):908–14.

[62] Kongsted AH, Husted SV, Thygesen MP, Christensen VG, Blache D, Tolver A., et al. Pre- and postnatal nutrition in sheep affects β-cell secretion and hypothalamic control. J Endocrinol 2013;219(2):159–71.

[63] Chen M, Zhang L. Epigenetic mechanisms in developmental programming of adult disease. Drug Discov Today 2011;16(23-24):1007–18.

[64] Heijmans BT, Tobi EW, Stein AD. Persistent epigenetic differences associated with prenatal exposure to famine in humans. PNAS 2008;105:17046–9.

[65] Weaver IC. Epigenetic effects of glucocorticoids. Semin Fetal Neonatal Med 2009;14(3):143–50.

[66] Dutriez-Casteloot I, Breton C, Coupé B, Hawchar O, Enache M, Dickes-Coopman A, et al. Tissue-specific programming expression of glucocorticoid receptors and 11 beta-HSDs by maternal perinatal undernutrition in the HPA axis of adult male rats. Horm Metab Res 2008;40(4):257–61.

[67] Hass BS. Effects of caloric restriction in animals on cellular function, oncogene expression, and DNA methylation *in vitro*. Mutat Res 1993;295:281–9.

[68] Stevens A. Epigenetic changes in the hypothalamic proopiomelanocortin and glucocorticoid receptor genes in the ovine fetus after periconceptional undernutrition. Endocrinology 2010;151:3652–64.

[69] Reynolds EH. The neurology of folic acid deficiency. Handb Clin Neurol 2014;120:927–43.

[70] Shim H, Harris ZL. Genetic defects in copper metabolism. J Nutr 2003;133(5):1527S–31S.

[71] Colleoni S, Galli C, Gaspar JA, Meganathan K, Jagtap S, Hescheler J, et al. Development of a neural teratogenicity test based on

human embryonic stem cells: response to retinoic acid exposure. Toxicol Sci 2011;124(2):370–7.

[72] Guadalupe M, Flores-Cruza B, Escobara A. Reduction of serotonergic neurons in the dorsal raphe due to chronic prenatal administration of a tryptophan-free diet. Int J Dev Neurosci 2012;30:63–7.

[73] Madore C, Nadjar A, Delpech JC, Sere A, Aubert A, Portal C, et al. Nutritional n-3 PUFAs deficiency during perinatal periods alters brain innate immune system and neuronal plasticity-associated genes. Brain Behav Immun 2014;14:S0889–1591.

[74] Church MW, Jen KL, Anumba JI. Excess omega-3 fatty acid consumption by mothers during pregnancy and lactation caused shorter life span and abnormal ABRs in old adult offspring. Neurotoxicol Teratol 2010;32(2):171–81.

[75] Liu NQ, Hewison M. Vitamin D, the placenta, and pregnancy. Arch Biochem Biophys 2012;523(1):37–47.

[76] Dror DK, Allen LH. Effect of vitamin B12 deficiency on neurodevelopment in infants: current knowledge and possible mechanisms. Nutr Rev 2008;66(5):250–5.

[77] Torres-Vega A, Pliego-Rivero BF, Otero-Ojeda GA, Gómez-Oliván LM, Vieyra-Reyes P. Limbic system pathologies associated with deficiencies and excesses of the trace elements iron, zinc, copper, and selenium. Nutri Rev 2012;70(12):679–92.

[78] Walsh CT, Sandstead HH, Prasad AS. Zinc: health effects and research priorities for the 1990s. Environ Health Perspect 1994;102(2):5–46.

[79] Gupta A, Lutsenko S. Human copper transporters: mechanism, role in human diseases, and therapeutic potential. Future Med Chem 2009;1:1125–42.

[80] Yang H, Brosel S, Acin-Perez R. Analysis of mouse models of cytochrome *c* oxidase deficiency owing to mutations in *Sco2*. Hum Mol Genet 2010;19:170–80.

[81] Robbins TW, Arnsten AF. The neuropsychopharmacology of fronto-executive function: monoaminergic modulation. Annu Rev Neurosci 2009;32:267–87.

[82] Morgane PJ, Austin-LaFrance R, Bronzino J, Tonkiss J, Díaz-Cintra S, Cintra L, et al. Prenatal malnutrition and development of the brain. Neurosci Biobehav Rev 1993;17(1):91–128.

[83] Levitsky DA, Strupp BJ. Malnutrition and the brain: changing concepts, changing concerns. J Nutr 1995;125(8):2212S–20S.

[84] Antonow SI, Schwab M, Cox LA, Li C, Stuchlik K, Witte OW, et al. Vulnerability of the fetal primate brain to moderate reduction in maternal global nutrient availability. Proc Natl Acad Sci USA 2011;108(7):3011–6.

[85] Rajanna B, Mascarenhas C, Desiraju T. Deviations in brain development due to caloric undernutrition and scope of their prevention by rehabilitation: alterations in the power spectra of the EEG of areas of the neocortex and limbic system. Brain Res 1987;465(1-2):97–113.

[86] Morgane PJ, Mokler DJ, Galler JR. Effects of prenatal protein malnutrition on the hippocampal formation. Neurosci Biobehav Rev 2002;26(4):471–83.

[87] Liu Y, Coresh J, Eustace JA, Longenecker JC, Jaar B, Fink NE, et al. Association between cholesterol level and mortality in dialysis patients: role of inflammation and malnutrition. JAMA 2004;291(4):451–9.

[88] Del Angel-Meza AR, Ramirez-Cortes L, Olver-Cortes E, Pérez-Vega MI, González-Burgos I. A tryptophan deficient corn-based diet induces plastic responses in cerebellar cortex cells of rat offspring. Int J Dev Neurosci 2001;19:447–53.

[89] Chopra JS, Sharma A. Protein energy malnutrition and the nervous system. J Neurol Sci 1992;110:8–20.

[90] Zamenhof S, Van Marthens E. Nutritional influences on prenatal brain development. In: Gottlicb G, editor. Studies on the development of behavior and the nervous system. New York: Academic Press; 1978. p. 149–86.

[91] Hatten ME, Alder J, Zimmerman K, Heintz N. Genes involved in cerebellar cell specification and differentiation. Curr Opin Neurobiol 1997;7:40–7.

[92] Bonatto F, Polydoro M, Andrades M, da Frota Júnior ML, Dal-Pizzol F, Rotta LN, et al. Effect of protein malnutrition on redox state of the hippocampus of rat. Brain Res 2005;1042(1):17–22.

[93] Gressens P, Muaku SM, Besse L, Nsegbe E, Gallego J, Delpech B, et al. Maternal protein restriction early in rat pregnancy alters brain development in the progeny. Dev Brain Res 1997;103(1):21–35.

[94] Montana-Rojas EA, Ferreira AA, Tenorio F, Barradas PC. Myelin basic protein accumulation is impaired in a model of protein deficiency during development. Nutr Neurosci 2005;8:49–56.

[95] Dobbing J, Hopewell JW. Permanent deficit of neurons in cerebral and cerebellar cortex following early mild undernutrition. Arch Dis Child 1971;46(249):736–7.

[96] Garcia-Ruiz M, Diaz-Cintra S, Cintra L, Corkidi G. Effect of protein malnutrition on CA3 hippocampal pyramidal cells in rats of three ages. Brain Res 1993;625(2):203–12.

[97] McConnell P, Berry M. The effects of undernutrition on Purkinje cell dendritic growth in the rat. J Comp Neurol 1978;177:159–72.

[98] Rotta LN, Leszczinski DN, Brusque AM, Pereira P, Brum LF, Nogueira CW, et al. Effects of undernutrition on glutamatergic parameters in the cerebral cortex of young rats. Physiol Behav 2008;94(4):580–5.

[99] Bedi KS. Spatial learning ability of rats undernourished during early postnatal life. Physiol Behav 1992;51(5):1001–7.

[100] Almeida SS, Tonkiss J, Galler JR. Prenatal protein malnutrition affects avoidance but not escape behavior in the elevated T-maze test. Physiol Behav 1996 b;60(1):191–5.

[101] Dongsheng Kai. Neuroinflammation and neurodegeneration in overnutrition induced diseases. Trends Endocrinol Metab 2013;24(1):40–7.

[102] Elmquist JK. Identifying hypothalamic pathways controlling food intake, body weight, and glucose homeostasis. J Comp Neurol 2005;493:63–71.

[103] Williams KW, Elmquist JK. From neuroanatomy to behavior: central integration of peripheral signals regulating feeding behavior. Nat Neurosci 2012;15:1350–5.

[104] Cai D, Liu T. Hypothalamic inflammation: a double-edged sword to nutritional diseases. Ann N Y Acad Sci 2011;1243:E1–39.

[105] Jaaro-Peled H, Hayashi-Takagi A, Seshadri S, Kamiya A, Brandon NJ, Sawa A. Neurodevelopmental mechanisms of schizophrenia: understanding disturbed postnatal brain maturation through neuregulin-1-ErbB4 and DISC1. Trends Neurosci 2009;32:485–95.

[106] Lewis DA, Levitt P. Schizophrenia as a disorder of neurodevelopment. Ann Rev Neurosci 2002;25:409–32.

[107] Galler JR, Bryce CP, Waber D, Hock RS, Exner N, Eaglesfield D, et al. Early childhood malnutrition predicts depressive symptoms at ages 11–17. J Child Psychol Psych 2010;51(7):789–98.

[108] Sanches M, Keshavan MS, Brambilla P, Soares JC. Neurodevelopmental basis of bipolar disorder: a critical appraisal. Prog Neuropsychopharmacol Biol Psychiatry 2008;32(7):1617–27.

[109] Brown AS, Susser ES. Prenatal nutritional deficiency and risk of adult schizophrenia. Schizophr Bull 2008;34(6):1054–63.

[110] Lahti J, Raikkonen K, Kajantie E, Heinonen K, Pesonen AK, Järvenpää AL, et al. Small body size at birth and behavioural symptoms of ADHD in children aged five to six years. J Child Psychol Psychiatr 2006;47:1167–74.

[111] Strang-Karlsson S, Raikkonen K, Pesonen AK. Very low birth weight and behavioral symptoms of attention deficit hyperactivity disorder in young adulthood: the Helsinki study of very-low-birth-weight adults. Am J Psychiatry 2008;165:1345–53.

[112] Vucetic Z, Totoki K, Schoch H, Whitaker KW, Hill-Smith T, Lucki I, et al. Early life protein restriction alters dopamine circuitry. Neuroscience 2010;68:359–70.

[113] Koot S, van den Bos R, Adriani W, Laviola G. Gender differences in delay-discounting under mild food restriction. Behav Brain Res 2009;200:134–43.

[114] Costello EJ, Worthman C, Erkanli A, Angold A. Prediction from low birth weight to female adolescent depression: a test of competing hypotheses. Arch Gen Psychiatry 2007;64:338–44.

[115] Thaler JP, Schwartz MW. Minireview: inflammation and obesity pathogenesis: the hypothalamus heats up. Endocrinology 2010;151:4109–15.

[116] Armstrong RA. What causes Alzheimer's disease? Folia Neuropathol 2013;51(3):169–88.

[117] Eyles DW, Burne TH, McGrath JJ. Vitamin D effects on brain development, adult brain function and the links between low levels of vitamin D and neuropsychiatric disease. Front Neuroendocrinol 2013;34(1):47–64.

[118] Partadiredja G, Worrall S, Bedi KS. Early life undernutrition alters the level of reduced glutathione but not the activity levels of reactive oxygen species enzymes or lipid peroxidation in mouse forebrain. Nutr Neurosci 2010;13(3):102–8.

[119] Chadio SE, Kotsampasi B, Papadomichelakis G, Deligeorgis S, Kalogiannis D, Menegatos I, et al. Impact of maternal undernutrition on the hypothalamic-pituitary-adrenal axis responsiveness in sheep at different ages postnatal. J Endocrinol 2007;192(3):495–503.

[120] Tofail F, Persson LA, Arifeen SE. Effects of prenatal food and micronutrient supplementation on infant development: a randomized trial from the Maternal and Infant Nutrition Interventions Matlab (MINIMat) study. Am J Clin Nutr 2008;87:704–11.

[121] Prado EL, Alcock KJ, Muadz H. Maternal multiple micronutrient supplements and child cognition: a randomized trial in Indonesia. Pediatrics 2012;130:e536–46.

[122] Winick M, Meyer KK, Harris RC. Malnutrition and environmental enrichment by early adoption. Science 1975;190:1173–5.

[123] Duffy PH, Leakey JE, Pipkin JL, Turturro A, Hart RW. The physiologic, neurologic, and behavioral effects of caloric restriction related to aging, disease, and environmental factors. Environ Res 1997;73:242–8.

[124] Chen YC, Chen QS, Lei JL. Physical training modifies the age-related decrease of GAP-43 and synaptophysin in the hippocampal formation in C57BL/6J mouse. Brain Res 1998;806:238–45.

23

Herbal Supplements in Pregnancy: Effects on Conceptions and Delivery

Fabio Facchinetti, MD, Giulia Dante, MD, Isabella Neri, MD

Policlinico Hospital – University of Modena and Reggio-Emilia, Modena, Italy

TRADITIONAL MEDICINE: INTRODUCTION

Traditional medicine (TM) is the sum total of the knowledge, skills, and practices based on the theories, beliefs, and experiences indigenous to different cultures, whether explicable or not, used in the maintenance of health, as well as in the prevention, diagnosis, improvement, or treatment of physical and mental illnesses. The terms complementary/alternative/nonconventional medicine (CAM) are used interchangeably with traditional medicine in some countries.

Traditional medication involves the use of herbal remedies (HR), animal parts, and minerals. Herbal medicines are the most widely used of the three and include herbs, herbal materials, herbal preparations, and finished herbal products that contain as active ingredients parts of plants or other plant materials, or combinations.

Despite its existence and continued use over many centuries, and its popularity and extensive use during the last decade, TM has not been officially recognized in most countries. Consequently, education, training, and research in this area have not been accorded due attention and support. The quantity and quality of the safety and efficacy data on TM are far from sufficient to meet the criteria needed to support its use worldwide. The reasons for the lack of research data are due not only to health care policies, but also to a lack of adequate or accepted research methodology for evaluating traditional medicine. It should also be noted that there are data on research in traditional medicine in various countries, but further research in safety and efficacy should be promoted, and the quality of the research should be improved [1].

Health care consumers have increasingly turned to TM, which they perceive as being more effective for chronic, emotional conditions and having fewer side effects than pharmaceuticals. Several data demonstrated an increase of CAM usage from 1990 through 2006 [2].

Use of HR has been documented among different patient groups and in the general population to promote health [3,4]. Multiple surveys have shown that women, especially those of white ethnicity, middle-aged, with high levels of education and income, are more likely to be users [5,6].

There has been an unprecedented explosion in the popularity of herbal preparations during the last few decades, especially in developed countries [7]. This phenomenon has stimulated considerable public health concern among physicians who are sometimes uncertain about the safety of herbs [8,9]. Despite these concerns, the global prevalence of use of medicinal herbs continues to rise as patients self-medicate with or without informing their physicians [1].

The suggested reasons for not reporting this use to doctors are that doctors themselves do not address this topic; patients may not think it necessary or be fearful of the doctor's reaction [10,11].

In this setting, the attitudes and knowledge of physicians would impact on the doctor–patient relationship and affect the overall quality of health care delivery, particularly with respect to issues such as possible adverse herb effects and herb–drug interactions [10,12].

TRADITIONAL CHINESE MEDICINE

A distinct chapter concerns Traditional Chinese medicine (TCM), a practice of medicine that originated in ancient China and evolved over thousands of years. TCM practitioners use herbal medicines and various mind and body practices, such as acupuncture and tai chi, to treat or prevent health problems.

The Chinese Materia Medica (a pharmacological reference book used by TCM practitioners) describes

Handbook of Fertility. http://dx.doi.org/10.1016/B978-0-12-800872-0.00023-8

thousands of medicinal substances, primarily plants, but also some minerals and animal products. Different parts of plants, such as the leaves, roots, stems, flowers, and seeds, are used. In TCM, herbs are often combined in formulas and given as teas, capsules, liquid extracts, granules, or powders.

The exact number of people who use TCM in the world is unknown. In the United States it was estimated in 1997 that some 10,000 practitioners served more than 1 million patients each year. According to the 2007 National Health Interview Survey (NHIS), which included a comprehensive survey on the use of complementary health approaches by Americans, an estimated 3.1 million US adults had used acupuncture in the previous year. The number of visits to acupuncturists tripled between 1997 and 2007.

An assessment of the research found that 41 of 70 systematic reviews of the scientific evidence (including 19 of 26 reviews on acupuncture for a variety of conditions and 22 of 42 reviews on Chinese herbal medicine) were unable to reach conclusions about whether the technique worked for the condition under investigation because there was not enough good-quality evidence. The other 29 systematic reviews (including 7 of 26 reviews on acupuncture and 20 of 42 reviews on Chinese herbal medicine) suggested possible benefits but could not reach definite conclusions because of the small quantity or poor quality of the studies.

Some may be safe, but others may not be. There have been reports of products being contaminated with drugs, toxins, or heavy metals or not containing the listed ingredients. Some of the herbs used in Chinese medicine can interact with drugs, can have serious side effects, or may be unsafe for people with certain medical conditions [13–15].

For all these reasons, and because we discuss a particular period such as pregnancy, we excluded Chinese herbal products from this chapter.

HERBAL REMEDIES AND FERTILITY

Before and during pregnancy women are apprehensive about the potential toxicity of conventional medicines, so they use HR to complement or to replace them, although much current practice is not evidence-based [16,17]. Indeed, there is evidence of negative effects associated with the use of some HR, and data on the safety for their use before and during pregnancy are limited [18,19].

Although HR contains active constituents with pharmacological properties and possible interactions with other compounds, they are considered by women to be natural and safer than conventional drugs [16,20].

Another issue is that HR are over-the-counter and offer women greater independence for their health care choices. Hence, the majority of consumers do not disclose their use to the doctor and rely on family/friends or web sites for information regarding such treatments [21,22]. Moreover, unfortunately, women's sources of information about herbal remedies are mainly "family and friends" and "health food shop" [23,24].

Infertility is a significant health problem [25,26].

Depending on the underlying cause, infertility can be treated by gynecologists, urologists, and reproductive endocrinologists using a range of medical options, including advice on the timing of intercourse, drugs to stimulate ovulation, surgery, intrauterine insemination, and assisted reproductive technology (ART). Based on data from the National Survey of Family Growth (NSFG), in the United States the percentage of women aged 15–44 who had used infertility services increased from 9% in 1982 to 15% in 1995, then declined to 12% in 2002, and remained at that level in 2006–2010.

Numerous previous analyses have shown that women who make use of medical help for fertility problems are a highly selective group among those who have fertility problems. Data from US surveys have shown that fertility-impaired women who use infertility services are significantly more likely to be married, non-Hispanic white, older, more highly educated, and more affluent than nonusers. Reasons for the disparities in the use of infertility services may include access barriers, such as the significant cost of medical services for infertility, and the lack of adequate health insurance to afford the necessary diagnostic or treatment services [25].

Several reports conducted, especially in developing countries, demonstrated that people seek any kind of medical or nonmedical therapy concordant with their culture and financial status [27]. As the number of couples seeking infertility care rises with increased age at conception, so do the costs and failed attempts at ART methods [28].

One of the rationales of infertile couples for trying an alternative method might be dissatisfaction with the outcome of ART and also its cost [27,29]. Moreover, depression and anxiety imposed on them by infertility might be another reason making this population vulnerable to nonorthodox medications such as herbal treatments (HT) and other CAMs [30].

The use of HT as part of CAM in infertile populations is investigated through limited studies [27,31,32], while data about the knowledge of HT in the population seeking infertility treatment are even more rare [33,34].

A nonrandomized controlled trial (RCT) on specific HR results was conducted from a literature search in the Medline electronic database about HR, their applications, and potential effects in female infertility over the period January 1990 to December 2013. Most of the published data were collected in Canada, Australia, the United States, and Europe, and analyzed mainly epidemiological data [35]. Estimates of frequency of the use of HR range from 3.5% reported in the studies conducted in the United Kingdom, Iran (3.5%), Canada (6%), and the United States (6.3%), to 29% described in a Turkish study. Such differences could be related to study designs, data collection methods, and cultural characteristics of investigated population.

A prospective cohort study conducted in the United States during 2010 identified the predictors of CAM use in couples seeking fertility care. The results of this study reported that CAM was used as a fertility treatment by 29% of the 428 infertile couples analyzed. Acupuncture and herbal therapy were the most commonly used modalities for the treatment of infertility (23 and 18%, respectively).

CAM was chosen most commonly by wealthier couples, those not achieving a pregnancy, and those with a baseline belief in the effectiveness of CAM treatments [36].

Interesting results were revealed from a European study that demonstrated less successful ongoing pregnancies and live births associated with cohorts using CAM during infertility treatments [37]. This was a prospective observational cohort study with a 12-month follow-up period. In this study several CAM interventions, including HR, were evaluated; the percentage of HR users in the group of CAM consumers was 50.2%.

The principal finding was that the concurrent use of CAM in ART was associated with a 30% lower ongoing pregnancy/live birth rate over a 12-month fertility treatment period despite the fact that CAM users underwent more *in vitro* fertilization/intracytoplasmic sperm injection (IVF/ICSI) cycles and had on average more embryos than did nonusers.

These results are important because they are the first to demonstrate in a large cohort of people undergoing fertility treatment that concurrent CAM use may have adverse effects on the chances of pregnancy. Future research needs, in particular RCTs, to focus on the effects of particular HR interventions used either exclusively or in combination and on the underlying behavioral or biological mechanisms that could account for the association between the use of HR and other CAMs and pregnancy failure in IVF/ICSI cycles.

The current stage of knowledge is still inadequate to sufficiently inform clinicians, researchers, and the public about the benefits or potential risks of the use of HR for female infertility treatments.

HERBAL REMEDIES DURING PREGNANCY

Epidemiological studies conducted in Europe, the United States, and Australia indicated that up to half of pregnant women take herbal drugs. The wide range in the prevalence may be explained by the use of different study methodologies, in addition to cultural and regional differences [38,39].

The extent of HR use in pregnancy has been widely researched throughout the world.

While some risks have been already documented, very few trials have been performed to demonstrate the potential benefits of such products. In spite of this, many women use HR during pregnancy and it is important to find out more about why [23,40,41].

The rationale for their use in pregnancy is threefold:

1. Pregnancy is a state that requires "more" in order to assure the support to the new, growing organism;
2. Pregnancy is associated with minor complaints (nausea, constipation, etc.) that require relief, possibly without drugs;
3. The best outcome in pregnancy is only reached with the removal of hazards like malformations, pregnancy-related disorders, and any other medical complications [42]. CAM intervention may prove helpful in this respect.

Pregnant women are inspired to choose this kind of treatment because they consider herb compounds as being natural, and therefore safer when compared to conventional drugs [43]. However, clear evidence of the negative effects exists in some cases while data on safety during pregnancy are limited [18].

The actual knowledge about HR, their applications, and potential effects in pregnancy are summarized in a recent systematic review published in 2013 [44].

This review performed a search over the period 1990–2013 concerning articles published in the English language. For inclusion, an article had to contain original data on either the prevalence of use or adverse effects of herbal treatments during pregnancy and only human studies were included; topical treatments were also included.

This review analyzed epidemiological data, use, and efficacy of single herbal treatments.

Out of 671 articles published during the last two decades (1990–2013), 20 were observational studies included in the review for the description of epidemiological data, and 26 were papers conducted on single HR. Of them, 15 were RCTs, 5 prospective observational, 1 retrospective observational, 4 cohort studies, 5 case reports, and a quasi-experimental study.

Most of the published data were collected in Europe or in the United States.

Most Popular Herbal Remedies

From the analysis of the observational studies, the frequency of the use of HR during pregnancy range from 0.9% [45] to 87% [46], and such differences could be related to study designs, data collection methods, and cultural characteristics of the investigated population.

TABLE 23.1 Herbs Most Frequently Used in Pregnant Women and Reasons for Their Use

Reasons for use	HR
Stretch marks	Almond oil
Digestive problems	Aloe, chamomile, peppermint, teas/green tea
Constipation	Aloe, teas/green tea
Anxiety	Chamomile, teas/green tea, valerian
Relax, sleep	Chamomile
Capillary frailty	Aloe
Treat and prevent urinary tract infections	Cranberry
Common cold, strengthen immune system	Echinacea
Fluid retention	Fennel
Nausea, vomiting	Ginger, peppermint
Heartburn	Peppermint
Induce and ease labor	Raspberry leaf
Depression	St. John's wort

However, some of these studies are limited by methodological flaws (lack of prestructured questionnaires/interviews, recall bias, etc.). Excluding them, it is reasonable to conclude that the consumers of herbal remedies during pregnancy ranges from 27 to 57% in Europe and from 10 to 73% in the United States.

According to the results from other reviews [22,47], women using HR were more likely to be Caucasian, middle-aged, not smokers, and with a high level of education.

The examination of these observational studies shows that 12 herbs are the most frequently consumed by pregnant women, and the most reason for their use were mood disorders such as anxiety and depression (Table 23.1). Indeed, there are at least three herbs claimed as anxiety relievers, that is, chamomile, teas, and valerian. None of them have a scientific demonstration of efficacy, apart from traditional beliefs.

Furthermore, it is necessary to highlight the adverse events associated with the prolonged use of almond oil. Despite the absence of studies devoted to this compound, in a survey performed in postpartum women, it was found that those who applied almond oil to their abdomen daily (to avoid stretch marks) were at higher risk for preterm delivery [48].

Most Investigated Herbal Remedies

The HR investigated during pregnancy, reasons for their use, and data on their efficacy and safety are summarized in Table 23.2.

Among the trials about single HR, ginger was the most investigated; indeed it was examined in 10 RCTs, 1 prospective observational study, and 1 cohort study.

Ginger, known scientifically as *Zingiber officinale*, is a perennial native to many Asian countries [49]. This is a popular herbal remedy that has been in use for thousands of years. It is an important part of the pharmacopoeia of Indian and Chinese Traditional Medicines for a number of actions, including anti-emetic effects, stimulation of digestion, relief of coughs and colds, and anti-inflammatory effect [50].

Recent clinical trials have demonstrated the anti-emetic effect of ginger in a number of contexts including chemotherapy [51], motion sickness [52,53], post-operative surgery [54,55], and women hospitalized for pregnancy-induced nausea. The mechanism of action of ginger on both nausea and vomiting has not been fully identified although several hypotheses have been proposed [56,57]. It has been reported that symptoms of nausea and vomiting during pregnancy improved in direct correlation to the improvement in pregnancy-induced gastric dysrhythmias. Therefore, ginger actions in pregnancy may be due to a direct effect of the drug on the gastrointestinal tract [58–60].

The primary objective of the RCTs was to investigate the effectiveness of ginger on nausea and vomiting during pregnancy. On the other hand, the primary outcome of the observational studies was to examine the safety of this product on congenital malformations and some pregnancy outcomes. Five RCTs reported the superiority of ginger compared to placebo, whereas four other trials found ginger to be equally effective when compared to vitamin B6 and dimenhydrinate.

There were no significant differences between ginger and the other treatments with respect to adverse events and no increased risk for major malformations, stillbirth/perinatal death, preterm birth, low birth weight, or low Apgar score [44].

St. John's wort is one of the most prevalent herb on the market. It has been used for its medicinal properties for hundreds of years in a variety of ailments [61]. Current evidence from studies in humans suggests that at least two constituents, hypericin and hyperforin, play a significant role in its pharmacologic effect for the treatment of depression [62,63]. In the folk medicine of various countries, this herb remedy has long been used orally and topically for healing of wounds, although there is limited clinical research supporting this practice [63–66].

The purpose of the observational studies was to determine whether exposure of this agent in pregnancy is associated with major fetal malformations or with infant adverse events. The use of St. John's wort was found to be safe in both conditions, whereas efficacy was not specifically reported.

TABLE 23.2 Single HR Investigated During Pregnancy, Reasons for Their Use, Efficacy and Safety

HR	Study	Reasons for use	Efficacy	Safety
Ginger	10 RCTs 1 observational 1 cohort	Nausea, vomiting	Yes	Yes
St. John's wort	4 observational	Depression For healing of wounds	Not described Yes	Yes Yes
Garlic	1 RCT 1 observational	Prevention of PE Prevention of PTD	No Yes	Yes Not described
Echinacea	1 observational	Common cold	Yes	Yes
Cranberry	1 RCT	Treat and prevent UTI	No	Yes
Blue cohosh	3 case report	Induce labor	Not described	No
Raspberry leaf	1 RCT 1 observational	Induce and ease labor	No	Yes
Castor oil	1 RCT 2 observational	Induce labor	Doubt	Not described
EPO	Retrospective quasi- experimental	Induce labor	No	Doubt

RCT, randomized controlled trial; PE, preeclampsia; UTI, urinary tract infection; PTD, preterm delivery; EPO, evening primrose oil.

The only RCT was performed to determine the effects of a topical preparation on cesarean wound healing. At the tenth day postpartum, St John's worth facilitated cesarean wound healing and minimized the formation of scars. In addition, significantly lower pain and pruritus were reported by the treatment group at the fortieth day postpartum [44].

Garlic is part of the allium, or onion, family. A hardy perennial herb, it probably originated in Central Asia. The traditional medicinal uses of garlic include prevention of infection and treatment of colds, influenza, bronchitis, whooping cough, gastroenteritis, dysentery, and skin problems [67–69]. More recently, it has been suggested that garlic may lower blood pressure, reduce oxidative stress, and/or inhibit platelet aggregation, which led to the hypothesis that garlic may have a role in the prevention of preeclampsia (PE) [69–72].

One RCT analyzed the effects of garlic on the prevention of PE in high-risk women. There was a reduction in the means of total cholesterol level, while neither hypertension nor PE was reduced. Minor adverse events, such as a foul odor and nausea, were reported in garlic users; no effect on neonates was found.

One observational cohort study demonstrated that garlic intake was associated with a lower risk of early and late preterm delivery. Maternal–fetal adverse events were not analyzed [44].

Echinacea is a perennial herb found in Eastern and Central United States and Southern Canada. Historically, Echinacea was the most commonly used herb among Native North Americans for a variety of conditions including wounds, insect bites, infections, toothache, joint pain, and as an antidote for rattlesnake bites. In the early twentieth century it was established as the remedy of choice for cold and flu and was commonly used as an anti-infective until the advent of modern antibiotics [73].

One prospective observational study evaluated the safety and the efficacy of Echinacea when used during the first trimester for upper respiratory tract ailments. No increased risk of major malformations was reported. Respiratory symptoms improved with respect to the nontreated group.

American cranberry, one of the few fruits native to Eastern North America, is also grown in Northern Europe. Cranberry's principal therapeutic value today continues to be the treatment and prevention of urinary tract infection (UTI). The currently accepted mechanism of action in treating and preventing UTI is through disabling *Escherichia coli*'s capacity to adhere to the urethra. The fruit contains two compounds, fructose and a proanthocyanidin, that adhere to proteins on the fimbriae of *E. coli*, effectively inhibiting the bacteria from sticking to the epithelial cell lining the urethra. Without the ability to establish a strong foothold via adherence, the infection is either attenuated or prevented at the outset [74].

In 2008, the Cochrane Collaboration reviewed cranberry use in the prevention and treatment of UTI [75]. Authors concluded that cranberry products significantly reduced the incidence of symptomatic UTIs, namely, in women with risk of recurrence.

On the contrary, in 2012, the Cochrane Collaboration published a new review and suggested that cranberry juice is less effective than previously indicated. Other preparations (such as powders) need to be quantified using standardized methods to ensure the potency, and must contain enough of the "active" ingredient before being evaluated in clinical studies or recommended [76].

Only one RCT compared cranberry extract with placebo in the prevention of UTI in pregnant women. A nonsignificant reduction in the frequency of both asymptomatic bacteriuria and urinary tract infections was reported in women receiving cranberry. The study, however, was not sufficiently powered to detect such a difference. Moreover, 38.8% of the participants withdrew, mostly due to gastrointestinal upset. There was no difference between groups with respect to obstetric and neonatal outcomes [44].

Blue cohosh is a small woodland perennial plant with large blue berries that is native to the American northeast. The medicinal effects of blue cohosh are derived from its root and rhizomes. Blue cohosh is also referred to as "papoose root" or "squaw root," which reflects the use of this herbal medicine by Native American women who brewed blue cohosh as a tea to relieve menstrual cramps and to ease the pains associated with childbirth.

In a 1999 survey of Certified Nurse Midwives in the United States, 64% claimed to use blue cohosh during labor. Currently, blue cohosh is used to induce labor, to help speed the process of labor, and generally, to help the mother through labor as quickly and painlessly as possible [77].

Only three case reports are available and they described cardiovascular side effects using blue cohosh at the time of delivery. In one case the neonate experienced acute myocardial infarction, profound congestive heart failure, and shock; in another case there was a severe multiorgan hypoxic injury, and in the last one, a perinatal stroke occurred.

No efficacy studies are available [44].

The use of raspberry to induce and ease labor was described in one RCT and one retrospective observational study.

The raspberry leaf plant has been used medicinally for centuries, certainly as early as the sixth century. There is a belief that this herb, taken in tablet, tea, or tincture form during pregnancy, shortens labor and makes it "easier"[78–82].

It is believed that the uterus is strengthened and toned by the regular ingestion of raspberry leaf throughout pregnancy and labor, assisting contractions and checking any hemorrhage during labor [78,79].

In both the included studies raspberry did not shorten the first stage of labor. The only clinically significant finding refers to the shortening of the second stage of labor with lower rate of forceps deliveries compared with the placebo. The use of raspberry was not associated with maternal–fetal adverse events [44].

Castor oil, a potent cathartic, is derived from the bean of the castor plant. Anecdotal reports, which date back to ancient Egypt, have suggested the use of castor oil to stimulate labor.

Castor oil has been widely used as a traditional method of initiating labor in midwifery practice. Its role in the initiation of labor is poorly understood and data examining its efficacy within a clinical trial are limited [83].

There were two observational studies and one RCT about the effect of castor oil on the induction of labor. The RCT showed a significant increase in labor initiation in the treated group compared with controls, and the same outcome was found in one prospective study. However, in the third study, castor oil showed no effect on the time of birth. Nausea was the most common maternal side effect reported. There were no data on neonatal mortality or morbidity [44].

Evening primrose oil (EPO) is obtained by cold expression or solvent extraction from the seeds of the evening primrose plant. It is a commonly used alternative therapy and a rich source of omega-6 essential fatty acids. It is best known for its use in the treatment of systemic diseases marked by chronic inflammation such as atopic dermatitis and rheumatoid arthritis. It is often used for several women's health conditions, including breast pain, menopausal and premenstrual symptoms, cervical ripening, and labor induction or augmentation. However, there is insufficient evidence to make a reliable assessment of its effectiveness for most clinical indications [84].

During pregnancy, EPO is a fatty acid used to trigger cervical ripening.

In a retrospective study of quasi-experimental design, EPO did not shorten gestation or decrease the overall length of labor; moreover, it increased the incidence of prolonged rupture of membranes, oxytocin augmentation, arrest of descent, and vacuum extraction.

There was one case of petechiae and ecchymosis in a newborn whose mother took primrose oil a week before giving birth [44].

For the HR employed in the induction of labor (raspberry, blue cohosh, castor oil, and evening primrose oil), very few scientific data are available to support such indication. Raspberry leaf as well as evening primrose oil has proven ineffective, the latter raising doubts about safety. Labor induction with castor oil seems promising and further studies will help to comprehend available contrasting data.

Of paramount importance is the alarm signal on the use of blue cohosh. Efficacy as a labor stimulant is lacking although a significant number of US midwives use it [41]. On the contrary, three case reports describe significant adverse events in neonates whose mother received the herb remedy [44].

In conclusion, despite the very large popular use of HT during pregnancy, there are very few studies that have been devoted to the specific evaluation of these treatments. With the exception of ginger supplementation for hyperemesis gravidarum, there is actually no clinical indication for the use of any other herbal treatment in pregnant women.

Vice versa, caution in the use of several compounds because of poor safety is available from case reports and epidemiological studies.

References

[1] General Guidelines for Methodologies on Research and Evaluation of Traditional Medicine. Geneve. WHO/EDM/TRM/2000.1.

[2] Frass M, Strassl RP, Friehs H, Müllner M, Kundi M, Kaye AD. Use and acceptance of complementary and alternativemedicine among the general population and medical personnel: a systematic review. Ochsner J 2012;12(1):45–56.

[3] Millen AE, Dodd KW, Subar AF. Use of vitamin, mineral, nonvitamin, and nonmineral supplements in the United States: the 1987, 1992, and 2000 National Health Interview Survey results. J Am Diet Assoc 2004;104:942–50.

[4] Lapi F, Vannacci A, Moschini M, Cipollini F, Morsuillo M, Gallo E, et al. Use, attitudes and knowledge of complementary and alternative drugs (CADs) among pregnant women: a preliminary survey in Tuscany. Alternat Med 2011;612159.

[5] Tesch BJ. Herbs commonly used by women. An evidence-based review. Am J Obstet Gynecol 2003;188:S44–5.

[6] Stevinson C, Ernst E. Complementary/alternative therapies for premenstrual syndrome: a systematic review of randomized controlled trials. Am J Obstet Gynecol 2001;185:227–35.

[7] Tindle HA, Davis RB, Phillips RS, Eisenberg DM. Trends in use of complementary and alternative medicine by US adults: 1997–2002. Altern Ther Health Med 2005;11:42–9.

[8] Risberg T, Kolstad A, Johansen A, Vingerhagen K. Opinions on and use of alternative medicine among physicians, nurses and clerks in northern Norway. In Vivo 1999;13:493–8.

[9] Hyodo I, Eguchi K, Nishina T, Endo H, Tanimizu M, Mikami I, et al. Perceptions and attitudes of clinical oncologists on complementary and alternative medicine, a nationwide survey in Japan. Cancer 2003;97:2861–8.

[10] Robison A, McGrail MR. Disclosure of CAM use to medical practitioners: a review of qualitative and quantitative studies. Complement Ther Med 2004;12:90–8.

[11] Alder SR, Fosket JR. Disclosing complementary and alternative medicine use in the medical encounter: a qualitative study in women with breast cancer. J Fam Pract 1991;48:453–8.

[12] Smith L, Ernst E, Ewings P, Myers P, Smith C. Co-ingestion of herbal medicines and warfarin. Br J Gen Pract 2004;54:439–41.

[13] Kaptchuk TJ. The web that has no weaver: understanding Chinese medicine. 2nd edn New York, NY: McGraw-Hill; 2000.

[14] Manheimer E, Wieland S, Kimbrough E, Cheng K, Berman BM. Evidence from the Cochrane Collaboration for traditional Chinese medicine therapies. J Altern Complement Med 2009;15(9):1001–14.

[15] Xue CC, O'Brien KA. Modalities of Chinese medicine. In: Leung P-C, Xue CC, Cheng Y-C, editors. A comprehensive guide to Chinese medicine. River Edge, NJ: World Scientific Publishing Co; 2003.

[16] Hall HG, Griffiths DL, McKenna LG. The use of complementary and alternative medicine by pregnant women: a literature review. Midwifery 2011;27:817–24.

[17] Dugoua JJ. Herbal medicines and pregnancy. J Popul Ther Clin Pharmacol 2010;17:e370–8.

[18] Ernst E. Herbal medicinal products during pregnancy: are they safe? Br J Obstet Gynecol 2002;109:227–35.

[19] Dante G, Pedrielli G, Annessi E, Facchinetti F. Herb remedies during pregnancy: a systematic review of controlled clinical trials. J Matern Fetal Neonatal Med 2013;26:306–12.

[20] Holst L, Wright D, Nordeng H, Haavik S. Use of herbal preparations during pregnancy: focus group discussion among expectant mothers attending a hospital antenatal clinic in Norwich, UK. Complement Ther Clin Pract 2009;15:225–9.

[21] Mitchell M. Risk, pregnancy and complementary and alternative medicine. Complement Ther Clin Pract 2010;16:109–13.

[22] Hall GH, McKenna LG, Griffiths DL. Midwives' support for complementary and alternative medicines: a literature review. Women Birth 2012;25:4–12.

[23] Nordeng H, Havnen GC. Impact of socio-demographic factors, knowledge and attitude on the use of herbal drugs in pregnancy. Acta Obstet Gynecol Scand 2005;84:26–33.

[24] Forster DA, Denning A, Wills G, Bolger M, McCarthy E. Herbal medicine use during pregnancy in a group of Australian women. BMC Pregnancy Childbirth 2006;19(6):21.

[25] Chandra A, Copen CE, Stephen EH. Infertility service use in the United States: data from the National Survey of Family Growth, 1982–2010. Nat Health Stat Report 2014;73:1–21.

[26] Oakley L, Doyle P, Maconochie N. Lifetime prevalence of infertility and infertility treatment in the UK: results from a population-based survey of reproduction. Hum Reprod 2008;23:447–50.

[27] Abbasi-Shavazi MJ, Inhorn MC, Rezeghi-Nasrabad HB, Toloo Gh. The "Iranian ART Revolution": infertility, assisted reproductive technology, and third-party donation in the Islamic Republic of Iran. J Middle East Women's Studies 2008;4:1–28.

[28] Rayner JA, Willis K, Burgess R. Women's use of complementary and alternative medicine for fertility enhancement: a review of the literature. J Altern Complement Med 2011;17:685–90.

[29] Onyiapat JL, Okoronkwo IL, Ogbonnaya NP. Complementary and alternative medicine use among adults in Enugu, Nigeria. BMC Complement Altern Med 2011;11:19.

[30] Freeman MP. Complementary and alternative medicine for perinatal depression. J Affect Disord 2009;112(1-3):1–10.

[31] Zini A, Fischer MA, Nam RK, Jarvi K. Use of alternative and hormonal therapies in male infertility. Urology 2004;63:141–3.

[32] Smith JF, Eisenberg ML, Millstein SG, Nachtigall RD, Shindel AW, Wing H, et al. The use of complementary and alternative fertility treatment in couples seeking fertility care: data from a prospective cohort in the United States. Fertil Steril 2010;93:2169–74.

[33] Tehrani Banihashemi SA, Asgharifard H, Haghdoost AA, Barghmadi M, Mohammadhosseini N. The use of complementary/alternative medicine among the general population in Tehran, Iran (in Persian). Payesh 2008;7:355–62.

[34] Bamidele JO, Adebimpe WO, Oladele EA. Knowledge, attitude and use of alternative medical therapy amongst urban residents of Osun State, southwestern Nigeria. Afr J Tradit Complement Altern Med 2009;6:281–8.

[35] Kashani L, Hassanzadeh E, Mirzabeighi A, Akhondzadeh S. Knowledge, attitude and practice of herbal remedies in a group of infertile couples. Acta Medica Iranica 2012;51:189–94.

[36] Smith JF, Eisenberg ML, Millstein SG, Nachtigall RD, Shindel AW, Wing H, et al. The use of complementary and alternative fertility treatment in couples seeking fertility care: data from a prospective cohort in the United States. Fertil Steril 2010;93:2169–74.

[37] Boivin J, Schmidt L. Use of complementary and alternative medicines associated with a 30% lower ongoing pregnancy/live birth rate during 12 months of fertility treatment. Hum Reprod 2009;24:1626–31.

[38] Hepner DL, Harnett M.J., Segal S, Camann W, Bader M, Tsen LC. Herbal medicinal products during pregnancy: are they safe? BJOG 2001;109:1425–6.

[39] Adams J, Sibbritt D, Lui CW. The use of complementary and alternative medicine during pregnancy: a longitudinal study of Australian women. Birth 2011;38:200–6.

[40] Forster DA, Denning A, Wills G, Bolger M, McCarthy E. Herbal medicine use during pregnancy in a group of Australian women. BMC Pregnancy Childbirth 2006;19(6):21.

[41] Glover GD, Amonkar M, Rybeck BF, Tracy TS. Prescription, over-the-counter, and herbal medicine use in a rural, obstetric population. Am J Obstet Gynecol 2003;188:1039–45.

[42] Facchinetti F, Dante G. . In: Chen S, editor. The ISDSP corner. San Antonio, TX: ISDSP; 2011.

[43] Adams C, Cannell S. Women's beliefs about "natural" hormones and natural hormone replacement therapy. Menopause 2001;8:433–40.

[44] Dante G, Bellei G, Neri I, Facchinetti F. Herbal therapies in pregnancy: what works? Curr Opin Obstet Gynecol 2014;26:83–91.

[45] Holst L, Nordeng H, Haavik S. Use of herbal drugs during early pregnancy in relation to maternal characteristics and pregnancy outcome. Pharmacoepidemiol Drug Saf 2008;17:151–9.

[46] Varga CA, Veale DJ. Isihlambezo: utilization patterns and potential health effects of pregnancy-related traditional herbal medicine. Soc Sci Med 1997;44:911–24.

[47] McKenna L, McIntyre M. What over-the-counter preparations are pregnant women taken? A literature review. J Adv Nurs 2006;56: 636–45.

[48] Facchinetti F, Pedrielli G, Benoni G, Joppi M, Verlato G, Dante G, et al. Herbal supplements in pregnancy: unexpected results from a multicentre study. Hum Reprod 2012;27:3161–7.

[49] Borrelli F, Capasso R, Aviello G, Pittler MH, Izzo AA. Effectiveness and safety of ginger in the treatment of pregnancy-induced nausea and vomiting. Obstet Gynecol 2005;105:849–56.

[50] Svoboda R, Lade A. Chinese medicine and Ayurveda. Delhi: Montilal Banarsidass Publishers Private Ltd; 1998. 126-127.

[51] Meyer K, Schwartz J, Crater D, Keyes B. *Zingiber officinale* (ginger) used to prevent 8-Mop associated nausea. Dermatol Nurs 1997;7:242–4.

[52] Grontved A, Brask T, Kambskard J, Hentzer E. Ginger root against seasickness. A controlled trial on the open sea. Acta Otolaryngol 1988;105:45–9.

[53] Mowrey DB, Clayson DE. Motion sickness, ginger and psychophysics. Lancet 1982;1:655–7.

[54] Phillips S, Ruggier R, Hutchinsons SE. *Zingiber officinale* (ginger) – an antiemetic for day case surgery. Anaesthesia 1990;45:669–71.

[55] Bone ME, WilkinsonDJ, Young JR, McNeil J, Charlton S. Ginger root – a new antiemetic: the effect of ginger root on postoperative nausea after major gynaecological surgery. Anaesthesia 1990;45:669–71.

[56] Matthews A, Dowswell T, Haas DM, Doyle M, O'Mathúna DP. Interventions for nausea and vomiting in early pregnancy. Cochrane Database Syst Rev 2010;8. CD007575.

[57] Chaiyakunapruk N, Kitikannakorn N, Nathisuwan S, Leeprakobboon K, Leelasettagool C. The efficacy of ginger for the prevention of postoperative nausea and vomiting: a meta-analysis. Am J Obstet Gynecol 2006;95–9.

[58] Ernst E, Pittler MH. Efficacy of ginger for nausea and vomiting: a systematic review of randomized clinical trials. Br J Anaesth 2000;84:367–71.

[59] Jednak MA, Shadigian EM, Kim MS, Woods ML, Hooper FG, Owyang C, et al. Protein meals reduce nausea and gastric slow wave dysrhythmic activity in first trimester pregnancy. Am J Physiol 1999;277:G855–61.

[60] Boon H, Smith M. The complete natural medicine guide to the 50 most common medicinal herbs. 2nd edn. Toronto: Robert Rose; 2004.

[61] Watkins RW. Herbal therapeutics: the top 12. Emerg Med 2002;34:16–9.

[62] Chatterjee SS, Noldner M, Koch E, Erdelmeier C. Antidepressant activity of *Hypericum perforatum* and hyperforin: the neglected possibility. Pharmacopsychiatry 1998,;31(Suppl. 1):7–15.

[63] Chatterjee SS, Bhattacharya SK, Wonnemann M, Singer A, Muller WE. Hyperforin as a possible antidepressant component of hypericum extracts. Life Sci 1998;63:499–510.

[64] Brown SS, Eisenberg L, editors. The best intentions: unintended pregnancy and the well-being of children and families. Washington, DC: National Academy Press; 1995.

[65] Moretti ME, Maxson A, Hanna F, Koren G. Evaluating the safety of St. John's wort in human pregnancy. Reprod Toxicol 2009;28: 96–9.

[66] Ozturk N, Korkmaz S, Ozturk Y. Wound-healing activity of St. John's wort (*Hypericum perforatum* L.) on chicken embryonic fibroblasts. J Ethnopharmacol 2007;111:33–9.

[67] Lissiman E, Bhasale AL, Cohen M. Garlic for the common cold. Cochrane Database Syst Rev 2012;3. CD006206.

[68] Hakimzadeh H, Ghazanfari T, Rahmati B, Naderimanesh H. Cytotoxic effect of garlic extract and its fractions on Sk-mel3 melanoma cell line. Immunopharmacol Immunotoxicol 2010;32:371–5.

[69] Borek C. Antioxidant health effects of aged garlic extract. J Nutr 2001;131:1010–5.

[70] Stevinson C, Pittler MH, Ernst E. Garlic for treating hypercholesterolemia: a meta-analysis of randomized clinical trials. Ann Intern Med 2000;133:420–9.

[71] Lau BHS. Suppression of LDL oxidation by garlic. J Nutr 2001;131:985–8.

[72] Amagase H, Petesch BL, Matsuura H, Kasuga S, Itakura Y. Intake of garlic and its bioactive components. J Nutr 2001;131:955S–62S.

[73] Blumenthal M. Market report. HerbalGram 2002;55:60.

[74] Dugoua JJ, Seely D, Perri D, Mills E, Koren G. Safety and efficacy of cranberry (*Vaccinium macrocarpon*) during pregnancy and lactation. Can J Clin Pharmacol 2008;15:e80–6.

[75] Jepson RG, Craig JC. Cranberries for preventing urinary tract infections. Cochrane Database Syst Rev 2008;23. CD001321.

[76] Schneeberger C, Geerlings SE, Middleton P, Crowther CA. Interventions for preventing recurrent urinary tract infection during pregnancy. Cochrane Database Syst Rev 2012;11. CD009279.

[77] DugouaJJ, Perri D, Seely D, Mills E, Koren G. Safety and efficacy of blue cohosh (*Caulophyllum thalictroides*) during pregnancy and lactation. Can J Clin Pharmacol 2008;15. e66-e73.

[78] Hoffman D. The new holistic herbal. Dorset: Element; 1990.

[79] Mills S. The a–z of modern herbalism. London: Thorsons; 1989.

[80] McFarlin BL, Gibson MH, O'Rear J, Harman P. A national survey of herbal preparation use by nurse-midwives for labor stimulation. Review of the literature and recommendations for practice. J Nurse Midwifery 1999;44:205–16.

[81] Below C. Herbs and the childbearing woman: guidelines for midwifes. J Nurse Midwifery 1999;44:231–52.

[82] Parsons M, Simpson M, Ponton T. Raspberry leaf and its effect on labor: safety and efficacy. Aust Coll Midwives Inc J 1999;12:20–5.

[83] Kelly AJ, Kavanagh J, Thomas J. Castor oil, bath and/or enema for cervical priming and induction of labor. Cochrane Database Syst Rev 2013;7. CD003099.

[84] Bayles B, Usatine R. Evening primrose oil. Am Fam Physician 2009;80:1405–8.

24

Selenium in Fertility and Reproduction

Hiten D. Mistry, PhD, Lesia O. Kurlak, PhD***

*University of Bern, Berne, Switzerland
**University of Nottingham, Nottingham, UK

SELENIUM BACKGROUND

Selenium is a trace element that occurs most naturally in carbonate rocks and volcanic and sedimentary soils. Although originally discovered in 1817 by Jöns Jacob Berzelius [1], it was not until 1957 that Schwarz and Foltz [2] proved that selenium is an essential nutrient for normal growth and reproduction in animals. This element can exist in organic forms, such as selenomethione, methylselenocysteine, and selenocysteine, as well as inorganic forms such as selenite, selenade, and selenide. Selenium is predominantly concentrated in the soils of drier regions of the world where the soil is usually more alkaline; in poorly aerated acidic soils, selenium is present mainly in selenite complexes, which are insoluble. The selenate form is taken up much more rapidly than selenite under most soil conditions. The highest concentrations of selenium are found in soils derived from seleniferous parent materials, which include shale, sandstone, limestone, slate, and coal. Therefore, seleniferous soils are widespread in parts of the United States, Canada, South America, China, and Russia [3]. Crop selenium uptake is influenced by the chemical species of the selenium in soil; inorganic selenium occurs in three soil-phases: fixed, adsorbed, and soluble, with only adsorbed and soluble forms available for plant uptake [3]. Availability of this element for incorporation into plants is affected by not only the actual selenium content of the soil, but also by pH, rainfall, use of high-sulfur fertilizers, as well as the presence of ions that can complex with selenium.

Selenoproteins

Selenium is integrated into enzymes, collectively referred to as selenoproteins, that have selenium in the form of the amino acid selenocysteine at their active center. In humans, 25 selenoproteins [4] have been identified having multiple actions; redox signaling [5] antioxidant protection against lipid peroxidation [6], thyroid hormone metabolism [7,8], T-cell immunity [5,9], and modulation of the inflammatory response [10] (Table 24.1).The three main classes of selenoproteins, the glutathione peroxidases (GPxs), thioredoxin reductases (TRs), and iodothyronine deiodinases (DIOs), were among the first eukaryotic selenoproteins discovered and are the most extensively studied [4,11]. Of particular importance to reproductive health are those having antioxidant effects, the GPxs, which play a pivotal role in reducing hydrogen peroxide and lipid peroxides to harmless products (water and alcohols) thereby mitigating damage caused by reactive oxygen species (ROS) [12]. As a result, these antioxidants maintain membrane and DNA integrity, ultimately protecting cell function. There is a form of GPx in virtually all cells and levels of this selenoenzyme respond rapidly to fluctations in selenium status. As a result, the nutritional status of selenium [13] is commonly assessed by direct measurement of its concentration in plasma or indirectly by measuring the activity of the GPx1 in erythrocytes [14], plasma GPx3, or whole blood GPx (GPx1 and 3). Selenoprotein P (SePP) [15], secreted from liver cells, is also important in the context of nutrition and reproductive health because it delivers selenium around the body maintaining selenium homeostasis. It contains 40–50% of the total selenium in the plasma. The tissues most affected by SePP deficiency are the testes and brain, with the kidney and heart less affected [15]. Under deficient conditions, plasma GPx activity decreases to approximately 1% of its activity under selenium-replete conditions; SePP, however, falls only to 5–10%. This shows that, even under severely deficient conditions, a considerable amount of the available selenium in animals and humans is used to synthesize SePP, thus emphasizing its importance among the hierarchy of selenoproteins and indicating it is a better index of selenium nutritional status [15]. Higher mammals have developed unique pathways to fine-tune the expression of all different selenoproteins according to developmental

Handbook of Fertility. http://dx.doi.org/10.1016/B978-0-12-800872-0.00024-X

TABLE 24.1 Known Mammalian Selenoproteins that Carry Out Functions of Selenium

Selenoprotein	Proposed function
Glutathione peroxidase (GPx)	
GPx1	Antioxidant in cell cytosol; selenium store?
GPx2	Antioxidant in gastrointestinal tract
GPx3	Antioxidant in extracellular space and plasma
GPx4	Antioxidant in membranes; structural protein in sperm; apoptosis?
GPx5	Unknown
GPx6	GPx1 homologue?
Thioredoxin reductase (TRs)	Multiple roles including dithiol-disulfide oxoreductase. Detoxifies peroxides, reduces thioredoxin (control of cell growth); maintains redox state of transcription factors
TR1	Mainly cytosolic, ubiquitous
TR2	Expressed by testes
TR3	Mitochondrial, ubiquitous
Iodothyronine deiodinases	
Type D1 and D2	Converts thyroxine (T4) to bioactive 3,5,3'-tri-iodothyronine (T3)
Type D1 and D3	Converts thyroxine (T4) to bio-inactive 3',3',5' reverse-tri-iodothyronine (T3)
Selenoprotein P	Selenium-transport protein. Antioxidant in endothelium
Selenoprotein W	Antioxidant in cardiac and skeletal muscle?
Selenophosphate synthetase (SPS2)	Synthesis of selenophosphate for selenoprotein synthesis
15 KDa Selenoprotein (Sep 15)	Protects against cancer?
Selenoproteins H,I,K,M,N,O,R,S,T,V	Role largely unknown

Adapted from Ref. [53].

stage, actual needs, and current availability of the trace element [16]. This means that not all tissues or all selenoproteins are equally well supplied when selenium becomes limiting. Selenium preferentially accumulates and is retained by the brain, testes, and adrenals [17–19], resulting in unequal biosynthesis of different selenoproteins; GPx4, TR1, and TR2 are all essential for life; GPx4 and GPx2 are ranked higher than GPx1 or GPx3; SePP and the deiodinase family are intermediate in the hierarchal positions [16].

These patterns also display differences between male and female individuals as seen in rat experiments, with females tending to retain the selenium more efficiently than males in most organs with the exception of the gonads [18], where there is an unparalleled high expression rate of GPx4.

Dietary Selenium

Dietary selenium is attained through a wide variety of food sources, which include bread and cereals, meat, fish, eggs, and milk/dairy products (Table 24.2) [20], with most forms of selenium being efficiently absorbed; 100% of selenate, >90% of ingested selenomethionine, and 80% of selenite [20]. The major form of selenium ingested by humans is selenomethionine (Fig. 24.1), which is predominant in cereals. A particularly rich source of selenium is Brazil nuts with selenium concentrations ranging from ~0.03 mg/kg to 512 mg/kg fresh weight depending on location; nuts from trees in central Brazil have ≤10 times more selenium than those from west Brazil.

Dietary intake of selenium will clearly vary between regions of the world largely due to the selenium content of the soil and hence where crops are cultivated, the soil/fodder to which animals are exposed, and the actual food types consumed. In the United States, a population study reported measured intake as an average of $105.5 \pm 50.7\ \mu g/day$ with resulting serum selenium levels of $136.7 \pm 18.9\ \mu g/L$ [21]. Selenium intakes in Europe are lower than in the United States with large variability [22] (see Table 24.3); Eastern Europe has the lowest intake at $7–30\ \mu g/day$, whereas Central and Western Europe have adequate levels of $30–90\ \mu g/day$ [23].

The health effects of selenium follow a U-shaped association with respect to selenium status [24]. Where nutritional intake is suboptimal, selenium supplementation can ameliorate conditions; however, supplementation where selenium intake is already adequate to high, can have adverse toxic effects, such as promotion of diabetes and insulin resistance [25]. Higher selenium status

TABLE 24.2 Selenium Content of General Food Types

Food	Typical selenium content (μg/g fresh weight)	Forms of selenium
Wheat	0.1–30	56–83% selenomethionine
Fish	0.1–5.0	50–70% selenomethionine
Meat	0.25–4.5	50–60% selenomethionine
Vegetables	0.5	Selenate/organic selenium
Garlic, selenium-enriched garlic	<0.5, 296	Selenate/organic selenium

Adapted from a Comprehensive Review [20].

is actually associated with adverse blood lipid profiles in the form of increased total and non-HDL cholesterol levels [21,26] and associated with increased cancer mortality [27,28].

In healthy adults, the World Health Organization (WHO) recommends a serum selenium concentration in the range 39.5–197.4 ng/mL [29]. However, since maximal GPx activity occurs at 70–90 ng/mL and SePP activity nearer 120 ng/mL, Fordyce suggests a range in the order of 60–105 ng/mL [30].

Selenium deficiency diseases have been identified in geographical areas notable for low selenium soil content; Keshan disease [31], first observed in Keshan County, Heilongjiang Province China; Kashin-Beck disease, mainly found in a diagonal belt from northwest to southwest China, in addition to Mongolia, Siberia, and North Korea. Keshan disease is a reversible endemic cardiomyopathy, characterized by focal myocardial necrosis associated with inflammatory infiltrates and calcification, which can lead to acute or chronic episodes of cardiogenic shock

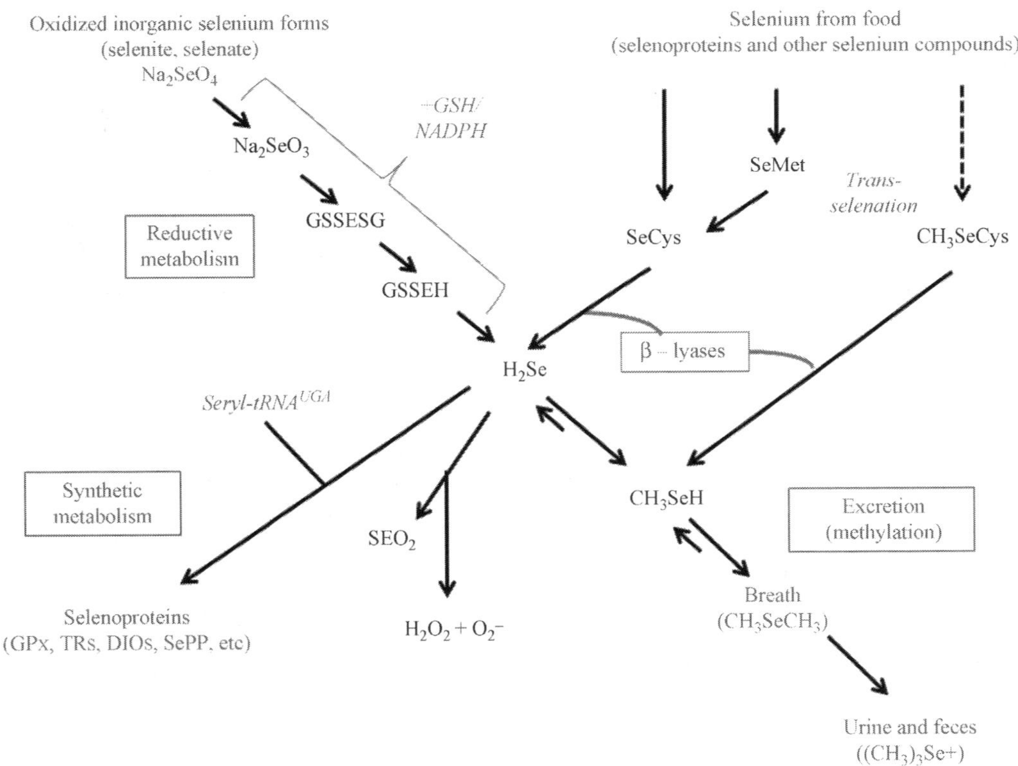

FIGURE 24.1 Metabolism of dietary selenium in humans. Oxidized inorganic form of selenium (selenate, selenite) undergo reductive metabolism, yielding hydrogen selenide (H_2Se), which is regarded as a precursor for supplying selenium in the active form for synthesis of selenoproteins (glutathione peroxidases (GPx, iodothyronine 5′-deiodinases (DIOs), thioredoxin reductases (TRs), selenoprotein P (SePP), and others) and cotranslationally through modification of tRNA-bound serinyl residues at certain loci encoded by specific UGA codons (Seryl-tRNA^{UGA}). Successive methylation of H_2Se detoxifies excess selenium, yielding methylselenol (CH_3SeH), dimethylselenide ($[CH_3]_2Se$), and trimethylselenonium ($(CH_3)_3Se^+$; the latter two metabolites are excreted in breath and urine, respectively. Food proteins can contain selenomethionine (SeMet), which can be incorporated nonspecifically into proteins in place of methionine, and selenocysteine (SeCys), which is a product of SeMet catabolism and is itself catabolized to H_2Se pool by a β-lyase. Another lyase releases CH_3SeH from Se-methylselenocysteine (CH_3SeCys) present in some foods (e.g., *Allium* vegetables). Oxidation of excess H_2Se leads to production of superoxide and other ROS; adapted from other sources [22,53].

D. NUTRITION AND REPRODUCTION

TABLE 24.3　Examples of Selenium Intake and Serum Selenium Concentrations in Different Regions of the World

Country	Selenium intake (μg/day)	Serum selenium (μg/L), mean ± SD/range
USA	60–220 [22]	136.7 ± 18.9 [21]
UK	29–39 [23]	40–158 [21]
Germany	35	63–95 [160]
Central Poland	30–40	50–135 [160]
China	7–4990 [23]	12–20

Source references are given with the values.

and congestive heart failure [32]. Kashin-Beck disease, on the other hand, is an endemic chronic degenerative osteoarthropathy with etiology involving the interaction of mycotoxins and selenium deficiency [33].

Aging

A relationship has been observed between estrogen and selenium status with erythrocyte selenium and GPx activity in women, coinciding with fluctuations in estrogen during the menstrual cycle [34]; this suggests that selenium status would be expected to fall following the decrease in estrogen after menopause. However, selenium status was maintained in healthy postmenopausal women and antioxidant capability, as measured by GPx activity, was preserved if dietary intake was greater than the recommended daily allowance (RDA) [34]. This has been substantiated by other groups, demonstrating selenium status in women around the age of menopause similar to those of younger women [35], except if contraceptive pills are used, in which case the selenium status is marginally higher in these younger women [35].

Trace elements, such as selenium, can affect the rate of biological aging because of their influence on modulating oxidative damage and DNA repair capacity, while levels of oxidized proteins increase with animal age [36]. It is difficult to establish whether or not selenium status decreases in elderly populations as there are conflicting data. A 9-year longitudinal Etude du Vieillissement Artériel (EVA) study of 1389 French subjects suggests that selenium does decrease with age [37]; however, the decrease is only slight in those independently living adults but may be more dramatic in subjects that are institutionalized [38] or ill [37]. In contrast, other studies have reported that selenium status increases with age, as in the Danish study [39], but this may reflect increased supplementation. Maybe some clearer insight is demonstrated by the InCHIANTI study, which measured selenium in a population between 60 and 110 years of age [40]; subjects were divided by 10 year intervals, there was no difference between any of the 60–90 year groups and only significant differences between the group containing nonagenarians and centenarians compared to all ages. The decrease after

90 years of age was the same for both genders. However, when this cohort was followed up after a 6-year period, the selenium level fell progressively with age [41], again with no differences between men and women.

SELENIUM AND REPRODUCTIVE HEALTH

At the physiological level, a certain amount of ROS is required for normal reproductive function [42]. However, when present in high concentrations, ROS can also have negative effects in fertility. Selenium is essential for both male and female fertility; these will be discussed next.

Male Fertility

Male fertility is often frustrating, with no indefinable cause being found in a significant number of cases due to the multifactorial disorder [43]. The damaging effects of oxidative stress is suggested to play a role in 30–80% of subfertile men [44]. Selenium is an important element for normal testicular development, spermatogenesis, and spermatozoa motility and function [45]. In animal studies, selenium has been shown to influence the gross and histological morphology of the testis [46]. SePP is responsible for the transport of selenium from the blood to the testis, while sperm formation requires GPx4 as a component of the midpiece sheath of the sperm tail [45], protecting the developing sperm cells from oxidative damage [47]. The cell membrane of spermatozoa contains large amounts of polyunsaturated fatty acids, which are susceptible to lipid peroxidation by oxygen radicals; ROS are being produced not only within the cytoplasm of the spermatozoa but also from leukocytes within the semen [47]. The part at which peroxidative damage to spermatozoa occurs is unknown, but there are clear effects of ROS on membrane integrity and morphology, possibly leading to cell death [48]. Antioxidant capacity in the semen is provided by a triad of enzymes: superoxide dismutase, catalase, and GPx4. Further in development, the GPx4 crosslinks with proteins in the midpiece to become a keratin-like structural part of the sperm that aids motility [45]. In healthy male subjects, a delicate balance between ROS and antioxidants is sustained in the reproductive tract. Studies have demonstrated that seminal oxidative stress negatively impacts on sperm motility, function, and concentration, ultimately affecting fusion events required for fertilization [49,50]. Decreased concentrations of GPx4 have been reported in spermatozoa of infertile males [51], particularly those classified as having oligoasthenozoospermia (i.e., where both mobility and concentration of spermatozoa are below normal).

In summary, serum selenium, semen total antioxidant capacity, and semen selenium are all lower in infertile men [52]; however, intervention studies supplementing dietary selenium have yielded mixed results [3] (and reviewed in Ref. [53]). The effects of selenium deficiency on spermatozoa have been demonstrated, including loss of motility, breakage at the midpiece level, and increased sperm shape abnormalities, mostly of the head [54]. In addition, positive associations were found between seminal plasma selenium and the proportion of normal sperm in semen samples [55] and sperm concentrations [56,57]. However, some trials have not corroborated these findings [58,59]. Several large randomized controlled trials (RCTs) in Scotland [60], Tunisa [61], and Iran [62] have indicated the benefits of selenium supplementation in improving male fertility. In contrast, others from the United States and Poland indicated no positive effects [63–65]. Such studies highlight that dosage of supplementation needs to be carefully considered, especially studies in the United States, in which basic selenium intake is generally higher.

A Cochrane review demonstrated that men taking antioxidants (including selenium) have statistically significant increases in both pregnancy rates and live birth rates [66]. Moreover, for those undergoing assisted reproduction, antioxidant supplementation improved both pregnancy rates (odds ratio (OR): 4.18; 95% confidence interval (CI), 4.18–6.59) and live birth rates (OR: 4.85; 95% CI, 1.92–12.24) [66]. However, it is unknown whether ROS production can be used as a criterion to select men for selenium supplementation, due to the many confounding factors, including among others, intracellular sperm antioxidant status, sperm counts, and abstinence time [42].

Female Fertility

Most studies into selenium and fertility establish a clear role for selenium in male fertility, but limited information exists as to its effects on female fertility. The emerging field of proteomics has enabled the identification of proteins involved in the cell cycle, as antioxidants, extracellular matrix, cytoskeleton, and their linkage to oxidative stress in female infertility related diseases [67]. Oxidative stress results in poor quality oocytes and embryos; significantly increased ROS, high lipid peroxidation, and decreased total antioxidant capacity in follicular fluid has been reported to correlate with lower fertilization rates [68]. Moreover, low concentrations of intrafollicular ROS are regarded as a promising marker for predicating *in vitro* fertilization successes [69]. Paszkowski et al. [70] completed a study of follicular fluid samples collected from 112 patients during transvaginal oocyte retrieval and showed that the follicular fluid GPx activity was significantly lower in follicles yielding oocytes, which failed to be fertilized. In addition, patients with unexplained infertility had significantly decreased selenium levels as compared with those with tubal infertility or male factor. These observations are supported by other groups [71], where supplementation was shown to normalize follicular fluid selenium levels [72]. Although no data were reported on the subsequent changes to pregnancy rates in this study, others suggest that selenium supplementation does have a positive effect on fertility [73]. Significantly raised serum concentrations of an autoantibody to selenium binding protein-1 have been demonstrated in women with unexplained infertility or premature ovarian failure and this appears to be unique to ovarian autoimmunity [74].

Another gynecological disorder, polycystic ovarian syndrome (PCOS) has also been associated with selenium. PCOS affects up to 10% of all females of reproductive age and is the most common endocrinopathy in women [75,76]. In the long term, PCOS, among other things, is considered a risk factor for infertility. The major characteristics of PCOS are chronic anovulation, menstrual abnormalities (oligomenorrhea or amenorrhea), hyperandrogenism, and polycystic ovary appearance on ultrasound [77]. PCOS has been linked with inflammation, oxidative stress and selenium (reviewed in Ref. [78]); in particular, studies have shown that PCOS patients experience increased oxidative stress and decreased antioxidant status [79, 80]. Furthermore, Coskun et al. reported reduced serum selenium concentrations in PCOS, compared to age- and BMI-matched control women [77], although this needs to be verified in further studies, as this is an isolated study.

Further work is warranted to fully evaluate these initial relationships between female reproductive disorders, infertility, and selenium deficiency.

SELENIUM AND PREGNANCY

Normal Pregnancy

During normal pregnancy, the selenium requirement is increased due to the demands of the growing fetus [81]; maternal blood selenium levels decrease from around 65 µg/L in the first trimester to 50 µg/L in the third trimester [82,83]. Umbilical cord values indicate that the fetus has lower selenium concentrations compared to the mother [84,85], reflecting the fact that selenium crosses the placenta by means of a concentration gradient via an anion exchange pathway shared with sulfate; both organic and inorganic forms of selenium are able to cross the placenta [86]. In the United States, the RDA of selenium in pregnancy, based on a fetal deposition of 4 μg/day throughout pregnancy, is 60 μg/day [87] compared to 55 μg/day in nonpregnant female adults. In the United Kingdom, where selenium intakes are lower, the reference nutrient intakes (RNI) are 75 μg/day [87], compared to 60 μg/day in nonpregnant female adults [88].

Type 3 iodothyronine deiodinase (DiO3) is a selenoprotein, which potentially protects the fetus from the elevated maternal thyroid hormones present in the placenta and amniotic fluid. In pregnancy, placental transfer of thyroid hormones from the maternal to the fetal circulation is essential for fetal development and in particular the central nervous system [89]. In early pregnancy, maternal triiodothyronine (T_3) and its prohormone thyroxine (T_4) are the only sources of these hormones for the fetus, which only starts to produce thyroid hormones at around 11–12 weeks, but does not produce any clinically significant amounts of T_4 until 17–19 weeks' gestation [89]. Although, transplacental thyroid hormone transport is modulated by plasma membrane transporters, DiO3 is the physiologic inactivator of thyroid hormone T_4, protecting fetal tissues against high toxic concentrations of this hormone. Uterine endometrium and the placenta are the only normal tissues known to express high levels of DiO3 activity in the mature human [90], whereas it is expressed in multiple fetal structures. Under conditions of low plasma selenium, as seen in the pregnancy-specific condition, preeclampsia, DiO3 concentrations and activity are maintained [91], in line with the selenoprotein hierarchy.

Pregnancy per se leads to an increased oxidative burden as high maternal and fetal oxygen demand increases oxygen metabolism [92,93]. Longitudinal studies of oxidative stress and antioxidant status in pregnancy [94,95] indicate that lipid peroxides rise in pregnancy to the highest levels in the second trimester, while total antioxidant capacity falls through pregnancy until 30 weeks' gestation and then rises again toward the end of the third trimester [96]. This is also mirrored by erythrocyte GPx concentrations [94].

Miscarriage

Miscarriage is defined as a pregnancy that ends spontaneously before 20 weeks of gestation and is estimated to occur in 12–15% of clinical pregnancies and 17–22% of all pregnancies (including early pregnancy losses) [97]. Genetic (chromosomal) abnormalities explain at least half of all miscarriages, although anatomical, endocrine, immune, infective, and thrombophilic conditions are other possible causes, most chromosomally normal miscarriages remain unexplained or idiopathic [98]. Some reports implicate selenium deficiency in miscarriage, although the evidence is inconsistent. A reduction in serum selenium normally occurs in the first trimester of pregnancies that progress to term; however, a further statistically highly significant decrease in serum selenium has been observed in women who have persistently miscarried [99–101]; one study reports higher selenium in women having recurrent spontaneous pregnancy loss [102]. Longer-term nutrient status can be measured in red cells and hair [13,103]. Studies in the United Kingdom [99] and Turkey [101,104] demonstrate lower hair selenium content in women suffering recurrent miscarriage, despite similar serum selenium levels. In contrast, a study in a South African population [105] did not observe such a difference although the control group also had low hair selenium content.

A few isolated studies also suggest that early pregnancy loss is associated with reduced antioxidant protection of biological membranes and DNA, potentially due to lower GPx concentrations. Further investigations are required to establish whether optimization of selenium status could be of benefit to women with recurrent pregnancy loss.

Preeclampsia

Preeclampsia occurs in approximately 2–8% of all pregnancies [106] and remains the second leading direct cause of maternal death, over 99% of which occurs in less developed countries. This condition is defined as *de novo* hypertension with proteinuria presenting after 20 weeks' gestation [107,108]. As both placental and systemic oxidative stress are components of the syndrome [109,110] and selenium status is associated with oxidative stress, it is not surprising that many studies have suggested that selenium deficiency may be linked to preeclampsia [85,111–113].

Maternal serum selenium concentrations correlate positively with plasma total GPx activity as well as placental tissue GPx activity [85], with lower values observed in the preeclamptic pregnancies [114]. This is supported by data obtained from different populations around the world [115–117].

Several genes that encode selenoproteins demonstrate functional polymorphisms. Selenoprotein S (SEPS1) is an anti-inflammatory protein, which acts primarily to limit the damaging consequences of endoplasmic reticulum stress. The A allele of the *SEPS1*-105G_ > A polymorphism is associated with impaired expression of *SEPS1* and is genetically associated with preeclampsia [118]. A single nucleotide polymorphism within the GPx4 gene

(*GPx4c18t*) affects GPx protein concentration and activity in females but not in males [119]. Interestingly, this polymorphism also exhibits differential effects on GPx3 and GPx1 when selenium supplementation is stopped in healthy volunteers but needs further investigation in preeclampsia.

Selenium supplementation studies are difficult as the optimal range of selenium intake to ensure biological benefit is narrow and still has not been determined with any certainty. A randomized, controlled pilot trial has looked at the effects of selenium on markers of risk in preeclampsia in pregnant women in the United Kingdom receiving 60 µg/day selenium-enriched yeast supplementation from 12–14 weeks' gestation until delivery [120]. It was not powered sufficiently to show any effect on preeclampsia risk factors, but clearly demonstrated an increase of 43% in whole blood selenium in the supplemented women compared to a fall of 12% in the placebo group. In addition, although not ideal, the odds of women having either preeclampsia or related pregnancy-induced hypertension, was significantly reduced in the selenium-treated group [120]. This finding needs to be validated in a larger, adequately powered RCT.

Small-for-Gestational-Age Deliveries

A small-for-gestational-age (SGA) infant is one whose individualized birth weight ratio falls below the 10th percentile and is associated with increased perinatal mortality and morbidity [121]. Some studies of SGA deliveries report reduced placental selenium concentrations (SGA: median [IQR]: 0.14 [0.1, 0.2] compared to 0.15 [0.1–0.24]) in placentae from appropriate-for-gestational-age (AGA) deliveries [122]. However, others report higher selenium in umbilical venous and arterial blood between SGA and AGA [123] and some, unchanged [124]. An investigation into a cohort of poor adolescent pregnant women from two inner cities in the United Kingdom found lower plasma selenium concentrations in mothers who delivered SGA infants compared to those who delivered AGA infants [125]. Geographical variations as well as differing markers of selenium status between these studies may account for some of the inconsistencies.

Preterm Birth

Preterm labor (<37 weeks) is a major cause of perinatal morbidity and mortality, occurring in around 6% of pregnancies in the developed world and as high as 25% of pregnancies in the undeveloped countries [126,127]. Several cross-sectional studies have reported lower plasma selenium in mothers delivering preterm as compared to term in Poland [128] (despite similar placental tissue selenium content), The Netherlands [129], and India [130], as well as in preterm infants compared

to those born at term in Germany [131] and Iran [132]. There is a significant association between umbilical cord selenium concentration and gestational age of newborns [133,134], with infants at 24–27 weeks' gestation having umbilical cord selenium concentrations in the range 29.4–84.5 µg/L and infants born 39–40 weeks' gestation having levels in the range 26.77–114.6 µg/L [133]. In addition, the lower the birth weight, the lower the cord selenium concentration in newborns [133,135].

Inflammation plays an important role in the etiology and pathophysiology of spontaneous preterm birth and SEPS1 is involved in regulating the inflammatory response [136]. Recently, the G-105A promoter polymorphism in *SEPS1* was shown to increase pro-inflammatory cytokine expression and may contribute to the risk of developing spontaneous premature delivery in a Chinese population [137].

Gestational Diabetes

Gestational diabetes mellitus (GDM) is a common medical disorder in pregnancy and is associated with adverse pregnancy outcomes, including fetal macrosomia, stillbirth, neonatal metabolic disturbances, and related problems [138]. The causes are not known but are closely related to a constitutional risk of type 2 diabetes in later life and strongly associated with obesity. Serum selenium appears to be lower in women with GDM or impaired glucose tolerance [139]; an inverse relationship between selenium concentrations and blood glucose concentrations has been observed even without changes in insulin [140], suggesting that selenium may affect glucose metabolism downstream from insulin or possibly through independent energy-regulating pathways such as thyroid hormones [140]. This relationship is unique to pregnancy, as diabetes in nonpregnant subjects is actually associated with higher, not lower, blood selenium concentrations [141,142].

SELENIUM AND EARLY CHILDHOOD

Pediatric allergic disease represents the largest category of chronic disease and this has increased in prevalence since the mid-twentieth century [143]. Airway development occurs predominately antenatally, commencing approximately 24 days after fertilization with the pre-acinar airway branching pattern being completed by about 17 weeks' gestation [144].

Fetal immune and airway development during a critical period of life can be potentially influenced by maternal diet during pregnancy, with long-term irreversible consequences, such as childhood asthma – a chronic disorder of the lung airways associated with increased airway responsiveness and variable airway obstruction. More specifically, a suboptimal fetal nutrient status antenatally adversely affects respiratory epithelial and mesenchymal development, resulting in suboptimal early-life airway function, which is associated with an increased risk of wheezing and asthma in later childhood [145]. Erythrocyte selenium is lower in the first few years of life [146].

Seaton et al. hypothesized that the increase in asthma had resulted from increasing population susceptibility rather than the air becoming more toxic or allergenic [147]. This hypothesis arose by the paralleled changes in the diet, resulting in deficiencies in associated antioxidants, potentially resulting in a decline in lung antioxidant defenses and thus increased oxidant-induced airway damage, airway inflammation, and asthma [147]. Associations between several antioxidant micronutrients concentrations during pregnancy with asthma, wheezing, and eczema during early childhood have been shown in a systematic review [148].

Numerous studies have observed significantly low mean plasma/serum selenium concentrations in very preterm (gestational age <32 weeks, birthweight <1500 g) compared to healthy term infants [132,149–151]. Several studies have suggested that these low selenium levels in preterm infants is associated with selenium accumulation during gestation [150,152,153]. Low selenium concentrations and GPx activities in the blood of preterm infants has been proposed to contribute to respiratory distress syndrome, retinopathy of prematurity, increased hemolysis or other prematurity related conditions [128]. Maternal plasma selenium concentrations during pregnancy are inversely associated with wheezing in children at 2 years of age [154]. In addition, low selenium concentrations in umbilical cord plasma have also been observed in children with wheezing compared to controls [154,155]. This has been further confirmed by measurements of selenium in nails from children at 12 years of age; those children with the highest quintile of nail selenium concentrations presented a fivefold decrease in the prevalence ratio of asthma, whereas those in the lowest selenium quintile presented with an almost 2.5-fold increase [156]. Similar findings have also been shown with lower plasma selenium concentrations found in children, also at 12 years of age with childhood asthma compared to healthy age-matched controls [157].

Low selenium status may exacerbate disease progression in conditions not otherwise associated with selenium deficiency (e.g., human immunodeficiency virus (HIV) infection) and has been associated with HIV-related mortality in children [158]. A recent study reported lower plasma selenium concentrations in prepubertal HIV-infected children compared with gender- and BMI-matched HIV-uninfected children [159]. The authors have suggested that selenium deficiency, could have a detrimental effect on the immune system and cardiac muscle of HIV-infected subjects [159].

CONCLUSIONS

Increased knowledge about the importance of selenium and its crucial part in maintaining healthy reproductive function, successful pregnancy, and determining both the long- and short-term health of the mother and baby needs to be addressed and be made a key focus for future health strategies in improving pregnancy outcomes.

An essential factor that needs to be highlighted when considering the health effects of selenium is the U-shaped link with status; whereas additional selenium intake may benefit people with low status, those with adequate to high status might be affected adversely and should not take selenium supplements. A definitive conclusion for many of these reproductive disorders linked to selenium deficiency cannot be drawn from the existing heterogeneous literature. Further adequately powered RCTs are required to reinforce or refute the argument of increasing selenium intake in selected populations. Finally, maternal selenium and antioxidant status should be assessed from periconceptional through postnatal periods. Furthermore, studies are required to evaluate the relationship between dietary and supplementary selenium intake before and during pregnancy in relation to adverse outcomes.

References

[1] Oldfield JE. The two faces of selenium. J Nutr 1987;117(12):2002–8.

[2] Schwarz K, Foltz CM. Selenium as an integral part of factor-3 against dietary necrotic liver degeneration. J Am Chem Soc 1957;79(12):3292–3.

[3] Fairweather-Tait SJ, Bao Y, Broadley MR, Collings R, Ford D, Hesketh JE, et al. Selenium in human health and disease. Antioxid Redox Signal 2011;14(7):1337–83.

[4] Reeves MA, Hoffmann PR. The human selenoproteome: recent insights into functions and regulation. Cell Mol Life Sci 2009;66(15):2457–78.

[5] Arner ES. Focus on mammalian thioredoxin reductases – important selenoproteins with versatile functions. Biochim Biophys Acta 2009;1790(6):495–526.

[6] Brigelius-Flohe R, Maiorino M. Glutathione peroxidases. Biochim Biophys Acta 2013;1830(5):3289–303.

[7] Kohrle J. Selenium and the thyroid. Current opinion in endocrinology, diabetes, and obesity 2013;20(5):441–8.

[8] Shrimali RK, Irons RD, Carlson BA, Sano Y, Gladyshev VN, Park JM, et al. Selenoproteins mediate T cell immunity through an antioxidant mechanism. J Biol Chem 2008;283(29):20181–5.

[9] Carlson BA, Yoo MH, Shrimali RK, Irons R, Gladyshev VN, Hatfield DL, et al. Role of selenium-containing proteins in T-cell and macrophage function. Proc Nutr Soc 2010;69(3):300–10.

[10] Huang Z, Rose AH, Hoffmann PR. The role of selenium in inflammation and immunity: from molecular mechanisms to therapeutic opportunities. Antioxid Redox Signal 2012;16(7):705–43.

[11] Bellinger FP, Raman AV, Reeves MA, Berry MJ. Regulation and function of selenoproteins in human disease. Biochem J 2009;422(1):11–22.

[12] Rotruck JT, Pope AL, Ganther HE, Swanson AB, Hafeman DG, Hoekstra WG. Selenium: biochemical role as a component of glutathione peroxidase. Science 1973;179(4073):588–90.

[13] Ashton K, Hooper L, Harvey LJ, Hurst R, Casgrain A, Fairweather-Tait SJ. Methods of assessment of selenium status in humans: a systematic review. Am J Clin Nutr 2009;89(6):2025S–39S.

[14] Stefanowicz FA, Talwar D, O'Reilly DS, Dickinson N, Atkinson J, Hursthouse AS, et al. Erythrocyte selenium concentration as a marker of selenium status. Clin Nutr 2013;32(5):837–42.

[15] Burk RF, Hill KE, Selenoprotein P. an extracellular protein with unique physical characteristics and a role in selenium homeostasis. Annu Rev Nutr 2005;25:215–35.

[16] Schomburg L, Schweizer U. Hierarchical regulation of selenoprotein expression and sex-specific effects of selenium. Biochim Biophys Acta 2009;1790(11):1453–62.

[17] Hopkins LL Jr, Pope AL, Baumann CA. Distribution of microgram quantities of selenium in the tissues of the rat, and effects of previous selenium intake. J Nutr 1966;88(1):61–5.

[18] Brown DG, Burk RF. Selenium retention in tissues and sperm of rats fed a Torula yeast diet. J Nutr 1973;103(1):102–8.

[19] Behne D, Hofer-Bosse T. Effects of a low selenium status on the distribution and retention of selenium in the rat. J Nutr 1984;114(7):1289–96.

[20] Fairweather-Tait SJ, Collings R, Hurst R. Selenium bioavailability: current knowledge and future research requirements. Am J Clin Nutr 2010;91(5):1484S–91S.

[21] Laclaustra M, Stranges S, Navas-Acien A, Ordovas JM, Guallar E. Serum selenium and serum lipids in US adults: National Health and Nutrition Examination Survey (NHANES) 2003–2004. Atherosclerosis 2010;210(2):643–8.

[22] Combs GF Jr. Selenium in global food systems. Br J Nutr 2001;85(5):517–47.

[23] Rayman MP. Food-chain selenium and human health: emphasis on intake. Br J Nutr 2008;100(2):254–68.

[24] Rayman MP. Selenium and human health. Lancet 2012;379(9822):1256–68.

[25] Rocourt CR, Cheng WH. Selenium supranutrition: are the potential benefits of chemoprevention outweighed by the promotion of diabetes and insulin resistance? Nutrients 2013;5(4):1349–65.

[26] Stranges S, Laclaustra M, Ji C, Cappuccio FP, Navas-Acien A, Ordovas JM, et al. Higher selenium status is associated with adverse blood lipid profile in British adults. J Nutr 2010;140(1):81–7.

[27] Bleys J, Navas-Acien A, Guallar E. Serum selenium levels and all-cause, cancer, and cardiovascular mortality among US adults. Arch Intern Med 2008;168(4):404–10.

[28] Bleys J, Navas-Acien A, Stranges S, Menke A, Miller ER 3rd, Guallar E. Serum selenium and serum lipids in US adults. Am J Clin Nutr 2008;88(2):416–23.

[29] World Health Organization. Vitamin and Mineral Requirements in Human Nutrition. Switzerland: World Health Organization; 2004.

[30] Fordyce F. Selenium deficiency and toxicity in the environment. In: Selinus O, Alloway B, Centeno J, Finkelman R, Fuge R, Lindh Y, editors. Essential of Medical Geology. London: Elsevier; 2005. p. 373–415.

[31] Chen J. An original discovery: selenium deficiency and Keshan disease (an endemic heart disease). Asia Pacific J Clin Nutr 2012;21(3):320–6.

[32] Li Q, Liu M, Hou J, Jiang C, Li S, Wang T. The prevalence of Keshan disease in China. Int J Cardiol 2013;168(2):1121–6.

[33] Yao Y, Pei F, Kang P. Selenium, iodine, and the relation with Kashin-Beck disease. Nutrition 2011;27(11–12):1095–100.

[34] Ha EJ, Smith AM. Selenium-dependent glutathione peroxidase activity is increased in healthy post-menopausal women. Biol Trace Elem Res 2009;131(1):90–5.

[35] Arnaud J, Arnault N, Roussel AM, Bertrais S, Ruffieux D, Galan P, et al. Relationships between selenium, lipids, iron status and hormonal therapy in women of the SU.VI.M.AX cohort. J Trace Elem Med Biol 2007;21(Suppl. 1):66–9.

[36] Stadtman ER. Protein oxidation and aging. Free Radic Res 2006;40(12):1250–8.

[37] Arnaud J, Akbaraly TN, Hininger I, Roussel AM, Berr C. Factors associated with longitudinal plasma selenium decline in the elderly: the EVA study. J Nutr Biochem 2007;18(7):482–7.

[38] Rambouskova J, Krskova A, Slavikova M, Cejchanova M, Wranova K, Prochazka B, et al. Trace elements in the blood of institutionalized elderly in the Czech Republic. Arch Gerontol Geriatr 2013;56(2):389–94.

[39] Rasmussen LB, Hollenbach B, Laurberg P, Carle A, Hog A, Jorgensen T, et al. Serum selenium and selenoprotein P status in adult Danes – 8-year followup. J Trace Elem Med Biol 2009;23(4):265–71.

[40] Savarino L, Granchi D, Ciapetti G, Cenni E, Ravaglia G, Forti P, et al. Serum concentrations of zinc and selenium in elderly people: results in healthy nonagenarians/centenarians. Exp Gerontol 2001;36(2):327–39.

[41] Lauretani F, Semba RD, Bandinelli S, Ray AL, Ruggiero C, Cherubini A, et al. Low plasma selenium concentrations and mortality among older community-dwelling adults: the InCHIANTI Study. Aging Clin Exp Res 2008;20(2):153–8.

[42] Mora-Esteves C, Shin D. Nutrient supplementation: improving male fertility fourfold. Semin Reprod Med 2013;31(4):293–300.

[43] Bonanomi M, Lucente G, Silvestrini B. Male fertility: core chemical structure in pharmacological research. Contraception 2002;65(4):317–20.

[44] Tremellen K. Oxidative stress and male infertility–a clinical perspective. Hum Reprod Update 2008;14(3):243–58.

[45] Ursini F, Heim S, Kiess M, Maiorino M, Roveri A, Wissing J, et al. Dual function of the selenoprotein PHGPx during sperm maturation. Science 1999;285(5432):1393–6.

[46] Ahsan U, Kamran Z, Raza I, Ahmad S, Babar W, Riaz MH, et al. Role of selenium in male reproduction – a review. Anim Reprod Sci 2014;146(1–2):55–62.

[47] Walczak-Jedrzejowska R, Wolski JK, Slowikowska-Hilczer J. The role of oxidative stress and antioxidants in male fertility. Cent European J Urol 2013;66(1):60–7.

[48] Aitken RJ. Free radicals, lipid peroxidation and sperm function. Reprod Fertil Dev 1995;7(4):659–68.

[49] Sharma RK, Agarwal A. Role of reactive oxygen species in male infertility. Urology 1996;48(6):835–50.

[50] Sikka SC. Relative impact of oxidative stress on male reproductive function. Curr Med Chem 2001;8(7):851–62.

[51] Imai H, Suzuki K, Ishizaka K, Ichinose S, Oshima H, Okayasu I, et al. Failure of the expression of phospholipid hydroperoxide glutathione peroxidase in the spermatozoa of human infertile males. Biol Reprod 2001;64(2):674–83.

[52] Eroglu M, Sahin S, Durukan B, Ozakpinar OB, Erdinc N, Turkgeldi L, et al. Blood serum and seminal plasma selenium, total antioxidant capacity and coenzyme q10 levels in relation to semen parameters in men with idiopathic infertility. Biol Trace Elem Res 2014;159(1–3):46–51.

[53] Mistry HD, Broughton Pipkin F, Redman CW, Poston L. Selenium in reproductive health. Am J Obstet Gynecol 2012;206(1):21–30.

[54] Watanabe T, Endo A. Effects of selenium deficiency on sperm morphology and spermatocyte chromosomes in mice. Mutat Res 1991;262(2):93–9.

[55] Noack-Fuller G, De Beer C, Seibert H. Cadmium, lead, selenium, and zinc in semen of occupationally unexposed men. Andrologia 1993;25(1):7–12.

[56] Bleau G, Lemarbre J, Faucher G, Roberts KD, Chapdelaine A. Semen selenium and human fertility. Fertil Steril 1984;42(6):890–4.

[57] Behne D, Gessner H, Wolters G, Brotherton J. Selenium, rubidium and zinc in human semen and semen fractions. Int J Androl 1988;11(5):415–23.

[58] Saaranen M, Suistomaa U, Kantola M, Saarikoski S, Vanha-Perttula T. Lead, magnesium, selenium and zinc in human seminal fluid: comparison with semen parameters and fertility. Hum Reprod 1987;2(6):475–9.

[59] Roy AC, Karunanithy R, Ratnam SS. Lack of correlation of selenium level in human semen with sperm count/motility. Arch Androl 1990;25(1):59–62.

[60] Scott R, MacPherson A, Yates RW, Hussain B, Dixon J. The effect of oral selenium supplementation on human sperm motility. Br J Urol 1998;82(1):76–80.

[61] Keskes-Ammar L, Feki-Chakroun N, Rebai T, Sahnoun Z, Ghozzi H, Hammami S, et al. Sperm oxidative stress and the effect of an oral vitamin E and selenium supplement on semen quality in infertile men. Arch Androl 2003;49(2):83–94.

[62] Safarinejad MR, Safarinejad S. Efficacy of selenium and/or N-acetyl-cysteine for improving semen parameters in infertile men: a double-blind, placebo controlled, randomized study. J Urol 2009;181(2):741–51.

[63] Hawkes WC, Turek PJ. Effects of dietary selenium on sperm motility in healthy men. J Androl 2001;22(5):764–72.

[64] Hawkes WC, Alkan Z, Wong K. Selenium supplementation does not affect testicular selenium status or semen quality in North American men. J Androl 2009;30(5):525–33.

[65] Iwanier K, Zachara BA. Selenium supplementation enhances the element concentration in blood and seminal fluid but does not change the spermatozoal quality characteristics in subfertile men. J Androl 1995;16(5):441–7.

[66] Showell MG, Brown J, Yazdani A, Stankiewicz MT, Hart RJ. Antioxidants for male subfertility. Cochrane Database Syst Rev 2011;2011 Jan 19(1). DOI 10.1002/14651858.CD007411.pub2.

[67] Gupta S, Ghulmiyyah J, Sharma R, Halabi J, Agarwal A. Power of proteomics in linking oxidative stress and female infertility. Biomed Res Int 2014;2014. DOI 10.1155/2014/916212.

[68] Jana SK, NB K, Chattopadhyay R, Chakravarty B, Chaudhury K. Upper control limit of reactive oxygen species in follicular fluid beyond which viable embryo formation is not favorable. Reprod Toxicol 2010;29(4):447–51.

[69] Attaran M, Pasqualotto E, Falcone T, Goldberg JM, Miller KF, Agarwal A, et al. The effect of follicular fluid reactive oxygen species on the outcome of in vitro fertilization. Int J Fertil Womens Med 2000;45(5):314–20.

[70] Paszkowski T, Traub AI, Robinson SY, McMaster D. Selenium dependent glutathione peroxidase activity in human follicular fluid. Clin Chim Acta 1995;236(2):173–80.

[71] Ozkaya MO, Naziroglu M. Multivitamin and mineral supplementation modulates oxidative stress and antioxidant vitamin levels in serum and follicular fluid of women undergoing in vitro fertilization. Fertil Steril 2010;94(6):2465–6.

[72] Ozkaya MO, Naziroglu M, Barak C, Berkkanoglu M. Effects of multivitamin/mineral supplementation on trace element levels in serum and follicular fluid of women undergoing in vitro fertilization (IVF). Biol Trace Elem Res 2011;139(1):1–9.

[73] Buhling KJ, Grajecki D. The effect of micronutrient supplements on female fertility. Curr Opin Obstet Gynecol 2013;25(3):173–80.

[74] Edassery SL, Shatavi SV, Kunkel JP, Hauer C, Brucker C, Penumatsa K, et al. Autoantigens in ovarian autoimmunity associated with unexplained infertility and premature ovarian failure. Fertil Steril 2010;94(7):2636–41.

[75] Franks S. Polycystic ovary syndrome. N Engl J Med 1995;333(13):853–61.

[76] Knochenhauer ES, Key TJ, Kahsar-Miller M, Waggoner W, Boots LR, Azziz R. Prevalence of the polycystic ovary syndrome in unselected black and white women of the southeastern United States: a prospective study. J Clin Endocrinol Metab 1998;83(9):3078–82.

[77] Coskun A, Arikan T, Kilinc M, Arikan DC, Ekerbicer HC. Plasma selenium levels in Turkish women with polycystic ovary syndrome. Eur J Obstet Gynecol Reprod Biol 2013;168(2):183–6.

[78] Brenneisen P, Steinbrenner H, Sies H. Selenium, oxidative stress, and health aspects. Mol Aspects Med 2005;26(4–5):256–67.

[79] Fenkci V, Fenkci S, Yilmazer M, Serteser M. Decreased total antioxidant status and increased oxidative stress in women with polycystic ovary syndrome may contribute to the risk of cardiovascular disease. Fertil Steril 2003;80(1):123–7.

[80] Yilmaz M, Bukan N, Ayvaz G, Karakoc A, Toruner F, Cakir N, et al. The effects of rosiglitazone and metformin on oxidative stress and homocysteine levels in lean patients with polycystic ovary syndrome. Hum Reprod 2005;20(12):3333–40.

[81] Smith AM, Picciano MF. Evidence for increased selenium requirement for the rat during pregnancy and lactation. J Nutr 1986;116(6):1068–79.

[82] Zachara BA, Wardak C, Didkowski W, Maciag A, Marchaluk E. Changes in blood selenium and glutathione concentrations and glutathione peroxidase activity in human pregnancy. Gynecol Obstet Invest 1993;35(1):12–7.

[83] Mihailovic M, Cvetkovic M, Ljubic A, Kosanovic M, Nedeljkovic S, Jovanovic I, et al. Selenium and malondialdehyde content and glutathione peroxidase activity in maternal and umbilical cord blood and amniotic fluid. Biol Trace Elem Res 2000;73(1):47–54.

[84] Gathwala G, Yadav OP, Singh I, Sangwan K. Maternal and cord plasma selenium levels in full-term neonates. Indian J Pediatr 2000;67(10):729–31.

[85] Mistry HD, Wilson V, Ramsay MM, Symonds ME, Broughton Pipkin F. Reduced selenium concentrations and glutathione peroxidase activity in preeclamptic pregnancies. Hypertension 2008;52(5):881–8.

[86] Shennan DB, Boyd CA. Ion transport by the placenta: a review of membrane transport systems. Biochim Biophys Acta 1987;906(3):437–57.

[87] Institute of Medicine Food and Nutrition Board. Dietary Reference Intakes for Vitamin C, Vitamin E, Selenium, and Carotenoids. Washington DC: National Academy Press; 2000. 284–324.

[88] Department of Health. Dietary reference values for food energy and nutrients for the United Kingdom. Report on social subjects no.41. London, UK; 1991.

[89] Chan SY, Vasilopoulou E, Kilby MD. The role of the placenta in thyroid hormone delivery to the fetus. Nat Clin Pract Endocrinol Metab 2009;5(1):45–54.

[90] Huang SA. Physiology and pathophysiology of type 3 deiodinase in humans. Thyroid 2005;15(8):875–81.

[91] Kurlak LO, Mistry HD, Kaptein E, Visser TJ, Broughton Pipkin F. Thyroid hormones and their placental deiodination in normal and pre-eclamptic pregnancy. Placenta 2013;34(5):395–400.

[92] Morris JM, Gopaul NK, Endresen MJ, Knight M, Linton EA, Dhir S, et al. Circulating markers of oxidative stress are raised in normal pregnancy and pre-eclampsia. Br J Obstet Gynaecol 1998;105(11):1195–9.

[93] Roes EM, Hendriks JC, Raijmakers MT, Steegers-Theunissen RP, Groenen P, Peters WH, et al. A longitudinal study of antioxidant status during uncomplicated and hypertensive pregnancies. Acta Obstet Gynecol Scand 2006;85(2):148–55.

[94] Hung TH, Lo LM, Chiu TH, Li MJ, Yeh YL, Chen SF, et al. A longitudinal study of oxidative stress and antioxidant status in women with uncomplicated pregnancies throughout gestation. Reprod Sci 2010;17(4):401–9.

[95] Little RE, Gladen BC. Levels of lipid peroxides in uncomplicated pregnancy: a review of the literature. Reprod Toxicol 1999;13(5):347–52.

[96] Patil SB, Kodliwadmath MV, Kodliwadmath SM. Study of oxidative stress and enzymatic antioxidants in normal pregnancy. Indian J Clin Biochem 2007;22(1):135–7.

[97] Garcia-Enguidanos A, Calle ME, Valero J, Luna S, Dominguez-Rojas V. Risk factors in miscarriage: a review. Eur J Obstet Gynecol Reprod Biol 2002;102(2):111–9.

[98] Hirschfeld AF, Jiang R, Robinson WP, McFadden DE, Turvey SE. Toll-like receptor 4 polymorphisms and idiopathic chromosomally normal miscarriage. Hum Reprod 2007;22(2):440–3.

[99] Barrington JW, Lindsay P, James D, Smith S, Roberts A. Selenium deficiency and miscarriage: a possible link? Br J Obstet Gynaecol 1996;103(2):130–2.

[100] Abdulah R, Noerjasin H, Septiani L, Mutakin, Defi IR, Suradji EW, et al. Reduced serum selenium concentration in miscarriage incidence of Indonesian subjects. Biol Trace Elem Res 2013;154(1):1–6.

[101] Kocak I, Aksoy E, Ustun C. Recurrent spontaneous abortion and selenium deficiency. Int J Gynaecol Obstet 1999;65(1):79–80.

[102] Ajayi OO, Charles-Davies MA, Arinola OG. Progesterone, selected heavy metals and micronutrients in pregnant Nigerian women with a history of recurrent spontaneous abortion. Afr Health Sci 2012;12(2):153–9.

[103] Sun Y, Li HZ. Determination of trace selenium in human plasma and hair with ternary inclusion compound-fluorescent spectrophotometry. Analyst 2000;125(12):2326–9.

[104] Al-Kunani AS, Knight R, Haswell SJ, Thompson JW, Lindow SW. The selenium status of women with a history of recurrent miscarriage. BJOG 2001;108(10):1094–7.

[105] Thomas VV, Knight R, Haswell SJ, Lindow SW, van der Spuy ZM. Maternal hair selenium levels as a possible long-term nutritional indicator of recurrent pregnancy loss. BMC Womens Health 2013;13:40.

[106] Steegers EA, von Dadelszen P, Duvekot JJ, Pijnenborg R. Pre-eclampsia. Lancet 2010;376(9741):631–44.

[107] Brown MA, Lindheimer MD, de Swiet M, Van Assche A, Moutquin JM. The classification and diagnosis of the hypertensive disorders of pregnancy: statement from the International Society for the Study of Hypertension in Pregnancy (ISSHP). Hypertens Pregnancy 2001;20(1):IX–XIV.

[108] Tranquilli AL, Brown MA, Zeeman GG, Dekker G, Sibai BM. The definition of severe and early-onset preeclampsia. Statements from the International Society for the Study of Hypertension in Pregnancy (ISSHP). Pregnancy Hyperten 2013;3:44–7.

[109] Poston L. The role of oxidative stress. In: Critchley H, MacLean A, Poston L, Walker J, editors. Pre-eclampsia. London: RCOG Press; 2004.

[110] Poston L, Igosheva N, Mistry HD, Seed PT, Shennan AH, Rana S, et al. Role of oxidative stress and antioxidant supplementation in pregnancy disorders. Am J Clin Nutr 2011;94(Suppl. 6):1980S–5S.

[111] Rayman MP, Bode P, Redman CW. Low selenium status is associated with the occurrence of the pregnancy disease preeclampsia in women from the United Kingdom. Am J Obstet Gynecol 2003;189(5):1343–9.

[112] Maleki A, Fard MK, Zadeh DH, Mamegani MA, Abasaizadeh S, Mazloomzadeh S. The relationship between plasma level of Se and preeclampsia. Hypertens Pregnancy 2011;30(2):180–7.

[113] Farzin L, Sajadi F. Comparison of serum trace element levels in patients with or without pre-eclampsia. J Res Med Sci 2012;17(10):938–41.

[114] Mistry HD, Kurlak LO, Williams PJ, Ramsay MM, Symonds ME, Broughton Pipkin F. Differential expression and distribution of placental glutathione peroxidases 1, 3 and 4 in normal and preeclamptic pregnancy. Placenta 2010;31(5):401–8.

[115] Atamer Y, Kocyigit Y, Yokus B, Atamer A, Erden AC. Lipid peroxidation, antioxidant defense, status of trace metals and leptin levels in preeclampsia. Eur J Obstet Gynecol Reprod Biol 2005;119(1):60–6.

[116] Wang Y, Walsh SW. Antioxidant activities and mRNA expression of superoxide dismutase, catalase, and glutathione peroxidase in normal and preeclamptic placentas. J Soc Gynecol Investig 1996;3(4):179–84.

[117] Vanderlelie J, Venardos K, Clifton VL, Gude NM, Clarke FM, Perkins AV. Increased biological oxidation and reduced antioxidant enzyme activity in pre-eclamptic placentae. Placenta 2005;26(1):53–8.

[118] Moses EK, Johnson MP, Tommerdal L, Forsmo S, Curran JE, Abraham LJ, et al. Genetic association of preeclampsia to the inflammatory response gene SEPS1. Am J Obstet Gynecol 2008;198(3):336 e1–5.

[119] Meplan C, Crosley LK, Nicol F, Horgan GW, Mathers JC, Arthur JR, et al. Functional effects of a common single-nucleotide polymorphism (GPX4c718t) in the glutathione peroxidase 4 gene: interaction with sex. Am J Clin Nutr 2008;87(4):1019–27.

[120] Rayman MP, Searle E, Kelly L, Johnsen S, Bodman-Smith K, Bath SC, et al. Effect of selenium on markers of risk of pre-eclampsia in UK pregnant women: a randomised, controlled pilot trial. Br J Nutr 2014;112(1):99–111.

[121] Cetin I, Foidart JM, Miozzo M, Raun T, Jansson T, Tsatsaris V, et al. Fetal growth restriction: a workshop report. Placenta 2004;25(8–9):753–7.

[122] Klapec T, Cavar S, Kasac Z, Rucevic S, Popinjac A. Selenium in placenta predicts birth weight in normal but not intrauterine growth restriction pregnancy. J Trace Elem Med Biol 2008;22(1):54–8.

[123] Osada H, Watanabe Y, Nishimura Y, Yukawa M, Seki K, Sekiya S. Profile of trace element concentrations in the feto-placental unit in relation to fetal growth. Acta Obstet Gynecol Scand 2002;81(10):931–7.

[124] Llanos MN, Ronco AM. Fetal growth restriction is related to placental levels of cadmium, lead and arsenic but not with antioxidant activities. Reprod Toxicol 2009;27(1):88–92.

[125] Mistry HD, Kurlak LO, Young SD, Briley AL, Broughton Pipkin F, Baker PN, et al. Maternal selenium, copper and zinc concentrations in pregnancy associated with small-for-gestational-age infants. Matern Child Nutr 2014;10(3):327–34.

[126] Steer P. The epidemiology of preterm labor–a global perspective. J Perinat Med 2005;33(4):273–6.

[127] Goldenberg RL, Culhane JF, Iams JD, Romero R. Epidemiology and causes of preterm birth. The Lancet. **371**(9606): 75–84.

[128] Dobrzynski W, Trafikowska U, Trafikowska A, Pilecki A, Szymanski W, Zachara BA. Decreased selenium concentration in maternal and cord blood in preterm compared with term delivery. Analyst 1998;123(1):93–7.

[129] Rayman MP, Wijnen H, Vader H, Kooistra L, Pop V. Maternal selenium status during early gestation and risk for preterm birth. CMAJ 2011;183(5):549–55.

[130] Gathwala G, Yadav OP, Sangwan K, Singh I, Yadav J. A study on plasma selenium level among pregnant women at Rohtak, Haryana. Indian J Public Health 2003;47(2):45–8.

[131] Sievers E, Arpe T, Schleyerbach U, Garbe-Schönberg D, Schaub J. Plasma selenium in preterm and term infants during the first 12 months of life. J Trace Elem Med Biol 2001;14(4):218–22.

[132] Iranpour R, Zandian A, Mohammadizadeh M, Mohammadzadeh A, Balali-Mood M, Hajiheydari M. Comparison of maternal and umbilical cord blood selenium levels in term and preterm infants. Zhongguo Dang Dai Er Ke Za Zhi 2009;11(7):513–6.

[133] Makhoul IR, Sammour RN, Diamond E, Shohat I, Tamir A, Shamir R. Selenium concentrations in maternal and umbilical cord blood at 24–42 weeks of gestation: basis for optimization of selenium supplementation to premature infants. Clin Nutr 2004;23(3):373–81.

[134] Galinier A, Périquet B, Lambert W, Garcia J, Assouline C, Rolland M, et al. Reference range for micronutrients and nutritional marker proteins in cord blood of neonates appropriated for gestational ages. Early Hum Dev 2005;81(7):583–93.

[135] Lockitch G, Jacobson B, Quigley G, Dison P, Pendray M. Selenium deficiency in low birth weight neonates: an unrecognized problem. J Pediatr 1989;114(5):865–70.

[136] Curran JE, Jowett JB, Elliott KS, Gao Y, Gluschenko K, Wang J, et al. Genetic variation in selenoprotein S influences inflammatory response. Nat Genet 2005;37(11):1234–41.

[137] Wang Y, Yang X, Zheng Y, Wu ZH, Zhang XA, Li QP, et al. The SEPS1 G-105A polymorphism is associated with risk of spontaneous preterm birth in a Chinese population. PloS One 2013;8(6):e65657.

[138] Coustan DR. Gestational diabetes mellitus. Clin Chem 2013;59(9):1310–21.

[139] Tan M, Sheng L, Qian Y, Ge Y, Wang Y, Zhang H, et al. Changes of serum selenium in pregnant women with gestational diabetes mellitus. Biol Trace Elem Res 2001;83(3):231–7.

[140] Hawkes WC, Alkan Z, Lang K, King JC. Plasma selenium decrease during pregnancy is associated with glucose intolerance. Biol Trace Elem Res 2004;100(1):19–29.

[141] Laclaustra M, Navas-Acien A, Stranges S, Ordovas JM, Guallar E. Serum selenium concentrations and diabetes in U.S. adults: National Health and Nutrition Examination Survey (NHANES) 2003–2004. Environ Health Perspect 2009;117(9):1409–13.

[142] Bleys J, Navas-Acien A, Guallar E. Serum selenium and diabetes in U.S. adults. Diabetes Care 2007;30(4):829–34.

[143] Jonsson U. The rise and fall of paradigms in world food and nutrition policy. World Pub Health Nutr 2010;1(3):128–58.

[144] Devereux G. Early life events in asthma–diet. Pediatr Pulmonol 2007;42(8):663–73.

[145] Menon P, Ruel MT, Loechl CU, Arimond M, Habicht JP, Pelto G, et al. Micronutrient Sprinkles reduce anemia among 9- to 24-mo-old children when delivered through an integrated health and nutrition program in rural Haiti. J Nutr 2007;137(4):1023–30.

[146] Chen Z, Myers R, Wei T, Bind E, Kassim P, Wang G, et al. Placental transfer and concentrations of cadmium, mercury, lead, and selenium in mothers, newborns, and young children. J Expo Sci Environ Epidemiol 2014;24(5):537–44.

[147] Seaton A, Godden DJ, Brown K. Increase in asthma: a more toxic environment or a more susceptible population? Thorax 1994;49(2):171–4.

[148] Patelarou E, Giourgouli G, Lykeridou A, Vrioni E, Fotos N, Siamaga E, et al. Association between biomarker-quantified antioxidant status during pregnancy and infancy and allergic disease during early childhood: a systematic review. Nutr Rev 2011;69(11):627–41.

[149] Sievers E, Arpe T, Schleyerbach U, Garbe-Schonberg D, Schaub J. Plasma selenium in preterm and term infants during the first 12 months of life. J Trace Elem Med Biol 2001;14(4):218–22.

[150] Mask G, Lane HW. Selected measures of selenium status in full-term and preterm neonates, their mothers and non-pregnant women. Nutr Res 1993;13:901–11.

[151] Freitas RG, Nogueira RJ, Antonio MA, Barros-Filho Ade A, Hessel G. Selenium deficiency and the effects of supplementation on preterm infants. Rev Paul Pediatr 2014;32(1):126–35.

[152] Smith AM, Chan GM, Moyer-Mileur LJ, Johnson CE, Gardner BR. Selenium status of preterm infants fed human milk, preterm formula, or selenium-supplemented preterm formula. J Pediatr 1991;119(3):429–33.

[153] Amin S, Chen SY, Collipp PJ, Castro-Magana M, Maddaiah VT, Klein SW. Selenium in premature infants. Nutr Metab 1980;24(5):331–40.

[154] Devereux G, McNeill G, Newman G, Turner S, Craig L, Martindale S, et al. Early childhood wheezing symptoms in relation to plasma selenium in pregnant mothers and neonates. Clin Exp Allergy 2007;37(7):1000–8.

[155] Shaheen SO, Newson RB, Henderson AJ, Emmett PM, Sherriff A, Cooke M. Umbilical cord trace elements and minerals and risk of early childhood wheezing and eczema. Eur Respir J 2004;24(2):292–7.

[156] Carneiro MF, Rhoden CR, Amantea SL, Barbosa F Jr. Low concentrations of selenium and zinc in nails are associated with childhood asthma. Biol Trace Elem Res 2011;144(1–3):244–52.

[157] Kocyigit A, Armutcu F, Gurel A, Ermis B. Alterations in plasma essential trace elements selenium, manganese, zinc, copper, and iron concentrations and the possible role of these elements on oxidative status in patients with childhood asthma. Biol Trace Elem Res 2004;97(1):31–41.

[158] Campa A, Shor-Posner G, Indacochea F, Zhang G, Lai H, Asthana D, et al. Mortality risk in selenium-deficient HIV-positive children. J Acquir Immune Defic Syndr Hum Retrovirol 1999;20(5): 508–13.

[159] Pugliese C, Patin RV, Palchetti CZ, Chiantelli CC, de Fatima Thome Barbosa Gouvea A, de Menezes Succi RC, et al. Assessment of antioxidants status and superoxide dismutase activity in HIV-infected children. Braz J Infect Dis 2014;18(5):481–6.

[160] Van Cauwenbergh R, Robberecht H, Van Vlaslaer V, Deelstra H. Comparison of the serum selenium content of healthy adults living in the Antwerp region (Belgium) with recent literature data. J Trace Elem Med Biol 2004;18(1):99–112.

The Role of Oxidative Stress in Endometriosis

Aditi Mulgund, MD,**, Sejal Doshi, MD*,**,*
*Ashok Agarwal, PhD***

*Northeastern Ohio Medical University, Rootstown, Ohio, USA
**Center for Reproductive Medicine, Cleveland Clinic, Cleveland, Ohio, USA

INTRODUCTION

Endometriosis is a disease that affects women of reproductive age primarily and is characterized by the presence of endometrial glands and stroma outside the endometrial lining and uterine muscle. This complex disease primarily affects 10% of all women, and is linked to infertility and chronic pelvic pain [1]. In the pathological state, the endometrial tissue is known to implant in other locations in the body, most commonly in the ovaries, peritoneal cavity, and even in the bladder, liver, kidneys, pleural cavity, and the gluteal muscles (Fig. 25.1) [2]. The etiology of endometriosis, though currently unclear, can be ascribed to one of five origins.

Endometriosis is a common gynecological disorder, affecting 10–15% of women in their reproductive years. Because surgical confirmation is necessary for the diagnosis, the true prevalence of the disease is underestimated [3]. Specifically, 10–70% of women presenting with pelvic pain have been found to have endometriosis and it has been shown to be the causal factor in 35–50% of those women diagnosed with infertility [4].

BACKGROUND OF DISEASE

Risk Factors

Women with first-degree relatives with the disease have an increased propensity (seven times more likely) to display symptoms of endometriosis [3]. One out of every 10 women with endometriosis has a first-degree relative with this disease [5]. Specifically, a new genetic linkage for the disease has been noted on chromosome 10q26 and chromosome 20p13, further showing that endometriosis has a familial relationship [6]. Endometriosis also shares a relationship with autoimmune disorders such as lupus.

Symptoms of Endometriosis

The classic presentation of endometriosis is cyclic pelvic pain, or dysmenorrhea, that peaks 1–2 days prior to the onset of menses, and then diminishing at the onset of flow. Symptoms of endometriosis vary depending on the area involved. Dysmenorrhea usually should be considered in women who develop pain after years of normal, pain-free cycles. Other symptoms associated with endometriosis include dysmenorrhea, dyspareunia, abnormal bleeding, and infertility. On physical examination, typically nodularity is felt on the uterosacral ligament, and on bimanual examination of the reproductive structures, the uterus is typically in fixed, retroverted position [3].

Diagnosis of Endometriosis

Unfortunately, diagnostic tools to detect endometriosis are few, invasive, and not ideal. The only way to diagnose endometriosis is through direct visualization of endometrial implants through laparoscopy or laparotomy [3]. These implants may appear as rust-colored or dark-brown powder burns or blue-colored raspberry lesions. Unfortunately, this is an invasive procedure that does not function ideally to detect subtle endometriosis and early preventive diagnosis. It is thereby essential to develop nonsurgical methods for early detection that would provide a more appealing option for women. Aside from laparoscopy, the most recent noninvasive alternatives for the diagnosis of endometriosis are via the

Handbook of Fertility. http://dx.doi.org/10.1016/B978-0-12-800872-0.00025-1

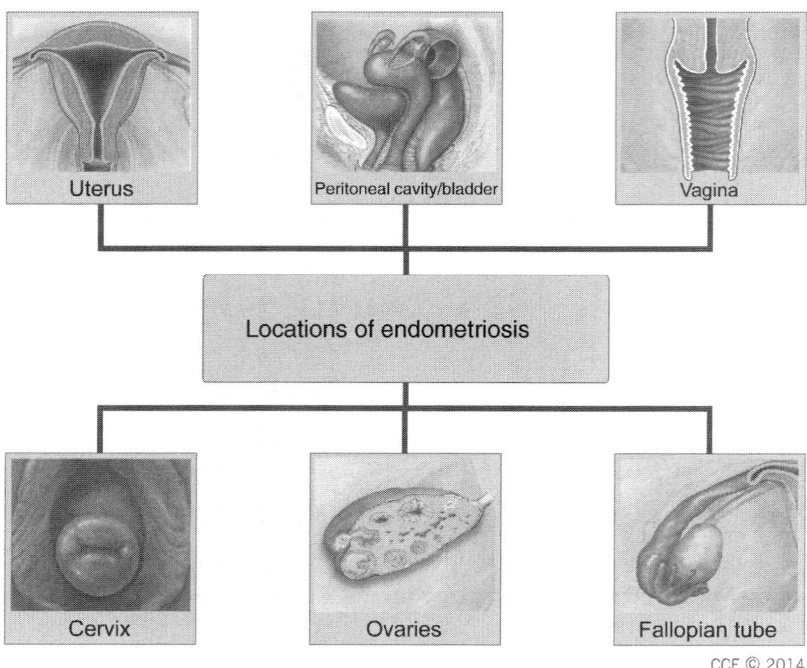

FIGURE 25.1   The figure demonstrates the female anatomy with areas that are commonly affected by endometriosis and the sources of ROS that contribute to the disease. Reprinted with permission, Cleveland Clinic Center for Medical Art & Artwork © 2014. All Rights Reserved.

analysis of serum and peritoneal markers. Such indirect markers include those that are indicated when levels of oxidative stress increase, and may be less accurate.

In this review, we will discuss the current markers of this disorder, which are mainly concentrated on oxidative stress elevation, and the various roles of different types of reactive oxygen species (ROS) such as nitric oxide, iron, and mediators of inflammation in endometriosis.

THE ORIGIN OF ENDOMETRIOSIS

The origin of endometriosis is yet unknown. However, multiple theories exist as to the causes and pathophysiology of the disease.

Sampson's Implantation Theory

Sampson's implantation theory is the most widely accepted theory for the cause of endometriosis. Sampson suggests that endometrial tissue gets transported back through the fallopian tubes during retrograde menstruation, leading to intra-abdominal pelvic implants [3]. This theory of retrograde reflux followed by peritoneal implantation is supported by the location of endometriotic lesions. The four main places that endometriotic lesions have been found include the pouch of Douglas at the rectosigmoid level, the cecum and ileocecal junction, the superior portion of the sigmoid mesocolon, and the right paracolic gutter [7]. These regions are those that experience increased contact with peritoneal fluid, as well as repetitive fluid flow [7].

Samson's theory also has three requirements for occurrence. These include retrograde menstruation through the fallopian tubes, which is present in 76–90% of women, the presence of viable refluxed cells in the peritoneal cavity, and the adherence of refluxed cells to the peritoneal cavity surface where they then implant and multiply [8].

Coelomic Metaplasia Theory

The coelomic metaplasia theory proposes that endometriosis comes from reversible changes of the epithelial lining in the peritoneal cavity. Specifically, coelomic epithelium is a derivative of endometrial and peritoneal cells and thus, this theory hypothesizes that transformation from one type of cell to the other in the form of metaplasia is perhaps triggered by inflammation seen in endometriosis. The idea that substances in menstrual fluid can induce peritoneal tissues to form endometrial cells suggests that there is a factor found in menstrual fluid that is possibly a precursor for the disease [8].

Embryonic Rest Theory

The embryonic rest theory implies that rest cells of mullerian origin differentiate into endometrium from certain stimuli [8]. Specifically, during the embryonic stage of development, certain endometrial cells that should grow in the uterus develop in the abdomen, which in turn, are activated during puberty under the influence of estrogen and progesterone and result in the vaginal bleeding and pelvic pain of endometriosis [9].

Vascular and Lymphatic Metastasis

Finally, the vascular and lymphatic metastasis theory states that endometrial cells spread through the body much like cancer, via the body's blood and lymphatic systems [8]. Moreover, approximately one third of the microvascular endothelium of ectopic endometrial tissue is derived from endothelial progenitor cells, which results in *de novo* formation of microvessels by the process of vasculogenesis seen in endometriosis [10].

RELATIONSHIP BETWEEN OXIDATIVE STRESS AND ENDOMETRIOSIS

Oxidative stress results from an imbalance of free radicals and antioxidant defense mechanisms in the body. Normally, antioxidants, enzymatic and nonenzymatic, scavenge free-radical species and protect the body from overexposure to oxidative stress [11]. An imbalance in the ratio of the antioxidants to ROS can result in a variety of pathological processes within the body. One such form of oxidative stress includes ROS. Specifically, ROS comprise a class of radical and nonradical oxygen derivatives that play a significant role in reproductive biology. Because they have an unpaired electron in their outer orbit, ROS are highly reactive and interact with a variety of lipids, proteins, and nucleic acids in the body [11]. Abnormal levels of ROS are not only harmful for reproductive potential, but they also generate more free radicals, thereby perpetuating a chain of reactions and creating high amounts of oxidative stress [12].

Present in the oviductal fluid, ROS has shown to affect a variety of female reproductive processes such as ovulation, fertilization, embryo development, and implantation. Specifically, this fluid has been found to contain cellular debris that act as a substrate for oxidative stress reactions in women with endometriosis. ROS are produced by activated neutrophils and macrophages, which are elevated in proinflammatory conditions such as endometriosis [13]. Despite the lack of substantial evidence of a causal relationship, multiple studies have shown a direct association between ROS and endometriosis, [14,15] while others have provided evidence to suggest the absence of the link between ROS and endometriosis [16,17].

Huge strides have been made to prove the detrimental effect of ROS to the reproductive potential, in general, but the purpose of this review is to address the impact of oxidative stress on endometriosis. A consolidation of the available information on this topic will allow the development of potential antioxidant treatments for conditions associated with this condition.

OXIDATIVE STRESS MARKERS IN ENDOMETRIOSIS

Oxidative stress markers present potential therapeutic targets for the treatment of endometriosis. By expanding and researching these markers, the diagnostics of endometriosis may progress from requiring surgery to a much less invasive method. The following sections will discuss the most relevant markers of ROS in endometriosis (Fig. 25.2).

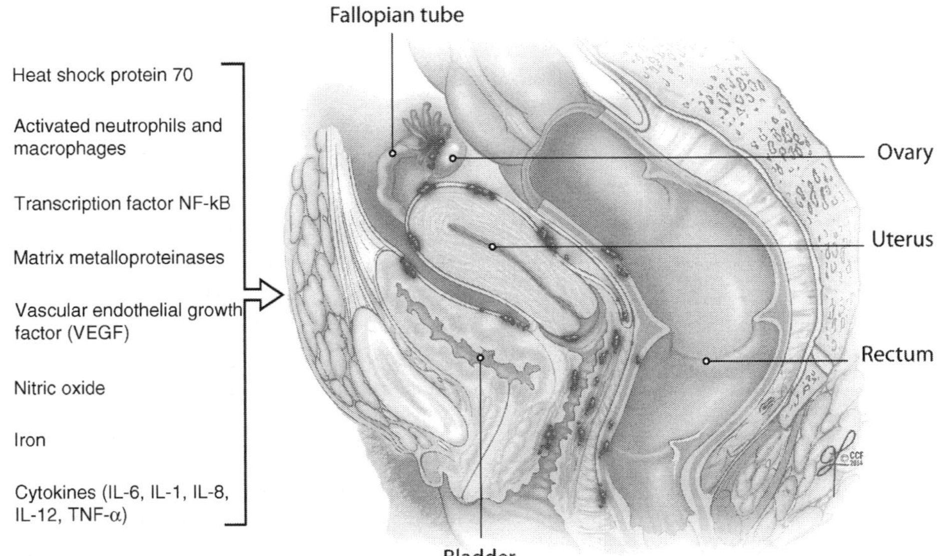

FIGURE 25.2 The figure illustrates the various contributors of oxidative stress in the female reproductive tract in the form of immune mediators, growth factors, and proteins. Reprinted with permission, Cleveland Clinic Center for Medical Art & Artwork © 2014. All Rights Reserved.

Heat Shock Protein 70

The normal function of heat shock proteins (HSP) is to regulate and guide protein synthesis and protect the proteins from undue stress. Previous studies have found that heat shock protein 70B' is increased in endometriosis [18]. Another study stated that HSP-70 stimulated vascular endothelial cell growth factor, interleukin-6 (IL-6), and tumor necrosis factor alpha (TNF-α) by macrophages in women with endometriosis when compared with healthy women [19]. Khan and colleagues also suspect that HSP-70 is implicated in the toll-like receptor 4 (TLR4) mediated growth of endometrial cells [19].

Macrophages

In endometriosis patients, macrophage number has been noted to increase in peritoneal fluid. This increase of phagocytic cells results in a higher amount of phagocytosis, which in turn releases more ROS [20]. Endometriosis results in the inflammation of the peritoneal cavity and other sites of implantation due to irritation from the refluxed blood, and results in the activation of neutrophils and macrophages – known producers of ROS [20].

Transcription Factor NF-kB

Nuclear factor kappa B (NF-kB) is a transcription factor that normally functions to control the transcription of DNA. It also plays a vital role in the regulation of immune factors during infection [21]. NF-kB is normally responsible for the expression of proinflammatory cytokines, growth factor, angiogenic factor, adhesion molecules, and inducible enzymes iNOS and COX-2 [22]. These products contribute to the development of endometriosis through the increase of endometrial fragment adhesion, proliferation, and neovascularization [23,24].

Matrix Metalloproteinases

The pathogenesis of endometriosis involves many different enzymes including matrix metalloproteinases (MMPs) that are produced in the endometrial stroma and function to degrade the extracellular matrix in the peritoneal mesothelium in order for ectopic cells to implant [25]. These MMPs may be necessary for the formation of endometriotic lesions [25].

Vascular Endothelial Growth Factor

Just as with metastatic tumors, endometriosis involves neovascularization for lesions to proliferate. Vascular endothelial growth factor (VEGF) is a known participant in endometriosis. It is hypothesized to be a participant in the angiogenesis that occurs in endometriotic lesions. Particularly, a study by Kim and colleagues determined that the VEGF polymorphism +405 C/G was associated with the increased risk of endometriosis in Korean women [26]. Additionally, a recent meta-analysis determined that certain VEGF polymorphisms are associated with an increased risk of endometriosis while others may be protective [27]. These studies indicate that VEGF has a strong role in the development of endometriosis, and potentially can function as a diagnostic factor or therapeutic target.

Destruction of Mesothelial Cells

During endometriosis and oxidative stress, the peritoneal mesothelium is particularly susceptible to damage [16,28]. In a healthy individual, the mesothelium serves as a protective barrier to ward off endometrial implants. However, in an individual with endometriosis, the mesothelium is fragile and susceptible to damage from oxidative stress. These damaged mesothelial cells function as adhesion sites for endometrial cells, which leads to progression of the disease [11,29].

NITRIC OXIDE AND ENDOMETRIOSIS

ROS not only includes oxygen radicals, such as the hydroxyl radical, superoxide radical, and hydrogen peroxide, but also a subclass of nitrogen-containing compounds collectively known as reactive nitrogen species (RNS). Examples of RNS include peroxynitrite anion, nitroxyl ion, nitrosyl-containing compounds, and nitric oxide [12]. All of these forms of RNS have been shown to play physiologic and pathologic roles in the female reproductive system; however, the role of nitric oxide in endometriosis has been most extensively reviewed in the literature.

Production of Nitric Oxide

Nitric oxide (NO) is produced from L-arginine via nitric oxide synthase (NOS). It requires oxygen and a number of cofactors, such as nicotinamide adenine dinucleotide phosphate (NADPH), flavin mononucleotide (FMN), flavin adenine dinucleotide (FAD), calmodulin, and calcium, resulting in the production of NO as well as a byproduct known as L-citrulline [30,31]. There are three forms of NOS that exert their effect through protein–protein interactions and catalyze the aforementioned reaction: (1) endothelial NOS (eNOS), (2) inducible NOS (iNOS), and (3) neuronal NOS (nNOS). Each isoform has a reductase domain that contains a compound known as tetrahydrobiopterin (BH4), which is essential for the efficient production of NO [32,33]. Pertinent to endometriosis, studies show that the peritoneal fluid in women with this condition contain increased concentrations of eNOS and iNOS, making the peritoneal fluid one of the richest sources of nitric oxide within the female reproductive tract. With regard to eNOS, it is located mainly in the endometrial glandular epithelium and elevated during the midluteal phase of the menstrual cycle. Moreover, studies report that the expression pattern of eNOS has an inverse relationship with an adhesive marker for uterine receptivity known as integrin alpha V beta 3. As levels of eNOS increase during the second half of the menstrual cycle, there is a respective decrease in integrin alpha V beta 3, potentially contributing to the implantation difficulties seen in endometriosis [13].

Aside from NOS, there are a variety of other compounds and biochemical reactions that produce NO in the body. Namely, studies have linked the generation of NO to the rate-limiting enzyme glucose-6-phosphate dehydrogenase as well as the NADPH-producing pentose phosphate pathway [34].

Additional pathways of nitric oxide transport can be generated via binding of NO to iron–sulfur clusters, formation of nitrotyrosines, and binding of NO to heme-containing proteins of the respiratory chain [35].

Furthermore, glucose has been shown to indirectly produce NO by stimulating the pentose phosphate pathway, as well as the conversion of L-arginine to L-citrulline. Other studies have indicated that NO can regulate its own activity via a feedback inhibition mechanism [32,33]. In particular, the reaction of NO with the superoxide anion results in the formation of a more noxious oxidant – peroxynitrite [36,37]. Specifically, the formation of peroxynitrite occurs only when NO has reached toxic levels and begins to compete with superoxide dismutase for the scavenging of superoxide [38]. While excessive levels of NO undoubtedly damage reproductive organs, it is nevertheless one of the least potent of the RNS [12].

Physiologic Actions of Nitric Oxide

Nitric oxide plays an important physiologic role within the female reproductive system. NO is known to regulate the endometrial microvasculature and is produced by eNOS, which is distributed in the glandular surface of epithelial cells of the endometrium.

Moreover, it regulates endometrial stromal edema production, which is an important step for endometrial growth during the menstrual cycle, embryo implantation, and uterine contraction. Furthermore, in healthy fertile women, NO promotes contractions in the subendometrial myometrium, necessary for a normal menstrual cycle [13].

Detrimental Actions of Nitric Oxide

In endometriosis, ROS and RNS are produced via the interaction of interferon-alpha and interferon-gamma with lipopolysaccharide (LPS). This reaction activates peritoneal fluid mononuclear cells and macrophages to produce nitric oxide [5,13]. Moreover, these peritoneal macrophages have the capacity to move from the peritoneal cavity to other parts of the female reproductive system including the fallopian tubes where fertilization takes place [11]. When the migration of these macrophages is left unchecked, such as in endometriosis, pathological concentrations of NO can accumulate within the fluid of the peritoneal cavity and inhibit fertility by altering the following reproductive processes: ovulation, gamete transport, sperm–oocyte interaction, peritoneal fluid environment, fertilization, and early embryonic development [39]. Moreover, increased NO production has been shown to inhibit tubal motility and uterine contractions, causing uterine dysperistalsis. Consequentially, this impairs implantation, and in turn leads to compromised fecundity [5,13].

Reproductive Hormones and Nitric Oxide Production

The levels of NOS and NO are also associated with the levels of the reproductive hormones estrogen and progesterone. In a study on endometriosis-related infertility, fasting blood samples procured from women with this condition demonstrated a positive correlation between the levels of estrogen and progesterone and eNOS protein levels [40]. Specifically, NO activates cyclooxygenase-2 (COX-2), which in turn increases the levels of prostaglandin E2, thereby causing the levels of aromatase, an enzyme necessary for estrogen production. The resultant estrogen elevation stimulates further eNOS gene expression in a positive feedback loop [41].

ROLE OF IRON

Retrograde menstruation includes red blood cells, which contain the factors of hemoglobin and heme. These factors are also proinflammatory and generate iron – a redox-generating molecule [42]. An overload of iron from these erythrocytes can result in damage mediated by the molecule, inflammation, and ultimately, oxidative stress. When women have endometriosis, the overload of iron facilitates the lysis of erythrocytes in the pelvic cavity. Normally, women have mechanisms to protect themselves from such lysis; however, two theories contribute to the lack of these protective mechanisms in women with endometriosis. These hypotheses include an abundance of reflux and defective auto-oxidative capacity [5]. Furthermore, as mentioned previously, the state of endometriosis contributes to high numbers of activated macrophages. As iron is continually delivered to these macrophages, ferritin is unable to store and sequester iron. This ultimately leads to iron's production of free radicals, and finally an imbalance between ROS and the body's natural antioxidant defense mechanisms. Furthermore, as iron is a generator of free radicals, it causes macromolecular oxidative stress and cellular damage to the normally protective mesothelial cells from the iron binding Hb protein, a factor prevalent in menstrual fluid [43].

Bilirubin is another antioxidant that is produced by heme oxidase, whose decrease further contributes to the formation of oxidative stress [44]. While heme oxygenase levels are elevated in the peritoneal fluid, macrophages, the main cells present in cases of endometriosis, do not express it [45]. Therefore, bilirubin cannot form and its antioxidant capacity is not present in women with endometriosis.

With iron playing such a fundamental role in endometriosis, it becomes a strong therapeutic target.

ROLE OF CYTOKINES

As cytokines get released into the peritoneal fluid, they begin to play a role in endometriosis. These cytokines contribute to tissue remodeling and implantation of cells or tissue within the endometrium [46].

Interleukin 1 (IL-1) is involved with the activation of T-cells and differentiation of B-cells; it has been implicated in the implantation of ectopic endometrium in the disease process of endometriosis [47].

Interleukin 6 (IL-6) is a regulator of other cytokines. Among other duties, it activates B-cells and is important in the folliculogenesis, steroid hormone synthesis, implantation, and growth of endometrial cells [48]. IL-6 was demonstrated to increase in the peritoneal fluid along with the increased severity of endometriosis, and levels of IL-6 were shown to be higher in women with minimal-moderate endometriosis when compared to controls [49,50]. Further studies have shown IL-6 to be upregulated in women with endometriosis, facilitating its use as a marker of the disease [51,52,53].

Interleukin 8 (IL-8) is an important factor involved in chemotaxis and angiogenesis. This cytokine was found to be elevated in the peritoneal fluid of women with endometriosis, and its levels appeared to increase as the disease progressed. The levels of IL-8 were also increased in women with endometriosis when compared to normal women during the proliferative phase of menarche [54,55].

TNF-α is crucial in the body, produced by multiple immunologic cells, and functions as a proinflammatory cytokine. TNF-α will disrupt glutathione, decreasing the protective mechanisms available in women. It has been found that with the addition of TNF-α in a time- and dose-dependent manner, spermatozoa quality decreased [56]. The levels of TNF-α in peritoneal fluid were found to be increased in patients with endometriosis [57,58,59].

TREATMENT OF ENDOMETRIOSIS

Current treatment for endometriosis involves symptomatic therapy and potentially curative therapy. Symptomatically, health care providers may give patients pain control, such as nonsteroidal anti-inflammatory drugs (NSAIDs) or opioids. These may accompany "watchful waiting" by the clinician, who will periodically examine the patient's symptoms and adjust the pain management regimen accordingly. Furthermore, the symptoms of endometriosis can be relieved by hormone therapy, which are aimed at inhibiting the production of estrogen and thus, prevent ovulation. Consequently, this inhibition helps to slow the growth and local activity of both the endometrium and the endometrial lesions. Treatment via hormone therapy also prevents the growth of new areas of endometriosis, but it will not make existing lesions go away [60]. Examples of such treatments includes gonadotropin-releasing hormone (GnRH) analogs, which have been found to diminish elevated levels of NO, and thus endometriosis-associated infertility. Specifically, this occurs via the reduction of eNOS by these analogs in the early and middle secretory and early proliferative stages of the menstrual cycle [40]. Peroxynitrate was also found to be diminished in eutopic and ectopic endometrium once treated with GnRH agonist [60]. Moreover, studies have also shown benefit in adding specific inhibitors of eNOS to the GnRH analog in order to diminish the positive feedback loop of estrogen and progesterone on eNOS, and in turn, reduce endometriotic implantation and improved pregnancy outcomes [11]. Danazol is another form of hormone therapy that prevents ovulation by suppressing the increase of luteinizing hormone during the middle of the menstrual cycle. Specifically, it promotes hypoestrogenic- and hyperandrogenic-like effects, causing atrophy of the endometrium, which in turn helps to alleviate the symptoms of endometriosis. Moreover, Danazol suppresses the production of interleukin-1 beta and tumor necrosis factor by human monocytes, which are key cytokines that contribute to ectopic implantation of the stromal and glandular endometrial tissue throughout the body [61].

Finally, surgical treatments can provide significant, albeit short-term, relief from the pain of endometriosis through laparoscopy, laparotomy, or surgery to sever pelvic nerves [62]. Although studies show improved pregnancy rates following this type of surgery, the success rate is not clear. Therefore, if pregnancy does not occur after laparoscopic treatment, *in vitro* fertilization (IVF) may be the best option to improve fertility [63]. Although treatments aimed at mitigating the symptoms and improving fertility outcomes caused by endometriosis are important, therapies that aid in the prevention of oxidative damage involved in this condition, such as antioxidants, have also proven to be beneficial.

Antioxidants

Antioxidants are a defense mechanism produced by the body to neutralize the effects of ROS. They can be enzymatic and nonenzymatic. Nonenzymatic sources of antioxidants include vitamin C, vitamin E, selenium, zinc, beta carotene, carotene, taurine, hypotaurine, and glutathione. Enzymatic antioxidants include SOD, catalase, glutaredoxin, and glutathione reductase [64]. However, as the body ages, antioxidant levels decline, resulting in a disruption in the balance between antioxidants and prooxidant molecules. This results in the generation of oxidative stress and in turn, overrides the scavenging capacity by antioxidants either due to the diminished availability of antioxidants or excessive generation of ROS. Therefore, supplementation with oral oxidants may help to alleviate oxidative stress and its contribution to the pathogenesis of obstetrical disease such as endometriosis [65]. Only the most relevant antioxidants beneficial to endometriosis will be discussed.

Vitamin E and Vitamin C

Two dietary vitamins, vitamin C (ascorbic acid) and E (α-tocopherol), can be used to thwart the oxidative damage caused by endometriosis via their ability to scavenge free radicals and neutralize oxidative stress [66]. In a study of infertile women, those with endometriosis were shown to have lower levels of vitamin C in their follicular fluid, compared to women who did not have endometriosis, implicating the important role antioxidants play in mitigating bodily harm. Moreover, in a randomized, placebo-controlled trial of antioxidant vitamins E and C in women with pelvic pain and endometriosis, patients in the treatment group were found to have decreased levels of oxidative stress cytokines, interleukin-6 in the peritoneal fluid, and improvement in the symptoms of dysmenorrhea and dyspareunia when compared to the placebo group [67]. Other studies report that lower intake of antioxidants, including vitamins E and C, in women with endometriosis correspond with increased disease severity [68]. Additionally, data have shown that women with endometriosis have lower levels of the endogenous antioxidant superoxide dismutase in their plasma, thereby suggesting the need for supplementation in these women. It is important to note, however, that at high doses, vitamin C and E have deleterious effects on the body. Specifically, large quantities of vitamin C (>2000 mg/day)

have been suggested to cause diarrhea, abdominal cramps, bloating, nausea, vomiting, and kidney stones [69]. Furthermore, high doses of vitamin E (>1000 mg/day) may increase the risk of bleeding by having an anticoagulant-like effect on the body and may also increase the risk of birth defects. Thus, when using vitamin C and E to quell the adverse effects of endometriosis, it is important that appropriate dosages be used [70].

Glutathione Peroxidase

Glutathione peroxidase is an antioxidant enzyme class with the capacity to scavenge free radicals. This is in turn helps to prevent lipid peroxidation and maintain intracellular homeostasis as well as redox balance [71]. Glutathione peroxidase is localized in the glandular epithelium of normal human endometrium and reaches a maximum level in the late proliferative and early secretory phases of the menstrual cycle [72]. A study on endometriosis-associated infertility demonstrated a lower mean activity of glutathione peroxidase and increased lipid peroxidation in infertile women with endometriosis compared to women without this disease. This suggests that low level of antioxidant enzymes in the peritoneal fluid plays an integral role in the development of endometrial pathology [73]. Furthermore, in women with endometriosis, abnormal expression of glutathione peroxidase in eutopic and ectopic endometrium has been reported [13]. Overall, this aberrant change in antioxidant enzyme level can be one of the many contributors of the oxidative damage seen in endometriosis.

CONCLUSIONS

Despite the lack of evidence of a causal relationship between the excess free-radical molecules and pathophysiology of endometriosis, it is clear that an association between oxidative stress and the pathogenesis of this condition exists. Therefore, investigation on the various oxidative markers implicated in endometriosis, such as iron, cytokines, nitric oxide, and other immune modulators, needs to be continued. Furthermore, measurement of these markers in the peritoneal fluid could possibly be used as noninvasive alternatives to diagnose endometriosis, but this remains to be investigated. Additionally, continued research on the role of antioxidant agents should be performed due to the proven benefit in both the prevention and treatment of endometriosis via the alleviation of endometriotic lesions and severity of symptoms. Overall, by understanding the role of oxidative stress in endometriosis, we can potentially stem the increasing rates of female infertility and improve the quality of life in women hindered by this condition.

References

[1] Eltabbakh GH, Bower NA. Laparoscopic surgery in endometriosis. Minerva Ginecol 2008;60(4):323–30.

[2] Pritts EA, Taylor RN. An evidence-based evaluation of endometriosis-associated infertility. Endocrinol Metab Clin North Am 2003;32(3):653–67.

[3] Callahan T, Caughey A. Endometriosis and adenomyosis. Blueprints: Obstetrics & Gynecology. 5th ed. Philadelphia: Lippincott Williams & Wilkins; 2009. p. 163–70.

[4] Lapp T. ACOG issues recommendations for the management of endometriosis. American College of Obstetricians and Gynecologists. Am Fam Physician 2000;62(6):1431–4.

[5] Gupta S, Chandra A, Kesavan S, Eapen D, Agarwal A. Oxidative stress and the pathogenesis of endometriosis. In: Garcia-Velasco J, Rizk B, editors. Endometriosis: Current Management and Future Trends. New Delhi: Jaypee Brothers Medical Publishers; 2011. p. 31–9.

[6] Treloar SA, Wicks J, Nyholt DR, Montgomery GW, Bahlo M, Smith V, et al. Genomewide linkage study in 1,176 affected sister pair families identifies a significant susceptibility locus for endometriosis on chromosome 10q26. Am J Hum Genet 2005;77(3):365–76.

[7] Chapron C, Chopin N, Borghese B, Foulot H, Dousset B, Vacher-Lavenu MC, et al. Deeply infiltrating endometriosis: pathogenetic implications of the anatomical distribution. Hum Reprod 2006;21(7):1839–45.

[8] Seli E, Berkkanoglu M, Arici A. Pathogenesis of endometriosis. Obstet Gynecol Clin North Am 2003;30(1):41–61.

[9] Worwood VA, Stonehouse J. The Endometriosis Natural Treatment Program: A Complete Self-Help Plan for Improving Health & Well-Being. California: New World Library; 2011.

[10] Laschke MW, Giebels C, Nickels RM, Scheuer C, Menger MD. Endothelial progenitor cells contribute to the vascularization of endometriotic lesions. Am J Pathol 2011;178(1):442–50.

[11] Gupta S, Shenoy R, Safi J, Gaglani A, Agarwal A. Oxidative stress and its role in endometriosis—mechanistic and therapeutic implications. In: Rizk B, Sallam H, editors. Clinical Infertility and Assisted Reproduction. New Delhi: Jaypee Brothers Medical Publishers; 2010. p. 316-25.

[12] Doshi SB, Khullar K, Sharma RK, Agarwal A. Role of reactive nitrogen species in male infertility. Reprod Biol Endocrinol 2012;10:109.

[13] Alpay Z, Saed GM, Diamond MP. Female infertility and free radicals: potential role in adhesions and endometriosis. J Soc Gynecol Investig 2006;13(6):390–8.

[14] Szczepańska M, Koźlik J, Skrzypczak J, Mikołajczyk M. Oxidative stress may be a piece in the endometriosis puzzle. Fertil Steril 2003;79(6):1288–93.

[15] Shanti A, Santanam N, Morales AJ, Parthasarathy S, Murphy AA. Autoantibodies to markers of oxidative stress are elevated in women with endometriosis. Fertil Steril 1999;71(6):1115–8.

[16] Arumugam K, Dip YC. Endometriosis and infertility: the role of exogenous lipid peroxides in the peritoneal fluid. Fertil Steril 1995;63(1):198–9.

[17] Wang Y, Sharma RK, Falcone T, Goldberg J, Agarwal A. Importance of reactive oxygen species in the peritoneal fluid of women with endometriosis or idiopathic infertility. Fertil Steril 1997;68(5):826–30.

[18] Lambrinoudaki IV, Augoulea A, Christodoulakos GE, Economou EV, Kaparos G, Kontoravdis A, et al. Measurable serum markers of oxidative stress response in women with endometriosis. Fertil Steril 2009;91(1):46–50.

[19] Khan KN, Kitajima M, Imamura T, Hiraki K, Fujishita A, Sekine I, et al. Toll-like receptor 4-mediated growth of endometriosis by human heat-shock protein 70. Hum Reprod 2008;23(10):2210–9.

[20] Zeller JM, Hening I, Radwanska E, Dmowski W. Enhancement of human monocyte and peritoneal macrophage chemiluminescence activities in women with endometriosis. Am J Reprod Immunol Microbiol 1987;13(3):78–82.

[21] Gilmore T. Introduction to NF-κB: players, pathways, perspectives. Oncogene 2006;25(51):6680–4.

[22] Viatour P, Merville MP, Bours V, Chariot A. Phosphorylation of NF-kappaB and IkappaB proteins: implications in cancer and inflammation. Trends Biochem Sci 2005;30(1):43–52.

[23] Lebovic DI, Mueller MD, Taylor RN. Immunobiology of endometriosis. Fertil Steril 2001;75(1):1–10.

[24] Gonzalez-Ramos R, Donnez J, Defrere S, Leclercq I, Squifflet J, Lousse JC, et al. Nuclear factor-kappa B is constitutively activated in peritoneal endometriosis. Mol Hum Reprod 2007;13(7):503–9.

[25] Pitsos M, Kanakas N. The role of matrix metalloproteinases in the pathogenesis of endometriosis. Reprod Sci 2009;16(8):717–26.

[26] Kim SH, Choi YM, Choung SH, Jun JK, Kim JG, Moon SY. Vascular endothelial growth factor gene +405 C/G polymorphism is associated with susceptibility to advanced stage endometriosis. Hum Reprod 2005;20(10):2904–8.

[27] Li Y, Wang L, Li X, Li SL, Wang JL, Wu ZH, et al. Vascular endothelial growth factor gene polymorphisms contribute to the risk of endometriosis: an updated systematic review and meta-analysis of 14 case-control studies. Genet Mol Res 2013;12(2):1035–44.

[28] Van Langendonckt A, Casanas-Roux F, Donnez J. Oxidative stress and peritoneal endometriosis. Fertil Steril 2002;77(5):861–70.

[29] Koks CA, Demir Weusten AY, Groothuis PG, Dunselman GA, de Goeij AF, Evers JL. Menstruum induces changes in mesothelial cell morphology. Gynecol Obstet Invest 2000;50(1):13–8.

[30] Sikka SC. Relative impact of oxidative stress on male reproductive function. Curr Med Chem 2001;8:851–62.

[31] Beckman JS, Conger K. Direct measurement of nitric oxide in solutions with an ozone based chemiluminescent detector. Methods 1995;7:35–9.

[32] Rosselli M, Keller PJ, Dubey RK. Role of nitric oxide in the biology, physiology, and pathophysiology of reproduction. Hum Reprod Update 1998;4:3–24.

[33] Archer S. Measurement of nitric oxide in biological models. FASB J 1993;7:349–60.

[34] Bolaños JP, Delgado-Esteban M, Herrero-Mendez A, Fernandez-Fernandez S, Almeida A. Regulation of glycolysis and pentose phosphate pathway by nitric oxide: impact on neuronal survival. Biochim Biophys Acta 2008;1777:789–93.

[35] Lee NP, Cheng CY. Nitric oxide/nitric oxide synthase, spermatogenesis, and tight junction dynamics. Biol Reprod 2004;70:267–76.

[36] Pacher P, Beckman JS, Liaudet L. Nitric oxide and peroxynitrite in health and disease. Physiol Rev 2007;87:315–424.

[37] Jourd'heuil D, Jourd'heuil FL, Kutchukian PS, Musah RA, Wink DA, Grisham MB. Reaction of superoxide and nitric oxide with peroxynitrite: implications for peroxynitrite-mediated oxidation reactions in vivo. J Biol Chem 2001;276:28799–805.

[38] de Lamirande E, Gagnon C. Capacitation-associated production of superoxide anion by human spermatozoa. Free Radic Biol Med 1995;18:487–95.

[39] Gupta S, Agarwal A, Krajcir N, Alvarez JG. Role of oxidative stress in endometriosis. Reprod Biomed Online 2006;13:126–34.

[40] Wang J, Zhou F, Dong M, Wu R, Qian Y. Prolonged gonadotropin-releasing hormone agonist therapy reduced expression of nitric oxide synthase in the endometrium of women with endometriosis and infertility. Fertil Steril 2006;85(4):1037–44.

[41] Kitawaki J, Kusuki I, Koshiba H, Tsukamoto K, Fushiki S, Honjo H. Detection of aromatase cytochrome P-450 in endometrial biopsy specimens as a diagnostic test for endometriosis. Fertil Steril 1999;72(6):1100–6.

[42] Reubinoff B, Shushan A. Preimplantation diagnosis in older patients, To biopsy or not to biopsy? Hum Reprod 1996;11:2071–5.

[43] Demir AY, Demol H, Puype M, de Goeij AF, Dunselman GA, Herrler A. Proteome analysis of human mesothelial cells during epithelial to mesenchymal transitions induced by shed menstrual effluent. Proteomics 2004;4:2608–23.

[44] Pelletier N, Boudreau F, Yu SJ, Zannoni S, Boulanger V, Asselin C. Activation of haptoglobin gene expression by cAMP involves CCAAT/enhancer-binding protein isoforms in intestinal epithelial cells. FEBS Lett 1998;439:275–80.

[45] Van Langendonckt A, Casanas-Roux F, Donnez J. Oxidative stress and peritoneal endometriosis. Fertil Steril 2002;77:861–70.

[46] Iwabe T, Harada T, Tsudo T, Nagano Y, Yoshida S, Tanikawa M. Tumor necrosis factor-alpha promotes proliferation of endometriotic stromal cells by inducing interleukin-8 gene and protein expression. J Clin Endocrinol Metab 2000;85:824–9.

[47] Simon C, Frances A, Piquette GN. Embryonic implantation in mice is blocked by interleukin-1 receptor antagonist. Endocrinology 1994;134(2):521–8.

[48] Barcz EP, Kaminski, Marianowski L. Role of cytokines in pathogenesis of endometriosis. Med Sci Monit 2000;6(5):1042–6.

[49] Rier SE, Parsons AK, Becker JL. Altered interleukin-6 production by peritoneal leukocytes from patients with endometriosis. Fertil Steril 1994;61(2):294–9.

[50] Martínez S, Garrido N, Coperias JL, Pardo F, Desco J, García-Velasco JA, et al. Serum interleukin-6 levels are elevated in women with minimal-mild endometriosis. Hum Reprod 2007;22(3):836–42.

[51] Harada T, Yoshioka H, Yoshida S, Iwabe T, Onohara Y, Tanikawa M, et al. Increased interleukin-6 levels in peritoneal fluid of infertile patients with active endometriosis. Am J Obstet Gynecol 1997;176(3):593–7.

[52] Umezawa M, Sakata C, Tanaka N, Kudo S, Tabata M, Takeda K, et al. Cytokine and chemokine expression in a rat endometriosis is similar to that in human endometriosis. Cytokine 2008;43(2):105–9.

[53] Othman Eel D, Hornung D, Salem HT, Khalifa EA, El-Metwally TH, Al-Hendy A. Serum cytokines as biomarkers for nonsurgical prediction of endometriosis. Eur J Obstet Gynecol Reprod Biol 2008;137(2):240–6.

[54] Arici A, Seli E, Zeyneloglu HB, Senturk LM, Oral E, Olive DL. Interleukin-8 induces proliferation of endometrial stromal cells: a potential autocrine growth factor. J Clin Endocrinol Metab 1998;83(4):1201–5.

[55] Ulukus M, Ulukus EC, Tavmergen Goker EN, Tavmergen E, Zheng W, Arici A. Expression of interleukin-8 and monocyte chemotactic protein 1 in women with endometriosis. Fertil Steril 2009;91(3):687–93.

[56] Said M, Agarwal A, Falcone T, Sharma RK, Bedaiwy MA, Li L. Infliximab may reverse the toxic effects induced by tumor necrosis factor alpha in human spermatozoa: an *in vitro* model. Fertil Steril 2005;83(6):1665–73.

[57] Eisermann J, Gast MJ, Pineda J, Odem RR, Collins JL. Tumor necrosis factor in peritoneal fluid of women undergoing laparoscopic surgery. Fertil Steril 1988;50(4):573–9.

[58] Overton C, Fernandez-Shaw S, Hicks B, Barlow D, Starkey P. Peritoneal fluid cytokines and the relationship with endometriosis and pain. Hum Reprod 1996;11:380–6.

[59] Taketani Y, Kuo TM, Mizuno M. Comparison of cytokine levels and embryo toxicity in peritoneal fluid in infertile women with untreated or treated endometriosis. Am J Obstet Gynecol 1992;167:265–70.

[60] Kamada Y, Nakatsuka M, Asagiri K, Noguchi S, Habara T, Takata M, et al. GnRH agonist-suppressed expression of nitric oxide synthases and generation of peroxynitrite in adenomyosis. Hum Reprod 2000;15(12):2512–9.

[61] Mori H, Nakagawa M, Itoh N, Wada K, Tamaya T. Danazol suppresses the production of interleukin-1 beta and tumor necrosis factor by human monocytes. Am J Reprod Immunol 1990;24(2):45–50.

[62] Yeung PP, Shwayder J, Pasic RP. Laparoscopic management of endometriosis: comprehensive review of best evidence. J Minim Inv Gynecol 2009;16(3):269–81.

[63] Giudice LC, Kao LC. Endometriosis. Lancet 2004;364:1789–99.

[64] Agarwal A, Gupta S, Sekhon L, Shah R. Redox considerations in female reproductive function and assisted reproduction: from molecular mechanisms to health implications. Antioxid Redox Signal 2008;10:1375–403.

[65] Sekhon HL, Gupta, Kim Y, Agarwal A. Female infertility and antioxidants. Curr Women's Health Rev 2010;6(2):84–95.

[66] Cook NR, Albert CM, Gaziano JM, Zaharris E, MacFadyen J, Danielson E. A randomized factorial trial of vitamins C and E and beta carotene in the secondary prevention of cardiovascular events in women: results from the women's antioxidant cardiovascular study. Arch Intern Med 2007;167:1610–8.

[67] Santanam N, Kavtaradze N, Murphy A, Dominguez C, Parthasarathy S. Antioxidant supplementation reduces endometriosis-related pelvic pain in humans. Transl Res 2013;161(3):189–95.

[68] Prieto L, Quesada JF, Cambero O, Pacheco A, Pellicer A, Codoceo R, et al. Analysis of follicular fluid and serum markers of oxidative stress in women with infertility related to endometriosis. Fertil Steril 2012;98(1):126–30.

[69] McSorely PT, Young IS, Bell PM, Fee JP, McCance DR. Vitamin C improves endothelial function in healthy estrogen-deficient postmenopausal women. Climacteric 2003;6:238–47.

[70] Kushi LH, Folsom AR, Prineas RJ, Mink PJ, Wu Y, Bostick RM. Dietary antioxidant vitamins and death from coronary heart disease in postmenopausal women. N Engl J Med 1996;334:1156–62.

[71] Gupta S, Agarwal A, Banerjee J, Alvarez JG. The role of oxidative stress in spontaneous abortion and recurrent pregnancy loss: a systematic review. Obstet Gynecol Surv 2007;62:335–47.

[72] Michitaka O, Mitsuaki S, Ikuo S, Hideo T, Keiichi W. Glutathione peroxidase activity in endometrium: effects of sex hormones and cancer. Gynecol Oncol 1996;60(2):277–82.

[73] Szczepańska M, Jacek K, Jana S, Mateusz M. Oxidative stress may be a piece in the endometriosis puzzle. Fertil Steril 2003;79(6):1288–93.

26

Oxidative Stress in Preeclampsia

Ashok Agarwal, PhD, Eva Tvrda, PhD*,**, Aditi Mulgund, MD*,†*

*Center for Reproductive Medicine, Cleveland Clinic, Cleveland, Ohio, USA
**Department of Animal Physiology, Slovak University of Agriculture, Nitra, Slovakia
†Northeastern Ohio Medical University, Rootstown, Ohio, USA

INTRODUCTION

The female reproductive system undergoes many changes throughout a woman's life. As women age, they are subjected to menarche and eventually menopause. Oxidative stress directly influences the pathophysiology of a woman's reproductive tract through mechanisms of lipid peroxidation, inhibition of protein synthesis, mitochondrial modifications, decrease of ATP levels, and damage to DNA. These processes are subsequently linked to embryo development and pregnancy outcomes, finally culminating in the adverse health of the baby and mother.

DEFINITION OF PREECLAMPSIA

Preeclampsia (PE) is one such process that can adversely affect a mother's and baby's health. During pregnancy, it is characterized by nondependent edema, hypertension, and proteinuria [1]. Classically, preeclampsia presents in a nulliparous woman during her third trimester. While the triad of fluid retention, high blood pressure, and protein in the urine is a classic presentation of preeclampsia, edema is no longer considered a component of the diagnosis.

Preeclampsia should be distinguished from chronic hypertension, a condition where high blood pressure is present before pregnancy. Furthermore, several conditions can be superimposed on preeclampsia. Liver injury is present in a small percentage of patients and is associated with HELLP (hemolysis, elevated liver enzymes, low platelets) syndrome and acute fatty liver of pregnancy. These conditions have high morbidity and mortality. Finally, preeclampsia may progress to eclampsia, in which seizures are present along with the triad of hypertension, edema, and proteinuria.

This disease process is present in 3.4% of all live births, with the rates and risk increasing in recent years [2].

Other studies claim that the rates of preeclampsia range from 3 to 7% in nulliparous women, and 1–3% in multiparous women [3]. Women delivering in 2003 were at a 6.7-fold increased risk compared to women delivering in 1980 [2]. Most commonly, preeclampsia occurs in the third trimester, typically after 20 weeks' gestation. If seen earlier, it may be associated with other conditions such as hydatidiform moles or chronic hypertension. After the diagnosis of preeclampsia is made, however, in 80% of cases, the diseases progress to HELLP syndrome; however, this is more common in severe preeclampsia [1].

PATHOPHYSIOLOGY OF PREECLAMPSIA

As a result of decreased vascular resistance during pregnancy, the blood pressure typically decreases in the pregnant woman during the first trimester. Subsequently, during the third trimester, the blood pressure will increase but should not rise higher than prepregnancy levels [1]. The pathophysiology of preeclampsia is not yet definitively determined; however, abnormal placentation is considered a primary cause [3]. Preeclampsia involves the process by which the cytotrophoblast invades into the uterus and differentiates causing fetal cells to become abnormal [4]. Furthermore, the involvement of reactive oxygen species (ROS), such as nitric oxide (NO), may prevent embryo implantation [5]. A generalized arteriolar constriction, or vasospasm, and intravascular depletion from transudative edema can produce these symptoms, which lead to placental ischemia, necrosis, and hemorrhage [1]. Fundamental theories for the development of preeclampsia include vascular damage leading to an imbalance of the concentrations of prostacyclin and thromboxane, as well as the release of free radicals, oxidized lipids, cytokines, and vascular endothelial growth factor, leading to endothelial dysfunction [1,6].

Handbook of Fertility. http://dx.doi.org/10.1016/B978-0-12-800872-0.00026-3

RISK FACTORS FOR PREECLAMPSIA

The risk factors for preeclampsia are generally related to either the hypertensive or renal associated symptoms or immunologic risk factors. Women with a family history of preeclampsia, especially on the paternal side of the family, are at a higher risk of developing this disease. Differences in parental ethnic background may also contribute to the development of preeclampsia. Risk factors associated to the disease include a history of chronic hypertension or chronic kidney disease, collagen vascular diseases, antiphospholipid antibody syndrome, pregestational diabetes, African-American ethnicity, and either very young or old maternal age (<20 or >35 years old) [1]. Immunogenic risk factors include nulliparity, previous history of preeclampsia, multiple gestation, abnormal placentation, or a new paternal partner [1].

EFFECTS ON MOTHER AND BABY

Most of the major complications of preeclampsia on the fetus manifest as premature delivery. The generalized vasoconstriction from this condition also results in decreased blood flow to the placenta, manifesting as uteroplacental insufficiency and subsequent placental abruption or fetal distress [1]. The uteroplacental insufficiency may be acute or chronic, leading to an intrauterine growth restricted (IUGR) fetus [1]. Additional fetal complications include placental infarct, intrapartum fetal distress, stillbirth, asymmetric and symmetric small-for-gestational-age (SGA) fetuses, IUGR, and oligohydramnios [1].

Maternal complications are typically due to generalized arteriolar vasoconstriction. These include seizure and cerebral hemorrhage in the brain, disseminated intravascular coagulation (DIC), thrombocytopenia from vasoconstriction on the small blood vessels, and renal failure and oliguria from kidney damage. These complications can also further affect the obstetric outcomes, leading to placental abruption, uteroplacental insufficiency, increased risk of premature delivery, and increased risk of emergency cesarean section [1].

TREATMENT OF PREECLAMPSIA

The ultimate treatment for preeclampsia is delivery. For pregnancies at term, along with unstable preterm pregnancies, or pregnancies with evidence of complete fetal lung maturity, induction of labor is the treatment of choice. In cases where fetal lung maturity is not yet reached, betamethasone is administered. Vaginal delivery is typically attempted along with the administration of prostaglandins, oxytocin, or amniotomy as needed [1].

Cesarean section is typically only performed as medically indicated.

The goals of treatment are to prevent eclampsia, control maternal blood pressure, and deliver the fetus. When delivery is not possible, patients with mild preeclampsia are given magnesium sulfate as prophylaxis for seizures. For control of blood pressure, hydralazine or labetalol are used in severe preeclampsia [1].

Following up with women who have preeclampsia, the risk of recurrence is between 25 and 33%. In patients with underlying issues of chronic hypertension as well as superimposed preeclampsia, the risk of recurrence is as high as 70%. Administering low doses of aspirin can decrease the risk of future preeclampsia, IUGR, and preterm birth [1].

PREECLAMPSIA-RELATED OXIDATIVE STRESS

Substantial evidence has connected oxidative stress with the etiology and pathophysiology of numerous pregnancy complications, preeclampsia being one of them.

Oxidative stress (OS) may be characterized as a state when the production of ROS overwhelms the intrinsic antioxidant defense systems of the organism [7]. ROS are defined as molecules containing one or more unpaired electrons, and it is this incomplete electron state that is responsible for their high reactivity [8]. ROS are generated by different pathways, in numerous cellular events, and under different conditions. The majority of ROS are considered to be essential by-products of cellular signaling and metabolism, located primarily in the mitochondria, or as specific products of enzymatic complexes such as NADPH oxidases, xanthine oxidases, and cytochrome P450 [9,10].

The primary generation of ROS is related to the reduction of molecular oxygen yielding the short half-life superoxide anion, $O_2^{\bullet-}$. Through enzymatic activity (superoxide dismutase, SOD) or spontaneous dismutation, $O_2^{\bullet-}$ conversion leads to the creation of hydrogen peroxide (H_2O_2) [11,12].

Among all ROS molecules, H_2O_2 is unique in its potential to be converted either to the highly damaging hydroxyl radical ($^{\bullet}OH$) via the Fenton and Haber–Weiss reaction, dependent upon the catalytic activities of free iron and copper. It may be catalyzed and excreted harmlessly as water through the action of catalase, or the conversion of reduced glutathione to its oxidized form [13].

Singlet oxygen is not considered to be a free radical, although it can be formed during radical chain reactions. Furthermore, it is rich in energy and is able to transfer this energy to a new molecule, acting as a catalyst for further ROS production [11,12].

The maternal–fetal interface of the developing placenta has emerged as intimately connected to the function of ROS in normal as well as pathological conditions [10]. Initially, the placenta has a hypoxic environment, but with further development, its vascularization leads to the creation of a strongly oxidative environment [14]. Furthermore, the placenta is rich in mitochondria, consumes approximately 1% of the basal metabolic rate of a pregnant woman, and is constantly exposed to high maternal oxygen partial pressure, leading to the potentially increased production of ROS [15]. Indeed, it is widely recognized that pregnant women are characterized by a higher state of oxidative stress when compared to nonpregnant women [16].

It is important to note that at physiological levels, ROS accomplish diverse essential functions related to cell homeostasis and communication during pregnancy. Furthermore, ROS stimulate cell proliferation and gene expression during placental and fetal development [7,8,15]. Furthermore, nitric oxide, a free radical belonging to the reactive nitrogen species (RNS), is an active vasodilator in the fetal placental vasculature [17] where it maintains basal vascular tone and attenuates the action of vasoconstrictors [18].

ROS production is observed to a much greater extent in preeclampsia [19]. The cause for the resulting oxidative stress is thought to be vascular, as early onset preeclampsia is associated with deficient conversion of the spiral arteries, especially the myometrial segments [20,21]. Another potential source of oxidative damage in preeclampsia is autoantibodies released against the angiotensin AT1 receptor [22]. These antibodies have the ability to stimulate the NADPH oxidase, leading to an increase in ROS production. Preeclampsia has also been linked to an increased release of free iron from iron sulfur clusters which, in reactive form, catalyze the production of highly aggressive hydroxyl radicals, mainly through the Haber–Weiss reaction [23]. Moreover, the placenta is rich in macrophages favoring the local placental overproduction of free radicals like reactive chlorine species (RClS) by autocatalysis in the presence of iron [24]. Finally, labor, in which the placenta is exposed to repeated episodes of ischemia–reperfusion, induces high ROS production [25]. This is associated with increased xanthine oxidase activity [26] and changes in the gene expression that mimic those described in preeclampsia [25].

A variety of complications resulting from ROS overproduction has been observed in preeclampsia patients [27]; these complications are presented next.

Lipid Peroxidation

Preeclampsia is accompanied by a significant elevation of serum triglycerides and serum free fatty acids [28], which are considered to be the prime targets for lipid peroxidation (LPO). In fact, hypertriglyceridemia

has been proposed to be a potential risk factor as well as a marker for early onset preeclampsia [29]. Furthermore, LPO products are considered to be the primary markers that may mediate the disturbance of the maternal vascular endothelium [30].

The placenta is the primary site of lipid peroxide generation [31]. Hydroxyl radicals, which are found in large amounts in the preeclamptic placenta, are considered to be extremely effective in causing LPO at the plasma membrane of any structure containing large quantities of polyunsaturated fatty acid side chains. By abstracting hydrogen from the hydrocarbon side-chain of a fatty acid, a carbon-centered radical, $C^{\bullet}$, is created. If oxygen is present, this unstable by-product may react to form peroxyl radicals ($-C-O-O^{\bullet}$), which in turn is capable of subtracting hydrogen from adjacent fatty acids, leading to further propagation of the reaction [27].

Increased generation of lipid peroxides has been widely reported in preeclamptic women [32–34]. Significant correlations have been found between increased diastolic pressure and LPO, indicating that the severity of hypertension is directly related to the extent of LPO [35,36].

Moreover, a variety of researchers agree that compared with normotensive pregnant women, women with preeclampsia have significantly higher mean plasma levels of malondialdehyde, an advanced lipoxidation end-product [31].

A different study of women undergoing C-section showed significantly higher concentrations of lipid hydroperoxides, phospholipids, and cholesterol in tissues from women with preeclampsia as compared with tissues from a normal pregnancy [37]. Removal of the decidual tissue has been shown to lead to a more rapid clinical recovery from preeclampsia [38]. This tissue has been later defined as a potential source of LPO products, such as isoprostane (8-iso-PGF2a), which may be released into maternal circulation and contribute to pregnancy complications [39]. Free 8-iso-PGF2a has furthermore potent activities relevant to preeclampsia, being a potent vasoconstrictor, platelet activator, and mitogen [40].

In vitro production of lipid hydroperoxides and thromboxane has been shown to be increased in trophoblast-derived cells and villous tissues from preeclamptic patients [41,42]. Production of 8-iso-PGF2a is also increased during an *in vitro* incubation of preeclamptic placental tissue in comparison with tissues from a normal pregnancy [43]. Unfortunately, a direct demonstration of the accumulation of LPO products in the maternal circulation has not been recorded yet.

Protein Alteration

Amino acids are free or bound in proteins, which are a frequent target for possible oxidative damage. A direct oxidation of the side chains may lead to the formation of carbonyl groups (aldehydes and ketones). Lysine, arginine, praline, and threonine are thought to be particularly vulnerable to oxidative insults [44].

Abstraction of hydrogen ions from the thiol group of cysteine can lead to the formation of disulfide bonds and abnormal protein folding, with a subsequent loss of function, protein aggregation, and apoptosis. Protein oxidation products, such as protein carbonyls or oxidized plasma proteins, are a common find in preeclamptic patients, followed by generally lower concentrations of proteins or lower enzymatic activity, proving the severity of oxidation level in this pathological condition [27].

Furthermore, peroxynitrite is capable of reacting with tyrosine residues to form 3-nitrotyrosine. While at physiological levels, protein nitration is defined as a selective and reversible process leading to molecular activation, higher levels have proven to be detrimental. Thus, protein nitration in the placenta may have diverse effects, with both gain and loss of function [45].

DNA Oxidation

OH$^{\bullet}$ radicals are the principal ROS to interact with the DNA structure, and a variety of products may be generated through reactions with either the DNA bases or deoxyribose [27]. OH$^{\bullet}$ can add on to guanine to produce 8-hydroxy-2′-deoxyguanosine, which has been found in high amounts in preeclamptic placentas [46]. At the same time, attacks on the sugar moieties may cause strand breakages, whereas those on histone proteins may lead to cross-linkages affecting chromatin folding, DNA repair, and transcription [47].

Mitochondrial DNA is particularly susceptible to ROS attacks based on the proximity to the $O_2^{\bullet-}$ generation site from the electron transport chain, as well as a lack of proper protective and repair mechanisms. As a result, damage to mitochondrial DNA is extensive and severe even under normal conditions, with a 5- to 10-fold higher increase of mutations in comparison with the nuclear DNA [48]. As mitochondrial DNA encodes several proteins, including enzymes of the electron transport chain, mutations may lead to impaired energy production and the risk of further electron leakage, supporting further oxidative pressure [27].

Moreover, investigators have observed elevated risks of preeclampsia in subjects with clinically diagnosed mitochondrial dysfunction [49]. Building upon these facts, other investigators [50] proposed that defects in the mitochondria of trophoblasts might actually be the initiating step in the pathophysiological cascade of preeclampsia.

Interactions with Cellular Signaling

Activation of redox-sensitive transcription factors, such as AP-1, p53, and NF-κB [51], may lead to an increased expression of pro-inflammatory molecules and cytokines, inhibition of cell differentiation, and apoptosis. NF-κB is normally inactive, bound to its inhibitory subunit IκB. However, due to ROS attacks, IκB may be phosphorylated and dissociate from NF-κB, leading to its translocation to the nucleus and activation of inflammation. Increased phosphorylation of IκB is frequently observed in placental explants subjected to hypoxia-reoxygenation *in vitro*, providing a model for malperfusion of the placenta *in vivo* [52].

ROS-induced activation of apoptosis-regulating signal kinase 1 (ASK1), ultimately leading to apoptosis, has been observed in placental explants exposed to either hypoxia-reoxygenation or hydrogen peroxide, and may be inhibited by the addition of vitamins C and E. At the same time, ASK1 activation may be associated with increased levels of the soluble receptor for vascular endothelial growth factor (sFlt-1), which has been implicated in the pathogenesis of preeclampsia. sFlt-1 levels can be reduced by vitamin C and E supplementation or treatment with inhibitors of the p38 pathway. Sulfasalazine has been shown to be equally effective, indicating considerable interactions and mutual reinforcements between the NF-κB and MAPK (mitogen-activated protein kinases) signaling pathways in the placenta [53].

Decrease of Antioxidant Power

Several antioxidant markers have been shown to be significantly decreased in preeclamptic women [54,55]. In the maternal circulation, the levels of enzymatic as well as nonenzymatic antioxidant markers, such as vitamins A, C, E, glutathione, iron binding capacity, and superoxide dismutase, may be altered [56].

Studies have also shown that women with preeclampsia have significantly reduced serum coenzyme Q10 [57] as well as serum and/or placental β-carotene, lycopene, and canthaxanthin [57,58].

One study focused on the evaluation of the total antioxidant response (TAR) determined by the redox status of the plasma, as well as the oxidative stress index (OSI), calculated from the percent ratio of the total plasma peroxide levels [59] in preeclamptic women.

Both parameters were found to be significantly higher in patients with preeclampsia. A twofold increase in the ratio between LPO and antioxidant capacity was observed in the antepartum period in women with preeclampsia [60]. At the same time, Hilari et al. [61] found that women whose pregnancies were complicated by mild preeclampsia had increased levels of total oxidant status (TOS) and a decreased total antioxidant status (TAS) level compared to women with normal pregnancies.

Superoxide dismutase is the primary antioxidant enzyme to be found modified in a preeclamptic placenta

[62]. Furthermore, Rosta et al. [63] suggested the role of SOD3 single nucleotide gene polymorphisms in the increased oxidative stress in preeclampsia.

Significantly lower levels of ascorbic acid were reported in the plasma of women with mild to severe preeclampsia [64]. Similarly, serum alpha-tocopherol levels were significantly reduced in pregnancies complicated by severe preeclampsia. As LPO is commonly observed in preeclamptic patients, the balance between lipid peroxides and antioxidant vitamin E will be tipped in favor of lipid peroxides in patients with mild and severe preeclampsia.

In a controlled clinical trial, significantly lower levels of vitamin C, E, and total thiols were seen in women with preeclampsia [65]. Furthermore, significantly reduced whole blood glutathione levels have been reported in women with preeclampsia and HELPP syndrome [66].

COMPLICATIONS IN THE FETUS AND NEWBORN

Preeclampsia is considered to be a major cause of fetal morbidity and mortality. Its pathogenesis involves the imbalance of angiogenic factors and oxidative stress [67].

In the fetus and newborn infant, infection and inflammation associated with preeclampsia cause the activation of polymorphonuclear leukocytes, which in turn can increase oxidative stress by generating ROS. At the same time, the antioxidant defense mechanisms are not fully developed in preterm infants, which further increase their vulnerability to oxidative damage.

Chamy et al. hypothesized that the antioxidant status changes observed in preeclamptic mothers had a similar pattern as in their newborns [68]. Also, Howlader et al. [69] as well as Namdev et al. [70] concluded that antioxidant levels in preeclamptic fetal circulation are low, accompanied by increased levels of lipid peroxides and protein carbonyls. Both studies suggest that oxidative stress and antioxidant status are altered toward proatherogenic level in cord blood of preeclamptic women, which may ultimately be responsible for different complications of newborn babies.

Low antioxidant protection and high oxidative stress status in preeclamptic infants is critical, as many illnesses in preterm infants, including bronchopulmonary dysplasia(BPD), retinopathy of prematurity (ROP), brain injury such as hypoxic/ischemic encephalopathy, and intraventricular hemorrhage (IVH) are thought to be related to the action of ROS [71]. Therefore, extensive research on such neonates, in a short- as well as long-term run, is crucially needed. The proposed role of oxidative stress in the pathogenesis of preeclampsia and the health consequences of this pregnancy-related disease on the mother and fetus is illustrated in Fig. 26.1.

INTERVENTIONS TO OVERCOME OXIDATIVE STRESS IN PREECLAMPSIA

Currently, no specific interventions are available for the prevention of preeclampsia-related oxidative stress. A variety of clinical trials investigating the benefits of antioxidant supplementation in patients with established preeclampsia have not yet shown any substantial beneficial results [72,73].

At the same time, it has been proposed that early (16–20 weeks' gestation) antioxidant supplementation may be helpful in preventing the incidence of preeclampsia in high-risk patients [72–77].

As significantly decreased levels of vitamins C and E were repeatedly reported in patients with preeclampsia [78], and these have mutually beneficial effects, most of the trials focused on the efficacy of the combination of these two antioxidants [79,80]. One study assessed the efficacy of any antioxidant and found no statistically significant difference in the assessed outcomes, except for the side effects [81].

Another study assessed only vitamin C as an antioxidant and showed a higher risk of preterm birth in women who took vitamin C compared to the placebo group, although the risk of preeclampsia was lower [82]. Other outcomes were not statistically significant [82]. Meanwhile, a study that analyzed only vitamin E also found a lower risk of preeclampsia among the group who took the vitamin versus placebo, however, with no statistically significant difference from other outcomes [83]. Despite the fact that these reviews analyzed the effects of vitamin C or E only, other antioxidants were included and the total number of included women was less than 1000 in both reviews [82,83].

A different study used a more complex mixture of up to six supplements including vitamin C, Halibut liver oil (vitamins A and D), as well as vitamin B1 in 1530 primigravida women [84]. Supplementation initiated before 24 weeks of gestation resulted in a lower incidence of hypertension and albuminuria. In a systematic review aimed at assessing the benefits of vitamin supplementation in improving various pregnancy outcomes, women who were taking any type of vitamin(s) were less likely to develop preeclampsia than control subjects [85].

The limitation of these trials includes a small patient number, which underscores the necessity of duplication and confirmation of the data by larger multicenter trials. The results of ongoing trials with sufficient patient numbers and power to demonstrate statistically significant results may definitely bring more insight to the investigation.

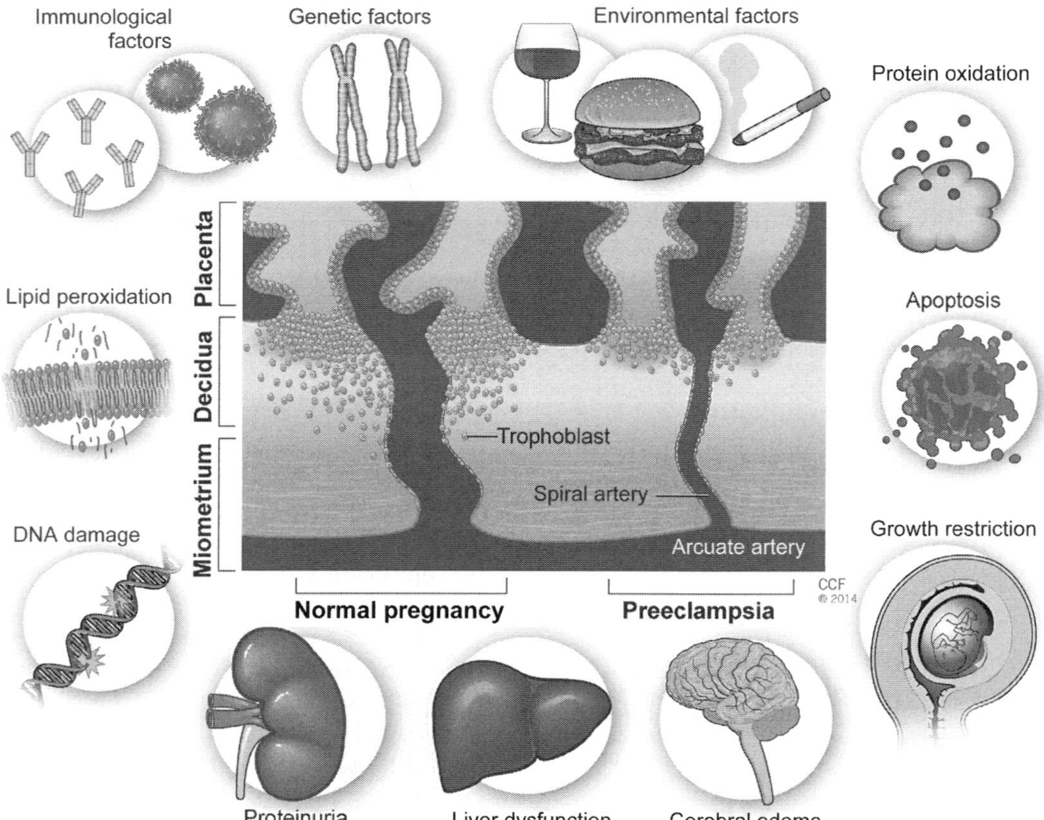

FIGURE 26.1   Placental dysfunction, triggered by a variety of mechanisms, plays the leading role in the pathogenesis of preeclampsia. Oxidative damage to vital biomolecules and the resulting cell death only contribute to further complications of this pregnancy-related pathology. The diseased placenta will ultimately release a variety of factors to the circulation, leading to maternal hypertension, proteinuria, abnormal liver function tests, elevated risk of seizures as well as intrauterine growth restriction of the fetus. Reprinted with permission, Cleveland Clinic Center for Medical Art & Photography © 2014. All Rights Reserved.

FUTURE DIRECTIONS

The past decade has brought many essential advances in the understanding of the health complications related to the pathogenesis of preeclampsia. Although the initiating events in preeclampsia are still not known, recent data suggest that excess free radicals link the placental disease and the systemic maternal as well as fetal manifestations. More work is needed to further define the regulation of placental vascular development and oxidative stress in normal pregnancy versus preeclampsia.

The biggest challenge, however, is related to the search for an effective treatment to prevent or at least reduce the damage resulting from preeclampsia-related oxidative stress, and unfortunately it is still unclear whether an eventual antioxidant supplementation would prove to be safe and effective.

CONCLUSIONS

The preeclamptic female can easily clinically decompensate, and requires evaluation, treatment, and close management. Preeclampsia is a disease process in pregnant women characterized by high blood pressure, proteinuria, and edema. It is a process that can adversely affect the mother and baby, and needs close management by obstetricians and health care providers. While the only true cure for preeclampsia is delivery, management is conducted with magnesium sulfate and blood pressure medications.

Oxidative stress plays an important role in the pathophysiology of preeclampsia, as a consequence of defective trophoblast invasion, reduction in placental perfusion, and placental hypoxia. The inability of endogenous antioxidant systems normally upregulated in normal pregnancy to control increased levels of oxidative damage is suggested as a possible factor in further overproduction of ROS and placental oxidative stress. This in turn may stimulate increased syncytiotrophoblast apoptosis, endothelial cell activation, and the hyperimmune response characteristic of preeclampsia. No specific management is available for the prevention of preeclampsia-related oxidative stress. Unfortunately, clinical trials focused on the possible benefits of antioxidant supplementation in preeclamptic subjects have not yet shown any substantial outcome.

References

[1] Callahan T, Caughey A. Hypertension and pregnancy. Blueprints: Obstetrics & Gynecology. 5th ed Philadelphia: Lippincott Williams & Wilkins; 2009. p. 92–8.

[2] Ananth CV, Keyes KM, Wapner RJ. Pre-eclampsia rates in the United States, 1980-2010: age-period-cohort analysis. BMJ 2013;347:f6564.

[3] Uzan J, Carbonnel M, Piconne O, Asmar R, Ayoubi JM. Pre-eclampsia: pathophysiology, diagnosis, and management. Vasc Health Risk Manag 2011;7:467–74.

[4] Fisher SJ, McMaster M, Roberts JM. The placenta in normal pregnancy and preeclampsia. Chesley's Hypertensive Disorders in Pregnancy. London: Elsevier; 2009. p. 73–86.

[5] Durán-Reyes G, Rocío Gómez-Meléndez Md, la Brena GM, Mercado-Pichardo E, Medina-Navarro R, Hicks-Gómez JJ. Nitric oxide synthesis inhibition suppresses implantation and decreases cGMP concentration and protein peroxidation. Life Sci 1999;65(21):2259–68.

[6] Roberts JM. Endothelial dysfunction in preeclampsia. Semin Reprod Endocrinol 1998;16:5–15.

[7] Agarwal A, Aponte-Mellado A, Premkumar BJ, Shaman A, Gupta S. The effects of oxidative stress on female reproduction: a review. Reprod Biol Endocrinol 2012;10:49.

[8] Agarwal A, Gupta S, Sharma RK. Role of oxidative stress in female reproduction. Reprod Biol Endocrinol 2005;3:28.

[9] Fruehauf JP, Meyskens FL Jr. Reactive oxygen species: a breath of life or death? Clin Cancer Res 2007;13:789–94.

[10] Bevilacqua E, Gomes SZ, Lorenzon AR, Hoshida MS, Amarante-Paffaro AM. NADPH oxidase as an important source of reactive oxygen species at the mouse maternal-fetal interface: putative biological roles. Reprod Biomed Online 2012;25(1):31–43.

[11] Halliwell B, Gutteridge JMC. Free Radicals in Biology and Medicine. 3rd ed Oxford: Oxford Science Publications; 1999.

[12] Thannickal VJ, Fanburg BL. Reactive oxygen species in cell signaling. Am J Physiol Lung Cell Mol Physiol 2000;279:L1005–28.

[13] Droge W. Free radicals in the physiological control of cell function. Physiol Rev 2002;82:47–95.

[14] Aris A, Benali S, Ouellet A, Moutquin JM, Leblanc S. Potential biomarkers of preeclampsia: inverse correlation between hydrogen peroxide and nitric oxide early in maternal circulation and at term in placenta of women with preeclampsia. Placenta 2009;30(4):342–7.

[15] Siddiqui IA, Jaleel A, Tamimi W, Al Kadri HM. Role of oxidative stress in the pathogenesis of preeclampsia. Arch Gynecol Obstet 2010;282(5):469–74.

[16] Siddiqui IA, Jaleel A, Al'Kadri HM, Akram S, Tamimi W. Biomarkers of oxidative stress in women with pre-eclampsia. Biomark Med 2013;7(2):229–34.

[17] Myatt L, Eis AL, Brockman DE, Greer IA, Lyall F. Endothelial nitric oxide synthase in placental villous tissue from normal, preeclamptic and intrauterine growth restricted pregnancies. Hum Reprod 1997;12(1):167–72.

[18] Lowe DT. Nitric oxide dysfunction in the pathophysiology of preeclampsia. Nitric Oxide 2000;4(4):441–58.

[19] Redman CW, Sargent IL. Placental stress and pre-eclampsia: a revised view. Placenta 2009;30(Suppl. A):S38–42.

[20] Gerretsen G, Huisjes HJ, Elema JD. Morphological changes of the spiral arteries in the placental bed in relation to pre-eclampsia and fetal growth retardation. Br J Obstet Gynaecol 1981;88:876–81.

[21] Meekins JW, Pijnenborg R, Hanssens M. A study of placental bed spiral arteries and trophoblast invasion in normal and severe pre-eclamptic pregnancies. Br J Obstet Gynaecol 1994;101:669–74.

[22] Dechend R, Viedt C, Muller DN. AT1 receptor agonistic antibodies from preeclamptic patients stimulate NADPH oxidase. Circulation 2003;107:1632–9.

[23] Dotsch J, Hogen N, Nyul Z, Hanze J, Knerr I, Kirschbaum M, et al. Increase of endothelial nitric oxide synthase and endothelin-1 mRNA expression in human placenta during gestation. Eur J Obstet Gynecol Reprod Biol 2001;97(2):163–7.

[24] Wisdom SJ, Wilson R, McKillop JH, Walker JJ. Antioxidant systems in normal pregnancy and in pregnancy-induced hypertension. Am J Obstet Gynecol 1991;165(6 Pt 1):1701–4.

[25] Cindrova-Davies T, Yung HW, Johns J. Oxidative stress, gene expression, and protein changes induced in the human placenta during labor. Am J Pathol 2007;171:1168–79.

[26] Many A, Roberts JM. Increased xanthine oxidase during labour-implications for oxidative stress. Placenta 1997;18:725–6.

[27] Burton GJ, Jauniaux E. Oxidative stress. Best Pract Res Clin Obstet Gynaecol 2011;25(3):287–99.

[28] Hubel CA, McLaughlin MK, Evans RW, Hauth BA, Sims CJ, Roberts JM. Fasting serum triglycerides, free fatty acids, and malondialdehyde are increased in preeclampsia, are positively correlated, and decrease within 48 hours post partum. Am J Obstet Gynecol 1996;174:975–82.

[29] Clausen T, Djurovic S, Henriksen T. Dyslipidemia in early second trimester is mainly a feature of women with early onset pre-eclampsia. BJOG 2001;108:1081–7.

[30] Hubel CA. Dyslipidemia, iron, and oxidative stress in preeclampsia: assessment of maternal and feto-placental interactions. Semin Reprod Endocrinol 1998;16:75–92.

[31] Gupta S, Agarwal A, Sharma RK. The role of placental oxidative stress and lipid peroxidation in preeclampsia. Obstet Gynecol Surv 2005;60(12):807–16.

[32] Mikhail MS, Anyaegbunam A, Garfinkel D, Palan PR, Basu J, Romney S.L. Preeclampsia and antioxidant nutrients: decreased plasma levels of reduced ascorbic acid, alpha-tocopherol, and beta-carotene in women with preeclampsia. Am J Obstet Gynecol 1994;171:150–7.

[33] Wang YP, Walsh SW, Guo JD, Zhang JY. The imbalance between thromboxane and prostacyclin in preeclampsia is associated with an imbalance between lipid peroxides and vitamin E in maternal blood. Am J Obstet Gynecol 1991;165:1695–700.

[34] Wang Y, Walsh SW, Kay HH. Placental lipid peroxides and thromboxane are increased and prostacyclin is decreased in women with preeclampsia. Am J Obstet Gynecol 1992;167:946–9.

[35] Aydin S, Benian A, Madazli R, Uludag S, Uzun H, Kaya S. Plasma malondialdehyde, superoxide dismutase, sE-selectin, fibronectin, endothelin-1 and nitric oxide levels in women with preeclampsia. Eur J Obstet Gynecol Reprod Biol 2004;113:21–5.

[36] Jain SK, Wise R. Relationship between elevated lipid peroxides, vitamin E deficiency and hypertension in preeclampsia. Mol Cell Biochem 1995;151:33–8.

[37] Staff AC, Ranheim T, Khoury J, Henriksen T. Increased contents of phospholipids, cholesterol, and lipid peroxides in *decidua basalis* in women with preeclampsia. Am J Obstet Gynecol 1999;180:587–92.

[38] Magann E, Martin J, Isaacs J, Perry K, Martin R, Meydrech E. Immediate postpartum curettage: accelerated recovery from severe preeclampsia. Obstet Gynecol 1993;81:502–6.

[39] Staff AC, Halvorsen B, Ranheim T, Henriksen T. Free 8-isoprostaglandin F2a is elevated in decidua basalis in women with preeclampsia. Am J Obstet Gynecol 1999;181:1211–5.

[40] Magann E, Martin J, Isaacs J, Perry K, Martin R, Meydrech E. Immediate postpartum curettage: accelerated recovery from severe preeclampsia. Obstet Gynecol 1993;81:502–6.

[41] Wang Y, Walsh SW. TNF alpha concentrations and mRNA expression are increased in preeclamptic placentas. J Reprod Immunol 1996;32:157–69.

[42] Raijmakers MT, Dechend R, Poston L. Oxidative stress and preeclampsia: rationale for antioxidant clinical trials. Hypertension 2004;44:374–80.

[43] Holmes VA, Young IS, Maresh MJ, Pearson DW, Walker JD, McCance D.R., et al. The Diabetes and Pre-eclampsia Intervention Trial. Int J Gynaecol Obstet 2004;87:66–71.

[44] Dalle-Donne I, Rossi R, Giustarini D. Protein carbonyl groups as biomarkers of oxidative stress. Clin Chim Acta 2003;329:23–38.

[45] Myatt L. Review: reactive oxygen and nitrogen species and functional adaptation of the placenta. Placenta 2010;31(Suppl.):S66–9.

[46] Fukushima K, Murata M, Tsukimori K, Eisuke K, Wake N. 8-Hydroxy-2-deoxyguanosine staining in placenta is associated with maternal serum uric acid levels and gestational age at diagnosis in pre-eclampsia. Am J Hypertens 2011;24(7):829–34.

[47] Tadesse S, Kidane D, Guller S, Luo T, Norwitz NG, Arcuri F, et al. In vivo and in vitro evidence for placental DNA damage in preeclampsia. PLoS One 2014;9(1):e86791.

[48] Richter C, Park J-W, Ames BN. Normal oxidative damage to mitochondrial and nuclear DNA is extensive. Proc Natl Acad Sci USA 1988;85:6465–7.

[49] Torbergsen T, Oian P, Mathiesen E, Borud O. Pre-eclampsia – a mitochondrial disease? Acta Obstet Gynecol Scand 1989;68:145–8.

[50] Widschwendter M, Schröcksnadel H, Mörtl MG. Pre-eclampsia: a disorder of placental mitochondria? Mol Med Today 1998;4:286–91.

[51] Bouhours-Nouet N, May-Panloup P, Coutant R, de Casson FB, Descamps P, Douay O, et al. Maternal smoking is associated with mitochondrial DNA depletion and respiratory chain complex III deficiency in placenta. Am J Physiol Endocrinol Metab 2005;288:E171–7.

[52] Masayesva BG, Mambo E, Taylor RJ, Goloubeva OG, Zhou S, Cohen Y, et al. Mitochondrial DNA content increase in response to cigarette smoking. Cancer Epidemiol Biomarkers Prev 2006;5:19–24.

[53] Clayton DA, Doda JN, Friedberg EC. The absence of a pyrimidine dimer repair mechanism in mammalian mitochondria. Proc Natl Acad Sci USA 1974;71:2777–81.

[54] Serdar Z, Gur E, Colakoethullary M, Develioethlu O, Sarandol E. Lipid and protein oxidation and antioxidant function in women with mild and severe preeclampsia. Arch Gynecol Obstet 2003;268(1):19–25.

[55] Orhan H, Onderoglu L, Yucel A, Sahin G. Circulating biomarkers of oxidative stress in complicated pregnancies. Arch Gynecol Obstet 2003;267(4):189–95.

[56] Fiore G, Capasso A. Effects of vitamin E and C on placental oxidative stress: an in vitro evidence for the potential therapeutic or prophylactic treatment of preeclampsia. Med Chem 2008;4(6):526–30.

[57] Palan PR, Shaban DW, Martino T, Mikhail MS. Lipid-soluble antioxidants and pregnancy: maternal serum levels of coenzyme Q10, alpha-tocopherol and gamma-tocopherol in preeclampsia and normal pregnancy. Gynecol Obstet Invest 2004;58:8–13.

[58] Sharma JB, Sharma A, Bahadur A, Vimala N, Satyam A, Mittal S. Oxidative stress markers and antioxidant levels in normal pregnancy and pre-eclampsia. Int J Gynaecol Obstet 2006;94(1):23–7.

[59] Harma M, Erel O. Measurement of the total antioxidant response in preeclampsia with a novel automated method. Eur J Obstet Gynecol Reprod Biol 2005;118:47–51.

[60] Davidge ST, Hubel CA, Brayden RD, Capeless EC, McLaughlin MK. Sera antioxidant activity in uncomplicated and preeclamptic pregnancies. Obstet Gynecol 1992;79:897–901.

[61] Hilali N, Kocyigit A, Demir M, Camuzcuoglu A, Incebiyik A, Camuzcuoglu H, et al. DNA damage and oxidative stress in patients with mild preeclampsia and offspring. Eur J Obstet Gynecol Reprod Biol 2013;170(2):377–80.

[62] Rosta K, Molvarec A, Enzsoly A, Nagy B, Ronai Z, Fekete A, et al. Association of extracellular superoxide dismutase (SOD3) Ala40Thr gene polymorphism with preeclampsia complicated by severe fetal growth restriction. Eur J Obstet Gynecol Reprod Biol 2009;142(2):134–8.

[63] Yoder SR, Thornburg LL, Bisognano JD. Hypertension in pregnancy and women of childbearing age. Am J Med 2009;122(10):890–5.

[64] Sagol S, Ozkinay E, Ozsener S. Impaired antioxidant activity in women with pre-eclampsia. Int J Gynaecol Obstet 1999;64:121–7.

[65] Kharb S. Total free radical trapping antioxidant potential in pre-eclampsia. Int J Gynaecol Obstet 2000;69:23–6.

[66] Knapen MF, Mulder TP, Van Rooij IA, Peters WH, Steegers EA. Low whole blood glutathione levels in pregnancies complicated by preeclampsia or the hemolysis, elevated liver enzymes, low platelets syndrome. Obstet Gynecol 1998;92:1012–5.

[67] Lapidus A. Effects of preeclampsia on the mother, fetus and child. Gyn Forum 1999;4(1):1.

[68] Chamy VM, Lepe J, Catalán A, Retamal D, Escobar JA, Madrid EM. Oxidative stress is closely related to clinical severity of pre-eclampsia. Biol Res 2006;39(2):229–36.

[69] Howlader MZ, Parveen S, Tamanna S, Khan TA, Begum F. Oxidative stress and antioxidant status in neonates born to pre-eclamptic mother. J Trop Pediatr 2009;55(6):363–7.

[70] Namdev S, Bhat V, Adhisivam B, Zachariah B. Oxidative stress and antioxidant status among neonates born to mothers with pre-eclampsia and their early outcome. J Matern Fetal Neonatal Med 2013;27(14):1481–4. Epub ahead of print.

[71] Dani C, Cecchi A, Bertini G. Role of oxidative stress as physiopathologic factor in the preterm infant. Minerva Pediatr 2004;56(4):381–94.

[72] Stratta P, Canavese C, Porcu M, Dogliani M, Todros T, Garbo E, et al. Vitamin E supplementation in preeclampsia. Gynecol Obstet Invest 1994;37:246–9.

[73] Gulmezoglu AM, Hofmeyr GJ, Oosthuisen MM. Antioxidants in the treatment of severe pre-eclampsia: an explanatory randomized controlled trial. BJOG 1997;104:689–96.

[74] Villar J, Say L, Shennan A, Lindheimer M, Duley L, Conde-Agudelo A, et al. Methodological and technical issues related to the diagnosis, screening, prevention, and treatment of pre-eclampsia and eclampsia. Int J Gynaecol Obstet 2004;85(Suppl. 1):S28–41.

[75] Steyn PS, Odendaal HJ, Schoeman J, Stander C, Fanie N, Grové D. A randomised, double-blind placebo-controlled trial of ascorbic acid supplementation for the prevention of preterm labour. J Obstet Gynaecol 2003;23:150–5.

[76] Lee BE, Hong YC, Lee KH, Kim YJ, Kim WK, Chang NS, et al. Influence of maternal serum levels of vitamins C and E during the second trimester on birth weight and length. Eur J Clin Nutr 2004;58:1365–71.

[77] Zhang C, Williams MA, King IB, Dashow EE, Sorensen TK, Frederick IO, et al. Vitamin C and the risk of preeclampsia—results from dietary questionnaire and plasma assay. Epidemiology 2002;13:409–16.

[78] Mikhail MS, Anyaegbunam A, Garfinkel D, Palan PR, Basu J, Romney SL. Preeclampsia and antioxidant nutrients: decreased plasma levels of reduced ascorbic acid, alpha-tocopherol, and beta-carotene in women with preeclampsia. Am J Obstet Gynecol 1994;171:150–7.

[79] Conde-Agudelo A, Romero R, Kusanovic JP, Hassan SS. Supplementation with vitamins C and E during pregnancy for the prevention of preeclampsia and other adverse maternal and perinatal outcomes: a systematic review and metaanalysis. Am J Obstet Gynecol 2011;204(6):503.e1–503.e12.

[80] Basaran A, Basaran M, Topatan B. Combined vitamin C and E supplementation for the prevention of preeclampsia: a systematic review and meta-analysis. Obstet Gynecol Surv 2010;65(10):653–67.

[81] Rumbold A, Duley L, Crowther CA, Haslam RR. Antioxidants for preventing pre-eclampsia. Cochrane Database Syst Rev 2008;1. CD004227.

[82] Rumbold A, Crowther CA. Vitamin C supplementation in pregnancy. Cochrane Database Syst Rev 2005;2. CD004072.

[83] Rumbold A, Crowther CA. Vitamin E supplementation in pregnancy. Cochrane Database Syst Rev 2005;2. CD004069.

[84] People's League of Health. Nutrition of nursing and expectant mothers. Lancet 1942;4:10–2.

[85] Rumbold A, Middleton P, Crowther C. Vitamin supplementation for preventing miscarriage. Cochrane Database Syst Rev 2005;2. CD004073.

NUTRITION AND LIFESTYLE IN *IN VITRO* FERTILIZATION

27

The Psychological Management of Infertility

Mandy J. Rodrigues, MA, Francisco Antonio Rodrigues, MBBCh

TUPS Stress, Medfem Fertility Clinic, Bryanston, Johannesburg, South Africa

INTRODUCTION

Thousands of couples long to have their own baby but are faced with the heartbreaking reality that they will never be able to do so without the help of medical science. There are others who might never have the opportunity to be parents, in spite of all the options available today like donor conception, surrogacy, and adoption. These are fortunately in the minority. However, all these women or couples go through a journey that is a lonely one with many casualties along the way. These include casualties to the woman as an individual, who describes immense feelings of helplessness and powerlessness sometimes causing depression and isolation [1,2]. There is an impact on one's marriage [2,3], and recent research shows that the reaction to infertility has been likened to a posttraumatic stress reaction [4]. Infertility has social implications and far-reaching consequences for the individual, the couple, and the extended family. This chapter will address these psychological issues and look into ways of managing them effectively.

STRESS AND INFERTILITY

We all have some plan of where our life is going and this often includes marriage and babies, or more recently, this might even include single parenthood in later life. Conception is often something we accept as a given, so when it does not happen, we are often totally unprepared for it. Suddenly, our life comes to a halt as we try to accommodate the various treatments and planning around ovulation. Even if the medical world is managing the process, there is a corresponding psychological management that is often lacking either due to ignorance or to a feeling that there is very little we can do to manage the emotions. We get bombarded by well-meaning advice suggesting that our lifestyles are contributing to our infertility, and that stress has an impact on fertility. But we feel helpless as to how to manage it. We often do not understand what came first: did the infertility cause the stress or did the stress cause the infertility? The answer is like the chicken and the egg.

THE CYCLE OF INFERTILITY

Once one is confronted with the diagnosis of infertility, one has probably been through months, even years, of trying unsuccessfully to conceive. The journey of infertility can be likened to a roller coaster ride. Each new cycle is met with excitement and anticipation, hopefulness and optimism. Ovulation is followed by a two-week wait fraught with anxiety and trepidation. Each symptom is interpreted and analyzed. With the availability of the Internet and social media, one is able to seek information at one's fingertips 24/7, and this information is not always accurate. We are bombarded with information from many sources, all trying to convince us not to give up hope. There is an eagerness to test for pregnancy, with a conflict that the results might be too early to be accurate. Once one receives a negative result, one is forced into a grief cycle. Kübler-Ross pioneered the work on the grief cycle as one consisting of a number of stages that need to be resolved, in the hope of coming to terms with a loss [5]. A negative pregnancy result can be regarded as one such loss [6]. Women seem to go through similar stages starting with shock and denial [7]. The first stage often consists of tearfulness alternating with numbness, followed by disbelief. This is followed by the second stage of bargaining where one tries to understand why this cycle has not worked. This is where questions about religion and blame might come into play. One is trying to grasp an understanding as to *why*. There is the third

Handbook of Fertility. http://dx.doi.org/10.1016/B978-0-12-800872-0.00027-5

stage of anger where one gets angry with those around her, like one's spouse or doctor, and then angry at the world. It is at this stage where the woman might be irritable and bitter. This is followed by a period of depression. This cycle is acutely experienced in a matter of days only to be replaced again by hope and anticipation instead of the final stage of acceptance as Kübler-Ross described. Acceptance allows one to adequately grieve the loss and move along. However, the infertile woman/couple do not get a reprieve as the cycle of hope begins again and seems to recur month after month.

THE CONSEQUENCES OF INFERTILITY

Along with this is a feeling of helplessness and lack of control. Infertility is something one feels that she or he has little control over. Even when medical science gets involved, the success rate is nowhere near 100%. Women describe the helplessness and hopelessness as despairing emotions as they feel they have very little control. They feel isolated and worry about where they fit in. They do not fit in with their single friends, and they do not feel they fit in with their friends with children. So they end up feeling isolated. The biggest buffer for stress during a life-changing event is social support [8,9]. People with higher social support cope better during stress. However, those going through infertility tend to isolate themselves either due to a lack of perceived understanding from others or due to feeling shame and being unable to communicate this to others. Isolation and lack of support impacts negatively on self-esteem, and further exacerbates the emotional effects of infertility [10].

Research has shown that two psychological consequences of infertility are depression and posttraumatic stress disorder [11,12]. A major cause of depression is helplessness, and a lack of being able to make an impact on the course of one's life as discussed earlier. The second reaction to infertility is that of a posttraumatic stress reaction [4,11]. In posttraumatic stress, we relive the trauma, dream about it, think of very little else, and there are stimuli or triggers reminding us of the trauma all the time. We start avoiding the stimuli that remind us of the trauma. We either think about it all the time or avoid it or do both; but it occupies so much energy. As with trauma, we struggle to make sense of it. We are reminded of it constantly when we see babies in trolleys, pregnant women, baby showers, adverts with families, and so on. We try to avoid situations that remind us of our infertility yet we cannot stop thinking about it. It starts to dominate our life and occupy our thoughts. A number of studies have shown that anxiety has an effect on fertility [13–18]. These studies also show that reducing anxiety results in increased pregnancy rates [19]. Depression appears to increase infertility rates [12]. There are a number of studies that indicate that psychological interventions lead to increased pregnancy rates [20–27]. All of these studies offer an explanation of the correlation between stress, immunity, and decreased fertility [28].

A further consequence is that our lives feel on hold. The very nature of infertility and the treatment processes themselves sometimes dictate that one has to live cycle to cycle. It becomes very difficult to plan into the future. It becomes difficult to predict future plans and goals, and one feels one is hanging in limbo. Suddenly one's life comes to a halt.

INFERTILITY AND ONE'S RELATIONSHIP

All of these lead to an impact on one's relationship. Our partners often seem to be on a different page when it comes to infertility [29]. Men generally seem more optimistic, and want to look for a solution. A book written by Pease and Pease looked at the way in which men and women interpret the world and react to it [30]. When talking to men going through infertility, their despair is primarily the helplessness they feel when they see their wives upset. In general, they battle more with the impact of the infertility on their wives and marriage than of the infertility itself. This ends up in couples doing independent coping, a phrase coined in infertility research that explains how each spouse begins to cope independently as they try to keep their own distress from each other. Petersen and his colleagues looked at couples and their reaction to treatment over a 5-year period [3]. They found that men and women started coping parallel to one another, and established a means of reacting to infertility, which estranged them from one another. This worsened across a 5-year period.

It is only natural to start stressing when one is battling to conceive. However, there comes a point where this stress is counter-productive and at this stage one needs to seek help in managing one's stress. We all get stressed, yet not all of us battle to conceive. In fact, during times of war, and in poverty-stricken communities, the pregnancy rate is often inordinately high. Surely people battling to meet their basic needs on a daily basis are stressed? We keep hearing that if we are stressed, we will not get pregnant. We hear of so many people who get pregnant after adopting a baby or giving up on their fertility treatment [31–33]. What is the link between fertility and stress?

PSYCHONEUROIMMUNOLOGY

First of all, we need to understand that there is a link between the mind and the body. There is an abundance of literature on this, and it has been proven over decades that our emotional state causes a corresponding biochemical or physical reaction [34–38]. Prior to the last two decades, the field of psychoneuroimmunology was that branch of medical science that was least known about. There have been huge strides in the last few decades [39,40]. Some of the studies are as follows: Haimovici and Hill looked at stress and depression and how these states altered specific cytokines that can effect fertility physiologically [22]. Stress and depression have a profound effect on the hypothalamic-pituitary-adrenal axis as well as increasing insulin levels [23,24]. This can, in turn, have a profound effect on ovarian function, fertilization, and endometrial implantation. Demyttenaere et al. found that women with functional chronic anovulation had higher serum cortisol and cerebrospinal fluid corticotrophin-releasing hormone concentrations than healthy controls [28]. A similar cortisol hypersecretion has been reported in women undergoing *in vitro* fertilization (IVF) and embryo transfer who fail to achieve implantation [41]. Studies have recently shown that follicular fluid obtained from women undergoing embryo transfer inhibits the proliferation of T-helper lymphocytes as well as the production of interleukin-1 and -2 [42]. However, cytokines, such as interleukin-2, -6, and -8, seem to play a role in follicular development through the local secretion of proteolytic enzymes [43]. Peritoneal immunosurveillance involving different cell types fulfills the function of clearing regurgitated endometrial cells from the peritoneal cavity [44]. Finally, endometriosis patients have been shown to have decreased T cells and natural killer cells as well as reduced macrophage activity [45,46]. They also have increased T suppressor activity, increased transforming growth factor and prostaglandin E2 [47]. Managing one's chronic stress appears to enhance cell-mediated immunity [48–50].

REAL AND SELF-INDUCED STRESS

Stress can be broadly divided into two types: real and self-induced. Real stresses are the triggers that make even relaxed people become stressed: poverty, crime, war, death, and even deadlines to a lesser degree. In these situations, the stress is externally induced, and adrenalin is secreted in order to cope. The flight-fight response or adrenalin-mediated stress appears to be less harmful on the immune system as regards fertility. The reaction to this stress is usually acute and short-term. However, it is chronic, prolonged stress that is the problem [51,52].

Before we explain this more chronic stress, let us look at the world today. Time has become emphasized in our modern world. There are reminders of time everywhere – the car, computer, the cell phone reminders that sync to our laptops. We are able to do things much faster today than ever before. We have access to the Internet, computers, cell phones – we are always in reach and also more readily available to others on a 24/7 basis. There is an abundance of information available. Years ago, the number of books published in a month worldwide was minimal, but nowadays, in a month, there are hundreds of thousands of new books available worldwide. There is information overload in a shorter period of time, in a quicker manner, which can occur at any time of day or night. No longer can you go home and be unreachable or leave work until the next day due to a lack of facilities. We also have a superabundance of things to do – we can travel more, we work more, we buy more, we play more – we have endless choices and all these require time.

Nowadays, more women have entered the corporate market than ever before [53,54]. This has meant becoming multiroled and multitasked, which means dividing time into various roles, which are each time-consuming in their own right as wife, mother, caregiver, or corporate woman. The working woman now enters the job market where she is expected to achieve as well as her counterparts regardless of whether she has a family or not. She struggles to juggle and then feels she is not good enough, so she pushes herself harder and makes herself work quicker in order to feel capable in her own eyes and in those of her colleagues.

Resources have also become scarcer in that we now compete more for jobs and keeping jobs more than ever before. So we try to be the best in what we do and being the best often requires us to work faster, more efficiently, and better than the other person.

TIME URGENCY PERFECTIONISM STRESS

All this has resulted in a more chronic, more harmful stress reaction. The construct of *time urgency perfectionism stress* was coined over 18 years ago [53,54]. Time urgency is where people are stressed for time, they are driven by time – they do things urgently, they are constantly aware of time, they try to attempt too much into a given time period, stressed in traffic, try to do more than one thing at a time, they hate being late, but often find they run late due to unrealistic assessments of time or procrastination or difficulties in prioritizing. These people expect a lot of themselves and others and expect others to be as responsible and goal-orientated as themselves. They often set unrealistically high standards for themselves and are very hard on themselves when they do not achieve these goals. When things are beyond their control, they tend to get stressed. They create extra stress in their lives by worrying about triggers that relaxed people would not worry about. This stress is inner driven – internally induced, maintained, and motivated. It is a learned stress, and it becomes a habitual stress after a period of time.

Every time one experiences this chronic stress, one's body releases cortisol and noradrenalin (norepinephrine). These serve to erode and weaken the immune system. The immune system becomes vulnerable and we are more susceptible to medical conditions like endometriosis [44–47] and thyroid and insulin problems [55], which all have an impact on our ability to conceive [56].

Research conducted at Medfem Clinic showed that the management of chronic stress improved IVF statistics to 67% and higher, whereas chronically stressed patients had an average success rate that was comparative to IVF results across the world [48].

Notwithstanding the emotional or psychological reactions of the general population when confronted with infertility, the time urgent perfectionist (TUP) has an even more difficult time when trying to cope with infertility. Time urgency perfectionism stress is compounded when situations are out of our control. Conception is supposedly something we accept as a given, so when it does not happen, we are often totally unprepared for this and do not have the coping skills to deal with it. And if we are the personality type prone to suffering from self-induced stress, we have a harder time to conceive and a harder time managing this event so out of our control. For TUPs, the case is even more disturbing due to a number of factors:

- They become urgent each month that the conception should happen immediately;
- Their perfectionism makes them doubt themselves and the processes they are undergoing, causing them either to personalize their infertility (internalize and blame themselves) or to externalize it (scarcity thinking whereby they might blame their spouses, or specialists/doctors assisting them or the inability of medical science, which has failed them thus far);
- As TUPs, they do not like the lack of control they have over the process of trying to conceive as well as the perceived lack of emotional control when they, once again, receive a negative result. The roller coaster of infertility makes it very hard for TUP individuals to foster some sort of prediction and control over their emotions. Take a look at the case study presented next:

Sandy was a 20-year-old woman who had always been a TUP but this had not caused any subjective distress in the past. She attributed her career success to her drive and competitiveness. She had always achieved well, received awards at school, and finally married and decided it was the right time to start a family. She visited her gynecologist, who stopped the contraceptive pill and said she would be pregnant in six months. However, Sandy did not like to think she had to wait six months, and decided to see a specialist in the field to "speed up the process," not yet believing there was a problem at all.

Suddenly, Sandy finds out that she has endometriosis and this might complicate the process of trying to conceive. She undergoes the necessary treatment, but still is not pregnant within three months. It is the first time in Sandy's life that she has little control over a process. She starts to think about pregnancy, and the failure to achieve one, constantly. Every month, the last two weeks of her cycle are fraught with despair and helplessness as she tries not to obsess about conception. However, as a perfectionist, she begins to structure her life and her goals, including her career, around gearing for a pregnancy. She stays in a dead-end job because she wants her maternity benefits. Nothing seems to make her joyful anymore.

For many TUP individuals, infertility is possibly the first obstacle they have faced where they have little control over the outcome. As perfectionists, they have probably always had a preconceived notion of what paths their lives would follow. They studied, entered a career, settled down in a stable relationship, and believed children would be the next logical step in this predetermined plan. However, despite having attained and achieved well in other areas of their lives, they are suddenly thrown into the possibility of their plans not working out as they were meant to. This exacerbates feelings of despair and helplessness. Society is sometimes insensitive to the needs and concerns of an infertile couple. These well-intentioned individuals do not realize the torment each of their harmless statements and suggestions actually creates for the woman/man who is struggling to conceive.

ADOPTION AND THE IMPACT ON INFERTILITY

Why is it that people who have just adopted or who have resigned themselves to the fact that they will not have children or have just attempted their third and last IVF, get pregnant afterward? [31–33]. We hear of these stories all the time. There is no conclusive research one can refer to, but could it be that these people have stopped that learned habitual worrying about getting pregnant? But it is easier said than done. The more someone tells you to relax, the harder it is. It is a task you cannot do all alone.

It is then very difficult – as an infertile couple – to hear that their time urgency and perfectionism probably played a major role in the cause of their infertility. This impacts on the bargaining stage of the grief cycle, where they begin to question the past or current behavior and its impact on their current situation, such as "if I had not focused on my career for so long" and "if only I had eaten properly and started trying for a baby sooner." The anger then becomes internalized as self-blame and recrimination. A perfectionist then tries to do everything possible to remedy the situation. The couple attempts to eat properly, exercise, manage their stress, and follow all the advice they can gather, but they do so almost obsessively, and want it done immediately with immediate effect. This input of time urgency causes more stress and perceived lack of control. Even though it is recommended and necessary to reassess certain lifestyle factors contributing to the infertility, the individual needs to also maintain a realistic outlook, by believing they are doing everything to enhance their chances of conceiving, and to attain a peace of mind about when the pregnancy will happen.

IDENTIFYING CHRONIC STRESS

Now that we understand the psychological consequences of infertility, how do we manage these? First, one needs to realize that the process of infertility in itself causes stress. Second, if one is chronically stressed or as the authors refer to as a TUP, then this needs to be ascertained and managed. Let us address the second construct first. The authors have developed an on-line stress management course that comprises of in-depth psychometric testing followed by a 10-session cognitive behavioral stress management course. The screening tests are available for free at www.timeurgency.com. Upon completion of the questionnaires, one is told whether one fits the category of the TUP. Let us address this concept of time urgency perfectionism stress a bit further.

TUPs often have low self-esteem, and they often feel they have to achieve more and more in order to feel acceptable to themselves. As perfectionists, they need to keep achieving to fulfill self-acceptance, and they tend to define achievement by their career, house, and tangible achievements. Many people who are TUPs tend to define themselves and their character by the job they do. They become their career. This becomes even more challenging when they decide to take time off to have a child or when they leave their careers temporarily in order to manage their infertility. They lose themselves as they have defined themselves for so long in a certain role that no longer exists and that they believe is detrimental to their pursuit of having a child. A word of caution here, if you have been told stress is impacting on your ability to have children, first manage your stress before you make major decisions about leaving your place of work. If you do not learn how to manage this chronic stress, it will emerge at home or in any other job you decide to pursue. It is a way of thinking and a way of life. It is not situation-specific.

An example is Julia who is 30. She presents with infertility to Medfem Clinic. She is a TUP personality, but denies it at first. She emphatically says she is not stressed by time because she is never late – so she has no need to be stressed by time. She also says she does not work so she does not have any need to worry as her time is her own. She also disagrees with being a perfectionist. Following further questioning, she is shown that she is in fact stressed by time. She spends much of her day looking at time and doing things urgently in order not to be late or to keep people waiting. When she drives, she races the clock by ensuring she is at

certain milestones at certain times. Of course, she says she is not usually stressed but should her perfectly planned drive go wrong or the traffic be beyond her control, then she gets aggressive with the other drivers and berates herself for not having left earlier. This negative self-talk and way of thinking merely exacerbates her anxiety. By the time she gets to her appointment, her neck muscles are stiff, she feels a headache starting, she is irritable, and she is anxious about the meeting that begins in 10 min. She is also hungry as she had no time for a breakfast, and quickly grabs a packet of crisps and a coke to boost her energy and her sugar levels.

In terms of her perfectionism, she is not an orderly person who ensures her house is in perfect order and neat and tidy and who washes the dishes perfectly after a meal, or who places her magazines in a certain order at home. However, she expects a lot of herself. She always checks her lists over again at least twice, in case there is an oversight, which merely aggravates her time constraints, and then – more stress! She is hard on others who do not do things the way she does, and struggles if she does not feel in control. She expects others to be as responsible as herself.

Julia is married to Alan who is the opposite. He is relaxed and enjoys watching television and having a beer. He works hard and is perfectionistic in his job, but he knows how to turn the traits off when he is not at work. Julia nags when they are due to go out, and nags for him to be on time. The more she nags, the slower he seems to be, which makes her angrier. When he drives, she moans at the route he takes, the speed at which he drives, and the way he drives. He has now offered that she drive when they go out. She hates a context such as driving when she is not in control.

Julia finds she is very passive with her husband's extended family as she does not like any conflict, but suppresses her irritation when she visits with them and feels tense later and cannot sleep. Even though her husband does not expect her to prepare a meal, she expects it of herself to be the perfect wife and mother. She demands of herself to fulfill the role of a traditional wife and homemaker, and modern career women when she was working, and she finds she cannot – given time constraints – so she oftentimes feels guilty.

First of all, Julia needed to be convinced that she needed the course. This is easier done when you have a motivation for getting less stressed. Fortunately, Julia wanted a baby, a huge motivating force. Others may want to feel more relaxed, or improve relationships, have less physical illnesses such as chronic colds or to focus on losing weight and maintaining health-promoting behaviors. She needed to be alerted to her tension. Oftentimes we are not aware of the anxiety or the tense shoulders and clenched jaw. We have grown accustomed to the habitual stress. Once Julia identified the tension, she had to look at what hooks caused her anxiety. That is, when did she feel stressed by an inner need to be time urgent and perfectionistic? Within a group setting, she was forced to look at herself as well as to learn from others with the same personality. Julia had to learn to talk to herself differently and to reassess her life goals. If a family was so important to her, why was she initially driven toward pursuing a career that was in conflict with her desperate desire to get pregnant?

Julia was taught to judge herself less harshly and to leave certain tasks that were less urgent and less important till the next day. In addition, she was taught that it was her that expected herself to be 110% at each role she played, not anyone else, and she was taught how to be more tolerant toward slowing down within herself. Julia was also taught how to tailor her behavior so that it did not vacillate between aggression and irritation, and passivity or between all-or-nothing thinking and to tolerate grey areas in her life. She was taught how to be more assertive and not compromise herself or others. This in itself reduced her stress levels. In addition, she learned how to relinquish control at times within her relationship and elsewhere such as when she could not control an external stressor like a traffic jam, a deadline, or her infertility. She realized that stressing was only permitted if it yielded a positive outcome, and this was rare. She learned instead how to maximize the use of her time and be efficient instead of wasting time worrying.

Ultimately, Julia stopped blaming herself for her infertility, and stopped obsessing every month. She relinquished control to her fertility specialist and trusted him to do what was in his expertise to do.

We all need a certain amount of mediocre stress to function; we need the motivation and the adrenaline (epinephrine) [57,58]. If we had no stress, we would get nothing done. As the stress gets more and more, we appear to cope better and better until the level of stress gets too much and our bodies (either psychologically or physically) signal us to stop, or stop us by producing some type of illness or emotional difficulty.

The effects of chronic stress can be reversed by eliminating the learned stress factors. That is, by removing the self-induced TUP stress. This decreases the "bad stress hormone" circulating, which in turn reduces physical symptomatology [48]. We know that the mind and body are intricately intertwined, so this will impact on the mind.

MANAGING ONE'S STRESS

So what do we do? We should learn to recognize when we are stressed. Chronic stress is a learned stress so it is often habitual and we become so used to the feelings and sensations that we take the stress as normal in our day. Learn to recognize stress either by physical discomfort, behavior (irritability, comfort eating, tearfulness, aggression), and/or negative self-talk. Begin looking at the hooks, that is, what context, situation, person elicited that stress? What about the situation made us feel stressed? Was it a time urgent or perfectionistic manner of thinking? Once we can identify what causes stress, we can analyze it into its components and start eliminating the stress. Remember, the hook or trigger is merely the start of the vicious circle. The hook sets off a chain reaction. It sets off a certain way of thinking about the situation. This then makes you physically tense, and you may start talking to yourself in a certain way. Finally, this evokes a certain behavior and once again this may make you feel muscle tension, and so the circle continues. For example, you get given a deadline of Thursday, and you immediately panic and say "I always leave things for the last minute, now I have no time to complete this deadline as I have too much else to do." This makes you feel a headache coming on, and you snap at the colleague who wants to ask you a question. You then feel guilty, which causes an internal recrimination of "why are you so nasty, this is what you get for never getting things done." By the time you realize the tension, you have almost forgotten the initial hook, and

your worrying has assumed huge proportions affecting behavior, physical comfort, and the manner in which you talk to yourself. You need to identify how you behave, think, and feel during stressful periods and focus on each of these. There are various techniques implemented that practically teach you how to eliminate these individual reactions. They can be located in detail at www.timeurgency.com with day-to-day instructions on how to manage your stress. Relaxation techniques, assertion training, and cognitive challenging are some of the cognitive behavioral techniques employed to manage the chronic stress [59].

In summary, identify your stress hooks then try to avoid them. If they are unavoidable, try to eliminate the negative stressful responses by minimizing their impact using one of the techniques presented earlier.

With practice, we learn how to manage this chronic TUP stress. As the stress decreases, we feel more relaxed, less anxious, and more affable. People like us more and we can accept ourselves and others easier. We become more efficient at work and at home. We suffer less from depression and we are less angry people. Our relationships improve. The symptoms we had before that could not be explained by conventional medicine, disappear or become more manageable. We have more peace of mind, and we like ourselves more. Life becomes more enjoyable again, instead of a task. Our goals become more meaningful. However, the most important change that happens is related to what we spoke about earlier, within the field of psychoneuroimmunology, namely, that the mind and body are closely interrelated and interdependent. As the mind relaxes, there is a corresponding change in the body. The initial focus of the research showed that female patients presenting with infertility had a 100% risk for having endometriosis, irrespective of the presence of the classical symptoms, if they scored 100 or more on the TUP screening test. In addition, management of TUP stress leads to reduced reoperation rates in endometriosis [48]. The value of TUP stress management relies on the modification of the immune system via psychoneuroimmunological factors [39–40,50]. Modifying the immune system controls endometriosis and immunological fertility factors. TUP stress management also enhances the ability of the patient to manage the stress related to the infertility process. These changes result in a dramatic increase in the pregnancy rates in women who have completed the 10-week course.

OTHER STRESS MANAGEMENT TECHNIQUES IN INFERTILITY

Once the TUP stress has been managed, one can manage the other stresses associated with infertility that is universal. Why is it that people who have just adopted or who have resigned themselves to the fact that they will not have children or have just attempted their third and last IVF, fall pregnant afterward? [31–33]. We hear of these stories all the time and we are starting to believe that the reason is that these people have stopped that learned habitual worrying about getting pregnant. But it is easier said than done. The more someone tells you to relax, the harder it is. It is a task you cannot do all alone. The problem with infertility is that we put our lives on hold and the last two weeks of every cycle are spent wondering whether or not you are pregnant and putting your goals aside. It is strongly recommended that one actively try to seek peace of mind about their fertility. How does one do this? The authors have found that by assisting individuals to write down their short-, medium-, and long-term goals in a list as well as their fertility plans or options on another list, they are forced to incorporate all life goals together. In this way, the fertility issues do not dominate or override other important goals. Neither do they negate the importance of other needs in the individual's life. They now are able to incorporate both infertility and other issues in their life plan. In this way, infertility gains perspective and a place, without replacing or neglecting other important areas. An important characteristic of writing down one's goals is that they should be flexible. The individual has a map forward. Psychologically it creates an amazing sense of peace and acceptance if the couple knows where they are going; they have negotiated this, and have a sense of direction. It places fertility somewhere in the back of their minds because they are moving forward in terms of their entire life. It also creates a sense of empowerment. The helplessness so often felt with infertility is lessened as they have direction and a way to manage their stress. We create a sense of distraction from the hooks that plague us every day in terms of trying to get pregnant.

SPOUSAL INVOLVEMENT

A further important factor is getting the husbands involved in this process. The women often feel their husbands are less involved in the process of fertility treatment. Get your spouse involved in the process so as to decrease alienation and withdrawal and to increase support. Couples should seek help from a psychologist specializing in the field. Men and women deal with the stress of fertility at different levels [29]. Men generally want solutions, are less likely to look at longer-term options until forced to, and feel helpless in the face of their wives' distress. Women do not generally want solutions or appeasements from their spouses. They just want to be heard and supported. They get angry and intolerant toward their spouses. The phenomenon of independent coping was discussed earlier in the chapter [3]. Therapy helps the couple establish the normal reaction to infertility within a marriage and how to manage this by bringing in communication skills and expectations. One wants to maintain a stable marriage in order to bring a child into a happy home.

CONCLUSIONS

As can be seen from all the discussions, the very nature of the infertility journey is one fraught with helplessness, grief, sadness, and depression. All this has an impact on the individual and the marriage. Research is still conflicted about what came first: the stress or the infertility? However, either way, the stress needs to be managed for success in the journey. The news is clear that depression and anxiety impact on infertility. The research also shows that chronic stress impacts on infertility due to the relationship between the mind and the body. The more stressed one is before the infertility journey, the harder it is to maneuver one's way through it and the more it impacts even further on the outcome, resulting in a cycle of stress and infertility. The authors recommend managing the chronic stress that possibly contributed to the problem while managing the acute stress for the best outcome.

References

[1] Domar AD, Broome A, Zuttermeister PC, Seibel MM, Friedman R. The prevalence and predictability of depression in infertile women. Fertil Steril 1992;58:1158–63.

[2] Berghuis JP, Stanton AL. Adjustment to a dyadic stressor: a longitudinal study of coping and depressive symptoms in infertile couples over an insemination attempt. J Consult Clin Psychol 2002;70(2):433–8.

[3] Peterson BD, Pirritano M, Christensen U, Boivin J, Block J, Schmidt L. The longitudinal impact of partner coping in couples following 5 years of unsuccessful fertility treatments. Hum Reprod 2009;24(7):1656–64.

[4] Bartlik B, Greene K, Graf M, Sharma G, Melnick H. Examining PTSD as a complication of infertility. Medscape Women's Health 1997;(2). available at http://www.medscape.com/viewarticle/408851.

[5] Kubler-Ross E. On Death and Dying. New York: First Scribner; 1969.

[6] Greenfeld D, Diamond M, DeCherney A. Grief reactions following in-vitro fertilization treatment. J Psychosom Obstet Gynecol 1988;8:169–74.

[7] Axelrod J. The 5 stages of loss and grief. Psych Central. Retrieved from: <http://psychcentral.com/lib/the-5-stages-of-loss-and-grief/000617> (05.06.14).

[8] Lechner L, Bolman C, van Dalen A. Definitive involuntary childlessness: associations between coping, social support, and psychological distress. Hum Reprod 2007;22:288–94.

[9] Cohen S, Wills TA. Stress, social support and the buffering hypothesis. Psychol Bull 1985;98(2):310–57.

[10] Amir M, Horesh N, Lin-Stein T. Infertility and adjustment in women: the effects of attachment style and social support. J Clin Psychol Med Settings 1999;6(4):463–79.

[11] Schwerdtfeger KL, Shreffler KM. Trauma of pregnancy loss and infertility for mothers and involuntary childless women in the contemporary United States. Retrieved on June 5, 2015 from http://www.ncbi.nlm.nih.gov/pmc/articles/PMC3113688/.

[12] Lapane LK, Zierler S, Lasatar TM, Stein M, Barbout MM, Hume AL. Is a history of depressive symptoms associated with an increased risk of infertility in women? Psychosom Med 1995;57: 509–13.

[13] Demyttenaere K, Nijs P, Steeno O, Koninckx PR, Evers-Kiebooms G. Anxiety and conception rates in donor insemination. J Psychosom Obstet Gynaecol 1988;8:175–81.

[14] Lynch CD, Sundaram R, Maisog JM, Sweeney AM, Buck Louis GM. Preconception stress increases the risk of infertility: results from a couple-based prospective cohort study – the LIFE study. Hum Reprod 2014;29(5):1067–75.

[15] Domar AD, Zuttermeister PC, Friedman R. The relationship between distress and conception in infertile women. J Am Med Womens Assoc 1999;54:196–8.

[16] Ferin M. Clinical review 105: stress and the reproductive cycle. J Clin Endocrinol Metab 1999;84:1768–74.

[17] Lynch CD, Sundaram R, Buck Louis GM, Lum KJ, Pyper C. Are increased levels of self-reported psychosocial stress, anxiety and depression associated with fecundity? Fertil Steril 2012;98: 453–8.

[18] Schenker JG, Meirow D, Schenker E. Stress and human reproduction. Eur J Obstet Gynecol Reprod Biol 1992;45:1–8.

[19] Domar AD, Clapp D, Slawsby E, Dusek J, Kessel B, Freizinger M. Impact of group psychological interventions on pregnancy rates in infertile women. Fertil Steril 2000;73:805–11.

[20] Domar AD, Seibel MM, Benson H. The mind/body program for infertility: a new behavioral treatment approach for women with infertility. Fertil Steril 1990;53:246–9.

[21] Domar AD, Zuttermeister PC, Seibel MM, Benson H. Psychological improvement in infertile women after behavioral treatment: a replication. Fertil Steril 1992;58:144–7.

[22] Haimovici F, Hill JA. The role of psycho-neuro-endocrine-immunology in reproduction. In: Hill JA, editor. Cytokines in Reproduction. Austin, TX: Landes Bioscience; 1998.

[23] Miller AH. Neuroendocrine and immune system interactions in stress and depression. Psychiatr Clin North Am 1998;21:443–63.

[24] Chrousos GP, Torpy DJ, Gold PW. Interactions between the hypothalamic-pituitary-axis and the female reproductive system: clinical implications. Ann Intern Med 1998;129:229–40.

[25] Domar AD, Rooney KL, Wiegand B, Orav EG, Alper MM, Berger BM, et al. Impact of a group mind/body intervention on pregnancy rates in IVF patients. Fertil Steril 2011;95:2269–73.

[26] Makrigiannakis A, Zoumakis E, Kalantaridou S, Coutifaris C, Margioris AN, Coukos G, et al. Corticotropin-releasing hormone promotes blastocyst implantation and early maternal tolerance. Nat Immunol 2001;2:1018–24.

[27] Ramezanzadeh F, Noorbala AA, Abedinia N, Rahimi FA, Naghizadeh MM. Psychiatric intervention improved pregnancy rates in infertile couples. Malays J Med Sci 2011;18:16–24.

[28] Demyttenaere K, Nijs P, Evers-Kiebooms G, Konnickx PR. Coping and ineffectiveness of coping influence the outcome of *in vitro* fertilization through stress responses. Psychoneuroendocrinology 1992;17:55–65.

[29] Jordan C, Revenson TA. Gender differences in coping with infertility: a meta-analysis. J Behav Med 1999;22(4):341–58.

[30] Pease A, Pease B. Why Men Don't Listen and Women Can't Read Maps. New York: Welcome Rain Publishers; 2000.

[31] Mai FM. Conception after adoption: an open question. Psychosom Med 1971;33:509–14.

[32] Rock J, Tietze C, McLaughlin HB. Effect of adoption on infertility. Fertil Steril 1965;16:305–12.

[33] Weir WC, Weir DR. Adoption and subsequent conceptions. Fertil Steril 1966;17:283–8.

[34] Kiecolt-Glaser J, McGuire L, Robles TF, Glaser R. Psychoneuroimmunology: psychological influences on immune function and health. J Consult Clin Psychol 2002;70:537–47.

[35] Herbert TB, Cohen S. Stress and immunity in humans: a meta-analytic review. Psychosom Med 1993;55:364–79.

[36] Esteling BA, Kiecolt-Glaser JK, Bodnar, Glaser R. Chronic stress, social support, and persistent alterations in the natural killer cell response to cytokines in older adults. Health Psychol 1994;13:291–9.

[37] Glaser R, Kiecolt-Glaser JK, editors. Handbook of Human Stress and Immunity. San Diego, CA: Academic Press; 1994.

[38] Segerstrom SC. Personality and the immune system: models, methods, and mechanisms. Ann Behav Med 2000;22:180–90.

[39] Kiecolt-Glaser J, McGuire L, Robles T, Glaser R. Emotions, morbidity and mortality: new perspectives from psychoneuroimmunology. In: Fiske ST,Schacter DL, Zahn-Waxler C, editors. Annual Review of Psychology: Annual Reviews. 2002. p. 83–107.

[40] Ader R, Felton D, Cohen N. Psychoneuroimmunology. 3rd ed San Diego: Academic Press; 2001.

[41] Facchinetti F, Matteo ML, Artini PG, Volpe A, Genazzani AR. An increased vulnerability to stress is associated with a poor outcome of *in vitro* fertilization–embryo transfer treatment. Fertil Steril 1997;67:309–14.

[42] Macciò A, Mantovani G, Turnu E, Artini P, Contu G, Serri FG, et al. Evidence that granulosa cells inhibit interleukin-1a and interleukin-2 production from follicular lymphomonocytes. J Assist Reprod Genet 1993;10:517–22.

[43] Fukuoka M, Yasuda K, Emi N, Fujiwara H, Iwai M, Takakura K, et al. Cytokine modulation of progesterone and estradiol secretion in cultures of luteinized human granulosa cells. J Clin Endocrinol Metab 1992;75:254–8.

[44] Braun DP, Dmowski WP. Endometriosis: abnormal endometrium and dysfunctional immune response. Curr Opin Obstet Gynecol 1998;10:365–9.

[45] Kupker W, Schultze-Mosgau A, Diedrich K. Paracrine changes in the peritoneal environment of women with endometriosis. Hum Reprod Update 1998;4:719–23.

[46] Ho HN, Wu MY, Yang YS. Peritoneal cell immunity and endometriosis. Am J Reprod Immunol 1997;38:400–12.

[47] Lebovic DI, Mueller M, Taylor RN. Immunobiology of endometriosis. Fertil Steril 2001;75:1–10.

[48] Wolff E, Rodrigues A, Wolff M. Stress – focused small group behavioral interventions and women's health. Presented at the XXX Congress of the European Association for Behavioral and Cognitive Therapies, Granada, Spain, September 26–28, 2000.

[49] Carmody J, Baer RA. Relationships between mindfulness practice and levels of mindfulness, medical and psychological symptoms and well-being in a mindfulness-based stress reduction program. J Behav Med 2008;31:23–33.

[50] Miller G, Cohen S. Psychological interventions and the immune system: a meta-analytic review and critique. Health Psychol 2001;20:47–63.

[51] Gallinelli A, Roncaglia R, Matteo ML, Ciaccio I, Volpe A, Facchinetti F. Immunological changes and stress are associated with different implantation rates in patients undergoing *in vitro* fertilization–embryo transfer. Fertil Steril 2001;76:85–91.

[52] Chrousos GP, Torpy DJ, Gold PW. Interactions between the hypothalamic-pituitary-adrenal axis and the female reproductive system: clinical implications. Ann Intern Med 1998;129:229–40.

[53] Rodrigues A, Wolff E, Wolff M. Changing health profiles in women's demographic and behavioral variables. Presented at the XXX Congress of the European Association for Behavioral and Cognitive Therapies, Granada, Spain, September 26–28, 2000.

[54] Wolff E, Rodrigues A, Wolff M. Stress, health and women: From Type-A to time urgency perfectionism. Presented at the XXX Congress of the European Association for Behavioral and Cognitive Therapies, Granada, Spain, September 26–28, 2000.

[55] Raikkonen K, Keltikangas-Jarvinen L, Adlercretz H, Hautanen A. Psychosocial stress and the insulin resistance syndrome. Metabolism 1996;45:1533–8.

[56] Rodriguez B, Bermudez L, Ponce de Leon E, Castro L. The relationship between infertility and anxiety: some preliminary findings. Presented at the Third World Congress of Behavior Therapy, Washington DC, December 8–11, 1983.

[57] Shellenbarger S. When stress is good for you. Wall Street Journal 2012;. Retrieved on June 5, 2014 from http://online.wsj.com/news/articles/SB10001424052970204301404577171192704005250.

[58] Mendes W, Gray H, Mendoza-Denton R, Major B, Epel E. Why egalitarianism might be good for your health: physiological thriving during stressful intergroup encounters. Psychol Sci 2007;18:991–8.

[59] Hofmann SG. An Introduction to Modern CBT: Psychological Solutions to Mental Health Problems. Oxford, UK: Wiley-Blackwell; 2011.

28

The Role of Body Mass Index on Assisted Reproductive Treatment Outcome

Kathryn Gebhardt, PhD,**, Deirdre Zander-Fox, PhD*,***

*Repromed, Adelaide, South Australia, Australia
**University of Adelaide, Adelaide, South Australia, Australia

INTRODUCTION: OBESITY AND FERTILITY

Obesity is an increasingly prevalent burden in Western society and is characterized by excessive lipid storage, which in many cases has a negative impact on health and can lead to a reduced life expectancy. Obesity is associated with various diseases, particularly cardiovascular diseases, diabetes mellitus type 2, obstructive sleep apnea, certain types of cancer, osteoarthritis, and asthma [1,2]. Maternal and paternal obesity compromises reproductive fitness. Currently, nearly half of the women of reproductive age are overweight or obese [3]. Previous research has identified numerous perturbed reproductive functions in overweight and obese adults of reproductive age, and is now beginning to disseminate the impact of parental obesity on subsequent offspring.

MATERNAL OBESITY AND FERTILITY

Female obesity is associated with menstrual irregularity, chronic oligoan ovulation, hirsutism, and infertility [4]. Interestingly, fertility in obese women has been shown to improve with weight loss, particularly anovulation [5,6]. Of interest is the impact of obesity on both embryo quality and pregnancy outcomes following *in vitro* fertilization (IVF) and other assisted reproductive technologies, which are being increasingly utilized by couples and individuals of reproductive age. Female reproductive function can be affected by obesity at various levels, including the control of ovulation, oocyte development, embryo development, endometrial development, embryo implantation, and fetal development [7]. In fact, obese women are three times more likely to suffer

infertility than women with a normal body mass index (BMI) [6] and have impaired fecundity both in natural and assisted conception cycles [8,9]. Obese women have been found to have a higher prevalence of anovulatory subfertility, and may take longer to conceive even if cycling regularly [10,11].

Maternal Obesity and Assisted Reproductive Treatment

The impact of obesity on assisted reproductive treatment (ART) outcome remains controversial as markers of ART success have varied in previous studies, but overall may include follicular development, number of oocytes retrieved, gonadotropin requirement, oocyte quality, embryo quality, endometrial alteration, implantation rates, ongoing pregnancy rates, miscarriage rates, and live birth rates [3]. The majority of studies conducted have been unable to identify the mechanism by which obesity negatively impacts on ART outcome, with suggestions of either ovarian or endometrial factors, or a combination of the two [3]. A systematic review on the effect of obesity on assisted reproduction treatment identified that women with a BMI >25 kg/m^2 have an overall lower chance of pregnancy following IVF, routinely require higher doses of gonadotropins, and even when pregnancy is achieved have increased rates of miscarriage [12]. In a further meta-analysis, the number of cancelled cycles per woman was found to be significantly higher in obese women, positive serum human chorionic gonadotropin (hCG) was decreased with increased BMI, and ongoing pregnancy was lower among obese versus normal-weight women [13]. The same study found that live-birth rate also appeared to be lower in obese women.

Handbook of Fertility. http://dx.doi.org/10.1016/B978-0-12-800872-0.00028-7

Maternal Obesity and the Ovary

While female obesity results in systemic changes to multiple organ systems, in regards to fertility, obesity may alter the nutrient and hormone status of the ovarian and uterine environments critical for oocyte and embryo development, resulting in detrimental changes to follicular metabolism [14]. *In vitro* animal studies have provided extensive evidence on the negative effects of an excess of certain metabolites, such as glucose [15,16], and a number of lipids [17–20] and hormones [21] on subsequent oocyte and embryo development. Negative environmental exposures may affect the developmental competence of the oocyte, resulting in poor oocyte and embryo quality. This is because the oocyte and ovarian environment is highly sensitive to systemic changes in both hormone and nutrient milieu. These changes are exacerbated in obesity in which multiple organ systems are affected due to large increases in the amount of adipose tissue and the stress this change places on the body.

The adipose tissue is an important site for steroid hormone production and metabolism, and its excess in obesity can alter the concentrations of steroid hormones [22]. Due to this, obese women frequently require greater amounts of gonadotropins for IVF, which can lead to perturbations in oocyte maturation and competence, and impact embryo development and subsequent implantation and pregnancy [23,24]. Obesity also leads to increased serum insulin concentrations and resistance to insulin action in cells [22]. Excess insulin can decrease sex hormone binding globulin (SHBG), which elevates testosterone, dihydrotestosterone, and androstenediol [7]. As peripheral insulin sensitivity is decreased, insulin production is increased and leads to further decreases in SHBG and increases in androgens [7]. In the theca and granulosa cells of the ovary, insulin stimulates steroidogenesis and upregulates luteinizing hormone (LH) receptor expression [25]. As a result of LH hypersecretion and an altered LH:FSH (follicle stimulating hormone) ratio, ovulation and the resumption of meiosis is often perturbed in obese women [7,26]. Obesity and insulin resistance can also result in type 2 diabetes, which is characterized by hyperglycemia. Many studies using mouse models of type 1 diabetes have shown that oocytes exhibit delays in maturation [27–29] accompanied by meiotic spindle defects and chromosome misalignment during oocyte maturation [27]. Human studies showed an impairment in oocyte maturation with increasing BMI [30,31]; however, there is much contradiction in the literature as others have found no difference in oocyte maturation between different BMI categories [32].

Similar contradictory results surround the impact of maternal obesity on embryo quality. Embryo quality is an integral factor in predicting the success of implantation and subsequent pregnancy [33]. Previous studies have investigated the correlation between embryo grade and pregnancy success; however, the use of embryo grading as a marker of viability is limited due to the various classification systems utilized and the highly subjective nature of embryo assessment [34]. As such, many studies have found no effect for maternal obesity on embryo grade [35–38], while others have found that mean embryo grade was significantly worse in the obese women compared with overweight and normal BMI women [32]. The evidence for this was further supported when looking at additional markers of embryo quality, namely, embryo utilization rate, number of embryos cryopreserved, and the number of embryos discarded, which were all found to be significantly worse in young obese women [32].

Maternal Obesity and Embryo Implantation

A number of studies have attempted to define the effect of female obesity on the endometrium. One of the best human models for discriminating between the effects of obesity on the oocyte/embryo and the endometrium and uterine receptivity is that of oocyte donation [7,39]. It is generally accepted that donor oocytes are obtained from young, healthy, normal BMI donors. In a retrospective study of 2656 oocyte donation cycles with transfer of good quality embryos, the effects of obesity in the recipient was observed. Recipients were appropriately allocated into BMI groups and statistical analysis demonstrated a negative trend in pregnancy rates with increasing BMI [40]. Interestingly, implantation rates were similar in all BMI groups, suggesting an increased pregnancy loss rate in obese women consistent with relevant literature. In a case-control study in women undergoing IVF with their own gametes, a reduction in implantation and pregnancy rates, with an increase in the rates of miscarriage, was observed in women with a BMI >25 kg/m^2 [41]. These studies, along with others, have identified a clear link between maternal obesity and decreased embryo implantation potential. However, the literature still remains controversial as to whether this reduction in embryo implantation seen in obese women is a result of oocyte/embryo contribution or the endometrial environment in which the embryo is transferred.

The histopathological and molecular effects of obesity on the endometrium may be a key contributor to the reductions in fertility seen in these women. Endometrial receptivity may be affected in obese women, leading to impaired implantation and miscarriage due to a hostile intrauterine milieu [40]. One such hypothesis centers on the balance of endometrial cytokines and inflammatory molecules. Leukemia inhibitory factor (LIF) has been implicated in the regulation of implantation [42]. In women with unexplained recurrent pregnancy loss, endometrial glandular LIF was examined in relation to

BMI. A significant negative correlation between glandular LIF and female BMI was observed although there were no differences in endometrial morphology or steroid receptor concentration [42,43]. Further to this, the state of relative hyperestrogenemia seen in obese women may have a negative effect on endometrial receptivity [23]. As previously discussed, visceral obesity alters insulin resistance, but also effects inflammatory mediators, coagulation, and fibrinolysis [7]. Elevated insulin levels are associated with a reduction in glycodelin (a major lipocalin of the reproductive axis with diverse actions in cell recognition and differentiation [44]) and a reduction in insulin-like growth factor binding protein 1 (IGFBP1). Low levels of glycodelin have been associated with recurrent pregnancy loss. IGFBP1 facilitates adhesion at the maternal–fetal interface; therefore, perturbations in these molecules may contribute to impaired fertility at an endometrial level [45,46]. Obese women also have elevated levels of acute phase proteins and pro-inflammatory cytokines (including IL-6, PAI1, and TNF), which are thought to exert a negative effect on implantation and early embryonic development [46–48].

Further evidence generated to suggest that maternal obesity impairs endometrial receptivity at the molecular level has demonstrated a different pattern of endometrial gene expression in obese women compared to normal-weight controls [49]. These changes were more pronounced when infertility was associated. As a result, the gene expression changes were seen to be higher in obese women with infertility when compared to the obese fertile group. Notably, the endometrial gene expression patterns identified were related to embryo implantation. This may exacerbate the epigenetic modifications to the embryo genome induced by pregravid or gravid female obesity, and has been suggested as a potential mechanism to explain some of the reproductive consequences seen in obese women, as well as the risk of chronic disease in their offspring [50].

Obesity and Pregnancy Outcomes

Obese women who do become pregnant are subject to a number of adverse maternal and fetal outcomes [51], including an increased risk of miscarriage, preeclampsia, gestational diabetes, and congenital defects in the offspring [52]. The impact of maternal obesity on pregnancy was especially distinct in one study examining embryo transfer cycles using donor oocytes [53]. The rate of spontaneous miscarriage was found to increase as recipient BMI increased; the risk of miscarriage was 38.1% for obese women and 15.5% for overweight women compared to 13.3% in women with a normal BMI. Several other studies have shown similar findings [54,55]. Prepregnancy BMI is a strong predictor of adverse obstetric and neonatal outcomes [56]. Obesity is strongly associated

with complications during pregnancy and increased rates of obstetric interventions [57]. In addition, maternal and neonatal morbidity and mortality are increased, and there are more neonatal hospital admissions [58,59].

Obese women have a significantly higher risk of obstetric complications including preeclampsia, gestational diabetes, prolonged duration of labor, unsuccessful induction of labor and resultant cesarean delivery, macrosomia, increased blood loss, and unexplained stillbirth [30,57,60–62]. In fact, the risk of preeclampsia has been shown to double with every 5–7 kg/m^2 increase in prepregnancy BMI [63]. Research also suggests that rapid fetal growth induced by hyperinsulinemia may be coupled with placental insufficiency, which may in turn result in intrauterine fetal demise in obese pregnant women [60,64]. In a study of approximately 350 obese and morbidly obese women who achieved pregnancy following ART, obstetric complications, such as preeclampsia, gestational diabetes, and cesarean delivery, were found to be significantly higher [30]. This increased risk is most noticeable in women with BMIs >35 kg/m^2 [57, 61].

Maternal Obesity and Offspring Health

It is not only conception that is affected by maternal obesity. During pregnancy, excess adipose tissue can lead to impaired fetal development and long-term health outcomes. The developing fetus relies upon a complex maternal–fetal interaction, which impacts on long-term neonatal, childhood, and adult health of the offspring. However, the exact mechanisms by which obesity mediates poor health outcomes for the mother and fetus are unclear [65]. Children of obese mothers are more at risk of overweight or obesity than children of obese fathers [66]. This demonstrates the pivotal role of intrauterine conditions on susceptibility to obesity, type 2 diabetes, and cardiovascular disease in the offspring [65]. Obesity is associated with chronic inflammation [67,68] and changes in the homeostasis of cytokines and adipokines. Adipose tissue is a source of pro-inflammatory cytokines (adipokines), such as adiponectin, which promotes insulin sensitivity and stimulates glucose uptake in skeletal muscles. In pregnancy, adiponectin influences the placental transfer of nutrients and placental insulin-stimulated amino acid uptake [65]. Obese pregnant women have low circulating levels of adiponectin, which is associated with increased fetal growth [69]. Maternal obesity may also lead to changes in placental development or functions, and is associated with a reduction in placental villous proliferation and apoptosis, which may increase susceptibility to adverse pregnancy outcomes [70]. Epidemiological studies have identified a correlation between maternal metabolic syndrome and placental dysfunction [71].

Due to difficulties in ultrasound imaging in obese women, the antenatal detection rate of congenital anomalies is lower in obese women, resulting in more affected liveborns and stillborns among obese mothers. Additionally, some congenital anomalies are more common in the offspring of obese women [65]. Due to this, antenatal care for obese women may require modification to better identify congenital anomalies during early pregnancy.

Maternal weight, weight gain during pregnancy, and neonatal birthweight are strongly correlated [72–75]. Obese women have an 18–26% increased chance of delivering large-for-gestational-age infants [57]. Excessive fetal growth carries significant perinatal morbidities, and macrosomic infants of obese mothers have a greater likelihood of operative delivery, shoulder dystocia, and brachial plexus injury [76]. Maternal obesity is associated with an overall increased rate of neonatal death, primarily related to pregnancy complications or preterm births [58]. Maternal overweight and obesity remain the most important potentially modifiable risk factors for stillbirth, which could be addressed during pre-conception care in more developed societies [65].

Maternal obesity has a life-long impact on fetal, childhood, and adult growth. Later in life, there is an inverse relationship between birth weight and health status [65]. Maternal prepregnancy overweight is an independent risk factor for infant and adolescent overweight and abdominal obesity [77] as excessive maternal body weight or weight gain during pregnancy disrupts the intrauterine environment and causes permanent changes in weight-regulating mechanisms in the offspring. Some studies also suggest that children from pregnancies complicated by metabolic conditions may be at an increased risk of neurodevelopmental disorders, autism spectrum disorders, and developmental delay [78].

PATERNAL OBESITY AND FERTILITY

Although the impact of maternal BMI on fertility is well documented in the literature, the impact of an increased paternal BMI has been less investigated and is not as well understood. Until only recently, the impact of male obesity has been primarily limited to semen parameter assessments, and it has only been in the last few years that the effect of increased paternal BMI on infertility and ART outcome has been investigated [79]. Interestingly, what is becoming more apparent is that paternal factors such as age, BMI, and lifestyle factors, such as smoking, drug usage, and toxin/occupational exposure, can have a profound effect on pregnancy establishment and offspring health. This has led to a renewed focus on the paternal contribution to infertility and ART outcome as it appears that the paternal contribution may be equally responsible for influencing ART success [80–82].

Paternal Obesity, Hormone Levels and Seminal Plasma

Spermatogenesis is a highly controlled process that relies heavily on gonadotropin-dependent mechanisms. Sex steroids, such as testosterone, FSH, LH, and estrogen, all play an intrinsic role in the regulation of spermatogenesis in conjunction with the hypothalamic–pituitary axis [83]. Several studies have demonstrated that altered hormone profiles in obese men and male obesity are associated with decreased levels of SHBG and testosterone coupled with an increase in plasma estrogen [84–86]. A recent meta-analysis has concluded that a negative relationship exists between BMI, testosterone, and SHBG [87]. Decreased levels of testosterone and increased levels of estrogen are associated with subfertility in men due to interference with the hypothalamic pituitary axis and disruption to the negative feedback loop [88]. Sertoli cell interaction with the developing germ cell is reliant on testosterone, and therefore decreased levels of testosterone can result in retention and phagocytosis of mature spermatids resulting in a reduced sperm count [89].

It is believed that excess adiposity in the obese male increases aromatization of androgens in adipose tissue leading to increased levels of circulating estrogen [90]. In addition, it has also been demonstrated that obese males have elevated circulating insulin [91,92]. Hyperinsulinemia can also reduce SHBG production in the liver and therefore decreases testosterone production due to decreased SHBG binding capacity.

Obese men have increased levels of insulin and leptin in their seminal plasma. [92] Insulin and leptin are hormones that are synthesized and secreted by sperm after ejaculation and are also synthesized by Sertoli cells. Studies have demonstrated that acute exposure of sperm in seminal plasma to exogenous insulin and leptin can increase sperm motility, acrosome reaction, and nitric oxide production; however, in obese men, increased insulin and leptin in seminal plasma is associated with decreased concentration and motility and it is hypothesized that in obesity, exposure to such chronic high levels of insulin may result in insulin resistance within the sperm themselves [93]. Therefore, the pathways that would normally be activated by acute exposure to insulin (PI3K/Akt), which mediate beneficial effects on motility, are suppressed due to insulin resistance thus decreasing sperm function [93]. This fact is further supported by the data obtained from men with hyperinsulinemia and hyperglycemia, as decreased sperm count and increased levels of DNA damage that are seen in men with diabetes are also seen in obese males [94,95].

Paternal Obesity and Semen Parameters

There are several studies that have investigated the impact of male obesity on traditional sperm parameters such as sperm motility, sperm concentration, and morphology utilizing guidelines mandated by the World Health Organization (WHO); however, a consensus has yet to be reached. It has been demonstrated that the majority of studies associate increasing BMI with decreased sperm concentration (15 out of 23 recent studies [96] and a recent meta-analysis has concluded that abnormal sperm count has a j-shaped association with BMI). Compared to men with a normal BMI, underweight men have an odds ratio (OR) of 1.15 (0.93–1.43) for oligizoospermia or azoospermia (as defined by <40 million/ejaculate), and as BMI increases to ≥ 25 kg/m^2, the likelihood of an abnormal sperm count increases: overweight men 1.11 (1.01–1.21), obese men 1.28 (1.06–1.55), and morbidly obese men 2.04 (1.59–2.62) [97]. A similar association was also seen with abnormal sperm concentrations of <15 million/ejaculate with ORs increasing in men who were overweight 1.06 (0.96–1.18), obese 1.31 (1.08–1.60), and morbidly obese 1.97 (1.27–3.07). In addition to overall BMI, it has also been shown that central adiposity can also influence semen parameters with an increased waist circumference (≥ 102 cm) being associated with decreased sperm concentration and total motile sperm count [98].

Sperm motility has also been assessed with contradicting findings as although some studies have shown decreased motility to be associated with increasing BMI [84,99,100], others have failed to identify any association [85,92,101,102]. In addition, a recent review concluded that only 7 out of 19 studies have demonstrated an association between BMI and sperm motility; therefore, it is unclear if this parameter is truly affected by obesity [96]. Sperm vitality, as measured by eosin-nigrosin staining, has also recently been assessed in relation to obesity, and it was discovered that obese men have a significantly decreased vitality (45.0 ± 26.1) compared to men with a normal BMI (62.6 ± 18.1) although progressive motility and total motility were not different [92]. A similar result has also been demonstrated in regard to morphology, with only a minimal amount of studies demonstrating an association between BMI and changes in sperm morphology further compounding whether obesity impacts on sperm morphological formation [96]. The difficulty with the interpretation and comparison of most of these studies is likely due to confounding factors that are unable to be controlled for. Lifestyle factors, such as smoking, alcohol consumption, heat and toxin exposure, as well as illness, can impact sperm function, are often not reported, or rely on self-reporting, which can lead to bias and may confound study outcomes.

Although conventional sperm morphological parameters are important markers of male fertility, the molecular and genetic structure within sperm also contributes to embryo development and more importantly pregnancy establishment and offspring health. Sperm DNA damage has been correlated to pregnancy outcome and is a diagnostic marker used clinically to determine the integrity of sperm DNA. Studies have demonstrated that elevated levels of DNA damage in sperm is associated with decreased fertilization rates [103–105], decreased embryo cleavage rates [103,106], and increased rates of failed implantation and pregnancy loss [105,107]. In regard to obesity, studies have correlated increased levels of DNA damage to elevated paternal BMI. Utilizing a sperm chromatin structure assay (SCSA), it was reported that obese men had a DNA fragmentation index of 27.0% ± 3.2, which was significantly higher than men with a normal BMI (19.9% ± 2.0) [108]. This was supported by a more recent study using terminal uridine nick-end labeling (TUNEL) assay that demonstrated that compared to men with a normal BMI, obese men have an increased DNA fragmentation rate with an OR of 2.5 (1.2–5.1) [109]. Other studies have also demonstrated correlations between paternal BMI and DNA integrity as well as alterations to metabolic markers such as mitochondrial membrane potential (MMP) and oxidative stress. La Vignera et al. demonstrated that obese men have sperm with significantly decreased MMP, increased levels of sperm with decondensed chromatin, increased rates of phosphatidylserine (PS) externalization (an early sign of apoptosis), and increased levels of fragmented DNA compared to men with a normal BMI [110]. Sperm mitochondrial activity has also been investigated in obese males, and in conjunction with increased levels of sperm DNA fragmentation, obese males also have sperm with decreased mitochondrial activity and increased levels of sperm with zero to low mitochondrial activity [100].

As well as altered mitochondrial parameters, a correlation has also been made between male obesity and oxidative stress. Reactive oxygen species (ROS) play an integral role in modulating essential sperm cellular functions including, capacitation, hyperactivation, and acrosomal reaction, as well as fertilization and subsequent embryonic cell division [100]. Although low levels of ROS are vital for fertility, imbalances in ROS production and antioxidant defense mechanism can result in increases in the level of ROS, which can significantly damage sperm. Studies have demonstrated that increased sperm ROS levels are associated with increased lipid peroxidation and decreased MMP as well as single and double strand DNA breaks [111–114]. In addition, there is an association between decreased sperm motility and increased ROS [115]. Due to the fact that sperm motility is reliant on ATP generation via mitochondria and that MMP is required to maintain ATP production, excessive ROS production may disrupt MMP and therefore inhibit efficient oxidative phosphorylation thus reducing ATP production and therefore sperm motility [116]. Interestingly, it has been demonstrated that obese men have a significantly increased level of seminal oxidative stress

TABLE 28.1   Summary of Studies Investigating the Impact of Paternal BMI on IVF Outcomes

	Fertilization	Embryo quality (day 3)	Blastocyst quality (day 5)	Clinical pregnancy	Implantation rate	Live birth rate
Keltz* et al. [117]	Not assessed	↔	Not assessed	↓	↓	Not assessed
Bakos et al. [79]**	↔	↔	↓	↓	↓	↓
Colaci et al. [118]	↔ (↑ for IVF)	↔	↓	↔ (↓ for ICSI)	Not assessed	↔ (↓ for ICSI)
Merhi et al. [119]*	Not assessed	↔	Not assessed	↓	↓	Not assessed
Anifandis et al. [120]*,**	Not assessed	↓	Not assessed	↓	↓	↓
Petersen et al. [121]**	Not assessed	Not assessed	Not assessed	Not assessed	Not assessed	↓

*Studies that compared a normal BMI (<25 kg/m^2) with overweight/obese BMI (>25 kg/m^2); all other studies utilized expanded BMI groups.
**Studies that assessed impact of both paternal and maternal BMI.

compared to men with a normal BMI, which may be a contributing factor to the sperm pathologies (such as decreased motility and MMP) seen in obese males and thus requires further investigation [84].

Paternal Obesity, Embryo Development, and IVF Outcome

Although the impact of paternal obesity on sperm function has been assessed in detail, the impact of paternal obesity on other aspects of *in vitro* technology (such as embryo quality and subsequent pregnancy and birth outcomes) has received limited attention with only a small number of studies undertaken (Table 28.1). In 2008, Keltz et al. demonstrated that male body mass index of >25 kg/m^2 (categorized as overweight according to WHO criteria) was associated with a significant reduction in clinical pregnancy rate after IVF compared to men with a normal BMI (33.6% vs. 53.2%, respectively). In addition increased paternal BMI was also associated with an increased average number of embryos transferred and therefore decreased implantation rate. Interestingly, multivariate logistic regression identified that this was only seen in the cohort that underwent IVF insemination and not in the ICSI (intracytoplasmic sperm injection) group. It was concluded that ICSI is able to overcome some of the aspects of obesity-related sperm dysfunction [117]. Following on from this, it has also been reported that increased BMI is also associated with decreased fetal heart rates as well as decreased live birth after ART. This study investigated fertilization rates, embryo development, and subsequent pregnancy and determined that although there was no impact of paternal BMI on cleavage stage (day 3) embryo development, there was a linear association with increasing paternal BMI and decreased blastocyst development rates. This is believed to be due to the fact that embryonic genome activation occurs between the 4–8 cell stage (prior to this the embryo is utilizing maternal transcripts only); therefore, the paternal contribution is primarily evident from day 3 onward. In addition, there was also a linear reduction in fetal heart rates with increasing paternal BMI (normal BMI 47.4%, overweight 33.9%, and obese 28.1%) and this was further decreased in men who were morbidly obese (16.7% fetal heart). Interestingly, it was discovered that this further reduction as seen in men with a BMI of ≥35 kg/m^2 was due to a combined effect of both paternal and maternal increased BMI. It was also noted that increased paternal BMI was associated with increased rates of pregnancy loss and decreased live birth rates [79]. This study was the first that determined the effect of paternal obesity on all aspects of ART including sperm parameters as well as fertilization, embryo development, and pregnancy/birth outcomes. Similar findings have also been reported showing decreased blastocyst development with increasing paternal BMI and decreased clinical pregnancy rates and live birth outcome where the father was obese [118,119].

Although emerging evidence indicates that paternal obesity in isolation can significantly impact embryo development and pregnancy establishment, it is also important to note that maternal obesity (as previously discussed) can also play a significant role in fertility and that in an IVF setting an overweight/obese women often presents for treatment with an overweight/obese male partner [79]. Therefore, two recent studies have investigated the impact of obesity in couples where overweight/obesity is present in one partner only or in combination (both male and female have a BMI ≥25 kg/m^2). In a study by Anifandis et al. the impact of paternal and maternal BMI was assessed using four groups: group 1 (both male and female partners had a normal BMI), group 2 (both male and female were overweight/obese), group 3 (the male was overweight/obese and female was in normal BMI range), and group 4 (female was overweight/obese and male had a BMI in the normal range) [120]. Interestingly, this study demonstrated that paternal BMI plays a more significant role in embryo grade and pregnancy establishment than female BMI. It was reported that group 2 and group 3 (where male BMI was ≥25 kg/m^2 with or without maternal BMI of $\geq 25\ kg/m^2$) had decreased embryo quality on day 2/day 3 of development. In addition, clinical pregnancy rates were also significantly decreased in group 3 (18.5%) compared to 26.7% in group 1. It has also been reported that increased maternal and paternal BMI ($\geq 25\ kg/m^2$) in combination significantly decreased the likelihood of a live birth following IVF (OR: 0.73, 95% CI) and that, similar to the study by Anifandis et al., when paternal BMI was $\geq 25\ kg/m^2$ and maternal BMI was in the normal range, the likelihood of pregnancy was similarly decreased (OR: 0.78 95% CI – although this subanalysis did not reach significance). A noteworthy limitation of this study was the missing information on paternal BMI and therefore subanalysis of the influence of paternal BMI only was limited; hence, a possible reason that paternal BMI in isolation did not reach significance [121]. Indeed this has been reported as a weakness in many studies on female obesity and ART outcome that paternal obesity has not be used as a possible confounding factor, as information on paternal BMI is often not routinely collected by ART clinics [79]. Due to the fact that often an overweight/obese female presents for treatment with an overweight/obese male, it is possible that a large proportion of data demonstrating the impact of female obesity on IVF outcome may in fact be at least in part due to the presence of an overweight/obese male.

Paternal Obesity and Offspring Health

It is very well understood that the environment surrounding the developing fetus can program offspring phenotype and susceptibility to disease later in life [122,123]. In addition, the environment surrounding the developing embryo (both in animal models and the human) can also program offspring developmental trajectory and can influence fetal weight and developmental milestones [124–126]. However, what is less understood is the role that paternal programming plays in determining offspring health. Studies have shown that some aspects of paternal physiology can program offspring health. It has been demonstrated that increased paternal age is linked to increased risk of autism spectrum disorder (ASD) [127], paternal smoking can increase the risk of childhood cancer, such as leukemia, and that paternal exposure to toxins, such as diesel engine exhaust, around the time of conception can increase the risk of childhood brain tumors [80,128]. Additionally, epidemiological studies demonstrate that obese fathers are more likely to father an obese child and that paternal obesity is linked to an increased risk of ASD [129,130]. Although the direct mechanism for this programming cannot be determined in the human and may be genetic, epigenetic, or environmental, animal models have demonstrated that paternal obesity (in the mouse) compromises offspring metabolic health with alterations to offspring weight, growth trajectories, and reproduction. In addition, female offspring

have altered pancreatic and islet cell function and that paternal obesity can also compromise second-generation health [131–133]. Due to the elimination of genetic and environmental factors through the use of animal models, this research suggests that paternal obesity at the time of conception directly influences offspring programming and therefore implicates sperm as the mediator. It is hypothesized that molecular or epigenetic changes occur due to the presence of paternal pathologies, such as obesity or exposure to environmental toxins, which then transmit into the embryo and thus offspring. Currently, the mechanism for transmission of programming is thought to be through alterations to the epigenetic code such as changes to acetylation or methylation or via the alterations to the noncoding microRNA (miRNA) content of sperm; however, this novel area of research is still emerging and has yet to be well understood [96].

CONCLUSIONS

The evidence suggests that maternal and paternal obesity has significant consequences for individuals of reproductive age, especially those considering pregnancy via spontaneous and assisted conception. From the period of preconception, through gestation and birth, extending to the long-term health of the offspring, the impact of obesity is compelling. Obesity impacts on all phases of reproductive health, yet remains one of the key modifiable complications. Weight loss improves reproductive outcomes and reduces the risk of pregnancy complication and neonatal morbidity and mortality, and therefore patients undergoing ART should be counseled as to the impact that increased BMI can have on not only IVF success but also on the future health of the child.

References

[1] Clark AM, Thornley B, Tomlinson L, Galletley C, Norman RJ. Weight loss in obese infertile women results in improvement in reproductive outcome for all forms of fertility treatment. Hum Reprod 1998;13(6):1502–5.

[2] Haslam DW, James WP. Obesity. Lancet 2005;366(9492):1197–209.

[3] Kumbak B, Oral E, Bukulmez O. Female obesity and assisted reproductive technologies. Semin Reprod Med 2012;30(6):507–16.

[4] Norman RJ, Noakes M, Wu R, Davies MJ, Moran L, Wang JX. Improving reproductive performance in overweight/obese women with effective weight management. Hum Reprod Update 2004;10(3):267–80.

[5] Clark AM, Ledger W, Galletly C, Tomlinson L, Blaney F, Wang X, Norman RJ. Weight loss results in significant improvement in pregnancy and ovulation rates in anovulatory obese women. Hum Reprod 1995;10(10):2705–12.

[6] Rich-Edwards JW, Goldman MB, Willett WC, Hunter DJ, Stampfer MJ, Colditz GA, Manson JE. Adolescent body mass index and infertility caused by ovulatory disorder. Am J Obstet Gynecol 1994;171(1):171–7.

[7] Brewer CJ, Balen AH. The adverse effects of obesity on conception and implantation. Reproduction 2010;140(3):347–64.

[8] Zaadstra BM, Seidell JC, Van Noord PA, te Velde ER, Habbema JD, Vrieswijk B, Karbaat J. Fat and female fecundity: prospective study of effect of body fat distribution on conception rates. BMJ 1993;306(6876):484–7.

[9] Crosignani PG, Ragni G, Parazzini F, Wyssling H, Lombroso G, Perotti L. Anthropometric indicators and response to gonadotrophin for ovulation induction. Hum Reprod 1994;9(3):420–3.

[10] Grodstein F, Goldman MB, Cramer DW. Body mass index and ovulatory infertility. Epidemiology 1994;5(2):247–50.

[11] Gesink Law DC, Maclehose RF, Longnecker MP. Obesity and time to pregnancy. Hum Reprod 2007;22(2):414–20.

[12] Maheshwari A, Stofberg L, Bhattacharya S. Effect of overweight and obesity on assisted reproductive technology--a systematic review. Hum Reprod Update 2007;13(5):433–44.

[13] Pinborg A, Gaarslev C, Hougaard CO, Nyboe Andersen A, Andersen PK, Boivin J, Schmidt L. Influence of female body-weight on IVF outcome: a longitudinal multicentre cohort study of 487 infertile couples. Reprod Biomed Online 2011;23(4):490–9.

[14] Robker RL, Akison LK, Bennett BD, Thrupp PN, Chura LR, Russell DL, Norman RJ. Obese women exhibit differences in ovarian metabolites, hormones, and gene expression compared with moderate-weight women. J Clin Endocrinol Metab 2009;94(5):1533–40.

[15] Moley KH, Chi MM, Knudson CM, Korsmeyer SJ, Mueckler MM. Hyperglycemia induces apoptosis in pre-implantation embryos through cell death effector pathways. Nat Med 1998;4(12):1421–4.

[16] Bermejo-Alvarez P, Roberts RM, Rosenfeld CS. Effect of glucose concentration during in vitro culture of mouse embryos on development to blastocyst, success of embryo transfer, and litter sex ratio. Mol Reprod Dev 2012;79(5):329–36.

[17] Nonogaki T, Noda Y, Goto Y, Kishi J, Mori T. Developmental blockage of mouse embryos caused by fatty acids. J Assist Reprod Genet 1994;11(9):482–8.

[18] Leroy JL, Van Hoeck V, Clemente M, Rizos D, Gutierrez-Adan A, Van Soom A, Bols PE. The effect of nutritionally induced hyperlipidaemia on in vitro bovine embryo quality. Hum Reprod 2010;25(3):768–78.

[19] Wonnacott KE, Kwong WY, Hughes J, Salter AM, Lea RG, Garnsworthy PC, Sinclair KD. Dietary omega-3 and -6 polyunsaturated fatty acids affect the composition and development of sheep granulosa cells, oocytes and embryos. Reproduction 2010;139(1):57–69.

[20] Van Hoeck V, Sturmey RG, Bermejo-Alvarez P, Rizos D, Gutierrez-Adan A, Leese HJ, Leroy JL. Elevated non-esterified fatty acid concentrations during bovine oocyte maturation compromise early embryo physiology. PLoS One 2011;6(8):e23183.

[21] Arias-Alvarez M, Bermejo-Alvarez P, Gutierrez-Adan A, Rizos D, Lorenzo PL, Lonergan P. Effect of leptin supplementation during in vitro oocyte maturation and embryo culture on bovine embryo development and gene expression patterns. Theriogenology 2011;75(5):887–96.

[22] Purcell SH, Moley KH. The impact of obesity on egg quality. J Assist Reprod Genet 2011;28(6):517–24.

[23] Tamer Erel C, Senturk LM. The impact of body mass index on assisted reproduction. Curr Opin Obstet Gynecol 2009;21(3):228–35.

[24] Farhi J, Ben-Haroush A, Sapir O, Fisch B, Ashkenazi J. High-quality embryos retain their implantation capability in overweight women. Reprod Biomed Online 2010;21(5):706–11.

[25] Poretsky L, Cataldo NA, Rosenwaks Z, Giudice LC. The insulin-related ovarian regulatory system in health and disease. Endocr Rev 1999;20(4):535–82.

[26] Jain A, Polotsky AJ, Rochester D, Berga SL, Loucks T, Zeitlian G, Santoro N. Pulsatile luteinizing hormone amplitude and progesterone metabolite excretion are reduced in obese women. J Clin Endocrinol Metab 2007;92(7):2468–73.

[27] Chang AS, Dale AN, Moley KH. Maternal diabetes adversely affects preovulatory oocyte maturation, development, and granulosa cell apoptosis. Endocrinology 2005;146(5):2445–53.

[28] Diamond MP, Moley KH, Pellicer A, Vaughn WK, DeCherney AH. Effects of streptozotocin- and alloxan-induced diabetes mellitus on mouse follicular and early embryo development. J Reprod Fertil 1989;86(1):1–10.

[29] Colton SA, Pieper GM, Downs SM. Altered meiotic regulation in oocytes from diabetic mice. Biol Reprod 2002;67(1):220–31.

[30] Dokras A, Baredziak L, Blaine J, Syrop C, Van Voorhis BJ, Sparks A. Obstetric outcomes after in vitro fertilization in obese and morbidly obese women. Obstet Gynecol 2006;108(1):61–9.

[31] Carrell DT, Jones KP, Peterson CM, Aoki V, Emery BR, Campbell BR. Body mass index is inversely related to intrafollicular HCG concentrations, embryo quality and IVF outcome. Reprod Biomed Online 2001;3(2):109–11.

[32] Metwally M, Cutting R, Tipton A, Skull J, Ledger WL, Li TC. Effect of increased body mass index on oocyte and embryo quality in IVF patients. Reprod Biomed Online 2007;15(5):532–8.

[33] Arce JC, Ziebe S, Lundin K, Janssens R, Helmgaard L, Sorensen P. Interobserver agreement and intraobserver reproducibility of embryo quality assessments. Hum Reprod 2006;21(8):2141–8.

[34] Baxter Bendus AE, Mayer JF, Shipley SK, Catherino WH. Interobserver and intraobserver variation in day 3 embryo grading. Fertil Steril 2006;86(6):1608–15.

[35] Fedorcsak P, Storeng R, Dale PO, Tanbo T, Abyholm T. Obesity is a risk factor for early pregnancy loss after IVF or ICSI. Acta Obstet Gynecol Scand 2000;79(1):43–8.

[36] Spandorfer SD, Kump L, Goldschlag D, Brodkin T, Davis OK, Rosenwaks Z. Obesity and in vitro fertilization: negative influences on outcome. J Reprod Med 2004;49(12):973–7.

[37] Dechaud H, Anahory T, Reyftmann L, Loup V, Hamamah S, Hedon B. Obesity does not adversely affect results in patients who are undergoing in vitro fertilization and embryo transfer. Eur J Obstet Gynecol Reprod Biol 2006;127(1):88–93.

[38] Zander-Fox DL, Henshaw R, Hamilton H, Lane M. Does obesity really matter? The impact of BMI on embryo quality and pregnancy outcomes after IVF in women aged </=38 years. Aust N Z J Obstet Gynaecol 2012;52(3):270–6.

[39] Bellver J, Rossal LP, Bosch E, Zuniga A, Corona JT, Melendez F, Pellicer A. Obesity and the risk of spontaneous abortion after oocyte donation. Fertil Steril 2003;79(5):1136–40.

[40] Bellver J, Melo MA, Bosch E, Serra V, Remohi J, Pellicer A. Obesity and poor reproductive outcome: the potential role of the endometrium. Fertil Steril 2007;88(2):446–51.

[41] Loveland JB, McClamrock HD, Malinow AM, Sharara FI. Increased body mass index has a deleterious effect on in vitro fertilization outcome. J Assist Reprod Genet 2001;18(7):382–6.

[42] Senturk LM, Arici A. Leukemia inhibitory factor in human reproduction. Am J Reprod Immunol 1998;39(2):144–51.

[43] Metwally M, Tuckerman EM, Laird SM, Ledger WL, Li TC. Impact of high body mass index on endometrial morphology and function in the peri-implantation period in women with recurrent miscarriage. Reprod Biomed Online 2007;14(3):328–34.

[44] Seppala M, Taylor RN, Koistinen H, Koistinen R, Milgrom E. Glycodelin: a major lipocalin protein of the reproductive axis with diverse actions in cell recognition and differentiation. Endocr Rev 2002;23(4):401–30.

[45] Carrington B, Sacks G, Regan L. Recurrent miscarriage: pathophysiology and outcome. Curr Opin Obstet Gynecol 2005;17(6): 591–7.

[46] Levens ED, Skarulis MC. Assessing the role of endometrial alteration among obese patients undergoing assisted reproduction. Fertil Steril 2008;89(6):1606–8.

[47] Gosman GG, Katcher HI, Legro RS. Obesity and the role of gut and adipose hormones in female reproduction. Hum Reprod Update 2006;12(5):585–601.

[48] Esinler I, Bozdag G, Yarali H. Impact of isolated obesity on ICSI outcome. Reprod Biomed Online 2008;17(4):583–7.

[49] Bellver J, Martinez-Conejero JA, Labarta E, Alama P, Melo MA, Remohi J, Horcajadas JA. Endometrial gene expression in the window of implantation is altered in obese women especially in association with polycystic ovary syndrome. Fertil Steril 2011;95(7):2335–41. 2341 e1-8.

[50] Symonds ME, Sebert SP, Hyatt MA, Budge H. Nutritional programming of the metabolic syndrome. Nat Rev Endocrinol 2009;5(11):604–10.

[51] Davies MJ. Evidence for effects of weight on reproduction in women. Reprod Biomed Online 2006;12(5):552–61.

[52] Jungheim ES, Moley KH. Current knowledge of obesity's effects in the pre- and periconceptional periods and avenues for future research. Am J Obstet Gynecol 2010;203(6):525–30.

[53] Bellver J, Busso C, Pellicer A, Remohi J, Simon C. Obesity and assisted reproductive technology outcomes. Reprod Biomed Online 2006;12(5):562–8.

[54] Bussen S, Sutterlin M, Steck T. Endocrine abnormalities during the follicular phase in women with recurrent spontaneous abortion. Hum Reprod 1999;14(1):18–20.

[55] Wang JX, Davies MJ, Norman RJ. Obesity increases the risk of spontaneous abortion during infertility treatment. Obes Res 2002;10(6):551–4.

[56] Abenhaim HA, Kinch RA, Morin L, Benjamin A, Usher R. Effect of prepregnancy body mass index categories on obstetrical and neonatal outcomes. Arch Gynecol Obstet 2007;275(1):39–43.

[57] Bhattacharya S, Campbell DM, Liston WA, Bhattacharya S. Effect of body mass index on pregnancy outcomes in nulliparous women delivering singleton babies. BMC Public Health 2007;7:168.

[58] Usha Kiran TS, Hemmadi S, Bethel J, Evans J. Outcome of pregnancy in a woman with an increased body mass index. BJOG 2005;112(6):768–72.

[59] Linne Y. Effects of obesity on women's reproduction and complications during pregnancy. Obes Rev 2004;5(3):137–43.

[60] Kristensen J, Vestergaard M, Wisborg K, Kesmodel U, Secher NJ. Pre-pregnancy weight and the risk of stillbirth and neonatal death. BJOG 2005;112(4):403–8.

[61] Maheshwari A, Scotland G, Bell J, McTavish A, Hamilton M, Bhattacharya S. The direct health services costs of providing assisted reproduction services in overweight or obese women: a retrospective cross-sectional analysis. Hum Reprod 2009;24(3):633–9.

[62] Weiss JL, Malone FD, Emig D, Ball RH, Nyberg DA, Comstock CH, D'Alton ME. Obesity, obstetric complications and cesarean delivery rate--a population-based screening study. Am J Obstet Gynecol 2004;190(4):1091–7.

[63] O'Brien TE, Ray JG, Chan WS. Maternal body mass index and the risk of preeclampsia: a systematic overview. Epidemiology 2003;14(3):368–74.

[64] Yu CK, Teoh TG, Robinson S. Obesity in pregnancy. BJOG 2006;113(10):1117–25.

[65] Tenenbaum-Gavish K, Hod M. Impact of maternal obesity on fetal health. Fetal Diagn Ther 2013;34(1):1–7.

[66] Whitaker KL, Jarvis MJ, Beeken RJ, Boniface D, Wardle J. Comparing maternal and paternal intergenerational transmission of obesity risk in a large population-based sample. Am J Clin Nutr 2010;91(6):1560–7.

[67] Schmatz M, Madan J, Marino T, Davis J. Maternal obesity: the interplay between inflammation, mother and fetus. J Perinatol 2010;30(7):441–6.

[68] Fujita K, Nishizawa H, Funahashi T, Shimomura I, Shimabukuro M. Systemic oxidative stress is associated with visceral fat accumulation and the metabolic syndrome. Circ J 2006;70(11): 1437–42.

[69] Aye IL, Powell TL, Jansson T. Review: Adiponectin--the missing link between maternal adiposity, placental transport and fetal growth? Placenta 2013;34(Suppl):S40–5.

[70] Higgins L, Mills TA, Greenwood SL, Cowley EJ, Sibley CP, Jones RL. Maternal obesity and its effect on placental cell turnover. J Matern Fetal Neonatal Med 2013;26(8):783–8.

[71] Ray JG, Vermeulen MJ, Schull MJ, McDonald S, Redelmeier DA. Metabolic syndrome and the risk of placental dysfunction. J Obstet Gynaecol Can 2005;27(12):1095–101.

[72] Abrams BF, Laros RK Jr. Prepregnancy weight, weight gain, and birth weight. Am J Obstet Gynecol 1986;154(3):503–9.

[73] Frentzen BH, Dimperio DL, Cruz AC. Maternal weight gain: effect on infant birth weight among overweight and average-weight low-income women. Am J Obstet Gynecol 1988;159(5): 1114–7.

[74] Johnson JW, Longmate JA, Frentzen B. Excessive maternal weight and pregnancy outcome. Am J Obstet Gynecol 1992;167(2):353–70. discussion 370-2.

[75] Mahony R, Foley M, McAuliffe F, O'Herlihy C. Maternal weight characteristics influence recurrence of fetal macrosomia in women with normal glucose tolerance. Aust N Z J Obstet Gynaecol 2007;47(5):399–401.

[76] Ecker JL, Greenberg JA, Norwitz ER, Nadel AS, Repke JT. Birth weight as a predictor of brachial plexus injury. Obstet Gynecol 1997;89(5 Pt 1):643–7.

[77] Pirkola J, Pouta A, Bloigu A, Hartikainen AL, Laitinen J, Jarvelin MR, Vaarasmaki M. Risks of overweight and abdominal obesity at age 16 years associated with prenatal exposures to maternal prepregnancy overweight and gestational diabetes mellitus. Diabetes Care 2010;33(5):1115–21.

[78] Krakowiak P, Walker CK, Bremer AA, Baker AS, Ozonoff S, Hansen RL, Hertz-Picciotto I. Maternal metabolic conditions and risk for autism and other neurodevelopmental disorders. Pediatrics 2012;129(5):pe1121–8.

[79] Bakos HW, Henshaw RC, Mitchell M, Lane M. Paternal body mass index is associated with decreased blastocyst development and reduced live birth rates following assisted reproductive technology. Fertil Steril 2011;95(5):1700–4.

[80] Chang JS. Parental smoking and childhood leukemia. Methods Mol Biol 2009;472:103–37.

[81] Sartorius GA, Nieschlag E. Paternal age and reproduction. Hum Reprod Update 2010;16(1):65–79.

[82] Lin CC, Wang JD, Hsieh GY, Chang YY, Chen PC. Increased risk of death with congenital anomalies in the offspring of male semiconductor workers. Int J Occup Environ Health 2008;14(2):112–6.

[83] Sofikitis N, Giotitsas N, Tsounapi P, Baltogiannis D, Giannakis D, Pardalidis N. Hormonal regulation of spermatogenesis and spermiogenesis. J Steroid Biochem Mol Biol 2008;109(3–5): 323–30.

[84] Tunc O, Bakos HW, Tremellen K. Impact of body mass index on seminal oxidative stress. Andrologia 2011;43(2):121–8.

[85] Chavarro JE, Toth TL, Wright DL, Meeker JD, Hauser R. Body mass index in relation to semen quality, sperm DNA integrity, and serum reproductive hormone levels among men attending an infertility clinic. Fertil Steril 2010;93(7):2222–31.

[86] Macdonald AA, Stewart AW, Farquhar CM. Body mass index in relation to semen quality and reproductive hormones in New Zealand men: a cross-sectional study in fertility clinics. Hum Reprod 2013;28(12):3178–87.

[87] MacDonald AA, Herbison GP, Showell M, Farquhar CM. The impact of body mass index on semen parameters and reproductive hormones in human males: a systematic review with meta-analysis. Hum Reprod Update 2010;16(3):293–311.

[88] Handelsman DJ, Swerdloff RS. Male gonadal dysfunction. Clin Endocrinol Metab 1985;14(1):89–124.

[89] Kerr JB, Millar M, Maddocks S, Sharpe RM. Stage-dependent changes in spermatogenesis and Sertoli cells in relation to the onset of spermatogenic failure following withdrawal of testosterone. Anat Rec 1993;235(4):547–59.

[90] Kley HK, Deselaers T, Peerenboom H, Kruskemper HL. Enhanced conversion of androstenedione to estrogens in obese males. J Clin Endocrinol Metab 1980;51(5):1128–32.

[91] Pasquali R, Patton L, Gambineri A. Obesity and infertility. Curr Opin Endocrinol Diabetes Obes 2007;14(6):482–7.

[92] Leisegang K, Bouic PJD, Menkveld R, Henkel RR. Obesity is associated with increased seminal insulin and leptin alongside reduced fertility parameters in a controlled male cohort. Reprod Biol Endocrin 2014;12(34).

E. NUTRITION AND LIFESTYLE IN *IN VITRO* FERTILIZATION

[93] Lampiao F, du Plessis SS. Insulin and leptin enhance human sperm motility, acrosome reaction and nitric oxide production. Asian J Androl 2008;10(5):799–807.

[94] La Vignera S, Condorelli R, Vicari E, D'Agata R, Calogero AE. Diabetes mellitus and sperm parameters. J Androl 2012;33(2):145–53.

[95] Amaral S, Oliveira PJ, Ramalho-Santos J. Diabetes and the impairment of reproductive function: possible role of mitochondria and reactive oxygen species. Curr Diabetes Rev 2008;4(1):46–54.

[96] Palmer NO, Bakos HW, Fullston T, Lane M. Impact of obesity on male fertility, sperm function and molecular composition. Spermatogenesis 2012;2(4):253–63.

[97] Sermondade N, Faure C, Fezeu L, Shayeb AG, Bonde JP, Jensen TK, Czernichow S. BMI in relation to sperm count: an updated systematic review and collaborative meta-analysis. Hum Reprod Update 2013;19(3):221–31.

[98] Hammiche F, Laven JS, Twigt JM, Boellaard WP, Steegers EA, Steegers-Theunissen RP. Body mass index and central adiposity are associated with sperm quality in men of subfertile couples. Hum Reprod 2012;27(8):2365–72.

[99] Sekhavat L, Moein MR. The effect of male body mass index on sperm parameters. Aging Male 2010;13(3):155–8.

[100] Fariello RM, Pariz JR, Spaine DM, Cedenho AP, Bertolla RP, Fraietta R. Association between obesity and alteration of sperm DNA integrity and mitochondrial activity. BJU Int 2012;110:863–7.

[101] Aggerholm AS, Thulstrup AM, Toft G, Ramlau-Hansen CH, Bonde JP. Is overweight a risk factor for reduced semen quality and altered serum sex hormone profile? Fertil Steril 2008;90(3):619–26.

[102] Shayeb AG, Harrild K, Mathers E, Bhattacharya S. An exploration of the association between male body mass index and semen quality. Reprod Biomed Online 2011;23(6):717–23.

[103] Sun JG, Jurisicova A, Casper RF. Detection of deoxyribonucleic acid fragmentation in human sperm: correlation with fertilization in vitro. Biol Reprod 1997;56(3):602–7.

[104] Lopes S, Sun JG, Jurisicova A, Meriano J, Casper RF. Sperm deoxyribonucleic acid fragmentation is increased in poor-quality semen samples and correlates with failed fertilization in intracytoplasmic sperm injection. Fertil Steril 1998;69(3):528–32.

[105] Bakos HW, Thompson JG, Feil D, Lane M. Sperm DNA damage is associated with assisted reproductive technology pregnancy. Int J Androl 2008;31(5):518–26.

[106] Saleh RA, Agarwal A, Nada EA, El-Tonsy MH, Sharma RK, Meyer A, Thomas AJ. Negative effects of increased sperm DNA damage in relation to seminal oxidative stress in men with idiopathic and male factor infertility. Fertil Steril 2003;79(Suppl 3):1597–605.

[107] Carrell DT, Liu L, Peterson CM, Jones KP, Hatasaka HH, Erickson L, Campbell B. Sperm DNA fragmentation is increased in couples with unexplained recurrent pregnancy loss. Arch Androl 2003;49(1):49–55.

[108] Kort HI, Massey JB, Elsner CW, Mitchell-Leef D, Shapiro DB, Witt MA, Roudebush WE. Impact of body mass index values on sperm quantity and quality. J Androl 2006;27(3):450–2.

[109] Dupont C, Faure C, Sermondade N, Boubaya M, Eustache F, Clement P, Levy R. Obesity leads to higher risk of sperm DNA damage in infertile patients. Asian J Androl 2013;15(5):622–5.

[110] La Vignera S, Condorelli RA, Vicari E, Calogero AE. Negative effect of increased body weight on sperm conventional and nonconventional flow cytometric sperm parameters. J Androl 2012;33(1):53–8.

[111] Sharma RK, Agarwal A. Role of reactive oxygen species in male infertility. Urology 1996;48(6):835–50.

[112] Alvarez JG, Storey BT. Differential incorporation of fatty acids into and peroxidative loss of fatty acids from phospholipids of human spermatozoa. Mol Reprod Dev 1995;42(3):334–46.

[113] Wang X, Sharma RK, Gupta A, George V, Thomas AJ, Falcone T, Agarwal A. Alterations in mitochondria membrane potential and oxidative stress in infertile men: a prospective observational study. Fertil Steril 2003;80(Suppl. 2):844–50.

[114] Kodama H, Yamaguchi R, Fukuda J, Kasai H, Tanaka T. Increased oxidative deoxyribonucleic acid damage in the spermatozoa of infertile male patients. Fertil Steril 1997;68(3):519–24.

[115] Agarwal A, Saleh RA, Bedaiwy MA. Role of reactive oxygen species in the pathophysiology of human reproduction. Fertil Steril 2003;79(4):829–43.

[116] Bourgeron T. Mitochondrial function and male infertility. Results Probl Cell Differ 2000;28:187–210.

[117] Keltz J, Zapantis A, Jindal SK, Lieman HJ, Santoro N, Polotsky AJ. Overweight men: clinical pregnancy after ART is decreased in IVF but not in ICSI cycles. J Assist Reprod Genet 2010;27(9–10):539–44.

[118] Colaci DS, Afeiche M, Gaskins AJ, Wright DL, Toth TL, Tanrikut C, Chavarro JE. Men's body mass index in relation to embryo quality and clinical outcomes in couples undergoing in vitro fertilization. Fertil Steril 2012;98(5):1193–9. e1.

[119] Merhi ZO, Keltz J, Zapantis A, Younger J, Berger D, Lieman HJ, Polotsky AJ. Male adiposity impairs clinical pregnancy rate by in vitro fertilization without affecting day 3 embryo quality. Obesity (Silver Spring) 2013;21(8):1608–12.

[120] Anifandis G, Dafopoulos K, Messini CI, Polyzos N, Messinis IE. The BMI of men and not sperm parameters impact on embryo quality and the IVF outcome. Andrology 2013;1(1):85–9.

[121] Petersen GL, Schmidt L, Pinborg A, Kamper-Jorgensen M. The influence of female and male body mass index on live births after assisted reproductive technology treatment: a nationwide register-based cohort study. Fertil Steril 2013;99(6):1654–62.

[122] Barker DJ, Osmond C. Infant mortality, childhood nutrition, and ischaemic heart disease in England and Wales. Lancet 1986;1(8489):1077–81.

[123] Barker DJ. The origins of the developmental origins theory. J Intern Med 2007;261(5):412–7.

[124] Dumoulin JC, Land JA, Van Montfoort AP, Nelissen EC, Coonen E, Derhaag JG, Evers JL. Effect of in vitro culture of human embryos on birthweight of newborns. Hum Reprod 2010;25(3):605–12.

[125] Eckert JJ, Porter R, Watkins AJ, Burt E, Brooks S, Leese HJ, Fleming TP. Metabolic induction and early responses of mouse blastocyst developmental programming following maternal low protein diet affecting life-long health. PLoS One 2012;7(12):e52791.

[126] Watkins AJ, Lucas ES, Torrens C, Cleal JK, Green L, Osmond C, Fleming TP. Maternal low-protein diet during mouse pre-implantation development induces vascular dysfunction and altered renin-angiotensin-system homeostasis in the offspring. Br J Nutr 2010;103(12):1762–70.

[127] Idring S, Magnusson C, Lundberg M, Ek M, Rai D, Svensson AC, Lee BK. Parental age and the risk of autism spectrum disorders: findings from a Swedish population-based cohort. Int J Epidemiol 2014;43(1):107–15.

[128] Peters S, Glass DC, Reid A, de Klerk N, Armstrong BK, Kellie S, Fritschi L. Parental occupational exposure to engine exhausts and childhood brain tumors. Int J Cancer 2013;132(12):2975–9.

[129] Danielzik S, Langnase K, Mast M, Spethmann C, Muller MJ. Impact of parental BMI on the manifestation of overweight 5–7 year old children. Eur J Nutr 2002;41(3):132–8.

[130] Suren P, Gunnes N, Roth C, Bresnahan M, Hornig M, Hirtz D, Stoltenberg C. Parental obesity and risk of autism spectrum disorder. Pediatrics 2014;133(5):e1128–38.

[131] Ng SF, Lin RC, Laybutt DR, Barres R, Owens JA, Morris MJ. Chronic high-fat diet in fathers programs beta-cell dysfunction in female rat offspring. Nature 2010;467(7318):963–6.

[132] Fullston T, Palmer NO, Owens JA, Mitchell M, Bakos HW, Lane M. Diet-induced paternal obesity in the absence of diabetes diminishes the reproductive health of two subsequent generations of mice. Hum Reprod 2012;27(5):1391–400.

[133] Mitchell M, Bakos HW, Lane M. Paternal diet-induced obesity impairs embryo development and implantation in the mouse. Fertil Steril 2011;95(4):1349–53.

Health Outcomes of Children Conceived Through Assisted Reproductive Technology

Fiona Langdon, MBBS, Abbie Laing, MBChB*, Roger Hart, MD**,†*

*King Edward Memorial Hospital, Perth, Western Australia, Australia
**School of Women's and Infants' Health, University of Western Australia, King Edward Memorial Hospital, Perth, Western Australia, Australia
†Fertility Specialists of Western Australia, Bethesda Hospital, Perth, Western Australia, Australia

INTRODUCTION

Thirty-five years since the first child was born from IVF, it is currently estimated that over 4 million children worldwide have been born as a result of (ART) [1]. Today, up to 4% of children born in the developed world are conceived using ART [2]. It is well-established that children born from IVF have an increased risk of congenital malformations at birth; however, it is less clear if there are any implications for the short- and long-term health outcomes of these children [3]. The majority of these children are yet to reach adulthood, let alone middle age, and thus we may not be fully aware of the possible long-term health implications.

Studying the health outcomes of children conceived through ART has been hampered by difficulties with controlling many confounding factors. The pathogenesis of infertility is not well-understood and for many couples the reason for their difficulty in conceiving is never elucidated. Thus, adverse health outcomes arising during pregnancy or manifesting in the offspring may well be due to the reason for infertility itself, rather than the process used to treat them. For research purposes we are unable to control these confounding factors and so creating a well-designed clinical trial to investigate the risks of ART is near impossible due to the difficulty in creating a control group. Some studies have used as controls those couples that have experienced subfertility (mostly defined as the inability to conceive within 12 months of trying) that conceived spontaneously. Even this is not an adequate control group, as couples requiring ART are likely to have more significant pathology or multifactorial pathology causing their infertility, and thus offer a potentially greater impact on the risks of complications in pregnancy and the long-term health of their offspring. Other studies have used sibling pairs, one who has been spontaneously conceived and the other through ART. These studies, for the most part, have smaller numbers and may not control for other factors, such as maternal age, birth order, or newly acquired maternal medical conditions, which may be the primary reason for requiring ART.

Most studies examining the complications of pregnancy and within the offspring use the general population as the control group and compare them with women who have required ART to conceive. To gain a large enough sample size to have strongly powered studies, patient cohorts have had to be drawn from multiple centers, often in different countries and over a large period of time. This creates further problems with analyzing and comparing data, as ART is a rapidly changing technology with multiple different methods and protocols used across institutions and by individual clinicians [4]. Studies have shown differences in outcomes in relation to fresh or frozen embryo transfer, the number of embryos transferred, the culture medium employed, the use of ICSI (intracytoplasmic sperm injection), and the type of ovarian hyperstimulation protocol followed. These may all play a role in the subsequent development of the fetus and the risk of complication in both the short and long term.

As described, ART has long been thought to play some role in creating a small but significant increased risk of health problems to the offspring born. This may be due to all or some of the steps it entails. Artificial hormonal

Handbook of Fertility. http://dx.doi.org/10.1016/B978-0-12-800872-0.00029-9

stimulation of the ovary to harvest the oocyte influences the oocyte maturation process and creates a supraphysiological hormonal environment within the endometrium. This may particularly impact the offspring produced in a fresh IVF cycle, where an embryo is transferred into the artificially stimulated endometrium. Furthermore, the culture medium used to culture the embryo potentially exposes the embryo to a hormonal and chemical environment different to that of the natural conception process. A lack of a natural selection process of first the egg and then, in particular with ICSI, the sperm, may result in genes that would otherwise not have been selected being passed on to the conceptus. ICSI also requires the zona pellucida and cytoplasmic membrane to be punctured, possibly damaging the oocyte and also introducing foreign material that may affect development of the conceptus, as in natural fertilization the sperm does not penetrate the oocyte. The process of, and chemicals used in, cryopreservation and thawing processes also leads to further steps that may impact negatively on the embryo.

CONGENITAL MALFORMATIONS

Since the first studies were conducted analyzing the outcomes of children conceived through ART, a definite link with an increased risk of major congenital malformation was identified [3,5–7]. The extent to which this link can be attributed to the technology itself and not to an underlying predisposition to congenital malformations in infertile couples is still difficult to determine. As discussed, the increased rate of malformation seen in ART children may be due to the medications used to hyperstimulate the ovary and provide luteal phase support, the media used for oocyte and embryo culture and storage, and in the case of ICSI, the selection and injection of the sperm in addition to a small amount of culture media into the egg. Added to this is the underlying cause of infertility that may increase the risk of malformation: the age of the couple, environmental toxin exposure, and the increased rate of chromosome abnormalities in ova and sperm of infertile couples, in particular Y chromosome abnormalities in the male. Studies aimed at determining the risk of major malformations in ART offspring lack homogeneity for a control group, as well as varying of criteria used to define a major malformation and at what age children need be assessed and by whom.

Multiple meta-analyses support an increased risk in ART of major malformation between 1.2 and 2 times over the background population risk [3,8–13]. Certain anatomical systems are overly represented in the data including cardiac malformations, in particular cardiac defects, gastrointestinal malformations including esophageal atresia and anorectal atresia, musculoskeletal anomalies, and genitourinary abnormalities [14].

Since the first successful pregnancy using ICSI in 1992, couples with infertility that have previously been unsuccessfully treated with routine IVF have increased the ability to conceive and have children. Now, despite its original use for severe male factor infertility, over two-thirds of IVF cycles currently performed in Australia and New Zealand employ ICSI [15]. ICSI, in theory, may carry a higher risk of congenital malformation due to the reduction in natural selection and due to the potential to transmit an increased rate of chromosomal or genetic abnormalities from subfertile males. Nearly 5% of men who suffer from oligozoospermia and over 13% of azoospermic men have a chromosomal abnormality, most commonly Y chromosome microdeletions and autosomal translocations [16]. Hence, providing these men with the ability to conceive now using a form of conception that circumvents the natural selection process of sperm may lead to the inheritance of these abnormalities in offspring. Analysis of the chromosomes of children conceived through ISCI has shown an increase in chromosomal abnormalities over spontaneously conceived children and children conceived through IVF [17]. However, this has not necessarily resulted in an increase in malformation rates. Studies that have examined the increased risk of malformation in ICSI over IVF have had differing results, some finding a small increased risk and others showing no significant risk between the two technologies [11,18,19]. There is, however, an increased risk of genitourinary malformations, in particular hypospadius, in the male offspring of ICSI pregnancies [20], thought to be linked to the increased genetic mutation within the male partner.

A large study that used women who suffered subfertility but subsequently conceived spontaneously as a control group to compare with children conceived through ART found a small but still significant increase of congenital malformation in the ART group, around 1.2 times the background population risk [20]. The authors went on to deduce from this that approximately 40% of major malformations in children conceived through ART were due to the technology itself, but that infertility plays a bigger role in the risk of major malformation. A subsequent meta-analysis extrapolated this 40% rate to other studies that used the general population as the control group and found that the relative risk of major congenital malformation was only 1.14 for ART pregnancies [21].

This was further demonstrated in a study of Australian birth registries that looked at the use of ART compared with ovulation stimulation only and also compared with a history of infertility but spontaneous conception. After controlling for maternal age, race, medical conditions, and parity it was found that IVF treatment had no greater risk than spontaneous conception in a subfertile couple. ICSI, however, was associated with a significantly greater risk of congenital malformation in this study (Table 29.1) [18].

TABLE 29.1 Odds Ratio for any Birth Defect According to Type of Assisted Conception

Type of assisted conception	Adjusted odds ratio*
Any	1.26 (1.14–1.40)
IVF: fresh or frozen-embryo cycle	1.07 (0.90–1.26)
IVF: fresh-embryo cycle	1.09 (0.89–1.33)
IVF: frozen-embryo cycle	1.02 (0.75–1.39)
ICSI: fresh or frozen-embryo cycle	1.57 (1.30–1.90)
ICSI: fresh-embryo cycle	1.66 (1.35–2.04)
ICSI: frozen-embryo cycle	1.28 (0.83–1.99)
Ovulation Induction	1.16 (0.76–1.75)
Spontaneous conception after previous birth from ART	1.25 (1.01–1.56)
Infertile but no history of ART	1.29 (0.99–1.68)

Adjusted for maternal age, parity, fetal sex, year of birth, maternal race, maternal conditions in pregnancy, maternal smoking, socio-economic status, and maternal and paternal occupation.
Data extracted from Davies et al. [22].

PERINATAL COMPLICATIONS

Complications that develop in pregnancy can have long-term implications for the health of the offspring. The concept of adult disease having its origins in early life events, especially in such diseases as coronary heart disease, type 2 diabetes, hypertension, and obesity, is well-recognized [23]. ART pregnancies have increased rates of many pregnancy complications, including multiple gestation, and this may well contribute to a potential subsequent increase in long-term health complications for the offspring [4].

Multifetal Gestation

The most significant and prevalent pregnancy complication affecting ART pregnancies is multiple gestations. The twinning and higher order multiple birth rate in ART pregnancies is dependent on national and institutional policies relating to the number of embryos permitted to be transferred per cycle. Also, countries with reimbursed and nationally funded IVF programs see much lower double embryo transfer (DET) rates and subsequently much lower multiple pregnancy rates. Worldwide the estimated total twinning rate in ART pregnancies is 25–38%, far higher than the 1% spontaneously conceived rate [24]. Australia and New Zealand have an enviable twin pregnancy rate of just under 7%, owing to strict protocols on number of embryos permitted to be transferred [25]. Monozygotic twinning is increased in ART pregnancies by a factor of three to five times over spontaneously conceived pregnancies [26], hence even a policy of elective single embryo transfer (eSET) will result in a higher multiple birth rate than the general population. This is particularly true when blastocyst transfer is used, which is associated with an up to 10 times increased risk of monozygotic twinning [27].

Multiple pregnancies are associated with an increased risk of preterm birth (PTB), congenital malformation, low birth weight, growth restriction, operative delivery, and perinatal mortality [4]. Even patients who undergo a multiple embryo transfer that results in a singleton pregnancy still have higher rates of pregnancy complications compared with elective single embryo transfer, thought to be due to the implication of the "vanishing twin" [28]. Elective single embryo transfer followed by a further frozen single embryo transfer, if unsuccessful, has no difference in the cumulative live birth rate compared with a DET, but with far fewer perinatal complications [29]. It is thus prudent to adhere to a policy of elective single embryo transfer wherever possible to reduce the risk of PTB, growth restriction, and perinatal mortality and possibly reduce the risk of long-term health complications.

Preterm Birth

The increased risk of preterm birth seen in ART technologies has largely been attributed to multifetal gestations; however, singleton pregnancies have also been associated with a 1.4–2 times increased risk of a PTB, and more concerning a threefold increased risk of a very PTB (less than 32 weeks' gestation) [28]. Other studies have placed this increased risk down to the "vanishing twin" phenomenon, stating there is no significant increased risk in PTB of true singleton ART pregnancies resulting from elective single embryo transfer [30,31]. This is supported by the reduction in PTB rates in ART pregnancies over time, with later studies having much lower PTB rates compared with earlier studies [28]. This has coincided with a move toward elective single embryo transfer.

Frozen/thawed embryos seem to have a protective role for PTB over fresh transfer embryos with a meta-analysis of nine studies giving a risk reduction of 0.85 [28]. Whether this is due to fresh cycles being associated with an abnormal hormonal environment secondary to the recent ovarian hyperstimulation or due to the freezing thawing cycle selecting out more robust embryos remains unclear.

ICSI also seems to be protective for PTB with a meta-analysis of five studies examining this giving an odds ratio of 0.8 over IVF pregnancies [28]. Other studies have found no significant difference in the PTB rate between the two technologies, but importantly no study has found an increased risk with ICSI [32].

To determine if the increased risk of PTB of ART pregnancies is related to ART or infertility itself, studies have looked at patients requiring ovulation induction only to conceive. It seems ovulation induction is associated with an increased risk of PTB, but less than the rate seen in ART pregnancies [33–35]. This may mean that infertility per se is a risk factor for PTB, but also that the abnormal hormonal environment ovulation induction produces, if multiple ovulations are triggered, may be the causative factor. In two studies that compared sibling pairs, one conceived through ART and one spontaneously conceived, a significantly higher rate of PTB was found in the ART siblings in one study [36] but not the other [37]; however, combining the two studies made the result significant with an increased risk of nearly 1.3 times for the ART sibling [28].

Birth Weight

ART pregnancies are associated with an increased incidence of low birth weight (<2500 g) and very low birth weight (<1500 g) infants, from 1.5 to 2 times and 2 to 3 times, respectively [9,13]. However, like PTB, birth weights have increased and continue to increase as IVF and ICSI treatment regimens evolve. PAPP-A levels, an inverse marker for the risk of low birth weight and growth restriction, have been found to be significantly lower in ART pregnancies [38]. This also has implications for aneuploidy screening tests that use PAPP-A as a marker. Meta-analysis of ICSI over IVF has shown no significant difference in birth weight between the two technologies [28].

Similar to PTB, a decreased rate of lower birth weight infants are seen resulting from frozen transfers, when compared to fresh cycles [28]. An increased rate of low birth weight infants is also seen in pregnancies conceived with ovulation induction [33]. Combining these two findings it is reasonable to think the exposure to gonadotropin stimulation and the subsequent altered hormonal environment may be detrimental to embryo and fetal growth.

Interestingly, studies have not consistently shown an increase in low birth weight infants from twin pregnancies conceived from ART treatment when compared to spontaneous conceived twin infants [39]. Whether this is due to confounding factors, in particular monozygocity, which has a much higher incidence in spontaneous twin pairs, is unclear [40,41].

The culture media used in ART has been shown to influence birth weight suggesting that ART, rather than just infertility, must indeed play a role in the risk of low birth weight [28]. This is further supported by the risk of low birth weight being greater with ART in studies comparing ART with subfertile couples requiring only ovulation induction [13]. Studies that compare sibling pairs conceived spontaneously and with ART have differing results with some having a significant increase in low birth weight with ART and others showing no significant difference [36]. It is likely that ART plays a small but significant role in low birth weight risk but that infertility itself also has a role.

IMPRINTING DISORDERS

Imprinting disorders, rare conditions that affect growth and neurodevelopment through aberrations in gene methylation arising during meiosis or fertilization, have been linked with ART conceptions [42]. Given that imprinting disorders occur so rarely, it is difficult to determine the exact risk ART presents and to what extent the risk is attributed to the underlying infertility and to what the technology.

The Beckwith–Weidemann syndrome (BWS) is an imprinting disorder associated with childhood overgrowth and an increased risk of certain malignancies. Of all the imprinting disorders, it has the most data supporting an association with ART. Epidemiological studies examining the risk ART poses to increased incidence of BWS give a risk of between 5 and 10 times that of the normal population [43]. Studies in animal models have found an increase in a similar condition in cows as a result of certain culture media used in IVF and thus a link between ART itself and BWS is being made. There is also some data to support infertility itself being the causative link with BWS as women requiring only ovulation induction may have an increased chance of bearing offspring with BWS [44,45].

The Silver–Russel syndrome is an imprinting disorder associated with growth restriction and abnormal facies. Epidemiological studies are all underpowered to show a significant association with ART; however, some studies have shown a trend toward an increased incidence in children born through IVF and ICSI [46]. There is some suggestion that the underlying mechanism behind the increased incidence comes from aberrant methylation during spermatogenesis, and it is these otherwise infertile sperm which are now, facilitated through predominantly ICSI, able to transfer their genome to the offspring [47].

The Angelman syndrome is a disorder of severe mental retardation, delayed motor development, and abnormal movements. Once again a possible link with ART has been shown in some studies but not in others, with a trend toward an increased incidence with ICSI pregnancies, but not IVF [44,48]. This trend has also been demonstrated in subfertile couples who did not require ART suggesting the causative link may be infertility itself [49].

More studies are needed before firm conclusions can be drawn regarding the risk of ART and imprinting disorders, but from the data available it can be concluded that there is a small but likely increased risk, especially for BWS. Given the very rare incidence of these conditions, the actual clinical impact any increased risk carries is debatable.

CARDIOVASCULAR AND METABOLIC EFFECTS

Body Mass Index

One of the first cohort studies of young IVF adults reported no significant differences when comparing body mass index (BMI) with a representative non-IVF population [50]. However, when body fat is measured more objectively by assessment of peripheral skin folds and use of dual energy X-ray absorptiometry, IVF children and adolescents appear to have more peripheral fat deposits despite minimal differences in BMI [51]. This adipose distribution seems to advance with pubertal stage in ICSI children [52] which is concerning given that a tendency for increased peripheral adiposity is known to have a more adverse metabolic profile.

Blood Pressure

There are concerns that IVF treatment might predispose offspring to early onset hypertension [53–56]. Higher systolic and diastolic blood pressure (BP) recordings were observed in an IVF cohort aged 8–18 years, and this could not be explained by BMI, early life factors, or parental characteristics [53]. Further work from this group suggested that a rapid growth during early childhood (1–3 years), but not late infancy (3 months to 1 year) was a significant risk factor for the development of hypertension in IVF children [54]. However, a later cross-sectional study found no such association in adolescents conceived via ICSI aged 14 years [57]. This study was adjusted for multiple confounders, although limited by its low response rate and cross-sectional design and so more research is needed.

Vascular Dysfunction

An interesting study reported generalized adverse vascular dysfunction among IVF and ICSI children despite no differences in their actual resting systolic and diastolic BP [58]. ART children had significantly smaller flow-mediated dilation of the brachial artery, faster mean carotid-femoral pulse wave velocity, greater average carotid intima thickness (410 μm vs. 370 μm), and a higher pulmonary vascular dysfunction (30% higher mean systolic pulmonary artery pressure at high altitude) [58]. The sample size was small in this study but all participants were term singletons without perinatal complications and multiple variables were analyzed.

Metabolic Profile

Subtle differences in the regulators of lipid metabolism have been demonstrated following ART [55,59,60]. One experimental study suggested that preconceptual manipulation by ICSI was an independent risk factor for metabolic disorders in later life after observing an altered expression of sterol-related genes in ICSI fetuses [59]. However, the maternal age of women in the study groups was not matched and was higher in the ICSI group. Furthermore, fetal tissue samples were not distinguished in each test. Another study suggested a favorable lipid profile in IVF and ICSI individuals, although this study was restricted to prepubertal children [61]. When participants at different stages of puberty are included in the analysis, higher triglyceride levels are reported in IVF subjects [55]. This last study, however, was not confined to singletons. Further large prospective trials are needed (Table 29.2).

MALIGNANCY

An increased risk of malignancy in IVF offspring has been suggested [62–64]. This concern is supported by findings of an altered epigenetic profile observed in human IVF embryos [65], which are known to have a role in human carcinogenesis [66]. However, the largest population-based study performed reported no increase in overall cancer risk [67]. When results were evaluated by cancer subtypes, increased risks were only found for hepatoblastoma and rhabdomyosarcoma. However, the absolute excess risks for both of these were low and hepatoblastoma was associated with low birth weight and rhabdomyosarcoma with older paternal age. Other smaller studies have shown similar results [68,69].

There have been concerns that an epigenetic mechanism involving hypermethylation in the RB1 promoter region may cause nonhereditary retinoblastoma in IVF children [70]. However, no evidence of this hypermethylation was

TABLE 29.2   Suggested Effects of ART on Cardiovascular and Metabolic Function

Cardiovascular and metabolic outcome	Effect
BMI	Increased peripheral fat deposits with advanced pubertal age Female predisposition to increased central, peripheral, and total body adiposity Male predisposition to higher total body adiposity
BP	Higher systolic and diastolic BP Growth during early childhood may program later systolic hypertension
Vascular dysfunction	Reduced flow-mediated dilation of the brachial artery Increased carotid-femoral pulse wave velocity Increased carotid intima-media thickness Pulmonary vascular dysfunction
Metabolic profile	Higher insulin induced gene expression Possible elevated triglyceride levels Elevated RBP-4 and NGAL levels Hyperglycemia

seen in seven tumors analyzed from IVF or ICSI children with nonhereditary retinoblastoma [70]. Furthermore, a review of nonfamilial retinoblastoma cases did not find a significantly increased risk of retinoblastoma with any type of IVF [71]. Instead retinoblastoma was significantly associated with increased maternal age and subfertility [71]. This study had excellent power to detect a difference; owing to its large sample of retinoblastoma cases and high frequency of infertility treatment in the reference population [71].

RESPIRATORY AND ALLERGY DISORDER: ASTHMA, ALLERGY, AND ATOPY

The prevalence of allergic disorders has been reported as similar between IVF and spontaneously conceived individuals [72]. One cross-sectional study even reported a favorable asthma profile for adults conceived by IVF [73]. Nonetheless, a large UK prospective study found that those conceived by reproductive assistance were more likely to experience asthma symptoms [74]. The association was also present, although reduced at 7 years of age. The results should be interpreted with caution as the study included a relatively small number of singletons. Furthermore, analysis of confounding factors suggests subfertility could be the putative risk factor for asthma not ART [74]. Other studies similarly report subfertility as the risk factor for asthma rather than ART per se [75,76].

ENDOCRINE

The limited studies available suggest IVF could confer a susceptibility to subclinical hypothyroidism. Higher thyroid stimulating hormone (TSH) levels, but no significant differences in T3 or T4, were demonstrated in one uncontrolled study (Table 29.3) [77], and another

cohort study demonstrated exaggerated TSH responses to thyrotropin-releasing hormone in IVF newborns [78], although the sample size in this was very small. Also, mild aberrations in insulin like growth factor secretion have been described among IVF subjects [61,79].

OPHTHALMOLOGICAL AND AUDITORY DISORDERS

Relatively little has been published on the auditory or visual impairments in IVF infants. A large follow-up study observed no difference in the hearing and vision of ICSI children compared with controls [81], although significantly more children in the control group were wearing glasses and despite close matching there was a higher paternal age and fewer siblings in the ICSI group. An increased risk for severe visual impairment was identified in another cohort study [82], but this effect lost significance after adjustment for parental subfertility. Other studies also reported no differences once adjustment is made for potential [83–86].

GROWTH AND PUBERTAL DEVELOPMENT

Growth

Several cohort studies report normal growth and development in children born from assisted techniques [87–89]. Measurements including height, weight, and head circumference recorded from birth up to the age of 12 years have been described as comparable to controls [85,90–92].

Two cohort studies have suggested that ICSI and IVF children are significantly smaller in weight up to and at 3 years of age but not thereafter [79,93]. However, both

TABLE 29.3 Descriptive Statistics for Height and Weight Measurements Between ART and non-ART Groups

Time of measurement	Mean height ± SD (cm)		
	IVF	ICSI	Non-ART
Time 1 (birth)	51.58 ± 2.90	50.91 ± 3.79	51.57 ± 2.88
Time 2 (59.61 months)	110.10 ± 3.36	110.40 ± 5.30	110.10 ± 5.42
Time 3 (106.86 months)	133.71 ± 5.69	136.90 ± 6.40	134.68 ± 5.22
Time 4 (131.86 months)	146.25 ± 5.90	148.76 ± 7.72	145.89 ± 6.81
Time of measurement	Mean weight ± SD (cm)		
	IVF	ICSI	Non-ART
Time 1 (birth)	3.32 ± 0.50	3.03 ± 0.82	3.44 ± 0.57
Time 2 (59.61 months)	19.21 ± 2.30	20.30 ± 4.52	20.01 ± 2.89
Time 3 (106.86 months)	29.56 ± 5.17	33.23 ± 9.08	31.67 ± 6.09
Time 4 (131.86 months)	39.33 ± 10.75	42.64 ± 9.64	40.51 ± 11.95

Data extracted from Basatemur et al. [80].

studies were not adjusted for PTB and low birth weights that might explain their results. Another cohort study reported IVF and ICSI children as significantly taller than their non-IVF controls, by a difference of approximately 3 cm [61]. This study had a small cohort size although it was corrected for age and parental height and only included term singletons in its analysis. They also reported higher serum concentrations of insulin growth factors and concluded that their findings might be due to epigenetic alterations. The underlying concern is whether this increased height could cause health risks in the future based on the evidence that rapid weight gain during early childhood is related subsequently to hypertension [54].

Puberty

A cross-sectional evaluation by Belva et al. found no evidence of delayed puberty in ICSI offspring as assessed by Tanner staging and onset of menarche, although breast development was less advanced in girls [57]. This last finding was unexpected given the authors' previous findings of increased adiposity in ICSI girls [52]. The authors suggested that less advanced breast development might be caused by a disturbed hypothalamic reactivation, but whether IVF technology can permanently affect the hypothalamic-pituitary axis is unknown. The Dutch OMEGA study also found no differences in pubertal stage between IVF and control children, but an advanced bone age and increased luteinizing hormone and dehydro-epiandosterone sulfate (DHEAS) levels were recorded in pubertal IVF girls [87]. It is possible that higher DHEAS concentrations occur following prenatal growth restriction [60]. But concentrations of DHEAS among IVF girls in the OMEGA study could not be explained by birth weight.

GENERAL HEALTH AND WELL-BEING

Higher levels of childhood illnesses have been described in IVF offspring [81,84,94–96]. One study group reported more respiratory diseases in ART children at age 3 and 7 years [96]. Another large European retrospective study reported higher usage of health care resources by ART children [84]. However, both these studies included premature children in their analysis, which is likely to be a major confounder influencing postnatal health. In a prospective controlled study of term ICSI children >37 weeks, no differences in acute or chronic childhood illnesses were found [81] and similar findings have been documented in other studies [86,88]. ICSI children were hospitalized more than controls but no specific reasons were given for this [81]. ICSI boys underwent more urogenital surgery because of a significantly increased risk of undescended testicles [81]; this was also previously described in the retrospective study by Bonduelle et al [84].

TESTICULAR FUNCTION

The gonadal development of male ART children is considered normal [97,98]. A study of prepubertal and pubertal ICSI-conceived boys found that the majority had normal testicular and penile size as well as normal serum anti-Mullerian hormone, inhibin B [97,98] and morning salivary testosterone levels [57].

CEREBRAL PALSY

Only a few population-based cohort studies have addressed the risk of cerebral palsy (CP) in IVF children and most of these are based on Scandinavian registry data [99–102]. The strongest association was found in a cohort study that used rehabilitation centers to identify children with CP [102]. CP was reported in 0.55% in the IVF group compared to 0.15% in the control group. For singletons, this risk remained significantly increased; however, after adjustment for prematurity and low birth weight, no difference in the rate of CP was recorded. A Danish population-based cohort study reported an increased risk of CP for IVF children, although in accordance with Stromberg and colleagues the risk attributable to IVF decreased when preterm delivery was included in the analyses [100].

In contrast, a large prospective cohort study recorded a more than twofold risk for CP in IVF and ICSI offspring even after adjusting for prematurity and multiple births [101]. Furthermore, Hvidtjørn et al. later performed a large systematic review and meta-analysis that described persuasive evidence of an increased risk of CP in IVF offspring based on the included 41 studies. Methodological problems were identified among included studies but the authors concluded that the risk for CP could be explained in part by an increased risk of preterm delivery and the concept of vanishing embryos in early pregnancy [100]. A Danish study showed that IVF singletons in which a coembryo had vanished before 22 weeks of gestational age had nearly twice the risk of CP [103]. More studies are needed for conclusions to be drawn particularly in countries outside of Scandinavia.

NEUROMOTOR DEVELOPMENT

Because most studies addressing the neurodevelopment of ART children have pitfalls in their methodology, two thorough systematic reviews have been performed [104,105]. Both of these tentatively conclude that the neurodevelopment of children born from ART is comparable to those born from spontaneous conception. Smaller previous studies might be limited by detection bias [105]; the use of health care resources by ART children has previously been reported as high [81,84] and thus diagnoses being made in this group are likely higher than in non-ART children. Furthermore, studies are often unmatched or unadjusted yet there is a multiplicity of factors potentially influencing child development. Recently, for example, developmental disorder has been associated with low fertility rather than ART per se [106].

COGNITIVE FUNCTION

One of the first prospective studies reported increased risks for intellectual delay in ICSI children aged 1 [107]. However, since this study no differences were reported in ICSI children at 2 years [108] or 5 years [109] when assessed by Bayley Scales and the Psychomotor Development Index, respectively. An age-matched study observed marginally but significantly lower IQ levels in ICSI children aged 5–8 years although mean IQ levels were still within normal range [110], and a follow-up study by Leunens et al. found that this difference in cognitive ability disappeared by 10 years [111]. A large international study found no significant differences for intelligences between ICSI, IVF, and naturally conceived children when assessed by verbal and performance tests [112], although older maternal age was significantly linked to lower IQ in the ART group [112]. In contrast a data analysis of 2.5 million birth records reported a small increased intellectual disability risk associated with the ICSI method [113]. This study benefits from its large size but it did not take into account influences of parental educational achievement or socioeconomic status and these demographic factors have been suggested as more important determinants for mental developmental outcome in ICSI children rather than ICSI per se [114].

SCHOOL PERFORMANCE

Children conceived by reproductive techniques appear to perform normally at school [53,115–117]. One study even reports higher results in standardized tests in IVF children [118]. As a follow-up on the Dutch OMEGA study [53], the school performance of IVF singletons was analyzed by means of requirement for educational support and by results from national tests. No evidence of educational limitations was obtained [115]. The same authors demonstrated no differences in cognitive function as assessed by visual motor function, perception, coordination, attention, and information processing [116].

SOCIO-EMOTIONAL DEVELOPMENT AND PSYCHOLOGICAL DISORDERS

The socio-emotional development of IVF children when assessed in terms of psychosocial adjustment and peer relationships appears normal [119–121]. A large collaborative study examined the behavior functioning of ICSI children and found minimal differences compared to matched controls [119]. Interestingly, parents of spontaneously conceived children reported higher levels of parental stress and dysfunction. It is possible that ICSI parents are more tolerant of stress associated with parenting after prolonged attempts at conception. Similar results were obtained in another large European study of 5 year olds [120] and in ART children aged 7–21 years [121]. There has only been one report of lower self-esteem in IVF adolescents compared to controls [122], but this result lost significance when young people who experienced difficulties at school were excluded from the study.

Despite normal socio-emotional development, psychological disorders appear more prevalent among offspring of children conceived by assisted techniques. Trends toward increasingly withdrawn behavior or depressive symptoms and fewer externalizing behaviors were described in IVF children when evaluated through parent and teachers assessments [115]. However, this was not the case when evaluated by the children themselves [123]. Similar symptoms of anxiety and depression were described in a recent cross-sectional analysis [124] and the follow-up cohort study by Beydoun et al. suggests that depressive symptoms continue into young adulthood [88].

AUTISM

Studies investigating the association between autism and reproductive techniques report inconsistent results and multiple methodological problems [86,105,113,125,126]. The heterogenic effect of autism means weak associations will only be apparent in very large samples [125]. Furthermore, the diagnosis and recognition of autism have increased in recent decades making an association with IVF difficult to derive [125]. One large population-based cohort study reported no significant risk for autism in ART offspring [125]. The study was large, well-adjusted, and assessed different types of conception separately although the follow-up time of some children was short (<6 years). Of interest, the adjusted analysis found an increased risk of autism following exposure to follicle stimulating hormone (FSH). The authors suggest interpreting these results with caution as they were unable to assess FSH use in detail or in the context of underlying subfertility. Another large prospective study has since suggested that autism risk was associated specifically with the techniques of ovulation induction/intrauterine insemination [105]. This was of interest given that these techniques include hormonal treatment alone. However, the study was not well adjusted for parental factors and no association between clomiphene citrate used for ovulation induction and autistic disorder was found. Additional studies with more detailed examination of the underlying causes of infertility and specific hormonal medications are needed to make conclusions.

ATTENTION DEFICIT DISORDER

The cross-sectional analysis by Beydoun et al. reported attention deficit disorder (ADD) in more than one-third of included IVF adolescents [88]. However, other studies found lower levels of ADD in IVF children with a

TABLE 29.4 Some Proposed Effects of IVF Treatment on Characteristics of Socio-Emotional Development

	Characteristic	Effect
1	Parental stress	$\downarrow\leftrightarrow$
2	Parent–child dysfunction	$\downarrow\leftrightarrow$
3	Child emotional behavior	$\leftrightarrow$
4	Child hyperactivity	$\leftrightarrow$
5	Peer problems among child	$\leftrightarrow$
6	Self-esteem	$\downarrow\leftrightarrow$

$\downarrow$, Likely reduced; $\leftrightarrow$, no effect; $\uparrow$, likely increased.

prevalence that is comparable to the background population [116,118] and a large data linkage did not find an association of ADD with IVF per se [127]. Instead, the last study identified many confounders that could explain a reported association; these included not cohabiting with a partner, smoking, subfertility, raised BMI, low birth weight, young maternal age, and maternal preeclampsia [127]. More studies are needed that take into account the complex confounding situation (Table 29.4).

CONCLUSIONS

The vast majority of children born through ART are born healthy and without complication. The data we have so far is reassuring regarding the long-term health outcomes of these children. There is some evidence that cardiovascular and metabolic health may be affected in children conceived through ART. Risk factors for these diseases including increased adiposity and higher BP are more apparent in ART-conceived children. As these children progress into later life this effect may become more evident. Whether this effect is a result of ART, or an indirect result of ART-associated obstetric complications, or a result of subfertile parents, remains unclear and requires further study. It is important that we continue to monitor the health outcomes of children conceived through ART as they enter middle age to better understand the impact this technology has on health.

References

[1] ESHRE. 2011. ESHRE ART fact sheet. In: Embryology ESoHRa, editor.

[2] Zegers-Hochschild F, Adamson GD, de Mouzon J. The International Committee for Monitoring Assisted Reproductive Technology and the World Health Organization revised glossary on ART terminology. Hum Reprod 2009;24:2683–7.

[3] Hansen M, Bower C, Milne E, de Klerk N, Kurinczuk JJ, Hansen M, et al. Assisted reproductive technologies and the risk of birth defects – a systematic review. Hum Reprod 2005;20(2):328–38.

[4] Hart R, Norman RJ, Hart R, Norman RJ. The longer-term health outcomes for children born as a result of IVF treatment: Part I. General health outcomes. Hum Reprod Update 2013;19(3):232–43.

[5] Lancaster PA. Congenital malformations after *in-vitro* fertilization. Lancet 1987;88:1392–3.

[6] Tan SL, Doyle P, Campbel S, Beral V, Rizk B, Brinsden P. Obstetric outcome of *in vitro* fertilization pregnancies compared with normally conceived pregnancies. Am J Obstet Gynecol 1992;167:778–84.

[7] Morin NC, Wirth FH, Johnson DH, Frank LM, PresbrgNJ, van der Water VL. Congenital malformations and psychosocial development in children conceived by *in vitro* fertilization. J Pediatr 1989;115:222–7.

[8] Hansen M, Bower C, Milne E, de Klerk N, Kurinczuk J. Assisted reproductive technologies and the risk of birth defects – a systematic review. Hum Reprod 2005;20:328–38.

[9] Allen VM, Wilson RD. Pregnancy outcomes after assisted reproductive technology. J Obstet Gynecol Can 2006;173:220–33.

[10] Bertelsmann H, de Carvalho Gomes H, Mund M, Bauer S, Matthlas K. The risk of malformation following assisted reproduction. Dtsch Arztebl Int 2008;105:11–7.

[11] Wen J, Jang J, Ding C, Dai J, Liu Y, Xia Y, et al. Birth defects in children conceived by *in vitro* fertilization and intracytoplasmic sperm injection: a meta-analysis. Fertil Steril 2012;97:1331–7.

[12] Buckett WM, Tan SL. Congenital abnormalities in children born after assisted reproductive techniques: how much is associated with the presence of infertility and how much with its treatment? Fertil Steril 2005;84:1318–9.

[13] Okun N, Sierra S. Pregnancy outcomes after assisted human reproduction. J Obstet Gynecol Can 2014;36:64–83.

[14] Reefhuis J, Honein MA, Schieye LA, Correa A, Hobbs CA, Rasmussen SA. Assisted reproductive technology and major structural birth defects in the United States. Hum Reprod 2009;24:360–6.

[15] Macaldowie A, Wang YA, Chambers GM, Sullivan EA. Assisted Reproductive Technology in Australia and New Zealand 2011. Sydney: National Perinatal Epidemiology and Statistics Unit, the University of New South Wales; 2013.

[16] Johnson MD. Genetic risks of intracytoplasmic sperm injection in the treatment of male infertility: recommendations for genetic counseling and screening. Fertil Steril 1998;70:397–411.

[17] Bonduelle M, Van Assche E, Joris H. Prenatal testing in ICSI pregnancies: incidence of chromosomal abnormalities in 1586 karyotypes and relation to sperm parameters. Hum Reprod 2002;17:2600–14.

[18] Davies MJ, Moore VM, Willson KJ, Van Essen P, Priest K, Scott H, et al. Reproductive technologies and the risk of birth defects. N Engl J Med 2012;366(19):1803–13.

[19] Alukal JP, Lamb DJ. Intracytoplasmic sperm injection (ICSI) – what are the risks? Urol Clin N Am 2008;35:277–88.

[20] Zhu JL, Basso O, Obel C. Infertility, infertility treatment, and congenital malformations: Danish national birth cohort. BMJ 2006;333:679.

[21] Rimm AA, Katayama AC, Katayama KP. A meta-analysis of the impact of IVF and ICSI on major malformations after adjusting for the effect of subfertility. J Assist Reprod Genet 2011;28:699–705.

[22] Davies MJ, Moore VM, Willson KJ, Van Essen P, Priest K, Scott H, et al. Reproductive technologies and the risk of birth defects. N Engl J Med 2012;366(19):1803–13.

[23] Hales CN, Barker DJ. Type II diabetes mellitus: the thrifty phenotype hypothesis. Diabetologia 1992;35:595–601.

[24] Dupont C, Sifer C. A review of outcome data concerning children born following assisted reproductive technologies. ISRN Obstet Gynecol 2012;.

[25] Hart, et al. Is there a place in infertility practice for the use of oil-based tubal flushing? ANZJOG 2014;54:1–2.

[26] Blickstein I. Estimation of iatrogenic monozygotic twinning rate following assisted reproduction: pitfalls and caveats. Am J Obstet Gynecol 2005;192:365–8.

[27] Miliki AA, Jun SH, Hinkley MD, Behr B, Guidice LC, Westphal LM. Incidence of monozygotic twinning after blastocyst transfer compared to cleavage-stage transfer. Fertil Steril 2003;79:503–6.

[28] Pinborg A, Wennerholm UB, Romundstad LB, Loft A, Aittomaki K, Soderstrom-Anttila V, et al. Why do singletons conceived after reproduction technology have adverse perinatal outcome? Systematic review and meta-analysis. Hum Reprod Update 2012;19:87–104.

[29] Pandian Z, Bhattacharya S, Ozturk O, Serour G, Templeton A. Number of embryos for transfer following *in-vitro* fertilization or intra-cytoplasmic sperm injection. Cochrane Database Syst Rev 2009;2. CD003416.

[30] Pinborg A. IVF/ICSI twin pregnancies: risks and prevention. Hum Reprod Update 2005;11:575–93.

[31] Pinborg A, Loft A, Nyboe AA. Neonatal outcome in a Danish national cohort of 8602 children born after *in vitro* fertilization or intracytoplasmic sperm injection: the role of twin pregnancy. Acta Obstet Gynecol Scand 2004;83:1071–8.

[32] Kailen B, Olausson PO, Nygren KG. Neonatal outcome in pregnancies from ovarian stimulation. Obstet Gynecol 2002;100:414–9.

[33] Kapiteijn K, de Bruijn CS, de Boer E, de Craen AJ, Burger CW, van Leeuwen FE. Does subfertility explain the risk of poor perinatal outcome after IVF and ovarian hyperstimulation? Hum Reprod 2006;21:3228–34.

[34] D'Angelo DV, Whitehad N, Helms K, Barfield W, Ahluwalia IB. Birth outcomes of intended pregnancies among women who used assisted reproductive technology, ovarian stimulation, or no treatment. Fertil Steril 2011;96:314–20.

[35] Klemetti R, Sevon T, Gissler M, Hemminki E. Health of children born after ovulation induction. Fertil Steril 2010;93:1157–68.

[36] Aaris Henningsen AK, Pinborg A, Lidegaard O, Vestergaard C, Forman JL, Andersen AN. Perinatal outcomes of singleton siblings born after assisted reproductive technology and spontaneous conception: Danish national sibling-cohort study. Fertil Steril 2011;95:959–63.

[37] Raatikainen K, Kuivasaari-Pinnen P, Hippelainen M, Heinonen S. Comparison of the pregnancy outcomes of subfertile women after fertility treatment and in naturally conceived pregnancies. Hum Reprod 2012;27:1162–9.

[38] Gjerris AC, Loft A, Pinborg A, Christiasen M, Tabor A. First-trimester screening markers are altered in pregnancies conceived after IVF/ICSI. Ultrasound Obstet Gynecol 2009;33:8–17.

[39] Schieve LA, Meikle SF, Ferre C. Low and very low birth weight in infants conceived with use of assisted reproductive technology. N Engl J Med 2002;326:731.

[40] Helmerhorst FM, Perquin DA, Donker D, Keirse MJ. Perinatal outcome of singletons and twins after assisted conception: a systematic review of controlled studies. BMJ 2004;328:261–7.

[41] Pinborg A, Loft A, Schmidt L. Maternal risks and perinatal outcome in a Danish national cohort of 1005 twin pregnancies: the role of *in vitro* fertilization. Acta Obstet Gynecol Scand 2004;83:75–9.

[42] Manipalviratn S, DeCherney A, Segars J. Imprinting disorders and assisted reproductive technology. Fertil Steril 2009;91(2):305–15.

[43] Vermeiden JP, Bernardus RE. Are imprinting disorders more prevalent after human *in vitro* fertilization or intracytoplasmic sperm injection? Fertil Steril 2013;99:642–51.

[44] Sutcliffe AG, Peters CJ, Bowdin S, Temple K, Reardon W, Wilson L. Assisted reproductive therapies and imprinting disorders – a preliminary British survey. Hum Reprod 2006;21:1009–11.

[45] Doornbos ME, Maas SM, McDonnell J, Vermeiden JP, Hennekam RC. Infertility, assisted reproduction technologies and imprinting disturbances: a Dutch study. Hum Reprod 2007;22:2476–80.

[46] Lammers TH, van Haelst MM, Alders M, Cobben JM. Silver–Russel syndrome in the Netherlands. Tijdschr Kindergeneeskd 2012;80:86–91.

[47] Vermeiden JP, Barnardus RE. Are imprinting disorders more prevalent after human *in vitro* fertilization or intracytoplasmic sperm injection? Fertil Steril 2013;99:642–51.

[48] Gnoth C, Godehardt D, Frank-Herrmann P, Freundi G. Time to pregnancy: results of the German Prospective Study and impact on the management of infertility. Hum Reprod 2003;18:1959–66.

[49] Doornbos ME, Maas SM, McDonnell J, Vermeiden JP, Hennekam RC. Infertility, assisted reproduction technologies and imprinting disturbances: a Dutch study. Hum Reprod 2007;22:2476–80.

[50] Beydoun HA, Sicignano N, Beydoun MA, Matson DO, Bocca S, Stadtmauer L, et al. A cross sectional evaluation of the first cohort of young adults conceived by *in vitro* fertilisation in the United States. Fertil Steril 2010;95:528–33.

[51] Ceelen M, van Weissenbruch MM, Roos JC, Vermeiden JP, van Leeuwin FE, Delemarre-van de Waal HA. Body composition in children and adolescents born after *in vitro* fertilisation or spontaneous conception. J Clin Endocrinol Metab 2007;92:3417–23.

[52] Belva F, Painter R, Bonduelle M, Devroey P, De Schepper J. Are ICSI adolescents at risk for increased adiposity? Hum Reprod 2012;27:257–64.

[53] Ceelen M, Weissenbruch MM, Vermeiden JP, van Leeuwen FE, Delemarre-van de waal HA. Cardiometabolic differences in children born after *in vitro* fertilisation: follow up study. J Clin Endocrinol Metab 2008;93:1682–8.

[54] Ceelen M, van Weissenbruch MM, Prein J, Smitt JJ, Vermeiden JP, Spreeuwenberg M, et al. Growth during infancy and early childhood in relation to blood pressure and body fat measures at aged 8–18 years of IVF children and spontaneously conceived controls born to subfertile parents. Hum Reprod 2009;24:2788–95.

[55] Sakka SD, Loutradis D, Kanaka-Gantenbein C, Margeli A, Papastamatakaki M, Papassotiriou I, et al. Absence of insulin resistance and low grade inflammation despite early metabolic syndrome manifestations in children born after *in vitro* fertilisation. Fertil Steril 2010;94:1693–9.

[56] Seggers J, Haadsma ML, La Bastide-Van Gemert S, Heineman MJ, Middelberg KJ, Roseboom TJ, et al. Is ovarian hyperstimulation associated with higher blood pressure in 4 year old IVF offspring? Part 1: multivariable regression analysis. Hum Reprod 2014;29:502–9.

[57] Belva F, Roelants M, Painter R, Bonduelle M, Devroey P, De Schepper J. Pubertal development in ICSI children. Hum Reprod 2012;27:1156–61.

[58] Scherrer U, Rimoldi SF, Rexhaj E, Stuber T, Duplain H, Garcin S, et al. Systemic and pulmonary vascular dysfunction in children conceived by assisted reproductive techniques. Circulation 2012;125:1890–6.

[59] Lou H, Le F, Zheng Y, Li L, Wang L, Wanf N, et al. Assisted reproductive technologies impair the expression and methylation of insulin induced gene-1 and sterol regulatory element binding factor 1 in the fetus and placenta. Fertil Steril 2014;101:974–80.

[60] Sakka SD, Margeli A, Loutradis D, Chrousos GP, Papassotiriou I, Kanaka-Gantenbein C. Gender dimorphic increase in RBP-4 and NGAL in children born after IVF: an epigenetic phenomenon? Eur J Clin Invest 2013;43:439–48.

[61] Miles HL, Hofman PL, Peek J, Harris M, Wilson D, Robinson EM, et al. *In vitro* fertilisation improves childhood growth and metabolism. J Clin Endocrinol Metab 2007;92:3441–5.

[62] Kallen B, Finnstrom O, Lindam A, Nilsson E, Nygren KG, Olausson PO. Cancer risk in children and young adults conceived by *in vitro* fertilisation. Pediatrics 2010;126:270–6.

[63] Petridou ET, Sergentanis TN, Panagopoulou P, Moschovi M, Polychronopoulou S, Baka M, et al. *In vitro* fertilisation and risk of childhood leukaemia in Greece and Sweden. Paediatr Blood Cancer 2012;58:930–6.

[64] Puumala SE, Ross JA, Feusner JH. Parental infertility, infertility treatment and hepatoblastoma: a report from the Children's Oncology Group. Hum Reprod 2012;27:1656–9.

[65] Santos F, Hyslop L, Stojkovic P. Evaluation of epigenetic marks in human embryos derived from IVF and ICSI. Hum Reprod 2010;2387–95.

[66] Dawson MA, Kouzarides T, Huntly BJP. Targeting epigenetics readers in cancer. N Engl J Med 2012;367:647–57.

[67] Carrie L, Williams MB, Kathryn J, Bunch MA, Charles A, Stiller MA, et al. Cancer risk among children born after assisted conception. N Engl J Med 2013;369:1819–27.

[68] Bruinsma F, Venn A, Lancaster P. Incidence of cancer in children born after *in vitro* fertilisation. Hum Reprod 2000;15:604–7.

[69] Raimondi S, Pedotti P, Taioli E. Meta analysis of cancer incidence in children born after assisted reproductive technologies. Br J Cancer 2005;93:1053–6.

[70] Dommering CJ, van der Hout AH, Meijers-Heijboer H, Marees T, Moll AC. IVF and retinoblastoma revisited. Fertil Steril 2012;97:79–81.

[71] Folix-L'Helias L, Aerts I, Marchand L, Lumbroso Le R, Gauthier-Villars M, Labrune P, et al. Are children born after infertility treatment at increased risk of retinoblastoma? Hum Reprod 2012;27:2186–92.

[72] Cetinkaya F, Gelen SA, Kervancioglu E, Oral E. Prevalence of asthma and other allergic diseases in children born after *in vitro* fertilisation. Allergo Immunopathol 2009;37:11–3.

[73] Sicignano N, Hind A, Russell H, Jones H, Oehninger S. A descriptive study of asthma in young adults conceived by IVF. Reprod Biomed Online 2010;21(6):812–8.

[74] Carson C, Sacker A, Kelly Y, Redshaw M, Kurinczuk J, Quigley MA. Asthma in children born after infertility treatment: findings from the UK Millennium Cohort Study. Hum Reprod 2012;28:471–9.

[75] Kallen B, Finnstrom O, Nygren KG, Otterblad Olausson P. Asthma in Swedish children conceived by *in vitro* fertilisation. Arch Dis Child 2012;98(2):92–6.

[76] Harju M, Keski-Nisula L, Raatikainen K, Pekkanen J, Heinonen S. Maternal fecundity and asthma among offspring – is the risk programmed preconceptionally? Retrospective observational study. Fertil Steril 2012;99:761–7.

[77] Sakka SD, Malamiitsi-Puchner A, Loutradis D, Chrousos GP, Kanaka-Gantenbein C. Euthyroid hyperthyrotropinemia in children born after *in vitro* fertilisation. J Clin Endocrinol Metab 2009;94:1338–41.

[78] Onal O, Ercan O, Adcal E, Ersen A, Onal Z. Subclinical hypothroidism in *in vitro* fertilisation babies. Acta Paediat 2012;101:248–52.

[79] Kai CM, Main KM, Andersen AM, Loft A, Chellakooty M, Skakkebaek NE, et al. Serum insulin like growth factor I (IGF-I) and growth in children born after assisted reproduction. J Clin Endocrinol Metab 2006;91:4352–60.

[80] Basatemur E, Shevlin M, Sutcliffe A. Growth of children conceived by IVF and ICSI up to 12 years of age. Reprod Biomed Online 2010;20:144–9.

[81] Ludwig AK, Katalinic A, Theyen U, Sutcliffe AG, Diedrich K, Ludwig M. Physical health at 5.5 years of age of term born singletons after intracytoplasmic sperm injection: results of a prospective, controlled single blind study. Fertil Steril 2009;91:115–24.

[82] Tornqvist K, Finnstrom O, Kallen B, Lindam A, Nilsson E, Nygren KG, et al. Ocular malformations or poor visual acuity in children born after *in vitro* fertilisation in Sweden. Am J Ophthalmol 2010;150:23–6.

[83] Sutcliffe AG, Taylor B, Saunders K, Thornton S, Lieberman BA, Grudzinskas JG. Outcome in the second year of life after *in vitro* fertilisation by intra cytoplasmic sperm injection: a UK case control study. Lancet 2001;357:2080–4.

[84] Bonduelle M, Wennerholm UB, Loft A, Tarlatzis BC, Peters C, Henriet S, et al. A multi centre cohort study of the physical health of 5 year old children conceived after intracytoplasmic sperm injection, *in vitro* fertilisation and natural conception. Hum Reprod 2005;20:413–9.

[85] Belva F, Henriet S, Liebaers I, Van Steireghem A, Cleestin-Westreich S, Bonduelle M. Medical outcome of 8 year old singleton ICSI children (born > or = 32 weeks gestation) and a spontaneously conceived comparison group. Hum Reprod 2007;22:506–15.

[86] Knoester M, Helmerhost FM, Vandenbroucke JP, van der Wester-laken LA, Walther FJ, Veen S. Perinatal outcome, health, growth and medical care utilisation of 5 to 8 year old intracytoplasmic sperm injection singletons. Fertil Steril 2008;89:1133–46.

[87] Ceelen M, van Weissenbruch MM, Vermeiden JP, van Leeuwen FE, Delemarre- van de waal HA. Pubertal development in chil-dren and adolescents born after IVF and spontaneous concep-tion. Hum Reprod 2008;23:2791–8.

[88] Beydoun HA, Sicignanon N, Beydoun MA, Matson DO, Bocca S, Stadtmauer L, et al. Pubertal development of the first cohort of young adults conceived by in vitro fertilisation in the United States. Fertil Steril 2011;95:528–33.

[89] Nakamura Y, Hirata K, Hattori H, Kuchiki M, Sakomoto E, Kyono K. Follow up of child growth regarding new technologies: testicular sperm extraction (TESE), in vitro maturation (IVM) and assisted oocytes activation (AOA). Hum Reprod 2013;28(287).

[90] Wennerholm F U.B., Albertsson-Wikland F K., Bergh C, Ham-berger L, Niklasson A, Nilsson L, et al. Postnatal growth and health in children born after cryopreservation as embryos. Lancet 1998;351:1085–90.

[91] Saunders K, Spensley J, Munro J, Halasz G. Growth and physical outcome of children conceived by in vitro fertilization. Pediatrics 1996;97:688–92.

[92] Basatemur E, Shevlin M, Sutcliffe A. Growth of children con-ceived by IVF and ICSI up to 12 years of age. Reprod Biomed Online 2010;20:144–9.

[93] Koivurova S, Hartikainen AL, Sovio U. Growth, psychomotor development and morbidity up to 3 years of age in children born after IVF. Hum Reprod 2003;18:2328–36.

[94] Ericson A, Nygren KG, Olausson PO, Kallen B. Hospital care uti-lization of infants born after IVF. Hum Reprod 2002;19:929–32.

[95] Chambers GM, Lee E, Hoang VP, Hansen M, Bower C, Sullivan EA. Hospital utilisation, costs and mortality rates during the first 5 years of life: a population study of ART and non ART single-tons. Hum Reprod 2013;29:601–10.

[96] Koivurova S, Hartikainen AL, Gissler M, Hemminki E, Jarvelin MR. Post neonatal hospitalisation and health care costs among IVF children: a 7 year follow up study. Hum Reprod 2007;22:2136–41.

[97] De Schepper J, Belva F, Schiettecatte J, Anckaert E, Tournaye H, Bonduelle M. Testicular growth and tubular function in prepu-bertal boys conceived by intracytoplasmic sperm injection. Horm Res 2009;71:359–63.

[98] Belva F, Bonduelle M, Painter RC, Schiettecatte J, Devroey P, De Schepper J. Serum inhibin B concentrations in pubertal boys conceived by ICSI: first results. Hum Reprod 2010;25:2811–4.

[99] Klemetti R, Sevon T, Gissler M. Health of children born as a result of in vitro fertilisation. Pediatrics 2006;118:1819–27.

[100] Hvidtjørn D, Grove J, Schendel DE, Vaeth M, Ernst E, Nielsen LF, et al. Cerebral palsy among children born after in vitro fertilization: the role of preterm delivery – a population-based, cohort study. Pediatrics 2006;118:475–82.

[101] Zhu JL, Hvidtjørn D, Basso O, Obel C, Thorsen P, Uldall P. Parental infertility and cerebral palsy in children. Hum Reprod 2010;25:3142–5.

[102] Strömberg B, Dahlquist G, Ericson A, Finnström O, Köster M, Stjernqvist K. Neurological sequelae in children born after in-vitro fertilization: a population-based study. Lancet 2002;359:461–5.

[103] Pinborg A, Lidegaard O, la Cour FN, Nyboe AA. Consequenc-es of vanishing twins in IVF/ICSI pregnancies. Hum Reprod 2005;20(10):2821–9.

[104] Middelburg KJ, Heineman MJ, Bos AF, Hadders-Algra M. Neuro-motor, cognitive, language and behavioural outcome in children born following IVK or ICSI – a systematic review. Hum Reprod Update 2008;14:19–31.

[105] Bay B, Mortensen E, Kesmodel US. Assisted reproduction and child neurodevelopmental outcomes: a systematic review. Fertil Steril 2013;100:844–53.

[106] Zhu JL, Obel C, Basso O, Olsen J. Parental infertility and de-velopmental co-ordination disorder in children. Hum Reprod 2010;25:908–13.

[107] Bowen JR, Gibson FL, Leslie GI, Saunders DM. Medical and developmental outcome at 1 year for children conceived by intracytoplasmic sperm injection. Lancet 1998;351:1529–34.

[108] Bonduelle M, Joris H, Hofmans K, Liebaers I, Van Steirteghem A. Mental development of 201 ICSI children at 2 years of age. Lancet 1998;351:1553.

[109] Leslie GI, Gibson FL, McMahon C, Cohen J, Saunders DM, Tennant C. Children conceived using ICSI do not have an in-creased risk of delayed mental development at 5 years of age. Hum Reprod 2003;18:2067–72.

[110] Knoester M, Helmerhorst FM, van der Westerlaken LA, Walther FJ, Veen S. Leiden Artificial Reproductive Techniques Follow-up Project. Matched follow-up study of 5–8-year-old ICSI single-tons: child behaviour, parenting stress and child (health-related) quality of life. Hum Reprod 2007;22:3098–107.

[111] Leunens L, Celestin-Westreich S, Bonduelle M, Liebaers I, Ponjaert-Kristoffersen I. Cognitive and motor development of 8 year old children born after ICSI compared to spontaneously conceived children. Hum Reprod 2006;21:2922–9.

[112] Ponjaert-Kristoffersen I, Bonduelle M, Barnes J, Nekkebroeck J, Loft A, Wennerholm UB, et al. An International collaborative study of ICSI conceived 5 year olds. Child outcomes: cognitive and motor assessment. Pediatrics 2005;115:283–9.

[113] Sandin S, Nygren KG, Lliadou A, Hultman CM, Reichenberg A. Autism and mental retardation among offspring born after in vitro fertilization. JAMA 2013;75–84.

[114] Leslie GI. Mental development of children conceived using in-tracytoplasmic sperm injection. The current evidence. Minerva Ginecol 2004;56:247–57.

[115] Wagenaar K, van Weissenbruch MM, Knol DL, Cohen-Kettenis PT, Delemarre-van de waal HA, Huisman J. Behaviour and socio-emotional functioning in 9-18 year old children born after in vitro fertilisation. Hum Reprod 2009;24:913–21.

[116] Wagenaar K, van Weissenbruch MM, Knol DL, Cohen-Kettenis PT, Delemarre-van de waal HA, Huisman J. Information process-ing, attention and visual motor function of adolescents born af-ter in vitro fertilisation compared with spontaneous conception. Hum Reprod 2009;24:913–21.

[117] Carson C, Redshaw M, Psychol C, Sacker A, Yvonne B, Kurinc-zuk J, et al. Effects of pregnancy planning, fertility, and assisted reproductive treatment on child behavioural problems at 5 and 7 years: evidence from the Millennium Cohort Study. Fertil Steril 2013;99:456–63.

[118] Mains L, Zimmerman M, Blaine J, Stegmann B, Sparks A, Ansley T, et al. Achievement test performance in children conceived by IVF. Hum Reprod 2010;25:2605–11.

[119] Ponjaert-Kristoffersen I, Tjus T, Nekkebroeck J, Squires J, Verte D, Heimann M, et al. Psychological follow up study of 5 year old ICSI children. Hum Reprod 2004;19:2791–7.

[120] Barnes J, Sutcliff AG, Kristoffersen I, Loft A, Wennerholm U, Tarlatzis BC, et al. The influence of assisted reproduction on fam-ily function and children's socio-emotional development: results from European study. Hum Reprod 2004;19:1480–7.

[121] Zhu JL, Obel C, Basso O, Henriksen TB, Bech BH, Hvidtjorn D, et al. Infertility, infertility treatment, and behavioural problems in the offspring. Paedaitr Perinat Epidemiol 2011;25:466–77.

[122] Golombok S, Owen L, Blake L, Murray C, Jadva V. Parent-child relationships and the psychological well-being of 18-year-old adolescents conceived by in vitro fertilisation. Hum Fertil 2009;12(2):63–72.

[123] Wagenaar K, van Weissenbruch MM, van Leeuwen FE, Cohen-Kettenis PT, Delemarre-van de Waal HA, Schats R, et al. Self-reported behavioural and socio-emotional functioning of 11- to 18-year-old adolescents conceived by *in vitro* fertilization. Fertil Steril 2011;(95):611–6.

[124] Ozbaran B. Psychiatric evaluation of children born with assisted reproductive technologies and their mothers: a clinical study. Noropsikiyatri Ars 2013;50:59–64.

[125] Hvidtjørn D, Grove J, Schendel D, Schieve LA, Svaerke C, Ernst E, et al. Risk of autism spectrum disorders in children born after assisted conception: a population-based follow-up study. J Epidemiol Community Health 2011;65(6):497–502.

[126] Zachor DA, Itzchak E. Assisted reproductive technology and risk for autism spectrum disorder. Res Dev Disab 2011;32(6): 2950–6.

[127] Kallen AJ, Finnstrom OO, Lindam AP, Nilsson EM, Nygren KG, Otterblad PM. Is there an increased risk for drug treated attention deficit hyperactivity disorder in children born after *in vitro* fertilisation? Eur J Paediatr Neurol 2011;15:247–53.

30

Influence of Male Hyperinsulinemia on IVF Outcome

Edolene Bosman, MSc, Aletta Dorothea Esterhuizen, PhD, Francisco Antonio Rodrigues, MBBCh

TUPS Stress, Medfem Fertility Clinic, Bryanston, Johannesburg, South Africa

INTRODUCTION

Insulin is a hormone released by the β-cells of the pancreatic islands of Langerhans in response to elevations in blood glucose. Insulin is a potent anabolic hormone that is essential in the transport of blood glucose across the largely impermeable cell membrane. Insulin binds to a receptor on the cell surface and causes a cascade of events that lead to the uptake of sugar. When receptors do not signal appropriately, the blood sugar will remain elevated unless the pancreas makes more insulin. The effect of insulin on the body is far reaching. Various processes are influenced by insulin. Hyperinsulinemia contributes to many autoimmune and degenerative disorders by stimulating the production of inflammatory chemicals and free radicals [1,2]. One of the many aspects in the body controlled by insulin is the maintenance of carbohydrate, fat, and protein metabolism. The prevalence of hyperinsulinemia and insulin resistance is on the rise in Western societies [3]. A sedentary lifestyle, unhealthy diet, and increased body mass index (BMI) predisposes a person to hyperinsulinemia and insulin resistance. Various hormonal mechanisms have been suggested as mediators of lifestyle effects on insulin resistance. Diets high in saturated fats and trans-fatty acids have been shown to decrease cell membrane fluidity and decrease insulin receptor binding and hormone responsiveness, thus promoting insulin resistance [4]. The combination of a high omega-6 essential fatty acid intake, as in the typical Western diet, with hyperinsulinemia is a potent stimulus to inflammation. This pro-inflammatory state increases the levels of cytokines, such as tumor necrosis factor, and free radical production [1,2]. Spermatozoa are particularly susceptible to damage by free radicals, due to their high content of unsaturated fatty acids

and relative lack of cytosolic antioxidant protection [5,6]. Micronutrients further affect cellular insulin balance, with some researchers stating that the primary defect of insulin-resistance is intracellular imbalance of calcium and magnesium [7]. Tobacco smoking has been found to increase insulin resistance [8,9] while chronic stress has been suggested to promote insulin resistance by increasing adrenalin (epinephrine) and cortisol levels [10]. At the cellular level insulin action is characterized by diverse effects, including changes in vesicle trafficking, stimulation of protein kinases and phosphatases, promotion of cellular growth and differentiation, and activation or repression of transcription [11]. Insulin is also a stimulator of growth factors, including insulin-growth factor 1 (IGF-I). Insulin and insulin growth factor II (IGF-II), as well as IGF-I, promote spermatogonial differentiation into primary spermatocytes [12].

Insulin plays a central role in the regulation of gonadal function; however, its significance in male fertility is poorly understood [13,14]. Glucose uptake and metabolism are essential for the proliferation and survival of sperm cells and is usually carried out through glucose transporters (GLUTS), also known as solute carrier family 2 (SLC2A). Insulin signaling might influence glucose transporter expression during spermatogenesis, steroidogenesis, and sperm maturation. The distribution of GLUTS suggests that they may play different roles not only in internal glucose movement but also in the trafficking of sugars for sperm maturation and motility [14,15]. Human sperm cells acquire the ability for fertilization through the process of capacitation and the acrosome reaction. Insulin plays an important role in the acrosome reaction and both the sperm plasma membrane and the acrosome represent cytological targets for insulin [16].

Handbook of Fertility. http://dx.doi.org/10.1016/B978-0-12-800872-0.00030-5

The insulin receptor is located within the cell membrane and any disturbances affecting the binding of the insulin to the receptors, or the receptor response to insulin, will cause reduced insulin activity or, with time, insulin resistance. Raised insulin levels and cellular insulin insensitivity is associated with poor glucose control. When there is a lack of correlation between carbohydrate uptake and insulin secretion, hyperglycemic periods will be experienced after meals. Chronic periods of hyperinsulinemia has an inhibitory effect on normal spermatogenesis, lowering the synthesis of sex hormone-binding globulin (SHBG), which results in lower testosterone levels and male infertility [17–19]. Production of competent sperm requires the control of hormonal and metabolic processes. Reactive oxygen species (ROS) are products of normal cellular metabolism and includes oxygen ions, free radicals, and peroxides. ROS-mediated damage to sperm is considered to be a major contributing pathology to male infertility [20].

Previous studies concentrated on the influence of insulin-dependent diabetes on semen parameters and the associated decline in male fertility [21–26].

Studies evaluating the relationship between hyperinsulinemia and semen parameters are currently lacking [27]. In the following study by Bosman et al. the negative effect of hyperinsulinemia on semen parameters, specifically CMA_3, hyaluronan binding assay, and DNA fragmentation (TUNEL), was investigated.

SEMEN PARAMETERS AND FUNCTIONAL ASSAYS IN HYPERINSULINEMIC PATIENTS

Semen Analysis and Male Participants

Semen samples were obtained from each research participant after the recommended 3–5 days of sexual abstinence. All the samples were assessed by determining the macroscopic variables such as volume, pH, coagulation, liquefaction, color, and viscosity. A conventional light microscope was used to ascertain sperm concentration and motility according to the World Health Organization recommendations [28]. Sperm morphology was assessed according to the Tygerberg strict criteria [29], while chromomycin A_3 (CMA_3) was assessed according to the method of Esterhuizen et al. [30]. Chromomycin A_3 is used to assess mature DNA packaging in the sperm head with a normal cut off value of $\leq 40\%$.

Fifty-eight men underwent semen analysis on the day of *in vitro* fertilization (IVF) treatment. Data from men with normal sperm counts and G-pattern morphology were included in the study. A hyaluronic binding assay (HBA) was performed (Biocoat, Inc., Horsham, USA) as indicated in the instruction sheet, and the percentage bound sperm calculated. DNA fragmentation was determined

with the ROCHE In Situ Cell Death Detection Kit, Fluorescein (Roche Diagnostics GmBH, Mannheim, Germany). A JC-1 mitochondrial membrane potential detection kit (Cell Technology, Inc. Minneapolis, USA) was used to identify mitochondrial activity. In this test the mitochondria in the neck of the spermatozoa were evaluated immediately after staining, using a Nikon Epi-fluorescence microscope by detecting the red or green fluorescent aggregates. J-aggregates were detected with a "dual-band pass" filter usually designed to detect Rhodamine/Texas Red and FITC. High mitochondrial membrane potential (MMP) was attributed to cells with an intense red fluorescence signal. Three categories of mitochondrial fluorescence activity were noted in each sample, that is, sperm with no mitochondrial activity (group A-green stain), sperm with minor mitochondrial activity in the midpiece (group B), and sperm with high (15 or more red aggregates of fluorescence) activity present in the midpiece (group C).

RESULTS

The mean semen parameter values and the CMA_3, DNA fragmentation, HBA, and MMP scores are shown in Table 30.1. The mean age for the male patients for the two study groups (normo- and hyperinsulinemic groups) were 34.3 and 33.4 years, respectively.

The mean count, motility, and morphology were similar for the normo- and hyperinsulinemic groups, and the values of these parameters fell in the same range. The patients in the hyperinsulinemic group showed a higher percentage of sperm with poor DNA chromatin packaging ($P = 0.04$) and DNA fragmentation ($P = 0.005$). The higher HBA scores found in the normoinsulinemic group ($P = 0.005$) was reflected in the higher IVF rate of 84.45% versus 78.81% for the study group. The MMP values between the two groups showed no difference. No correlation was found between DNA fragmentation and MMP, CMA_3, HBA, or any of the other semen parameters. Results from the correlation analysis performed are reported in Table 30.2. No relationship could be found between DNA fragmentation, morphology, and CMA_3 values in a multiple regression analysis. Only 2.3% of variation in DNA fragmentation could be explained by variation in morphology and CMA_3 ($P = 0.275$).

No correlation was found between insulin and CMA_3, or between insulin and morphology values ($P = 0.9$). A statistically significant, negative correlation was found between morphology and CMA_3 values ($r = -0.82$; $P = 0.0001$).

Hyperinsulinemic conditions had a negative effect on chromatin packaging as measured with the CMA_3 test. The hyperinsulinemic group had a mean CMA_3 value of 52.07% versus 43.16% of the normoinsulinemic group ($P = 0.04$). The replacement of somatic histones by protamines is important for sperm nuclear chromatin compaction, sperm maturation, and fertility [31]. Testing

TABLE 30.1 Semen Parameters on the Day of IVF Aspiration for Normo- and Hyperinsulinemic Patients

Parameter	Normoinsulinemic males (N = 33)	Hyperinsulinemic males (N = 25)
Age	34.33 ± 3.4 (28–42)	33.48 ± 3.1 (28–39)
Sperm count (10^6 mL^{-1})	47.69 ± 27.39 (18–132)	49.48 ± 21.35 (14–107)
Motility (%)	55.15 ± 9.72 (35–75)	53.96 ± 9.39 (40–70)
Morphology (%)	8.44 ± 4.74 (2–21)	9.92 ± 7.34 (3–40)
CMA$_3$ (%)	43.16 ± 8.57 (30–58)*	52.07 ± 3.82 (33–70)*
MMP		
Group A	26.15 ± 9.13 (11–49)	24.8 ± 12.13 (6–50)
Group B	56.2 ± 11.63 (36–77)	56.16 ± 17.02 (21–82)
Group C	17.58 ± 12.83 (2–45)	19.04 ± 13.25 (1–47)
DNA fragmentation	15.3 ± 3.58*	23.3 ± 7.71*
HBA	81.73 ± 11.91*	75.07 ± 12.78*
% Fertilization	84.45 ± 14.84	78.81 ± 23.09
Fasting insulin (μIE/mL)	6.03 ± 1.50	14.61 ± 4.64
% Clinical pregnancies	14/33 = 42.4%	9/25 = 36%

** P < 0.05 statistical significant difference.*

for sperm DNA integrity may help in selecting spermatozoa with the least amount of damage for the use in ART – this in turn may alleviate the financial, social, and emotional problems associated with failed ART attempts. Based on correlations with normal sperm morphology, the hyaluronan binding assay level of differentiating higher from lower expectation of fertility is estimated to be approximately 80% [32,33]. The mean HBA value for the hyperinsulinemic group was 75.07% and below the recommended threshold level. The mean oocyte fertilization rate of 78% in the raised insulin group was also lower than 84% obtained in the normal insulin group.

A number of studies have shown that the MMP ($\Delta\psi_m$) can be used as an indicator of sperm quality. MMP is seen as a strong indicator of sperm function, partly due to its inverse relationship with ROS generation [34]. Although the factors responsible for stimulating increased free radical release from the mitochondria of spermatozoa

are still unresolved, there is evidence for the role of mitochondrial ROS in the etiology of male infertility in human spermatozoa. A possible cause–effect relationship exists between ROS and DNA damage [35,36], but no correlation could be found between DNA damage and MMP values in this study. MMP measurements were made by fluorescent microscopy and not by flow cytometry, with the aim of finding a practical test that can easily be performed in an IVF laboratory without expensive equipment. The fact that no correlation was found between sperm, DNA damage and MMP could possibly be due to the use of fluorescent microscopy as an evaluation technique, which is less sensitive and accurate than flow cytometry.

Clinically, the importance of oxidative stress-mediated male infertility has given rise to a number of commercial antioxidant treatments available. It may prove beneficial to use antioxidant therapies that specially target mitochondria, the source of ROS production [34]. Little effort has been

TABLE 30.2 Correlation (P-Value) of Apoptosis with Different Semen Parameters

Different parameters tested	Correlation analysis	
	Pearson	Spearman
Fasting insulin values	0.175 (0.518)	−0.132 (0.626)
Sperm count	0.135 (0.618)	0.160 (0.554)
Sperm motility	−0.176 (0.513)	−0.315 (0.235)
Sperm morphology	0.336 (0.203)	0.193 (0.474)
CMA$_3$ values	−0.252 (0.346)	−0.145 (0.592)
HBA	0.184 (0.496)	0.176 (0.514)

P-value in parenthesis.

made with improving sperm handling media so that antioxidants and other protectants could improve or maintain the overall functional parameters of spermatozoa [37]. The effect of such sperm preparation techniques still needs to be investigated for hyperglycemic males as it can improve IVF results. Hyperinsulinemic males ended up with poorer IVF results as illustrated in the next two studies.

IVF OUTCOME OF HYPERINSULINEMIC SPERM DONORS

In this study, the prevalence of hyperinsulinemia in a normozoospermic group of sperm donors was determined and the influence of insulin levels on embryo quality was measured in an IVF program. Standard insulin sensitivity and secretion tests are labor intensive and difficult to perform in a large number of patients. Therefore, this study investigated the use of fasting and 2 h postmeal insulin levels as simple indices of insulin sensitivity. The mean concentration of fasting and 2 h postmeal serum insulin levels was measured.

Initial Assessment of Research Subjects

A hundred and fifty-one potential donors were screened over a period of 2 years. Each participant was requested to complete a self-administered lifestyle questionnaire. The aim of the questionnaire was to investigate environmental/lifestyle factors (e.g., tobacco use, alcohol consumption, physical exercise, chronic disease, emotional stress, work place, and medication) that might influence semen parameters and/or hormonal levels. Thereafter, each participant's BMI was calculated. A BMI over 25 kg/m^2 is defined as overweight and a BMI of over 30 kg/m^2 as obese [38]. Only 34 normozoospermic men [28] with a healthy lifestyle met the minimum requirements for participating as sperm donors. The majority of aspirant donors (77%) were disqualified on the fact that their sperm count and morphology were suboptimal [28].

Insulin and Hormonal Testing

A hormonal profile (FSH, LH, SHBG, GH, and testosterone) was performed on each of the research participants to exclude any non-insulin-related abnormal hormonal influences. The required period of fasting (2200 the previous evening until 0800 the next morning) for the fasting insulin blood test was explained to the participants, as well as 3–5 days of sexual abstinence required for the standard semen analysis. A fasting/basal insulin level (I_0) was measured on the morning of testing (before 1000) and then repeated 120 min (I_{120}) after they had a normal breakfast. Insulin, FSH, LH, SHBG, GH, and testosterone assays were performed on the fasting blood sample using an Immulite® Automated Analyser (Siemens Healthcare Diagnostics Inc., Flanders, NJ, USA). The research participants were divided into three groups according to their I_0 and I_{120} insulin levels. Group GG (Good, Good) consisted of participants with normal fasting and I_{120} levels, group GB (Good, Bad) had normal I_0 but abnormal I_{120}, and group BB (Bad, Bad) had abnormal I_0 and I_{120}. The cut-off value for normal basal insulin levels is ≤9.2 µIU/mL and for 120 min posteating 6–20 µIU/mL [39]. The insulin ranges for the different groups are indicated in Table 30.3.

Only normozoospermic males were allowed to participate in the study; their samples were frozen and quarantined according to prescribed rules for donor banks [40]. An analysis of the completed lifestyle questionnaires indicates that all of the participants followed a healthy lifestyle. Two (22.2%) of the participants in group BB with abnormal insulin levels changed their lifestyle after they were informed of their abnormal fasting insulin levels. They started to eat regular meals and took up sports activities. Their basal insulin levels were normal 3 months after adopting a healthy lifestyle and the sperm of both achieved pregnancies.

In Vitro Fertilization

The controlled ovarian hyperstimulation of IVF patients was achieved after GnRH suppression with Lucrin, using recombinant FSH (Gonal F®, Serono-Merck, Gauteng, South Africa) and Luveris®. The patients were triggered for aspiration after 36 h with Ovidrel® (Gonal F, Serono-Merck, Gauteng, South Africa) when two or more follicles were >18 mm. The frozen sperm samples were thawed on the day of aspiration and prepared using a one-step wash and swim-up technique [41]. The oocytes were inseminated with 100,000 motile sperm 3–4 h after aspiration and checked for fertilization 19 h later. The embryos were cultured to the blastocyst stage in Cook Medium [42]. A maximum of two blastocysts were transferred on day 5.

TABLE 30.3   Insulin Ranges for Donor Groups

Insulin groups	Fasting insulin µIU/mL (I_0)	Post-eating insulin µIU/mL (I_{120})
Group GG ($n = 13$)	0–9.2	6–20
Group GB ($n = 10$)	0–9.2	>20
Group BB ($n = 11$)	>9.2	>20

Embryo development and quality was noted for each patient. Day 5 blastocysts were evaluated using a scoring system similar to the one described by Gardner and Schoolcraft [43]. In cases of positive serum HCG levels, a clinical pregnancy was confirmed with ultrasound 2 weeks later.

Results

Approximately 50% of the participants ($n = 19$) had normal basal insulin levels (range 2–23 μIU/mL). The prevalence of raised insulin (>9 μIU/mL) was 44% with a 95% confidence interval of (27.2%; 62.1%). This indicates that the point estimated prevalence is large despite the small sample size and the wide interval estimate for the prevalence. The prevalence for abnormal postmeal insulin (I_{120}) ($\geq$ 20 μIU/mL) was 55.9%. Thirteen participants had normal I_0 and normal I_{120}; they were allocated to group GG. Ten participants were allocated to group GB with normal I_0 but abnormal I_{120}, while group BB consisted of 11 participants with abnormal I_0 and I_{120}.

All the participants had normal LH values. The prevalence of abnormal FSH values and SHBG values were very low for both parameters, namely 2.94%. Five of the participants (14.7%) had abnormally high (>94.8) free index testosterone values. The sperm of one of these five participants achieved a pregnancy in the IVF program. Also, four of these five participants had lower than normal GH levels (<0.1 ng/mL). In the case of one participant this coincided with a low SHBG value (11 nmol/l) and a high BMI of 31.2. The sperm of this particular participant did not achieve a pregnancy in the IVF program. A summary of the data for the different hormonal parameters is presented in Tables 30.4 and 30.5.

The different insulin groups were compared with respect to embryo development. The incidence of 2–4 cell embryo development on day 2, 6–8 cell embryos on day 3, and blastocyst formation on day 5 of culture were assessed. No statistically significant differences were found between the groups for day 2 development ($\chi^2 = 0.55$; $P = 0.58$), day 3 ($\chi^2 = 0.61$; $P = 0.54$), or day 5 development ($\chi^2 = 0.14$; $P = 0.86$). The research participants with normal I_0 levels (groups GG and GB) had a slightly higher percentage of top quality blastocysts (A and B scores) on day 5. Group GG (37.0%) and group GB (32.7%) displayed a higher percentage of good quality blastocysts compared to the abnormal insulin group, group BB (31.3%) (Fig. 30.1).

Groups BB and GB were comparable for the outcome of the biochemical pregnancy rate but Group GG showed a significant increase ($P = 0.04$). The clinical pregnancy rate in group GG (30.7%) and GB (26.3%) were higher than that of group BB (18.1%). However, these clinical differences were not statistically significant ($P = 0.54$) [44].

A high prevalence of increased BMI and low GH levels was noted in the group BB. Some studies suggest that low GH levels may have an influence on insulin sensitivity [45,46]. An increase in BMI and prevalence to obesity has been reported in the Western world [47]. Overweight and obesity in men have been associated with poorer semen quality [48–50], decreased chromatin integrity [51], and infertility [52,53]. The positive effects of a healthy diet, exercise, and lifestyle changes to decrease high insulin levels are well known, and were also demonstrated in two participants of this study who adjusted their lifestyles.

A higher number of good quality blastocysts and pregnancies were obtained in this study from donors with normal fasting insulin levels, suggesting an association between hyperinsulinemia and IVF outcome.

TABLE 30.4 Hormonal and BMI Values of the Donors

Variable	Normal range	Mean, standard deviation, minimum and maximum values	Abnormal hormonal prevalence (%)	95% Confidence interval
Basal insulin (I_0)	0–9.2 μIU/mL	8.76 ± 5.08 (2–23)	44.10	27.2; 62.1
Post-eating insulin (I_{120})	6–20 μIU/mL	23.41 ± 14.16 (5.1–61)	55.90	37.9; 72.8
FSH	0.7–11.1 mIU/mL	3.56 ± 2.03 (0.6–8.6)	2.94	0.07; 15.33
LH	0.8–7.6 mIU/mL	3.66 ± 1.65 (1.1–7.6)	0	N/A
Free testosterone	14.9–94.8	69.36 ± 24.83 (32–127)	14.7*	31.1; 49.5
GH	0.1–1 ng/mL	0.61 ± 0.86 (0.01–2.5)	20.60	8.7; 37.9
SHBG	13–71 nmol/l	27.61 ± 10.63 (11–61)	2.94	0.07; 15.33
BMI	18.5–25.0 kg/m²	25.93 ± 5.68 (19.9–42.5)	35.3**	19.8; 53.5

Four of these five participants also presented with low GH values, while the semen of one of the participants achieved pregnancy.
**Six of these participants fall in group Bad, Bad N/A, not applicable.*
Adapted from Ref. [44].

TABLE 30.5 Summary of IVF Data

Donor group	Group GG (normal I_0 fasting and I_{120} levels)	Group GB (normal I_0 but abnormal I_{120})	Group BB (abnormal I_0 and I_{120})
Number of donors	8	7	7
Number female patients with transfers	26	19	22
Female's mean age	35.4 ± 2.4	34.3 ± 4.2	33.6 ± 4.3
Number mature ova	200	134	174
Number of fertilized ova	168 (84.0%)	93 (69.4%)	120 (68.9%)
Cleavage rate	161 (95.8%)	90 (96.7%)	118 (98.3%)
Good quality embryos	60 (37.3%)	29 (32.2%)	37 (31.3%)
Mean number embryos transferred	1.96	1.73	1.73
Biochemical pregnancy	13* (50.0%)	5 (26.3%)	5 (22.7%)
Clinical pregnancy	8** (30.7%)	5 (26.3%)	4 (18.1%)
Multiple pregnancy per transfer	3 (11.5%)	0 (0%)	2 (9.1%)

** The biochemical pregnancy in group GG was significantly higher than group GB and BB (P = 0.04).*
*** The clinical pregnancy in the three groups were not statistically significant different (P = 0.54).*
Adapted from Ref. [44].

Infertile males with hyperinsulinemia should be advised to take their BMI, diet, and lifestyle into consideration as healthy adaptations may enhance their IVF outcome.

Further investigations on the relationship between male hyperinsulinemia, infertility, and IVF outcome is discussed in the next section.

INFLUENCE OF MALE HYPERINSULINEMIA ON IVF OUTCOME

The study group consisted of 82 (n = 82) couples participating in the IVF program at Medfem Clinic, Johannesburg. Female patients younger than 38 years

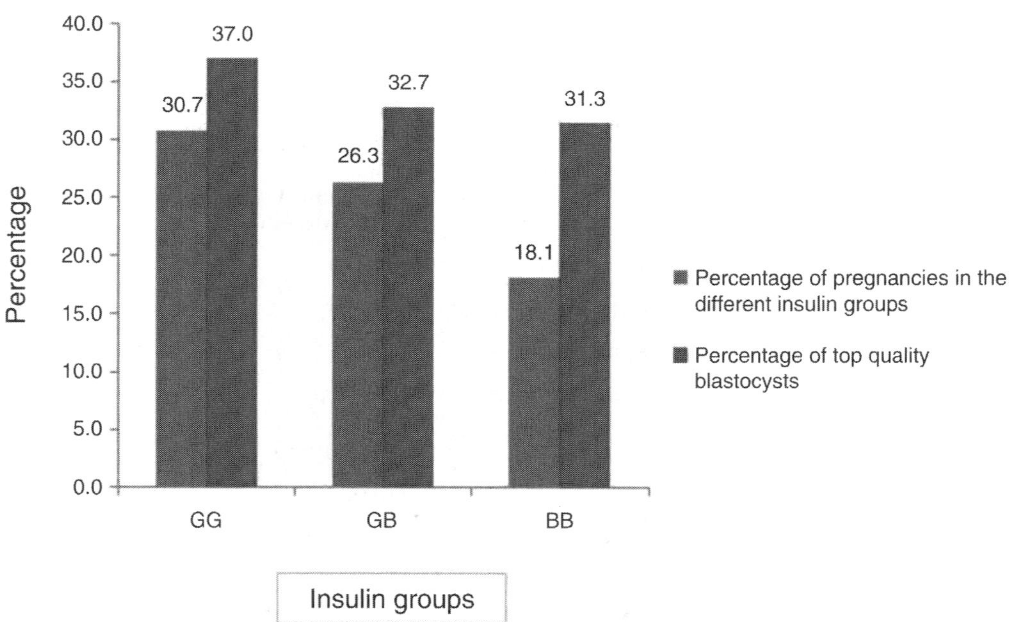

FIGURE 30.1 Pregnancy and blastocyst outcome for the different insulin groups.

TABLE 30.6 The Mean (±SD) Sperm Parameters for Groups A and B

Parameter	Normal insulin group A ($n = 38$)	Hyperinsulinemic group B ($n = 44$)
Mean fasting insulin (μIU/mL)	6.52 ± 1.50	16.18 ± 6.80
Mean sperm count (million/mL)	48.41 ± 26.20	47.65 ± 16.00
Mean sperm motility (%)	54.74 ± 8.05	55.56 ± 7.33
Mean sperm morphology (%)	7.27 ± 3.40	6.74 ± 2.67
Mean DNA packaging (CMA_3) (%)	45.73 ± 8.32*	48.98 ± 8.47*

** The mean DNA packaging of groups A and B differed marginally ($0.05 < P \leq 0.1$).*
Adapted from Ref. [27].

of age with normal insulin levels, blocked Fallopian tubes, and/or endometriosis, which responded well to hormonal stimulation and produced five or more ova, were included in the study. The male study group was selected for normal sperm counts and motility [28] and G-pattern morphology according to Tygerberg strict criteria [29]. The age range for the males was between 32 and 45 years. The weight or BMI of the respective

The IVF couples were divided into two groups; the men in group A had normal fasting insulin levels ($\leq$9.2 μIU/mL) and those in group B were hyperinsulinemic (>9.2 μIU/mL). Data on age, sperm parameters, number of ova, fertilization rate, embryo development, and pregnancy outcome were collected. Semen analysis and IVF methods were as discussed in the mentioned study.

Eighty two couples undergoing IVF treatment at Medfem Clinic were included in this study. The 38 normal insulinemic men with fasting insulin levels <9.2 μIU/mL were allocated to group A, and 44 males with fasting insulin levels >9.2 μIU/mL were included in group B. The mean age of the females in groups A and B were 32.3 ± 3.6 and 32.9 ± 3.0 years, respectively.

Results

The fasting insulin levels for the males and their sperm parameters are presented in Table 30.6. The mean sperm count, motility, and morphology of the two groups were similar. The morphology of the two groups fell in the moderate teratozoospermia or G-pattern category [54], and coincided with high fertilization rates. The two groups did not differ significantly for any of the tested parameters except for DNA packaging and maturity where the mean CMA_3 value for group B was higher ($0.05 < P \leq 0.1$) than for group A.

Group B presented with a higher mean CMA_3 value of 48.98% versus 45.73% for group A. CMA_3 is a guanine cytosine-specific fluorochrome that reveals poorly packed chromatin. DNA compaction is important to safeguard the delivery of stable genetic information to the oocyte [55,56]. High CMA_3 fluorescence (>40%) is a strong indicator of poorly packed chromatin, low protamination, and immature sperm. Recent research has highlighted the importance of sperm DNA integrity during fertilization and embryogenesis. Embryo development and quality appear to be influenced by semen parameters and CMA_3 values, and the influence of sperm quality continues into the blastocyst stage [57,58]. The normal insulin group (group A) included 13 males (34.2%) with immature CMA_3 values above 50% and in the hyperinsulinemic group (group B) there were 21 males (47.7%) with CMA_3 values above 50%. Previously, we have reported significantly decreased IVF pregnancy rates per embryo transfer (15.1% vs. 39%) for couples in the category $CMA_3 \geq 50\%$ [59].

Group A (normal insulinemic) presented with marginally higher fertilization and superior embryo quality rate on day 2 and 3; however, it did not differ statistically significantly from group B for these parameters (Table 30.5). However, the pregnancy rate of group A was significantly higher than that of group B, namely, 57.9 versus 31.8%. The clinical pregnancy outcomes of the two groups are presented in Table 30.7.

Euglycemia is very difficult to maintain in patients with abnormal insulin levels. Patients often experience periods of hyper- and hypoglycemia. These fluctuations in insulin and glucose concentrations can lead to several health problems and have unpredictable effects on the overall metabolism of testicular cells [60]. Although the exact etiology of male hyperinsulinemia is not known, glucose availability and insulin dysfunction can result in increased oxidative stress that is often reflected in mitochondrial and nuclear fragmentation [60]. Immature spermatozoa have high levels of DNA damage and ROS production, and are likely to have alterations in protamination resulting in poor chromatin packaging [61,62]. Patients should be advised about the negative impact of male hyperinsulinemia on IVF pregnancy rates. The use of antioxidants can be beneficial to patients with a CMA_3 above 50%, improving both CMA_3 values and IVF outcome [59]. The effect of drug and antioxidant treatment on the semen parameters of hyperinsulinemic males is presented in the next section.

TABLE 30.7   Embryo Quality and IVF Outcome for Groups A and B

Parameter	Normal insulin groups A	Hyperinsulinemic group B
Fertilization rate (%)	79.3	74.6
Embryos on day 2 with good grading (%)	48.1	42.6
Embryos on day 3 with good grading (%)	55. 8	50.4
Embryos on day 5 with good grading (%)	16.9	16.3
Number of patients with embryo transfer	38	44
Singleton pregnancy (%)	39.47* ($n = 15$)	15.91* ($n = 7$)
Twin pregnancy (%)	18.42* ($n = 7$)	11.36* ($n = 5$)
Triplet pregnancy (%)	0	4.55 ($n = 2$)
Total % pregnancy per embryo transfer	57.9* ($n = 22$)	31.8* ($n = 14$)

Pregnancy rates in groups A and B statistically significantly different (*$P \leq 0.05$).
Adapted from Ref. [27].

THE EFFECT OF METFORMIN THERAPY AND DIETARY SUPPLEMENTS ON SEMEN PARAMETERS IN HYPERINSULINEMIC MALES

In this study male patients with raised insulin levels were treated with metformin and given the option to use the nutritional supplement StaminoGro containing antioxidant components. The aim of this study was to investigate the effect of metformin and antioxidant treatment on the semen parameters of hyperinsulinemic men. The use of antioxidants, like vitamins C and E, has been proven to be beneficial in the treatment of oxidative stress and DNA fragmentation [63].

Fasting insulin levels were determined with a blood test on the day of semen analysis. Hyperinsulinemic males were followed up and treated with Glucophage® (metformin) and/or with the supplement StaminoGro®. Patients were retested 3 months after treatment and semen parameters were noted. Males that presented with varicocele or hyperthyroidism were not included in the study. Routine semen analysis was done as discussed in the previous sections. A fluorescent microscope with a 460-nm filter was used to read CMA$_3$ slides and at least 200 spermatozoa were scored under the 100× magnification with oil. Spermatozoa with poor chromatin packaging and stain brightly yellow were expressed as the percentage of sperm with immature packaging DNA. Nonreacted spermatozoa appear faintly yellowish-green under the fluorescence microscope and were counted as mature forms (Fig. 30.2).

A total of 34 patients met the inclusion criteria for hyperinsulinemia after initial testing and semen analysis. 19 patients were treated with metformin alone (group A) and 15 patients used metformin in combination with StaminoGro (group B). The ages of the males ranged between 33 and 45 years. The fasting insulin levels for the groups ranged from 10.2 to 38.3 μIU/mL with an average of 16.9 ±5.3 μIU/mL. The pre- and post-treatment semen parameters for the two different treatment groups are presented in Table 30.8.

A comparison of the findings for the patients in groups A and B indicate that the improvement in sperm morphology was similar for the two groups. The mean percentage normal morphology of the metformin group improved from 3.9 to 5.5% normal forms, while the group that used a combination of metformin and the supplement improved from 4.2 to 5.5%. The improvement in CMA$_3$ levels was higher in group B than in group A. The chromatin packaging quality in group B improved with approximately 12% from 64.2 to 52.3%, while it only improved with approximately 6% from 57.3 to 50.5% in the metformin treatment group. However, when these values were statistically adjusted for baseline variables, the changes were 10 and 8.3%, respectively, and not statistically significant ($P = 0.5929$).

Combined data of the two different hyperinsulinemic treatment groups demonstrated that there was no statistically significant difference between the mean pretreatment versus post treatment sperm counts, percentage motile sperm, or sperm vitality (Table 30.9). The overall posttreatment morphology and CMA$_3$ values (data of both treatments pooled together) differed statistically significantly from pretreated values ($P = 0.0003$; $P < 0.0001$).

There is a growing body of evidence that illustrates the important role of sperm DNA damage in male infertility and the clinical relevance thereof in the outcome in ART. Sperm DNA integrity has been associated with decreased embryo quality and an increase in miscarriage rates [64]. According to Paldi [65], a proper chromatin structure is critical for correct methylation of imprinted genes. DNA methylation and histone modifications are the two major mechanisms involved in genetic imprinting

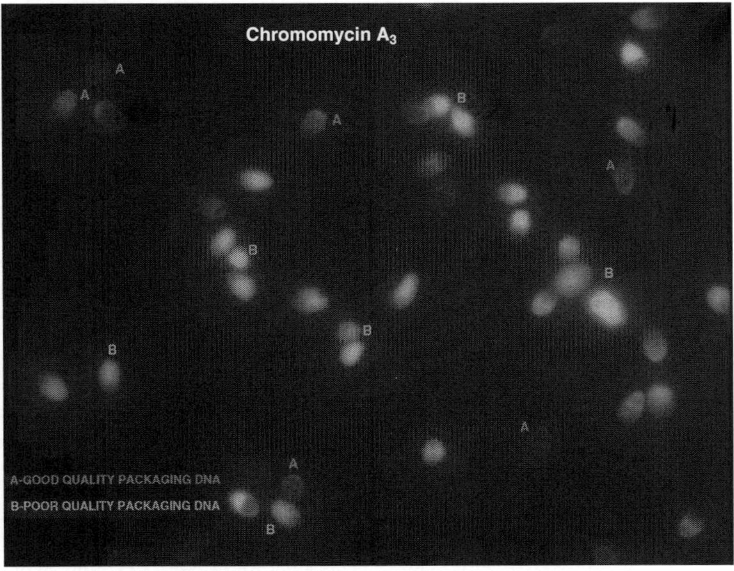

A = good quality DNA packaging in the sperm head

B = poor quality DNA packaging in the sperm head

FIGURE 30.2   Spermatozoa with poor chromatin package showing bright yellow under fluorescence. A, good quality DNA packaging in the sperm head; B, poor quality DNA packaging in the sperm head. Source: *Bosman et al. (2013) [71].*

in humans [66,67]. The primary role of the sperm protamines is to ensure proper sperm chromatin structure. It has been suggested that protamine and chromatin structural defects may leave sperm susceptible to improper imprinting patterns in critical genes [68]. Recently, some concern has been raised over a possible relation between assisted reproductive technologies (ART) and genomic imprinting disorders [69].

Spermatogenesis is an extremely intricate process in which an abundance of nutritional supplies are crucial during spermatogenesis. Antioxidant supplementation is one of the strategies that can be employed to enhance DNA integrity [62]. Tunc and Tremellen [70] found that three months of supplementation with antioxidant supplement resulted in a significant decrease in seminal ROS and sperm DNA fragmentation of infertile males, while increasing sperm DNA methylation. Similarly, the use of the mineral and vitamin supplement in combination with metformin in this study resulted in a greater improvement of DNA integrity than the use of metformin alone [71]. The significantly positive effect of metformin/supplement treatment on sperm morphology and CMA$_3$ values needs

TABLE 30.8   The Pretreatment and Posttreatment Sperm Parameters of Group A (Glucophage Treatment; $n = 19$) and Group B (Glucophage in Combination with StaminoGro Treatment; $n = 15$)

	Group A (metformin)		Group B (metformin and supplement)	
Sperm parameters	Pretreatment (mean ± SD)	Posttreatment (mean ± SD)	Pretreatment (mean ± SD)	Posttreatment (mean ± SD)
Sperm concentration ($\times 10^6$ mL^{-1})	32.6 ± 21.2	42.7 ± 32.9	36.0 ± 32.3	34.0 ± 18.4
% Motility	53.9 ± 16.3	49.3 ± 11.9	53.1 ± 11.5	53.0 ± 11.5
% Vitality	62.3 ± 13.7	59.3 ± 10.6	61.3 ± 9.6	63.1 ± 11.7
% Normal morphology	3.9 ± 2.2*	5.5 ± 2.9*	4.2 ± 1.6 **	5.5 ± 2.8**
CMA$_3$ (% immature forms)	57.7 ± 13.8†	50.5 ± 10.9†	64.3 ± 11.6‡	52.3 ± 6.8‡

Statistically significant difference between pre- and posttreatment values.
* *P = 0.002.*
** *P = 0.006.*
† *P = 0.040.*
‡ *P = 0.016.*
Adapted from Ref. [71].

TABLE 30.9   The Pretreatment and Posttreatment Sperm Parameters of the Hyperinsulinemic Group ($n = 34$)

Semen parameter	Pretreatment (mean ± SD)	Posttreatment (mean ± SD)	*P*-value
Sperm count ($\times 10^6$/mL)	34.1 ± 26.2	38.8 ± 27.4	0.3099
% Motility	53.6 ± 14.3	50.9 ± 11.7	0.2287
% Vitality	61.8 ± 11.8	61.0 ± 11.1	0.6161
% Normal morphology	4.1 ± 1.9*	5.5 ± 2.8*	0.0003
% Immature forms (CMA$_3$)	60.4 ± 13.2*	51.3 ± 9.2*	<0.0001

Statistically significant difference.
Adapted from Ref. [71].

to be further investigated. Early diagnosis and treatment of hyperinsulinemia is beneficial prior to ART as improved morphology and CMA$_3$ values can result in healthier embryos with higher DNA integrity and increased pregnancy outcome. The significant improvement of the CMA$_3$ values seen after treatment with supplements suggests that ROS is responsible for the poorer chromatin quality seen in the sperm of hyperinsulinemic men.

The association between hyperinsulinemia and high immature CMA$_3$ values has been apparent in all the mentioned studies. Infertile hyperinsulinemic men can benefit from metformin and antioxidant treatment and further studies are needed to validate the positive effect in IVF programs.

References

[1] Ceriello A, Giugliano D, Quatraro A, Lefebvre PJ. Antioxidants show an antihypertensive effect in diabetic and hypertensive subjects. Clin Sci 1991;81(6):739–42.

[2] Paolisso G. Pharmacological doses of vitamin E improve insulin action in healthy subjects and non-insulin dependent diabetic patients. Am J Clin Nutr 1993;57:650–6.

[3] Li C, Ford ES, McGuire LC, Mokdad AH, Little RR, Reaven GM. Trends in hyperinsulinaemia among non-diabetic adults in the US. Diabetes Care 2006;29:2396–402.

[4] Field CJ, Ryan EA, Thomson ABR, Clandininin MT. Dietary fat and the diabetic state alter insulin binding and the fatty acid composition of the adipocyte plasma membrane. Biochem J 1988;253:417–24.

[5] Aitken RJ, Baker MA, Sawyer D. Oxidative stress in the male germ line and its role in the aetiology of male infertility and genetic disease. Reprod Biomed Online 2003;7:65–70.

[6] Aitken RJ, Ryan AL, Curry BJ, Baker MA. Multiple forms of redox activity in populations of human spermatozoa. Mol Hum Reprod 2003;9:645–61.

[7] Anderson HG. Sugars and health: a review. Nutr Res 1997;18(9):1485–98.

[8] Kong C, Nimmo L, Elatrozy T, Anyaoku V, Hughes C, Robinson S, et al. Smoking is associated with increased hepatic lipase activity, insulin resistance, dyslipidemia and early atherosclerosis in Type II diabetes. Atherosclerosis 2001;156:373–8.

[9] Facchini FS, Hollenbeck CB, Jeppesen J, Chen YD, Reaven GM. Insulin resistance and cigarette smoking. Lancet 1992;339:1128–30.

[10] Sapolsky RM. Why stress is bad for your brain? Science 1996;273:749–50.

[11] Pessin JE, Saltiel AR. Signalling pathways in insulin action: molecular targets of insulin resistance. J Clin Invest 2000;106:165–9.

[12] Nakayama Y, Yamamoto T, Abe SI. IGF-I, IGF-II and insulin promote differentiation of spermatogonia to primary spermatocytes in organ culture of newt testes. Int J Dev Biol 1999;43:343–7.

[13] Acquila S, Gentile M, Middea E, Calatano S, Andò S. Autocrine regulation of insulin secretion in human ejaculated spermatozoa. Endocrinology 2005;146:552–7.

[14] Kim ST, Moley KH. The expression of GLUT8, GLUT9a and GLUT9b in the mouse testis and sperm. Reprod Sci 2007;14:445–55.

[15] Lampiao F, Du Plessis SS. Insulin stimulates GLUT8 expression in human spermatozoa. J Biosci Technol 2010;1(2):90–3.

[16] Silvestroni L, Modesti A, Sartori C. Insulin-sperm interaction: effects on plasma membrane and binding to acrosome. Arch Androl 1992;28:201–11.

[17] Jensen TK, Andersson A, Jørgensen N, Andersen A, Carlsen E, Petersen JH, et al. Body mass index in relation to semen quality and reproductive hormones among 1,558 Danish men. Fertil Steril 2004;82(4):863–70.

[18] Pitteloud N, Hardin M, Dwyer AA, Valassi E, Yialamas M, Elahi D, et al. Increasing insulin resistance is associated with decrease in Leydig cell testosterone secretion in men. J Clin Endocrinol Metab 2005;90:2636–41.

[19] Aboua G, Oguntibeju OO, Du Plessis S. Can lifestyle factors of Diabetes Mellitus patients affect their fertility. In: Trounson AO, Gardner DK, editors. Handbook of *In Vitro* Fertilization. 2nd ed. Boca Raton, FL: CRC Press; 2013. p. 205–264.

[20] Tremellen K. Oxidative stress and male infertility – a clinical perspective. Hum Reprod Update 2008;14:243–58.

[21] Bacetti B, La Marca A, Piomboni P, Capitani S, Bruni E, Petraglia F, et al. Insulin-dependent diabetes in men is associated with hypothalamo-pituitary derangement and with impairment in semen quality. Hum Reprod 2002;17:2673–7.

[22] Handelsman DJ, Conway AJ, Boylan LM, Yue DK, Turtle JR. Testicular function and glycaemic control in diabetic men. A controlled study. Andrologia 1985;17:288–496.

[23] Vignon F, Le Faou A, Montagnon D, Pradignac A, Cranz C, Winiszewsky P, et al. Comparative study of semen in diabetic and healthy men. Diabetes Metab 1991;17:350–4.

[24] Agbaje IM, Rogers DA, Mcvicar CM, Mcclure N, Atkinson AB, Mallidis C, et al. Insulin dependent diabetes mellitus: implications for male reproductive function. Hum Reprod 2007;22(7):1871–7.

[25] Agbaje IM, Mcvicar CM, Schock BC, Mcclure N, Atkinson AB, Rogers D, et al. Increased concentrations of oxidative DNA adduct 7,8-dihydro-8-oxo-2-deoxyguanosine in the germ-line of men with type 1 diabetes. Reprod Biomed Online 2008;16(3):401–9.

[26] La Vignera S, Condorelli R, Vicari E, D' Agata R, Calogera AE. Diabetes and sperm parameters: a brief review. J Androl 2012;33:145–53.

[27] Bosman E, Esterhuizen ADE, Rodrigues FA, Becker P, Hoffmann WA. Influence of male hyperinsulinaemia on IVF outcome. Andrologia 2014;. doi: 10.1111/and.12227.

[28] World Health Organization. WHO Laboratory Manual for the Examination of Human Semen and Sperm-Cervical Mucus Interaction. Cambridge, UK: Cambridge University Press; 1999.

[29] Menkveld R, Stander FSH, Kotze TJ, Kruger TF, Van Zyl JA. The evaluation of morphological characteristics of human spermatozoa according to stricter criteria. Hum Reprod 1990;5(5):586–92.

[30] Esterhuizen AD, Franken DR, Lourens JGH, Prinsloo E, Van Rooyen LH. Sperm chromatin packaging as an indicator of *in-vitro* fertilization rates. Hum Reprod 2000;15(3):657–61.

[31] Carrell DT, Hammoud SS. The human sperm epigenome and its potential role in embryonic development. Mol Hum Reprod 2010;16:37–47.

[32] Huszar G, Cayli S, Jakab A, Ovari L, Delpiano E, Celik-Ozenci C, et al. Biochemical markers of sperm function: male fertility and sperm selection for ICSI. Reprod Biomed Online 2003;7:1–7.

[33] Huszar G, Ozenci CC, Cayli S, Zavaczki Z, Hansch E, Vigue L. Hyaluronic acid binding by human sperm indicates cellular maturity, viability, and unreacted acrosomal status. Fertil Steril 2003;79:1616–24. (b).

[34] Koppers AJ. Mitochondria as a source of ROS in mammalian spermatozoa. Studies on Men's Health and Fertility, Oxidative Stress in Applied Basic Research and Clinical Practice. Springer Science + Business media; 2012. Chapter 2 p 34.

[35] Oehninger S, Morshedi M, Weng S-L, Taylor S, Duran H, Beebe S. Presence and significance of somatic cell apoptosis markers in human ejaculated spermatozoa. Reprod Biomed Online 2003;7(4):469–76.

[36] Mahfouz R, Sharma R, Thiyagarajan A, Kale V, Gupta S, Sabanegh E, et al. Semen characteristics and sperm DNA fragmentation in infertile men with low and high levels of seminal reactive oxygen species. Fertil Steril 2010;94(6):2141–6.

[37] Sakkas D. Novel technologies for selecting the best sperm for *in vitro* fertilization and intracytoplasmic sperm injection. Fertil Steril 2013;99(4):1023–9.

[38] World Health Organization. Obesity: Preventing and Managing the Global Epidemic. WHO Technical Report Series, vol. 894. Geneva: World Health Organization; 2000. p. 1–253.

[39] DPC. Immulite Insulin, Package Insert. Los Angeles, USA: Diagnostic Products Corporation (DPC); 2009. (ISO 8601).

[40] McLaughlin EA. British Andrology Society guidelines for the screening of semen donors for donor insemination. Hum Reprod 1999;14:1823–6.

[41] Mahadevan MM, Baker HWG. Assessment and preparation of semen for *in vitro* fertilization. In: Wood C, Trounson AA, editors. EBT Clinical *In Vitro* Fertilization. New York: Springer-Verlag; 1984. p. 83.

[42] Cook Medical. Embryo Culture, Suggested Laboratory Protocols. Sydney, Australia: Cook Medical Incorporated; 2007.

[43] Gardner DK, Schoolcraft WB. *In-vitro* culture of human blastocysts. In: Jansen R, Mortimer D, editors. Towards Reproductive Certainty: Infertility and Genetics Beyond. Carnforth: Parthenon Press; 1999. p. 378–88.

[44] Bosman E, Esterhuizen AD, Rodrigues FA, Becker PJ, Hoffmann WA. Prevalence of hyperinsulinaemia among normozoospermic donors at Medfem Clinic, South Africa. Andrologia 2013;45(1):18–25.

[45] Johannsson G, Albertson-Wikland K, Bengtsson BA. Discontinuation of growth hormone (GH) treatment; metabolic effects in GH deficient and GH sufficient adolescent patients compared with control subjects Swedish Study Group for GH treatment in children. J Clin Endocrinol Metab 1999;84:1416–24.

[46] Larson Z, Avitzur Y, Klinger B. Carbohydrate metabolism in primary growth hormone resistance (Laron Syndrome) before and during insulin-like growth factor-1 treatment. Metabolism 1995;44:113–8.

[47] Mokdad AH, Bowman BA, Ford ES, Vinicor F, Marks JS, Koplan JP. The continuing epidemics of obesity and diabetes in the United States. JAMA 2001;286:1195–200.

[48] Hammoud AO, Gibson M, Peterson CM, Meikle AW, Carrell DT. Impact of male obesity on infertility: a critical review of the current literature. Fertil Steril 2008;90:897–904.

[49] Hammoud AO, Wilde N, Gibson M, Parks A, Carrell DT, Meikle AW. Male obesity and alteration in sperm parameters. Fertil Steril 2008;90:2222–5.

[50] Sermondade N, Faure C, Fezeu L, Levy R, Czernichow S. Obesity and increased risk for oligozoospermia and azoospermia. Arch Intern Med 2012;172:440–2.

[51] Kort HI, Massey JB, Elsner CW, Mitchell-Leef D, Shapiro DB, Witt MA, et al. Impact of body mass index values on sperm quantity and quality. J Androl 2006;27:450–2.

[52] Sallmén M, Neto M, Mayan ON. Reduced fertility among shoe manufacturing workers. Occup Environ Med 2008;65:518–24.

[53] Colaci DS, Afeiche M, Gaskins AJ, Wright DL, Toth TL, Tanrikut C, et al. Men's body mass index in relation to embryo quality and clinical outcomes in couples undergoing *in vitro* fertilization. Fertil Steril 2012;98(5):1193–9. e1.

[54] Kruger TF, Acosta AA, Simmons KF, Swanson RJ, Matta JF, Oehninger S. Predictive value of abnormal sperm morphology in IVF. Fertil Steril 1988;49:112–7.

[55] Sakkas D, Mariethoz E, Manicardi G, Bizzaro D, Bianchi PG, Bianchi U. Origin of DNA damage in ejaculated human spermatozoa. Rev Reprod 1999;4:31–7.

[56] Carrell DT, Emery BR, Hammoud S. Altered protamine expression and diminished spermatogenesis: what is the link? Hum Reprod Update 2007;13:313–27.

[57] Shoukir Y, Chardonnes Campana A, Sakkas D. Blastocyst development from supernumerary embryos after intracytoplasmic sperm injection: a paternal influence? Hum Reprod 1998;13:1632–7.

[58] Tomsu M, Sharma V, Miller D. Embryo quality and IVF treatment outcomes may correlate with different sperm comet assay parameters. Hum Reprod 2002;17(7):1856–62.

[59] Bosman E, Rodrigues FA, Esterhuizen AD. Poster: Improvement of DNA packaging in sperm of IVF patients after nutrient supplement treatment. IFFS Congress, XIX World Congress on Fertility and Sterility, Conference Proceedings 2007;48.

[60] Alves MG, Martins AD, Rato L, Moreira PI, Socorro S, Oliviera PF. Molecular mechanisms beyond glucose transport in diabetes-related male infertility. Biochim Biophys Acta 2013;1832:626–35.

[61] Bianchi PG, Manicardi GC, Urner F, Campana A, Sakkas D. Chromatin packaging and morphology in ejaculated human spermatozoa: evidence of hidden anomalies in normal spermatozoa. Mol Hum Reprod 1996;2:139–44.

[62] Sharma RK, Said T, Agarwal A. Sperm DNA damage and its clinical relevance in assessing reproductive outcome. Asian J Androl 2004;6:139–48.

[63] Koca Y, Ozdal OL, Celik M, Unal S, Balaban N. Antioxidant activity of seminal plasma in fertile and infertile men. Arch Androl 2003;49:355–9.

[64] Morris ID, Ilott S, Dixon L, Brison DR. The spectrum of DNA damage in human sperm assessed by single cell gel electrophoresis (Comet assay) and its relationship to fertilization and embryo development. Hum Reprod 2002;17(4):990–8.

[65] Paldi A. Genomic imprinting: could the chromatin structure be the driving force? Curr Top Dev Biol 2003;53:115–38.

[66] Manipalviratn S, DeCherney A, Segars J. Imprinting disorders and assisted reproductive technology. Fertil Steril 2009;91(2):305–15.

[67] Kobayashi H, Sato A, Otsu E. Aberrant DNA methylation of imprinted loci in sperm from oligospermic patients. Hum Mol Genet 2007;16:2542–51.

[68] Aoki VW, Liu L, Carrell TC. Identification and evaluation of a novel sperm protamine abnormality in a population of infertile males. Hum Reprod 2005;20(5):1298–306.

[69] Lawrence LT, Moley KH. Epigenetics and assisted reproductive technologies: human imprinting syndromes. Semin Reprod Med 2008;26(2):143–52.

[70] Tunc O, Tremellen K. Oxidative DNA damage impairs global sperm DNA methylation in infertile men. J Assist Reprod Genet 2009;26:537–44.

[71] Bosman E, Esterhuizen AD, Rodrigues FA, Becker PJ, Hoffmann WA. The effect of metformin therapy and dietary supplements on sperm parameters in hyperinsulinaemic males. Andrologia 2014; doi: 10.1111/and.12366.

NUTRITION, LIFESTYLE, AND MALE FERTILITY

31

Dietary Zinc Deficiency and Testicular Apoptosis

Deepa Kumari, MSc, PhD, Neena Nair, MSc, PhD,
Ranveer Singh Bedwal, MSc, PhD

Cell Biology Laboratory, Department of Zoology, University of Rajasthan, Jaipur, India

ZINC BIOLOGY

Zinc, an essential trace element and a member of one of the major subgroups of the micronutrients, has attained prominence in human nutrition and health especially for the growth and development of infants and young children [1–4]. It is ubiquitously present in all biological systems and has unique and extensive functions within these systems [5–7]. The adult human (~70 kg weight) contains about 2–3 g of zinc in the body, with higher average in men than in women, 0.1% of which is replenished daily through dietary intake [8]. It is found in all organs, secretions, fluids, and tissues of the body [9], with 85% of the whole body zinc in muscle and bone, 11% in the skin and liver, and the remaining in all other tissues. In multicellular organisms, virtually all zinc is intracellular, 30–40% is located in the nucleus, 50% in the cytoplasm, organelles and specialized vesicles (for digestive enzymes or hormone storage), and the remainder in the cell membrane [8,10]. It serves as a catalytic, cocatalytic, structural and protein interface component for diverse biological functions [11]. These include carbohydrates, lipids, and nucleic acid synthesis, sequence recognition, transcriptional regulation, cytoskeletal and membrane integrity, apoptosis, synaptic signaling, gene expression, cellular energy metabolism, cell division, hormonal storage and release, neurotransmission, memory, and visual process [2,8,12–14]. It also has a specific role in enzyme functions – some of them are alkaline phosphatase, alcohol-dehydrogenases, Cu–Zn superoxide dismutase, 5α-reductase, dipeptidyl carboxypeptidase, DNA polymerase, RNA polymerase, and reverse transcriptase [5,9,10], involved in most of the major metabolic pathways and consequently is necessary for a wide range of biochemical, immunological, clinical, and antioxidant functions [8,11]. The role of zinc as an antioxidant has gained importance as it prevents free radical formation, protects biological structures from damage, and functions against oxidative stress [1,6,10,15,16]. This mineral stabilizes the quaternary structures of enzymes and structures of RNA, DNA, and ribosomes [17]. Zinc finger proteins, the largest class of DNA binding proteins, provide a scaffold that organizes protein subdomains for the interaction with either DNA or other proteins [10,11]. From prokarya to eukarya, between 4–10% of the genome encodes zinc proteins, and in human cells this percentage corresponds to the remarkable number of about 3000 zinc proteins [18,19]. Zinc plays an important role in regulating the production and secretion of proteins at transcriptional and translational level [20]. Zinc mimics the actions of hormones, growth factors, and cytokines, which suggest that zinc may act on the intracellular signaling molecule [1,21,22]. It is vital for a variety of hormonal activities including glucagon, insulin, growth hormone, as well as sex hormones [23]. Its role in immune function is evident from thymic hormone-thymulin, which is zinc dependent and is required for T-cell maturation and differentiation [24,25]. It is also known for its antiinflammatory (antiviral, antibacterial, and antifungal) and anticancerous properties and has been found to protect animals against otherwise lethal irradiation by neutrons [16,26]. The therapeutic roles of zinc in acute inflammation, diarrhea, acrodermatitis enteropathica, cognitive impairment, type 2 diabetes [13,27,28], prevention of blindness in patients with age related macular degeneration, and treatment of common cold [5] have also been reported.

ZINC ABSORPTION

Zinc-dependent proteins are found in the cytoplasm and within many organelles of eukaryotic cell including the nucleus, the endoplasmic reticulum, Golgi, secretory

Handbook of Fertility. http://dx.doi.org/10.1016/B978-0-12-800872-0.00031-7

vesicles, and mitochondria [29]. Regulation of intracellular zinc is likely to be mediated by a complex network of uptake, efflux, and sequestration methods by zinc transport proteins [30–31], all of which have transmembrane domains and are encoded by two solute-linked carrier (SLC) gene families: ZnT/CDF/SLC30 and ZIP/SLC39 [32]. In humans, 9 ZnTs (ZnT1–ZnT9) and 15 Zip transporters have been reported, which are antagonistic to each other in the maintenance of cellular zinc homeostasis [30]. ZnT/SLC30 transporters reduce intracellular zinc (Zn_i) availability by promoting zinc efflux from cells or into intracellular vesicles, while Zip/SLC39 transporters increase intracellular zinc (Zn_i) availability by promoting Zn_i uptake and perhaps vesicular zinc release into the cytoplasm [30–32]. The ZnTs and Zip transporter families exhibit unique tissue-specific expression and differential responsiveness to dietary zinc deficiency and excess. Out of nine ZnT transporters, ZnT1, ZnT2, and ZnT3 are known to be expressed in testes [10,33]. ZnT1, a ubiquitously expressed zinc transporter although present on the plasma membrane of neurons and glia cells [34,35], is also present in the cytoplasm of Sertoli cells, late-stage spermatids, spermatozoa, and Leydig cells in mouse testes [33], suggesting its importance in spermatogenesis. ZnT1 mRNA is highly expressed in tissues involved in zinc recycling or accumulation such as small intestine, renal tubular epithelium, testes, and placenta [36]. ZnT2 mRNA is present in only a few tissues of rodents viz. small intestine, kidney, placenta, pancreas, testes, seminal vesicle, prostate, and mammary gland [36,37]. Relatively high levels of ZnT2 mRNA in lateral and ventral prostate correlates with its high zinc concentration [38]. Zn T-3 mRNA is expressed in rodent brain, testes, and in human breast epithelial cell line (PMC2) [39,40]. Metallothioneins (MTs) also play an important role in buffering Zn_i [41]. MT-I and MT-II, two distinct isoforms, bind zinc accumulates when animals are exposed to excess zinc and gets depleted under conditions of zinc deficiency, thus protecting cells under conditions of zinc deficiency and zinc toxicity [42]. MT-I/II proteins are expressed exclusively in germ cells, notably in spermatocytes and absent in Sertoli and interstitial cells regulating Zn_i in testes [33,43]. Therefore, the cellular distribution of zinc is precisely managed to provide appropriate zinc concentrations required by each cell compartment for maintaining the structural and functional integrity as well as its multifunctional biological roles [31,44,45]. Factors recognized to affect zinc absorption at the whole body level have been reviewed and include the amount and form of zinc consumed; dietary promoters such as animal protein and low molecular weight organic compounds; dietary inhibitors such as phytate and possibly iron and calcium when consumed as supplements; and physiological states such as pregnancy, lactation, early infancy, and tissue repair, all of which increase the demand for absorbed zinc [46–48].

ZINC DEFICIENCY

Zinc deficiency has become a global health problem than is generally recognized [49–50] and can be due to inadequate dietary intake, decreased absorption, increased requirements, decreased utilization, increased loss, or genetic disease [9,51,52]. The ubiquitous nature of zinc in the human biological system indicates the widespread consequences and complexity of inadequate dietary supply of zinc and zinc depletion [53]. It has been related to various dysfunctions such as liver disease, sickle cell anemia, elevated blood pressure, malabsorption syndrome, Down's syndrome, Crohn's disease, protein-energy malnutrition (PEM), increased risk of cardiovascular diseases, acute respiratory infections, autoimmune disorders, renal disease, Alzheimer's disease, malignancy, gastrointestinal disorders, tobacco smoking, excessive intake of alcohol, skin lesions, and other chronic debilitating diseases [8,14,54–56]. Situations of stress, acute trauma, and infection can lead to lower plasma zinc [57]. The concentration of zinc in plasma is approximately 15 μM/L of which 84% is bound to albumin, 15% is tightly bound to α-2 macroglobulin, and 1% to amino acids [10]. Zinc deficiency is associated with a wide variety of clinical features such as growth impairment, alopecia, hypogeusia, hyposmia, diarrhea, dementia, delayed sexual development and impotency, eyes and skin lesions, emotional disorders, immune dysfunction, susceptibility to infections, impaired appetite, abnormal pregnancy, reduced glucose tolerance, and increased carcinogenesis [5,28,52]. Thymic-atrophy, lymphopenia, and compromised cell and antibody mediated response that cause increased rates of infections of longer duration are the immunological hallmarks of zinc deficiency in humans and higher animals [10]. Zinc deficiency in human beings may be manifested as severe, moderate, or mild and it is estimated to affect one-third of the world's population [8]. Although severe zinc deficiency is rare, it is estimated that mild to moderate deficiency is common in developing countries [5,58]. The lack of awareness of the importance of zinc in human nutrition, paucity of zinc, and phytate food composition values and difficulties in identifying zinc deficiency are some of the major reasons why it is still not included in the UN micronutrient list [59]. Recent intervention trials in lower-income countries have found that zinc supplementation decreases the rates of diarrhea and lower respiratory infections – two major causes of child mortality [54]. Several studies have detected reduced mortality rates among zinc-supplemented children [60–62]. Maternal zinc deficiency may cause adverse pregnancy outcomes for the mother and fetus including infertility, embryo/fetal death, intrauterine growth retardation and teratogenesis, behavioral abnormalities, impaired immunocompetence, and an elevated risk for high blood pressure in the offspring [63,64].

ZINC IN MALE REPRODUCTION

Zinc affects different aspects of mammalian reproduction [14,65]. It is vital for spermatogenesis and for the development of primary and secondary sexual characteristics [66], being directly involved in spermatozoa maturation and preservation of germinative epithelium [67]. Low zinc intake by young males of several species, including humans, interferes with normal sexual development [33,65,68,69], leads to gonadal dysfunction, decreases testicular weight, and causes shrinkage of seminiferous tubules [70,71]. Mitochondrial disarray in spermatids, acrosomal deformities, incomplete formation or disorganization of the axonemal complex, and dense fibers of spermatozoa tail and other defects, such as decapitation, disorganization, and redundant tail elements with superfluous cytoplasm, have been frequently observed in zinc-deficient rats [72–74].

Zinc content (μg/g wet weight) is high in the adult testis (165.90) as compared to immature animals (68.9) or efferent duct ligated animals (104.2) [68]. In men, its concentration increases at puberty and reaches to maximum at the age of 30–40 years, when the functional activity of the organ is at its peak, while in rats with maldescended testicle, the ectopic testis has a decreased zinc content whereas that of the other testicle is normal. In addition, severe zinc deficiency leads to necrosis of germ cell precursors that may lead to tubular atrophy and abnormal differentiation of spermatids [75]. Zinc deficiency impairs the action of the Müllerian Inhibitory Factor (MIF), which is essential for testicular differentiation [76]. The high concentration of zinc in organs like the prostate (for secretion into the seminal fluid), testicles, and in the ejaculated sperm suggests its importance in the process of spermatogenesis [9,77–78]. Zinc appears to be an indispensable element in reproduction for another reason too. The gonads are the most rapidly growing tissues in the body and vital enzymes involved in the nucleic acid and protein synthesis are zinc metalloenzymes [79]. Zinc may contribute to fertility through its significant effects on various semen parameters [80]. Zinc is also known to provide protection against the toxic effects of cadmium [81,82] and lead [83]. Appropriate zinc can antagonize male reproductive toxicity of fluorine on the molecular level by antagonizing lipid peroxidation, influencing reproduction, activity of enzyme, and Fas expression [84]. All this indicates an important role of the element in spermatogenesis and its probable site of action is spermatogonia, primary spermatocytes, spermatids, spermatozoa, Sertoli cells, and Leydig cells [77]. Zinc is also involved in histone displacement during nuclear condensation in spermiogenesis [85,86]. It protects the testes against degenerative changes [83] and may play a regulatory role in the process of capacitation and acrosome reaction [87]. Kvist [88] concluded from *in vitro* experiments on human spermatozoa that one of the functions of zinc is to preserve the ability of nuclear chromatin to undergo decondensation at the stage of male genome transfer and thus it plays a crucial role in fertilization [69]. Spermatozoal zinc is suggested to protect an inherent capacity of decondensation thereby helping to extend the functional life span of the ejaculated sperm [89]. Zinc is related to the structural integrity of DNA and prevents destruction of DNA by inhibiting degrading enzymes [90,91]. Zinc also plays an essential role in the synthesis and secretion of luteinizing hormone (LH), follicle-stimulating hormone (FSH) and prolactin, gonadal differentiation, testicular growth and development of seminiferous tubules, testicular steroidogenesis, androgen metabolism, and genetic expression of steroid hormone receptors [33,68,70,80]. In zinc deficiency, testicular cells are able to take up cholesterol and neutral lipids, which are precursors of sex steroids but are incapable of converting them into sex steroids, leading to the arrest of spermatogenesis and impairment of fertilization [92].

APOPTOSIS AND APOPTOTIC GENES

Apoptosis (from the Greek words, apo = from and ptosis = falling from, introduced by Kerr et al. [93]), a cell death program mediated by the caspase family of cysteine proteases, is essential for the appropriate removal of excess, unwanted, crippled, and damaged cells in many developmental and physiological states [94,95]. It can be triggered by intrinsic signals, such as physiological messenger molecule (e.g., cytokine or hormone), genotoxic stress, radiations or chemicals, growth factor withdrawal, or by extrinsic signals, such as the binding to cell surface death receptors [96]. Various upstream signaling pathways can modulate apoptosis by converging on, and thereby altering the activity of the bcl-2 family proteins, caspases, p53, death factors (TNFR-1, Fas, TRAIL etc), NF-κB and inhibitors of apoptosis proteins (IAPs) [97,98]. At the molecular level, the changes during apoptosis are thought to be accompanied by the activation of Ca^{2+}/Mg^{2+} dependent endonuclease, which cuts the internucleosomal regions into double-stranded DNA fragments of 180–200 base-pairs [99,100]. Bcl-2 encodes a 26 kDa protein that has a transmembrane domain found on the cytoplasmic face of the outer mitochondrial membrane, the endoplasmic reticulum, the nuclear membrane, and even in mitotic chromosomes, and its relatives reside on one or more of these membranes or assemble there during apoptosis [101]. The bcl-2 family members can be divided into three groups according to their effect on cell death [102,103]. The first group includes bcl-2, ced-9, BHRF (gene of Epstein–Barr virus), Mcl-1, and bcl-x$_l$, which inhibit cell death by preventing cytochrome c release from mitochondria [104]. The second group includes bax, bak, and bok, which inhibit bcl-2's ability to block growth factor induced by cell death and thereby promote apoptosis [105] and are also essential for organelle perturbation [106]. A third divergent group of BH-3-only proteins (Bid, Bim, Bik, Bad, Bmf, Hrk, Noxa, Puma, Bik, BNIP-3, and Spike) have only a short BH-3 domain that is both necessary and sufficient for their killing action [107]. It primarily antagonizes antiapoptotic bcl-2 proteins to promote apoptosis [108]. Bax and bak are crucial for mitochondrial outer membrane permeabilization (MOMP) and subsequent release of apoptogenic molecules [such as cytochrome c and DIABLO, also known as second mitochondrial activation of caspases (SMAC)], which leads to caspase activation [109]. At the same time as cytochrome c release, bax and bak also induce mitochondria to fragment into numerous smaller units, which suggests a connection between mitochondrial division processes and functions of the bcl-2 family [110]. Caspases fall into three categories: (1) Initiator caspases or apoptotic initiators (caspases 2, 8, 9, 10, and adaptor proteins); (2) Executioner caspases or apoptotic executioners (caspases 3, 6, and 7); and (3) Cytokine processors (caspases 1, 4, 5, 11, 12, 13, and 14). The activated executioners cleave key proteins required for the maintenance of homeostasis leading to collapse and death of the cell [111]. All the initiator and executioner caspases have either a direct or indirect role in the processing, propagation, and amplification of apoptotic signals that results in the destruction of cellular structures [112]. The intrinsic pathway for apoptosis involves the release of cytochrome c from mitochondria into the cytosol [113] where it binds to apoptotic protease activating factor-1 (Apaf-1) resulting in the activation of initiator caspase 9 and the subsequent proteolytic activation of executioner caspases 3, 6, and 7 resulting in amplification of death signals and eventually in the execution of cell death [114]. Death factors induce apoptosis through activation of specific death receptors that belong to the growing superfamily of tumor necrosis factor/nerve growth factor receptors [115,116]. Ligation of death receptors results in the rapid formation of an intracellular death inducing signaling complex (DISC) through an amino acid stretch within the carboxy terminus of the receptor called the death domain (DD) [117], which is responsible for coupling the death receptor to either caspases cascade, leading to induction of apoptosis [114], or to activation of kinase-signaling pathways resulting in gene expression through NF-κB or activator protein1 (AP-1), which suppresses apoptosis [118]. NF-κB, a redox-sensitive transcription factor [119], is involved in a broad range of physiological and pathological processes [120]. NF-κB may function as a central regulator of stress responses, since numerous stressful stimuli, including cytokines (TNFα), free radicals, UV irradiation, bacterial and viral infections, exposure to certain chemicals, and heavy metals, can lead to its activation [121]. In unstimulated cells, the NF-κB-family of proteins exist as heterodimers or homodimers that are sequestered in the cytoplasm by the IκB family of inhibitory proteins (inhibitor of κB) [122], but traverses to the nucleus when cells encounter oxidative stress or other apoptotic stimuli [96]. The tumor suppressor gene p53 has a key role in DNA damage recognition and repair, differentiation, cell cycle regulation, triggering apoptosis after genetic injury, and carcinogenesis [123]. The ability of p53 to eliminate excess, damaged or infected cells by apoptosis [93] is vital for the proper regulation of cell proliferation in multicellular organisms [124].

ZINC AND APOPTOSIS

Increased cell death is in part responsible for the effects of zinc deficiency in the thymus, prostate, testes, intestinal and retinal pigment epithelium, skin, pancreatic acinar cells [125,126], as well as the neuroepithelium of fetal rats [127–130]. The earliest report that explicitly linked zinc to cell death or apoptosis was published by Elmes [127], who observed a massive increased number of apoptotic bodies in the crypt region of the mucosa of the small intestine of zinc-deficient rats as compared to pair-fed controls that received a zinc-sufficient diet, suggesting that zinc deficiency induces apoptosis [131]. Cell death as assessed by TUNEL analysis was shown to be increased in peri-implantation embryos that were cultured in low-zinc medium [132]. Similar findings in testes have been reported by Kumari et al. [74]. Embryogenesis was significantly altered in zinc-deficient mice due to increased apoptosis in the neural crest, optic, and head regions. Insufficient zinc initiated programmed cell death in hepatocytes, glioma, kidney, monocytes, fibroblasts, and testicular cells [131,133,134]. Zinc is also essential for retaining the balance between proliferation of the prostatic cells and their physiological death [135,136]. Addition of zinc has been reported to inhibit various models of apoptosis induced by different agents: mastocytoma cells exposed to cytotoxic-T lymphocytes, thymocytes exposed to dexamethasone, fibrosarcoma cells treated by actinomycin, HL-60 cells exposed to UV-B, and thymocytes treated by NO donors [137–139].

Addition of zinc in culture medium prevents the rapid strand breaks of DNA induced by exposure to UV-B [140] and partially prevents apoptosis as measured by the increase in histone-associated DNA fragments inside the cell [141]. Zinc was also very efficient in protecting fibroblasts against the effect of UV-A, decreasing double chain breaks lipid peroxidation, cytotoxicity, and apoptosis [142]. In mice, zinc deficiency caused a 300% increase in the amount of apoptosis among pre-T-cells, which was a major cause of thymic atrophy that alters host defense [133]. Low-zinc cultures substantially increased apoptosis among Jurkat T-cells, 3T3 fibroblasts, and neuroblastoma cells [143]. A human monocytic cell line THP-1, when maintained in zinc chelated cultures, exhibited reduced viability followed by apoptosis [144]. Similar findings were reported by Fernandes et al. [145] who studied three strains of mice (viz. A/Jax, C57BL/K3, and CBA (H)) that were fed a zinc-deficient diet at age 6–8 weeks, in comparison to pair-fed controls on a zinc-sufficient diet. Many of the zinc-deficient mice developed thymic atrophy and signs of acrodermatitis enteropathica (i.e., diarrhea, skin lesions on paws and tail, and stunted growth). Apoptosis is also markedly increased in conjunction with congenital abnormalities in fetal rats born by zinc-deficient dams as emphasized in the neuroepithelium, where excessive apoptosis interfered with neural tube closure [146]. Rogers et al. [129] concluded that apoptotic embryonic cell death, particularly in neural crest cells (NCC), could arise within 4 days of maternal zinc deficiency, suggesting that NCC may be particularly more sensitive to adverse effects of zinc deficiency [134].

There are numerous potential targets of zinc in the apoptotic cascade and by analogy with multiple sites of action of zinc in the mitotic cycle; one might expect zinc to act at more than one site in the apoptotic cascade. The effects of zinc in the regulation of apoptosis appear to be cell type specific [147]. Zinc deprivation by starvation or by zinc chelation leads cells to death and the mode of cell death in many cell types is apoptosis [1,148–150]. Zinc deficiency may lead to mitochondrial-dependent apoptosis, modulating ROS and apoptotic proteins (p53, p21, and fas/fas ligand) in H4IIE cells [151]. On the other hand, zinc may also become cytotoxic when its extracellular concentration gets elevated (150–200 μM), which exceeds the capacity of the zinc homeostatic system [152–154]. The mechanism of zinc influence on apoptosis is complex. It might directly prevent apoptosis by blocking the activation or inhibiting the activity of Ca^{2+}/Mg^{2+}-dependent endonuclease [155,156]. Giannakis et al. [157] reported the inhibition of Ca^{2+}/Mg^{2+} endonuclease activity and DNA fragmentation in intact splenocytes, liver cells, and chronic lymphatic leukemia (CLL) cells treated with micromolar concentrations of zinc plus zinc ionophore. Endonuclease activity was inhibited regardless of whether chromatin or isolated DNA was used as a substrate; the action of zinc must be directed, at least primarily, at the enzyme level rather than the substrate [125].

Zinc influences apoptosis by acting on several molecular regulators of PCD including caspases and proteins from the bcl-2 and bax families [49]. Zinc finger proteins, which interact with apoptotic cascade, include glucocorticoid receptors, protein kinase-C, and poly-ADP-ribose polymerase [158]. An upstream event in apoptosis is the activation of ICE/CED-3 proteases, which is commonly observed as proteolysis of a substrate protein, poly(ADP-ribose) polymerase (PARP) [159]. The ICE/CED-3 (caspase-3) proteases are themselves activated by proteolysis, and this was detected by cleavage of one family member, CPP32 [160]. Zinc prevented cleavage of CPP32 and PARP [160]. Pretreatment of intact Molt-4 leukemia cells with micromolar concentrations of Zn^{2+} caused an inhibition of PARP proteolysis induced by the chemotherapeutic agent etopside [148]. Thus, caspase-3 acts as a novel target of Zn^{2+} inhibition in apoptosis [12,131,161]. Chai et al. [162] also reported that zinc-deficiency-induced apoptosis is dependent on caspase-3 activation, since cytosolic caspase-3 activity is increased in zinc-deficient cells while its specific inhibitor z-VAD-fmk (N-benzyloxycarbonyl-Val-Ala-Asp-fluromethylketone) can partially suppress apoptosis. Increased expression of caspase-3 in HaCaT keratinocyte cell line after zinc deficiency has also been observed by Wilson et al. [163]. Zinc deficiency also induces apoptotic neuronal death through the intrinsic (mitochondrial) pathway triggered by the activation of zinc-regulated enzyme caspase-3 and as a consequence of abnormal regulation of prosurvival signals (ERK1/2 and NF-κB) [63]. Zinc regulates upstream signaling pathways leading to the activation [164] as well as potential regulation of NF-κB [26]. There is a large number of zinc finger proteins, such as A20, Gfi1, IKK-γ and ZAS3, that regulates (inhibits/increases) the expression of NF-κB [165,166]. In zinc deficiency, a decreased NF-κB binding activity has been described in the nuclear extracts from rat testes [15], 3T3 fibroblasts [167], C6 rat glioma cells [168], T-lymphoblastoid cell line (HUT-78) [122], and human neuroblastoma IMR-32 cells [169]. A decreased NF-κB-dependent gene expression in zinc deficiency could participate in increased apoptotic cell death due to general antiapoptotic function of this transcription factor [170]. Zinc depletion not only increases the rate of apoptosis but there is a potent synergy in the induction of apoptosis, between zinc depletion and other apoptotic inducers such as colchicine, tumor necrosis factor, and HIV-1 Tat protein [171,172]. Peri-implantation and midgestation embryos from zinc-deficient dams are characterized by increased cell death and caspase-3-like activity [132]. Zinc deficiency increases the susceptibility of respiratory epithelial cells and human premonocytic U937 cells to H_2O_2-induced apoptosis [12]. Another sensitive molecular target of zinc is caspase-6, which cleaves and activates the proenzyme form of caspase-3 and is also responsible for the cleavage of lamins and therefore is directly involved in nuclear membrane dissolution [173]. Zinc blocks both these events [161]. Zinc acts as a selective inhibitor of caspase-6/Mch 2α, where complete inhibition was observed at 10 μM zinc [173,174]. *In vitro* studies of zinc supplementation to human premonocytic

U937 cells and enterocyte-like Caco-2 cells increased the Bcl2/Bax ratio and reduced caspase-3 activity, indicating increased resistance of the cells to apoptosis [175]. In zinc-deficient neuronal cells, cytosolic cytochrome *c*, mitochondrial BAD concentrations, DNA fragmentation, and percentage of apoptotic cells were higher, whereas levels of antiapoptotic proteins, like cIAP1, XIAP, Bcl-Xl, and bcl-2, were low compared to zinc adequate cells [63]. Zinc shares features in parallel to antiapoptotic protein bcl-2, both of which have synergistic functions in the control of apoptosis at multiple overlapping sites [150]. Like zinc, bcl-2 protects cells from oxidative stress [176] and directly suppresses caspase processing [177]. Bcl-2 knockout mice [176] and zinc-deficient rodents [9,152] were growth retarded, had severe immunodeficiency associated with massive apoptotic involution of the thymus and spleen, depletion of CD4[+] T-cells, and exhibited hair hypopigmentation [178]. All these findings indicate the role of zinc in protecting cells at early steps in the apoptotic pathway and regulating the upstream apoptotic machinery.

Zinc has potential mechanisms to protect against apoptosis. It functions against oxidative stress [9,12,16,179–181], an essential requirement for the occurrence of programmed cell death [182,183]. Reactive oxygen species (ROS) represent a broad category of molecules, including a collection of radical (hydroxyl ion, superoxide, nitric oxide, peroxyl, etc.) and nonradical (ozone, singlet oxygen, lipid peroxide, hydrogen peroxide) oxygen derivatives [184], and are commonly produced during normal cellular metabolism. However, under certain conditions of stress, such as poor nutrition, increase in production of ROS results in oxidative damage. All cellular components, including lipids, proteins, nucleic acids, and sugars, are potential targets of oxidative stress [184,185]. Oxidative stress-mediated damage is a consequence of an imbalance between the production of ROS and the body's antioxidant defense mechanisms. Reactive nitrogen species (RNS) (nitrous oxide, peroxynitrite, and nitroxyl ion) are also a class of free radicals derived from nitrogen and considered a subclass of ROS [185]. Clegg et al. [96] reported that low zinc concentration provides an environment that contributes to the production of ROS/RNS, caspase activation and NF-κB inhibition, leading to cell death. Increased ROS generation disrupts the inner and outer mitochondrial membranes inducing the release of the cytochrome *c* protein and activating caspases and apoptosis [186]. High levels of lipid, protein, and DNA oxidation have been described in different models of zinc deficiency [15,187–190]. Zinc deficiency not only causes oxidative stress and DNA damage, but also compromises the cell's ability to repair the damage suggesting the role of zinc in the maintenance of DNA integrity [91,191,192]. Song et al. [191] also reported interactions among zinc deficiency, DNA integrity, oxidative stress, and DNA repair. The role of zinc with redox properties in

cellular protection in relation to metallothionein expression and apoptotic processes is evident by an efficient antioxidant system [193]. Zinc protects protein sulfhydryl groups from oxidation by forming readily reversible thiolate complexes, thereby inhibiting enzymatic activity [96,150]. Affinity of zinc for sulfhydryl groups, coupled with the lack of redox activity, enables it to reversibly suppress cysteine-dependent enzymes (including caspases) that contain a sulfhydryl group [194], susceptible to oxidation [195] and S-nitrosation [196]. It is also an essential component of copper–zinc superoxide dismutase (CuZnSOD), one of the cell's first line of defense against ROS [150,180,197], which dismutates superoxide anion into hydrogen peroxide. Zinc protects nuclear factors from oxidation, mainly by preventing iron or toxic metals to bind to cysteine and to oxidize the protein or DNA binding element, a toxic effect observed in the nucleus by metal chelating protein metallothionein [198,199]. Such an effect is very important for the apoptotic antioncogenic protein p53 that becomes inactive after oxidation [140]. Under zinc-deficient conditions, cells and tissues are characterized by iron accumulation in part due to increased expression of proteins involved in iron transport, storage, and regulation [200]. Marginal zinc deficiency also sensitizes the prostate to oxidative stress and DNA damage by increasing p53 and PARP expression in rats [201]. Ho and Ames [168] reported that DNA repair is compromised although there is an increase in p53 expression after zinc deficiency in rat glioma C6 cells but this p53 is dysfunctional [63,201]. The specific DNA binding domain of p53 has a complex tertiary structure that is stabilized by zinc [64,96,202]. Depending on the duration of the zinc deficiency, p53 can cause cell cycle arrest to allow DNA repair or trigger apoptosis in response to DNA damage [203]. Rogers et al. [129] revealed that cell cycle distribution (the percentage of cells in the G0/G1, S, and G2 M phases of cell cycle) in embryos was not adversely affected by short-term maternal zinc deficiency, but cells cultured in low-zinc medium were characterized by growth arrest in the G0/G1 phase and reduced progression to S phase [204,205]. In addition, levels of proteins involved in the G1 checkpoint control mechanism, such as p21 and p53, were higher and levels of proteins that have a major role in cell cycle progression, such as cyclin D1 and cyclin E, were lower in zinc-deficient cells compared to zinc-adequate cells [205]. Modulation of binding of p53 to DNA by physiological concentrations of zinc might represent a pathway that regulates p53 activity *in vivo* [206]. Fanzo et al. [207] reported modulations of p53 protein and p21 mRNA levels by providing zinc-normal control (ZN), zinc-adequate (ZA), and zinc-supplemented (ZS) treatment with 3.0, 16.0, and 32.0 μM zinc, respectively, and serum-reduced low-zinc medium (ZD) in human aortic endothelial cells (HAECS). p53 nuclear protein levels were

>100% higher in the ZD and ZS cells and approximately 200% higher in the ZA cells than in ZN controls, while p21 mRNA was increased in ZS cells compared with the ZN control and ZD cells [207]. The transcriptional process of p21 is downregulated by depressed zinc status in human hepatoblastoma (HepG2) cells [204]. Other potential targets of zinc are the microtubular cytoskeleton, which is disrupted in zinc deficiency and apoptosis [208–211]. Zinc protects tubulin from oxidation where sulfhydryls are required for polymerization into microtubules [212].

Increased apoptosis *in vivo* may occur as a direct or indirect consequence of a decrease in intracellular zinc concentrations. Therefore, cellular zinc is described as a potent inhibitor of apoptosis while its depletion induces death in many cell lines [171,213,214]. The balance between life and cell death, that is, zinc homeostasis inside the cells is regulated by means of zinc transporters (ZIP/ZRT) and several zinc channels, controlling the intracellular zinc movements and the free amount of the metal [214]. Cao et al. [215] reported that the depletion of intracellular zinc (on treatment with >5 μM TPEN-N,N,N',N' tetrakis (2-pyridylmethylethylenediamine) in cultured THP-1 cells and human peripheral blood mononuclear cells decreases metallothionein expression, induces apoptosis, and increases ZIP 2 mRNA expression suggesting the homeostatic adjustment of cells to zinc depletion. The percentage of DNA fragmentation is strictly related to the decrease in cellular zinc [171]. Treves et al. [216] reported intracellular zinc as a regulator of apoptosis in lymphocytes. During the process of apoptosis, intracellular zinc (Zn^{2+}) rises as shown by an intense reaction with the fluorescent probe, Zinquin [217]. Zinquin-reactive zinc arises as a result of a change in the redox state of the cell that releases zinc bound to protein via zinc-S thiolate bonds [217]. Because apoptosis is a process by which the dying cell dismantles itself and zinc is a structural building block in many cellular components (e.g., membrane cytoskeleton and chromatin) [9], its release during apoptosis is perhaps not surprising. After phagocytosis, zinc may be recycled or excreted from the body. Whether the release is simply an effect of apoptosis or further accelerates the process (by destabilization of the microtubules and other structures) thereby facilitating the action of caspases and endonucleases is not known [12]. Alternatively, an increase in fluorescence may result from an influx of extracellular zinc due to loss of membrane integrity or decreased membrane efflux [218].

SPERMATOGENESIS AND GERM CELL APOPTOSIS

Spermatogenesis is a dynamic process and a fundamental function of the testes, in which a cohort of undifferentiated diploid cells (spermatogonial stem cells) proceeds throughout a sequence of mitotic and meiotic cell division and differentiation to generate spermatozoa [219–221]. It involves four basic processes: (1) spermatogonial development (stem cells and subsequent cell mitotic divisions), (2) meiosis (DNA synthesis and two meiotic divisions to yield haploid spermatids), (3) spermiogenesis (spermatid development involving differentiation of head and tail structures), and (4) spermiation (the process of release of mature sperm into the tubule lumen). These occur along similar lines in all mammals and are well described at the morphological level [222,223]. The production of a normal number of spermatozoa depends on the highly specific regulation of gene expression in the germ cells, the paracrine and hormonal control of germ cell proliferation, differentiation and survival, and the structural and functional support of the germ cells provided by the Sertoli cells [219,224,225].

For maintenance of normal spermatogenesis in rodents, humans, and various mammalian species, physiological germ cell apoptosis occurs spontaneously at various maturational stages continuously throughout life [223,226–229]. In order to limit the number of germ cells, surplus and genetically abnormal cells are all eliminated by apoptosis to match the supportive capacity of terminally differentiated Sertoli cells [230]. In adult testes, germ cell depletion during normal spermatogenesis results in the loss of approximately 75% of the potential number of mature spermatozoa [231] and has been extensively investigated in spermatogonia (mainly Type-A), zygotene, both early and late pachytene spermatocytes, and round spermatids [230–234].

Abnormally accelerated apoptosis of germ cells may lead to an imbalance of cell proliferation and death resulting in spermatogenic impairment as reported in the case of testes of infertile men with hypospermatogenesis or maturational arrest [235,236], cryptorchidism, and testes torsions [232,237,238]. Massive testicular germ cell loss is known to result from toxicant exposure [239], depletion of growth factors [240], alterations of hormonal support (testosterone and pituitary hormones including FSH and LH) [70,241], heat exposure [242,243], radiation or treatment with chemotherapeutic agents [244,245], and during zinc deficiency [74,246,247]. Defective spermatogenesis and male fertility is observed when the expression of apoptosis-related genes is disrupted or inappropriately controlled [224,231,232]. High levels of proapoptotic bcl-2 family members, bax and bak, in relation to antiapoptotic bcl-2 family members are associated with germ cells during development and their imbalance results in infertility [232,248,249]. Mice misexpressing bcl-2 in spermatogonia displayed an accumulation of spermatogonia before puberty, but during adulthood exhibited loss of germ cells in the majority of tubules [250]. However, the antiapoptotic bcl-2 member bcl-w appears to be an important survival factor for Sertoli cells, spermatogonia, and spermatocytes and its deficiency leads to increased postpubertal Sertoli and germ cell death and finally disruption of spermatogenesis [251]. Bax, a multidomain proapoptotic member of the bcl-2 family, is required for the normal spermatogenesis in mice [252], and its deficiency results in increased apoptosis and testicular atrophy, large accumulation of premeiotic germ cells in mature animals, and near complete absence of spermatocytes and mature sperms [238,248]. Hikim et al. [253] reported redistribution of bax from cytoplasmic to paranuclear localization in heat-susceptible germ cells of rat testes. The tumor suppressor p53 apoptotic gene is involved in radiation and cryptorchidism and is found to be overexpressed in germ cells [254,256] resulting in increased germ cell apoptosis and decreased production of spermatozoa [257]. Fas ligand expression, a tumor necrosis factor-related type II transmembrane protein [258], is shown to be upregulated in rat testes after exposure to Sertoli cell toxicants, mono-2-ethylhexyl phthalate (MEHP), and 2,5-hexanedione (2,5-HD) [259]. In addition, mice with disruption of various other genes, including those encoding the transcriptional activator CREM (cAMP-responsive element modulator), heat shock protein HSP70-2, and several other proteins involved in DNA repair and cell cycle control, exhibit defective spermatogenesis and increased germ cell apoptosis [224,231,260].

Apoptosis, a physiological process essential for eliminating cells after signaling molecules (TNF family members, fas, etc.) targets it and has been associated with intracellular events involving caspases and Bcl-2 family proteins. Association of the essential trace element zinc with apoptosis has been reviewed. Evidences have shown that depletion of zinc either through diet or by any other factors (disease state, environmental contaminants) activates the extrinsic as well as intrinsic pathways in testes leading to apoptosis. Studies confirm the critical role of zinc in gonad (testes) although further molecular analysis of pathways would provide more insight in testes undergoing apoptosis.

References

[1] John E, Laskow TC, Buchser WJ. Zinc in innate and adaptive tumor immunity. J Transl Med 2010;8:118–24.

[2] Bhowmik D, Chiranjib, Kumar KPS. A potential medical importance of zinc in human health and chronic disease. Int J Pharm Biomed Sci 2010;1:5–11.

[3] Hambidge KM, Miller LV, Westcott JE, Sheng X, Krebs NF. Zinc bioavailabilty and homeostasis. Am J Clin Nutr 2010;91:1478S–83S.

[4] Kaur K, Gupta R, Saraf SA, Saraf SK. Zinc: The metal of life. Compr Rev Food Sci Food Saf 2014;13:358–76.

[5] Stefanidou M, Maravelias C, Dona A, Spiliopoulou C. Zinc: a multipurpose trace element. Arch Toxicol 2006;80:1–9.

[6] Prasad AS. Zinc in human health: effect of zinc on immune cells. Mol Med 2008;14:353–7.

[7] Song Y, Elias V, Wong CP, Scrimgeour AG, Ho E. Zinc transporters expression profiles in the rat prostate following alterations in the dietary zinc. Biometals 2010;23:51–8.

[8] Hotz C, Brown KH. IZiNCG Technical Document: assessment of the risk of zinc deficiency in populations and options for its control. Food Nutr Bull 2004;25:S99–S103.

[9] Vallee BL, Falchuk KF. The biochemical basis of zinc physiology. Physiol Rev 1993;73:79–118.

[10] Tapiero H, Tew KD. Trace elements in human physiology and pathology: zinc and metallothioneins. Biomed Pharmacother 2003;57:399–411.

[11] Maret W. Molecular aspects of human cellular zinc homeostasis: redox control of zinc potentials and zinc signals. Biometals 2009;22:149–57.

[12] Truong-Tran AQ, Ho LH, Chai F, Zalewski PD. Cellular zinc fluxes and the regulation of Apoptosis/gene directed cell death. J Nutr 2000;130:1459S–S1466.

[13] Ackland ML, Michalczyk A. Zinc deficiency and its inherited disorders – a review. Genes Nutr 2006;1:41–50.

[14] Kumari D, Nair N, Bedwal RS. Effect of dietary zinc deficiency on testes of Wistar rats: morphometric and cell quantification studies. J Trace Elem Med Biol 2011;25:47–53.

[15] Oteiza PI, Clegg MS, Keen CL. Short term zinc deficiency affects nuclear factor κB nuclear binding activity in rat testes. J Nutr 2001;131:121–6.

[16] Prasad AS, Bao B, Beck FWJ, Kucuk O, Sarkar FH. Antioxidant effect of zinc in humans. Free Radic Biol Med 2004;37:1182–90.

[17] Prask JA, Plocke DJ. A role for zinc in structural integrity of the cytoplasmic ribosomes of *Lugiena gacilis*. Plant Physiol 1971;48:150–5.

[18] Andreini C, Banci L, Bertini I. Counting the zinc proteins encoded in the human genome. J Proteome Res 2006;5:196–201.

[19] Maret W. Zinc biochemistry: from a single zinc enzyme to a key element of life. Am Soc Nutr Adv Nutr 2013;4:82–91.

[20] Bahuguna A, Bedwal RS. Testicular protein profile (SDS-PAGE) study of zinc deficient Wistar albino rat. Indian J Exp Biol 2008;46:27–34.

[21] Beyersmann D, Haase H. Functions of zinc in signaling, proliferation and differentiation of mammalian cells. Biometals 2001;14:331–41.

[22] Yamasaki S, Sagata-Sagawa K, Hasegawa A, Suzuki T, Kabu K, Sato E, et al. Zinc is a novel intracellular second messenger. J Cell Biol 2007;177:637–45.

[23] Bryce-Smith D. Zinc deficiency – the neglected factor. Chem Br 1989;25:783–6.

[24] Bray TM, Bettger WJ. The physiological role of zinc as an antioxidant. Free Radic Biol Med 1990;8:281–91.

[25] Mocchegiani E, Muzzioli M, Giacconi R. Zinc and immunoresistance to infections in ageing: new biological tools. Trends Pharmacol Sci 2000;21:205–8.

[26] Bao B, Prasad AS, Beck FWJ, Fitzgerald JT, Snell D, Bao GW, et al. Zinc decreases C-reactive protein, lipid peroxidation, and inflammatory cytokines in elderly subjects: a potential implication of zinc as an atheroprotective agent. Am J Clin Nutr 2010;91:1634–41.

[27] Soinio M, Marnieni J, Laasko M, Pyörata K, Lehto S, Riönemaa T. Serum zinc levels and coronary heart disease in patients with type 2 diabetes. Diabetes Care 2007;30:523–8.

[28] Tuerk MJ, Fazel N. Zinc deficiency. Curr Opin Gastroenterol 2009;25:136–43.

[29] Eide DJ. Zinc transporters and the cellular trafficking of zinc. Biochim Biophys Acta 2006;1763:711–22.

[30] Liuzzi JP, Cousins RJ. Mammalian zinc transporters. Annu Rev Nutr 2004;24:151–72.

[31] Lichten LA, Cousins RJ. Mammalian zinc transporters: nutritional and physiologic regulation. Annu Rev Nutr 2009;29:153–76.

[32] Kelleher SL, McCormick NH, Velasquez V, Lopez V. Zinc in specialized secretory tissues: roles in the pancreas, prostate, and mammary gland. Adv Nutr 2011;2:101–11.

[33] Elgazar V, Razanov V, Stolenberg M, Hershfinkel M, Huleihel M, Nitzan YB, et al. Zinc regulating proteins ZnT-1 and metallothionein I/II are present in different cell populations in the mouse testis. J Histochem Cytochem 2005;53:905–12.

[34] Palmiter RD. Protection against zinc-toxicity by metallothionein and zinc transporter 1. Proc Natl Acad Sci USA 2004;101:4918–23.

[35] Nolte C, Gore A, Sekler I, Kresse W, Hershfinkel M, Hoffmann A, et al. ZnT-1 expression in astroglial accumulation of intracellular zinc. Glia 2004;48:145–55.

[36] Liuzzi JP, Bobo JA, Cui L, McMohan RJ, Cousins RJ. Zinc transporters 1, 2 and 4 are differentially expressed and localized during pregnancy and lactation. J Nutr 2003;133:342–51.

[37] Falcon-Prez JM, Dell'angelica EC. Zinc transporter 2 (SLC 30 A2) can suppress the vesicular zinc defect of adaptor protein 3-depleted fibroblasts by promoting zinc accumulation in lysosomes. Exp Cell Res 2007;313:1473–83.

[38] Iguchi K, Usui S, Inoue T, Sugimura Y, Tatematsu M, Hirano K. High level expression of zinc transporter-2 in the rat lateral and dorsal prostate. J Androl 2002;23:819–24.

[39] Michalczyk A, Allen J, Blomeley R, Ackland L. Constitutive expression of hZnT4 zinc transporter in human breast epithelial cells. Biochem J 2002;364:105–13.

[40] Friedlich AL, Lu JY, van Groen T, Cherny RA, Volitakis I, Cole TB, et al. Neuronal zinc exchange with the blood vessel wall promotes cerebral amyloid angiopathy in an animal model of Alzheimer's disease. J Neurosci 2004;24:3453–9.

[41] Bremner I. Involvement of metallothionein in the regulation of mineral metabolism. In: Sujuki KT, Imura N, Kimura M, editors. Metallothionein III, Biological Roles and Medical Implications. Bäsel, Switzerland: Birkhäuser Verlag; 1993. p. 111–24.

[42] Palmiter RD. The elusive function of metallothioneins. Proc Natl Acad Sci USA 1998;95:8230–48.

[43] Tohyama C, Nishimura N, Suzuki JS, Karasawa M, Nishimura H. Metallothionein mRNA in the testes and prostate of the rat detected by digoxigenin-labelled riboprobe. Histochemistry 1994;101:341–6.

[44] Tubek S. Zinc supplementation or regulation of its homeostasis, advantages and threats. Biol Trace Elem Res 2007;119:1–9.

[45] Haase H, Rink L. Functional significance of zinc-related signaling pathways in immune cells. Annu Rev Nutr 2009;29:133–52.

[46] Pathak P, Kapil U. Role of trace elements zinc, copper and magnesium during pregnancy and its outcome. Indian J Pediatr 2004;71:1003–5.

[47] Hess SY, King JC. Effects of maternal zinc supplementation on pregnancy and lactation outcomes. Food Nutr Bull 2009;30:S60–78.

[48] Walingo MK. Indigenous food processing methods that improve zinc absorption and bioavailability of plant diets consumed by the Kenyan population. Afr J Food Agric Nutr Dev 2009;9:524–35.

[49] Plum LM, Rink L, Haase H. The essential toxin: impact of zinc on human health. Int J Environ Res Public Health 2010;7:1342–65.

[50] Prasad AS. Zinc: a miracle element. Its discovery and impact on human health. JSM Clin Oncol Res 2014;2:1030.

[51] Agte V, Chiplonkar SA, Tarwadi KV. Factors influencing zinc status of apparently healthy Indians. J Am Coll Nutr 2005;24:334–41.

[52] Yanagisawa H. Zinc deficiency and clinical practice – validity of zinc preparations. Yakugaku Zasshi 2008;128:333–9.

[53] Lowe NM, Fekete K, Decsi T. Methods of assessment of zinc status in humans: a systematic review. Am J Clin Nutr 2009;89:2040S–51S.

[54] Brown KH, Peerson JM, Baker SK, Hess SY. Preventive zinc supplementation among infants, preschoolers and older prepubertal children. Food Nutr Bull 2009;30:S12–40.

[55] Prasad AS. Impact of the discovery of human zinc deficiency on health. J Am Coll Nutr 2009;28:257–65.

[56] Deshpande JD, Joshi MM, Giri PA. Zinc: the trace element of major importance in human nutrition. Int J Med Sci Public Health 2013;2:1–6.

[57] Nutrient Reference values for Australia and New Zealand including recommended dietary intakes, Commonwealth of Australia. 2006 nhmrc publications @nhmrc.gov.au.

[58] Fesharakinia A, Zarban A, Sharifzadeh GR. Prevalence of zinc deficiency in elementary school children of south Khorasan province (East Iran). Iranian J Pediatr 2009;19:249–54.

[59] Gibson RS. Zinc: the missing link in combating micronutrient malnutrition in developing countries. Proc Nutr Soc 2006;65:51–60.

[60] Sazawal S, Black RE, Menon VP. Zinc supplementation in infants born small for gestational age reduces mortality: a prospective, randomized, controlled trial. Pediatrics 2001;108:1280–6.

[61] Walker CL, Black RE. Zinc for the treatment of diarrhoea: effect on diarrhoea morbidity, mortality and incidence of future episodes. Int J Epidemiol 2010;39:63–9.

[62] Roth DE, Richard SA, Black RE. Zinc supplementation for the prevention of acute lower respiratory infection in children in developing countries: meta-analysis and meta-regression of randomized trials. Int J Epidemiol 2010;39:795–8.

[63] Adamo AM, Oteiza PI. Zinc deficiency and neurodevelopment: the case of neurons. Bio Factors 2010;36:117–24.

[64] Uriu-Adams JY, Keen CL. Zinc and reproduction: effects of zinc deficiency on prenatal and early postnatal development. Birth Defects Res 2010;89:313–25.

[65] Dissanayake DMAB, Wijesinghe PS, Ratnasooriya WD, Wimalasena S. Effects of zinc supplementation on sexual behavior of male rats. J Hum Reprod Sci 2009;2:57–61.

[66] Davies S. Zinc, nutrition and health. In: Bland J, editor. Yearbook of Nutritional Medicine. New Canaan, CT: Keats Publishing; 1985. p. 113–52.

[67] Hurley IR, Hershlag A, Benoff S. Mannose receptors, zinc and fertilization success. Arch Sexually Transmit Dis HIV 1997;11:241–53.

[68] Bedwal RS, Bahuguna A. Zinc, copper and selenium in reproduction. Experientia 1994;50:626–40.

[69] Kaji M. Zinc in endocrinology. Int Pediatr 2001;16:1–7.

[70] Kumari D, Nair N, Bedwal RS. Protein carbonyl, 3β- and 17β-hydroxysteroid dehydrogenases in testes and serum FSH, LH and testosterone levels in zinc deficient Wistar rats. Biofactors 2012;38:234–9.

[71] Saxena R, Bedwal RS, Mathur RS. Histopathology of epididymis and prostate of zinc deficient Sprague Dawley rats. Trace Elem Med 1993;10:55–9.

[72] Wallace E, Calvin HI, Salgo MP, Dennis DE, Ploetz K. Normal levels of zinc and sulphydryls in morphologically abnormal populations of spermatozoa from moderately zinc deficient rats. Gamete Res 1984;9:375–86.

[73] Bahuguna A, Bedwal RS. Effects of zinc deficiency on boundary layers of seminiferous tubules of testes of Wistar rats. Curr Sci 1998;75:611–6.

[74] Kumari D, Nair N, Bedwal RS. Testicular apoptosis after dietary zinc deficiency: ultrastructural and TUNEL studies. Syst Biol Rep Med 2011;57:233–43.

[75] Hamdi SA, Nassif OI, Ardawi MS. Effect of marginal or severe dietary zinc deficiency on testicular development and functions of the rat. Arch Androl 1997;38:243–53.

[76] Budzik GP, Huston JM, Ikawa H, Donahoe PK. The role of zinc in Mullerian duct regression. Endocrinology 1982;110:1521–5.

[77] Sorensen MB, Stoltenberg M, Henriksen K, Ernst E, Dansher G, Parvinen M. Histochemical tracing of zinc ions in the rat testis. Mol Hum Reprod 1998;4:423–8.

[78] Franklin RB, Costello LC. Zinc as an anti-tumor agent in prostate cancer and in other cancers. Arch Biochem Biophys 2007;463:211–7.

[79] Bedwal RS, Nair N, Mathur RS. Effects of zinc deficiency and toxicity on reproductive organs, pregnancy and lactation – a review. Trace Elem Med 1991;8:89–100.

[80] Ali H, Ahmed M, Baig M, Ali M. Relationship of zinc concentrations in blood and seminal plasma with various semen parameters in infertile subjects. Pak J Med Sci 2007;23:111–4.

[81] Gunnarsson D, Sevensson M, Selstam G, Nordberg G. Pronounced induction of testicular PGF (2α) and suppression of testosterone by cadmium-prevention by zinc. Toxicology 2004;2004:49–58.

[82] Burukoğlu D, Bayçu C. Protective effects of zinc on testes of cadmium-treated rats. Bull Environ Contam Toxicol 2008;81:521–4.

[83] Batra N, Nehru B, Bansal MP. Reproductive potential of male Portan rats exposed to various levels of lead with regard to zinc status. Br J Nutr 2004;91:387–91.

[84] Li Y, Jiang C, Cheng X, Cui L. Antagonistic mechanisms of zinc on male reproductive toxicity induced by excessive fluorine in rats. Life Sci J 2006;3:32–4.

[85] Marushige Y, Marushige K. Transformation of sperm histone during formation and maturation of rat spermatozoa. J Biochem 1975;250:39–45.

[86] Hunt CD, Johnson PE. The effects of dietary zinc on human sperm morphology and seminal mineral loss. In: Hurley LS, Keen CL, Lönnerdal B, Rucker RB, editors. Trace Element Metabolism in Man and Animals. New York: Plenum Press; 1990.

[87] Andrews JC, Nolan JP, Hammerstedt RH, Bavister BD. Role of zinc during hamster sperm capacitation. Biol Reprod 1994;51:1238–47.

[88] Kvist U. Spermatozoal thiol disulphide interaction, a possible event underlying physiological sperm nuclear chromatin decondensation. Acta Physiol (Scand) 1982;115:503–5.

[89] Kvist U, Björndhal L, Kjelberg L. Sperm nuclear zinc, chromatin stability and male fertility. Scanning Microsc 1987;1:1241–7.

[90] Devenson DP, Enerik RJ, Jost LK, Kayongo-Male H, Stewart SR. Zinc/silicon interactions influencing sperm chromatin integrity and testicular cell development in the rat as measured by flow cytometry. J Anim Sci 1993;71:955–62.

[91] Ho E, Courtemanche C, Ames BN. Zinc deficiency induces oxidative DNA damage and increases p53 expression in human lung fibroblasts. J Nutr 2003;133:2543–8.

[92] Lei KY, Abbasi A, Prasad AS. Function of the pituitary-gonadal axis in zinc deficient rats. Am J Physiol 1976;230:1730–2.

[93] Kerr JFR, Wyllie AH, Currie AR. Apoptosis: basic biological phenomenon with wide range of implications in tissue kinetics. Br J Cancer 1972;26:239–57.

[94] Kinchen JM. A model to die for: signaling to apoptotic cell removal in worm, fly and mouse. Apoptosis 2010;15:998–1006.

[95] Nakanishi Y, Nagaosa K, Shiratsuchi A. Phagocytic removal of cells that have become unwanted: implications for animal development and tissue homeostasis. Dev Growth Differ 2011;53:149–60.

[96] Clegg MS, Hanna LA, Niles BJ, Momma TY, Keen CL. Zinc deficiency-induced cell death. IUBMB Life 2005;57:661–9.

[97] Zhang A, Wu Y, Lai HWL, Yewc DT. Apoptosis – a brief review. Neuroembryology 2005;3:47–59.

[98] Lamkanfi M, Declercq W, Berghe TV, Vandenabeele P. Caspases leave the beaten track caspase-mediated activation of NF-κB. J Cell Biol 2006;173:165–71.

[99] Wyllie AH, Kerr JFR, Curie AR. Cell death: the significance of apoptosis. Int Rev Cytol 1980;68:251–6.

[100] Saraste A, Pulkki K. Morphological and biochemical hallmarks of apoptosis. Cardiovasc Res 2000;45:528–37.

[101] Kaufmann T, Schlipf S, Sanz J, Neubert K, Stein R, Borner C. Characterization of signal that directs Bcl-Xl but not Bcl-2 to the mitochondrial outer membrane. J Cell Biol 2003;160:53–64.

[102] Youle RJ, Strasser A. The bcl-2 protein family: opposing activities that mediate cell death. Nat Rev Mol Cell Biol 2008;9:47–59.

[103] Dewson G, Kluck RM. Bcl-2 family-regulated apoptosis in health and disease. Cell Health Cytoskeleton 2010;2:9–22.

[104] Cuconati A, White E. Viral homologues of Bcl-2: role of apoptosis in the regulation of virus infection. Genes Dev 2002;16:2465–78.

[105] Kiefer MC, Brauer MJ, Powers VC, Wu JJ, Umansky SR, Tomei LD, et al. Modulation of apoptosis by the widely distributed bcl-2 homologue bak. Nature 1995;374:736–9.

[106] Wei MC, Zong WX, Cheng FH, Lindsten T, Panoutsakopoulou V, Ross AJ, et al. Proapoptotic Bax and Bak: a requisite gateway to mitochondrial dysfunction and death. Science 2001;292:727–30.

[107] Mund T, Gewies A, Schoenfeld N, Bauer MK, Grimm S. Spike a novel BH3 only protein regulates apoptosis at the endoplasmic reticulum. FASEB J 2005;17:696–8.

[108] Willis SN, Fletcher JI, Kaufmann T, Delft MFV. Apoptosis initiated when BH3 ligands engage multiple Bcl-2 homologues not Bax or Bak. Science 2007;315:856–9.

[109] Newmeyer DD, Ferguson Miller S. Mitochondria releasing power for life and unleashing the machineries of death. Cell 2003;112:481–90.

[110] Martinou JC, Youle RJ. Which came first, the cytochrome-C release or the mitochondrial fission? Cell Death Differ 2006;13:1291–5.

[111] Thornberry NA. The caspase family of cysteine proteases. Br Med Bull 1997;53:478–90.

[112] Lavrik IN, Golks A, Krammer PH. Caspases: pharmacological manipulations of cell death. J Clin Invest 2005;115:2665–72.

[113] Riedl SJ, Salvesen GS. The apoptosome: signalling platform of cell death. Nat Rev Mol Cell Biol 2007;8:405–13.

[114] Denault JB, Salvesen GS. Caspases: keys in the ignition of cell death. Chem Rev 2002;102:4489–90.

[115] Nagata S, Golstein P. The Fas death factor. Science 1995;267:1449–56.

[116] Schmitz I, Kirchhoff S, Krammer PH. Regulation of death receptor-mediated apoptosis pathways. Int J Biochem Cell Biol 2000;32:1123–36.

[117] Huang B, Eberstadt M, Olejniczak ET, Meadows RP, Fesik SW. NMR structure and mutagenesis of the Fas (APO-1/CD95) death domain. Nature 1996;384:638–41.

[118] Li X, Stark GR. NF kappaB-dependent signaling pathways. Exp Hematol 2002;30:285–96.

[119] Herbein A, Varin, Fulop T. NF-kappaB, AP-1, zinc-deficiency and aging. Biogerontology 2006;7:409–19.

[120] Karin M, Lin A. NF-kappaB at the crossroads of life and death. Nat Immunol 2002;3:221–7.

[121] Gilmore TD. Introduction to Nf-κB: players, pathways, perspectives. Oncogene 2006;25:6680–4.

[122] Prasad AS, Bao B, Beck FW, Sarkar FH. Zinc enhances the expression of interleukin-2 and interleukin-2 receptors in HUT-78 cells by way of NF-kappaB activation. J Lab Clin Med 2002;140:272–89.

[123] Jin S, Levine AJ. The p53 functional circuit. J Cell Sci 2001;114:4120–39.

[124] Zabkiewicz J, Clarke AR. DNA damage-induced apoptosis: insights from the mouse. Biochim Biophys Acta 2004;1705:17–25.

[125] Zalewski PD, Forbes IJ. Intracellular zinc and the regulation of apoptosis. In: Lavin M, Watters D, editors. The Cellular and Molecular Biology of Apoptosis. Melbourne: Harwood Academic Publishers; 1993. p. 73–86.

[126] Sunderman FW. The influence of zinc on apoptosis. Ann Clin Lab Sci 1995;25:134–42.

[127] Elmes ME. Apoptosis in small intestine of zinc deficient and fasted rats. J Pathol 1977;123:219–24.

[128] Elmes ME, Jones JG. Ultrastructural changes in the small intestine of zinc deficient rats. J Pathol 1980;130:37–43.

[129] Rogers JM, Taubeneck MW, Daston GP, Sulik KK, Zucker RM, Elstein KH, et al. Zinc deficiency causes apoptosis but not cell cycle alterations in organogenesis-stage rat embryos: effect of varying duration of deficiency. Teratology 1995;52:149–59.

[130] Nodera M, Yanagisawa N, Wada O. Increased apoptosis in a variety of tissues of zinc deficient rats. Life Sci 2001;69:1639–49.

[131] Henning JM, Clegg M, Daston G, Rogers J, Keen C. Zinc deficient rat embryos have increased caspase 3-like activity and apoptosis. Biochem Res Commun 2000;271:250–6.

[132] Hanna LA, Clegg MS, Momma TY, Daston GP, Rogers JM, Keen CL. Zinc influences the *in vitro* development of peri-implantation mouse embryos. Birth Defects Res A Clin Mol Teratol 2003;67:414–20.

[133] Fraker PJ. Roles for cell death in zinc deficiency. J Nutr 2005;135:359–62.

[134] Lopez V, Keen CL, Lanoue L. Prenatal zinc deficiency: influence on heart morphology and distribution of key heart proteins in a rat model. Biol Trace Elem Res 2008;122:238–55.

[135] Costello LC, Franklin RB, Feng P. Mitochondrial function, zinc and intermediary metabolism relationships in normal prostate and prostate cancer. Mitochondrion 2005;5:143–53.

[136] Franklin RB, Costello LC. Zinc as an anti-tumor agent in prostate cancer and in other cancers. Arch Biochem Biophys 2007;463:211–7.

[137] Cohen J. Programmed cell death in the immune system. Adv Immunol 1991;50:55–85.

[138] McGowan AJ, Fernandes RS, Verhaegen S, Cotter TG. Zinc inhibits UV-radiation induced apoptosis but fails to prevent subsequent cell death. Int J Radiat Biol 1994;66:343–9.

[139] Fehsel K, Kroncke KD, Meyer KL, Huber H, Wahn V, Koeb-Bachofen V. Nitric oxide induces apoptosis in mouse thymocytes. J Immunol 1995;155:2858–65.

[140] Favier A. Is zinc a cellular mediator in the regulation of apoptosis? In: Bratter P, de Bratter VN, Khasanova L, Etienne JC, editors. Metal Ions Biol Med, Paris: John Libbey Eurotext;1998;5:164–7.

[141] Parat MO, Richard MJ, Pollet S, Hadjur C, Favier A, Blani JC. Zinc and DNA fragmentation in keratinocyte apoptosis: its inhibitory effect in UVB irradiated cells. J Photochem Photobiol, B 1997;37:101–6.

[142] Leccia MT, Richard MJ, Favier A, Beans JC. Zinc protects against UVA-1 induced DNA damage and apoptosis in cultured human fibroblasts. Biol Trace Elem Res 1999;69:177–90.

[143] Verstraeten SV, Zago MP, Mackenzie GG, Keen CL, Oteiza PI. Influence of zinc deficiency on cell membrane fluidity in Jurkat 3T3 and IMR-32 cells. Biochem J 2004;378:579–87.

[144] Cao J, Bobo J, Liuzzi J, Cousins R. Effects of intracellular zinc depletion on metallothionein and ZIP2 transporter expression and apoptosis. J Leukoc Biol 2001;70:559–66.

[145] Fernandes G, Nair N, Onoe K, Tanaka T, Floyd R, Good RA. Impairment of cell mediated immunity functions by dietary zinc deficiency in mice. Proc Natl Acad Sci USA 1979;76:457–61.

[146] Record IR, Tulsi RS, Dreosti IE, Fraser FJ. Cellular necrosis in zinc deficient rat embryos. Teratology 1985;32:397–405.

[147] Franklin RB, Costello LC. The important role of the apoptotic effects of zinc in the development of cancers. J Cell Biochem 2009;106:750–7.

[148] Perry DK, Smyth MJ, Stennicke HR, Salvesen GS, Duriez P, Poirier GG, et al. Zinc is a potent inhibitor of the apoptotic protease caspase-3: a novel target for zinc in the inhibition of apoptosis. J Biol Chem 1997;25:18530–3.

[149] Duffy JY, Miller CM, Rutschilling GL, Ridder GM, Clegg MS, Keen CL, et al. A decrease in intracellular zinc level precedes the detection of early indicators of apoptosis in HL-60 cells. Apoptosis 2001;6:161–72.

[150] Truong-Tran AQ, Carter J, Ruffin RE, Zalewski PD. The role of zinc in caspase activation and apoptotic cell death. Biometals 2001;14:315–30.

[151] Sanka Varam K, Chong L, Bruno RS, Freake HC. Zinc deficiency induces apoptosis but zinc induces necrosis in rat hepatoma cells. FASEB J 2010;24(Meeting Abstract Suppl.):7183.

[152] Fraker PJ, Telford WG. A reappraisal of the role of zinc in life and death decisions of cells. Proc Soc Exp Biol Med 1997;215:229–36.

[153] Wätzen W, Haase H, Biagioli M, Beyersmann D. Induction of apoptosis in mammalian cells by cadmium and zinc. Environ Health Perspect 2002;110:865–7.

[154] Chimienti F, Aouffen M, Favier A, Seve M. Zinc homeostasis-regulating proteins: new drug targets for triggering cell fate. Curr Drug Targets 2003;4:323–38.

[155] Ribeiro JM, Carson DA. Ca^{2+}/Mg^{2+} dependent endonuclease from human spleen: purification properties and role in apoptosis. Biochemistry 1993;32:9129–36.

[156] Hughes FM, Cidloweski JA. Apoptotic DNA fragmentation: evidence for normal enzymes. Cell Death Differ 1994;1:11–7.

[157] Giannakis C, Forbes IJ, Zalewski PD. Ca^{2+}/Mg^{2+} dependent nuclease tissue distribution relationship to inter-nucleosomal DNA fragmentation and inhibition by Zn^{2+}. Biochem Biophys Res Commun 1991;181:915–20.

[158] Gradwohl G, Menissier de-Murcia JM, Molinete M, Simonin F, Koken M, Hoeijmakers JH, et al. The second zinc-finger domain of poly (ADP-ribose) polymerase determines specificity for single-stranded breaks in DNA. Proc Natl Acad Sci USA 1990;87:2990–4.

[159] Lazebnik YA, Kaufmann SH, Disnoyers S, Povier GG, Earnshaw WC. Cleavage of poly (ADP-ribose) polymerase by a proteinase with properties like ICE. Nature 1994;371:346–7.

[160] Wolf CM, Morana SJ, Eastman A. Zinc inhibits apoptosis upstream of ICE/CED-3 protease rather than at the level of an endonuclease. Cell Death Differ 1997;4:125–9.

[161] Aiuchi T, Miharr S, Nakaya M, Masuda Y, Nakajo S, Nakaya K. Zinc ions prevent processing of caspase-3 during apoptosis induced by geranyl-geraniol in HL-60 cells. J Biochem 1998;124:300–3.

[162] Chai F, Truong-Tran AQ, Evdokiou A, Young GP, Zalewski PD. Intracellular zinc depletion induces caspase activation and p21Waf1/Cip1 cleavage in human epithelial cell lines. J Infect Dis 2000;182:S85–92.

[163] Wilson D, Varigos G, Ackland ML. Apoptosis may underlie the pathology of zinc-deficient skin. Immunol Cell Biol 2006;84:28–37.

[164] Haase H, Ober-Blobaum JL, Engelhardt G, Hebel S, Heit A, Heine H, et al. Zinc signals are essential for lipopolysaccharide-induced signal transduction in monocytes. J Immunol 2008;181:6491–2.

[165] Shifera AS, Horwitz MS. Mutations in the zinc finger domain of IKK gamma block the activation of NF-kappa B and the induction of IL-2 in stimulated T lymphocytes. Mol Immunol 2008;45(6):1633–45.

[166] Sharif-Askari E, Vassen L, Kosan C, Khandanpour C, Gaudreau MC, Heyd F, et al. Zinc finger protein Gfi1 controls the endotoxin-mediated Toll-like receptor inflammatory response by antagonizing NF-kappaB p65. Mol Cell Biol 2010;30(16):3929–42.

[167] Oteiza PI, Clegg MS, Zago MP, Keen CL. Zinc deficiency induces oxidative stress and AP-1 activation in 3T3 cells. Free Radic Biol Med 2000;28:1091–9.

[168] Ho E, Ames BN. Low intracellular zinc induces oxidative DNA damage disrupts p53, NF-κB and AP1 DNA binding and affects DNA repair in a rat glioma cell line. PNAS 2002;99:16770–5.

[169] Mackenzie GG, Zago MP, Keen CL, Oteiza PI. Low intracellular zinc impairs the translocation of activated NF-kappa B to the nuclei in human neuroblastoma IMR-32 cells. J Biol Chem 2002;277:34610–7.

[170] Herbein G, Varin A, Fulop T. NFκB, AP-1, zinc deficiency and aging. Biogerontology 2006;7:409–19.

[171] Zalewski PD, Forbes IJ, Betts WH. Correlation of apoptosis with change in intracellular labile Zinc (II) using zinquin [(2-methyl-8-*p*-toluene sulphonamido-6-quindyloxy) acetic acid] a new specific fluorescent probe for Zn(II). Biochem J 1993;296:403–8.

[172] Meerarani P, Ramdass P, Toborek M, Bauer HC, Bauer H, Hennig B. Zinc protects against apoptosis of endothelial cells induced by linoleic acid and TNF-α. Am J Clin Nutr 2000;71:81–7.

[173] Takahashi A, Alnemri ES, Lazebnik YA, Fernandes-Alnemri T, Litwack G, Moir RD, et al. Cleavage of lamin A by Mch 2α but not CPP32 multiple interleukin 1β-converting enzymes related

proteases with distinct substrate recognition properties are active in apoptosis. Proc Natl Acad Sci USA 1996;93:8395–400.

[174] Stennicke HR, Salvesen GS. Biochemical characteristics of caspase-3-6-7 and -8. J Biol Chem 1997;272:25719–23.

[175] Kilari S, Pullakhandam R, Nair KM. Zinc inhibits oxidative stress-induced iron signaling and apoptosis in Caco-2 cells. Free Radic Biol Med 2010;48(7):961–8.

[176] Korsmeyer SJ, Shutter JR, Veis DJ, Merry DE, Oltvai ZN. A rheostat that regulates an anti-oxidant pathway and cell death. Semin Cancer Biol 1993;4:327–32.

[177] Esposti MD, Hatzinisiriou I, McLennan H, Ralph S. Bcl-2 and mitochondrial oxygen radicals: new approaches with reactive oxygen species-sensitive probes. J Biol Chem 1999;274:29831–7.

[178] Chai F, Truong-Tran AQ, Ho LH, Zalewski PD. Regulation of caspase activation and apoptosis by cellular zinc fluxes and zinc deprivation: a review. Immunol Cell Biol 1999;77:272–8.

[179] Nair N, Bedwal RS, Prasad S, Saini MR, Bedwal RS. Short term zinc deficiency in diet induces increased oxidative stress in testes and epididymis of rats. Indian J Exp Biol 2005;43:786–94.

[180] Oteiza PI, Mackenzie GG. Zinc, oxidant-triggered cell signaling, and human health. Mol Aspects Med 2005;26:245–55.

[181] Bruno RS, Song Y, Leonard SW, Mustacich DJ, Taylor AW, Traber MG, et al. Dietary zinc restriction in rat alters antioxidant status and increases plasma F2 isoprostanes. J Nutr Biochem 2007;18:509–18.

[182] Buttke TM, Sandström PA. Oxidative stress as a mediator of apoptosis. Immunol Today 1994;15:7–10.

[183] Slater AFG, Nobel CSI, Orrenius S. The role of intracellular oxidants in apoptosis. Biochim Biophys Acta 1995;1271:59–62.

[184] Agarwal A, Prabakaran SA. Mechanism, measurement, and prevention of oxidative stress in male reproductive physiology. Indian J Exp Biol 2005;43:963–74.

[185] Sikka SC. Relative impact of oxidative stress on male reproductive function. Curr Med Chem 2001;8:851–62.

[186] Makker K, Agarwal A, Sharma R. Oxidative stress and male infertility. Ind J Med Res 2009;129:357–67.

[187] Oteiza PI, Olin KL, Fraga CG, Keen CL. Oxidant defense systems in testes from zinc deficient rats. Proc Soc Exp Biol Med 1996;213:85–91.

[188] Oteiza PI, Adonaylo VN, Keen CL. Cadmium induced testes oxidative damage in rats can be influenced by dietary zinc intake. Toxicology 1999;137:13–22.

[189] Oteiza PI, Olin KL, Fraga CG, Keen CL. Zinc deficiency causes oxidative damage to proteins lipids and DNA in rat testes. J Nutr 1995;125:823–9.

[190] Carter JE, Truong-Tran AQ, Grosser D, Ho L, Ruffin RE, Zalewski PD. Involvement of redox events in caspase activation in zinc depleted airway epithelial cells. Biochem Biophys Res Commun 2002;297:1062–70.

[191] Song Y, Leonard SW, Bruno RS, Traber MG, Ho E. Zinc deficiency affects DNA damage, oxidative stress, antioxidant defenses and DNA repair in rats. J Nutr 2009;139:1626–31.

[192] Silwinski T, Czechowska A, Kolodziejcjak M, Jajte J, Wisniewska-Jarosinska M, Blasiak J. Zinc salts differentially modulate DNA damage in normal and cancer cells. Cell Biol Int 2009;33:542–7.

[193] Formigari AQ, Irato P, Santon A. Zinc antioxidant systems and metallothionein in metal-mediated apoptosis: biochemical and cytochemical aspects. Comp Biochem Physiol C Toxicol Pharmacol 2007;46:443–59.

[194] Nicholson DW, Ali A, Thornberry NA, Vaillancourt JP, Ding CK, Gallant M, et al. Identification and inhibition of the ICE/CED-3 protease necessary for mammalian apoptosis. Nature 1995;376:37–43.

[195] Sen CK, Sashwati R, Packer L. Fas mediated apoptosis of human Jurkat T-cells intracellular events and potentiation by redox-active alpha-lipoic acid. Cell Death Differ 1999;481–91.

[196] Rossig L, Fichtlscherer B, Breitschopf K, Haendeler J, Zeiher AM, Mülsch A, et al. Nitric oxide inhibits caspase-3 by S-nitrosation *in vivo*. J Biol Chem 1999;74:6823–6.

[197] Ho E. Zinc deficiency, DNA damage and cancer risk. J Nutr Biochem 2004;5:572–8.

[198] Maret W. The function of zinc metallothionein: a link between cellular zinc and redox state. J Nutr 2000;130:1455S–S1458.

[199] Wasowicz W, Reszka E, Gromadzinska J, Rydzynski K. The role of essential elements in oxidative stress. Comments Toxicol 2003;9:39–48.

[200] Niles BJ, Clegg MS, Hanna LA, Chou SS, Momma TY, Hong H, et al. Zinc deficiency-induced iron accumulation, a consequence of alterations in iron regulatory protein-binding activity, iron transporters, and iron storage proteins. J Biol Chem 2008;283:5168–77.

[201] Song Y, Elias V, Loban A, Scrimgeour AG, Ho E. Zinc deficiency increases oxidative DNA damage in the prostate after chronic exercise. Free Radic Biol Med 2010;48:82–8.

[202] Verhaegh GW, Parat MO, Richard MJ, Hainaut P. Modulation of p53 protein conformation and DNA- binding activity by intracellular chelation of zinc. Mol Carcinog 1998;21:205–14.

[203] Corniola RS, Tassabehji NM, Hare J, Sharma G, Levenson CW. Zinc deficiency impairs neuronal precursor cell proliferation and induces apoptosis via p53-mediated mechanisms. Brain Res 2008;1237:52–61.

[204] Wong SH, Zhao Y, Schoene NW, Han CT, Shih RS, Lei KY. Zinc deficiency depresses p21 gene expression: inhibition of cell cycle progression is independent of the decrease in p21 protein level in HepG2 cells. Am J Physiol Cell Physiol 2007;292:C2175–84.

[205] Adamo AM, Zago MP, Mackenzie GG, Aimo L, Keen CL, Keenan A, et al. The role of zinc in the modulation of neuronal proliferation and apoptosis. Neurotox Res 2010;17:1–14.

[206] Palecek E, Brazdova M, Cernocka H, Vik D, Brazda V, Vojtesek B. Effect of transition metals on binding of p53 protein to supercoiled DNA and to consensus sequence in DNA fragments. Oncogene 1999;17:3617–25.

[207] Fanzo JC, Reaves SK, Cui L, Zhu L, Lei KY. p53 protein and p21 mRNA levels and caspase-3 activity are altered by zinc status in aortic endothelial cells. Am J Physiol Cell Physiol 2002;283:C631–8.

[208] Nickolson VJ, Veldstra H. The influence of various cations on the binding of colchicine by rat brain homogenates: stabilization of intact microtubules by zinc and cadmium ions. FEBS Lett 1972;23:309–14.

[209] Hesketh JE. Microtubule assembly in rat brain extracts: further characterization of the effects of zinc on assembly and cold stability. Int J Biochem 1984;16:1331–9.

[210] Martin SJ, Cotter TG. Specific loss of microtubules in HL-60 cells leads to programmed cell death (apoptosis). Biochem Soc Trans 1990;18:299–301.

[211] Wyllie AH. Apoptosis: an overview. Br Med Bull 1997;53:451–65.

[212] Roychowdhury M, Sarkar N, Manna T, Bhattacharyya S, Sarkar T, Basusarkar P, et al. Sulfhydryls of tubulin: a probe to detect conformational changes of tubulin. Eur J Biochem 2000;267: 3469–76.

[213] Lizard G, Moisant M, Cordelet C, Monier S, Gambert P, Lagrost L, et al. Induction of similar features of apoptosis in human and bovine vascular endothelial cells treated by 7- ketocholesterol. J Pathol 1997;183:330–8.

[214] Seve M, Chimienti F, Favier A. Role of intracellular zinc in programmed cell death. Pathol Biol 2002;50:212–21.

[215] Cao J, Bobo J, Liuzzi J, Cousins R. Effects of intracellular zinc depletion on metallothionein and ZIP2 transporter expression and apoptosis. J Leukoc Biol 2001;70:559–66.

[216] Treves S, Trentin PL, Ascanelli M, Bucci G, Di Virgilo F. Apoptosis is dependent on intracellular zinc and independent of intracellular calcium in lymphocytes. Exp Cell Res 1994;211:339–43.

[217] Zalewski PD, Forbes IJ, Seamark RF, Borlinghaus R, Betts WH, Lincoln SF, et al. Flux of intracellular labile zinc during apoptosis (gene-directed cell death) revealed by a specific chemical probe, Zinquin. Chem Biol 1994;1:153–61.

[218] Cuajungao MP, Lees GJ. Zinc metabolism in the brain: relevance to human neurodegenerative disorders. Neurobiol Dis 1997;4:137–69.

[219] De Krester DM, Loveland KL, Meinhardt A, Simorangkir D, Wreford N. Spermatogenesis. Hum Reprod 1998;1:1–8.

[220] McLachlan RI, Donell LO, Meachem SJ, Stanton PG, De Krester DM, Pratis K, et al. Identification of specific sites of hormonal regulation in spermatogenesis in rats, monkeys and man. Recent Prog Horm Res 2002;57:149–79.

[221] Kerr JB, Loveland KLO, Bryan MK, deKrester DM. Cytology of the testis and intrinsic control mechanisms. In: Neill JD, editor. The Physiology of Reproduction. Waltham, MA: Elsevier Academic Press; 2006. p. 827–947.

[222] Clermont Y. Kinetics of spermatogenesis in mammals, seminiferous epithelium cycle and spermatogonial renewal. Physiol Rev 1972;52:198–236.

[223] Russell LD, Etllin RA, Hikim APS, Clegg ED. Histological and histopathological evaluation of the testis. Int J Androl 2008;16:33–83.

[224] Sassone-Corsi P. Transcriptional checkpoints determining the fate of male germ cell. Cell 1997;88:163–6.

[225] Grootegoed JA, Slip M, Baarends WM. Molecular and cellular mechanisms in spermatogenesis. Baillieres Best Pract Res Clin Endocrinol Metab 2000;14:331–43.

[226] Clermont Y. Quantitative analysis of spermatogenesis of the rat, a revised model for the removal of spermatogonia. Am J Anat 1962;111:111–29.

[227] Hikim APS, Wang C, Lue Y, Johnson Wang XH, Swerdloff RS. Spontaneous germ cell apoptosis in humans: evidence for ethnic differences in the susceptibility of germ cells to programmed cell death. J Clin Endocrinol Metab 1998;83:152–6.

[228] Ricci G, Perticarari S, Fragonas C, Giolo E, Canova S, Pozzobon C, et al. Apoptosis in human sperm: its correlation with semen quality and the presence of leukocytes. Hum Rep 2002;17:2665–72.

[229] Chaki SP, Misro MM, Ghosh D, Gautam DK, Srinivas M. Apoptosis and cell removal in the cryptorchid rat testis. Apoptosis 2005;10:395–405.

[230] Jahnukainen K, Chrysis D, Hou M, Parvinen M, Eksborg S, Soder O. Increased apoptosis occurring during first wave of spermatogenesis is stage-specific and primarily affects mid-pachytene spermatocytes in the rat testis. Biol Reprod 2004;70:290–6.

[231] Hikim APS, Swerdloff RS. Hormonal and genetic control of germ cell apoptosis in the testis. Rev Reprod 1999;4:38–47.

[232] Blanco-Rodriguez I, Ody C, Araki K, Garcia I, Vassalli P. An early and massive wave of germinal cell apoptosis is required for the development of functional spermatogenesis. EMBO J 1997;16:2262–70.

[233] Fan X, Robaire B. Orchidectomy induces a wave of apoptotic cell death in the epididymis. Endocrinology 1998;139:42128–36.

[234] Pentikainen V, Erkkilä K, Suonalainen L, Otala M, Pentikäinen MO, Parvinen M, et al. TNF-down regulates the fas ligand and inhibits germ cell apoptosis in the human testis. J Clin Endocrinol Metab 2001;86:4480–8.

[235] Takagi S, Itoh N, Kimura M, Sasao T, Tsukamato T. Spermatogonial proliferation and apoptosis in hypospermatogenesis associated with non-obstructive azoospermia. Fertil Steril 2001;76:901–7.

[236] Hassan A, El-Nashar EM, Mostafa I. Programmed cell death in variocele-bearing testes. Andrology 2009;41:39–45.

[237] Kocak I, Dunder M, Hekimgil M, Okyay P. Assessment of germ cell apoptosis in cryptorchid rats. Asian J Androl 2002;4:183–6.

[238] Dündar M, Kocak I, Culhaci N, Erol H. Determination of apoptosis through bax expression in cryptorchid testis, an experimental study. Pathol Oncol Res 2005;11:170–3.

[239] Allard EK, Boekelheide K. Fate of germ cells in 2,5-hexadione-induced testicular injury. Toxicol Appl Pharmacol 1996;137:149–56.

[240] Yoshinaga K, Nishikawa S, Ogawa M, Hayashi S, Kunisada T, Fujimoto T. Role of c-kit in mouse spermatogenesis, identification of spermatogonia as a specific site of c-kit expression and function. Development 1991;113:689–99.

[241] Hikim APS, Wang C, Leung A, Swerdloff RS. Involvement of apoptosis in the induction of germ cell degeneration in adult rats after gonadotropin-releasing hormone antagonist treatment. Endocrinology 1995;136:2770–5.

[242] Sakallioglu AE, Ozdemir BH, Basaran O, Nacar A, Suren D, Haberal MA. Ultrastructural study of severe testicular damage following acute scrotal thermal injury. Burns 2007;33:328–33.

[243] Paul C, Teng S, Saunders PTK. A single, mild, transient scrotal heat stress causes hypoxia and oxidative stress in mouse testes, which induces germ cell death. Biol Reprod 2009;80:913–9.

[244] Meistrich ML. Effects of chemotherapy and radiotherapy on spermatogenesis. Eur Urol 1993;23:136–41.

[245] Henriksen K, Kulmala J, Toppari J, Mehrotra K, Parvinen. Stage-specific apoptosis in the rat seminiferous epithelium: quantification of irradiation effects. J Androl 1996;17:394–402.

[246] Li J, Xu P, He Z. Effect of zinc deficiency on apoptosis of spermatogenic cells testis. Zhonghua Yi Xue Za Zhi 1998;78:91–3.

[247] Nodera M, Yanagisawa H, Wada O. Increased apoptosis in a variety of tissues of zinc deficient rats. Life Sci 2001;69:1639–49.

[248] Knudson CM, Tung KSK, Tourtellotte WG, Brown GAJ, Korsmeyer SJ. Bax-deficient mice with lymphoid hyperplasia and male germ cell death. Science 1995;270:96–9.

[249] Yan W, Samson M, Jégou B, Toppari J. Bcl-w forms complexes with Bax and Bak, and elevated ratios of Bax/Bcl-w and Bak/Bcl-w correspond to spermatogonial and spermatocyte apoptosis in the testis. Mol Endocrinol 2000;14:682–99.

[250] Furuchi T, Masuko K, Nishimune Y, Obinata M, Matsui Y. Inhibition of testicular germ cell apoptosis and differentiation in mice misexpressing Bcl-2 in spermatogonia. Development 1996;122:1703–9.

[251] Print CG, Loveland KL, Gibson L, Meehan T, Stylianou A, Wreford N, et al. Apoptosis regulator Bcl-w is essential for spermatogenesis but appears otherwise redundant. Proc Natl Acad Sci USA 1998;95:12424–31.

[252] Russell LD, Chiarini-Garcia H, Korsmeyer SJ, Knudson CM. BAX-dependent spermatogonia apoptosis is required for testicular development and spermatogenesis. Biol Reprod 2002;66:950–8.

[253] Sinha Hikim AP, Lue Y, Diaz-Romero M, Yen PH, Wang C, Swerdloff RS. Deciphering the pathways of germ cell apoptosis in the testis. J Steroid Biochem Mol Biol 2003;85:175–82.

[254] Rogel A, Pepliker M, Webb CG, Oren M. p53 cellular tumour antigen: analysis of mRNA levels in normal adult tissues, embryos and tumours. Mol Cell Biol 1985;5:2851–5.

[255] Almon E, Goldfinger N, Kapon A, Schwartz D, Levine AJ, Rotter V. Testicular tissue-specific expression of the p53 suppressor gene. Dev Biol 1993;156:107–16.

[256] Yin Y, Dewolf WC, Morgentaler A. Experimental cryptorchidism induces testicular germ cell apoptosis by p53 dependent and independent pathways in mice. Biol Reprod 1998;58:492–6.

[257] Allemand I, Anglo A, Jeantet AY, Cerutti I, May E. Testicular wild type p53 expression in transgenic mice induces spermiogenesis alterations ranging from differentiation defects to apoptosis. Oncogene 1999;18:6521–30.

[258] Suda T, Takahashi T, Golstein P, Nagata S. Molecular cloning and expression of the fas ligand, a novel member of the tumor necrosis factor family. Cell 1993;75:1169–78.

[259] Lee J, Richburg JH, Youkin SC, Boekelheide K. The fas system is a key regulator of germ cell apoptosis in the testis. Endocrinology 1997;138:2081–8.

[260] Eddy EM. Role of heat shock protein hsp-70 in spermatogenesis. Rev Reprod 1999;44:23–30.

32

Environmental Factors, Food Intake, and Social Habits in Male Patients and its Relationship to Intracytoplasmic Sperm Injection Outcomes

Daniela Paes de Almeida Ferreira Braga, DVM, MSc,**,*
Edson Borges Jr., MD, PhD,***

*Fertility – Assisted Fertilization Centre, São Paulo, SP – Brazil
**Sapientiae Institute – Educational and Research Centre in Assisted Reproduction,
São Paulo, SP – Brazil

INTRODUCTION

Infertility affects 8–16% of reproductive-aged couples [1]. Over the past two decades, the use of assisted reproductive technology (ART) has increased dramatically worldwide and has made pregnancy possible for many infertile couples. However, although high-quality embryos may be available for transfer, most *in vitro* produced embryos fail to implant [2].

The identification of factors that may influence the implantation of *in vitro* produced embryos is one of the most studied areas in ART. Special attention has been given to the effects of the patient's lifestyle on the outcomes of ART. In addition, there is growing concern on sperm quality since it was reported that semen quality has declined in the latter half of the twentieth century [1–5]. It has been suggested that this decline is associated with many external agents, such as environmental factors, as well as eating habits and lifestyle.

Lifestyle factors are behaviors and circumstances that are, or were once, modifiable and can be a contributing factor to the overall well-being and they are ultimately under one's own control. They play a key role in determining reproductive health and can positively or negatively influence fertility.

Since there is some evidence that certain lifestyle habits may have an adverse impact on fertility, it is important to understand which behaviors have the highest negative impact, so that appropriate recommendations can be made to patients. Therefore, this chapter aims to discuss the different aspects of food intake and social habits in male patients that may affect their reproductive health.

WEIGHT AND EXERCISE

Overweight

Body mass index (BMI) is reported as a number. If it is below 18.5 it is considered underweight, between 18.5 and 24.9 is normal, above 25 is overweight, and over 30 is considered obese.

A combination of reduced exercise, changes in dietary composition, and increased energy intake have contributed to a growing worldwide epidemic in obesity, with serious impacts on several aspects of health [3]. Obesity is associated with many health conditions, such as diabetes, hypertension, and heart disease, and although not often discussed, it is linked to infertility as well [4]. The interaction between obesity and fertility has received increased attention owing to the recent and rapid increase in the prevalence of obesity in the developed world [5,6].

The deleterious effects of obesity on reproductive health include menstrual disorders and infertility. Such disorders are probably related to multiple factors,

Handbook of Fertility. http://dx.doi.org/10.1016/B978-0-12-800872-0.00032-9

such as endocrine and metabolic functions, including the balance of sex steroids, insulin, and leptin, which, in turn, may directly or indirectly affect ovarian function, follicular growth, implantation, and the development of a clinical pregnancy [7,8].

Although it is well documented in women, the relationship between BMI and male fertility is less known. Previous reports showed that the accumulation of fatty tissue in men is associated with a significantly reduced testosterone-to-estradiol ratio, suggesting that high BMI may influence the hypothalamo-pituitary system and decrease the sperm quality [9].

Research into the relationship between BMI and semen parameters has reported conflicting results. While an early systematic review with meta-analysis did not find a relationship between BMI and semen quality [10], more recent meta-analysis concluded that obesity did significantly increase the risk of oligozoospermia and azoospermia [11]. Some studies have found a significant negative relationship between BMI and sperm concentration or sperm count [12–14], while others have not found such an association [15–19].

Reports have suggested that the distribution of body fat, as assessed by waist circumference (WC), may provide a more robust measure of the adverse metabolic implications of excess body size rather than BMI [20]. Some investigators have examined the relationship between WC and semen parameters and identified a negative correlation between sperm counts and WC [21,22]. A recent article suggests an inverse relationship between WC and sperm production. In this study, body size was assessed by BMI and WC, which gave similar findings [23].

The exact mechanism by which increased BMI causes decline of semen quality is to be elucidated. Excess weight has been related directly or indirectly to biologic changes that could reduce male fertility. As described before, previous studies have reported reductions in testosterone with obesity [12,24]. Another hypothesis would be that overweight and obesity may reduce sperm motility due to an increased sperm DNA fragmentation [25].

The correlation between male obesity and ART success rates was analyzed in literature reviewed by Hammoud et al. [26]. Their observational findings indicate that obesity at a population level results in reduced male fertility among infertile couples. However, the effect of obesity on ART outcomes seems to be modest. Keltz et al. [27] describe a decreased chance of clinical pregnancy after conventional *in vitro* fertilization (IVF) but not after ICSI, suggesting that ICSI overcomes some as yet uncharacterized aspect of obesity-related sperm dysfunction. Indeed, Wegner et al. [28] showed a negative correlation between BMI and the hyaluronic acid binding capacity. However, when the zona pelucida competence was tested, no significant association was found between BMI and the results of the zona pelucida binding sperm ability [29].

Interestingly, studies have suggested that the quality of human semen has been declining over recent decades, presumably because of lifestyle or environmental factors. Previous studies; however, have reported that the increased risk of subfertility observed in the last decades is independently associated with overweight and obesity in male partners [30]. It has therefore been suggested that overweight and obesity may represent a significant potentially reversible cause of male subfertility.

Underweight

Eating disorders leading to weight loss are also associated with a reduced frequency or the cessation of ovulation. Food is used as a source of energy for a variety of essential and nonessential functions. In times of deprivation, it is necessary to ration available oxidizable substrates in favor of essential functions that are required to sustain life. Reproduction is expendable at least in the short term and can be deferred until times are more favorable. For example, menstrual cycles return in some female athletes when energy expenditure is reduced, such as after an injury [31].

Previous studies found a positive relationship between BMI and sperm concentration or total sperm count [14,16]. A Danish study of young men completing their military physical examination found a higher rate of abnormal sperm among men with either low or high BMI (<20 or ≥25) [12]. Men with low BMI may present with an abnormal semen analysis due to lower circulating testosterone levels [32]. However, this may be attributed to physical factors of training, since these individuals are typically athletes [33].

In a recent study from our group [34], the fertilization and the pregnancy and implantation rates after ICSI were negatively affected in patients who reported being on a weight loss diet. It is well known that the reproductive system is extremely sensitive to influences from the external environment and the mechanisms responsible for the adjustment of reproductive function involve the availability of calories [35,36]. Women with anovulation associated with strenuous exercise or who are underweight have low levels of leptin, LH, and estradiol. The frequency of gonadotropin pulses is too low to sustain development of antral follicles to the point of ovulation [37]. However, whether the mechanism by which reproductive function is impaired in men with food deprivation also involves the endocrinologic system is still unknown. In our study, the effect of being in a weight loss diet on implantation and pregnancy rates were dependent on the female BMI. It could be argued that partners usually share the same habits, and if one is undergoing a weight loss diet, it is likely that the other is also experiencing food deprivation, especially when they are overweight and the decreased chance of assisted fertilization treatment success in these couples is due to the diminished quality oocyte rather than the sperm.

Excessive Exercising

A healthy amount of exercise in men can be beneficial. However, not having an adequate knowledge of how to perform these activities might lead to the appearance of negative side effects. Exhaustive workout has been considered as a factor affecting the hypothalamic-pituitary-gonadal axis in men and women [38]. Decreased testosterone levels have the potential to interfere with the androgenic and reproductive processes [39–42] and therefore a trend toward a decline in reproductive function has been reported not only in female but also in male athletes [43–45].

Vaamonde et al. [40] described the existence of differences in the seminal profiles of individuals exercising in different modalities. The differences are more marked as intensity and volume of exercise increase, especially for morphology. In fact, early investigations suggest that the exercise volume affects reproduction, thus hypothesizing a volume threshold for reproductive disorders [46,47]. Other authors have suggested that exercise intensity is equally deleterious to, or even more so than, volume on reproductive function [48,49]. Both parameters may have a harmful effect on reproductive function.

In a study of young wrestlers, significant fluctuations in testosterone levels coincided with changes in dietary intake and body composition [38]. Similarly, in a prospective study of male soldiers [50], training intervals that coincided with severe energy restriction resulted in a decline in testosterone to levels well below the normal range for men, whereas training intervals with normal feeding produced a prompt recovery in testosterone levels even though the physiological stressors persisted. Therefore, it could be argued that energy deficit rather that the exercise per se, is associated with the decline in testosterone and decline in reproductive function.

There is growing evidence in animal studies indicating that athletes may be able to prevent or reverse infertility by dietary reform without any moderation of their exercise regimen [51]. In other words, in cases of energy imbalance, reproductive function in men and women can be restored with simple changes in lifestyle habits.

Sedentary

Although negative effects have been reported in the excessive practice of physical exercise, practicing regular physical activity has been shown to improve overall health [52]. It has been observed that, when comparing physically active patients with other groups that undergo a more intense exercise routine, the less demanding physical activity does not promote any semen alterations, while the other two do. Physically active men who exercised at least three times a week for 1 h typically have a better sperm quality in comparison to men who participated in more frequent and rigorous exercise [40].

In fact, a more anabolic environment has been reported by some authors for people who regularly exercise [53]; furthermore, exercise has been proposed as a treatment option in reproduction-related diseases such as polycystic ovarian syndrome, hypogonadism, and erectile dysfunction [54,55]. So, the beneficial effect of sports practice seems to also have an effect on the reproductive system, and also on semen and hormone values.

Vaamonde et al. [56] found a significant difference in semen and hormonal parameters between physically active and sedentary patients. Physically active patients show improved values for several sperm and hormonal parameters, which suggests a more anabolic microenvironment and better maintenance of homeostasis for suitable spermatogenesis.

Therefore, it seems clear that while high intensity stimuli are detrimental for hormones and semen, not having any exercise stimulus may also result in detrimental effects. Besides being effective in preventing or ameliorating several chronic diseases, moderate exercise has been reported to favor a more anabolic state improving the overall hormonal milieu, even enabling restoration of gonadotropin secretion [56]. Some studies have also reported a decrease in stress-related hormones as a result of moderate exercise [57,58]. Thus, it seems plausible that, as a consequence, moderate exercise might exert a positive effect on semen production as well since semen production is greatly dependent on an adequate hormonal environment.

RECREATIONAL SUBSTANCES AND ILLICIT DRUGS

Cigarette Smoking

It is well-documented that cigarette smoke contains over 4000 chemicals and is associated with a number of potential health complications. It is estimated that 35% of reproductive-aged males smoke [59].

Large epidemiological studies have demonstrated decreased fecundity in unassisted reproduction, due to smoking by women and/or their male partners [60–62]. Within the controlled settings of assisted reproduction cycles using IVF, these observations were confirmed, not only in smoking women but also in smoking male partners – cigarette smoking by either or both partners undergoing IVF significantly reduces success rates in terms of clinical pregnancies [63].

Concerning the effects of smoking on the paternal factor of sperm functions, cigarette chemicals seem to reduce sperm motility and/or morphology. Also, sperm membrane functions and the ability to undergo capacitation, the process essential for oocyte penetration, were reported to be impaired by smoking [64]. Therefore, it has been suggested that chances for pregnancies in smokers might be increased by ICSI, because the initial step of fertilization are bypassed through this technique. In fact Zitzmann et al. [65] did not find any association between oocyte fertilization by ICSI and male smoking. Conversely, in a previous report from Braga et al. [66], cigarette smoking was shown to be involved in sperm quality and fertilization function impairment.

Smoking habits may be associated with decreased levels of seminal plasma antioxidants, which places their sperm at additional risk of oxidative damage [67,68]. In recent years, oxidative stress and the role of reactive oxygen species (ROS) in the pathophysiology of human sperm function and male infertility have been explored intensively. Indeed, spermatozoa, from the moment that they are produced in the testes to when they are ejaculated into the female reproductive tract, are constantly exposed to oxidizing environments, and oxidative stress has been recognized as one of the most important causes of male infertility [69].

Alcohol Consumption

Alcohol consumption is one of the fastest growing health problems in the world. It has been reported that alcoholic beverages are consumed by nearly 90% of the population in England, which is equivalent to 36 million people (adults aged 16 years or over) [70]. Drinking alcohol is widely socially accepted and associated with relaxation and pleasure, and some people drink alcohol without experiencing harmful effects. However, a growing number of people experience physical, social, and psychological harmful effects of alcohol.

A negative effect of chronic alcoholism on male fertility has been previously described. Increased impotence has been reported in subjects suffering from chronic alcoholism, as compared with the case of nondrinkers [71,72]. Additionally, testicular atrophy, gynecomastia, and loss of sexual interest are often associated with alcoholism in men.

In adults, a possible effect of chronic alcohol exposure on spermatogenesis has been assessed with conflicting results. Although some studies did not observe any statistically significant differences in semen parameters between alcohol consumers versus nonconsumers [62,73,74], or even a protective effect of moderate amounts of alcohol [75], most of them suggest that chronic alcohol intake may affect male fertility by decreasing volume, sperm count [76,77], sperm motility

[76–78] and/or morphology [77,78]. Dunphy et al. [73] evaluated the relationship between male alcohol intake and fertility in 258 couples attending an infertility clinic. No association between the amount of alcohol and sperm parameters was observed in this study. In addition, there was no significant association between the amount of alcohol consumed per week and the fertility outcome. A multicenter prospective study evaluated whether the amount and the timing of female and male alcohol use during IVF programs affected the reproductive outcome. The risk of not achieving a live birth increased by 2.28 to 8.32 times, depending on the time period, in men who drank one additional drink per day [35].

Experimental and clinical studies suggest that alcohol consumption may alter both testosterone secretion and spermatogenesis. It is well known that alcohol consumption produces significant spermatozoon morphological changes, which include breakage of the sperm head, distention of the midsection, and tail curling [35]. In addition, seminiferous tubules in alcohol users mostly contain degenerated spermatids with a consequent azoospermia [36]. These effects may be due to alteration of the endocrine system controlling the hypothalamic-pituitary-gonadal axis function and/or to a direct effect on testis and/or male accessory glands [36].

In particular, experimental evidence suggests that ethanol is a Leydig cell toxin, although dose-dependent effects of alcohol on human spermatogenesis are not well known.

Although an effect of chronic alcohol exposure on spermatogenesis has been demonstrated, a published meta-analysis demonstrated that alcohol withdrawal allows a very fast and drastic improvement of semen characteristics; strictly normal semen parameters were observed after no more than 3 months. This suggests that patients should be effectively questioned about their alcohol intoxication before any semen analysis, especially when undergoing ART, and be strictly informed about the adverse effect of alcohol for spermatogenesis and about the efficacy of abstinence [37].

Marijuana

Many of the commonly used drugs of abuse have a profound effect upon male fertility, and it is important that the use of these agents is identified from the history in men presenting to an infertility clinic.

Marijuana is the most commonly used recreational drug worldwide [79]. The active ingredients of cannabis are the cannabinoids that act in the central and the peripheral nervous system and interfere with reproductive function at multiple levels.

In the 1990s, it was found that cannabinoid compounds are also naturally synthesized by the human body from fatty acid derivatives, termed endogenous cannabinoids or endocannabinoids [80,81]. Endocannabinoid receptors are found on numerous cells, including those in the testes [82] and the head and middle pieces of sperm [83].

Human studies consistently conclude that cannabinoids negatively affects male reproductive physiology [82–87]. There are observed disruptions in the hypothalamic-pituitary-testicular axis, with marijuana users having decreased levels of LH [84,85]. In addition, cannabinoids negatively influence spermatogenesis by modulating the apoptosis of Sertolli cells and by reducing testosterone production by Leydig cells [79].

Other adverse effects of cannabinoids on the fertilizing capacity of sperm have been reported [88]. Activation of the endocannabinoid receptors in sperm by cannabis reduces sperm motility in a dose-dependent manner and inhibit the capacitation-induced acrosomal reaction [82,83,86]. However, Schuel et al. [82] found a biphasic effect at different concentrations of cannabinoids. Sperm was inhibited at higher levels but hyperactivated at lower levels of cannabinoids.

The binding of cannabinoid to sperm receptors also impairs acrosome reaction and mitochondrial function [83,89]. Mitochondria are the principal suppliers of sperm energy, and as reported by Mahoney and Harris [90], the polycyclic structure of cannabinoids imposes a strong adverse effect on membrane-dependent processes, such as the inner mitochondrial membrane, and it was described that cannabinoids are potent inhibitors of sperm respiration [87].

Cannabinoids have a long half-life ranging from 24 to 36 h because it is lipid-soluble, which allows it to enter fatty tissue that then acts as a reservoir to slowly release cannabinoids back into the blood. Studies with rats have shown that after cannabinoids administration, approximately 0.06 and 0.02% of the administered dose concentrates in the brain and testis, respectively [91]. The presence of such doses in the testis may overstimulate endocannabinoid receptors and contribute to alter sperm motility and male infertility [92].

Cocaine

Evidence demonstrates the teratogenic effects of cocaine use on fetal development when abused by women during pregnancy, but very few articles have investigated the effects of cocaine on males because of the inability to prospectively study. Studies are further complicated by probable concurrent use of other illicit substances.

The association of cocaine use with sperm concentration, morphology, and motility has been previously suggested. It was described that cocaine use for five or more years was more common in men with low sperm motility, low concentration, or poor morphology. Cocaine use within the previous two years was twice as frequent in men with oligozoospermia [93].

An understanding of the specific effects of cocaine on male fertility is best estimated through animal studies. Cocaine receptors were observed in testicular tissue in rats and was found to bind cocaine in a saturable and specific manner [94]. Rats that were chronically exposed to cocaine at a level comparable to that of a heavy cocaine user had respective pregnancy rates of 33 and 50% for 100 and 150 days of cocaine exposure, compared with 86 and 100% of controls, respectively, for the same exposure duration [95]. In another study cocaine exposure in rats resulted in negative effects on spermatogenesis due to cellular degeneration, cell sloughing, and abnormal cell structures into the seminiferous tubules [96].

Li et al. reported that chronic exposure to peripubertal rats leads to significant apoptosis in the testes and the mechanism of cocaine-induced testicular injury may be related to the induction of apoptosis [97]. The same group also report that blood flow to rat testes is reduced after subcutaneous injections of high-dose cocaine [97]. They conclude that adverse effects of cocaine on the testes may be in part due to ischemic and postischemic reperfusion injury [98].

OCCUPATIONAL AND ENVIRONMENTAL FACTORS

Recently, the effect of pollutants and occupational factors on reproductive health has been a matter of debate. Environmental pollutants, such as methyl mercury, pesticides, lead, welding smoke, organic solvents, radiation, xenoestrogens, and household glues, have been shown to have a negative effect on fertility [3,99–103].

Organic Solvents

Although the effect of specific substances on semen quality is in expanding evidence, the relationship between environmental chemical exposure and male infertility is not always available and few studies have focused on the effects of pollutants on reproductive health in an infertile population [104–106].

The negative effect of exposure to organic solvents on reproductive health has been demonstrated. Tielemans et al. [107] evaluated patients from two infertility clinics and found an association between aromatic solvent exposure and abnormal semen parameters. Moreover, a published meta-analysis confirmed that occupational exposure to organic solvents in fathers is associated with increased risk of central nervous system malformations, in particular neural tube defects including anencephaly [108].

Alteration of the male reproductive health may be due to effects on the endocrine control of reproductive system [109,110] or by the direct effect on spermatogenesis [111,112]. However, one possible mechanism associated with adverse pregnancy outcomes due to paternal exposure to organic solvents is a direct effect on sperm DNA, producing mutations or chromosomal abnormalities [108].

Heavy Metals

Lead, cadmium, mercury, and arsenic, often referred to as "heavy metals," are toxic for wildlife, experimental animals, and humans. Experimental animal and human occupational studies support an adverse role for these metals in human reproductive outcomes [113]. An association has been found between welding and reduced semen quality (sperm count and motility) [104,114].

Lead exposure levels have been associated with adverse human male reproductive effects in several studies. Blood lead level has been associated with reduced sperm count, poor sperm motility, and abnormal sperm morphology [115,116]. However, in another trial, no associations were observed between blood lead and sperm concentration, sperm motility, or sperm morphology in models adjusted for age and current smoking [117].

The association of lead exposure and IVF outcomes has been previously suggested. Benoff et al. [118] indicated that blood plasma lead levels were elevated in some IVF patients and were inversely correlated with the rate of fertilization [119] and IVF outcomes [120].

Pillai et al. [121] found that lead may affect pituitary membrane function and cause alterations in receptor binding and secretory mechanism(s) of pituitary hormones. This may be an important factor in the pathogenesis of infertility.

Previous studies have examined the male reproductive effects of exposure to low levels of cadmium, and have provided some evidence in support of decreased semen quality and/or altered level of reproductive hormones. A negative correlations between blood cadmium level and sperm characteristics have been reported [122,123]. Buck Louis et al. [124] suggest that environmentally relevant concentrations of blood lead and, possibly, cadmium are associated with a longer time to pregnancy [124]. Men with low sperm motility had significantly higher blood cadmium levels than did men with normal sperm motility [125] and cadmium exposure among male partners may affect oocyte fertilization during IVF [126].

Mercury has been detected in semen and its levels have correlated with reproductive outcomes [127]. Patients with abnormal sperm parameters had significantly higher blood mercury levels than did fertile men [128]

and the seminal fluid mercury level was significantly positively correlated with abnormal sperm morphology and lower motility [129].

Although early studies [130,131] reported conflicting results regarding the arsenic effects on fertility in males and females, more recent reports suggest that exposure alters female reproductive physiology [132]. Laboratory animals exposed to arsenic exhibited impaired steroidogenesis possibly leading to infertility [133]. It was also suggested that chronic arsenic exposure may contribute to male infertility [134] and has a negative impact on erectile function [135]. Recently, Xu et al. [136] described that arsenic exposure may be associated with reduced human semen quality.

Endocrine-Disruptor Compounds

Endocrine-disrupting compounds (EDCs) are synthetic and naturally occurring chemicals that are characterized by their ability to mimic the effects of endogenous hormones. A number of chemicals commonly encountered in the environment have been associated with altered endocrine function in animals and humans, and exposure to some EDCs may result in adverse effects on reproduction, metabolism, fetal/child development, neurological function, and other vital processes [137]. Recently, the role of EDCs in adverse reproductive outcomes has received attention because of their persistence and ability to mimic natural hormones.

An EDC can affect many target sites in the male reproductive tract. The most important being the testes, the sites of spermatogenesis and androgen production. There are paracrine and autocrine regulations in various compartments of the testis that are under endocrine influences from the pituitary and hypothalamus [138]. Because EDCs mimic natural hormones, they inhibit the action of hormones, or alter the normal regulatory function of the endocrine system and thus have potential hazardous effects on male reproductive axis causing infertility.

Populations are exposed to EDCs in air, water, food, and in a variety of industrial products, including personal care goods. Many EDCs accumulate in the environment and in animals higher up on the food chain [137,139]. For example, EDCs that were banned decades ago, namely, DDT and PCB, are still found in human fluids [140]. This is due to their lipophilicity and resistance to biodegradation [141].

Although there is chronic exposure to EDCs through inhalation and skin contact [142], the major source of human exposure is ingestion of food, as well as plain water and other beverages, meat, fish, dairy food, and so on. EDC-contaminated food and water may contain environmental pollutants, such as pesticide residues [143] and heavy metals [144], in addition to processing aids and anabolic steroids used in food production. Most individuals have traceable amounts of these substances in their serum or urine [144,145].

The use of xenoestrogens in the food industry may result in an increased level of sex steroids in processed foods, such as meat or milk, whose intake contributes significantly to daily exposures [146,147]. Recently, we have suggested that successful pregnancy and implantation outcomes are decreased in patients reporting a more frequent intake of meat [66]. Xenoestrogens, which are highly lipophilic substances that can also accumulate in fat-rich foods, may be suspected as partially responsible for the decline in semen quality. In a study by Mendiola et al. [148], an association between semen quality and the consumption of foods containing processed meat (sausages and others) was observed.

CAFFEINE

Caffeine has become an integral part of society, since its stimulant properties were first discovered in beverages (coffee, tea, and soft drinks) and some foods such as chocolate. Its consumption has been reported to prolong the time to pregnancy, although the mechanism for this is unclear.

There are few reports on the effect of caffeine on the ART population with inconclusive results. While some have reported a negative influence of caffeine on ART outcomes [149], others have reported no effect or even a positive effect [150].

In previous studies, caffeine was added in culture medium to stimulate hamster sperm motility. The addition of caffeine to the medium increased motility of cryopreserved sperm, reduced percentage of penetrated oocytes, and decreased fertilizing ability and embryo development (for review see Ref. [151]).

There are several biological pathways by which caffeine could affect reproductive health. It could affect the hypothalamic-pituitary-gonadal axis through alterations in hormone levels. Caffeine consumption is inversely correlated with levels of estradiol in pregnant women and positively correlated with levels of sex-hormone-binding globulin [152].

In a previous report, we have demonstrated that the consumption of coffee is related to decreased fertilization capacity [66]; however, the pregnancy, implantation, and miscarriage rates were not affected, suggesting that if there is a harmful effect of caffeine on sperm fertilization potential, once fertilized, the embryo development is not impaired. In addition, Klonoff-Cohen et al. [149] showed a significant association between female but not male caffeine consumption and live births.

FOOD INTAKE

Many studies have suggested that semen quality may be influenced by food intake. It has been described that men with poor semen quality have a more frequent intake of some food items that may adversely affect semen quality or that act as carriers of deleterious products to the reproductive system.

Dairy Food

Emerging evidence suggests that environmental estrogens may be related to lower semen quality [153]. A particularly prevalent exposure route to environmental estrogens is by consumption of dairy foods [154]. Because commercial milk is a mixture of milk from cows at different stages of pregnancy, dairy products contain detectable amounts of estrogens and other hormones that increase during pregnancy [155,156], and it has been pointed to as the main source of estrogens from foods (for a review see Ref. [157]).

Intake of milk and other dairy products has been related to lower semen quality in some studies [148,158,159], but not others [160]. Afeiche et al. [159] found an inverse correlation between full-fat dairy intake and sperm motility and morphology. The authors stated that these associations are driven primarily by intake of cheese and are independent of overall dietary patterns. Other articles reported an increased consumption of dairy foods among oligoasthenoteratospermic [148] and asthenospermic [158] patients.

Besides being a source of estrogens, full-fat dairy foods are an important source of saturated fat, which has been previously related to low sperm counts [161]. In fact, it was found that low-fat dairy foods, especially low-fat milk, are positively associated with sperm concentration and progressive motility. However, cheese intake is associated with lower sperm concentration [157].

The literature regarding the effect of excessive intake of dairy products on the ART outcomes is scarce. In a previous report from our group, no association was found between the frequency of dairy products intake and sperm quality or ICSI outcomes [66]; therefore, it could be argued that the effect of dairy products on male fertility, if there is any, is bypassed by the use of ART.

Vegetables, Legumes, and Fruits

As a reflection of global changes in dietary behavior, the prevalence of unhealthy diets, characterized by low intakes of fruits and vegetables and high intakes of foods rich in saturated fats, has increased in women and men within the reproductive age range [162], and has been pointed to as the potential cause for the decline in human fertility observed over the past decades.

Mendiola et al. [148] compared dietary habits in normospermic and oligoasthenoteratospermic patients attending an ART clinic and found that normospermic patients have a higher intake of lettuce and tomatoes, and some fruits. This is in line with previous reports suggesting a lower intake of some antioxidant nutrients, such as vitamins A, C, and E, carnitines, folate, zinc, and selenium, among infertile patients [163–165].

It is known that spermatozoa are susceptible to oxidative damage. ROS levels are higher and levels of seminal plasma antioxidants are significantly lower in subfertile patients than in fertile subjects [166,167], and many published studies have demonstrated that antioxidant therapy may improve certain seminal parameters in male factor infertility patients (for a review see Ref. [163]).

Mínguez-Alarcón et al. [168] found a positive association between the dietary intake of several antioxidant nutrients, such as cryptoxanthin, vitamin C, lycopene, and β-carotene, and the total motile sperm count. In this study, the semen volume also increased with higher intakes of vitamin C, lycopene, and β-carotene.

In a Dutch trial, subfertile couples with high adherence to a "Mediterranean" dietary pattern have a 40% increased probability of achieving pregnancy after IVF/ICSI treatment [169]. The adherence to this diet is reflected by relatively high concentrations of folate and vitamin B6 in blood and follicular fluid. The principal aspect of the "Mediterranean" diet [23] is the high consumption of vegetable oils, vegetables, fruits, nuts, fish, and legumes, low dairy intake, and moderate intake of alcohol.

Although vegetables, legumes, and fruits consumption is beneficial to reproductive health, these items could show a large presence of xenoestrogens like pesticides [170], but it has been suggested that their beneficial effects would outweigh the negative consequences [148].

Sweet Foods and Beverages

Sugar-sweetened foods and beverages intake has been found to increase insulin resistance [171]. Insulin resistance is known to increase oxidative stress [172], which in turn can negatively influence sperm quality. In a report [173], the intake of sugar-sweetened beverages was related to lower sperm motility among young healthy men. This relation was independent of a large number of potential confounders, but was confined to lean men. There was also a suggestion of an inverse relation between sugar-sweetened beverages intake and FSH levels. This finding is consistent with animal experimental data showing that comparatively low levels of added sugar consumption have substantial negative effects on mouse survival, competitive ability, and reproduction [174]. However, in one of our published articles, no relationship among the consumption of sugar-sweetened foods and semen quality or ICSI outcomes was noted [66].

Red Meat

As discussed before, previous studies have reported an association between poor semen quality and an increased consumption of products that may incorporate xenobiotics, mainly xenoestrogens or certain anabolic steroids [175,176], such as red meat.

Mendiola et al. [148] found an increased intake of meat processed foods with especially high saturated fat content in patients with poor semen quality. In fact, Swan et al. [175] reported a significant association between a mother's reported beef consumption while pregnant and her son's sperm concentration. Moreover, the proportion of men with a sperm concentration below the World Health Organization (WHO) threshold and of men with a history of subfertility is increased among higher beef consumers. The authors suggest that these associations may be related to the presence of anabolic steroids and other xenobiotics in beef.

This is in agreement with a previous study from our group suggesting that successful pregnancy and implantation outcomes after ICSI cycles are decreased in patients reporting a more frequent intake of meat.

On the other side, a positive effect of the consumption of meat has also been observed. In a trial [160] evaluating the effect of different dietary patterns on semen quality in men undergoing IVF/ICSI treatment, it was observed that the "Traditional Dutch" diet, which comprises high intakes of meat products, potatoes, and whole grains, is positively correlated with sperm concentration. It was suggested that the high intake of meat, a natural source of zinc elements, influences the bioavailability of dietary folate. The zinc-dependent enzyme γ-glutamylhydrolase in the jejunum efficiently converts dietary folate as polyglutamates into monoglutamates, which are the only absorbable form of folate. Furthermore, zinc is a cofactor of methionine synthase involved in the remethylation of tHcy into methionine thereby reducing tHcy. The beneficial effects of zinc are supported by a randomized, placebo-controlled intervention study in which administration of zinc and/or folic acid to subfertile patients led to a significant increase in semen concentration [177,178].

CONCLUSIONS

While there has been much attention on the potential effects of environmental factors on male reproduction, literature on its effects on male patients undergoing ART is still scarce.

Nevertheless, during the past few decades, a number of reports have raised serious concerns about the decrease in semen quality and development of reproductive problems. Simultaneously with the growing evidence for decreasing quality of semen during the past 50 years, the use of ART has increased radically.

In spite of more than 2 million children having been born as a consequence of ART, most *in vitro* produced embryos fail to implant and around two-thirds of IVF cycles fail to result in pregnancy.

On this scenario, devising innovative strategies in order to increase the efficiency of ART have been a major concern in our area. This highlights the detrimental effects of specific lifestyle factors on reproductive health and ART outcomes.

In fact, there is strong evidence of the harmful effects of factors related to lifestyle, such as cigarette smoking, obesity or occupational risk factors, including heavy metals and solvents exposure, and even diet, on men's reproductive health.

However, the interpretation of studies on the exposure of couples to environmental factors and the association with IVF success is difficult because many confounding variables are present and it is difficult to isolate the effect of a single factor on the results. Moreover, there is little consistency among studies and the lack of literature makes it difficult to draw any strong conclusions.

Although the evidence linking environmental factors and impaired human fertility is weak, there is still some evidence demonstrating that semen quality and ICSI outcomes may be influenced by environmental factors, food intake, and social habits. Therefore, couples seeking ARTs must be advised about the adverse effects of the male and female lifestyles on the treatment success.

References

[1] Stephen EH, Chandra A. Declining estimates of infertility in the United States: 1982–2002. Fertil Steril 2006;86(3):516–23.

[2] de Mouzon J, Goossens V, Bhattacharya S, Castilla JA, Ferraretti AP, Korsak V, et al. Assisted reproductive technology in Europe, 2007: results generated from European registers by ESHRE. Hum Reprod 2012;27(4):954–66.

[3] Homan GF, Davies M, Norman R. The impact of lifestyle factors on reproductive performance in the general population and those undergoing infertility treatment: a review. Hum Reprod Update 2007;13(3):209–23.

[4] Brannian JD. Obesity and fertility. S D Med 2011;64(7):251–4.

[5] Popkin BM. Recent dynamics suggest selected countries catching up to US obesity. Am J Clin Nutr 2010;91(1):284S–8S.

[6] Finucane MM, Stevens GA, Cowan MJ, Danaei G, Lin JK, Paciorek CJ, et al. National, regional, and global trends in body-mass index since 1980: systematic analysis of health examination surveys and epidemiological studies with 960 country-years and 9.1 million participants. Lancet 2011;377(9765):557–67.

[7] Pasquali R, Pelusi C, Genghini S, Cacciari M, Gambineri A. Obesity and reproductive disorders in women. Hum Reprod Update 2003;9(4):359–72.

[8] Pasquali R, Gambineri A. Metabolic effects of obesity on reproduction. Reprod Biomed Online 2006;12(5):542–51.

[9] Macdonald AA, Stewart AW, Farquhar CM. Body mass index in relation to semen quality and reproductive hormones in New Zealand men: a cross-sectional study in fertility clinics. Hum Reprod 2013;28(12):3178–87.

[10] MacDonald AA, Herbison GP, Showell M, Farquhar CM. The impact of body mass index on semen parameters and reproductive hormones in human males: a systematic review with meta-analysis. Hum Reprod Update 2010;16(3):293–311.

[11] Sermondade N, Faure C, Fezeu L, Shayeb AG, Bonde JP, Jensen TK, et al. BMI in relation to sperm count: an updated systematic review and collaborative meta-analysis. Hum Reprod Update 2013;19(3):221–31.

[12] Jensen TK, Andersson AM, Jorgensen N, Andersen AG, Carlsen E, Petersen JH, et al. Body mass index in relation to semen quality and reproductive hormones among 1,558 Danish men. Fertil Steril 2004;82(4):863–70.

[13] Koloszar S, Fejes I, Zavaczki Z, Daru J, Szollosi J, Pal A. Effect of body weight on sperm concentration in normozoospermic males. Arch Androl 2005;51(4):299–304.

[14] Qin DD, Yuan W, Zhou WJ, Cui YQ, Wu JQ, Gao ES. Do reproductive hormones explain the association between body mass index and semen quality? Asian J Androl 2007;9(6):827–34.

[15] Aggerholm AS, Thulstrup AM, Toft G, Ramlau-Hansen CH, Bonde JP. Is overweight a risk factor for reduced semen quality and altered serum sex hormone profile? Fertil Steril 2008;90(3):619–26.

[16] Chavarro JE, Toth TL, Wright DL, Meeker JD, Hauser R. Body mass index in relation to semen quality, sperm DNA integrity, and serum reproductive hormone levels among men attending an infertility clinic. Fertil Steril 2010;93(7):2222–31.

[17] Duits FH, van Wely M, van der Veen F, Gianotten J. Healthy overweight male partners of subfertile couples should not worry about their semen quality. Fertil Steril 2010;94(4):1356–9.

[18] Martini AC, Tissera A, Estofan D, Molina RI, Mangeaud A, de Cuneo MF, et al. Overweight and seminal quality: a study of 794 patients. Fertil Steril 2010;94(5):1739–43.

[19] Eskandar M, Al-Asmari M, Babu Chaduvula S, Al-Shahrani M, Al-Sunaidi M, Almushait M, et al. Impact of male obesity on semen quality and serum sex hormones. Adv Urol 2013;2012:407601.

[20] Janssen I, Katzmarzyk PT, Ross R. Waist circumference and not body mass index explains obesity-related health risk. Am J Clin Nutr 2004;79(3):379–84.

[21] Fejes I, Koloszar S, Szollosi J, Zavaczki Z, Pal A. Is semen quality affected by male body fat distribution? Andrologia 2005;37(5):155–9.

[22] Hammiche F, Laven JS, Twigt JM, Boellaard WP, Steegers EA, Steegers-Theunissen RP. Body mass index and central adiposity are associated with sperm quality in men of subfertile couples. Hum Reprod 2012;27(8):2365–72.

[23] Eisenberg ML, Kim S, Chen Z, Sundaram R, Schisterman EF, Buck Louis GM. The relationship between male BMI and waist circumference on semen quality: data from the LIFE study. Hum Reprod 2013;29(2):193–200.

[24] Fejes I, Koloszar S, Zavaczki Z, Daru J, Szollosi J, Pal A. Effect of body weight on testosterone/estradiol ratio in oligozoospermic patients. Arch Androl 2006;52(2):97–102.

[25] Kort HI, Massey JB, Elsner CW, Mitchell-Leef D, Shapiro DB, Witt MA, et al. Impact of body mass index values on sperm quantity and quality. J Androl 2006;27(3):450–2.

[26] Hammoud AO, Gibson M, Peterson CM, Meikle AW, Carrell DT. Impact of male obesity on infertility: a critical review of the current literature. Fertil Steril 2008;90(4):897–904.

[27] Keltz J, Zapantis A, Jindal SK, Lieman HJ, Santoro N, Polotsky AJ. Overweight men: clinical pregnancy after ART is decreased in IVF but not in ICSI cycles. J Assist Reprod Genet 2010;27(9–10):539–44.

[28] Wegner CC, Clifford AL, Jilbert PM, Henry MA, Gentry WL. Abnormally high body mass index and tobacco use are associated with poor sperm quality as revealed by reduced sperm binding to hyaluronan-coated slides. Fertil Steril 2010;93(1):332–4.

[29] Sermondade N, Dupont C, Faure C, Boubaya M, Cedrin-Durnerin I, Chavatte-Palmer P, et al. Body mass index is not associated with sperm-zona pellucida binding ability in subfertile males. Asian J Androl 2013;15(5):626–9.

[30] Bonde JP, Ramlau-Hansen CH, Olsen J. Trends in sperm counts: the saga continues. Epidemiology 2011;22(5):617–9.

[31] Loucks AB. Energy availability, not body fatness, regulates reproductive function in women. Exerc Sport Sci Rev 2003;31(3):144–8.

[32] Ayers JW, Komesu Y, Romani T, Ansbacher R. Anthropomorphic, hormonal, and psychologic correlates of semen quality in endurance-trained male athletes. Fertil Steril 1985;43(6):917–21.

[33] Lucia A, Chicharro JL, Perez M, Serratosa L, Bandres F, Legido JC. Reproductive function in male endurance athletes: sperm analysis and hormonal profile. J Appl Physiol (1985) 1996;81(6):2627–36.

[34] Braga DP, Halpern G, Figueira Rde C, Setti AS, Iaconelli A Jr, Borges E Jr. Food intake and social habits in male patients and its relationship to intracytoplasmic sperm injection outcomes. Fertil Steril 2011;97(1):53–9.

[35] Klonoff-Cohen H, Lam-Kruglick P, Gonzalez C. Effects of maternal and paternal alcohol consumption on the success rates of *in vitro* fertilization and gamete intrafallopian transfer. Fertil Steril 2003;79(2):330–9.

[36] Hadi HA, Hill JA, Castillo RA. Alcohol and reproductive function: a review. Obstet Gynecol Surv 1987;42(2):69–74.

[37] Sermondade N, Elloumi H, Berthaut I, Mathieu E, Delarouziere V, Ravel C, et al. Progressive alcohol-induced sperm alterations leading to spermatogenic arrest, which was reversed after alcohol withdrawal. Reprod Biomed Online 2010;20(3):324–7.

[38] Roemmich JN, Sinning WE. Weight loss and wrestling training: effects on growth-related hormones. J Appl Physiol (1985) 1997;82(6):1760–4.

[39] Safarinejad MR, Azma K, Kolahi AA. The effects of intensive, long-term treadmill running on reproductive hormones, hypothalamus-pituitary-testis axis, and semen quality: a randomized controlled study. J Endocrinol 2009;200(3):259–71.

[40] Vaamonde D, Da Silva-Grigoletto ME, Garcia-Manso JM, Vaamonde-Lemos R, Swanson RJ, Oehninger SC. Response of semen parameters to three training modalities. Fertil Steril 2009;92(6):1941–6.

[41] Hill EE, Zack E, Battaglini C, Viru M, Viru A, Hackney AC. Exercise and circulating cortisol levels: the intensity threshold effect. J Endocrinol Invest 2008;31(7):587–91.

[42] Hackney AC. Effects of endurance exercise on the reproductive system of men: the "exercise-hypogonadal male condition". J Endocrinol Invest 2008;31(10):932–8.

[43] De Souza MJ, Arce JC, Pescatello LS, Scherzer HS, Luciano AA. Gonadal hormones and semen quality in male runners. A volume threshold effect of endurance training. Int J Sports Med 1994;15(7):383–91.

[44] Di Luigi L, Gentile V, Pigozzi F, Parisi A, Giannetti D, Romanelli F. Physical activity as a possible aggravating factor for athletes with varicocele: impact on the semen profile. Hum Reprod 2001;16(6):1180–4.

[45] Gebreegziabher Y, Marcos E, McKinon W, Rogers G. Sperm characteristics of endurance trained cyclists. Int J Sports Med 2004;25(4):247–51.

[46] Arce JC, De Souza MJ, Pescatello LS, Luciano AA. Subclinical alterations in hormone and semen profile in athletes. Fertil Steril 1993;59(2):398–404.

[47] De Souza MJ, Miller BE. The effect of endurance training on reproductive function in male runners. A 'volume threshold' hypothesis. Sports Med 1997;23(6):357–74.

[48] Jensen CE, Wiswedel K, McLoughlin J, van der Spuy Z. Prospective study of hormonal and semen profiles in marathon runners. Fertil Steril 1995;64(6):1189–96.

[49] Vaamonde D, Da Silva ME, Poblador MS, Lancho JL. Reproductive profile of physically active men after exhaustive endurance exercise. Int J Sports Med 2006;27(9):680–9.

[50] Friedl KE, Moore RJ, Hoyt RW, Marchitelli LJ, Martinez-Lopez LE, Askew EW. Endocrine markers of semistarvation in healthy lean men in a multistressor environment. J Appl Physiol (1985) 2000;88(5):1820–30.

[51] Williams NI, Helmreich DL, Parfitt DB, Caston-Balderrama A, Cameron JL. Evidence for a causal role of low energy availability in the induction of menstrual cycle disturbances during strenuous exercise training. J Clin Endocrinol Metab 2001;86(11):5184–93.

[52] Pedersen BK, Saltin B. Evidence for prescribing exercise as therapy in chronic disease. Scand J Med Sci Sports 2006;16(Suppl. 1):3–63.

[53] Grandys M, Majerczak J, Duda K, Zapart-Bukowska J, Kulpa J, Zoladz JA. Endurance training of moderate intensity increases testosterone concentration in young, healthy men. Int J Sports Med 2009;30(7):489–95.

[54] Alessio HM, Hagerman AE, Nagy S, Philip B, Byrnes RN, Woodward JL, et al. Exercise improves biomarkers of health and stress in animals fed ad libitum. Physiol Behav 2005;84(1):65–72.

[55] Badawy A, Elnashar A. Treatment options for polycystic ovary syndrome. Int J Womens Health 2011;3:25–35.

[56] Vaamonde D, Da Silva-Grigoletto ME, Garcia-Manso JM, Barrera N, Vaamonde-Lemos R. Physically active men show better semen parameters and hormone values than sedentary men. Eur J Appl Physiol 2012;112(9):3267–73.

[57] Adlard PA, Cotman CW. Voluntary exercise protects against stress-induced decreases in brain-derived neurotrophic factor protein expression. Neuroscience 2004;124(4):985–92.

[58] Duclos M, Corcuff JB, Rashedi M, Fougere V, Manier G. Trained versus untrained men: different immediate post-exercise responses of pituitary adrenal axis. A preliminary study. Eur J Appl Physiol Occup Physiol 1997;75(4):343–50.

[59] ASRM. Smoking and infertility. Fertil Steril 2008;90(5):S254–9.

[60] Bolumar F, Olsen J, Boldsen J. Smoking reduces fecundity: a European multicenter study on infertility and subfecundity. The European Study Group on Infertility and Subfecundity. Am J Epidemiol 1996;143(6):578–87.

[61] Chia SE, Lim ST, Tay SK, Lim ST. Factors associated with male infertility: a case-control study of 218 infertile and 240 fertile men. BJOG 2000;107(1):55–61.

[62] Curtis KM, Savitz DA, Arbuckle TE. Effects of cigarette smoking, caffeine consumption, and alcohol intake on fecundability. Am J Epidemiol 1997;146(1):32–41.

[63] Klonoff-Cohen H, Natarajan L, Marrs R, Yee B. Effects of female and male smoking on success rates of IVF and gamete intra-fallopian transfer. Hum Reprod 2001;16(7):1382–90.

[64] Sofikitis N, Takenaka M, Kanakas N, Papadopoulos H, Yamamoto Y, Drakakis P, et al. Effects of cotinine on sperm motility, membrane function, and fertilizing capacity *in vitro*. Urol Res 2000;28(6):370–5.

[65] Zitzmann M, Rolf C, Nordhoff V, Schrader G, Rickert-Fohring M, Gassner P, et al. Male smokers have a decreased success rate for *in vitro* fertilization and intracytoplasmic sperm injection. Fertil Steril 2003;79(Suppl. 3):1550–4.

[66] Braga DP, Halpern G, Figueira Rde C, Setti AS, Iaconelli A Jr, Borges E Jr. Food intake and social habits in male patients and its relationship to intracytoplasmic sperm injection outcomes. Fertil Steril 2012;97(1):53–9.

[67] Braga DP, Figueira Rde C, Rodrigues D, Madaschi C, Pasqualotto FF, Iaconelli A Jr, et al. Prognostic value of meiotic spindle imaging on fertilization rate and embryo development in *in vitro*-matured human oocytes. Fertil Steril 2008;90(2):429–33.

[68] Mostafa T, Tawadrous G, Roaia MM, Amer MK, Kader RA, Aziz A. Effect of smoking on seminal plasma ascorbic acid in infertile and fertile males. Andrologia 2006;38(6):221–4.

[69] Lanzafame FM, La Vignera S, Vicari E, Calogero AE. Oxidative stress and medical antioxidant treatment in male infertility. Reprod Biomed Online 2009;19(5):638–59.

[70] Clemens S, Jotangia D, Lynch S, Nicholson S, Pigott S, Fuller E, editors. Drug Use, Smoking and Drinking Among Young People in England in 2007. London: National Centre for Social Research, National Foundation for Educational Research; 2009.

[71] Oldereid NB, Rui H, Purvis K. Lifestyles of men in barren couples and their relationships to sperm quality. Eur J Obstet Gynecol Reprod Biol 1992;43(1):51–7.

[72] Marshburn PB, Sloan CS, Hammond MG. Semen quality and association with coffee drinking, cigarette smoking, and ethanol consumption. Fertil Steril 1989;52(1):162–5.

[73] Dunphy BC, Barratt CL, Cooke ID. Male alcohol consumption and fecundity in couples attending an infertility clinic. Andrologia 1991;23(3):219–21.

[74] Lopez Teijon M, Garcia F, Serra O, Moragas M, Rabanal A, Olivares R, et al. Semen quality in a population of volunteers from the province of Barcelona. Reprod Biomed Online 2007;15(4):434–44.

[75] Marinelli D, Gaspari L, Pedotti P, Taioli E. Mini-review of studies on the effect of smoking and drinking habits on semen parameters. Int J Hyg Environ Health 2004;207(3):185–92.

[76] Brzek A. Alcohol and male fertility (preliminary report). Andrologia 1987;19(1):32–6.

[77] Muthusami KR, Chinnaswamy P. Effect of chronic alcoholism on male fertility hormones and semen quality. Fertil Steril 2005;84(4):919–24.

[78] Donnelly GP, McClure N, Kennedy MS, Lewis SE. Direct effect of alcohol on the motility and morphology of human spermatozoa. Andrologia 1999;31(1):43–7.

[79] Battista N, Pasquariello N, Di Tommaso M, Maccarrone M. Interplay between endocannabinoids, steroids and cytokines in the control of human reproduction. J Neuroendocrinol 2008;20(Suppl. 1):82–9.

[80] Devane WA, Hanus L, Breuer A, Pertwee RG, Stevenson LA, Griffin G, et al. Isolation and structure of a brain constituent that binds to the cannabinoid receptor. Science 1992;258(5090):1946–9.

[81] Felder CC, Veluz JS, Williams HL, Briley EM, Matsuda LA. Cannabinoid agonists stimulate both receptor- and non-receptor-mediated signal transduction pathways in cells transfected with and expressing cannabinoid receptor clones. Mol Pharmacol 1992;42(5):838–45.

[82] Schuel H, Burkman LJ, Lippes J, Crickard K, Mahony MC, Giuffrida A, et al. Evidence that anandamide-signaling regulates human sperm functions required for fertilization. Mol Reprod Dev 2002;63(3):376–87.

[83] Rossato M, Ion Popa F, Ferigo M, Clari G, Foresta C. Human sperm express cannabinoid receptor Cb1, the activation of which inhibits motility, acrosome reaction, and mitochondrial function. J Clin Endocrinol Metab 2005;90(2):984–91.

[84] Cone EJ, Johnson RE, Moore JD, Roache JD. Acute effects of smoking marijuana on hormones, subjective effects and performance in male human subjects. Pharmacol Biochem Behav 1986;24(6):1749–54.

[85] Vescovi PP, Pedrazzoni M, Michelini M, Maninetti L, Bernardelli F, Passeri M. Chronic effects of marihuana smoking on luteinizing hormone, follicle-stimulating hormone and prolactin levels in human males. Drug Alcohol Depend 1992;30(1):59–63.

[86] Whan LB, West MC, McClure N, Lewis SE. Effects of delta-9-tetrahydrocannabinol, the primary psychoactive cannabinoid in marijuana, on human sperm function *in vitro*. Fertil Steril 2006;85(3):653–60.

[87] Badawy ZS, Chohan KR, Whyte DA, Penefsky HS, Brown OM, Souid AK. Cannabinoids inhibit the respiration of human sperm. Fertil Steril 2009;91(6):2471–6.

[88] Schuel H, Chang MC, Berkery D, Schuel R, Zimmerman AM, Zimmerman S. Cannabinoids inhibit fertilization in sea urchins by reducing the fertilizing capacity of sperm. Pharmacol Biochem Behav 1991;40(3):609–15.

[89] Makler A, David R, Blumenfeld Z, Better OS. Factors affecting sperm motility. VII. Sperm viability as affected by change of pH and osmolarity of semen and urine specimens. Fertil Steril 1981;36(4):507–11.

[90] Mahoney JM, Harris RA. Effect of 9-tetrahydrocannabinol on mitochondrial precesses. Biochem Pharmacol 1972;21(9):1217–26.

[91] Nahas GG, Frick HC, Lattimer JK, Latour C, Harvey D. Pharmacokinetics of THC in brain and testis, male gametotoxicity and premature apoptosis of spermatozoa. Hum Psychopharmacol 2002;17(2):103–13.

[92] Fronczak CM, Kim ED, Barqawi AB. The insults of illicit drug use on male fertility. J Androl 2012;33(4):515–28.

[93] Bracken MB, Eskenazi B, Sachse K, McSharry JE, Hellenbrand K, Leo-Summers L. Association of cocaine use with sperm concentration, motility, and morphology. Fertil Steril 1990;53(2):315–22.

[94] Li H, George VK, Crossland WJ, Anderson GF, Dhabuwala CB. Characterization of cocaine binding sites in the rat testes. J Urol 1997;158(3 Pt 1):962–5.

[95] George VK, Li H, Teloken C, Grignon DJ, Lawrence WD, Dhabuwala CB. Effects of long-term cocaine exposure on spermatogenesis and fertility in peripubertal male rats. J Urol 1996;155(1):327–31.

[96] Rodriguez MC, Sanchez-Yague J, Paniagua R. Effects of cocaine on testicular structure in the rat. Reprod Toxicol 1992;6(1):51–5.

[97] Li H, Jiang Y, Rajpurkar A, Dunbar JC, Dhabuwala CB. Cocaine induced apoptosis in rat testes. J Urol 1999;162(1):213–6.

[98] Li H, George VK, Crawford SC, Dhabuwala CB. Effect of cocaine on testicular blood flow in rats: evaluation by percutaneous injection of xenon-133. J Environ Pathol Toxicol Oncol 1999;18(1):73–7.

[99] McDiarmid MA, Gardiner PM, Jack BW. The clinical content of preconception care: environmental exposures. Am J Obstet Gynecol 2008;199(6 Suppl. 2):S357–61.

[100] Foster WG, Neal MS, Han MS, Dominguez MM. Environmental contaminants and human infertility: hypothesis or cause for concern? J Toxicol Environ Health B Crit Rev 2008;11(3–4):162–76.

[101] Mendola P, Messer LC, Rappazzo K. Science linking environmental contaminant exposures with fertility and reproductive health impacts in the adult female. Fertil Steril 2008;89(2 Suppl.): pe81–94.

[102] Bretveld R, Kik S, Hooiveld M, van Rooij I, Zielhuis G, Roeleveld N. Time-to-pregnancy among male greenhouse workers. Occup Environ Med 2008;65(3):185–90.

[103] Mendiola J, Torres-Cantero AM, Moreno-Grau JM, Ten J, Roca M, Moreno-Grau S, et al. Exposure to environmental toxins in males seeking infertility treatment: a case-controlled study. Reprod Biomed Online 2008;16(6):842–50.

[104] Irgens A, Kruger K, Ulstein M. The effect of male occupational exposure in infertile couples in Norway. J Occup Environ Med 1999;41(12):1116–20.

[105] Bigelow PL, Jarrell J, Young MR, Keefe TJ, Love EJ. Association of semen quality and occupational factors: comparison of case-control analysis and analysis of continuous variables. Fertil Steril 1998;69(1):11–8.

[106] Cherry N, Labreche F, Collins J, Tulandi T. Occupational exposure to solvents and male infertility. Occup Environ Med 2001;58(10):635–40.

[107] Tielemans E, Burdorf A, te Velde ER, Weber RF, van Kooij RJ, Veulemans H, et al. Occupationally related exposures and reduced semen quality: a case-control study. Fertil Steril 1999;71(4):690–6.

[108] Logman JF, de Vries LE, Hemels ME, Khattak S, Einarson TR. Paternal organic solvent exposure and adverse pregnancy outcomes: a meta-analysis. Am J Ind Med 2005;47(1):37–44.

[109] Skakkebaek NE, Rajpert-De Meyts E, Main KM. Testicular dysgenesis syndrome: an increasingly common developmental disorder with environmental aspects. Hum Reprod 2001;16(5):972–8.

[110] Sharpe RM, Irvine DS. How strong is the evidence of a link between environmental chemicals and adverse effects on human reproductive health? BMJ 2004;328(7437):447–51.

[111] Robins TG, Bornman MS, Ehrlich RI, Cantrell AC, Pienaar E, Vallabh J, et al. Semen quality and fertility of men employed in a South African lead acid battery plant. Am J Ind Med 1997;32(4):369–76.

[112] Wyrobek AJ, Schrader SM, Perreault SD, Fenster L, Huszar G, Katz DF, et al. Assessment of reproductive disorders and birth defects in communities near hazardous chemical sites. III. Guidelines for field studies of male reproductive disorders. Reprod Toxicol 1997;11(2–3):243–59.

[113] Wirth JJ, Mijal RS. Adverse effects of low level heavy metal exposure on male reproductive function. Syst Biol Reprod Med 2010;56(2):147–67.

[114] Bonde JP. The risk of male subfecundity attributable to welding of metals. Studies of semen quality, infertility, fertility, adverse pregnancy outcome and childhood malignancy. Int J Androl 1993;16(Suppl. 1):1–29.

[115] Jurasovic J, Cvitkovic P, Pizent A, Colak B, Telisman S. Semen quality and reproductive endocrine function with regard to blood cadmium in Croatian male subjects. Biometals 2004;17(6):735–43.

[116] Telisman S, Colak B, Pizent A, Jurasovic J, Cvitkovic P. Reproductive toxicity of low-level lead exposure in men. Environ Res 2007;105(2):256–66.

[117] Meeker JD, Rossano MG, Protas B, Diamond MP, Puscheck E, Daly D, et al. Cadmium, lead, and other metals in relation to semen quality: human evidence for molybdenum as a male reproductive toxicant. Environ Health Perspect 2008;116(11): 1473–9.

[118] Benoff S, Jacob A, Hurley IR. Male infertility and environmental exposure to lead and cadmium. Hum Reprod Update 2000;6(2):107–21.

[119] Benoff S. Receptors and channels regulating acrosome reactions. Hum Fertil (Camb) 1999;2(1):42–55.

[120] Benoff S, Centola GM, Millan C, Napolitano B, Marmar JL, Hurley IR. Increased seminal plasma lead levels adversely affect the fertility potential of sperm in IVF. Hum Reprod 2003;18(2):374–83.

[121] Pillai A, Laxmi Priya PN, Gupta S. Effects of combined exposure to lead and cadmium on pituitary membrane of female rats. Arch Toxicol 2002;76(12):671–5.

[122] Xu B, Chia SE, Tsakok M, Ong CN. Trace elements in blood and seminal plasma and their relationship to sperm quality. Reprod Toxicol 1993;7(6):613–8.

[123] Akinloye O, Arowojolu AO, Shittu OB, Anetor JI. Cadmium toxicity: a possible cause of male infertility in Nigeria. Reprod Biol 2006;6(1):17–30.

[124] Buck Louis GM, Sundaram R, Schisterman EF, Sweeney AM, Lynch CD, Gore-Langton RE, et al. Heavy metals and couple fecundity, the LIFE Study. Chemosphere 2012;87(11):1201–7.

[125] Chia SE, Ong CN, Lee ST, Tsakok FH. Blood concentrations of lead, cadmium, mercury, zinc, and copper and human semen parameters. Arch Androl 1992;29(2):177–83.

[126] Kim K, Fujimoto VY, Parsons PJ, Steuerwald AJ, Browne RW, Bloom MS. Recent cadmium exposure among male partners may affect oocyte fertilization during in vitro fertilization (IVF). J Assist Reprod Genet 2010;27(8):463–8.

[127] Rignell-Hydbom A, Axmon A, Lundh T, Jonsson BA, Tiido T, Spano M. Dietary exposure to methyl mercury and PCB and the associations with semen parameters among Swedish fishermen. Environ Health 2007;6:14.

[128] Choy CM, Lam CW, Cheung LT, Briton-Jones CM, Cheung LP, Haines CJ. Infertility, blood mercury concentrations and dietary seafood consumption: a case-control study. BJOG 2002;109(10):1121–5.

[129] Choy CM, Yeung QS, Briton-Jones CM, Cheung CK, Lam CW, Haines CJ. Relationship between semen parameters and mercury concentrations in blood and in seminal fluid from subfertile males in Hong Kong. Fertil Steril 2002;78(2):426–68.

[130] Golub MS, Macintosh MS, Baumrind N. Developmental and reproductive toxicity of inorganic arsenic: animal studies and human concerns. J Toxicol Environ Health B Crit Rev 1998;1(3):199–241.

[131] Shukla JP, Pandey K. Impaired spermatogenesis in arsenic treated freshwater fish, Colisa fasciatus (Bl. and Sch.). Toxicol Lett 1984;21(2):191–5.

[132] Chatterjee A, Chatterji U. Arsenic abrogates the estrogen-signaling pathway in the rat uterus. Reprod Biol Endocrinol 2010;8:80.

[133] Biswas R, Poddar S, Mukherjee A. Investigation on the genotoxic effects of long-term administration of sodium arsenite in bone marrow and testicular cells in vivo using the comet assay. J Environ Pathol Toxicol Oncol 2007;26(1):29–37.

[134] Leke RJ, Oduma JA, Bassol-Mayagoitia S, Bacha AM, Grigor KM. Regional and geographical variations in infertility: effects of environmental, cultural, and socioeconomic factors. Environ Health Perspect 1993;101(Suppl 2):73–80.

[135] Hsieh FI, Hwang TS, Hsieh YC, Lo HC, Su CT, Hsu HS, et al. Risk of erectile dysfunction induced by arsenic exposure through well water consumption in Taiwan. Environ Health Perspect 2008;116(4):532–6.

[136] Xu W, Bao H, Liu F, Liu L, Zhu YG, She J, et al. Environmental exposure to arsenic may reduce human semen quality: associations derived from a Chinese cross-sectional study. Environ Health 2012;11:46.

[137] Diamanti-Kandarakis E, Bourguignon JP, Giudice LC, Hauser R, Prins GS, Soto AM, et al. Endocrine-disrupting chemicals: an endocrine society scientific statement. Endocr Rev 2009;30(4):293–342.

[138] Schlatt S, Ehmcke J. Regulation of spermatogenesis: An evolutionary biologist's perspective. Semin Cell Dev Biol 2014;29C:2–16.

[139] Letcher RJ, Bustnes JO, Dietz R, Jenssen BM, Jorgensen EH, Sonne C, et al. Exposure and effects assessment of persistent organohalogen contaminants in arctic wildlife and fish. Sci Total Environ 2010;408(15):2995–3043.

[140] Younglai EV, Foster WG, Hughes EG, Trim K, Jarrell JF. Levels of environmental contaminants in human follicular fluid, serum, and seminal plasma of couples undergoing in vitro fertilization. Arch Environ Contam Toxicol 2002;43(1):121–6.

[141] Darnerud PO, Risberg S. Tissue localisation of tetra- and pentabromodiphenyl ether congeners (BDE-47, -85 and -99) in perinatal and adult C57BL mice. Chemosphere 2006;62(3):485–93.

[142] Stahlhut RW, Welshons WV, Swan SH, Bisphenol A. Data in NHANES suggest longer than expected half-life, substantial nonfood exposure, or both. Environ Health Perspect 2009;117(5):784–9.

[143] Schiliro T, Gorrasi I, Longo A, Coluccia S, Gilli G. Endocrine disrupting activity in fruits and vegetables evaluated with the E-screen assay in relation to pesticide residues. J Steroid Biochem Mol Biol 2011;127(1-2):139–46.

[144] Iavicoli I, Fontana L, Bergamaschi A. The effects of metals as endocrine disruptors. J Toxicol Environ Health B Crit Rev 2009;12(3):206–23.

[145] Calafat AM, Ye X, Wong LY, Reidy JA, Needham LL. Exposure of the U.S. population to bisphenol A and 4-tertiary-octylphenol: 2003–2004. Environ Health Perspect 2008;116(1):39–44.

[146] Milnes MR, Bermudez DS, Bryan TA, Edwards TM, Gunderson MP, Larkin IL, et al. Contaminant-induced feminization and demasculinization of nonmammalian vertebrate males in aquatic environments. Environ Res 2006;100(1):3–17.

[147] Fusani L, Della Seta D, Dessi-Fulgheri F, Farabollini F. Altered reproductive success in rat pairs after environmental-like exposure to xenoestrogen. Proc Biol Sci 2007;274(1618):1631–6.

[148] Mendiola J, Torres-Cantero AM, Moreno-Grau JM, Ten J, Roca M, Moreno-Grau S, et al. Food intake and its relationship with semen quality: a case-control study. Fertil Steril 2009;91(3):812–8.

[149] Klonoff-Cohen H, Bleha J, Lam-Kruglick P. A prospective study of the effects of female and male caffeine consumption on the reproductive endpoints of IVF and gamete intra-fallopian transfer. Hum Reprod 2002;17(7):1746–54.

[150] Bolumar F, Olsen J, Rebagliato M, Bisanti L. Caffeine intake and delayed conception: a European multicenter study on infertility and subfecundity. European Study Group on Infertility Subfecundity. Am J Epidemiol 1997;145(4):324–34.

[151] Klonoff-Cohen H. Female and male lifestyle habits and IVF: what is known and unknown. Hum Reprod Update 2005;11(2):180–204.

[152] Hatch EE, Bracken MB. Association of delayed conception with caffeine consumption. Am J Epidemiol 1993;138(12):1082–92.

[153] Sharpe RM. The 'oestrogen hypothesis' – where do we stand now? Int J Androl 2003;26(1):2–15.

[154] Ganmaa D, Wang PY, Qin LQ, Hoshi K, Sato A. Is milk responsible for male reproductive disorders? Med Hypotheses 2001;57(4):510–4.

[155] Garcia-Pelaez B, Ferrer-Lorente R, Gomez-Olles S, Fernandez-Lopez JA, Remesar X, Alemany M. Technical note: measurement of total estrone content in foods. Application to dairy products. J Dairy Sci 2004;87(8):2331–6.

[156] Pape-Zambito DA, Roberts RF, Kensinger RS. Estrone and 17beta-estradiol concentrations in pasteurized-homogenized milk and commercial dairy products. J Dairy Sci 2010;93(6):2533–40.

[157] Afeiche MC, Bridges ND, Williams PL, Gaskins AJ, Tanrikut C, Petrozza JC, et al. Dairy intake and semen quality among men attending a fertility clinic. Fertil Steril 2014;101(5):1280–7.

[158] Eslamian G, Amirjannati N, Rashidkhani B, Sadeghi MR, Hekmatdoost A. Intake of food groups and idiopathic asthenozoospermia: a case-control study. Hum Reprod 2012;27(11):3328–36.

[159] Afeiche M, Williams PL, Mendiola J, Gaskins AJ, Jorgensen N, Swan SH, et al. Dairy food intake in relation to semen quality and reproductive hormone levels among physically active young men. Hum Reprod 2013;28(8):2265–75.

[160] Vujkovic M, de Vries JH, Dohle GR, Bonsel GJ, Lindemans J, Macklon NS, et al. Associations between dietary patterns and semen quality in men undergoing IVF/ICSI treatment. Hum Reprod 2009;24(6):1304–12.

[161] Attaman JA, Toth TL, Furtado J, Campos H, Hauser R, Chavarro JE. Dietary fat and semen quality among men attending a fertility clinic. Hum Reprod 2012;27(5):1466–74.

[162] Vujkovic M, Ocke MC, van der Spek PJ, Yazdanpanah N, Steegers EA, Steegers-Theunissen RP. Maternal Western dietary patterns and the risk of developing a cleft lip with or without a cleft palate. Obstet Gynecol 2007;110(2 Pt 1):378–84.

[163] Agarwal A, Nallella KP, Allamaneni SS, Said TM. Role of antioxidants in treatment of male infertility: an overview of the literature. Reprod Biomed Online 2004;8(6):616–27.

[164] Agarwal A, Said TM. Carnitines and male infertility. Reprod Biomed Online 2004;8(4):376–84.

[165] Comhaire FH, Mahmoud A. The role of food supplements in the treatment of the infertile man. Reprod Biomed Online 2003;7(4):385–91.

[166] Abd-Elmoaty MA, Saleh R, Sharma R, Agarwal A. Increased levels of oxidants and reduced antioxidants in semen of infertile men with varicocele. Fertil Steril 2010;94(4):1531–4.

[167] Agarwal A, Sharma RK, Nallella KP, Thomas AJ Jr, Alvarez JG, Sikka SC. Reactive oxygen species as an independent marker of male factor infertility. Fertil Steril 2006;86(4):878–85.

[168] Mínguez-Alarcón L, Mendiola J, Lopez-Espin JJ, Sarabia-Cos L, Vivero-Salmeron G, Vioque J, et al. Dietary intake of antioxidant nutrients is associated with semen quality in young university students. Hum Reprod 2012;27(9):2807–14.

[169] Vujkovic M, de Vries JH, Lindemans J, Macklon NS, van der Spek PJ, Steegers EA, et al. The preconception Mediterranean dietary pattern in couples undergoing *in vitro* fertilization/intracytoplasmic sperm injection treatment increases the chance of pregnancy. Fertil Steril 2009;94(6):2096–101.

[170] Rozati R, Reddy PP, Reddanna P, Mujtaba R. Role of environmental estrogens in the deterioration of male factor fertility. Fertil Steril 2002;78(6):1187–94.

[171] Stanhope KL, Schwarz JM, Keim NL, Griffen SC, Bremer AA, Graham JL, et al. Consuming fructose-sweetened, not glucose-sweetened, beverages increases visceral adiposity and lipids and decreases insulin sensitivity in overweight/obese humans. J Clin Invest 2009;119(5):1322–34.

[172] Park K, Gross M, Lee DH, Holvoet P, Himes JH, Shikany JM, et al. Oxidative stress and insulin resistance: the coronary artery risk development in young adults study. Diabetes Care 2009;32(7):1302–7.

[173] Chiu YH, Afeiche MC, Gaskins AJ, Williams PL, Mendiola J, Jorgensen N, et al. Sugar-sweetened beverage intake in relation to semen quality and reproductive hormone levels in young men. Hum Reprod 2014;29:1575–84.

[174] Ruff JS, Suchy AK, Hugentobler SA, Sosa MM, Schwartz BL, Morrison LC, et al. Human-relevant levels of added sugar consumption increase female mortality and lower male fitness in mice. Nat Commun 2013;4:2245.

[175] Swan SH, Liu F, Overstreet JW, Brazil C, Skakkebaek NE. Semen quality of fertile US males in relation to their mothers' beef consumption during pregnancy. Hum Reprod 2007;22(6):1497–502.

[176] Santti R, Makela S, Strauss L, Korkman J, Kostian ML. Phytoestrogens: potential endocrine disruptors in males. Toxicol Ind Health 1998;14(1–2):223–37.

[177] Ebisch IM, Peters WH, Thomas CM, Wetzels AM, Peer PG, Steegers-Theunissen RP. Homocysteine, glutathione and related thiols affect fertility parameters in the (sub)fertile couple. Hum Reprod 2006;21(7):1725–33.

[178] Ebisch IM, Pierik FH, FH DEJ, Thomas CM, Steegers-Theunissen RP. Does folic acid and zinc sulphate intervention affect endocrine parameters and sperm characteristics in men? Int J Androl 2006;29(2):339–45.

33

The Role of Over-the-Counter Supplements in Male Infertility

Alan Scott Polackwich Jr., MD, Edmund S. Sabanegh, MD

Glickman Urological and Kidney Institute, Cleveland Clinic Foundation, Cleveland, Ohio, USA

INTRODUCTION

Over-the-counter supplements are a significant part of many Americans' approach to healthcare and daily life. A recent study has shown that 47% of adult men use at least one daily supplement or vitamin [1], and the average number of over-the-counter drugs taken by Americans is nearly 1.8 drugs per person [2]. With such widespread use of supplements, it is likely that patients are utilizing over-the-counter supplements either concurrently or as an adjunct to medical treatments. It is therefore important to know the effectiveness of these agents, and whether incorporating them into patients' treatment is useful.

The field of male infertility is no different than the rest of medicine with respect to the use of supplements. In the treatment of the infertile male, there is an increasing volume of data suggesting the efficacy of antioxidants and micronutrients to maximize fertility potential [3]. By manipulating DNA synthesis or lipid peroxidation pathways, these supplements are theoretically able to decrease spermatogenic susceptibility to exogenous stressors and counteract deficiencies or errors in metabolism that may contribute to idiopathic male infertility [3,4]. With the high cost and personal investment in assisted reproductive techniques (ART), it is no wonder that patients and practitioners are interested in the usefulness of these supplements.

OXIDATIVE STRESS AND MALE INFERTILITY

Many supplements used for fertility address reactive oxygen species (ROS) and limiting their detrimental effects on sperm. ROS are a family of molecules within the body that contain unpaired electrons, most commonly the superoxide anion and the hydroxyl radical. They occur as a by-product of aerobic metabolism, and as their name suggests, are highly reactive, which allows them to interfere with the normal metabolic functions of cells [5].

Sperm is highly sensitive to oxidation and ROS, due to interference with normal sperm function and development through the induction of lipid peroxidation and impairment of DNA replication and repair mechanisms [6]. The process of lipid peroxidation affects membrane fluidity, which is crucial to the movement of the sperm as well as its development; this involves significant manipulation of the cellular membrane [7]. This interference with the normal cellular machinery leads to the clinical findings of oligo-, astheno-, and teratospermia, as well as increased DNA fragmentation [5,6,8]. Correlations between increased by-products of ROS and abnormal semen parameters are well documented as well as increases in ROS in men with idiopathic infertility [9,10]. Not all individuals are as affected by environmental changes, and it may be that genetic polymorphisms between individuals explain these differences in response to environmental stressors [11].

There are multiple sources of oxidative stress in the modern world, intrinsic and extrinsic, some of which have been directly linked to changes in seminal parameters [12]. Animal studies and some human study suggest that electromagnetic radiation within the frequencies used for cellular phones and WiFi may affect seminal parameters by increasing ROS [13]. Due to the nature of the exposure, it is difficult to study this in a definitive way in humans, and conflicting evidence exists at this point [14].

Smoking has been found to be a significant factor in increasing oxidative stress within sperm and seminal fluid. Men who smoke have lower activity of known antioxidant pathways within the semen [15–17], lower serum levels of antioxidants [18], higher levels of MDA,

Handbook of Fertility. http://dx.doi.org/10.1016/B978-0-12-800872-0.00033-0

and higher sperm DNA fragmentation indexes than nonsmokers [19,20].

Varicoceles are also associated with oxidative stress with increased testicular temperatures and associated changes in metabolism [21,22]. Use of antioxidants in patients with low-grade varicoceles and abnormal semen parameters has been shown to improve these parameters [23,24]. Varicocelectomy has also been able to improve seminal oxidation status, which may be one of the mechanisms that improve seminal parameters after surgery [22].

SUPPLEMENTS

Most over-the-counter supplements are either antioxidants or micronutrients; they decrease ROS and reduce lipid peroxidation, or improve seminal maturation and division by increasing seminal concentration of necessary micronutrients [3]. There is good evidence that suggests that vitamins C and E and carnitines are helpful at improving relevant seminal parameters and possibly pregnancy rates, but there is also evidence for the use of selenium, vitamin A, zinc, glutathione, coenzyme Q10, polyunsaturated fatty acids (PUFAs), folate, and arginine [3,25].

Sperm has high concentrations of PUFAs, a trait that aids with motility and is integral to sperm production, but also makes them susceptible to reactive oxidative species [26,27]. ROS are able to initiate lipid peroxidation that, due to the chemical structure of PUFAs, propagates the reaction resulting in changes in membrane structure and interference with sperm function [7]. In addition to lipid peroxidation pathways, ROS directly interfere with DNA replication and repair, leading to increased DNA fragmentation [6,8]. Oral antioxidant therapy can protect against these changes to DNA [28].

While there are many studies that have been done to determine the efficacy of over-the-counter supplements, there are issues with many of these studies. Many of these studies have small patient numbers, and few are placebo controlled. Therefore, many of the studies are vulnerable to regression bias. Additionally, while many of these studies report statistically significant findings, it is unclear whether these findings are clinically relevant, as few studies evaluate pregnancy outcomes. Those that do report pregnancy outcomes are rarely designed to do so and therefore the applicability of this information is not known.

Over-the-counter supplements are also rarely given in isolation, with many profertility supplements being a combination of many, if not all the vitamins and micronutrients that we will discuss in this chapter. Some supplements have also been found to work synergistically, while others seem to act independently. This complex interaction makes studying the individual impact of each the supplements more difficult [25]. From a clinical perspective though, evaluating the effect of a host of these vitamins and minerals simultaneously may be more relevant as patients rarely take only one or two (Table 33.1).

Vitamin E

Vitamin E (alpha-tocopherol) is a fat-soluble vitamin that acts throughout the body as a free-radical scavenger [54]. It functions in the male reproductive system in this capacity as well, improving germ cell integrity and function [12,55]. Indeed, seminal levels of vitamin E are reduced by up to 65% in oligospermic and azoospermic men [56].

TABLE 33.1 Common Dosages for Supplements Used Within Human Studies

Supplement	Dose	References
Vitamin E	400 (300–1000) mg/d PO	[28–34]
Vitamin C	1000 mg/d PO	[28,33]
Vitamin A	1 mg/d PO	[35]
Selenium	200 μg/d PO	[29,36]
Zinc sulfate	66 (66–200) mg/d PO	[37,38]
Glutathione	600 mg/OD IM	[39]
Polyunsaturated fatty acids	400–1840 mg/d PO of EPA/DHA	[40,41]
Coenzyme Q10	200 (200–300) mg/d PO	[42–44]
Folic acid	5 mg/d PO	[38,45]
Lycopene	20 (2–20) mg/d PO	[46,47]
Carnitines	2 g/d L-carnitine, 1 g/d L-acetyl-carnitine	[24,48–52]
Arginine	500 mg/d PO	[53]

Doses listed are those used in highest level of evidence available. Ranges are included when more than one dosage has been used.

The mechanism for this effect is hypothesized to be due to its antioxidant effect and subsequent decrease in oxidative damage and DNA fragmentation. This has been demonstrated in animal models with decreased apoptosis and necrosis within rat testicular tissue of stressed rats supplemented with vitamin E [57]. In human studies, men with infertility have improvements in DNA fragmentation [28] as well as chromatin stability [58].

Vitamin E can also ameliorate the effects of external stressors on spermatogenesis. Animal studies with vitamin E supplementation have shown that vitamin E can be protective against environmental and reproductive toxins that are known to decrease the effect of glutathione related enzymes [59]. Vitamin E similarly improves motility and DNA fragmentation after thawing when used in cryopreservation medium, another source of oxidative stress [60].

Increasing the intake of vitamin E, even solely in a patient's diet, is associated with improved semen parameters [61,62] and DNA fragmentation on Comet assay [63]. A large prospective study with 690 infertile men found an increase of >5% in motility or morphology in 52.6% of patients, and a 10.8% spontaneous pregnancy rate [29]. Oral supplementation has been found to increase seminal plasma concentrations of vitamin E in a randomized trial [64]. This subsequent increase in vitamin E levels within the seminal fluid significantly decreases markers of lipid peroxidation [65]. *In vitro* studies with vitamin E have shown that supplementation reduces ROS-induced increase in lipid peroxidation in a dose-dependent fashion [66].

Vitamin E is one of the most studied of the antioxidants, and as such, there are multiple randomized controlled trials (RCT) evaluating its efficacy, though study sizes are at times small and initially, results were mixed. Early studies found no difference in semen parameters as was seen in a study of 31 patients randomized to high-dose vitamins C and E or placebo [67]. However, significant improvements in binding to the zona pellucida were seen in a randomized trial of 30 patients [68].

More recently, there are larger RCTs that demonstrate improvement in semen analysis and two that suggest an improvement in the pregnancy rate. One trial with 60 patients randomized to either vitamin E and clomiphene citrate versus placebo found increases in semen parameters as well as 36.7% pregnancy rate in the treatment arm versus 13.3% in the placebo arm [30]. Another larger RCT, evaluating 106 infertile men, found an increase in the live pregnancy rate with five pregnancies in the control group versus nine in the treatment group, though this study was not specifically designed to evaluate pregnancy rate [31].

In addition to spontaneous pregnancy rates, IVF-ICSI outcomes are also improved with supplementation as demonstrated in a randomized trial of 60 patients [32]. Because of these studies, it is now believed that vitamin E supplementation does improve seminal parameters and pregnancy rates, and is therefore a common supplement given to patients. There is published data by Kristal et al. suggesting that supplementation with vitamin E at a dose of 400 IU/d increases the risk of prostate cancer by as much as 17% in men with low baseline levels. Because of this, men should be counseled about this possible risk and supplementation should not be at high doses [69].

Vitamin C

Vitamin C is a water-soluble vitamin involved in the creation of extracellular matrices, and acts in concert with vitamin E as an antioxidant [12,70]. Its role in spermatogenesis has been known for some time, though the mechanism of action has not been completely elucidated [71,72].

Seminal vitamin C levels are decreased in patients presenting with oligo- and asthenospermia when compared to controls [73,74], as well as patients who smoke, which has been associated with an increase in seminal oxidation status [75]. Seminal vitamin C levels have also been correlated with reduced DNA fragmentation indices [76].

Animal studies have found some evidence that vitamin C supplementation may increase epididymal sperm concentration and testosterone levels. While no difference in motility was seen in this study of 24 rats, significant improvements in levels of lipid peroxidation were seen [77]. Other animal studies have demonstrated increases in serum and seminal antioxidant capacity in addition to improvements in sperm count [78] as well as preservation of fertilization capacity in the presence of environmental stressors [79].

Vitamin C supplementation has been found to decrease sperm agglutination in those who were vitamin C deficient [75]. A study of 13 infertile men with oligospermia saw significant increases in concentration and motility after only 2 months of treatment [80].

There are few trials evaluating vitamin C in isolation, as trials typically combine it with vitamin E as vitamin C is felt to act synergistically with vitamin E [12]. In a RCT, combination therapy with vitamins E and C found improvements in DNA fragmentation levels after only 2 months of treatment [28]. Another RCT of combination vitamins C and E therapy found no difference after 56 days of supplementation [33]. Vitamin C is a common supplement, and while it may not be helpful for someone to take it in isolation to maximize fertility, it is likely a useful adjunct to other supplements, especially vitamin E.

Vitamin A

Vitamin A is a fat-soluble vitamin found in most of the mucosal surfaces within the body, originally discovered for its role in embryological development of the visual system

[12,81]. Its action within the male reproductive tract itself was discovered later when fertility could be restored in vitamin-A-deficient rats with supplementation [81].

Animal models of vitamin A deficiency have found that this causes early cessation of spermatogenesis and increased apoptosis within the testicle [82], and vitamin A supplementation is able to maintain mouse germ line stem cells differentiation ability [83]. Mice who have administered retinoic acid receptor antagonist also exhibit a reversible arrest of spermatogenesis [84,85]. This interference in sperm maturation is likely related to the role of vitamin A in supporting the blood–testis barrier [86] by maintaining Sertoli cell tight junctions [87].

In humans, patients with severe oligospermia have been found to have nearly half the levels of seminal fluid vitamin A compared to normospermic men [88]. Similarly, men presenting with infertility have been found to have decreased levels of vitamin A synthesis enzymes in testis tissue [89], and decreased concentrations of vitamin A metabolites in testis biopsy samples [90]. Infertility related to alcoholism is also believed to be due to the inhibition of vitamin A metabolism by alcohol [91]. Interestingly though, one study of 40 subfertile men found no significant difference in serum levels of vitamin A when compared to controls, suggesting that the relationship between overall vitamin A levels and intratesticular vitamin A activity may be more complex than previously thought [92].

There is only one RCT on vitamin A, a study that was actually designed to evaluate the efficacy of selenium. This study determined that there was significant improvement in motility when combined with selenium and vitamins C and E, though the effect of vitamin A alone was not elucidated [35]. While there is a need for more studies on the effect of vitamin A, most research is currently evaluating other vitamins and antioxidants.

Selenium

Selenium is a trace element that is incorporated into a host of selenoproteins, which act as powerful antioxidants [93]. In these selenoproteins, selenium acts within the active site to aid in their catalytic activity [94]. It is commonly combined with vitamin E, though a few studies have evaluated selenium's activity alone [35].

Decreasing levels of dietary selenium corresponds with increasing oxidative stress and decrease in the fertility of male mice [95]. This is linked to decreasing levels of heat shock protein 70-2 and increasing p53 activity with increased cell death [96]. Selenium-deficient rats are found to have significant infertility that persists despite vitamin E, or other antioxidant supplementation, though it is recoverable with selenium supplementation suggesting that selenium is necessary in spermatogenesis independent of other vitamins or micronutrients [97].

Motility is also affected as flagellar dysfunction has been found in selenium-deficient rats [98].

In humans, selenium has been found in both sperm and seminal fluid [99]. Selenium levels are reduced in patients presenting to infertility clinics [100,101]. Supplementation has been found to successfully increase serum and seminal concentrations [102]. This has been linked to improving seminal mitochondrial function, increasing the testicular activity of selenoprotein glutathione peroxidase within the testicle, and improving seminal parameters [103,104]. This activity has also been associated with a decrease in malondialdehyde, a marker of oxidative stress, and subsequent improvement in motility [105].

Studies on selenium have demonstrated improvement in seminal parameters and some suggestion of improvements in pregnancy rates. A study of 690 patients evaluating a combination therapy of selenium and vitamin E did find increases in semen parameters in 52.6% of cases. In this group of previously infertile patients, 10.8% of patients did achieve a pregnancy, though the impact of supplementation is unknown [29]. A RCT of 69 patients given placebo versus selenium alone versus selenium in addition to vitamins A, C, and E found improvement in sperm motility with selenium supplementation with no significant difference between selenium alone and selenium with additional vitamin supplementation. Eleven percent of the patients were able to father a child compared to no patients in the control arm [35]. Another RCT by Safarinejad and Safarinejad, involving 468 men with OAT, studied not only semen parameters but also reproductive hormones. They found a decrease in FSH and increases in testosterone and inhibin B, which corresponded with improvements in all aspects of semen analysis; these are the expected hormonal effects of a profertility agent [36].

Zinc

Zinc is a microelement involved in the regulation and synthesis of many important enzymes. In the male reproductive system, it is a cofactor involved in embryogenesis, sperm maturation, and even hormonal production [106]. Its presence within sperm during development then subsequent elimination from the flagellum is an important part of spermatic generation of motility [107]. It is also found in significant quantities in prostatic fluid and is felt to have important antioxidant properties as well [108].

Zinc's presence within the male reproductive tract has been shown in multiple studies to be essential for normal spermatogenesis. Compared with fertile controls, patients with abnormal semen parameters have lower levels of seminal and serum zinc, while patients with diets high in zinc and those with high seminal zinc levels have better antioxidant status and decreased

DNA damage. This is the basis for zinc supplementation for male infertility [63,100,109]. The role of supplementation seems to be best for men with low concentration and motility as morphology has not been correlated to zinc levels [110]. This is especially true in men with leukospermia for whom this link between severe oligospermia and zinc levels is strongest [111]. Interestingly, serum zinc has also been inversely related to FSH levels suggesting a further relationship between zinc status and healthy spermatogenesis [45].

Oral zinc supplementation has been shown to successfully restore seminal zinc binding proteins back to normal levels [112]. In healthy men, this increase in zinc intake decreases sperm aneuploidy [113]. Even after varicocelectomy, zinc administration may improve sperm recovery, as measured by increasing levels of inhibin B, though the clinical relevance of this is likely low [114]. In animal models zinc supplementation is protective against environmental-stress-induced oxidative damage when compared to nonsupplemented controls [115]. There is an upper level to the beneficial effects of zinc as high levels of seminal zinc have been linked to increased time to pregnancy in fertile couples, though the mechanism of this is not understood [116].

There have been multiple RCTs evaluating zinc supplementation in the treatment of male infertility. Each has shown improvements in seminal parameters. One study in 45 men with asthenospermia and oligospermia demonstrated significant improvements in oxidative stress and DNA fragmentation along with semen parameters [37]. Another comparing 108 fertile men with 103 subfertile men found a 74% increase in total sperm count with supplementation among the infertile men. There was no difference in the fertile men when compared to placebo. Interestingly, while there was an improvement with supplementation in the infertile men, there was no difference seen in pretreatment seminal plasma zinc levels between the two groups [117].

In trying to elucidate the mechanism by which zinc is able to improve seminal parameters, Ebisch et al. conducted an RCT evaluating not only the seminal parameters, but also the hormonal effects of supplementation. Pretreatment serum zinc was correlated to sperm concentration and inhibin B and an inverse relationship to FSH. After treatment with zinc, improvements with hormone levels were not seen, though semen parameters did improve, suggesting improvement in spermatogenesis without involvement of the hormonal pathways [45].

This improvement in parameters is not only relevant in patients hoping for spontaneous pregnancy, but those undergoing IVF as well. An RCT of 60 couples having IVF for male factor infertility found that the administration of multiple antioxidants including zinc was associated with increased viable pregnancy rate of 38.5% versus 16%, though there was no change in the oocyte fertilization rate and the effect of live birth rate was not reported [32].

Glutathione

Glutathione, and the family of glutathione peroxidases, is the archetype free-radical scavenger, reducing oxidative damage within all cells of the body [118]. It is one of the most prevalent antioxidants in the body, and is present within all parts of the male reproductive system. It is most abundant in the proximal aspects of the system – testis, proximal epididymis, and seminal fluid – where it is involved in protecting sperm during development and after ejaculation [119,120]. One of the difficulties with glutathione supplementation is that it is given as an intramuscular injection, which may not be reasonable to many patients [39].

In animal models of increased oxidative stress there are improvements in motility with intramuscular injections of GSH (the reduced form of glutathione) [121]. Rats demonstrate age-related changes in the functionality of glutathione within the epididymis and seminal vesicles, which are associated with decreased semen parameters such as motility [120].

Glutathione levels are reduced in couples presenting with male factor infertility [122], and strict morphology, motility, and count are correlated with intracellular GSH levels [123,124]. Errors in genes affecting glutathione metabolism have been discovered in patients with previously "idiopathic" infertility [125]. Glutathione levels are modifiable by external events, and during times of psychological stress, associated reduction in glutathione levels are accompanied by worsening of motility and morphology [126] – an effect likely attributed to increased oxidative stress and decreased antioxidant levels in seminal fluid [127].

Because supplementation is not oral, there are fewer RCTs on the utilization of glutathione. Lenzi et al. conducted an RCT of glutathione treatment in 20 patients with infertility and presumed increased oxidative stress (varicocele or "germ-free genital tract inflammation"). They demonstrated improvements in motility, velocity, and morphology [39].

Increased glutathione levels are important in ART as they protect against oxidative stress after cryopreservation, with improvements in membrane status, DNA integrity, and motility post-thaw [128,129]. Treated sperm had higher sperm penetration rates and a decreased rate of premature chromosome condensation [130]. It appears that these improvements in sperm DNA integrity may improve subsequent embryo development as these higher glutathione levels were associated with improved day 3 embryo morphology [131]. Whether or not these possible benefits are worth the difficulty in supplementation though is yet to be seen.

Polyunsaturated Fatty Acids

Sperm membranes contain a high concentration of PUFAs, which are integral to the proper functioning of sperm, from spermatogenesis and motility to capacitation and acrosomal activity [121]. Preventing lipid peroxidation and preserving membrane integrity is considered a main pathway through which antioxidants exert their effect on fertility. There is evidence that PUFAs, specifically omega-3 fatty acids, can reduce membrane lipid peroxidation and subsequently improve sperm form and function [4].

In normospermic men, there are positive correlations between superoxide dismutase activity and PUFA composition. In these men, seminal plasma glutathione peroxidase and superoxide dismutase protect sperm against lipid peroxidation [132]. Men with OAT and asthenozoospermia have been found to have increased levels of PUFAs. This makes these patients more susceptible to lipid peroxidation, which may explain patients' varying responses to exposure to oxidative stressors [26].

The two main types of PUFAs, omega-3 and omega-6, function differently within the body, and also affect fertility differently due to their effects on membrane fluidity [133]. Omega-3 fatty acid concentrations are higher in spermatozoa in fertile men. Docosahexaenoic acid (DHA), an omega-3 fatty acid, comprises more than 60% of total PUFA in normospermic males [133]. A high omega-6 to omega-3 ratio has a negative correlation to total sperm count, motility, and morphology [134]. In a study of dietary patterns and sperm parameters, men with the highest dietary levels of omega-3 fatty acids had improved sperm morphology compared to those with the lowest. Additionally, increased intake of saturated fatty acids was correlated with lower sperm concentration and counts [135].

Evidence on supplementation suggests that intake of increased levels of PUFAs allows for restoration of membranes that have been impaired by oxidative stress [136]. In a study by Safarinejad, 238 patients were randomized to EPA and DHA or placebo. There were significant increases in serum, seminal plasma, and spermatozoa omega-3 levels. These increases were associated with improvements in semen parameters with sperm concentration improving from 15.6 mil/mL to 28.7 mil/mL ($P = 0.001$) [40]. Not all studies show a benefit as a much smaller randomized, double-blind, placebo controlled trial of 28 men evaluating the effect of DHA on fertility determined that while levels of DHA increased in blood and semen, incorporation of DHA into sperm phospholipids was not seen and there was no improvement in sperm motility [41].

Coenzyme Q10

Coenzyme Q10 (CoQ10) is a protein found within all tissues of the body, so much so that it is also referred to as ubiquinone. Its main role within human cells is as part of the mitochondrial respiratory chain, though today, it is also known to have antioxidant properties [137]. Due to its important role in cellular energy production and antioxidant capabilities, it has been suggested to be important in the pathology of idiopathic male infertility, and even has a role in seminal derangements in those with varicocele [138].

With its central role in energy production, it would follow that healthy CoQ10 levels might correlate with improved forward motility. Indeed, higher levels of CoQ10 are seen in patients with better semen parameters, especially motility [139]. Seminal CoQ10 is correlated with other seminal parameters as well, including count, in men with abnormal semen analysis when compared to controls. Those with varicoceles and oligospermia have decreased seminal-to-serum-CoQ10 ratio, though this decrease in CoQ10 is more pronounced in those with idiopathic asthenozoospermia compared to those with varicocele [140,141]. While there might not be as significant an association between CoQ10 and semen parameters in men with varicocele, CoQ10 supplementation in these patients has been shown to improve concentration and motility, which was strongly correlated with improvements in oxidation status [23]. A trial of 287 patients treated for 12 months confirmed these improvements in concentration, morphology, and motility. These patients, who previously were diagnosed as infertile, had a 34.1% pregnancy rate during the study time, though there were no fertile controls and the effect of CoQ10 on pregnancy rate is not known [42].

Coenzyme Q10 is a supplement that has been fairly well studied with multiple RCTs demonstrating its effectiveness. Nadjarzadeh et al. conducted an RCT of 3 months of supplementation with 60 patients, 47 of which completed the trial. Those in the treatment arm experienced a modest improvement of 5.78% (+/− 15.6%) in sperm motility [43]. In another RCT of 212 infertile men with OAT, greater improvement was seen with prolonged supplementation. They were given 300 mg CoQ10 BID for 26 weeks. There was a 30.7, 24.3, and 33% improvement above baseline in sperm concentration, motility, and morphology, respectively. These differences decreased over the next 26 weeks without supplementation, but were still elevated over those in the placebo arm by the end of the study [142]. Similar results were found in a follow-up study of 228 men by the same author with improvements in motility, morphology, and concentration [44]. Comparing the modest improvements of shorter supplementation with the longer studies, it appears that periods of supplementation longer than 12 weeks are required to see significant improvements [43,142]. A meta-analysis of these studies continued to show improvements in motility and concentration with unknown effect on pregnancy [143].

Folic Acid

Folate, through their conversion to active metabolites, is a necessary cofactor in DNA synthesis and DNA methylation as well as other cellular functions. Because of this role, folate is important within rapidly dividing cells including germ cells [144]. Additionally, folate has antioxidant properties with significant ability to inhibit lipid peroxidation [145].

Men who have mutations in folate metabolism, specifically the methylenetetrahydrofolate reductase (*MTHFR*) and methionine synthase (*MTR*) gene mutations, have increased risk of infertility. Decreased folate is associated with reductions in sperm concentration, progressive motility, and normal morphology [146]. In a study of 133 men with either nonobstructive azoospermia (NOA) or severe oligospermia compared to controls, polymorphisms in the *MTHFR* C677T and *MTR* A2756G genes were associated with worse semen parameters [147]. Eloualid et al. found a similar correlation between severe oligospermia and A1298C polymorphism of the *MTHFR* gene [148]. There is also evidence that actual changes to the gene are unnecessary and that epigenetic changes can silence the gene leading to NOA [149].

Increased dietary folate intake improves spermatogenesis in healthy men. A study on dietary folate and seminal DNA quality found that those with the highest folate intake have the lowest rate of errors in DNA replication and meiosis (i.e., aneuploidy) [113], and also less DNA damage, measured by Comet assay [63]. There is evidence that folate supplementation can improve semen analysis in infertile men as well.

The effects of folate supplementation have been studied with two RCTs, though not in isolation as both studies evaluated its effects with zinc supplementation. One study with 40 subfertile and 47 fertile men treated for 26 weeks demonstrated a significant increase in sperm concentration [45]. Another RCT by Wong et al. evaluated 103 subfertile men and 108 fertile men treated for 26 weeks with combinations of placebo, zinc, folate, or a combination of zinc and folate. In this study there was an increase in total normal sperm concentration from 5.1 mil/mL to 8.9 mil/mL with combination therapy. While this was a statistically significant increase of 74%, it is not clear whether this is clinically relevant, and this was not related to any data on pregnancy rates [38].

Lycopene

Lycopene is a carotenoid antioxidant within the same family as vitamin A. It is lipophilic and found throughout the body. Its action is important in many disease states; cardiovascular, osteoporosis, hypertension, and male infertility [150]. Like vitamin A and other antioxidants, its role as a free-radical scavenger makes it important to normal spermatogenesis and sperm function [150].

Two studies evaluating antioxidant intake showed that in 250 men, semen parameters improved with increased dietary intake of lycopene [151,152]. Dietary supplementation with lycopene has been found to successfully raise seminal and serum lycopene levels [153]. In animal studies evaluating spermatic response to toxic (cisplatin) and oxidative (ischemia/reperfusion) insults, Lycopene provided an element of protection against spermatic damage [154,155].

As previously discussed, men with infertility have elevated ratios of omega-6 (arachidonic acid) to omega-3 (docosahexaenoic acid), which impairs sperm function. Three months of treatment with lycopene has been found to significantly improve this ratio in a study of 44 infertile men and 13 fertile controls. Interestingly, in the infertile men, 16% of patients had a spontaneous pregnancy and 42% underwent successful IVF, though this was not a study controlled to look at fertility rates [46]. In a noncontrolled study involving 30 men with OAT treated with lycopene for 3 months, improvements were noted in motility and morphology (25 and 10%, respectively) and concentration increased by 22 mil/mL. Those with significant oligospermia (defined as less than 5 mil/mL) had no improvement [156].

There are no RCTs evaluating the effect of lycopene on either semen analysis or pregnancy rate, though Oborna et al. performed an RCT with 15 fertile men and 13 normospermic men who have been unable to conceive (no female factor identified). In this trial, they were able to determine that lycopene was able to significantly lower oxidative stress, which in other studies has been shown to improve fertility outcomes [47].

Carnitines

Carnitines are integral in regulating cellular metabolism and fatty acid transportation within mitochondria in addition to having antioxidant effects [12,157]. It is through this action within the mitochondria that carnitines may aid spermatogenesis and improve semen parameters [157]. L-carnitine and acetyl-L-carnitine are found in the epididymis and are believed to be important in sperm maturation [158]. A word of caution in the use of carnitines in the treatment of male infertility though; L-carnitine has been shown to be converted to trimethylamine-*N*-oxide by intestinal microflora, a compound that has been shown to increase atherosclerosis in mouse models [159].

Seminal plasma carnitine levels are lower in infertile men, especially in those with oligospermia and oligoasthenozoospermia [160]. Normal seminal morphology has also been related to seminal carnitine levels [161].

In animal models, carnitines protect against oxidative stress and damage when compared to placebo [162].

Carnitine supplementation is effective in many different clinical scenarios. In patients with abnormal semen parameters and varicocele, supplementation improves seminal parameters except in patients with the largest varicoceles [24]. Vicari and Calogero found that in patients with chronic inflammation of the prostate, seminal vesicles and epididymis could experience improvements in seminal ROS and motility and viability, though the improvements were decreased in those with elevated WBC in the seminal fluid. Among patients without leukocytospermia, the pregnancy rate was 11.7% [163]. These effects of carnitines on semen parameters may be enhanced through the use of prostaglandins [164]. Another study by Vicari et al. with 98 patients with infertility and inflammation within the reproductive tract found that those with leukocytospermia had improved sperm motility and viability with sequential use of carnitine supplementation and nonsteroidal anti-inflamatory drugs [165].

Multiple RCTs in men with abnormal semen parameters randomized to either carnitine supplementation or placebo found improvements in motility and concentration [24,48–50,166–169]. A nonrandomized study of 114 infertile men found improvements in motility in addition to 16 of the patients obtaining pregnancy [170]. Lenzi et al. found the improvement most pronounced in those with severe asthenozoospermia [48]. Improvements in ROS seen in these studies were attributed to this effect [50], though one study evaluating 30 asthenospermic patients only found improvements in those patients with normal levels of PHGPx [49]. Interestingly, the study by Morgante et al. also reported subjective improvements in sexual satisfaction in patients [168]. There is one RCT that showed no effect of carnitine supplementation, though this trial only had 21 total patients randomized [171]. A systematic review of these papers found that carnitine supplementation was most effective with improving morphology, and less so, motility. This review also found an improvement in the pregnancy rate with an OR of 4.1 (2.08–8.08) [172].

Arginine

Arginine is a lesser-studied amino acid used for supplementation for male infertility. It has many roles within the body, including ammonia metabolism, and as part of the nitric oxide pathway for endothelial relaxation [53]. Arginine's role in nitric oxide synthesis is important for male infertility as nitric oxide plays an important role in capacitation as well as preventing lipid peroxidation [173]. There are few modern studies evaluating arginine in isolation. *In vitro* animal studies have found that arginine supplementation improves spermatic acrosomal function [173]. *In vivo* studies have found improvements in spermatozoa metabolism with arginine supplementation, though large doses can be harmful to sperm function [174].

In vitro studies in humans have found improvements in sperm motility with incubation in arginine-containing solutions [175]. Arginine supplementation is associated with increased seminal metabolic activity with increases in motility through the nitric oxide pathway [176]. Clinically, this correlates with improvements in motility in multiple studies [177]. In an RCT of 180 patients with asthenospermia, Morgante et al. evaluated arginine in combination with carnitine supplementation and found an improvement in progressive forward motility [168].

CONCLUSIONS

Oxidative stress and antioxidant status are a rising concern relevant to many health issues today. Due to sperm's unique sensitivity to the action of ROS, environmental stressors and toxins can adversely affect sperm form and function. By improving antioxidant status and DNA synthesis/repair mechanisms, over-the-counter supplements have been shown to have a positive effect on male fertility parameters – motility, concentration, abnormal forms, as well as oxidation status and DNA fragmentation.

A recent Cochrane review that included 34 trials and 2876 patients found that while not ideal, there was sufficient evidence to determine that the use of oral antioxidant therapy can increase pregnancy rates, especially in patients who are utilizing ARTs [25]. These data should be discussed with patients, both the positive effects, and the potential costs. For many patients, this could be a useful adjunct to other medical treatments for male factor infertility.

References

[1] Radimer K, Bindewald B, Hughes J, Ervin B, Swanson C, Picciano MF. Dietary supplement use by US adults: data from the National Health and Nutrition Examination Survey, 1999–2000. Am J Epidemiol 2004;160:339–49.

[2] Hanlon JT, Fillenbaum GG, Ruby CM, Gray S, Bohannon A. Epidemiology of over-the-counter drug use in community dwelling elderly: United States perspective. Drugs Aging 2001;18:123–31.

[3] Agarwal A, Sekhon LH. The role of antioxidant therapy in the treatment of male infertility. Hum Fertil (Camb) 2010;13:217–25.

[4] Tavilani H, Goodarzi MT, Doosti M, Vaisi-Raygani A, Hassanzadeh T, Salimi S, et al. Relationship between seminal antioxidant enzymes and the phospholipid and fatty acid composition of spermatozoa. Reprod Biomed Online 2008;16:649–56.

[5] Sharma RK, Agarwal A. Role of reactive oxygen species in male infertility. Urology 1996;48:835–50.

[6] Saleh RA, Agarwal A, Nada EA, El-Tonsy MH, Sharma RK, Meyer A, et al. Negative effects of increased sperm DNA damage in relation to seminal oxidative stress in men with idiopathic and male factor infertility. Fertil Steril 2003;79(Suppl. 3):1597–605.

[7] Aitken RJ. The Amoroso Lecture. The human spermatozoon – a cell in crisis? J Reprod Fertil 1999;115:1–7.

[8] Menezo Y, Evenson D, Cohen M, Dale B. Effect of antioxidants on sperm genetic damage. Adv Exp Med Biol 2014;791:173–89.

[9] Walczak-Jedrzejowska R, Wolski JK, Slowikowska-Hilc3er J. The role of oxidative stress and antioxidants in male fertility. Cent European J Urol 2013;66:60–7.

[10] Benedetti S, Tagliamonte MC, Catalani S, Primiterra M, Canestrari F, De Stefani S, et al. Differences in blood and semen oxidative status in fertile and infertile men, and their relationship with sperm quality. Reprod Biomed Online 2012;25:300–6.

[11] Rubes J, Rybar R, Prinosilova P, Veznik Z, Chvatalova I, Solansky I, et al. Genetic polymorphisms influence the susceptibility of men to sperm DNA damage associated with exposure to air pollution. Mutat Res 2010;683:9–15.

[12] Ko EY, Sabanegh ES. The role of nutraceuticals in male fertility. Urol Clin North Am 2014;41:181–93.

[13] Nazıroğlu M, Yüksel M, Köse SA, Özkaya MO. Recent reports of Wi-Fi and mobile phone-induced radiation on oxidative stress and reproductive signaling pathways in females and males. J Membr Biol 2013;246:869–75.

[14] Agarwal A, Singh A, Hamada A, Kesari K. Cell phones and male infertility: a review of recent innovations in technology and consequences. Int Braz J Urol 2011;37:432–54.

[15] Wong WY, Thomas CM, Merkus HM, Zielhuis GA, Doesburg WH, Steegers-Theunissen RP. Cigarette smoking and the risk of male factor subfertility: minor association between cotinine in seminal plasma and semen morphology. Fertil Steril 2000;74:930–5.

[16] Viloria T, Meseguer M, Martínez-Conejero JA, O'Connor JE, Remohí J, Pellicer A, et al. Cigarette smoking affects specific sperm oxidative defenses but does not cause oxidative DNA damage in infertile men. Fertil Steril 2010;94:631–7.

[17] Mostafa T, Tawadrous G, Roaia MM, Amer MK, Kader RA, Aziz A. Effect of smoking on seminal plasma ascorbic acid in infertile and fertile males. Andrologia 2006;38:221–4.

[18] Galan P, Viteri FE, Bertrais S, Czernichow S, Faure H, Arnaud J, et al. Serum concentrations of beta-carotene, vitamins C and E, zinc and selenium are influenced by sex, age, diet, smoking status, alcohol consumption and corpulence in a general French adult population. Eur J Clin Nutr 2005;59:1181–90.

[19] Kiziler AR, Aydemir B, Onaran I, Alici B, Ozkara H, Gulyasar T, et al. High levels of cadmium and lead in seminal fluid and blood of smoking men are associated with high oxidative stress and damage in infertile subjects. Biol Trace Elem Res 2007;120:82–91.

[20] Sepaniak S, Forges T, Gerard H, Foliguet B, Bene MC, Monnier-Barbarino P. The influence of cigarette smoking on human sperm quality and DNA fragmentation. Toxicology 2006;223:54–60.

[21] Agarwal A, Prabakaran SA, Sikka S. Current opinions on varicocele and oxidative stress and its role in male infertility. Business briefing. US Genito-Urinary Dis 2006;2006:62–8.

[22] Baazeem A, Belzile E, Ciampi A, Dohle G, Jarvi K, Salonia A, et al. Varicocele and male factor infertility treatment: a new meta-analysis and review of the role of varicocele repair. Eur Urol 2011;60:796–808.

[23] Festa R, Giacchi E, Raimondo S, Tiano L, Zuccarelli P, Silvestrini A, et al. Coenzyme Q10 supplementation in infertile men with low-grade varicocele: an open, uncontrolled pilot study. Andrologia 2013.

[24] Cavallini G, Ferraretti AP, Gianaroli L, Biagiotti G, Vitali G. Cinnoxicam and L-carnitine/acetyl-L-carnitine treatment for idiopathic and varicocele-associated oligoasthenospermia. J Androl 2004;25:761–70; discussion 771–772.

[25] Showell MG, Brown J, Yazdani A, Stankiewicz MT, Hart RJ. Antioxidants for male subfertility. Cochrane Database Syst Rev 2011. CD007411.

[26] Khosrowbeygi A, Zarghami N. Fatty acid composition of human spermatozoa and seminal plasma levels of oxidative stress biomarkers in subfertile males. Prostaglandins Leukot Essent Fatty Acids 2007;77:117–21.

[27] Wathes DC, Abayasekara DR, Aitken RJ. Polyunsaturated fatty acids in male and female reproduction. Biol Reprod 2007;77:190–201.

[28] Greco E, Iacobelli M, Rienzi L, Ubaldi F, Ferrero S, Tesarik J. Reduction of the incidence of sperm DNA fragmentation by oral antioxidant treatment. J Androl 2005;26:349–53.

[29] Moslemi MK, Tavanbakhsh S. Selenium-vitamin E supplementation in infertile men: effects on semen parameters and pregnancy rate. Int J Gen Med 2011;4:99–104.

[30] Ghanem H, Shaeer O, El-Segini A. Combination clomiphene citrate and antioxidant therapy for idiopathic male infertility: a randomized controlled trial. Fertil Steril 2010;93:2232–5.

[31] Chen XF, Li Z, Ping P, Dai JC, Zhang FB, Shang XJ. Efficacy of natural vitamin E on oligospermia and asthenospermia: a prospective multi-centered randomized controlled study of 106 cases. Zhonghua Nan Ke Xue 2012;18:428–31.

[32] Tremellen K, Miari G, Froiland D, Thompson J. A randomised control trial examining the effect of an antioxidant (Menevit) on pregnancy outcome during IVF-ICSI treatment. Aust N Z J Obstet Gynaecol 2007;47:216–21.

[33] Rolf C, Cooper TG, Yeung CH, Nieschlag E. Antioxidant treatment of patients with asthenozoospermia or moderate oligoasthenozoospermia with high-dose vitamin C and vitamin E: a randomized, placebo-controlled, double-blind study. Hum Reprod 1999;14:1028–33.

[34] Kessopoulou E, Powers HJ, Sharma KK, Pearson MJ, Russell JM, Cooke ID, et al. A double-blind randomized placebo cross-over controlled trial using the antioxidant vitamin E to treat reactive oxygen species associated male infertility. Fertil Steril 1995;64:825–31.

[35] Scott R, MacPherson A, Yates RW, Hussain B, Dixon J. The effect of oral selenium supplementation on human sperm motility. Br J Urol 1998;82:76–80.

[36] Safarinejad MR, Safarinejad S. Efficacy of selenium and/or N-acetyl-cysteine for improving semen parameters in infertile men: a double-blind, placebo controlled, randomized study. J Urol 2009;181:741–51.

[37] Omu AE, Al-Azemi MK, Kehinde EO, Anim JT, Oriowo MA, Mathew TC. Indications of the mechanisms involved in improved sperm parameters by zinc therapy. Med Princ Pract 2008;17:108–16.

[38] Wong WY, Merkus HM, Thomas CM, Menkveld R, Zielhuis GA, Steegers-Theunissen RP. Effects of folic acid and zinc sulfate on male factor subfertility: a double-blind, randomized, placebo-controlled trial. Fertil Steril 2002;77:491–8.

[39] Lenzi A, Culasso F, Gandini L, Lombardo F, Dondero F. Andrology: placebo-controlled, double-blind, cross-over trial of glutathione therapy in male infertility. Hum Reprod 1993;8:1657–62.

[40] Safarinejad MR. Effect of omega-3 polyunsaturated fatty acid supplementation on semen profile and enzymatic anti-oxidant capacity of seminal plasma in infertile men with idiopathic oligoasthenoteratospermia: a double-blind, placebo-controlled, randomised study. Andrologia 2011;43:38–47.

[41] Conquer JA, Martin JB, Tummon I, Watson L, Tekpetey F. Effect of DHA supplementation on DHA status and sperm motility in asthenozoospermic males. Lipids 2000;35:149–54.

[42] Safarinejad MR. The effect of coenzyme Q_{10} supplementation on partner pregnancy rate in infertile men with idiopathic oligoasthenoteratozoospermia: an open-label prospective study. Int Urol Nephrol 2012;44:689–700.

[43] Nadjarzadeh A, Shidfar F, Amirjannati N, Vafa MR, Motevalian SA, Gohari MR, et al. Effect of Coenzyme Q10 supplementation on antioxidant enzymes activity and oxidative stress of seminal plasma: a double-blind randomised clinical trial. Andrologia 2014;46:177–83.

[44] Safarinejad MR, Safarinejad S, Shafiei N, Safarinejad S. Effects of the reduced form of coenzyme Q10 (ubiquinol) on semen parameters in men with idiopathic infertility: a double-blind, placebo controlled, randomized study. J Urol 2012;188:526–31.

[45] Ebisch IM, Pierik FH, Jong FH DE, Thomas CM, Steegers-Theunissen RP. Does folic acid and zinc sulphate intervention affect endocrine parameters and sperm characteristics in men? Int J Androl 2006;29:339–45.

[46] Filipcikova R, Oborna I, Brezinova J, Novotny J, Wojewodka G, De Sanctis JB, et al. Lycopene improves the distorted ratio between AA/DHA in the seminal plasma of infertile males and increases the likelihood of successful pregnancy. Biomed Pap Med Fac Univ Palacky Olomouc Czech Repub 2013;157–163.

[47] Oborna I, Malickova K, Fingerova H, Brezinova J, Horka P, Novotny J, et al. A randomized controlled trial of lycopene treatment on soluble receptor for advanced glycation end products in seminal and blood plasma of normospermic men. Am J Reprod Immunol 2011;66:179–84.

[48] Lenzi A, Sgrò P, Salacone P, Paoli D, Gilio B, Lombardo F, et al. A placebo-controlled double-blind randomized trial of the use of combined L-carnitine and L-acetyl-carnitine treatment in men with asthenozoospermia. Fertil Steril 2004;81:1578–84.

[49] Garolla A, Maiorino M, Roverato A, Roveri A, Ursini F, Foresta C. Oral carnitine supplementation increases sperm motility in asthenozoospermic men with normal sperm phospholipid hydroperoxide glutathione peroxidase levels. Fertil Steril 2005;83:355–61.

[50] Balercia G, Regoli F, Armeni T, Koverech A, Mantero F, Boscaro M. Placebo-controlled double-blind randomized trial on the use of L-carnitine, L-acetylcarnitine, or combined L-carnitine and L-acetylcarnitine in men with idiopathic asthenozoospermia. Fertil Steril 2005;84:662–71.

[51] Costa M, Canale D, Filicori M, D'Iddio S, Lenzi A. L-carnitine in idiopathic asthenozoospermia: a multicenter study. Italian Study Group on Carnitine and Male Infertility. Andrologia 1994;26:155–9.

[52] Wang YX, Yang SW, Qu CB, Huo HX, Li W, Li JD, et al. L-carnitine: safe and effective for asthenozoospermia. Zhonghua Nan Ke Xue 2010;16:420–2.

[53] Appleton J. Arginine: clinical potential of a semi-essential amino acid. Altern Med Rev 2002;7:512–22.

[54] Niki E, Traber MG. A history of vitamin E. Ann Nutr Metab 2012;61:207–12.

[55] Bolle P, Evandri MG, Saso L. The controversial efficacy of vitamin E for human male infertility. Contraception 2002;65:313–5.

[56] Bhardwaj A, Verma A, Majumdar S, Khanduja KL. Status of vitamin E and reduced glutathione in semen of oligozoospermic and azoospermic patients. Asian J Androl 2000;2:225–8.

[57] Hemadi M, Saki G, Rajabzadeh A, Khodadadi A, Sarkaki A. The effects of honey and vitamin E administration on apoptosis in testes of rat exposed to noise stress. J Hum Reprod Sci 2013;6:54–8.

[58] Song B, He XJ, Jiang HH, Peng YW, Wu H, Cao YX. Compound Xuanju Capsule combined with vitamin E improves sperm chromatin integrity. Zhonghua Nan Ke Xue 2012;18:1105–7.

[59] Yousef MI. Vitamin E modulates reproductive toxicity of pyrethroid lambda-cyhalothrin in male rabbits. Food Chem Toxicol 2010;48:1152–9.

[60] Kalthur G, Raj S, Thiyagarajan A, Kumar S, Kumar P, Adiga SK. Vitamin E supplementation in semen-freezing medium improves the motility and protects sperm from freeze-thaw-induced DNA damage. Fertil Steril 2011;95:1149–51.

[61] Vézina D, Mauffette F, Roberts KD, Bleau G. Selenium-vitamin E supplementation in infertile men. Biol Trace Elem Res 1996;53:65–83.

[62] Eskenazi B, Kidd SA, Marks AR, Sloter E, Block G, Wyrobek AJ. Antioxidant intake is associated with semen quality in healthy men. Hum Reprod 2005;20:1006–12.

[63] Schmid TE, Eskenazi B, Marchetti F, Young S, Weldon RH, Baumgartner A, et al. Micronutrients intake is associated with improved sperm DNA quality in older men. Fertil Steril 2012;98:1130–1137e1.

[64] Moilanen J, Hovatta O, Lindroth L. Vitamin E levels in seminal plasma can be elevated by oral administration of vitamin E in infertile men. Int J Androl 1993;16:165–6.

[65] Suleiman SA, Ali ME, Zaki ZM, el-Malik EM, Nasr MA. Lipid peroxidation and human sperm motility: protective role of vitamin E. J Androl 1996;17:530–7.

[66] Verma A, Kanwar KC. Effect of vitamin E on human sperm motility and lipid peroxidation in vitro. Asian J Androl 1999;1:151–4.

[67] Rolf C, Cooper TG, Yeung CH, Nieschlag E. Antioxidant treatment of patients with asthenozoospermia or moderate oligoasthenozoospermia with high-dose vitamin C and vitamin E: a randomized, placebo-controlled, double-blind study. Hum Reprod 1999;14:1028–33.

[68] Kessopoulou E, Powers HJ, Sharma KK, Pearson MJ, Russell JM, Cooke ID, et al. A double-blind randomized placebo cross-over controlled trial using the antioxidant vitamin E to treat reactive oxygen species associated male infertility. Fertil Steril 1995;64:825–31.

[69] Kristal AR, Darke AK, Morris JS, Tangen CM, Goodman PJ, Thompson IM, et al. Baseline selenium status and effects of selenium and vitamin E supplementation on prostate cancer risk. J Natl Cancer Inst 2014;. 106:djt456.

[70] Luck MR, Jeyaseelan I, Scholes RA. Ascorbic acid and fertility. Biol Reprod 1995;52:262–6.

[71] Yazama F, Furuta K, Fujimoto M, Sonoda T, Shigetomi H, Horiuchi T, et al. Abnormal spermatogenesis in mice unable to synthesize ascorbic acid. Anat Sci Int 2006;81:115–25.

[72] Gonzalez E. Sperm swim singly after Vitamin C therapy. JAMA 1983;249:2747–8.

[73] Ebesunun MO, Solademi BA, Shittu OB, Anetor JI, Onuegbu JA, Olisekodiaka JM, et al. Plasma and semen ascorbic levels in spermatogenesis. West Afr J Med 2004;23:290–3.

[74] Colagar AH, Marzony ET. Ascorbic Acid in human seminal plasma: determination and its relationship to sperm quality. J Clin Biochem Nutr 2009;45:144–9.

[75] Colagar AH, Marzony ET. Ascorbic acid in human seminal plasma: determination and its relationship to sperm quality. J Clin Biochem Nutr 2009;45:144.

[76] Song GJ, Norkus EP, Lewis V. Relationship between seminal ascorbic acid and sperm DNA integrity in infertile men. Int J Androl 2006;29:569–75.

[77] Sönmez M, Türk G, Yüce A. The effect of ascorbic acid supplementation on sperm quality, lipid peroxidation and testosterone levels of male Wistar rats. Theriogenology 2005;63:2063–72.

[78] Ashamu E, Salawu E, Oyewo O, Alhassan A, Alamu O, Adegoke A. Efficacy of vitamin C and ethanolic extract of Sesamum indicum in promoting fertility in male Wistar rats. J Hum Reprod Sci 2010;3:11–4.

[79] Saki G, Jasemi M, Sarkaki AR, Fathollahi A. Effect of administration of vitamins C and E on fertilization capacity of rats exposed to noise stress. Noise Health 2013;15:194–8.

[80] Akmal M, Qadri JQ, Al-Waili NS, Thangal S, Haq A, Saloom KY. Improvement in human semen quality after oral supplementation of vitamin C. J Med Food 2006;9:440–2.

[81] Kamal-Eldin A, Appelqvist LA. The chemistry and antioxidant properties of tocopherols and tocotrienols. Lipids 1996;31:671–701.

[82] Boucheron-Houston C, Canterel-Thouennon L, Lee TL, Baxendale V, Nagrani S, Chan WY, et al. Long-term vitamin A deficiency induces alteration of adult mouse spermatogenesis and spermatogonial differentiation: direct effect on spermatogonial gene expression and indirect effects via somatic cells. J Nutr Biochem 2013;24:1123–35.

[83] Zhang S, Sun J, Pan S, Zhu H, Wang L, Hu Y, et al. Retinol (vitamin A) maintains self-renewal of pluripotent male germline stem cells (mGSCs) from adult mouse testis. J Cell Biochem 2011;112:1009–21.

[84] Chung SS, Wang X, Roberts SS, Griffey SM, Reczek PR, Wolgemuth DJ. Oral administration of a retinoic Acid receptor antagonist reversibly inhibits spermatogenesis in mice. Endocrinology 2011;152:2492–502.

[85] Howell JM, Thompson JN, Pitt GA. Histology of the lesions produced in the reproductive tract of animals fed a diet deficient in vitamin A alcohol but containing vitamin A acid. I. The male rat. J Reprod Fertil 1963;5:159–67.

[86] Morales A, Cavicchia JC. Spermatogenesis and blood–testis barrier in rats after long-term Vitamin A deprivation. Tissue Cell 2002;34:349–55.

[87] Huang HF, Yang CS, Meyenhofer M, Gould S, Boccabella AV. Disruption of sustentacular (Sertoli) cell tight junctions and regression of spermatogenesis in vitamin-A-deficient rats. Acta Anat (Basel) 1988;133:10–5.

[88] Singer R, Barnet M, Sagiv M, Allalouf D, Landau B, Segenreich E, et al. Vitamin A and beta-carotene content of seminal fluid and sperm of normospermic and oligozoospermic men. Arch Androl 1982;8:61–4.

[89] Amory JK, Arnold S, Lardone MC, Piottante A, Ebensperger M, Isoherranen N, et al. Levels of the retinoic acid synthesizing enzyme ALDH1A2 are lower in testicular tissue from men with infertility. Fertil Steril 2014.

[90] Nya-Ngatchou JJ, Arnold SL, Walsh TJ, Muller CH, Page ST, Isoherranen N, et al. Intratesticular 13-cis retinoic acid is lower in men with abnormal semen analyses: a pilot study. Andrology 2013;1:325–31.

[91] Van Thiel DH, Gavaler J, Lester R. Ethanol inhibition of vitamin A metabolism in the testes: possible mechanism for sterility in alcoholics. Science 1974;186:941–2.

[92] Omu AE, Fatinikun T, Mannazhath N, Abraham S. Significance of simultaneous determination of serum and seminal plasma alpha-tocopherol and retinol in infertile men by high-performance liquid chromatography. Andrologia 1999;31:347–54.

[93] Beckett GJ, Arthur JR. Selenium and endocrine systems. J Endocrinol 2005;184:455–65.

[94] Lu J, Holmgren A. Selenoproteins. J Biol Chem 2009;284:723–7.

[95] Kaur P, Bansal MP. Effect of selenium-induced oxidative stress on the cell kinetics in testis and reproductive ability of male mice. Nutrition 2005;21:351–7.

[96] Kaushal N, Bansal MP. Selenium variation induced oxidative stress regulates p53 dependent germ cell apoptosis: plausible involvement of HSP70-2. Eur J Nutr 2009;48:221–7.

[97] Wu SH, Oldfield JE, Whanger PD, Weswig PH. Effect of selenium, vitamin E, and antioxidants on testicular function in rats. Biol Reprod 1973;8:625–9.

[98] Olson GE, Winfrey VP, Hill KE, Burk RF. Sequential development of flagellar defects in spermatids and epididymal spermatozoa of selenium-deficient rats. Reproduction 2004;127:335–42.

[99] Shinohara A, Chiba M, Takeuchi H, Kinoshita K, Inaba Y. Trace elements and sperm parameters in semen of male partners of infertile couples. Nihon Eiseigaku Zasshi 2005;60:418–25.

[100] Türk S, Mändar R, Mahlapuu R, Viitak A, Punab M, Kullisaar T. Male infertility: decreased levels of selenium, zinc and antioxidants. J Trace Elem Med Biol 2013;.

[101] Oluboyo AO, Adijeh RU, Onyenekwe CC, Oluboyo BO, Mbaeri TC, Odiegwu CN, et al. Relationship between serum levels of testosterone, zinc and selenium in infertile males attending fertility clinic in Nnewi, South East Nigeria. Afr J Med Med Sci 2012;41(Suppl):51–4.

[102] Iwanier K, Zachara BA. Selenium supplementation enhances the element concentration in blood and seminal fluid but does not change the spermatozoal quality characteristics in subfertile men. J Androl 1995;16:441–7.

[103] Ursini F, Heim S, Kiess M, Maiorino M, Roveri A, Wissing J, et al. Dual function of the selenoprotein PHGPx during sperm maturation. Science 1999;285:1393–6.

[104] Michaelis M, Gralla O, Behrends T, Scharpf M, Endermann T, Rijntjes E, et al. Selenoprotein P in seminal fluid is a novel biomarker of sperm quality. Biochem Biophys Res Commun 2014;443:905–10.

[105] Keskes-Ammar L, Feki-Chakroun N, Rebai T, Sahnoun Z, Ghozzi H, Hammami S, et al. Sperm oxidative stress and the effect of an oral vitamin E and selenium supplement on semen quality in infertile men. Arch Androl 2003;49:83–94.

[106] Ebisch IM, Thomas CM, Peters WH, Braat DD, Steegers-Theunissen RP. The importance of folate, zinc and antioxidants in the pathogenesis and prevention of subfertility. Hum Reprod Update 2007;13:163–74.

[107] Henkel R, Bittner J, Weber R, Hüther F, Miska W. Relevance of zinc in human sperm flagella and its relation to motility. Fertil Steril 1999;71:1138–43.

[108] Frenette G, Tremblay RR, Dubé JY. Zinc binding to major human seminal coagulum proteins. Arch Androl 1989;23:155–63.

[109] Chia SE, Ong CN, Chua LH, Ho LM, Tay SK. Comparison of zinc concentrations in blood and seminal plasma and the various sperm parameters between fertile and infertile men. J Androl 2000;21:53–7.

[110] Fuse H, Kazama T, Ohta S, Fujiuchi Y. Relationship between zinc concentrations in seminal plasma and various sperm parameters. Int Urol Nephrol 1999;31:401–8.

[111] Awadallah SM, Salem NM, Saleh SA, Mubarak MS, Elkarmi AZ. Zinc, magnesium and gammaglutamyltransferase levels in human seminal fluid. Bahrain Med Bull 2003;25:1–8.

[112] Hadwan MH, Almashhedy LA, Alsalman AR. Oral zinc supplementation restores high molecular weight seminal zinc binding protein to normal value in Iraqi infertile men. BMC Urol 2012;12:32.

[113] Young SS, Eskenazi B, Marchetti FM, Block G, Wyrobek AJ. The association of folate, zinc and antioxidant intake with sperm aneuploidy in healthy non-smoking men. Hum Reprod 2008;23:1014–22.

[114] Nematollahi-Mahani SN, Azizollahi GH, Baneshi MR, Safari Z, Azizollahi S. Effect of folic acid and zinc sulphate on endocrine parameters and seminal antioxidant level after varicocelectomy. Andrologia 2014;46:240–5.

[115] Jana K, Samanta PK, Manna I, Ghosh P, Singh N, Khetan RP, et al. Protective effect of sodium selenite and zinc sulfate on intensive swimming-induced testicular gamatogenic and steroidogenic disorders in mature male rats. Appl Physiol Nutr Metab 2008;33:903–14.

[116] Sørensen MB, Bergdahl IA, Hjøllund NH, Bonde JP, Stoltenberg M, Ernst E. Zinc, magnesium and calcium in human seminal fluid: relations to other semen parameters and fertility. Mol Hum Reprod 1999;5:331–7.

[117] Wong WY, Merkus HM, Thomas CM, Menkveld R, Zielhuis GA, Steegers-Theunissen RP. Effects of folic acid and zinc sulfate on male factor subfertility: a double-blind, randomized, placebo-controlled trial. Fertil Steril 2002;77:491–8.

[118] Pompella A, Visvikis A, Paolicchi A, Tata VD, Casini AF. The changing faces of glutathione, a cellular protagonist. Biochem Pharmacol 2003;66:1499–503.

[119] Agrawal YP, Vanha-Perttula T. Glutathione, L-glutamic acid and gamma-glutamyl transpeptidase in the bull reproductive tissues. Int J Androl 1988;11:123–31.

[120] Zubkova EV, Robaire B. Effect of glutathione depletion on antioxidant enzymes in the epididymis, seminal vesicles, and liver and on spermatozoa motility in the aging brown Norway rat. Biol Reprod 2004;71:1002–8.

[121] Tripodi L, Tripodi A, Mammí C, Pullé C, Cremonesi F. Pharmacological action and therapeutic effects of glutathione on hypokinetic spermatozoa for enzymatic-dependent pathologies and correlated genetic aspects. Clin Exp Obstet Gynecol 2003;30:130–6.

[122] Ebisch IM, Peters WH, Thomas CM, Wetzels AM, Peer PG, Steegers-Theunissen RP. Homocysteine, glutathione and related thiols affect fertility parameters in the (sub)fertile couple. Hum Reprod 2006;21:1725–33.

[123] Garrido N, Meseguer M, Alvarez J, Simón C, Pellicer A, Remohí J. Relationship among standard semen parameters, glutathione peroxidase/glutathione reductase activity, and mRNA expression and reduced glutathione content in ejaculated spermatozoa from fertile and infertile men. Fertil Steril 2004;82(Suppl. 3):1059–66.

[124] Atig F, Raffa M, Habib BA, Kerkeni A, Saad A, Ajina M. Impact of seminal trace element and glutathione levels on semen quality of Tunisian infertile men. BMC Urol 2012;12:6.

[125] Aydemir B, Onaran I, Kiziler AR, Alici B, Akyolcu MC. Increased oxidative damage of sperm and seminal plasma in men with idiopathic infertility is higher in patients with glutathione S-transferase Mu-1 null genotype. Asian J Androl 2007;9:108–15.

[126] Eskiocak S, Gozen AS, Yapar SB, Tavas F, Kilic AS, Eskiocak M. Glutathione and free sulphydryl content of seminal plasma in healthy medical students during and after exam stress. Hum Reprod 2005;20:2595–600.

[127] Raijmakers MT, Roelofs HM, Steegers EA, Ré Steegers-Theunissen R, Mulder TP, Knapen MF, et al. Glutathione and glutathione S-transferases A1-1 and P1-1 in seminal plasma may play a role in protecting against oxidative damage to spermatozoa. Fertil Steril 2003;79:169–72.

[128] Meseguer M, Garrido N, Simón C, Pellicer A, Remohí J. Concentration of glutathione and expression of glutathione peroxidases 1 and 4 in fresh sperm provide a forecast of the outcome of cryopreservation of human spermatozoa. J Androl 2004;25:773–80.

[129] Anel-López L, Alvarez-Rodríguez M, García-Álvarez O, Alvarez M, Maroto-Morales A, Anel L, et al. Reduced glutathione and Trolox (vitamin E) as extender supplements in cryopreservation of red deer epididymal spermatozoa. Anim Reprod Sci 2012;135:37–46.

[130] Meybodi AM, Mozdarani H, Moradi ShZ, Akhoond MR. Importance of sperm gluthatione treatment in ART. J Assist Reprod Genet 2012;29:625–30.

[131] Meseguer M, de los Santos MJ, Simón C, Pellicer A, Remohí J, Garrido N. Effect of sperm glutathione peroxidases 1 and 4 on embryo asymmetry and blastocyst quality in oocyte donation cycles. Fertil Steril 2006;86:1376–85.

[132] Calamera J, Buffone M, Ollero M, Alvarez J, Doncel GF. Superoxide dismutase content and fatty acid composition in subsets of human spermatozoa from normozoospermic, asthenozoospermic, and polyzoospermic semen samples. Mol Reprod Dev 2003;66:422–30.

[133] Zalata AA, Christophe AB, Depuydt CE, Schoonjans F, Comhaire FH. The fatty acid composition of phospholipids of spermatozoa from infertile patients. Mol Hum Reprod 1998;4:111–8.

[134] Safarinejad MR, Hosseini SY, Dadkhah F, Asgari MA. Relationship of omega-3 and omega-6 fatty acids with semen characteristics, and anti-oxidant status of seminal plasma: a comparison between fertile and infertile men. Clin Nutr 2010;29:100–5.

[135] Attaman JA, Toth TL, Furtado J, Campos H, Hauser R, Chavarro JE. Dietary fat and semen quality among men attending a fertility clinic. Hum Reprod 2012;27:1466–74.

[136] Lenzi A, Gandini L, Picardo M, Tramer F, Sandri G, Panfili E. Lipoperoxidation damage of spermatozoa polyunsaturated fatty acids (PUFA): scavenger mechanisms and possible scavenger therapies. Front Biosci 2000;5:E1–E15.

[137] Littarru GP, Tiano L. Clinical aspects of coenzyme Q10: an update. Nutrition 2010;26:250–4.

[138] Mancini A, De Marinis L, Littarru GP, Balercia G. An update of Coenzyme Q10 implications in male infertility: biochemical and therapeutic aspects. Biofactors 2005;25:165–74.

[139] Gvozdjakova A, Kucharska J, Lipkova J, Bartolcicova B, Dubravicky J, Vorakova M, et al. Importance of the assessment of coenzyme Q10, alpha-tocopherol and oxidative stress for the diagnosis and therapy of infertility in men. Bratisl Lek Listy 2013;114:607–9.

[140] Mancini A, De Marinis L, Oradei A, Hallgass ME, Conte G, Pozza D, et al. Coenzyme Q10 concentrations in normal and pathological human seminal fluid. J Androl 1994;15:591–4.

[141] Balercia G, Arnaldi G, Fazioli F, Serresi M, Alleva R, Mancini A, et al. Coenzyme Q10 levels in idiopathic and varicocele-associated asthenozoospermia. Andrologia 2002;34:107–11.

[142] Safarinejad MR. Efficacy of coenzyme Q10 on semen parameters, sperm function and reproductive hormones in infertile men. J Urol 2009;182:237–48.

[143] Lafuente R, González-Comadrán M, Solà I, López G, Brassesco M, Carreras R, et al. Coenzyme Q10 and male infertility: a meta-analysis. J Assist Reprod Genet 2013;30:1147–56.

[144] Singh K, Jaiswal D. One-carbon metabolism, spermatogenesis, and male infertility. Reprod Sci 2013;20:622–30.

[145] Joshi R, Adhikari S, Patro BS, Chattopadhyay S, Mukherjee T. Free radical scavenging behavior of folic acid: evidence for possible antioxidant activity. Free Radic Biol Med 2001;30:1390–9.

[146] Safarinejad MR, Shafiei N, Safarinejad S. Relationship between genetic polymorphisms of methylenetetrahydrofolate reductase (C677T, A1298C, and G1793A) as risk factors for idiopathic male infertility. Reprod Sci 2011;18:304–15.

[147] Gava MM, Kayaki EA, Bianco B, Teles JS, Christofolini DM, Pompeo AC, et al. Polymorphisms in folate-related enzyme genes in idiopathic infertile Brazilian men. Reprod Sci 2011;18:1267–72.

[148] Eloualid A, Abidi O, Charif M, El Houate B, Benrahma H, Louanjli N, et al. Association of the *MTHFR* A1298C variant with unexplained severe male infertility. PLoS One 2012;7:e34111.

[149] Khazamipour N, Noruzinia M, Fatehmanesh P, Keyhanee M, Pujol P. *MTHFR* promoter hypermethylation in testicular biopsies of patients with non-obstructive azoospermia: the role of epigenetics in male infertility. Hum Reprod 2009;24:2361–4.

[150] Rao AV, Ray MR, Rao LG. Lycopene. Adv Food Nutr Res 2006;51:99–164.

[151] Mendiola J, Torres-Cantero AM, Vioque J, Moreno-Grau JM, Ten J, Roca M, et al. A low intake of antioxidant nutrients is associated with poor semen quality in patients attending fertility clinics. Fertil Steril 2010;93:1128–33.

[152] Zareba P, Colaci DS, Afeiche M, Gaskins AJ, Jørgensen N, Mendiola J, et al. Semen quality in relation to antioxidant intake in a healthy male population. Fertil Steril 2013;100:1572–9.

[153] Goyal A, Chopra M, Lwaleed BA, Birch B, Cooper AJ. The effects of dietary lycopene supplementation on human seminal plasma. BJU Int 2007;99:1456–60.

[154] Hekimoglu A, Kurcer Z, Aral F, Baba F, Sahna E, Atessahin A. Lycopene, an antioxidant carotenoid, attenuates testicular injury caused by ischemia/reperfusion in rats. Tohoku J Exp Med 2009;218:141–7.

[155] Ateşşahin A, Karahan I, Türk G, Gür S, Yilmaz S, Ceribaşi AO. Protective role of lycopene on cisplatin-induced changes in sperm characteristics, testicular damage and oxidative stress in rats. Reprod Toxicol 2006;21:42–7.

[156] Gupta NP, Kumar R. Lycopene therapy in idiopathic male infertility – a preliminary report. Int Urol Nephrol 2002;34:369–72.

[157] Savitha S, Tamilselvan J, Anusuyadevi M, Panneerselvam C. Oxidative stress on mitochondrial antioxidant defense system in the aging process: role of DL-alpha-lipoic acid and L-carnitine. Clin Chim Acta 2005;355:173–80.

[158] Agarwal A, Said TM. Carnitines and male infertility. Reprod Biomed Online 2004;8:376–84.

[159] Koeth RA, Wang Z, Levison BS, Buffa JA, Org E, Sheehy BT, et al. Intestinal microbiota metabolism of l-carnitine, a nutrient in red meat, promotes atherosclerosis. Nat Med 2013;19:576–85.

[160] Li K, Li W, Huang Y. Determination of free L-carnitine in human seminal plasma by high performance liquid chromatography with pre-column ultraviolet derivatization and its clinical application in male infertility. Clin Chim Acta 2007;378:159–63.

[161] Gürbüz B, Yalti S, Fiçicioğlu C, Zehir K. Relationship between semen quality and seminal plasma total carnitine in infertile men. J Obstet Gynaecol 2003;23:653–6.

[162] Ramadan LA, Abd-Allah AR, Aly HA, Saad-el-Din AA. Testicular toxicity effects of magnetic field exposure and prophylactic role of coenzyme Q10 and L-carnitine in mice. Pharmacol Res 2002;46:363–70.

[163] Vicari E, Calogero AE. Effects of treatment with carnitines in infertile patients with prostato-vesiculo-epididymitis. Hum Reprod 2001;16:2338–42.

[164] Biagiotti G, Cavallini G, Modenini F, et al. Prostaglandins pulsed down-regulation enhances carnitine therapy performance in severe idiopathic oligoasthenospermia. Presented at: 19th Annual Meeting of the ESHRE; 2003. Madrid, Spain.

[165] Vicari E, La Vignera S, Calogero AE. Antioxidant treatment with carnitines is effective in infertile patients with prostatovesiculoepididymitis and elevated seminal leukocyte concentrations after treatment with nonsteroidal anti-inflammatory compounds. Fertil Steril 2002;78:1203–8.

[166] Lenzi A, Lombardo F, Sgrò P, Salacone P, Caponecchia L, Dondero F, et al. Use of carnitine therapy in selected cases of male factor infertility: a double-blind crossover trial. Fertil Steril 2003;79:292–300.

[167] Peivandi S, Karimpour A, Moslemizadeh N. Effects of L-carnitine on infertile men's spermogram; a randomized clinical trial. J Reprod Infertil 2010;10 (4):331.

[168] Morgante G, Scolaro V, Tosti C, Di Sabatino A, Piomboni P, De Leo V. Treatment with carnitine, acetyl carnitine, L-arginine and ginseng improves sperm motility and sexual health in men with asthenopermia. Minerva Urol Nefrol 2010;62:213–8.

[169] Li Z, Chen GW, Shang XJ, Bai WJ, Han YF, Chen B, et al. A controlled randomized trial of the use of combined L-carnitine and acetyl-L-carnitine treatment in men with oligoasthenozoospermia. Zhonghua Nan Ke Xue 2005;11:761–4.

[170] Busetto GM, Koverech A, Messano M, Antonini G, De Berardinis E, Gentile V. Prospective open-label study on the efficacy and tolerability of a combination of nutritional supplements in primary infertile patients with idiopathic astenoteratozoospermia. Arch Ital Urol Androl 2012;84:137–40.

[171] Sigman M, Glass S, Campagnone J, Pryor JL. Carnitine for the treatment of idiopathic asthenospermia: a randomized, double-blind, placebo-controlled trial. Fertil Steril 2006;85:1409–14.

[172] Zhou X, Liu F, Zhai S. Effect of L-carnitine and/or L-acetyl-carnitine in nutrition treatment for male infertility: a systematic review. Asia Pac J Clin Nutr 2007;16(Suppl. 1):383–90.

[173] O'Flaherty C, Rodriguez P, Srivastava S. L-arginine promotes capacitation and acrosome reaction in cryopreserved bovine spermatozoa. Biochim Biophys Acta 2004;1674:215–21.

[174] Patel AB, Srivastava S, Phadke RS, Govil G. Arginine activates glycolysis of goat epididymal spermatozoa: an NMR study. Biophys J 1998;75:1522–8.

[175] Keller DW, Polakoski KL. L-arginine stimulation of human sperm motility in vitro. Biol Reprod 1975;13:154–7.

[176] Srivastava S, Agarwal A. Effect of anion channel blockers on L-arginine action in spermatozoa from asthenospermic men. Andrologia 2010;42:76–82.

[177] Scibona M, Meschini P, Capparelli S, Pecori C, Rossi P, Menchini Fabris GF. L-arginine and male infertility. Minerva Urol Nefrol 1994;46:251–3.

34

Sperm Physiology and Assessment of Spermatogenesis Kinetics *In Vivo*

Sandro C. Esteves, MD, PhD, Ricardo Miyaoska, MD, PhD*,**,†*

*ANDROFERT, Referral Center for Male Reproduction, Campinas SP, Brazil
**University of Campinas, Campinas SP, Brazil
†Sumaré State Hospital, Sumaré SP, Brazil

INTRODUCTION

It has been recently proposed that spermatogenesis and steroidogenesis are controlled by a master switch (gonadotropin releasing hormone (GnRH) pulse generator) that controls two separate and independent feedback systems: androgen production (LH-testosterone) and sperm production (FSH-inhibin) [1]. The testes require stimulation by the pituitary gonadotropins, luteinizing hormone (LH), and follicle-stimulating hormone (FSH), which are secreted in response to hypothalamic GnRH. Their action on germ cell development is affected by androgen and FSH receptors on Leydig and Sertoli cells, respectively. While FSH acts directly on the germinative epithelium, LH stimulates secretion of testosterone by the Leydig cells. Testosterone stimulates sperm production and virilization, and also feeds back the hypothalamus and pituitary to regulate GnRH secretion. FSH stimulates Sertoli cells to support spermatogenesis and to secrete inhibin B that negatively feedback FSH secretion.

When spermatozoa leave the testis, they are neither fully motile nor able to recognize or fertilize oocytes. Human spermatozoa must migrate through the epididymis and undergo a specific maturation process in order to become a functional gamete. The epididymis is a dynamic organ that promotes sperm maturation under the influence of androgens. It also provides a place for sperm storage and plays a role in the transport of the spermatozoa from the testis to the ejaculatory duct. In addition, the epididymis protects the male gametes from harmful substances and reabsorbs both fluids and products of sperm breakdown, thus enabling the sperm to fertilize the ovum and to contribute to the formation of a healthy embryo.

THE TESTIS: STRUCTURE AND FUNCTION

The human testis is an ovoid mass that lies within the scrotum. Since the process of sperm production is optimal at temperatures about 2 °C lower than that of core body temperature, the scrotum has mechanisms that facilitate the dissipation of heat, including a thin and wrinkled skin, minimal subcutaneous fat, sparse distribution of hair, a large number of sweat glands, and the cremaster and dartos muscles.

The testes in all mammals are paired encapsulated organs consisting of seminiferous tubules, which correspond to approximately 90% of testicular volume, separated by interstitial tissue. The testis weight and volume increases at puberty, reaching an average of 20 cm³ in young men, and then decreases slightly with age [2]. Normal longitudinal length of the testis is approximately 4.5–5.1 cm. The average weight of the human testis is 15–19 g with a specific gravity of 1.038 g/mL [3]. The right testis is usually 10% larger than the left. Spermatogenesis probably decreases concomitantly to the decline in the overall testicular size [4].

The seminiferous tubules are long V-shaped tubules; both ends drain toward the central superior and posterior regions of the testis, the rete testis, which has a flat cuboidal epithelium. These cells appear to form a valve or plug that may prevent the passage of fluid from the rete into the tubule. The rete lies along the epididymal edge of the testis and coalesces in the superior portion of the testis, just anterior to the testicular vessels, to form 5–10 efferent ductules. The ductules leave the testis and travel a short distance to enter the head or caput region providing a connecting conduit for sperm transport to the epididymis. The seminiferous tubules

Handbook of Fertility. http://dx.doi.org/10.1016/B978-0-12-800872-0.00034-2

are arranged in about 300 lobules each containing one to four tubules. The rete testis is located in close proximity to the testicular artery that has part of its course on the surface of the testis. From the testis surface, the main artery branches turn back into the parenchyma and begin to ramify. The interstitial tissue fills up the spaces between the seminiferous tubules and contains all the blood and lymphatic vessels and nerves of the testicular parenchyma.

A layer of Sertoli cells coated by a lamina propria lines the seminiferous tubules, containing germ cells. The lamina propria consists of the basal membrane covered by peritubular cells (fibroblasts). The main component of the interstitial space is the Leydig or interstitial cells, but it also contains macrophages, lymphocytes, loose connective tissue, and neurovascular bundles.

The peritubular cells are distributed concentrically in layers around the seminiferous tubules, separated by collagen fibers. These cells produce extracellular matrix, connective tissue proteins (collagen, laminin, vimentin, fibronectin), and proteins related to cellular contractility such as smooth muscle myosin and actin. They also synthetize adhesion molecules including nerve growth factor (NGF) and monocyte chemoattractant protein 1 (MCP-1) [5]. Secretion of these factors is regulated by tubular necrosis factor-α (TNF-α), which in turn is produced by mast cells; as such, an interaction between peritubular and mast cells is suggested. It has also been shown that the number of mast cells increases in the testis in some infertility cases [6]. Peritubular cells have contractility properties that aid in the transport of sperm through the seminiferous tubules. Peritubular contractility is regulated by oxytocin, prostaglandins, androgens, and endothelin [7–9]. Endothelin is, in turn, modulated by the relaxant peptide adrenomedullin produced by Sertoli cells [7]. Peritubular cells also secrete insulin-like growth factor-1 (IGF-1) and cytokines that modulate the function of Sertoli cells, particularly the secretion of transferring, inhibin, and androgen-binding protein [8]. Due to the complex interactions between peritubular cells and other cellular elements, it has been suggested that these cells have a role in fertility. In fact, loss of contractility markers, tubular fibrosis, and sclerosis, as well as an increased number of mast cells, are seen in some derangements of spermatogenesis leading to subfertility [6,10,11]. Peritubular and interstitial fibrosis, in association with spermatogenic damage, have also been demonstrated in the testis of vasectomized men [12].

Androgen Production

The differentiation of Leydig cells is determined, at least in part, by peritubular and Sertoli cells, which secrete leukemia inhibitory factor (LIF), platelet-derived growth factor-α (PDGF-α), and other factors that trigger Leydig stem cells to proliferate and migrate into the interstitial compartment of the testis, where they differentiate in the so-called progenitor Leydig cells. After that, growth factors and hormones (LH, IGF-1, PDGF-α, and others) transform them into immature Leydig cells and, finally, into the adult Leydig cell population, which is primarily responsible for the production and secretion of testosterone [13]. Adult Leydig cells exhibit endoplasmatic reticulum and mitochondria, typical of a steroid-producing cell. The major substrate for androgen synthesis is cholesterol. Mitochondrial enzymes cytochrome P450SCC (side chain cleavage) or CYP11A1 (cytochrome P450, family 11, subfamily A, polypeptide 1) transform cholesterol into pregnenolone, a process of androgen synthesis limited by the availability of cholesterol substrate. In the so called delta-4 pathway, pregnenolone is converted to progesterone by 3β-hydroxysteroid dehydrogenase, which in turn is converted to 17α-hydroxy-progesterone and androstenedione by 17α-hydroxylase or CYP17A; androstenedione is finally converted to testosterone by cytochrome P450c17 (Fig. 34.1). In the delta-5 pathway, pregnenolone is hydroxylated to 17α-hydroxypregnenolone and dehydroepiandrosterone by 17α-hydroxylase or CYP17A, which in turn are converted to androstenediol by cytochrome P450c17; finally, androstenediol is transformed into testosterone by 3β-hydroxysteroid dehydrogenase. Testosterone can be converted to estradiol by aromatase or to dihydrotestosterone by 5α-reductase. LH stimulates the transcription of genes that encode the enzymes involved in the steroidogenic pathways to testosterone.

Sperm Production

Sertoli cells form the structure of the seminiferous tubules; their base rest on the basal membrane and their apex is oriented toward the lumen of the tubule (Fig. 34.2). Tight junctions between adjacent cells create a basal compartment that acts as a blood–testis barrier. Spermatogonia and early preleptotene primary spermatocytes are enclosed in the basal compartment while spermatocytes and spermatids are confined to the adluminal compartment [14]. Spermatocytes and spermatids first appear at puberty and therefore after the development of the immune system. The blood–testis barrier separates spermatocytes and spermatids from the immune system, thus avoiding the formation of autoantibodies. A continuous remodeling of the tight junctions, mediated by proteases and protease inhibitors self-secreted by Sertoli cells, occurs as the germ cells are transferred from the basal to the adluminal compartment [15]. Sertoli cells synthesize and secrete a large number of proteins, under the influence of testosterone and FSH [16,17], such as transport proteins (transferrin, ceruloplasmin, androgen binding protein), proteins involved

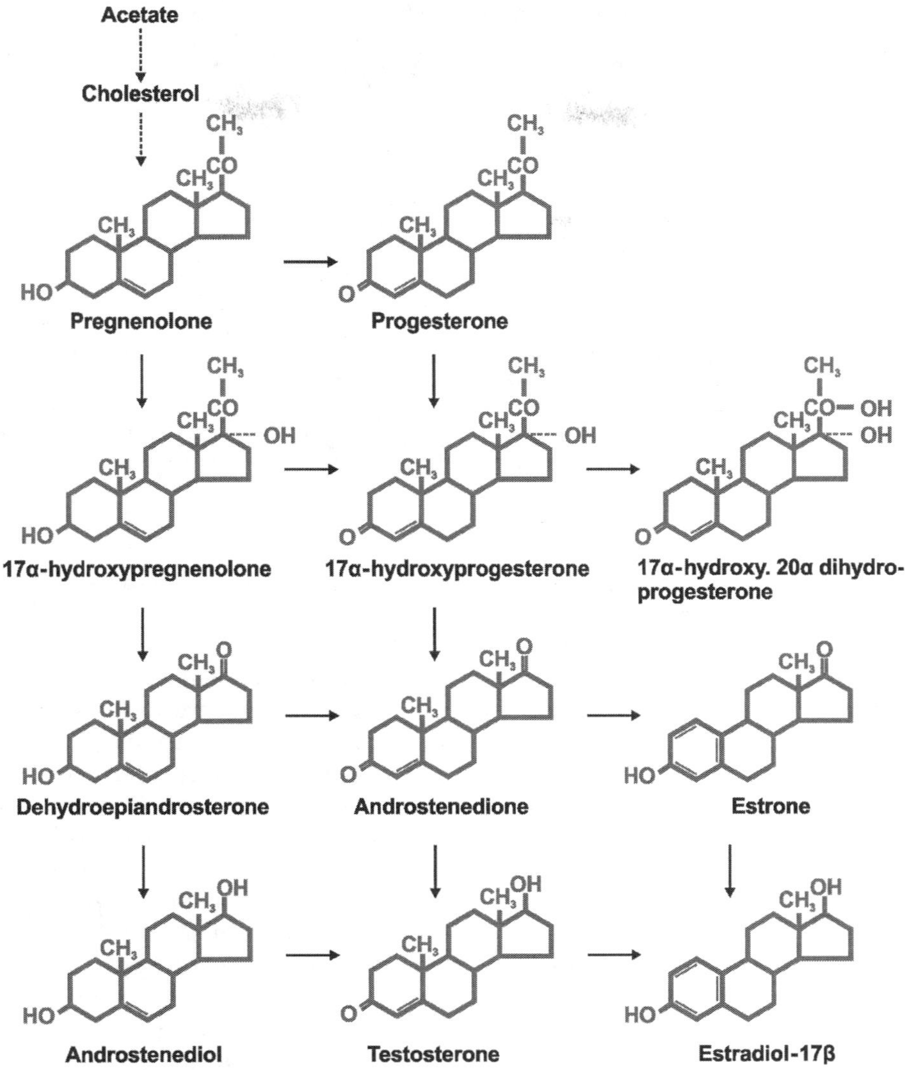

Acetate

Cholesterol

Pregnenolone → Progesterone

17α-hydroxypregnenolone → 17α-hydroxyprogesterone → 17α-hydroxy. 20α dihydro-progesterone

Dehydroepiandrosterone → Androstenedione → Estrone

Androstenediol → Testosterone → Estradiol-17β

FIGURE 34.1 Steroidogenic pathways to testosterone production. *Reprinted from Ref. [102], with permission from Jaypee Brothers Medical Publisher.*

in tight junction remodeling (cadherins, connexins, laminins), and regulatory proteins (antimullerian hormone, seminiferous grown factor), which are crucial for their interaction with germ cells. The functions of Sertoli cells include the regulation of spermatogenesis (by providing support and nutrition to germ cells) and the release of spermatids and spermiation.

The germinative cells are enclosed within the compartments created by the Sertoli cells and the tubular lumen (Fig. 34.3). Three types of spermatogonia are found in the basal compartment: (1) type A dark (considered as testicular stem cells), (2) type A pale (replicate by mitosis), and (3) type B (the cell type that will progress into meiosis) [1,18]. Cohorts of type B spermatogonia initiate meiosis by entering the leptotene stage of the first meiotic prophase. Meiosis is initiated after mitotic proliferation of spermatogonia by DNA synthesis that accomplishes precise replication of each chromosome to form two chromatids. Thus, the DNA content doubles but the number of chromosomes remains the same, that is, diploid. These cells are the primary spermatocytes; they progressively show the nuclear features that identify meiosis I stages of leptotene, zygotene, pachytene, and diplotene. During meiosis I, homologous chromosome pair, forming bivalents, and undergo reciprocal recombination, resulting in a new combination of gene alleles. The first meiotic division is reductional, separating the members of each homologous pair and reducing the chromosome number from 2N to 1N. The result is two haploid cells, secondary spermatocytes, each with 23 chromosomes, but with each chromosome still comprised of two chromatids. The meiosis II is an equational division that separates the chromatids to separate cells, each containing the haploid number of chromosome and DNA content. The products of these meiotic divisions are four spermatids (Fig. 34.4).

The arrangement of germ cells along the basal compartment of the seminiferous tubules differs between

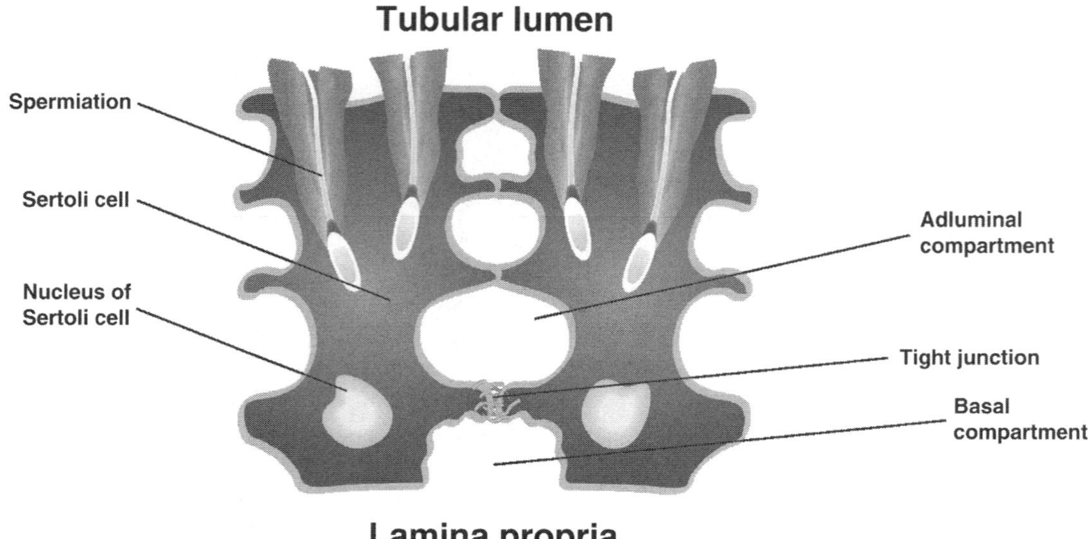

FIGURE 34.2 Schematic representation of Sertoli cells and their relation to the compartments that enclose the germ cells. *Reprinted from Ref. [102], with permission from Jaypee Brothers Medical Publisher.*

species [1]. While in the mouse all germ cells in a cross-section of a seminiferous tubule are in developmental synchrony, the situation in man is different as germ cells at different developmental stages can be identified in each tubular cross-section. In addition, the longitudinal extension of the premeiotic germ cell cohorts attached to the basement membrane area also differs between species. In man the cohorts of germ cells are much shorter than in mice (Fig. 34.5).

When meiosis is completed, the haploid round spermatids are conjoined in a syncytium as they initiate the final differentiation stage of spermatogenesis, termed spermiogenesis. It involves no cell division, representing a complex series of cytological changes leading to the transformation of the round spermatids to spermatozoa. In humans, there are eight different stages (S_{a-1}, S_{a-2}, S_{b-1}, S_{b-2}, S_{c-1}, S_{c-2}, S_{d-1}, and S_{d-2}) involved in the maturation of spermatids to spermatozoa. This process includes (1) changes in the position of the nucleus from a central to an eccentric location, together with reduction in the nuclear size and condensation of the nuclear DNA (chromatin compaction); (2) formation of the acrosome from the Golgi complex, which interposes between the nucleus and the cell membrane; (3) tail formation from a pair of centrioles lying adjacent to the Golgi complex and the aggregation of mitochondria; (4) elimination of most of the cytoplasm, which are phagocytosed by Sertoli cells. Once formed, sperm is released into the tubular lumen (spermiation). In summary, spermatogenesis and spermiogenesis involve a series of extremely intricate processes of cell differentiation, which begin in the basal compartment and end in the apical compartment, and that result in the production of highly specialized spermatozoa with fully compacted chromatin.

Spermatozoon Structure

A morphologically normal sperm cell is about 45–50 μm in length and consists of a head and tail (Fig. 34.6). A morphologically normal head is smooth and symmetrically oval in shape with a broad base and tapering apex. The sperm head measures between 4.0–5.5 μm in length and 2.5–3.5 μm in width, with a length-to-width ratio of between 1.50 and 1.70 [19]. The head is the most important part of the mature male gamete as it contains a nucleus, which is composed of packed chromosomal paternal genetic material (mostly DNA) containing 23 chromosomes. The nucleus comprises about 65% of the head, but like most somatic cells, lacks a large cytoplasm in contrast to somatic cells [20]. The head also contains a well-defined acrosome region, a cap-like covering of the anterior two thirds of the head (40–70% of the apex). The acrosome is represented by the Golgi complex and contains a number

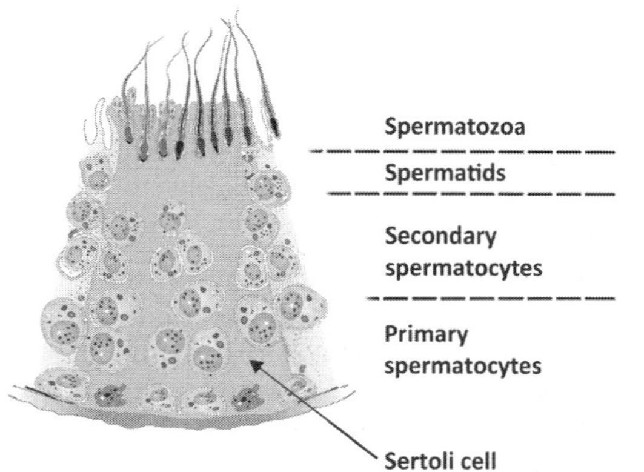

FIGURE 34.3 Schematic representation of a histologic cross-section of a single seminiferous tubule depicting its cell components.

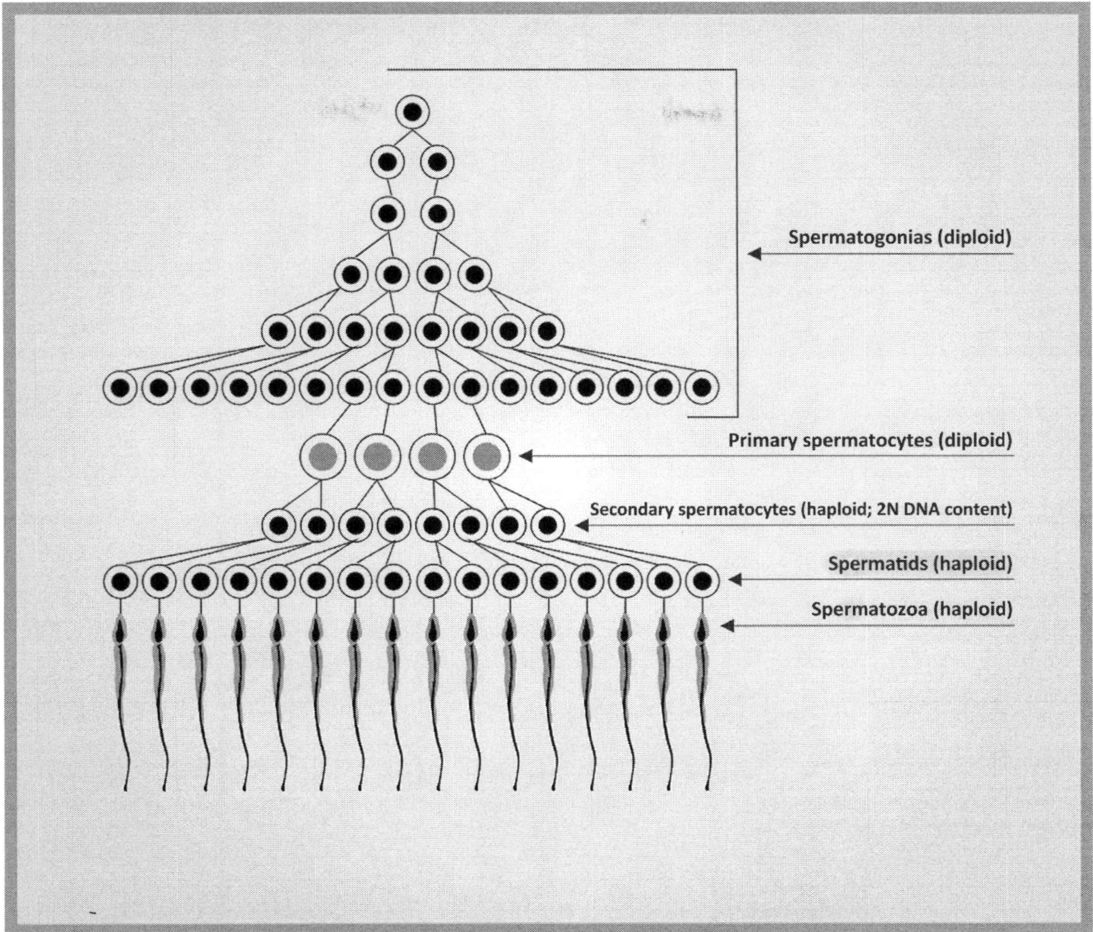

FIGURE 34.4 Schematic representation of cell division stages during spermatogenesis.

of hydrolytic enzymes, such as hyaluronidase and acrosin, which are required for fertilization [20].

The sperm tail measures 40–50 μm in length (nearly 10 times the length of the head) and between 0.4 μm and 0.5 μm in diameter. It is divided into the midpiece, principal piece, and endpiece. The midpiece supports the head at exactly the center position. It is slender (maximum width of 1 μm), yet thicker than the rest of the tail and measures between 7.0 μm and 8.0 μm in length. The midpiece consists of tightly packed mitochondria surrounded by a sheath. The mitochondria in the midpiece supply energy in the form of ATP for tail movement. The principal piece is the longest part of the tail and comprises most of the propellant machinery. The normal tail has a well-defined endpiece, without any coiling or abnormal bending.

THE EPIDIDYMIS: STRUCTURE AND FUNCTION

The epididymis is a single highly convoluted duct extending from the superior to the inferior pole of the testis. It is closely attached to the testis surface by connective tissue and the tunica vaginalis, which surrounds the testis and the epididymis except for its posterior aspect. The posterior surface is attached to the scrotum and spermatic cord by a fibrofatty connective tissue [21]. The epididymis consists of the efferent ductules and the epididymal duct. Between 10 and 15 efferent ductules arise from the rete testis. These ductules come together to form the epididymal duct that varies from 3 m to 4 m [22]. This convoluted tube folds repeatedly upon itself and forms the main bulk of the organ. The epididymis is conventionally divided in the caput, the corpus, and cauda. An alternative subdivision has been proposed by Glover and Nicander [23], which is based on histologic and functional criteria, and divides the epididymis into three regions: initial, middle, and terminal segments. The initial and middle segments are primarily concerned with sperm maturation, while the terminal segment coincides with the region where mature sperm are stored before ejaculation [24,25]. The initial segment comprises the region where the efferent ductules void, while the terminal portion disappears into the epididymal fat making it seem that the epididymis has a minimal cauda region. In fact, the still-coiled human

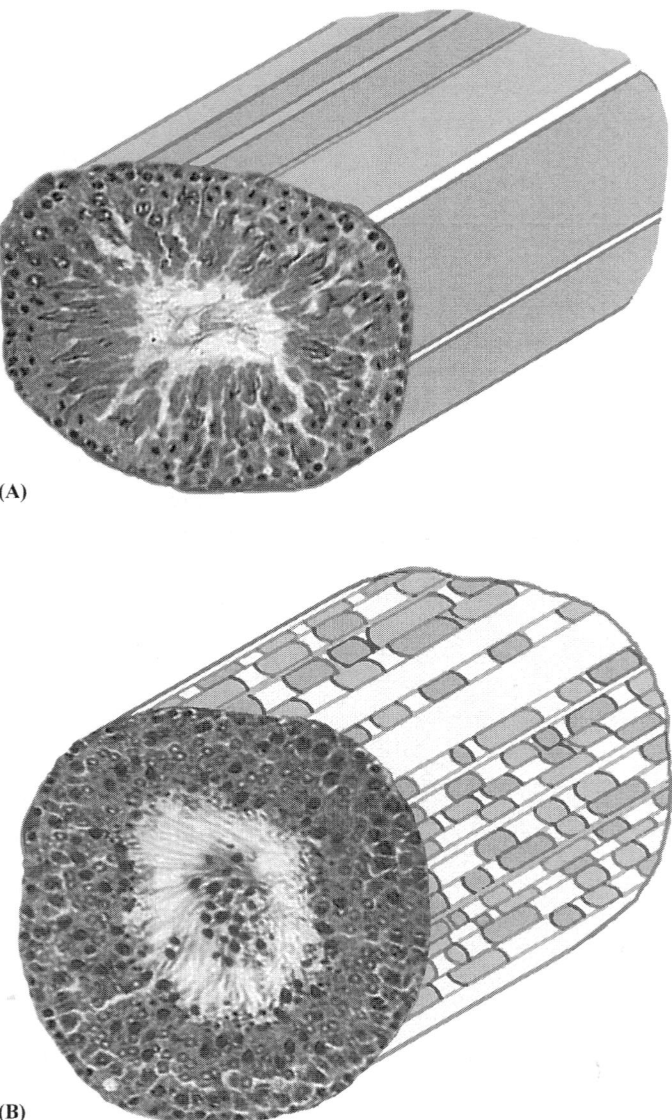

FIGURE 34.5 Schematic arrangement of germ cells along the basement membrane of the seminiferous tubules. (A) The mouse is an example of a species in which a single seminiferous epithelial stage arrangement is seen at a given cross-section of a seminiferous tubule. (B) In man, there is a mixed seminiferous epithelial stage at a given cross-section of a seminiferous tubule. For both species the longitudinal extension of the premeiotic germ cell cohorts attached to the basement membrane are depicted. The cohorts of germ cells are shorter in humans compared with mice. *Adapted from Ref. [1], copyright 2014, with permission from Elsevier.*

epididymis is 10–12 cm long before becoming the convoluted vas, and the convoluted vas extends for another 7–8 cm [26].

The epididymal epithelium is composed of several cell types that include the principal, basal, apical, halo, clear, and narrow cells. The distribution of these cells varies in number and size at different points along the epididymal duct [26,27]. The primary cell type throughout the tubule is the principal cell, which constitutes approximately 80% of the epithelium and is, by far, the most studied since it is responsible for the bulk of the proteins that are secreted into the lumen [28]. The infranuclear compartments of the principal cells are rich in rough endoplasmic reticulum, and the supranuclear compartments of these cells have numerous mitochondria and highly developed Golgi complexes. The principal cells secrete carnitine, glycerylphosphorylcholine and sialic acid, inositol, and a variety of glycoproteins [29]. Narrow, apical, and clear cells contain the vacuolar H^+-ATPase and secrete protons into the lumen, therefore participating in its acidification [30,31]. Clear cells are endocytic cells that seem to be responsible for the clearance of proteins from the epididymal lumen. Basal cells are in close association with the overlying principal cells, as indicated by the presence of cytoplasmic extensions between these cell types, which suggest

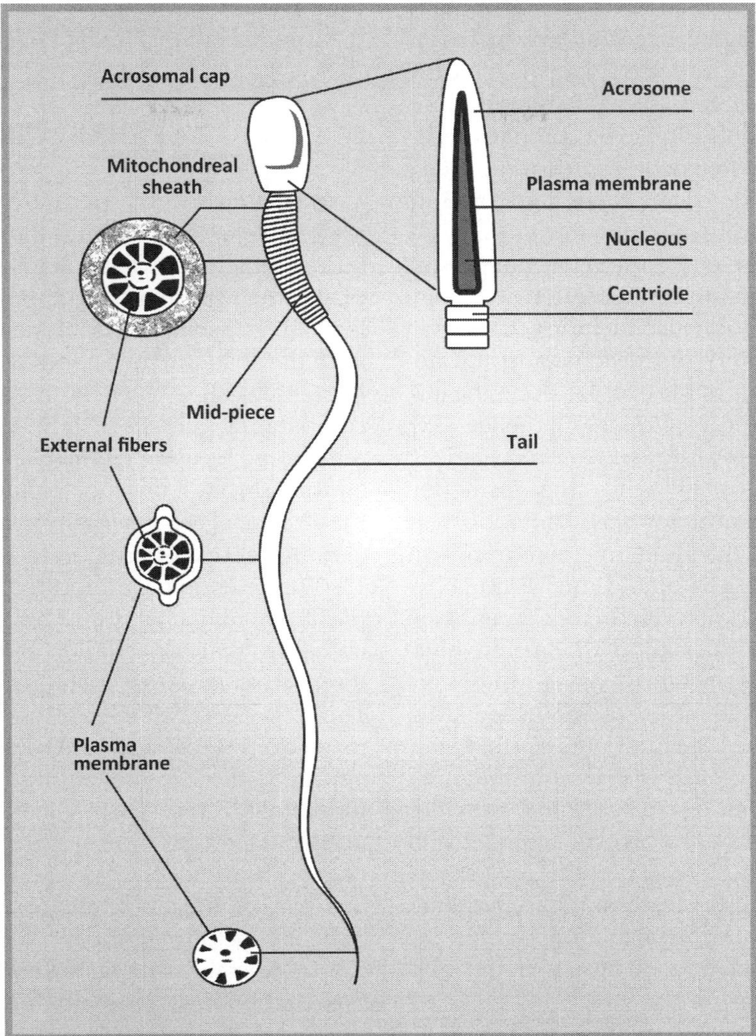

FIGURE 34.6 Schematic representation of a mature human spermatozoon showing its components, the head, midpiece, and tail.

that the basal cells regulate the function of the principal cells [32,33]. Halo cells appear to be the primary immune cell type within the epididymis, while apical cells are involved in endocytosis of luminal components. The most detailed study of the human epididymis was carried out by Yeung et al. [34]. The authors described at least seven types of tubules, connected by at least eight types of junctions to form a network, each one characterized by a different epithelium. The differences in the cellular architecture are primarily due to the functional roles of each epithelium within each epididymal region. In the proximal region there is considerable absorption of water, hence the epithelium takes on the classical appearance of a water-absorbing surface with long stereocilia and many mitochondria in the basal aspects. The cells at the distal epididymis are much smaller and are specialized in removing cellular debris.

Sperm Maturation

Spermatozoa mature during epididymal transit and acquire functional competence. Although the mechanisms by which the epididymis perform its functions of sperm maturation, transport, and storage are not completely understood, it is believed that these functions are affected by the cells and fluid milieu within the epididymal lumen [26]. During their development, spermatozoa are continually bathed in fluid secretions provided by the seminiferous tubules and epididymal ducts [27]. Microanalytic studies revealed that several biochemical changes critical for sperm maturation and survival occur within the epididymal duct [35,36]. The fluid in the proximal epididymis is quite acidic with a pH around 6.5, which increases to approximately 6.8 in the distal region. The epididymal lumen contains the most complex fluid found in any exocrine gland. It results from the continuous changes in composition as well as the presence of many components in unusually high concentrations, such as L-carnitine, glutamate, inositol, sialic acid, taurine, glycerophosphorylcholine, and lactate [28]. Epididymal tubule segments represent unique physiological compartments, each one possessing distinctive gene expression profiles that dictate segment-specific secretion of proteins into the luminal fluid thus affecting sperm maturation.

Sperm maturation involves alterations in the plasma membrane, as well as chromatin condensation and stabilization [37]. Sperm maturation also includes changes in the dimension and appearance of the acrosome and nucleus, migration of the cytoplasmic droplet along the tail, and structural changes in intracellular organelles [38]. Spatially separated lipids and proteins are reorganized during maturation possibly allowing the formation of signaling complexes critical for fertilization. These changes are promoted by the fluid microenvironment within the epididymis [39]. The epididymal fluid is hyperosmotic and as aforesaid, its major constituents are L-carnitine, glutamate, inositol, sialic acid, taurine, glycerophosphorylcholine, and lactate. Concentrations of these substances vary from 20 mM to 90 mM depending upon the epididymal region. Sodium, potassium, bicarbonate, and chloride are also present in the luminal fluid [39]. The organic substances, electrolytes and enzymes are involved in the acquisition of sperm motility, and epididymal cell metabolism. Several proteins, such as albumin, transferrin, immobilin, clusterin (SGP-2), metalloproteins, and proenkephalin, are also found within the epididymal lumen and are claimed to be associated with sperm maturation [22].

Sperm Transport

Spermatozoa released into the seminiferous tubule lumen transit approximately 6 m before they leave the urethral meatus [26]. These cells move in part by hydrostatic pressure originating from fluids secreted in the seminiferous tubules and by tubular peristaltic-like contractions. Contractions of the tunica albuginea play a role in the generation of positive fluid pressure onto the head of the epididymis [22,23]. Transport through the proximal epididymis is due to peristaltic contractions of the smooth muscles surrounding the epididymal duct. The mechanisms responsible for driving the spermatozoa through the epididymal lumen include fluid currents established by action of cilia along the walls of the efferent ducts, hydrostatic pressure gradients, and spontaneous rhythmic contractions of the contractile cells surrounding the epididymal duct [23]. Adrenergic and cholinergic mechanisms and vasopressin have been proposed as regulating factors for the epididymal duct peristaltic activity. Transport rates are estimated to be more rapid in the efferent ducts and proximal epididymis where the

fluid is nonviscous and water is rapidly absorbed from the luminal compartment [22].

SPERM FUNCTION

In vivo, ejaculated sperm from all eutherian mammals are unable to fertilize until they have undergone capacitation, which allows the acrosome reaction (AR) to take place when spermatozoa approach or contact the oocyte [40–43]. Capacitation is a time-dependent phenomenon, with the absolute time course being species-specific [44]. It prepares the sperm to undergo the AR with the accompanying release of lytic enzymes and exposure of membrane receptors, which are required for sperm penetration through the zona pellucida (ZP) and fusion with the oolema [44]. Sperm transport through the female genital tract can occur quite rapidly (times as short as 15–30 min have been reported in humans), whereas capacitation may take from 3 h to 24 h [44]. It is speculated; therefore, that capacitation is not completed until after the spermatozoa have entered the cumulus oophorus. This delay is physiologically beneficial because spermatozoa do not respond to AR-inducing signals until they approach the ZP, preventing premature ARs that lead ultimately to the sperm's inability to penetrate the egg vestments [44,45]. Sperm capacitation is a postejaculatory modification of the sperm plasma membrane, which involves mobilization and/or removal of surface components, including glycoproteins, decapacitation factor, acrosome-stabilizing factor, and acrosin inhibitor. Sperm capacitation involves major biochemical and biophysical changes in the membrane complex and energy metabolism. The presence of high concentrations of cholesterol in the seminal plasma, which maintains a high cholesterol concentration in sperm membranes, seems to be the most important factor for inhibiting capacitation [46]. Capacitation is associated with an increased membrane fluidity, caused by removal of cholesterol from the sperm plasma membrane via sterol acceptors present in the female tract secretions [47,48]. A marked change in sperm motility, named hyperactivation, is also associated with capacitation. Hyperactivated spermatozoa exhibit an extremely vigorous but nonprogressive motility pattern, as a result of Ca^{2+} influx, which causes increased flagellar curvature [49] and extreme lateral movement of the sperm head [47]. Proteasome participates in activating calcium channels that also leads to an increased membrane fluidity, and permeability [50–53]. These events are followed by, or occur simultaneously with, (1) a decrease in net surface charge, (2) devoided area of intramembrane protein and sterols, and (3) increased concentrations of anionic phospholipids [53,54]. Hyperactivated motility is essential for sperm penetration into the intact oocyte–cumulus complexes both *in vitro* and *in vivo* [55,56].

The spermatozoon binds to the ZP with its intact plasma membrane after penetrating into the cumulus oophorus. Sperm binding occurs via specific receptors to ZP glycoproteins located over the anterior sperm head [57]. Glycosylation of ZP glycoproteins is an important aspect of sperm–ZP interaction. It is believed that human ZP glycoprotein-3 (ZP3) has a central role in initiating the AR [58]; however, it has been recently demonstrated that human ZP1 and ZP4 are also implicated in the process [59,60]. The AR is a stimulus-secretion coupled exocytotic event in which the acrosome fuses with the overlying plasma membrane [61,62]. The multiple fusions between the outer acrosomal membrane and the plasma membrane result in the release of hydrolytic enzymes (mostly acrosin) and exposure of new membrane domains, both of which are essential if fertilization is to proceed further. The hydrolytic enzymes released from the acrosome digest the ZP, allowing the spermatozoon to penetrate the oocyte [57]. The AR seems to be physiologically induced by natural stimulants such as the follicular fluid (FF), progestin, progesterone, and hydroxyprogesterone [54]. Follicular fluid and cumulus cells contain protein-bound progesterone that has been identified as one of the most important acrosome-reaction-inducing agents [62,63]. Follicular fluid stimulates the AR in a dose-dependent manner [63,64]. A link between locally produced estradiol by ejaculated spermatozoa, AR, and sperm capacitation was described [65]. Moreover, evidence indicates that environmental estrogens can significantly stimulate mammalian sperm capacitation and AR [66]. Hence, AR is probably initiated when ligands produced by the oocyte bind to sperm receptors. This signal is transduced intracellularly via second messengers, ultimately leading to exocytosis [67]. A number of second messenger pathways have been identified in human spermatozoa, including those that result in the activation of cyclic adenosine monophosphate (cAMP), cyclic guanosine monophosphate (cGMP), and phospholipid-dependent protein kinases [53,68–70]. These kinases are called, respectively, protein kinase A, protein kinase G, and protein kinase C. It is possible that these pathways interact to ensure an optimal response at the correct place and time during the fertilization process. Although the concentration of cGMP in ejaculated human semen is almost seven times lower than cAMP, it is speculated that both nucleotides have a similar role in AR since their dependent protein kinases are closely related [71–73]. The AR can be also induced by artificial stimulants that cause and increase in sperm intracellular calcium [53,74].

Acrosome integrity is crucial for fertilization. A high proportion of sperm with intact acrosomes is seen in ejaculates of normal men. In such individuals, however, ~5–20% sperm cells may exhibit spontaneous ARs that are of no clinical significance [74]. Conversely, several abnormal conditions affecting the sperm acrosome may lead to a decreased fertilization ability. Acrosomeless round-headed spermatozoa (globozoospermic spermatozoa) are unable to fertilize the oocytes, and increased percentages of morphologically abnormal acrosomes were related to fertilization failure in assisted reproductive technology (ART) using conventional *in vitro* fertilization (IVF) [75]. For AR is a time-dependent phenomenon, it cannot take place prematurely or too late [76]. Both premature AR and the inability of spermatozoa to release the acrosomal contents in response to proper stimuli (known as AR insufficiency) have been associated with unexplained male infertility [77]. Although the cause of premature AR is unknown, premature (stimulus-independent) initiation of acrosomal exocytosis seems to be related to a perturbation of plasma membrane stability. In this situation, AR may not involve a premature activation of the receptor-mediated process, but rather reflect an inherent fragility of the sperm membrane, leading to a receptor-independent acrosomal loss [78]. Antisperm antibodies (ASA) may adversely affect the ability of sperm to undergo capacitation and AR [79]. Chang et al. [80] reported a reduction in fertilization rate either by IgG directly bound to sperm or IgM present in female serum. The combination of IgG and IgA may have a synergistic negative effect on fertilization [81–84]. Toxic substances to sperm can also affect AR. High concentrations of dietary phytochemicals, such as genistein, isoflavone, and β-lapachone, were shown to suppress AR in a dose- and time-dependent manner in the rat model [46]. Inhibition of AR by genistein seems to involve the protein kinase C pathway while β-lapachone has a direct cytotoxic effect on sperm cell membrane. It is suggested that genistein and β-lapachone might have an impact on male fertility via AR suppression in high doses and AR induction in low doses [85]. Calcium channel blockers may also interfere with the AR exocytotic event. Sperm incubation with calcium blockers, such as trifluoperazine (calmodulin inhibitor), verapamil (Ca^{2+} channel inhibitor), and nifedipine (voltage-dependent Ca^{2+} channel inhibitor) significantly reduced the ability of hamster sperm to undergo AR [86].

Recent publications on AR have focused on the biochemical and functional aspects of sperm–oocyte fusion. It has been demonstrated that in humans, ZP1 in addition to ZP3 and ZP4 binds to capacitated spermatozoa and induces acrosomal exocytosis [59,87–89]. The ZP3-induced AR involves the activation of T-type voltage-operated calcium channels (VOCCs), whereas ZP1- and ZP4 involve T- and L-type VOCCs. Chiu et al. [90] reported that glycodelin-A, a glycoprotein present in the female reproductive tract, sensitizes spermatozoa to the ZP-induced AR in a glycosylation-specific mechanism involving the activation of the adenylyl cyclase/PKA pathway, suppression of extracellular signal-regulated kinase activation, and upregulation of ZP-induced calcium influx. These findings suggest that glycodelin-A may be important, *in vivo*, to ensure full responsiveness of human spermatozoa to the ZP.

ASSESSMENT OF SPERMATOGENESIS KINETICS *IN VIVO*

The Past

In the past century, one of the most important discoveries in the field of spermatogenesis relates to the duration of the germinative cycle in mammals [91,92]. The experimental approach initially used consisted of an analysis of the rate of disappearance of germ cells from the seminiferous tubules after testis irradiation. When administered in proper doses, X-rays destroy a large proportion of spermatogonia, which results in a progressive disappearance of spermatocytes and spermatids from the seminiferous tubules. Obviously, this method could not be applied to humans.

With the advent of radioactive tracers and the development of radioautography, it became possible to determine the duration of the cycle in intact animals. This method involved radiographic analysis of sequential biopsies after the testes have been injected with tracers. Tritiated thyamidine, which is selectively incorporated in the nuclei of cells preparing for mitosis or meiosis, became the label of choice and was extensively used to time the cycle. Quantitative analysis of the tubular cross-sections containing labeled cells yielded consistent results for the duration of the cycle.

In 1963, a breakthrough study was conducted by Heller and Clermont in which a group of seven men who had undergone vasectomy received an intratesticular injection of tritiated thiamidine, and was followed with serial testicular biopsies [93]. Based on these data, the production of spermatocytes from spermatogonia was estimated to be 16 days, and the duration of the three phases of spermatogenesis, namely, (1) the proliferation of spermatogonia to give rise to diploid spermatocytes; (2) meiotic division, which gives rise to spermatids; and (3) cytological transformation, which leads to mature spermatozoa, was estimated to be 64 days, a value exclusive of epididymal transit time. The human epididymal transit time had been estimated to be 5.5 days, based on mammalian studies mainly involving pigs [94]. Our knowledge that human spermatogenesis requires approximately 3 months to complete comes from these data, and they have been used for the last five decades to guide the time lines proposed in the clinical management of male infertility [95].

Due to its toxicity and invasiveness, the aforementioned studies could never be repeated in men. Nevertheless, their findings have helped investigators to understand the different cellular events that occur during spermatogenesis. For instance, although it has been difficult to describe a clear-cut germinative cycle in humans due to an apparent mixing of germ cells, the analysis of serial sections in well-fixed biopsies revealed typical cell associations into six stages. A mixing of germ cells at the interface of adjacent cell associations was present, including a frequent absence of one or more germ cell generations. Of note, the number of germ cells at the same developmental stage was relatively small, and occupied restricted tubular areas [92].

The Present

A noninvasive and nontoxic method for accurately measuring spermatogenesis kinetics *in vivo* would be ideal for basic research and clinical applications. Such a method might allow the assessment of the effects of infertility treatments in the form of surgery or medication, and the effects of environmental agents or gonadotoxins on testicular function.

In 2006, Misell et al. described a nontoxic and noninvasive method for measuring spermatogenesis *in vivo* [96]. It involved the use of a stable isotope labeling with 70% enriched heavy water (^2H_2O) and analysis of DNA isotopic enrichment in ejaculated sperm by gas chromatography/mass spectrometry (GC/MS). The authors also characterized the kinetics of human spermatogenesis *in vivo* in a group of healthy men with normal sperm production.

Briefly, the method described by Misell et al. involved the daily intake of ^2H_2O for 3 weeks in order to achieve and maintain a body water enrichment of approximately 1.5%. This level of body water enrichment has been previously shown to allow adequate label incorporation for subsequent analysis of DNA synthesis (Fig. 34.7) [97–99]. In their experiment, a total of 11 healthy men with normal sperm concentrations ingested deuterated (heavy) water (^2H_2O) daily and semen samples were collected every 2 weeks for up to 90 days. Label incorporation into sperm DNA was quantified by gas chromatography/mass spectrometry, allowing calculation of the percent of new cells in ejaculates. The overall mean time to detection of labeled sperm in the ejaculate was 64 ± 8 days (Fig. 34.8).

The aforementioned study represents the first time in five decades that a noninvasive direct kinematic measurement of spermatogenesis *in vivo* has been done in humans. In contrast to the study of Heller and Clermont, in which spermatogenesis was estimated to take 64 days excluding epididymal transit time, Misell et al. showed that spermatogenesis duration was much shorter. The authors showed that the appearance of new sperm in

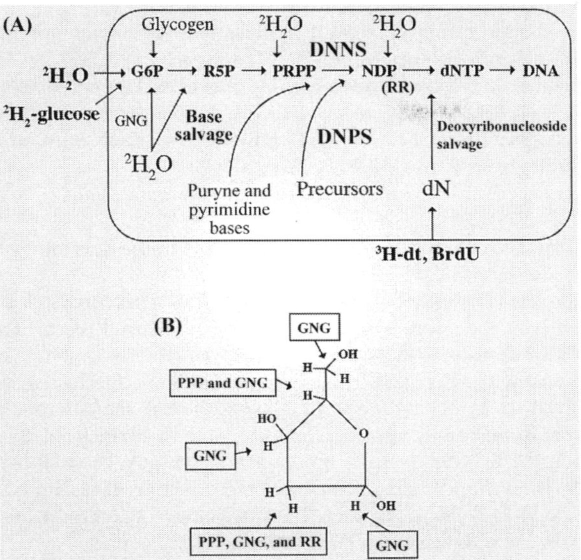

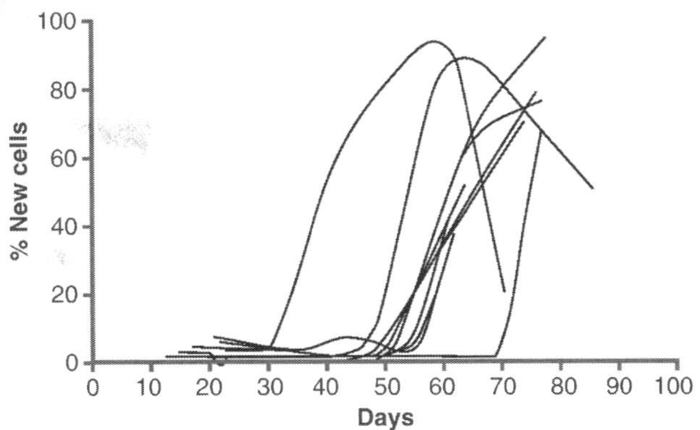

FIGURE 34.8 Spermatocyte labeling curves for 11 subjects with normal semen analyses. Cases were labeled with 50 mL 70% deuterated (heavy) water (^2H_2O) twice daily for 3 weeks. Semen samples were collected every 2 weeks for 90 days from start on ^2H_2O. Spermatocyte DNA enrichment was measured by gas chromatography/mass spectrometry and compared to that of fully turned over cell (monocyte) to calculate the percentage of new cells present. A considerable interindividual variability between normal subjects was observed. *Reprinted from Ref. [96], copyright 2006, with permission from Elsevier.*

FIGURE 34.7 (A) Labeling pathways for measuring DNA synthesis and thus cell proliferation. (B) Sites of 2H incorporation from ^2H_2O into C–H bonds of dR in replicating DNA. GNG, gluconeogenesis/glycolysis; PPP, pentose–phosphatase pathway; RR, ribonucleotide reductase; DNPS, *de novo* purine/pyrimidine synthesis pathway; DNNS, *de novo* nucleotide synthesis pathway. *Reprinted from Ref. [98], copyright (2002), National Academy of Sciences, USA.*

ejaculates of normal men occurred at a mean of 64 days, but this value included epididymal transit time.

The time to new sperm in the ejaculates ranging from 42 to 76 days, as observed by Misell et al., indicates a significant interindividual variation, which contradicts the current belief that spermatogenesis duration is fixed among individuals. While in one subject the time lag was 42 days to detect greater than 33% of new sperm at this time point, in all others it took at least 60 days. All subjects achieved greater than 70% new sperm in the ejaculate by day 90, but plateau labeling was not attained in most, suggesting rapid washout of old sperm in the epididymal reservoir. Although these findings can be related to a variation in spermatogenesis per se, it could also be possible that they were influenced by epididymal transit time, since a separate analysis of spermatogenesis and epididymal transit duration was not possible using their methods. Indeed, it has been speculated that epididymal transit time varies due to the rate of passage through the epididymis cauda, which in turn are influenced by ejaculatory frequency [100]. Also, it has been suggested that men with high testicular sperm production have shorter epididymal transit time than men with lower testicular sperm output [101]. This difference might be explained by a direct association between the production of sperm and fluid, since testes that produce more sperm also produce more fluid, so the movement of spermatozoa along the epididymal duct could be more rapid [102].

The Misell et al. data also suggested that in normal men sperm released from the seminiferous epithelium enter the epididymis in a coordinated manner with little mixing of old and new sperm before subsequent ejaculation. It is a novel concept since it had been suggested that because of mixing, in any segment of the epididymal duct, the population of sperm would be heterogeneous in age and biological status. Their kinetic data showed a sharp increase and subsequent steep decrease in sperm enrichment in the men with complete labeling curves, thus indicating that sperm age is not heterogeneous (Fig. 34.8). If significant mixing of young and old sperm had occurred, the slope of the delabeling curve would have been far more gradual. These kinetic data suggest that the epididymal reservoir is purged of old sperm fairly rapidly and completely in normal men.

CONCLUSIONS

Spermatogenesis is a highly organized and complex sequence of differentiation events that yields genetically distinct male gametes for fertilization. Sperm production is a continuous process, initiated at puberty and continuing throughout life, which occurs in the seminiferous tubules within an immune privileged site. Spermatozoa released from the seminiferous tubules into the epididymis undergo post-testicular maturation. Before fertilization can occur, spermatozoa must undergo further biochemical changes via capacitation and acrosome reaction, both of which occur after ejaculation. Recent knowledge originated from a novel direct measurement of human spermatogenesis kinetics *in vivo* indicates that the entire sperm production process is shorter than previously believed. Based on this new method involving a stable isotope labeling with enriched heavy water and analysis of DNA isotopic enrichment in ejaculated sperm by gas chromatography/mass spectrometry, it has been also suggested that there is a large individual biological variability in the duration of spermatogenesis. This method may become a novel tool for characterizing the relationship between spermatogenesis and semen quality in male infertility, including the measurement of the effects of gonadotoxic exposure as well as medical and surgical interventions.

References

[1] Schlatt S, Ehmcke J. Regulation of spermatogenesis: an evolutionary biologist's perspective. Semin Cell Dev Biol 2014; 29:2–16.

[2] Johnson L, Petty CS, Neaves WB. Age-related variation in seminiferous tubules in men: a stereologic evaluation. J Androl 1986;7(5):316–22.

[3] Handelsman DJ, Staraj S. Testicular size: the effects of aging, malnutrition and illness. J Androl 1985;6(3):144–51.

[4] Kothari LK, Gupta AS. Effect of aging on the volume, structure and total Leydig cell content of the human testis. Int J Fertil 1974;19(3):140–6.

[5] Schell C, Albrecht M, Mayer C, Shwarzer JU, Frungieri MB, Mayerhofer A. Exploring human testicular peritubular cells: identification of secretory products and regulation by tumor necrosis factor-alpha. Endocrinology 2008;149(4):1678–86.

[6] Apa DD, Cayan S, Polat A, Akbay E. Mast cells and fibrosis on testicular biopsies in male infertility. Arch Androl 2002;48(5):337–44.

[7] Romano F, Tripiciano A, Muciaccia B, De Cesaris P, Ziparo E, Palombi F, et al. The contractile phenotype of peritubular smooth muscle cells is locally controlled: possible implications in male infertility. Contraception 2005;72(4):294–7.

[8] Verhoeven G, Hoeben E, De Gendt K. Peritubular cell-Sertoli cell interactions: factors involved in PmodS activity. Andrologia 2000;32(1):42–5.

[9] Zhang JQ, Qin DN, Cui HY. Relationship between ferm cell apoptosis and Sertoli cell vimentin in prepubertal rats induced by local exposure to heat. Zhonghua Nan Ke Xue 2006;12(3):202–6.

[10] Gulkesen KH, Erdogru T, Sargin CF, Karpuzoglu G. Expression of extracellular matrix proteins and vimentin in testes of azoospermic man: an immunohistochemical andmorphometric study. Asian J Androl 2002;4(1):55–60.

[11] Roaiah MM, Khatab H, Mostafa T. Mast cells in testicular biopsies of azoospermic men. Andrologia 2007;39(5):185–9.

[12] Raleigh D, O'Donnell L, Southwick GJ, de Kretser DM, McLachlan RI. Stereological analysis of the human testis after vasectomy indicates impairment of spermatogenic efficiency with increasing obstructive interval. Fertil Steril 2004;81(6):1595–603.

[13] Svechnikov K, Izzo G, Landreh L, Weisser J, Söder O. Endocrine disruptors and Leydig cell function. J Biomed Biotechnol 2010;. 684504.

[14] Main SJ, Waites GM. The blood–testis barrier and temperature damage to the testis of the rat. J Reprod Fertil 1977;51(2):439–50.

[15] Griswold MD. The central role of Sertoli cells in spermatogenesis. Semin Cell Dev Biol 1998;9(4):411–6.

[16] Verhoeven G. A Sertoli cell-specific knock-out of the androgen receptor. Androgen 2005;37(6):207–8.

[17] Gromoll J, Simoni M. Genetic complexity of FSH receptor function. Trends Endocrinol Metab 2005;16(8):368–73.

[18] Ehmcke J, Schlatt S. A revised model for spermatogonial expansion in man: lessons from non-human primates. Reproduction 2006;132(5):673–80.

[19] Menkveld R, Stander FS, Kotze TJ, Kruger TF, van Zyl JA. The evaluation of morphological characteristics of human spermatozoa according to stricter criteria. Hum Reprod 1990;5(5):586–92.

[20] Esteves SC. Fisiologia dos testículos, anexos e da espermatogênese. In: Reis RB, Zequi SC, Zerati Filho M, editors. Urologia Moderna. 1ed. São Paulo: MBC & ATD Editora, Sociedade Brasileira de Urologia Seção São Paulo; 2013. p. 102–9.

[21] Turner TT. De Graaf's thread: the human epididymis. J Androl 2008;29(3):237–50.

[22] Turner TT. On the epididymis and its role in the development of the fertile ejaculate. J Androl 1995;16(4):292–8.

[23] Glover TD, Nicander L. Some aspects of structure and function in the mammalian epididymis. J Reprod Fertil Suppl 1971;13(Suppl):39–50.

[24] Yamamoto M, Nagai T, Miyake K. Different mechanisms are involved in 3H-androgen uptake by the rat seminiferous and epididymal tubules *in vivo*. Nagoya J Med Sci 1991;53(1–4):33–41.

[25] Setchell BP, Maddocks S, Brooks IDE. Anatomy, vasculature, innervations and fluids of the male reproductive tract. 2nd ed. In: Knobil E, Neill JD, editors. The Physiology of Reproduction, 1. New York: Raven Press; 1994. p. 1063–1175.

[26] Yamamoto M, Turner TT. Epididymis, sperm maturation and capacitation. In: Lipshultz LI, Howards SS, editors. Infertility in the Male. 2nd ed. St. Louis: Mosby Year Book; 1991. p. 103–23.

[27] Hinton BT, Lan ZT, Rudolph DB, Labus JC, Lye RJ. Testicular regulation of epididymal gene expression. J Reprod Fertil Suppl 1998;53(suppl):47–57.

[28] Cornwall GA. New insights of epididymal biology and function. Hum Reprod Update 2009;15(2):213–27.

[29] Flickinger CJ. Synthesis and secretion of glycoprotein by the epididymal epithelium. J Androl 1983;4(2):157–61.

[30] Pietrement C, Sun-Wada GH, Silva ND, McKee M, Marshansky V, Brown D, et al. Distinct expression patterns of different subunit isoforms of the V-ATPase in the rat epididymis. Biol Reprod 2006;74(1):185–94.

[31] Kujala M, Hihnala S, Tienari J, Kaunisto K, Hästbacka J, Holmberg C, et al. Expression of ion transport-associated proteins in human efferent and epididymal ducts. Reproduction 2007;133(4):775–84.

[32] Veri JP, Hermo L, Robaire B. Immunocytochemical localization of the Yf subunit of glutathione S-transferase P shows regional variation in the staining of epithelial cells of the testis, efferent ducts, and epididymis of the male rat. J Androl 1993;14(1):23–44.

[33] Seiler P, Cooper TG, Yeung CH, Nieschlag E. Regional variation in macrophage antigen expression by murine epididymal basal cells and their regulation by testicular factors. J Androl 1999;20(6):738–46.

[34] Yeung CH, Cooper TG, Bergmann M, Schulze H. Organization of tubules in the human caput epididymis and the ultrastructure of their epithelia. Am J Anat 1991;191(3):261–79.

[35] Howards SS, Johnson A, Jessee S. Micropuncture and microanalytic studies of the rat testis and epididymis. Fert Steril 1975;26(1): 13–9.

[36] Hilton BT. The epididymal microenvironment: a site of attack for a male contraceptive? Invest Urol 1980;18(1):1–10.

[37] Mortimer ST. Essentials of sperm biology. In: Patton PE, Battaglia DE, editors. Office Andrology. New Jersey: Humana Press; 2005. p. 1–9.

[38] Olson GE, NagDas SK, Winfrey VP. Structural differentiation of spermatozoa during post-testicular maturation. In: Robaire B, Hinton BT, editors. The Epididymis: From Molecules to Clinical Practice. New York: Kluwer Academic/Plenum Publishers; 2002. p. 371–87.

[39] Toshimori K. Biology of spermatozoa maturation: an overview with an introduction to this issue. Microsc Res Tech 2003;61(1):1–6.

[40] Toshimori K, Ito C. Formation and organization of the mammalian sperm head. Arch Histol Cytol 2003;66(5):383–96.

[41] WHO Laboratory Manual for the Examination of Human Semen and Semen-Cervical Mucus Interaction. 4th ed. Cambridge: Cambridge University Press; 1999. p. 19.

[42] Florman HM, Jungnickel MK, Sutton KA. Regulating the acrosome reaction. Int J Dev Biol 2008;52(5–6):503–10.

[43] Henkel R, Muller C, Miska W, Gips H, Schill WB. Determination of the acrosome reaction in human spermatozoa is predictive of fertilization *in vitro*. Hum Reprod 1993;8(12):2128–32.

[44] Mortimer D. From the semen to oocyte: the long route in vivo and in vitro short cut. In: Testart J, Frydman R, editors. Human *In-Vitro* Fertilization: Actual Problems and Prospects. Amsterdam: Elsevier Science Publishers; 1985. p. 93.

[45] Fraser LR. Mechanisms controlling mammalian fertilization. In: Clarke JR, editor. Oxford Reviews of Reproductive Biology. Oxford: Oxford University Press; 1984. p. 173.

[46] Cross NL. Effect of cholesterol and other sterols on human sperm acrosomal responsiveness. Mol Reprod Dev 1996;45(2):212–7.

[47] Mortimer D. Practical Laboratory Andrology. New York: Oxford University Press; 1994.

[48] Lamirande E, Leclerc P, Gagnon C. Capacitation as a regulatory event that primes spermatozoa for the acrosome reaction and fertilization. Mol Hum Reprod 1997;3(3):175–94.

[49] Lindemann CB, Kanous KS. Regulation of mammalian sperm motility. Arch Androl 1989;23(1):1–22.

[50] Matsumura K, Aketa K. Proteasome (multicatalytic proteinase) of sea urchin sperm and its possible participation in the acrosome reaction. Mol Reprod Dev 1991;29(2):189–99.

[51] Inaba K, Akazome Y, Morisawa M. Two high molecular mass proteases from sea urchin sperm. Biochem Biophys Res Commun 1992;182(2):667–74.

[52] Morales P, Kong M, Pizarro E, Pasten C. Participation of the sperm proteasome in human fertilization. Hum Reprod 2003;18(5):1010–17.

[53] Zaneveld LJ, De Jonge CJ, Anderson RA, Mack SR. Human sperm capacitation and the acrosome reaction. Hum Reprod 1991;6(9):1265–74.

[54] Mansour RT, Serour MG, Abbas AM, Nevine AK, Tawab A, Aboulghar MA, et al. The impact of spermatozoa preincubation time and spontaneous acrosome reaction in intracytoplasmic sperm injection: a controlled randomized study. Fertil Steril 2008;90(3):584–91.

[55] Yanagimachi R, Bhattacharyya A. Acrosome-reacted guinea pig spermatozoa become fusion competent in the presence of extracellular potassium ions. J Exp Zool 1988;248(3):354–60.

[56] Katz DF, Drobnis EA, Overstreet JW. Factors regulating mammalian sperm migration through the female reproductive tract and oocyte vestments. Gamete Res 1989;22:443.

[57] Breitbart H, Spungin B. The biochemistry of the acrosome reaction. Mol Hum Reprod 1997;3(3):195–202.

[58] Kopf GS, Gerton GL. The mammalian sperm acrosome and the acrosome reaction. In: Wasserman PM, editor. Elements of Mammalian Fertilization. Boston: CRC Press; 1991. p. 153–203.

[59] Chiu PCN, Wong BST, Chung MK, Lam KKW, Pang RTK, Lee KF, et al. Effects of native human zona pellucida glycoproteins 3 and 4 on acrosome reaction and zona pellucida binding of human spermatozoa. Biol Reprod 2008;79(5):869–77.

[60] Ganguly A, Bukovsky A, Sharma RK, Bansal P, Bhandari B, Gupta SK. In humans, zona pellucida glycoprotein-1 binds to spermatozoa induces acrosomal exocytosis. Hum Reprod 2010;25(7):1643–56.

[61] Yanagimachi R. Mammalian fertilization. In: Knobil E, Neill JD, Ewing LL, Markert CL, Greenwald GS, Pfall DW, editors. The Physiology of Reproduction. New York: Raven Press; 1994. p. 189.

[62] Brucker C, Lipford GB. The human sperm acrosome reaction: physiology and regulatory mechanisms. An update. Hum Reprod Update 1995;1(1):51–62.

[63] Thomas P, Meizel S. Phosphatidylinositol 4,5-bisphosphate hydrolysis in human sperm stimulated with follicular fluid or progesterone is dependent upon Ca2+ influx. Biochem J 1989;264(2):539–46.

[64] Burrello N, Vicari E, D'Amico L, Satta A, D'Agata R, Calogero AE. Human follicular fluid stimulates the sperm acrosome reaction by interacting with the gamma-aminobutyric acid receptors [abstract]. Fertil Steril 2004;82(Suppl. 3):1086–90.

[65] Aquila S, Sisci D, Gentile M, Carpino A, Middea E, Catalano S, et al. Towards a physiological role for cytochrome P450 aromatase in ejaculated human sperm. Hum Reprod 2003;18(8):1650–9.

[66] Adeoya-Osiguwa SA, Markoulaki S, Pocock V, Milligan SR, Fraser LR. 17b-Estradiol and environmental estrogens significantly affect mammalian sperm function. Hum Reprod 2003;18(1):100–7.

[67] Baldi E, Luconi M, Bonaccorsi L, Muratori M, Forti G. Intracellular events and signaling pathways involved in sperm acquisition of fertilizing capacity and acrosome reaction. Front Biosci 2000;1(5):110–23.

[68] Bielfeld P, Anderson RA, Mack SR, De Jonge CJ, Zaneveld LJ. Are capacitation or calcium ion influx required for the human sperm acrosome reaction? Fertil Steril 1994;62(6):1255–61.

[69] De Jonge CJ, Han HL, Lawrie H, Mack SR, Zaneveld LJ. Modulation of the human sperm acrosome reaction by effectors of the adenylate cyclase/cyclic AMP second-messenger pathway. J Exp Zool 1991;258(1):113–25.

[70] Doherty CM, Tarchala SM, Radwanska E, De Jonge CJ. Characterization of two second messenger pathways and their interactions in eliciting the human sperm acrosome reaction. J Androl 1995;16(1):36–46.

[71] Robison GA, Sutherland EW. Cyclic AMP and the function of eukaryotic cells: an introduction. Ann NY Acad Sci 1971;185(M2):5–9.

[72] Gray JP. Cyclic AMP and cyclic GMP in gametes. Ph.D. Dissertation. Nashville (TN): Vanderbilt University; 1971.

[73] Birnabaumer L, Pohl SL, Rodbell M. Adenyl cyclase in fat cells. 1. Properties and the effects of adrenocorticotropin and fluoride. J Bioch Chem 1969;244(13):3468–76.

[74] Esteves SC, Sharma RK, Thomas AJ Jr, Agarwal A. Cryopreservation of human spermatozoa with pentoxifylline improves the post-thaw agonist-induced acrosome reaction rate. Hum Reprod 1998;13(12):3384–9.

[75] Kilani Z, Ismail R, Ghunaim S, Mohamed H, Hughes D, Brewis I, et al. Evaluation and treatment of familial globozoospermia in five brothers. Fertil Steril 2004;82(5):1436–9.

[76] Cummins JM, Fleming AD, Crozet N, Kuehl TJ, Kosower NS, Yanagimachi R. Labelling of living mammalian spermatozoa with the fluorescent thiol alkylating agent, monobromobimane (MB): immobilization upon exposure to ultraviolet light and analysis of acrosomal status. J Exp Zool 1986;237(3):375–82.

[77] Tesarik J, Mendoza C. Alleviation of acrosome reaction prematurity by sperm treatment with egg yolk. Fertil Steril 1995;63(1):153–7.

[78] Esteves SC, Sharma RK, Thomas AJ Jr, Agarwal A. Effect of *in vitro* incubation on spontaneous acrosome reaction in fresh and cryopreserved human spermatozoa. Int J Fertil Womens Med 1998;43(5):235–42.

[79] Chiu WW, Chamley LW. Clinical associations and mechanisms of action of antisperm antibodies. Fertil Steril 2004;82(2):529–35.

[80] Chang TH, Jih MH, Wu TC. Relationship of sperm antibodies in women and men to human *in vitro* fertilization, cleavage, and pregnancy rate. Am J Reprod Immunol 1993;30(2–3):108–12.

[81] Junk SM, Matson PL, Yovich JM, Bootsma B, Yovich JL. The fertilization of human oocytes by spermatozoa from men with antispermatozoal antibodies in semen. J In Vitro Fert Embryo Transf 1986;3(6):350–2.

[82] Meinertz H, Linnet L, Fogh-Andersen P, Hjort T. Antisperm antibodies and fertility after vasovasostomy: a follow-up study of 216 men. Fertil Steril 1990;54(2):315–21.

[83] De Almeida M, Gazagne I, Jeulin C, Herry M, Belaisch-Allart J, Frydman R, et al. *In-vitro* processing of sperm with autoantibodies and *in-vitro* fertilization results. Hum Reprod 1989;4(1):49–53.

[84] Matson PL, Junk SM, Spittle JW, Yovich JL. Effect of antispermatozoal antibodies in seminal plasma upon spermatozoal function. Int J Androl 1988;11(2):101–6.

[85] Kumi-Diaka J, Townsend J. Toxic potential of dietary genistein isoflavone and beta-lapachone on capacitation and acrosome reaction of epididymal spermatozoa. J Med Food 2003;6(3):201–8.

[86] Feng HL, Han YB, Hershlag A, Zheng LJ. Impact of Ca2+ flux inhibitors on acrosome reaction of hamster spermatozoa. J Androl 2007;28(4):561–4.

[87] Gupta SK, Chakravarty S, Suraj K, Bansal P, Ganguly A, Jain MK, et al. Structural and functional attributes of zona pellucida glycoproteins [abstract]. Soc Reprod Fertil 2007;63(Suppl.): 203–16.

[88] Gupta SK, Bansal P, Ganguly A, Bhandari B, Chakrabarti K. Human zona pellucida glycoproteins: functional relevance during fertilization. J Reprod Immunol 2009;(1–2):50–5.

[89] Ganguly A, Bukovsky A, Sharma RK, Bansal P, Bhandari B, Gupta SK. In humans, zona pellucida glycoprotein-1 binds to spermatozoa and induces acrosomal exocytosis. Hum Reprod 2010;25(7):1643–56.

[90] Chiu PC, Wong BS, Lee CL, Lam KK, Chung MK, Lee KF, et al. Zona pellucida-induced acrosome reaction in human spermatozoa is potentiated by glycodelin-A via down-regulation of extracellular signal-regulated kinases and up-regulation of zona pellucida-induced calcium influx. Hum Reprod 2010;25(11):2721–33.

[91] LeBlond C, Clermont Y. Definition of the stages of the cycle of the seminiferous epithelium in the rat. Ann NY Acad Sci 1952;55:548–73.

[92] Clermont Y. Kinetics of spermatogenesis in mammals: seminiferous epithelium cycle and spermatogonial renewal. Physiol Rev 1972;52(1):198–236.

[93] Heller CG, Clermont Y. Spermatogenesis in man: an estimate of its duration. Science 1963;140:184–6.

[94] França LR, Avelar GF, Almeida FF. Spermatogenesis and sperm transit through the epididymis in mammals with emphasis on pigs. Theriogenology 2005;63(2):300–18.

[95] Esteves SC, Agarwal A. Novel concepts in male infertility. Int Braz J Urol 2011;37(1):5–15.

[96] Misell LM, Holochwost D, Boban D, Santi N, Shefi S, Hellerstein MK, et al. A stable isotope-mass spectrometric method for measuring human spermatogenesis kinetics *in vivo*. J Urol 2006;175(1):242–6.

[97] Macallan DC, Fullerton CA, Neese RA, Haddock K, Park SS, Hellerstein MK. Measurement of cell proliferation by labeling of DNA with stable isotope-labeled glucose: studies *in vitro*, in animals, and in humans. Proc Natl Acad Sci USA 1998;95:708–13.

[98] Neese RA, Misell LM, Turner S, Chu A, Kim J, Cesar D, et al. Measurement *in vivo* of proliferation rates of slow turnover cells by ^2H_2O labeling of the deoxyribose moiety of DNA. Proc Natl Acad Sci USA 2002;99:15345–50.

[99] Neese RA, Siler SQ, Cesar D, Antelo F, Lee D, Misell L, et al. Advances in the stable isotope-mass spectro photometric measurement of DNA synthesis and cell proliferation. Anal Biochem 2001;298:189–95.

[100] Amann RP. A critical review of methods for evaluation of spermatogenesis from seminal characteristics. J Androl 1981;2(1): 37–58.

[101] Johnson L, Varner DD. Effect of daily sperm production but not age on the transit times of spermatozoa through the human epididymis. Biol Reprod 1988;39(4):812–7.

[102] Haddad FH, Spaine DM, Esteves SC. The Testis: Function and hormonal control. In: Sabanegh E Jr, Agarwal A, Aziz N, Rizk B, editors. Male Infertility Practice. 1st ed. New Delhi: Jaypee Brothers Medical Publisher; 2013. p. 8–13.

35

Antioxidant Treatment and Prevention of Human Sperm DNA Fragmentation: Role in Health and Fertility

C. Abad Gairín, MD Bs*, J. Gual Frau, MD Bs*,
N. Hannaoui Hadi, MD Bs*, A. García Peiró, PhD**

*Urology Department, Corporació Sanitària Parc Taulí, Institut Universitari Parc Taulí, Sabadell, Spain
**Centro de Infertilidad Masculina y Análisis de Barcelona (CIMAB), Edifici Eureka, Bellaterra, Spain

INTRODUCTION

Male fertility problems are present in 25–50% of infertile couples. This is a situation that although not new, has gained importance in recent decades especially in developed countries where a higher quality of life has been associated with better health care in the broadest sense, which therefore also includes reproductive health.

Numerous pathologies have been detected that can cause male infertility: genetic, infectious, hormonal disorders, obstruction of the seminal duct, varicocele, and abnormal ejaculation. In a significant number of male patients, we cannot define the precise etiology of infertility and we have a lack of precise diagnosis and treatment. Many of us andrologists that study infertile patients daily are aware of the frequency in which we are dealing with individuals who do not show any apparent cause of infertility and that usually appear in our consultation with a semen analysis showing an astenoteratozoospermia that is more or less severe or even normal.

The mission of the spermatozoa is simply to transport the genetic material in the best possible conditions to the ovule; lesions of this material will therefore be a source of problems in both fertilization and the offspring. Elevated levels of DNA fragmentation have been shown to be a robust indicator of male infertility. A fragmentation DNA index greater than 30% is significantly associated with reduced fertility [1]. It is important that there may be an increased sperm DNA fragmentation in infertile patients with normozoospermia; this may account for some cases of infertility of hitherto unknown origin [2].

In the absence of clearly effective treatments directed at the cause of infertility, medical efforts have been oriented to the use of assisted reproductive technology (ART). Progress in laboratory techniques have allowed the fertilization of ovules with sperm of poor quality and even testicular sperm directly obtained. ART has achieved pregnancies in patients with severe sperm abnormalities, but in many cases do not act on the cause of infertility. ART facilitates fertilization of the egg, but keep in mind that when we actively fertilize with intracytoplasmic seminal microinjection (ICSI), we are not able to ensure a proper legacy genetic embryo. Several authors have shown results of a higher frequency of abnormalities in children who have been conceived using this technique. Hansen et al. found that the incidence of birth defects is doubled in those conceived by *in vitro* fertilization (IVF)/ICSI than in naturally conceived children [3]. Other studies showed an increase of imprinting disorders [4], congenital malformations [5], and an eightfold increase in the frequency of undescended testicles in children conceived by ICSI [6]. IVF/ICSI generally have good results and there is no evidence to qualify it as unsafe.

ART is useful and has helped couples who had virtually lost all hope of having children. However, it should not be the ultimate solution. We must deepen our knowledge of the biology of reproduction to know the origin of sperm damage and what should be the appropriate treatment to avoid or repair such damage.

The classical books of andrology have always cited the existence of a number of toxic factors that can affect spermatogenesis. The way by which sperm damage occurs is not known but are associated with poor semen

Handbook of Fertility. http://dx.doi.org/10.1016/B978-0-12-800872-0.00035-4

quality. These factors include heat, exposure to heavy metals, pesticides, ionizing radiation, alcohol, smoking, and obesity [7]. Advances in the study of the biology of sperm have allowed us to relate certain cases of infertility not justified by the presence of sperm DNA fragmentation. The sperm DNA fragmentation is today considered an important factor in the etiology of male infertility [8]. This relationship between fragmented DNA and infertility has helped to clarify the mechanism that causes infertility in a fair amount of infertile men, and today it has been documented that breakage of sperm DNA may be a factor favoring apoptosis, poor fertilization rate, high frequency of miscarriage, and morbidity in offspring or cancer [9–15].

An increase in sperm DNA fragmentation has been found in men with varicocele or leukospermia, and in men who smoke and are exposed to various toxic substances in the environment. On the other hand, age is also able to increase the percentage of breaks in the DNA of sperm [16,17]. DNA fragmentation can be caused by different reasons, but the most common cause is the excess of production of free radicals. These substances are the superoxide anion and hydroxyl radical, which are typically very reactive species and which have one or more unpaired electrons in its valence shell [18]. Another highly oxidizing substance is hydrogen peroxide, but it is not, strictly speaking, a radical. There are as well free radicals that are not directly connected to oxygen but dependent on nitrogen, like nitric oxide (NO), an important compound found in Leydig cells, which contributes in controlling tight junctions in the testis. The reaction of NO with superoxide results in highly toxic molecules, such as peroxynitrate, which are capable of inducing peroxidation of tyrosine molecules that facilitate signal transduction [19].

Mac Leod in 1943 observed that human sperm is capable of producing reactive oxygen species (ROS) [20]. Subsequently, it has been demonstrated that oxidative stress is a major cause of sperm DNA damage at the mitochondrial and nuclear level [21], and that this damage is manifested primarily by fragmentation. The spermatozoon is a cell that is very sensitive to oxidative stress. It has scant cytoplasm and little chance of fighting free radicals due to its lack of intrinsic antioxidative protection by ROS scavengers. This means that an increase of free radicals causes injuries predominantly caused in the sperm membrane and fragmentation of its DNA [22] at the mitochondrial and nuclear level. In the first case, it affects the respiratory chain and cellular metabolism, and in the second, it may lead to defects in the sperm's structure, which could be responsible for failure of fertilization, abortions, or various alterations in the offspring.

The sperm membrane has a high content of unsaturated fatty acids susceptible to attack by ROS (lipid peroxidation) causing a decrease in sperm motility [23].

Several authors have recorded an increase in ROS in the semen of infertile patients compared to fertile men, even suggesting that it is a significant contributing factor in 30–80% of all cases of male infertility [24–26]. Also, some have described higher levels of ROS in the case of smokers and the elderly, who also are more likely to have spontaneous genetic mutations, such as dominant achondroplasia [27], and therefore present in the offspring a greater risk of congenital defects and diseases such as epilepsy, schizophrenia, bipolar disease, and autism. Smoking and age increased the levels of DNA damage in spermatozoa [28]. Besides genetic reasons, an oxidative stress situation has been considered as a major contributory factor in reduced male fertility [29].

The primary sources of ROS are the immature sperm [30] and leukocytes [31]. In the first case, these radicals come from the oxygen consumption during cell metabolism. The production of ROS in leukocytes is a necessity because it is needed in leukocytes' action against pathogens [22,32]. Leukocytes are found in the seminal tract although there is no infection and it has been suggested that moderate levels of leukocytes improve fertilization rate and pregnancy results probably by acrosome reaction. Only high levels of leukocytes are harmful; infection will occur with a significant inflammatory process having clinical significance when leukocyte concentration is higher than 2 million/mL [33]. Infection is a known cause of infertility and its prevalence is believed to be between 10 and 20%, but reached 35% in the study of over 4000 patients by Moretti et al. [34]. In these cases, the excessive production of ROS is an important factor in the impoverishment of semen quality. Other external factors that have been implicated as a source of ROS are environmental contaminants such as heavy metals, phthalate, pesticides, alcohol, smoke, varicocele, and spinal cord injury [35].

However, all is not negative in relation to free radicals. Certain levels of ROS are required for sperm capacitation, hyperactivation, acrosome reaction, sperm binding, and for the fusion zone with the oocyte. There is therefore a relationship between the formation and destruction of ROS so that an increased production of ROS and reduced efficacy of the natural antioxidants can cause cell impairment by these radicals, but in turn, an excess antioxidant activity might inhibit the necessary action of ROS in sperm maturation and acrosome reaction [36]. This would also hinder fertilization and is called the "antiox paradox" [37]. It should also be taken into account when applying an uncontrolled antioxidant treatment at high doses or subjecting sperm to a high concentration of antioxidants before the use of assisted reproductive techniques.

Sperm protection from ROS during spermatogenesis depends mainly on the action of glutathione that comes from Sertoli cells. In the epididymis, enzyme systems, such as glutathione peroxidase, catalase, superoxide dismutase, and indoleamine dioxygenase, are responsible for removing excess free radicals [38]. Other substances responsible for the protection of spermatozoa against

ROS are found in the seminal plasma and are primarily vitamins C and E, uric acid, and glutathione.

Therefore, we can summarize that free radicals are produced by sperm cell metabolism and that the higher or lower presence of leukocytes in the semen will affect the sperm in two ways: a positive when leukocyte levels are not very high, and a negative when the activity of natural antioxidants look overwhelmed by factors that favor the proliferation of ROS such as infection, environmental toxins, or varicocele. When this occurs, the free radicals can affect the sperm cell membrane itself causing a decrease in sperm motility and an increase in DNA fragmentation, both of which are related to infertility.

It seems clear that in an ideal situation, to provide a correct treatment to infertile couples, a study of sperm DNA fragmentation and consequently a measure of free-radical levels and total antioxidant activity (TAC) in semen should be performed. Some authors, such as Omran et al. [39], relate the decline in TACs with increased DNA fragmentation and abnormal semen analysis in infertile men. DNA fragmentation is not used routinely in semen analysis but is still being debated in scientific societies if this test should be included [40]. We think that given the high incidence of fragmented DNA in infertile men, its assessment should be a parameter that should be included in the seminal study.

ANTIOXIDANTS AND GENERAL HEALTH

It has been suggested that in oxidative stress produced by the interaction of cellular proteins, structural damage may be the origin of chronic diseases including diabetes, cardiovascular diseases, and cancer. There are lots of reports that multivitamins are the most common dietary supplement, regularly taken by at least 1/3 of US adults [41].

In diabetes, it has been observed that there is an increase in lipid peroxidation mediated by free radicals and a decrease in the antioxidant levels. Cuerda et al. [42] published an interesting meta-analysis of the work performed after the additional treatment with antioxidants (mainly vitamins E and C) in patients with diabetes or with cardiovascular disorders. They concluded that the intervention studies including different antioxidants have not demonstrated any beneficial effect on cardiovascular and overall morbimortality in different populations, including diabetic patients. None of these studies has demonstrated a beneficial effect of antioxidant supplementation on the prevention of diabetes. According to these studies, these substances can decrease lipid peroxidation, LDL-cholesterol oxidation particles, and improve endothelial function and endothelial-dependent vasodilatation, without significant improvement in the metabolic monitoring of these patients.

Among the most important studies to observe the role of antioxidant therapy on overall health and disease prevention is the SUVIMAX [43], where the effect of a daily combination of 6 mg β-carotene, 120 mg of vitamin C, 30 mg of vitamin E, 100 μg of selenium, and 25 mg of Zn was evaluated in a population of 13,017 French adults 35–60 years old in a randomized, double-blind, placebo-controlled study for an average of 7.5 years. No significant differences between groups in the overall incidence of cancer, incidence of cardiovascular disease or all-cause mortality were detected. However, a protective effect of antioxidants against cancer and all-cause mortality in men but not in women was observed and attributed to lower basal levels of β-carotene in men.

Another important large study was published by Gaziano et al. [44], which included 14,641 US physicians, male and over 50 years old (1312 with a history of cancer), in a randomized, double-blind, placebo-controlled, 2×2×2×2 factorial trial evaluating the balance of risks and benefits of a multivitamin (Centrum Silver vitamin E 400 IU, vitamin C 500 mg and 50 mg β-carotene, or its placebo) daily between 1997 and 2011. They observed that daily multivitamin supplementation modestly reduced the risk of total cancer during a mean of 11 years of treatment and follow-up. These data provide support for the potential use of multivitamin supplements in the prevention of cancer in middle-aged and older men.

Oxidative stress plays an important role in atherogenesis by promoting the oxidation of lipids and proteins in the vascular wall and the proliferation and migration of smooth muscle cells to the intima. Hypertensive patients exhibit a significantly higher production of H_2O_2 plasma than normotensive subjects.

Other authors, such as Schiffrin [45], have contributed their work on the role of antioxidants in hypertension and cardiovascular disease concluding that oxidative stress in hypertension and other cardiovascular conditions contributes to vascular injury by promoting vascular smooth muscle cell growth, endothelial dysfunction, inflammation, increased vascular tone, and matrix metalloproteinase activation. These processes lead to altered vascular contractility and structural remodeling, characteristic features of vessels in hypertension and other forms of cardiovascular disease. Foods such as dark chocolate, along with flavonoids in red wine, and inhibitors of the renin-angiotensin-aldosterone system, suggest that antioxidant treatments may be vasculoprotective.

These and other studies highlight the concern in the relationship of oxidative stress and disease, but a conclusion has not yet been reached. Sometimes the methodology and dosage are criticized. The U.S. Preventive Services Task Force [46] concludes that the current evidence is insufficient to assess the balance of the benefits and harms of multivitamins for the prevention of cardiovascular disease or cancer, and recommends against β-carotene or vitamin E supplements for the prevention of cardiovascular disease or cancer (D recommendation).

OXIDATIVE STRESS AND DNA FRAGMENTATION PREVENTION IN INFERTILE MAN

The first step to avoid fragmentation of sperm DNA would be to know what are the factors that can injure sperm and avoid these factors as much as possible.

Modifiable Factors that Suppose an Alternative Source of Free Radicals

Leucocytes

In addition to the abnormal and immature sperm, leukocytes are the major source of free radicals and oxidative stress. The leukospermia usually coexists with increase in sperm DNA damage [32].

Varicocele

The varicocele is the largest known cause of male infertility with an incidence of 15% in the general population and 40% in infertile men [47]. The oxidative stress is a major factor in its pathophysiology and high levels of DNA fragmentation have been shown in numerous studies. Increased concentrations of ROS and decreased antioxidant capacity coexist in most patients [48]. Although it has led to many debates, there exists a general consensus that symptomatic varicocele or the one that causes infertility has to be treated, and the best considered treatment is surgical and the gold standard is microsurgical varicocelectomy [49].

Smoke

Smoke contains large amounts of ROS involved in DNA damage affecting sperm with high levels of oxidative stress. Nicotine is oxidative and induces a breakup of the double-stranded DNA, associating to dramatic increases of ROS, number of leucocytes in semen, and decreased antioxidant concentration, demonstrating the importance of smoke in the etiology of idiopathic infertility [50].

Xenobiotics and Toxic Metals

Xenobiotics and toxic metals cause DNA damage and infertile patients with high levels of oxidative stress may be more susceptible. The occupational and environmental exposure to certain pollutants that can sometimes act as hormone disruptors could be a cause to take into account of what we define today as unexplained infertility.

The U.S. Environmental Protection Agency now list some 80,000 chemicals in human use, but only a small percentage has been tested for long-term safety. There is a wide variety of xenobiotics with aromatic rings or conjugated bonds that can be enzymatically reduced to form free radicals. Such substances that have been identified as Paraquat, Metadione, aromatic amines, organochlorine or organophosphate, and pesticides (1,2-dibromo-3-chloropropane) have shown DNA fragmentation rates >30% [51]. Bisphenol A present in some foods, plastics, and packaging has been studied as a probable toxic element [52,53]; the same has happened with vinclozolin [54]. Some medications, such as selective serotonin reuptakes, which act on the hormonal system [55], cancer chemotherapy [56], or opioids, have been implicated in sperm DNA damage [57]. The chemical industry works with concentrations considered safe but the damage that many xenobiotics, alone or in combination, can perform at these concentrations is unknown [58,59]. In the absence of further studies we should avoid possible overexposure to these substances.

Heat

Increased scrotal temperature is associated with infertility and impaired spermatogenesis [60]. The factors can include very tight underwear, cycling (tight clothing and position), regular use of saunas or hot tubs, and use of laptops on the legs. All of these have shown a significant increase in sperm DNA fragmentation [61,62].

Mobile Phone Radiation

A study has shown that DNA fragmentation was the only parameter altered with mobile phone use (>4 h/day) and keeping the terminal in the pocket [63].

Wi-Fi Radiation

Some studies have shown a decrease in sperm motility and an increase in sperm DNA fragmentation with Wi-Fi radiation [64,65].

Radiation due to Brachytherapy and Ionizing Radiations

Brachytherapy as radical treatment of prostate cancer was shown in a recent study as a possible cause of infertility resulting in decreased sperm motility and increased sperm DNA fragmentation [66]. Kumar et al. [67] reviewed the relationship between sperm DNA fragmentation and occupational exposure to ionizing radiation. They found a positive correlation with sperm DNA integrity in exposed subjects. To conclude, their study shows distinctly altered sperm chromatin integrity that in radiation health workers is associated with an increase in seminal plasma antioxidant level. Furthermore, the increased seminal plasma glutathione and total antioxidant capacity could be an adaptive measure to tackle the oxidative stress to protect genetic and functional sperm deformities in radiation health workers.

Obesity

A various number of studies have directly and significantly related positively the body mass index (BMI) and percentage of sperm DNA fragmentation. It has also been shown that exercise and weight loss improve semen parameters, which were altered in obese patients [68].

Ethanol

Alcohol abuse is considered one of the problems associated with loss of semen quality. Alcohol alters sperm motility and sperm DNA integrity due to alterations in secondary sperm maturation and reduces the condensation of chromatin in spermatozoa [69].

All these factors represent possible causes of male infertility that are modifiable through changes in lifestyle without risk. They should be investigated and correction should be recommended, especially in cases of idiopathic male infertility.

Some studies have linked low levels of dietary antioxidants with poor semen quality [70]; other studies have found lower sperm aneuploidy with high frequency in men with a higher dietary intake of antioxidants than in those with a lower intake [71]. As popular wisdom says "we are what we eat" and at this point we should do some reflection.

A balanced diet, rich in fruits and vegetables, should provide us a sufficient source of antioxidants that should not necessitate any additional treatment. A clear example of a healthy diet that has been considered for years is the "Mediterranean diet," where legumes, cereals, fruits, vegetables, and vegetable oils have a predominant role. However, some Western countries have new eating habits that focus on fast food and thus are losing the habit of variety, abusing animal fats, and preparing food by adding preservatives and other chemicals to improve taste and visual aspect to stimulate consumption.

The concern that some individuals have for this type of diet leads them not to change it, but to complete their potential shortcomings with mineral and vitamin supplements. These substances tend to be self-administered by these consumers more or less empirically according to trade publications that have fallen into their hands, often without control and without taking into account the actual requirements or the possible side effects from excessive intake.

The so-called nutraceuticals are "food or food ingredients that provide medical or health benefits, including the prevention and/or treatment of a disease" [72]. Regarding their antioxidant action, these substances are mainly (Table 35.1) the vitamins A, C, and E, folic acid, selenium, zinc, carnitine, arginine, coenzyme Q10, glutathione, lycopene, and omega fatty acids. Sufficient and regular intake of food containing them is the most natural way to maintain the proper level of antioxidants so that they can perform all their functions in a physiological manner.

TABLE 35.1 Main Antioxidants Used in the Infertile Male

	Recommended daily allowance	Maximum dose	Dosages used *in vivo* in infertile men	References
Arginine	20 g/day	30 g/day	20 g/day	[107–109]
Carnitine	Not established	4 g/day	1–4 g/day	[77–79]
Coenzyme Q10	Not established	200 mg/day	20–200 mg/day	[123–126]
Folic acid	400 μg	1000 μg/day	50–475 g/day	[95,96,133]
Glutathione	Not established	3 g/day	50–600 mg/day	[113,114]
Lycopene	5–7 mg/day	Not established	400 mg/day	[105]
N-Acetylcysteine	Not established	Not established	600 mg/day	[87,117,118]
Pentoxifylline	400 mg/12 h	1200 mg/day	800–200 mg/day	[112]
Selenium	55 μg/day	300–400 μg/day	200 μg/day	[97–100]
Vitamin A	900 μg/day	3000 μg/day	2100–2400 μg/day	[74]
Vitamin C	90 mg/day	2000 mg/day	200–1000 mg/day	[81,82]
Vitamin E	15 mg/day	400 mg/day	200–400 mg/day	[86–88]
Zinc	11 mg/day	200 mg/day	66–400 mg ZnSO$_4$	[95,96]

ANTIOXIDANT TREATMENT

Based on the current understanding of the action of free radicals in DNA fragmentation and the harmful effects they have on fertility, it seems clear that a treatment leading to antioxidant effect should be seen as an important option in patients with high percentage of sperm DNA fragmentation. Several articles support this theory; it seems that semen from fertile men has a more effective antioxidant capacity than that from infertile men [25]. There are many publications that support the positive role of antioxidant treatment in infertile patients in relation to semen parameters and pregnancy rates after ART [73–75].

Therefore, it does not seem unreasonable to use antioxidants in the treatment of patients with increased DNA fragmentation, and in fact many studies have been conducted to assess the efficacy and function of many of them, which we will now try to summarize.

Carnitine

Carnitines are synthesized from the amino acids lysine and methionine. Its major forms are L-carnitine and L-acetylcarnitine. Carnitine plays an important role in sperm energy metabolism, accumulates in the epididymis, and protects sperm DNA and cell membranes from free-radical-induced damage and apoptosis. It also, protects against peroxidation of the lipid membranes [76]. Carnitine has been used at doses of 3 g/day for 4 months, 4 g/day for 2 months, and 2 g/day, appreciating increased motility, concentration, and morphology [77–79]. Other authors have found seminal improvement after treatment with carnitine although Sigman et al. [80] found no improvement in asthenospermia after 6 months of treatment with 2 and 1 g of L-carnitine and L-acetyl-carnitine, respectively.

Dietary sources of carnitine are meats, poultry, and fish. The recommended daily allowance is not well established. Excess intake greater than 4 g/day can result in gastrointestinal distress, seizures, and malodorous body secretions [72].

Vitamin C

Vitamin C or ascorbic acid is a cofactor for hydroxylation and amidation reactions. It is a high-potency water-soluble ROS scavenger and is an important antioxidant. Its seminal plasma levels have been associated with the percentage of normal sperm [81]; in heavy smokers, an intake of at least 200 mg of vitamin C improved sperm count, motility, and vitality [82]. The best results were obtained with a dose of 1000 mg/day.

Moreover, Abel et al. [83] and Hargreave et al. [84] found no improvement in pregnancy rate with doses of 200 mg/day for 6 months compared with other treatments such as clomiphene citrate and mesterolone.

Vitamin C is abundant in fruits and vegetables such as citrus, kiwi, papaya, strawberry, broccoli, bell pepper, and kale. The minimum daily dose to fill the reservoirs is 90 mg. The intake of over 2000 mg/day can cause dyspepsia, headache, and increased risk of nephrolithiasis [85].

Vitamin E

Vitamin E is one of the most important fat-soluble antioxidants. It inhibits the damage level of free radicals in cell membrane, prevents lipid peroxidation, and improves the activity of other antioxidants. Combination therapy with vitamin C has demonstrated a significant improvement in DNA fragmentation [86]. Comhaire et al. [87] reported a reduction of ROS in a group of 27 infertile men treated using 180 mg/day of vitamin E and 30 mg/day of β-carotene, but no improvement in semen parameters was observed. Suleiman et al. [88] administered 300 mg of vitamin E/day for 6 months and observed a reduction of lipid peroxidation and an increase in motility and pregnancy rate in the control group. Up to 18 studies have found seminal improvement *in vitro* and *in vivo* using vitamin E alone or in combination. On the other hand, other authors, such as Rolf et al. [89], Kessopoulou et al. [90], and Moilanen and Hovatta [91], found no improvement in semen quality.

Vitamin E is found in vegetables, grains, vegetable oils, wheat germ, eggs, meat, and poultry. The recommended daily allowance is 15 mg. Doses above 400 mg/day have been associated with increased cardiovascular risk, especially heart failure, and a dose above 800 mg/day may increase the risk of bleeding for antiplatelet effect. Patients presenting with bleeding disorders should be informed of possible complications with the intake of this vitamin. Other side effects are fatigue, headache, rash, gastrointestinal distress, muscle weakness, and vision changes.

Zinc

Zinc is a cofactor for many enzymes used in DNA transcription and protein synthesis. It has antioxidant properties by its action over superoxide dismutase. It is antiapoptotic, immunological, and is involved in testicular development and steroidogenesis. Zinc deficiency has been associated with flagellar and adapter [92] anomalies. It is found in high concentrations in the prostate and its deficiency has been associated with oligospermia and decreased testosterone [72, 93]. Combining vitamins C and E has been shown to decrease ROS, apoptosis, and DNA fragmentation as well as an increase in sperm motility [94]. Wong et al. [95] observed an improvement in sperm concentration in a study of 103 infertile men treated with 5 mg of folic acid and 66 mg of zinc sulfate per day, while Young et al. [96] observed a significant decrease in X disomy in 89 healthy men with sperm aneuploidy using a combination of the same substances and antioxidants.

We can find zinc in plants, seeds, wheat, and beef. Its deficiency is rare because in industrialized countries it is often an additive in many prepared foods. The recommended daily allowance is 11 mg and more than 200 mg intake can cause gastric ulcers, loss of appetite, headaches, rash, and dehydration. A greater intake than 450 mg/day may affect the function of iron, decrease copper levels, reduce immune function, and cause sideroblastic anemia [72].

Selenium

Selenium is an important element in testicular development, spermatogenesis, and sperm motility training. It plays an important antioxidant role and is a component of glutathione peroxidase, being also involved in the reduction of antioxidant enzymes.

Selenium deficiency causes a decrease in motility, breakage at the intermediate piece, and increased teratozoospermia. Some studies have shown a relationship between the sperm concentration and selenium levels in seminal plasma [97] and increased motility after administration of selenium, alone or in combination with other antioxidants [98]. Moslemi and Tavanbakhsh found an increase in pregnancy rate in 690 infertile men treated although their study is not blinded, placebo-controlled, or randomized control trial [99]. In conjunction with vitamin E, it has an antiperoxidative effect [100,101]. Moreover, other authors have found no seminal improvement after selenium supplementation [102,103].

Plant foods are the major dietary sources of selenium: nuts, Brazil nuts, cereals, eggs, meat, and seafood products. The recommended daily allowance is 55 μg and tolerance limit is 400 μg/day according to the Institute of Medicine (United States) or 300 μg/day according to the European Commission and the World Health Organization. The content in food depends on the selenium content of the soil in which plants are grown. Selenium deficiency can occur in volcanic regions and is endemic in certain rural areas of China causing Keshan disease (reversible cardiomyopathy with selenium supplementation). Moreover, selenium toxicity or seleniosis causes brittleness or loss of hair and nails, gastrointestinal problems, rashes, garlic breath odor, muscle tenderness, tremors, metallic taste, hematologic derangements, and hepatic and renal insufficiency.

Carotenoids and Vitamin A

There is a series of compounds related to vitamin A that show a protective effect of the epithelia and an antioxidant action. Carotenoids from fruits and vegetables are the main dietary source of vitamin A. The β-carotene molecule is split into two molecules of vitamin A (retinol), in a process that occurs in the intestine. Vitamin A deficiency has been experimentally shown to have a harmful effect on spermatogenesis, operation of the Sertoli cell, and the blood–testis barrier [104]. Moreover, a positive effect with a combined treatment of vitamin A and other antioxidants has been shown in relation to sperm motility and concentration [74].

Vitamin A is found in carrots, pumpkin, eggs, oily saltwater fish, and meat products. The recommended daily allowance is 900 μg; a daily dose of 3000 μg should not be exceeded. Being a fat-soluble vitamin, it can accumulate and manifest signs of toxicity such as fatigue, papilledema, dizziness, anorexia, fever, peeling skin, liver toxicity, hypoplastic anemia, joint aches, visual changes, and mental status changes.

Lycopene is a carotenoid that is involved in the regulation of cell growth, modulation of gene expression, immune response, and protection from lipid peroxidation. It is abundant in adrenal glands, breast, liver, prostate, and the testes. Dosed at 200 mg twice daily for 3 months has been shown to improve sperm concentration in 66% of patients, 53% motility and morphology [105]. Lycopene is abundant in tomatoes and cooked products so a supplementary treatment is not usually necessary.

Within this group of substances, Comhaire et al. posted a double-blind, randomized, astaxanthin-controlled trial with 16 m/day for 3 months in which an increase in pregnancy rate compared with placebo was observed [106].

Arginine

Arginine, a precursor of nitric oxide related to inflammatory response, acts by promoting cell protection against oxidative damage. It is essential for sperm motility, training, metabolism, and acrosome reaction. Several studies have shown an improvement in sperm concentration and motility with arginine supplementation [107], while others have not [108]. Srivastava and Agarwal have reported that an excessive concentration of arginine may worsen sperm function [109].

The recommended daily allowance is 20 g arginine without exceeding 30 g [72], and can be obtained from brown rice, barley, nuts, cereal, chocolate, coconut, seeds, and various meat products. The side effects from excess consumption can be electrolyte abnormalities, bleeding risk, increased glucose levels, renal insufficiency, gastrointestinal distress, worsening symptoms of sickle disease, and asthma.

Pentoxifylline

Pentoxifylline is a competitive nonselective phosphodiesterase inhibitor that raises intracellular cyclic adenosine monophosphate. It reduces free radicals [110], preserve sperm motility *in vitro* [111], and has been used orally to improve sperm parameters, in particular mobility. Okada et al. have only found improvement with high doses like 1200 mg/day [112].

Glutathione and *N*-acetylcysteine

Glutathione is one of the most abundant endogenous antioxidant. It protects lipids, proteins, and nucleic acids. It combines with vitamin E and selenium to form glutathione peroxidase. Glutathione is synthesized in the liver from cysteine, glutamic acid, and glycine. Among its properties is its detoxification action of carcinogenic substances and harmful external agents. Glutathione deficiency may cause sperm instability of the midpiece causing disturbances in motility and morphology.

Oral glutathione supplements have improved motility in patients with varicocele [113] and also when administered in combination with vitamins C and E [114]. *In vitro* studies have described a protective effect of glutathione against DNA damage caused by free radicals [115], even against the action produced by leukocytes [116].

As a result of endogenous synthesis, glutathione has no established recommended daily allowance, although up to 3 g supplementation has been proven safe. It should be administered intramuscularly because of its poor intestinal absorption.

N-Acetylcysteine (derived from the amino acid L-cysteine) as a precursor of glutathione is responsible for the increasing levels of endogenous antioxidants. According to Comhaire et al. [87], intake of *N*-acetylcysteine does not improve semen parameters, while Ciftci et al. [117] reported improved motility, viscosity, and seminal oxidative status, and it seems that the results may be better if administered with selenium [118]. *In vitro* beneficial effects have been observed in the work of Baker et al. [116] and Oeda et al. [119] among others.

Superoxide Dismutase

Some studies have reported beneficial effects of superoxide dismutase against lipid peroxidation *in vitro* [120,121] appreciating an improvement in sperm motility. However, Twigg et al. did not find the same benefit [122].

Coenzyme Q10

Coenzyme Q10 is an electron carrier in the mitochondrial respiratory chain and is used to produce energy in the sperm midpiece. The sperm membrane contains high levels of polyunsaturated fatty acids, and a liposoluble antioxidant is able to protect against oxidative stress. Coenzyme Q10 can be found in sperm and seminal plasma, which passes through an active process that can be altered in processes such as varicocele. We hypothesize that the increase of free radicals in the cell may cause consumption of CoQ10, which induces a decrease of its energy role [123].

Several studies have shown that CoQ10 levels in seminal plasma correlate with sperm count and motility; some authors have reported an increase in motility between 4.5 and 6% [123–125]. In a meta-analysis, Lafuente et al. [126] concluded that there is no evidence in the literature that CoQ10 increases either live birth or pregnancy rates, but there is a global improvement in sperm parameters.

CoQ10 can be obtained in whole grains, rice bran, soybeans, nuts, carrots, onions, potatoes, spinach, oily fish, and cabbage, and can be synthesized in the body so its recommended daily allowance has not been established. Based on previous studies, it is estimated that the optimal intake is 200–300 mg/day. Daily needs are about 12 mg/kg/day. With proper diet, it is not necessary to take supplements. Excessive intake of CoQ10 may cause headache, loss of appetite, skin rash, and gastrointestinal distress [72].

Omega Fatty Acids

Polyunsaturated fatty acids (PUFAs) are primarily omega-3 (α-linolenic acid, ALA; docosahexanoic acid, DHA; eicosapentanoic acid, EPA) and omega-6 (arachidonic acid, linoleic acid, Y-linolenic acid) fatty acids. The PUFAs are essential because they cannot be synthesized by the human body. As already discussed, one structural constituent of the cell membrane is to protect the sperm from free radicals. The successful fertilization of spermatozoa depends on the lipids of the membrane [127].

Several studies have documented the deleterious health effects of increased dietary omega-6/omega-3 ratio. In recent decades the increased consumption of omega-6 acids in Western countries has changed the ideal ratio (1:1) in the intake of these two types of PUFAs from 25:1 to 40:1 [128]. Decreased concentrations of DHA in spermatozoa of oligozoospermic asthenozoospermic men have been described [129]. Safarinejad et al. found a correlation between blood plasma and spermatozoa levels and concentration of omega-3 fatty acid, and omega-3 fatty acid concentration and sperm concentration, motility, and morphology, concluding that fertile men had higher omega-3 and lower omega-6 fatty acids [130].

Other studies have shown an improvement on the morphology and total sperm count after treatment with omega-3 [131,132].

Dietary sources for omega-3 fatty acids include oils derived from fish, plants, and nuts. DHA and EPA are found in cold-water fish (anchovy, salmon, tuna, halibut, herring, mackerel, and sardines) and in fish oil supplements. ALA is found in flaxseed, soybeans, pumpkin seeds, purslane, perilla seeds, walnuts, and their respective oils.

A daily intake of 1.6 g/d of omega-3 is acceptable, although there is no recommended daily allowance. The omega-3 fatty acids are used to enrich many processed foods (bakery products, eggs, margarines, meat, milk, oils, pasta); this makes their deficiency unlikely in industrialized countries. Excessive intake can increase bleeding risk. Please note in patients receiving anticoagulant or antiplatelet therapy treatment.

Folic Acid

Folic acid plays an important role in DNA synthesis. It is a powerful free-radical scavenger that provides methyl donors; it is a water-soluble vitamin (B_9). The intake of folic acid has been linked to a lower frequency of abnormalities in the DNA [96]. In the study of Wong et al., sperm count increased from 74% with folic acid and zinc [95] intake. Other studies have found no improvement in semen parameters [133].

The recommended daily allowance of folic acid is 400 μg. Excess intake can result on sleep disturbances, irritability, confusion, and gastrointestinal distress. In patients with a cardiac history, even lower doses of folate have been associated with increased risk of myocardial infarction [134]. Folic acid can be found in avocados, beans, enriched cereals, citrus fruits, eggs, dark green leafy vegetables, and meat products.

Phytoestrogens

Phytoestrogens are nonsteroidal plant-derived substances with a structure similar to estradiol, and can bind or activate estrogen receptors. A group of these phytoestrogens are isoflavones, substances found mainly in soy products and whose metabolites have antioxidant properties.

Song et al. [135] have reported an improvement of sperm count and motility with soy supplements, although Chavarro et al. [136] found that intake of soy foods and isoflavones was inversely related to sperm concentration in a study of 99 men.

Although a meta-analysis showed that isoflavone intake does not alter testosterone levels [137], there is no randomized placebo-controlled study recommending its routine use in infertile men because of the estrogenic effect of these substances.

CONCLUSIONS

The discovery of free radicals in the sperm and seminal fluid, the relationship of oxidative stress to sperm DNA fragmentation, and the testing that semen has antioxidants that can reduce aggression caused by these agents, has opened a new door for the study of infertile men.

If we ignore the assisted reproductive technologies, we must recognize that to date there are few treatments that have proven effective in the management of male infertility. It is for that reason that any possibility of improving the fertility of our patients, within the field of andrology, should be received with enthusiasm, especially if, as in the case of antioxidants, side effects do not usually present gravity.

So many authors have reported on the antioxidant treatment of infertile man and published many works, some of which have been collected in these pages. Many of these publications show a positive effect of this treatment and probably rightly so in many cases. We have also checked the beneficial effects of a combined treatment of antioxidants in patients with asthenoteratozoospermia, appreciating an improvement not only in semen parameters, but also decreased levels of DNA fragmentation of sperm [138]. However, according to the comprehensive reviews conducted by Lombardo et al. [76], Ko and Sabanegh [72], and Mora-Esteves and Shin [92], they are of the opinion of having more randomized, double-blind, placebo-controlled trials with enough patients so that we can reach more conclusive results.

In practice, we can summarize that the antioxidant treatment appears effective *in vitro* [139] (level B evidence) and *in vivo*. The combination of antioxidants has been performed empirically and does not appear to increase the side effects of the treatment. The duration of the published treatments ranges from 2 to 6 months (the spermatic cycle is about 64 days [140]). In any case, in the absence of more conclusive studies, every physician should use these treatments as closely as possible to the recommendations so far in the literature, and the combination that seems more appropriate for a time sufficient to cover the entire sperm cycle. Care should be taken with the side effects especially with substances that tend to accumulate in the body (fat-soluble vitamins and selenium).

Different considerations make it difficult to make a fair assessment of the results of antioxidant treatment. First, a consensus has not been reached for each antioxidant therapeutic dose so that they are strong enough to eliminate the negative effects of free radicals but not enough to suppress their favorable action in sperm capacitation and fertilization process. Also, we do not have objective data to discriminate which patient will benefit from antioxidant treatment and use many times the percentage of fragmented DNA. We should also quantify the oxidizing power of semen or the TAC. Menezo et al. [141] recommend a study of the spermatic nucleus (DNA fragmentation or lack of nuclear condensation) before starting treatment.

Finally, the intrapersonal variability of semen, the ability of the oocyte to repair sperm DNA damage, and the presence of a possible unknown or underrated female factor are added to other difficulties in assessing the success of treatment and pregnancy rates achieved in infertile couples with male infertility.

In 2011, the Cochrane Collaboration conducted a review of antioxidant treatments used in male infertility, asking whether oral antioxidant supplementation improved the outcomes of assisted reproductive abnormalities in couples with male factor or unexplained infertility and whether these different antioxidants offered different results when using these techniques.

Thirty-four randomized controlled studies involving 2876 couples were reviewed. The primary outcome was live birth rate per couple. The secondary outcome included pregnancy rate, miscarriage rate, stillbirth rate, level of sperm DNA damage, sperm parameters, and side effects of treatment.

The results showed that there is a significant increase in live birth rates when compared with men taking a placebo. The authors concluded that supplementation with antioxidants may improve pregnancy rate and live birth outcomes for subfertile couples in assisted reproduction. No evidence of harmful side effects of the antioxidant therapy used was reported. Further head to head comparisons are necessary to identify the antioxidant superiority of one over another [142].

All these considerations indicate how difficult it will be until we reach definitive conclusions, but today we must recognize that at least antioxidant treatment is a promising challenge in the yet disappointing treatment of infertile man.

References

[1] Evenson DP, Jost LK, Marshall D, Zinaman MJ, Clegg E, Purvis K, et al. Utility of the sperm chromatin structure assay as a diagnostic and prognostic tool in the human fertility clinic. Hum Reprod 1999;14:1039–49.

[2] Oleszczuk K, Augustinsson L, Bayat N, Giwercman A, Bungum M. Prevalence of high DNA fragmentation index in male partners of unexplained infertile couples. Andrology 2013;1:357–60.

[3] Hansen M, Kurinczuk JJ, Bower C, Webb S. The risk of major birth defects after intracytoplasmic sperm injection and *in vitro* fertilization. N Engl J Med 2002;346:725–30.

[4] Shiota K, Yamada S. Intrauterine environment-genome interaction and children's development (3): assisted reproductive technologies and development disorders. J Toxicol Sci 2009;34(Suppl 2): SP287–91.

[5] Bunduelle M, Liebaers I, Deketelaere V, Derde MP, Camus M, Devroey P, et al. Neonatal data on a cohort of 2889 infants born after ICSI (1991–1999) and of 2995 infants born after IVF (1883–1999). Hum Reprod 2002;17:671–94.

[6] Ludwig AK, Katalinic A, Thyen U, Sutcliffe AG, Diedrich K, Ludwig M. Physical health at 5.5 years of age of term-born singletons after intracytoplasmatic sperm injection results of a prospective, controlled, single-blinded study. Fertil Steril 2009;91: 115–24.

[7] Marinelli D, Gaspari L, Pedotti P, Taioli E. Mini-review of studies on the effect of smoking and drinking habits on semen parameters. Int J Hyg Environ Health 2004;207:185–92.

[8] Zhang Y, Wang H, Wang L, Zhou Z, Sha J, Mao Y, et al. The clinical significance of sperm DNA damage detection combined with routine semen testing in assisted reproduction. Mol Med Rep 2008;1:617–24.

[9] Aitken RJ, Koppers AJ. Apoptosis and DNA damage in human spermatozoa. Asian J Androl 2011;13(1):36–42.

[10] Agarwal A, Said TM. Role of sperm chromatin abnormalities and DNA damage in male infertility. Hum Reprod Update 2003;9:331–45.

[11] Bungum M, Humaidan P, Spano M, Jepson K, Bungum L, Giwercman A. The predictive value of sperm chromatin structure assay (SCSA) parameters for the outcome of intrauterine insemination, IVF and ICSI. Hum Reprod 2004;19:1401–8.

[12] Carrell DT, Liu L, Peterson CM, Jones KP, Hatasaka HH, Erikson L, et al. Sperm DNA fragmentation is increased in couples with unexplained recurrent pregnancy loss. Arch Androl 2003;49: 49–55.

[13] Matés JM, Segura JA, Alonso FJ, Márquez J. Oxidative stress in apoptosis and cancer: an update. Arch Toxicol 2012;86(11): 1649–65.

[14] Perrin A, Nguyen MH, Bujan L, Vialard F, Amice V, Guéganic N, et al. DNA fragmentation is higher in spermatozoa with chromosomally unbalanced content in men with a structural chromosomal rearrangement. Andrology 2013;1(4):632–8.

[15] Zini A, Boman JM, Belzile E, Ciampi A. Sperm DNA damage is associates with an increase of pregnancy loss after IVF and ICSI: systematic review and meta-analysis. Hum Reprod 2008;23: 2663–8.

[16] Humm KC, Sakkas D. Role of increased male age in IVF and egg donation: is sperm DNA fragmentation responsible? Fertil Steril 2013;99(1):30–6.

[17] Rybar R, Kopecka V, Prinosilova P, Markova P, Rubes J. Male obesity and age in relationship to semen parameters and sperm chromatin integrity. Andrologia 2011;43(4):286–91.

[18] Aitken RJ, De Iuliis GN. On the possible origins of DNA damage in human spermatozoa. Mol Hum Reprod 2010;16:3–13.

[19] Doshi SB, Khullar K, Sharma RK, Agarwal A. Role of reactive nitrogen species in male infertility. Reprod Biol Endocrinol 2012;10:109.

[20] Mac Leod J. The role of oxygen in the metabolism and motility of human spermatozoa. Am J Physiol 1943;138:512–8.

[21] Sawyer DE, Mercer BG, Wiklendt AM, Aitken RJ. Analysis of gene-specific DNA damage and single-strand DNA breaks induced by pro-oxidant treatment of human spermatozoa *in vitro*. Mutat Res 2003;529:21–34.

[22] Henkel RR. Leukocytes and oxidative stress: dilemma for sperm function and male fertility. Asian J of Androl 2011;13:43–52.

[23] Tavilani H, Doosti M, Abdi K, Vaisiraygani A, Joshaghani HR. Decreased polyunsaturated and increased saturated fatty acid concentration in spermatozoa from ashenozoospermic males as compared with normozoospermic males. Andrologia 2006;38:173–8.

[24] Auger J, Eustache F, Andersen AG, Irvine DS, Jorgensten N, Skakkebaek NE, et al. Sperm morphological defects related to environment, lifestyle and medical history of 1001 male partners of pregnant women from four European cities. Hum Reprod 2001;16:2710–7.

[25] Tremellen K. Oxidative stress and male infertility – a clinical prospective. Hum Reprod Update 2008;14:243–58.

[26] Showell MG, Brown J, Yardazi A, Stankiewicz MT, Hart RJ. Antioxidants for male subfertility. Cochrane Database Syst Rev 2011;Jan(1). Art No.:CD007411.

[27] Crow JF. The origins, patterns and implications of human spontaneous mutation. Nat Rev Genet 2000;1:40–7.

[28] Aitken RJ, de Luliis GN. On the possible origins of DNA damage in human spermatozoa. Mol Hum Reprod 2010;16:3–13.

[29] Agarwal A, Allamaneni SS. Free radicals and male reproduction. J Indian Med Assoc 2011;109:184–7.

[30] Gil-Guzman E, Ollero M, Lopez MC, Sharma RK, Alvarez JG, Thomas AJ Jr, et al. Differential production of reactive oxygen species by subsets of human spermatozoa at different stages of maturation. Hum Reprod 2001;16:1922–30.

[31] Henkel R, Kierspel E, Stalf T, Mehnert C, Menkveld R, Tinneberg HR, et al. Effect of reactive oxygen species produced by spermatozoa and leukocytes on sperm functions in non-leukocitospermic patients. Fertil Steril 2005;83:635–42.

[32] Alvarez JG, Sharma RK, Ollero M, Saleh RA, Lopez MC, Thomas AJ Jr, et al. Increased DNA damage in sperm from leukocytospermic semen samples as determined by the sperm chromatin structure assay. Fertil Steril 2002;78:319–29.

[33] Yanushpolsky EH, Politich JA, Hill JA, Anderson DJ. Is leukocytospermia clinically relevant? Fertil Steril 1996;66(5):822–5.

[34] Moretti E, Capitani S, Figura N, Pammolli A, Federico MG, Pammolli A, et al. The presence of bacteria species in semen and sperm quality. J Assist Reprod Genet 2009;26:47–56.

[35] Agarwal DK, Maronpont RR, Lamb JC, Kluwe WM. Adverse effects of butyl benzyl phthalate on the reproductive and hematopoietic systems of male rats. Toxicology 1985;35:189–206.

[36] De Lamirande E, Lamothe G. Reactive oxygen-induced reactive oxygen formation during human sperm capacitation. Free Radic Biol Med 2009;46:502–10.

[37] Halliwell B. The antioxidant paradox. Lancet 2000;355:1179–80.

[38] Vernet P, Aitken RJ, Drevel JR. Antioxidant strategies in the epididymis. Mol Cell Endocrinol 2004;216:31–9.

[39] Omran HM, Bakhiet M, Dashti MG. DNA integrity is a critical molecular indicator for the assessment of male infertility. Mol Med Rep 2013;7(5):1631–5.

[40] Practice Committee of the American Society for Reproductive Medicine. The clinical utility of sperm DNA integrity testing: a guideline. Fertil Steril 2013;99:673–7.

[41] Bailey RL, Gahche JJ, Lentino CV, Dwyer JT, Engel JS, Thomas PR, et al. Dietary supplement use in the United States 2003–2006. J Nutr 2011;141(2):261–6.

[42] Cuerda C, Luengo LM, Valero MA, Vidal A, Burgos R, Calvo FL, et al. Antioxidants and diabetes mellitus: review of the evidence. Nutr Hosp 2011;26(1):68–78.

[43] Hercberg S, Galan P, Preziosi P, Bertrais S, Mennen L, Malvy D, et al. The SU.VI.MAX Study: a randomized, placebo-controlled trial of the health effects of antioxidant vitamins and minerals. Arch Intern Med 2004;164(21):2335–42.

[44] Gaziano JM, Sesso HD, Christen WG, Bubes V, Smith JP, Mac Fadyen J, et al. Multivitamins in the prevention of cancer in men: the Physicians' Health Study II randomized controlled trial. JAMA 2012;308(18):1871–80.

[45] Schiffrin EL. Antioxidants in hypertension and cardiovascular disease. Mol Interv December 2010;10(6):354–62.

[46] Moyer VA. U.S. Preventive Services Task Force. Vitamin, mineral, and multivitamin supplements for the primary prevention of cardiovascular disease and cancer: U.S. Preventive services Task Force recommendation statement. Ann Intern Med 2014;160(8):558–64.

[47] Practice Committee of the American Society for Reproductive Medicine. Report on varicocele and infertility. Fertil Steril 2004;82:S142–5.

[48] Smith R, Kaune H, Parodi D, Madariaga M, Rios R, Morales I, et al. Increased sperm DNA damage in patients with varicocele: relationship with seminal oxidative stress. Hum Reprod 2006;21:986–93.

[49] Ficarra V, Crestani A, Novara G, Mirone V. Varicocele repair for infertility: what is the evidence? Curr Opin Urol 2012;22:489–94.

[50] Valavanidis A, Vlachogianni T, Fiotakis K. Tobacco smoke: involvement of reactive oxygen species and stable free radicals in mechanisms of oxidative damage, carcinogenesis and synergistic effects with other respirable particles. Int J Environ Res Public Health 2009;6:445–62.

[51] Miranda-Contreras L, Gomez-Perez R, Rojas G, Cruz I, Berrueta L, Salmen S, et al. Occupational exposure to organophosphate and carbamate pesticides affects sperm chromatin integrity and reproductive hormone levels among Venezuelan farm workers. J Occup Health 2013;55:195–203.

[52] Hulak M, Gazo I, Shaliutina A, Linhartova P. *In vitro* effects of bisphenol A on the quality parameters, oxidative stress, DNA integrity and adenosine triphosphate content in sterlet (*Acipenser ruthenus*) spermatozoa. Comp Biochem Physiol C Toxicol Pharmacol 2013;158(2):64–71.

[53] Meeker JD, Ehrlich S, Toth TL, Wright DL, Calafat AM, Trisini AT, et al. Semen quality and sperm DNA damage in relation to urinary bisphenol A among men from an infertility clinic. Reprod Toxicol 2010;30:532–9.

[54] Gazo I, Linhartova P, Shaliutina A, Hulak M. Influence of environmentally relevant concentrations of vinclozolin on quality, DNA integrity, and antioxidant responses of sterlet *Acipenser ruthenus* spermatozoa. Chem Biol Interact 2013;203(2):377–85.

[55] Tu HY, Zini A. Finasteride-induced secondary infertility associated with sperm DNA damage. Fertil Steril 2011;95(6). 2125. e13–e14.

[56] Romerius P, Ståhl O, Möell C, Relander T, et al. Sperm integrity in men treated for childhood cancer. Clin Cancer Res 2010;16(15):3843–50. 1.

[57] Safarinejad MR, Asgari SA, Farshi A, Ghaedi G, Kolahi AA, Iravani S, et al. The effects of opiate consumption on serum reproductive hormone levels, sperm parameters, seminal plasma antioxidant capacity and sperm DNA integrity. Reprod Toxicol 2013;36:18–23.

[58] Gharagozloo P, Aitken RJ. The role of sperm oxidative stress in male infertility and the significance of oral antioxidant therapy. Hum Reprod 2011;26(7):1628–40.

[59] Comhaire F, Mahmoud A, Schoonjans F. Sperm quality, birth rates and environment in Flandes (Belgium). Reprod Toxicol 2007;23:133–7.

[60] Mieusset R, Bujan L, Mondinat C, Mansat A, Pontonnier F, Grandjean H. Association of scrotal hyperthermia with impaired spermatogenesis in infertile men. Fertil Steril 1987;48:1006–11.

[61] Ahmad G, Moinard N, Esquerré-Lamare C, Mieusset R, Bujan L. Mild induced testicular and epididymal hyperthermia alters sperm chromatin integrity in men. Fertil Steril 2012;97(3):546–53.

[62] Sheynkin Y, Welliver R, Winer A, Hajimirzaee F, Ahn H, Lee K. Protection from scrotal hyperthermia in laptop computer users. Fertil Steril 2011;95:647–51.

[63] Rago R, Salacone P, Caponecchia L, Sebastianelli L, Marcucci I, Calogero AE, et al. The semen quality of the mobile phone users. J Endocrinol Invest 2013;36:970–4.

[64] Avendaño C, Mata A, Sanchez Sarmiento CA, Doncel GF. Use of laptop computers connected to internet through Wi-Fi decreases

human sperm motility and increases DNA fragmentation. Fertil Steril 2012;97(1):39–45.

[65] Freour T, Barriere P. Wi-Fi decreases human sperm motility and increases sperm DNA fragmentation. Fertil Steril 2012;97doi: 10.1016/j.fertnstert.2012.02.004.

[66] Singh DK, Hersey K, Perlis N, Crook J, Jarvi K, Fleshner N. The effect of radiation on semen quality and fertility in men treated with brachytherapy for early stage prostate cancer. J Urol 2012;187:987–9.

[67] Kumar D, Salian SR, Kalthur G, Uppangala S, Kumari S, Challapalli S, et al. Association between sperm DNA integrity and seminal plasma antioxidant levels in health workers occupationally exposed to ionizing radiation. Environ Res 2014;132C: 297–304.

[68] Fariello RM, Pariz JR, Spaine DM, Cedenho AP, Bertolla RP, Fraietta R. Association between obesity and alteration of sperm DNA integrity and mitochondrial activity. BJU Int 2012;110: 863–7.

[69] Talebi AR, Sarcheshmeh AA, Khalili MA, Tabibnejad N. Effects of ethanol consumption on chromatin condensation and DNA integrity of epididymal spermatozoa in rat. Alcohol 2011;45: 403–9.

[70] Mendiola J, Torres Cantero AM, Vioque J, Moreno-Grau JM, Ten J, Roca M, et al. A low intake of antioxidant nutrients is associated with poor semen quality in patients attending fertility clinics. Fertil Steril 2010;93:1128–33.

[71] Lewis SE, Sterling ES, Young IS, Thompson W. Comparison of individual antioxidants of sperm and seminal plasma in fertile and infertile men. Fertil Steril 1997;67:142–7.

[72] Ko EY, Sabanegh ES. The role of nutraceuticals in male fertility. Urol Clin N Am 2014;41(1):181–93.

[73] Keskes-Ammar L, Feki-Chakroun N, Rebai T, Sahnoun Z, Ghozzi H, Hammami S, et al. Sperm oxidative stress and the effect of an oral vitamin E and selenium supplement on semen quality in infertile men. Arch Androl 2003;49:53.

[74] Scott R, MacPherson A, Yates RW, Hussain B, Dixon J. The effect of oral selenium supplementation on human sperm motility. Br J Urol 1998;82:76–80.

[75] Ross C, Morris A, Khairy M, Khalal Y, Braude P, Coomarasamy A, et al. A systematic review of the effect of oral antioxidants on male infertility. Reprod Biomed Online 2010;20:711–23.

[76] Lombardo F, Sansone A, Romanelli F, Paoli D, Gandini L, Lenzi A. The role of antioxidant therapy in the treatment of male infertility: an overview. Asian J Androl 2011;13:690–7.

[77] Lenzi A, Sgrò P, Salacone P, Paoli D, Gilio B, Lombardo F, et al. A placebo-controlled double-blind randomized trial of the use of combined L-carnitine and L-acetyl-carnitine treatment in men with asthenozoospermia. Fertil Steril 2004;81(6):1578–84.

[78] Lenzi A, Lombardo F, Sgrò P, Salacone P, Caponecchia L, Dondero F, et al. Use of carnitine therapy in selected cases of male factor infertility: a double-blind crossover trial. Fertil Steril 2003;79(2):292–300.

[79] Zhou X, Liu F, Zhai S. Effects of L-carnitine and/or L-acetyl-carnitine in nutrition treatment for male infertility: a systematic review. Asia Pac J Clin Nutr 2007;16:383–90.

[80] Sigman M, Glass S, Campagnone J, Pryor JL. Carnitine for the treatment of idiopathic asthenospermia: a randomized, double-blind, placebo-controlled trial. Fertil Steril 2006;85:1409–14.

[81] Thiele JJ, Friesleben HJ, Fuchs J, Ochsendorf FR. Ascorbic acid and urate in human seminal plasma: determination and inter-relationships with chemiluminescence in washed semen. Hum Reprod 1995;10(1):110–5.

[82] Dawson EB, Harris WA, Teter MC, Powell LC. Effect of ascorbic acid supplementation on the sperm quality of smokers. Fertil Steril 1992;58:1034–9.

[83] Abel BJ, Carswell G, Eiton R, Hargreave TB, Kyle K, Orr S, et al. Randomised trial of clomiphene citrate treatment and vitamin C for male infertility. Br J Urol 1982;54:780–4.

[84] Hargreave TB, Kyle KF, Baxby K, Rogers AC, Scott R, Tolley D, et al. Randomised trial of mesterolona versus vitamin C for male infertility. Scottish Infertility Group. Br J Urol 1984;56:740–4.

[85] Alpers DH, Stenson WF, Taylor BE, editors. Manual of nutritional therapeutics. 5th ed. Philadelphia: Lippincott Williams & Wilkins; 2008.

[86] Greco E, Iacobelli M, Rienzi L, Ubaldi F, Ferrero S, Tesarik J, et al. Reduction of the incidence of sperm DNA fragmentation by oral antioxidant treatment. J Androl 2005;26:349–53.

[87] Comhaire FH, Christophe AB, Zalata AA, Dhooge Ws, Mahmoud AM, Depuydt CE. The effects of combined conventional treatment, oral antioxidants and essential fatty acids on sperm biology in subfertile men. Prostaglandins Leukot Essent Fatty Acids 2000;63:159–65.

[88] Suleiman SA, Ali ME, Zaki ZM, el Mallik EM, Nasr MA. Lipid peroxidation and human sperm motility protective role of vitamin E. J Androl 1996;17:530–7.

[89] Rolf C, Cooper TG, Yeung CH, Nieschlag E. Antioxidant treatment of patients with asthenozoospermia or moderate oligoasthenozoospermia with high-dose vitamin C and vitamin E: a randomized, placebo-controlled, double-blind study. Hum Reprod 1999;14:1028–33.

[90] Kessopoulou E, Powers HJ, Sharma KK, Pearson MJ, Russell JM, Cooke ID, et al. A double-blind randomized placebo crossover controlled trial using the antioxidant vitamin E to treat reactive oxygen species associated male infertility. Fertil Steril 1995;64:825–31.

[91] Moilanen J, Hovatta O. Excretion of alpha-tocopherol into human seminal plasma after oral administration. Andrologia 1995;27:133–1336.

[92] Mora-Esteves C, Shin D. Nutrient supplementation: improving male fertility fourfold. Semin Reprod Med 2013;31:293–300.

[93] Prasad AS. Zinc in human health: effect of zinc on immune cells. Mol Med Biol 2006;20:3–18.

[94] Omu AE, Al Azemi MK, Kehinde EO, Anim JT, Oriowo MA, Mathew TC. Indications of the mechanisms involved in improved sperm parameters by zinc therapy. Med Princ Pract 2008;17:108–16.

[95] Wong WY, Merkus HM, Thomas CM, Menkveld R, Zielhuis GA, Steegers-Theunissen RP. Effects of folic acid and zinc sulfate on male factor subfertility: a double-blind, randomized, placebo-controlled trial. Fertil Steril 2002;77:491–8.

[96] Young SS, Eskenazi B, Marchetti FM, Block G, Wyrobek AJ. The association of folate, zinc and antioxidant intake with sperm aneuploidy in healthy non-smoking men. Hum Reprod 2008;23:1014–22.

[97] BehneD, Gessner H, Wolters G, Brotherton J. Selenium, rubidium and zinc in human semen and semen fractions. Int J Androl 1988;11(5):415–23.

[98] Vezina D, Mauffette F, Roberts KD, Bleau G. Selenium-vitamin E supplementation in infertile men. Effects in semen parameters and micronutrient levels and distribution. Biol Trace Elem Res 1996;53:65–83.

[99] Moslemi MK, Tavanbakhsh S. Selenium-vitamin E supplementation in infertile men: effects on semen parameters and pregnancy rate. Int J Gen Med 2011;4:99–104.

[100] Maiorino M, Coassin M, Roveri A, Ursini F. Microsomal lipid reoxidation: effect of vitamin E and its functional interaction with phospholipid hydroperoxide glutathione peroxidase. Lipids 1989;24(8):721–6.

[101] Keskes-Ammar L, Feki-chakroun N, Rebai T, Sahnoun Z, Ghozzi H, Hammami S, et al. Sperm oxidative stress and the effect of

an oral vitamin E and selenium supplement on semen quality in infertile men. Arch Androl 2003;49(2):84–94.

[102] Iwanier K, Zachara BA. Selenium supplementation enhances the element concentration in blood and seminal fluids but does not change the spermatozoal quality characteristics in subfertile men. J Androl 1995;16:441–7.

[103] Hawkes WC, Alkan Z, Wong K. Selenium supplementation does not affect testicular selenium status or semen quality in North American men. J Androl 2009;30:525–33.

[104] Morales A, Cavicchia JC. Spermatogenesis and blood–testis barrier in rats after long-term Vitamin A deprivation. Tissue Cell 2002;34(5):349–55.

[105] Gupta NP, Kumar R. Lycopene therapy in idiopathic male infertility – a preliminary report. Int Urol Nephrol 2002;34(3):369–72.

[106] Comhaire FH, El Garem Y, Mahmoud A, Eertmans F, Schoonjans F. Combined conventional/antioxidant "Astaxanthin" treatment for male infertility: a double blind, randomized trial. Asian J Androl 2005;7(3):257–62.

[107] Schibona M, Mseschini P, Capparelli S, Pecori C, Rossi P, Menchini Fabris GF. L-Arginine and male infertility. Minerva Urol Nefrol 1994;46(4):251–3.

[108] Pryor JP, Blandy JP, Evans P, Chaput De Saintonge DM, Usherwood M. Controlled clinical trial of arginine for infertile men with oligozoospermia. Br J Urol 1978;50:47–50.

[109] Srivastava S, Agarwal A. Effect of anion channel blockers on L-arginine action in spermatozoa from asthenospermic men. Andrologia 2010;42(2):76–82.

[110] Gavella M, Lipovac V. Pentoxifylline-mediated reduction of superoxide anion production by human spermatozoa. Andrologia 1992;24(1):37–9.

[111] Pang SC, ChanPJ, Lu A. Effects of pentoxifylline on sperm motility and hyperactivation in normozoospermic and normokinetic semen. Fertil Steril 1993;60(2):336–43.

[112] Okada H, Tatsumi N, Kanzaki M, Fujisawa M, Arakawa S, Kamidoro S. Formation of reactive oxygen species by spermatozoa from asthenospermic patients: response to treatment with pentoxifylline. J Urol 1997;157(6):2140–6.

[113] Lenzi A, Culasso F, Gandini L, Lombardo F, Dondero F. Placebo-controlled, double blind, cross-over trial of glutathione therapy in male infertility. Hum Reprod 1993;8:1657–62.

[114] Kodam H, Yamaguchi R, Fukuda J, Kasai H, Tanaka T. Increased oxidative deoxyribonucleic acid damage in the spermatozoa of infertile male patients. Fertil Steril 1997;68:519–24.

[115] Griveau JF, Le Lannou D. Effects of antioxidants on human sperm preparation techniques. Int J Androl 1994;17:225–31.

[116] Baker HW, Brindle J, Irvine DS, Aitken RJ. Protective effect of antioxidants on the impairment of sperm motility by activated polymorphonuclear leukocytes. Fertil Steril 1996;65:411–9.

[117] Ciftci H, Verit A, Savas M, Yeni E, Erel O. Effects of N-acetylcysteine on semen parameters and oxidative/antioxidant status. Urology 2009;74:73–6.

[118] Safarinejad MR, Safarinejad S. Efficacy of selenium and/or N-acetylcysteine for improving semen parameters in infertile men: a double-blind, placebo controlled, randomized study. J Urol 2009;181(2):741–51.

[119] Oeda T., Henkel R., Ohmori H., Schill WB. Scavenging effect of N-acetyl-L-cysteine against reactive oxygen species in human semen: a possible therapeutic modality for male factor infertility? Andrologia 1997;29:125–31.

[120] Kobayashi T, Miyazaki T, Natori M, Nozawa S. Protective role of superoxide dismutase in human sperm motility: superoxide dismutase activity and lipid peroxide in human seminal plasma and spermatozoa. Hum Reprod 1991;6(7):987–91.

[121] Aitken RJ, Buckingham D, Harkiss DJ. Use of a xanthine oxidase free radical generating system to investigate the cytotoxic effects of reactive oxygen species on human spermatozoa. Reprod Fertil 1993;97(2):441–50.

[122] Twigg J, Fulton N, Gomez E, Irvine DS, Aitken RJ. Analysis of the impact of intracellular reactive oxygen species generation on the structural and functional integrity of human spermatozoa: lipid peroxidation, DNA fragmentation and effectiveness of antioxidants. Hum Reprod 1998;13(6):1429–36.

[123] Mancini A, Balercia G. Coenzyme Q(10) in male infertility: physiopathology and therapy. Biofactors 2011;37(5):374–80.

[124] Safarinejad MR. Efficacy of coenzyme Q10 on semen parameters, sperm function and reproductive hormones in infertile men. J Urol 2009;182:237–48.

[125] Balercia G, Buldreghini E, Vignini A, Tiano L, Paggi F, Amoroso S, et al. Coenzyme Q10 treatment in infertile men with idiopathic asthenozoospermia: a placebo-controlled, double-blind randomized trial. Fertil Steril 2009;91(5):1785–92.

[126] Lafuente R, González-Comadrán M, Solà I, López G, Brassesco M, Carreras R, et al. Coenzyme Q10 and male infertility: a meta-analysis. J Assist Reprod Genet 2013;30(9):1147–56.

[127] Lenzi A, Gandini L, Maresca V, Rago R, Sgrò P, Dondero F, et al. Fatty acid composition of spermatozoa and immature germ cells. Mol Hum Reprod 2000;6(3):226–31.

[128] Weaver KL, Ivester P, Seeds M, Case LD, Arm JP, Chilton FH. Effect of dietary fatty acids on inflammatory gene expression in healthy humans. J Biol Chem 2009;284(23):15400–7.

[129] Conquer JA, Martin JB, Tummon I, Watson L, Tekpetey F. Fatty acid analysis of blood serum, seminal plasma, and spermatozoa of normozoospermic vs. asthenozoospermic males. Lipids 1999;34(8):793–9.

[130] Safarinejad MR, Hosseini SY, Dadkhah F, Asgari MA. Relationship of omega-3 and omega-6 fatty acids with semen characteristics, and anti-oxidant status of seminal plasma: a comparison between fertile and infertile men. Clin Nutr 2010;29(1):100–5.

[131] Attaman JA, Toth TL, Furtado J, Campos H, Hauser R, Chavarro JE. Dietary fat and semen quality among men attending a fertility clinic. Hum Reprod 2012;27(5):1466–74.

[132] Safarinejad MR. Effect of omega-3 polyunsaturated fatty acid supplementation on semen profile and enzymatic anti-oxidant capacity of seminal plasma in infertile men with idiopathic oligoasthenoteratospermia: a double-blind, placebo-controlled, randomised study. Andrologia 2011;43(1):38–47.

[133] Landau B, Singer R, Klein T, Segenreich E. Folic acid levels in blood and seminal plasma of normo- and oligospermic patients prior and following folic acid treatment. Experientia 1978;34(10):1301–2.

[134] Lange H1, Suryapranata H, De Luca G, Börner C, Dille J, Kallmayer K, et al. Folate therapy and in-stent restenosis after coronary stenting. N Engl J Med 2004;350(26):2673–81.

[135] Song G, Kochman L, Andolina E, Herko RC, Brewer KJ, Lewis V. Beneficial effects of dietary intake of plant phytoestrogens on semen parameters and sperm DNA integrity in infertile men. Fertil Steril 2006;86:S49.

[136] Chavarro JE, Toth TL, Sadio SM, Hauser R. Soy food and isoflavone intake in relation to semen quality parameters among men from an infertility clinic. Hum Reprod 2008;23(11):2584–90.

[137] Hamilton-Reeves JM, Vazquez G, Duval SJ, Phipps WR, Kurzer MS, Messina MJ. Clinical studies show no effects of soy protein or isoflavones on reproductive hormones in men: results of a meta-analysis. Fertil Steril 2010;94(3):997–1007.

[138] Abad C, Amengual MJ, Gosálvez J, Coward K, Hannaoui N, Benet J, et al. Effects of oral antioxidant treatment upon the dynamics of human sperm DNA fragmentation and subpopulations of sperm with highly degraded DNA. Andrologia 2013;45(3):211–6.

[139] Greco E, Romano S, Iacobelli M, Ferrero S, Baroni E, Minasi MG, et al. ICSI in cases of sperm DNA damage: beneficial effect of oral antioxidant treatment. Hum Reprod 2005;20(9):2590–4.

[140] Misell LM, Holochwost D, Boban D, Santi N, Shefi S, Hellerstein MK, et al. A stable isotope-mass spectrometric method for measuring human spermatogenesis kinetics *in vivo*. J Urol 2006;175(1):242–6.

[141] Menezo Y, Evenson D, Cohen M, Dale B. Effect of antioxidants on sperm genetic damage. Adv Exp Med Biol 2014;791:173–89.

[142] Showell MG, Brown J, Yazdani A, Stankiewicz MT, Hart RJ. Antioxidants for male subfertility. Cochrane Database Syst Rev 2011;Jan(1). CD007411.

36

Epigenetics and its Role in Male Infertility

Eva Tvrda, PhD,**, Jaime Gosalvez, PhD[†], Ashok Agarwal, PhD**

**Center for Reproductive Medicine, Cleveland Clinic, Cleveland, Ohio, USA*
***Department of Animal Physiology, Slovak University of Agriculture, Nitra, Slovakia*
[†]Universidad Autonoma de Madrid, Madrid, Spain

INTRODUCTION

The innovation that stemmed from the Human Genome Project has made it possible to investigate the protein coding genes, as well as the intergenic regulatory sequences. Additionally, the recent completion of the human protesome has added to a wealth of knowledge about the complex functionality of proteins. Genes are therefore a static and permanent imprint of information deciding upon an individual's phenotype [1,2].

However, the original belief that inheritance via the genetic route (transfer of DNA sequence from a parent to the offspring) is the exclusive way of transgenerational transmission of information is changing, as a variety of theories focused on a nongenetic mode of transfer of information is emerging [3].

Epigenetics is defined as the study of heritable changes, not related to actual alterations in the DNA sequence. As opposed to traditional genetics, epigenetic variations in gene expression or phenotyping are not based on the actual changes within the DNA sequence. The term is also related to the functionally relevant changes within the genome independent of changes in the nucleotide sequence [4].

The concept of an epigenome was introduced and defined by Waddington, to explain the complexity of developmental mechanisms that cannot be explained solely by genotype and phenotype. According to his landscape theory, the sperm and egg are highly specialized cells. Once these germ cells fuse, they give rise to what Waddington considers the very top of his landscape theory – a very unspecialized and totipotent zygote that has the ability to differentiate into all tissues of the organism. A process known as epigenetic reprogramming can actually return the highly differentiated sperm and egg nuclei to this nascent state in the early embryo. Furthermore, Waddington explains that differences in cells and tissues derive from changes in gene expression as cells differentiate. Waddington describes the route of a cell toward its final differentiation as a ball traveling downward along branching valleys; once it has entered its final valley, it is not able to easily cross the mountain into a neighboring valley or return back to the beginning [5,6].

EPIGENETICS OF THE MALE GAMETE

The paternal contribution to the growing embryo represents an essential role in our comprehension of early developmental processes as well as the impact on the health state of a newborn. Spermatozoa are highly organized, differentiated, and specialized cells, whose characteristics are intimately connected to specific structural and functional integrity. While in the past, spermatozoa have been regarded as simple vectors carrying the paternal genetic information, the importance of sperm contribution to embryogenesis has recently been revived, with substantial evidence pointing to the active participation of spermatozoa structures in early development [7,8].

Sperm DNA is structurally very specific when compared to a somatic cell. The major portion of sperm DNA is coiled into highly condensed toroids due to the incorporation of protamines. A smaller amount of DNA is bound to histones in a much looser form, and the remaining DNA is attached to the sperm nuclear matrix within matrix attachment regions (MARs) [9].

During fertilization, the sperm transmits more than nuclear DNA to the egg: the oocyte activation factor (essential for fertilization), centrosomes (crucial for cell division) [10], and messenger RNA [7,11]. Studies focused

Handbook of Fertility. http://dx.doi.org/10.1016/B978-0-12-800872-0.00036-6

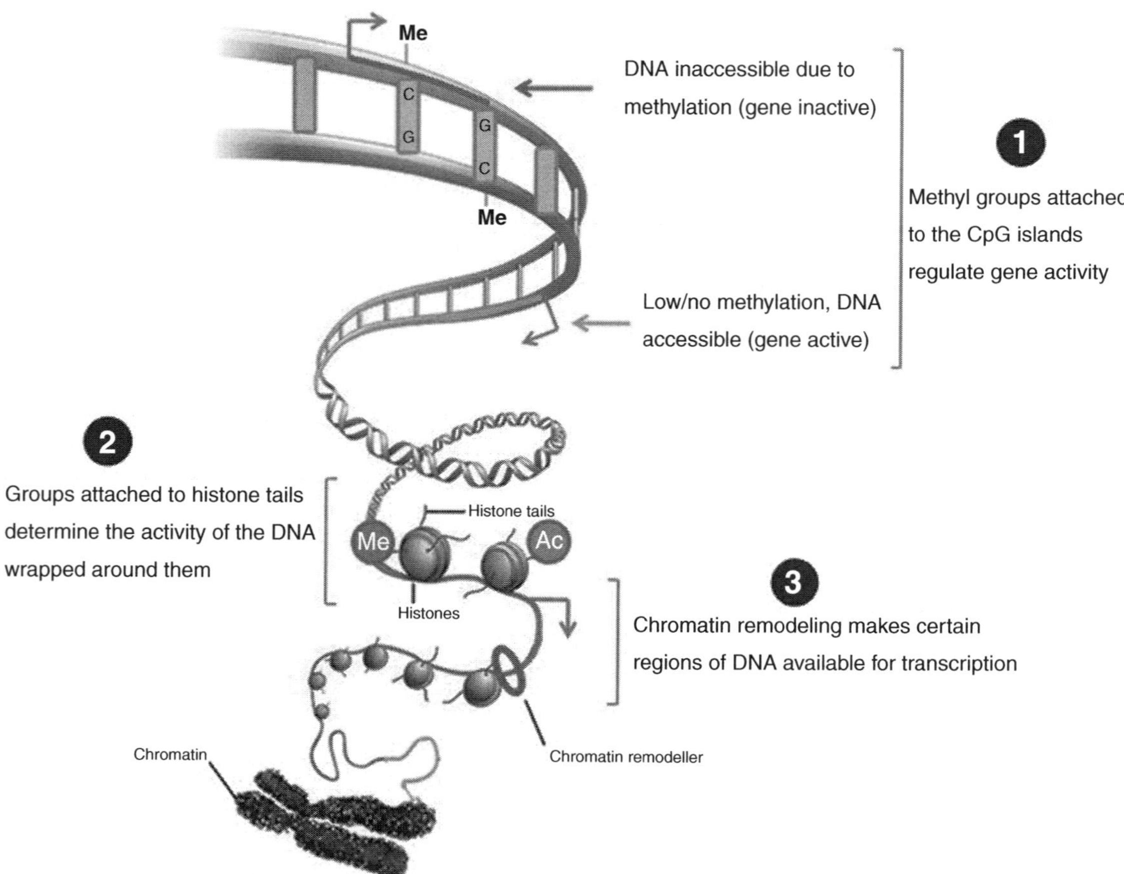

FIGURE 36.1   Essential epigenetic changes occurring in the sperm cell. *Reprinted with permission from Elsevier [2].*

on the epigenetic modifications within the developing spermatozoan have provided new insights that are of critical significance to the sperm epigenome in the developing embryo. These epigenetic modifications include DNA methylation, histone tail modifications, histone retention, and protamine incorporation into the chromatin. All of these adjustments are critical to protect the DNA molecule throughout spermatogenesis along with a defined contribution through the sperm, to the development of the prospect embryo (Fig. 36.1) [11].

DNA METHYLATION

DNA methylation is a powerful epigenetic mechanism linked to genetic silencing, imprinting, and inactivation of the X chromosome [12]. This process is based on the addition of a methyl group from *S*-adenosyl-methionine to the fifth position of the cytosine ring (5meC) found in the CpG dinucleotides. It has been suggested that 3–5% of the cytosine residues within mammalian DNA contain 5meC [13]. These islands are usually 500 bp long [14] and are systematically distributed throughout the genome. Strong CpG promoters are usually unmethylated; however, CpG islands directly located in the promoter

or in the regulatory regions of transposable elements are methylated. This specific modification helps to prevent the replication of parasitic transposons and other repetitive elements. At the same time, CpG-poor promoters are generally found to be hypermethylated in somatic cells [14].

The basic role of DNA methylation is to decrease or completely block the access of polymerases to hypermethylated promoter regions within specific genes. Inversely, the lack of methylation marks (hypomethylation) effectively places the gene in an inducible state for activation and subsequent transcription [15].

DNA methylation patterns within regulatory regions are defined as stable; however, dynamic *de novo* methylation of regulatory regions is possible in order to prevent reexpression [16]. Furthermore, it has been hypothesized that DNA methylation may occur in embryonic cells at cytosine residues in CpA or CpT dinucleotides, although the concrete functions and mechanisms of non-CpG methylation in sperm function remain unclear [17].

This regulatory mechanism relies on the DNA methyltransferase (DNMT) protein family [18], which includes both catalytic members responsible for the deposition of new methylation marks, as well as proteins responsible for the maintenance of these marks. DNMT3A,

DNMT3B, and DNMT1 work in harmony in order to mediate *de novo* methylation within the mammalian sperm, with DNMT3A and DNMT3B being responsible for the actual catalysis of the reaction. DNMT1 lacks a catalytic domain, yet this enzyme is crucial for the methylation establishment, as it facilitates the activities of DNMT3A and DNMT3B and coordinates the correct placement of the methylation marks [19]. DNMT1 is the predominant methyltransferase in somatic tissues. Even though this protein lacks catalytic capabilities, it is vital to maintain the previously determined methylation patterns during DNA replication [20]. Although, it is possible for the actual fertilization to occur without the methylation patterns, the resulting embryo is unable to develop properly [21].

Methylation in male germ cells is a unique process crucial for proper spermatogenesis [17]. It was shown that the demethylation–remethylation process during spermatogenesis involves massive erasure of somatic-like patterns of DNA methylation followed by *de novo* creation of sex-specific patterns [22]. In a developing embryo, a second round of demethylation occurs between 8 and 14 days *post coitum*, erasing all methylation markers [23,24]. Subsequently, specific remethylation begins at around 15 days *post coitum* [25,26]. This specific erasure process is intimately linked to parental imprinting, enabling monoallelic and specific expression of genes [27]. Most of the methylation occurs during fetal life, although complete DNA methylation is reached around the pachytene spermatocyte stage, taking place postnatally, however, prior to meiosis [28]. Only four imprinted loci are methylated in the male germ line: Igf2/H19, Rasgrf1, Dlk1-Gtl2, and Zdbf2 [29].

Contrary to other mammals, a complete explanation about the precise kinetics of erasure and remethylation in humans is still missing. Although it has been confirmed that the methylation markers are acquired before meiosis, it is still questionable if the process takes place during fetal life, in the perinatal period, and/or at the onset of puberty [30].

CHROMATIN REORGANIZATION

Mammalian fertilization is a very complex and demanding process during which spermatozoa are required to exhibit a variety of activities eventually leading to the fusion with the oocyte [31]. To accomplish all required steps related to successful fertilization, the spermatozoan has a distinctive architecture, largely due to the dramatic changes within the chromatin structure based on the replacement of the majority (90–95%) of histones with protamines [32]. Spermatozoa undergo this specific and extensive chromatin remodeling soon after their speciation and primarily during the differentiation

process in order to become mature and functionally active [22].

Chromatin of a mature spermatozoan is a very firm and compact structure that results from substantial rearrangement during spermatogenesis [22]. Protamination of the sperm chromatin enables proper nuclear condensation, which is essential for sperm motility and protects the nuclear genome against oxidative damage or harmful effects potentially caused by molecules from the female reproductive tract [32]. At the same time, high DNA packaging following protamination serves to inhibit any transcriptional activity, making this a specialized epigenetic mechanism to the spermatozoa.

Protamination is considered a stepwise process. Early exchange events include the replacement of target histones with specific histone variants that are expressed solely during spermatogenesis, such as the testis-specific histone 2B, the most abundant histone variant found in structurally and functionally mature spermatozoan [33]. Along with histone replacement, an increased acetylation of other histones has been often recorded. Hyperacetylation is considered to be the starting point and ultimately several additional events lead to the final protamine replacement [34]. Hyperacetylation is regulated by interactions with acetylases as well as deacetylases leading to a "relaxed" chromatin structure, which allows topoisomerase-induced strand breaks, subsequently followed by the removal of histones and their replacement with transition proteins [34,35].

The transition proteins 1 and 2 (TP1 and TP2) are molecules of intermediate basic nature present in Stages 12 and 13 of spermatogenesis, and eventually removed at Stage 14. When bound to DNA, TP1 and TP2 enable histone removal, and appear to be crucial for future protamine transition [36]. The absence of either of the proteins presents with a high risk for infertility, although several animal models have been shown to develop a partial compensation from the remaining TP [37]. In the following step, the transition proteins are completely replaced by protamines [38]. Sometimes an overlap is noted in the genetic expression for both types of proteins, although both are unique within their timing and function [39]. Phosphorylation is the principal posttranslational modification, considered crucial for proper chromatin condensation [40]. Successful protamine replacement results in tight chromatin packaging, essential for a functionally compact sperm nucleus. This tight structure comes from disulfide bonds established between the protamines [41].

It is crucial to note that 5–10% of histones, including the testis-specific histone variants along with canonical histones, will not be replaced [42]. Their retention was originally thought to result from an inefficient machinery or a different regulatory mechanism, indicative of infertility, although their presence was later confirmed to be present in fertile men too. Thus, it has been suggested that nucleosome-bound DNA is retained at specific coding regions in a demethylated state to enable transcription and recruitment of signaling factors required for early embryogenesis and to ensure proper delivery of the parental genome to the egg as well [43]. Expectedly, abnormal protamine replacement of histones is connected to decreased semen and *in vitro* fertilization (IVF) embryo quality [44,45].

HISTONE MODIFICATIONS

DNA protamination is the defining epigenetic feature, specific to only the spermatozoan [42]. Multiple sperm histone variants including the H2A and H2B, H3, H4, together with the testis-specific variant tH2B, are essential for proper spermatozoa production, structure, and function [46,47].

Modifications at histone tails are essential for driving epigenetic changes. The histones that persisted through the protamination process are presumed to carry out regulatory functions. Specifically, modifications at the histone tails on lysine or serine residues facilitate gene activation or silencing [48]. The main histone modifications in spermatozoa include methylation, acetylation, ubiquitination, and phosphorylation which, either alone or in concert, can ensure activation or suppression of a specific gene or promoter. H3K4 methylation, H3 and H4 acetylation, and H2B ubiquitination are responsible for gene activation, while H3K9 and H3K27 methylation, deacetylation at H3 and H4, and H2A ubiquitination enrichment lead to gene silencing [49,50]. These modifications are, in turn, regulated by a variety of enzymes. The histone methyltransferases and demethylases catalyze methylation and demethylation, while acetylation and deacetylation are regulated by histone acetyltransferases and deacetylases, respectively [51]. These histone modifications function to relax the chromatin structure and to ensure chromatin accessibility to transcription factors [52]. Acetylation of H3 and H4 lysine residues is high in male germ cells, but is completely removed during meiosis, while the reacetylation of H4K in elongating spermatids is a crucial prerequisite for histone-to-protamine exchange [51].

Depending on the context, methylation of H3 or H4 lysine residues can promote gene activation or repression. H3K4 methylation is linked with gene expression, whereas H3K9 and H3K27 methylation are related to gene silencing and heterochromatin. The synchronization of the establishment and removal of methylation markers is vital to spermatogenesis: mono-, di-, and trimethylation of H3K4, H3K9, and H3K27 are essential for proper progression through spermatogenesis [49,50].

Recently, a novel histone modification called crotonylation was discovered. This process is based on the

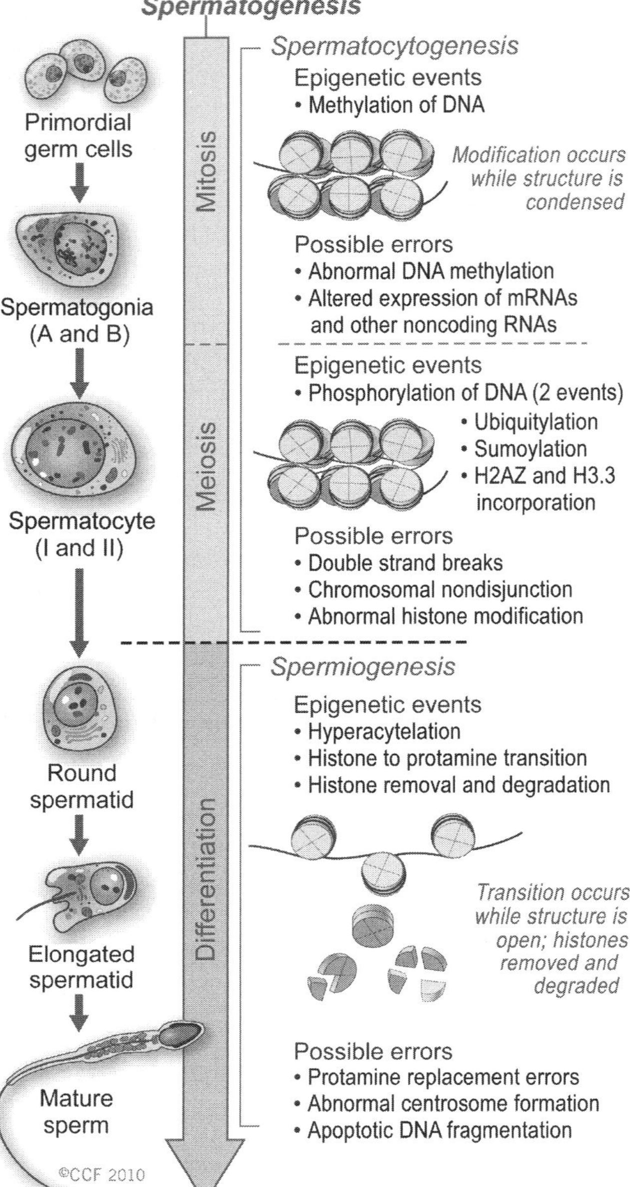

FIGURE 36.2   Epigenetic events taking place during different steps of spermatogenesis. *Reprinted with permission, Cleveland Clinic Center for Medical Art & Photography © 2010-2014. All Rights Reserved.*

addition of a crotonyl group (C_4H_5O) on the lysine residue of all core histones [53] and marks sex chromosomes and gonosomes in order to transmit genetic resistance to transcription repressors (Fig. 36.2) [54].

SPERM RNAs

The spermatozoa nucleus has been previously defined as inert based on the extraordinarily compact structure of the chromatin. Nevertheless, the spermatozoan does contain a variety of specific RNAs, mRNAs, miRNAs, and piwi-interacting RNAs (piRNAs) that might be useful to embryo development by modifying gene expression following fertilization [55]. Furthermore, it has been proposed that spermatozoal RNA is capable of preserving nucleosome-bound DNA and prevents protamination in specific chromatin regions [56]. Specific mRNAs have been detected in mature human spermatozoa, some of them encoding protamines and hormonal receptors [57]. Furthermore, abnormally high protamine mRNA retention in the ejaculated sperm is associated with alterations in the protamine translation, protamine deficiency in sperm, and infertility [44]. At the same time, it has been proposed that the sperm RNAs could contribute to the nuclear envelope stabilization, and the interactions between DNA and histones during the protamine transition as well. Furthermore, they could play a significant role in marking the DNA sequences that stay bound to histones [56]. It is interesting to note that different mRNA patterns have been detected between fertile and infertile men, which were associated with assisted reproduction techniques (ART)-related pregnancy outcomes in ART [57,58]. The *IGF-2* gene, crucial for embryonic development, provides an excellent example for RNA involvement in the maintenance of nucleosome-retained regions within chromatin, as this gene and its promoter are located close to the nuclear envelope, a major region of histone retention [58,59].

MALE INFERTILITY

According to the World Health Organization (WHO), infertility is defined as the inability of a sexually active couple to achieve pregnancy in one year, in the absence of contraception. Generally speaking, around 25% of couples do not achieve successful pregnancy within 1 year, of which 15% seek medical treatment. Male causes for infertility are found in 50% of the cases [60].

Genetic Causes of Infertility

Genetic causes of male infertility may be of different origins and etiologies. Chromosome abnormalities can be numerical, such as trisomy, or structural, including inversions or translocations. According to Johnson [61], the incidence of chromosomal abnormalities in infertile men is estimated to be around 5.8%, with 4.2% related to the sex chromosome and 1.5% linked to autosomal abnormalities.

Klinefelter's syndrome and variants (47,XXY; 46,XY; 47,XXY mosaicism) are the most frequent sex chromosome abnormalities [62], while the Kallmann's syndrome is a common X-linked disorder linked to infertility. Adult men with Klinefelter's syndrome have small firm testicles lacking germ cells and an impaired Leydig cell function. Patients with Kallmann's syndrome are diagnosed with hypogonadotrophic hypogonadism.

Although it is often possible to stimulate spermatogenesis with replacement therapy [63], the inheritance risk is high in both diseases.

Alterations linked to the Y chromosome may be considered the flagship of genetics-related male infertility. Although it has been shown that Y chromosome microdeletions naturally occur in fertile men, their prevalence is much higher in infertile men [64]. Microdeletions have been found in three nonoverlapping regions within the Y chromosome, AZF a,b,c [65]. A fourth region, AZFd, overlaps with AZFc, but is defined as a separate area [64]. The most frequently recorded Y chromosome-associated microdeletion is located in the AZFc region, encompassing the *DAZ* gene [41]. Although poor correlations have been found between the AZFc or the *DAZ* gene deletions and spermatogenesis, different subjects with similar microdeletions may have different degrees of alterations to the sperm cell production or maturation [64–66]. AZFa and AZFb microdeletions are rare; however, if the AZFa microdeletion is large enough to have both the USP9Y (DFFRY) and the DDX3Y (DBY) genes deleted, azoospermia is a traditional consequence [67]. Azoospermia is also frequently observed in men with larger AZFb Y microdeletions [68].

The majority of microdeletions are defined as a subset of rearrangements within the long arm of the Y chromosome; others may include duplications and inversions. Their exact biological significance has yet to be determined [65], but it is possible that some of them are associated with reduced male fertility [69].

Epigenetic Factors and Infertility

As in other complex diseases, epigenetic alterations may be an important factor contributing to inducing male fertility. A variety of studies have shown that epigenetic modifications within the spermatozoa stabilize the paternal genome and its contribution to early embryo development.

Oligozoospermia has been linked to altered DNA methylation or improper histone to protamine replacement [42–44]. These observations subsequently led scientists to question if methylation defects of imprinted or nonimprinted genes, as well as other epigenetic abnormalities (histone localization or modifications in mature sperm), could play an important role in the development and growth of ART offspring [70]. Given the importance of epigenetic factors in sperm function, it could be expected that alterations in epigenetic patterns of infertile men would provide answers to an increased risk for IVF pregnancy-related complications [71]. This necessitates an in-depth understanding of the impact of epigenetics on early gametes in order to estimate accurately the factors responsible for a constantly diminishing reproduction potential in the general population [72].

Studies linking altered DNA methylation patterns in paternal imprinted loci and severe oligozoospermia have been recently observed [70,71]. These reports hypothesized that improper methylation may extend beyond severely oligozoospermic patients to men with relatively normal sperm concentration in the presence of abnormal chromatin packaging.

Furthermore, Hammoud et al. [73] reported that the methylation patterns actually varied between the specific etiologies of infertility. Oligozoospermic patients were characterized by predominant hypermethylation at the mesoderm specific transcript (*MEST*) gene, an imprinted gene associated with the Silver–Russell Syndrome (SRS) – a growth disorder linked to intrauterine growth restriction. Patients exhibiting abnormal protamine ratios have significant changes at the *LIT1* and small nuclear ribonucleoprotein polypeptide N (*SNRPN*) genes, which have been associated with neonatal diabetes mellitus. This study concluded that the risk of transmitting epigenetic alterations vary depending on the classification of infertility.

Houshdaran et al. [74] observed that poor semen quality was associated with extensive DNA hypermethylation at various unique regions, as well as at several repetitive elements. This excessive methylation was suggested to be a consequence of deficient erasures of previously established methylation marks rather than aberrations in *de novo* postreprogramming methylation.

Moreover, altered methylation has been shown to have a significant impact on the expression patterns of selective paternal loci in several groups of patients suffering from infertility. The most commonly altered loci included *PLAG1*, *DIRAS3*, and *MEST*. Similarly, Boissonnas et al. [75] found that many subjects diagnosed with teratozoospermia and oligoasthenoteratozoospermia showed hypomethylation at the CpG islands located within the H19 locus. Inversely, hypermethylation of *MEST* locus was observed in patients diagnosed with idiopathic infertility, where they exhibited low spermatozoa motility and morphology.

The exchange of protamines to histones is another essential epigenetic step in the process of spermatozoa production. The P1/P2 ratio representing the balance between Protamine 1 and Protamine 2 has been defined as 0.8–1.2 during normal spermatogenesis. Any deviation will most likely lead to infertility as it adversely affects general spermatozoa quality and DNA integrity. Subjects diagnosed with an altered P1/P2 ratio will subsequently present with decreased spermatozoa concentration, motility, morphology, as well as decreased fertilization capacity. Last, insufficient DNA packaging and condensation have been associated with infertility with variable phenotypic presentations in different male animal models [22].

EPIGENETICS AND TESTICULAR CANCER

Alterations in major epigenetic pathways effective in the male germline have been reported to be associated with a variety of transgenerational adult onset diseases, including testicular cancer [76].

Testicular cancer (TC) is a relatively rare solid tumor accounting for 1% of cancers diagnosed in men. However, it is the prevalent cancer type affecting men in their reproductive age (20–40 years). The most common type of TC are germ cell tumors (TGCTs), which represent 95% of this malignancy [77].

Potential involvement of epigenetics in TGCT development has been shown in numerous studies reporting modified epigenetic profiles of cancer cells as opposed to the "normal" ones. At the same time, malignant cells and spermatogonia have similar epigenetic patterns, possibly related to high metabolic and proliferative rate [78].

The essential role of epigenetic alterations in carcinogenesis has been documented through the repression of tumor suppressor genes and has been recognized as a major mechanism during TGCT formation and progression [79]. Furthermore, DNA methylation patterns seem to be highly associated with histological features of a specific TGCT type. Undifferentiated TGCTs, such as seminomas or gonadoblastomas, have been found to be hypomethylated, while differentiated TGCTs, including teratomas and choriocarcinomas, have shown hypermethylated profiles. It may be suggested that such epigenetic parameters could be used as diagnostic markers to discriminate between seminomas and nonseminomas [78].

Consistent with the cellular origin of TGCT, the DNMT enzymes were expressed differently in various cancer types when compared to normal cells. It is known that these enzymes are primarily expressed in fetal testis and later in spermatogonia during spermatogenesis. On analyzing different cancer types, it was found that DNMT1 was not expressed in seminoma, but overexpressed in embryonal carcinoma [80]. Inversely, DNMT3A was found to be upregulated in TGCTs compared to nontumor testicular tissues [81]. DNMT3B is seen to be overexpressed in relapsed Stage I seminomas [82] and DNMT3L was overexpressed in nonseminoma malignancies [83].

Histone methylation has been implicated in the deregulation of gene expression in cancer. Histone modifications within chromatin have been associated with high transcriptional and proliferative activities of the cell. High-levels of histone methylation, especially in the case of EZH2, are typical for a normal testicular cell development. Inverse correlations between the expression levels of histone methyltransferases and the severity of spermatogenic failure link to their potential value as molecular markers for spermatogenic defects [84].

Furthermore, it has been suggested that H2A and H4A3 dimethylation could be the key mechanisms by which seminomas maintain their undifferentiated state, while a decrease or complete absence of these modifications could be involved in the somatic differentiation reported in nonseminomas. Finally, the H4 hyperacetylation during spermatogenesis [85] plays a critical role for the removal of histones and their replacement by protamines, leading to nuclear condensation of the spermatozoan. Choriocarcinomas have shown high expression patterns for all three classes of histone deacetylase isoforms [86].

NUTRITION AND EPIGENETICS

Extensive data from human epidemiologic studies as well as animal experiments have shown that prenatal and early postnatal nutrition have an impact on susceptibility to a variety of diet-related chronic diseases including cancer, obesity or diabetes, at a later time in life [87,88].

Focusing on male fertility, numerous studies have reported on the associations between dietary nutrients and DNA methylation. It has been shown that dietary components may serve to regulate the availability of methyl donors for the subsequent methylation reactions. The B vitamin family serves as a coenzyme for one carbon (C1) reaction network, where a C1 unit is received from a methyl donor and subsequently used in the methylation of biomolecules. At the same time, agents affecting the C1 metabolism have an impact on S-adenosyl methionine, which is the primary donor for the DNA methylation [89].

The genome of the preimplantation embryo undergoes widespread demethylation, followed by a postimplantation cytosine remethylation [90]. These patterns must be conserved over numerous rounds of cellular proliferation during embryogenesis. Availability of dietary methyl donors and cofactors during gestation might therefore have an impact on DNA methylation patterns [91]. Thus, methyl donor over- or under-nutrition might lead to an elevated susceptibility to decreased fertility in later life.

A variety of clinical studies have reported positive associations between folate intake and fertility [92]. Furthermore, it has been shown that mutations in the 5,10-methylenetetrahydrofolate reductase (MTHFR), a key enzyme of the folate metabolism, have led to infertility [93]. Thus, the availability of folate will most likely alter the extent of DNA methylation in spermatogenesis, with a consequent impact on the sperm epigenome and possible pregnancy outcomes.

Waterland and Jirtle focused on the impact of nutrients on epigenetic profiles following cytosine methylation, taking place on both strands of palindromic CpG dinucleotide islands. The availability of nutrients essential for CpG methylation affects transcription either directly by controlling the binding of methyl-sensitive DNA-binding proteins, or indirectly by affecting regional chromatin conformation [87]. Specific profiles of CpG methylation are created in early development and expanded during DNA replication by DNMT1 [90]. Early nutrition may have an impact on the establishment and maintenance of cytosine methylation, and it has been demonstrated that the transposons [92] and genomically imprinted genes [93,94] are the two genomic elements most susceptible to nutritional dysregulation.

THE IMPACT OF ENVIRONMENT ON EPIGENETICS

Numerous possible causes for epigenetic alterations in sperm and embryogenesis currently exist, the most prominent of them being environmental toxins. Although many studies have evaluated different environmental agents and their possible effects on general male fertility, only few have focused on specific epigenetic alterations that may be present as a consequence of prior environmental exposures [15].

Epigenetic alterations have been demonstrated with heavy metals [95,96], polycyclic aromatic hydrocarbons [97], drugs, such as cocaine [98], endocrine disruptors [99,100], phytoestrogens [101], and fungicides or pesticides [102]. Recent data provide convincing evidence on the effects of selected lead and arsenic on sperm nuclear proteins, especially protamine 2. Quintanilla-Vega et al. [95] reported that heavy metals have the ability to inhibit protamine 2 binding to DNA, leading to abnormal chromatin compaction. Yauk et al. [103] noted that mice exposed to particulate ambient air pollution had DNA hypermethylation compared to controls housed with high-efficiency air particulate (HEPA) filtered ambient air.

Endocrine-disrupting agents are however considered to be the prime environmental contaminants with a pronounced impact on spermatozoa epigenetic patterns and subsequently male infertility. McLachlan [104] was the first to propose that exposure to environmental endocrine-disrupting chemicals during early development significantly affects adults, potentially involving gene imprinting and leading to persistent alterations of DNA methylation. Endocrine disruptors have the ability to alter the DNA methylation patterns of key genes in charge of significant transcriptional changes [11,101,105]. Administration of the endocrine-disrupting phytoestrogens, coumestrol, and equol, to newborn mice upregulates DNA methylation that functions to inactivate the proto-oncogene H-ras [106]. DNA methylation patterns are further modified in young mice supplemented with high doses of the phytoestrogen genistein [107]. Environmental compounds with endocrine-disrupting activity studied for possible epigenetic effects include the fungicide vinclozolin, plasticizer Bisphenol (BPA), and the pharmacological molecule Diethylstilbestrol (DES) [106]. Exposure to environmentally relevant doses of BPA during the neonatal development in rats caused DNA methylation changes associated with carcinogenesis [99,100]. Changes in methylation also explain the increased susceptibility for tumor formation in the F2 generation after developmental exposure to DES [106].

Furthermore, it is important to note the emerging role of environmental compounds on epigenetic mechanisms from an evolutionary and ecological perspective [108].

Based on the available data, it may be concluded that the environmental toxins must be targeted in order to analyze the correct etiology of related epigenetic alterations in sperm and to aid in improving diagnostic and treatment options in men who present with infertility [15].

EPIGENETICS AND ART

The majority of genes are expressed equally from maternal and paternal alleles; however, imprinted genes are expressed exclusively from only one allele in a parent-of-origin dependent manner. Only a few imprinted genes have been defined in the male germline, three of which have been described in detail: *IGF2/H19*, Rasgrf1126, and Gtl2. Genomic imprinting may be defined as a temporal balance between DNA demethylation, establishment of epigenetic marks, fertilization, and embryogenesis. Therefore, misregulation of one or more of these processes is associated with pathologies related to infertility or malformations during early embryogenesis [109].

In vitro manipulation of gametes and/or early embryos during ART might alter a normal distribution of epigenetic marks before, during, or after fertilization, resulting in imprinting errors of variable severity [109]. This theory has been proven by several studies reporting a link between ART and the loss of imprinting, which results in epigenetic alteration of specific genes clustered together with the loss of parent-of-origin specific expression of imprinted genes [109]. It is therefore of crucial importance to recognize and understand the epigenetic mechanisms that take place during every step of the ART procedure, including *in vitro* growth and maturation of oocytes [110], use of immature spermatids [111], specific composition of each cell culture media [112], and blastocyst culture [113].

Current evidence from studies in spermatozoa suggests that male infertility may contribute to epigenetic alterations in pregnancies through ART. Aberrant DNA methylation of several imprinting genes associated with male infertility might be related to imprinting disorders taking place after ART. Aberrant methylation has been reported in case of *MEST* and ICR1 of the *IGF2/H19* gene in subfertile men, and it has been hypothesized that male germ cells could be a potential source of aberrant epigenetic features in children conceived via ART. Alterations of several imprinted genes including *LIT1*, *MEST*, *SNRPN*, *H19*, and *IGF2* have been reported in infertile subjects diagnosed with an abnormal histone-to-protamine replacement [109,114].

However, the differences in the statistical significance between the infertile groups suggest that the risk of transmission of epigenetic alterations may actually differ depending on the underlying cause of male infertility. Furthermore, the methylation imprinting patterns of two oppositely imprinted genes, *H19* and *MEST/PEG1*, were assessed in testicular spermatozoa from azoospermic patients. It was found that spermatozoa from subjects with abnormal spermatogenesis display methylation defects in the *H19* imprinted gene, affecting the CCCTC-binding factor-binding site, further confirming an association between the occurrence of imprinting errors and disruptive spermatogenesis [42,73].

Studies have also shown that epigenetic alterations are associated with vitrification, a technique based on an instant cryopreservation of gametes or embryos. Wang et al. [115] showed that vitrification aggravated the loss of methylation in the *H19/Igf2* DMD (differentially methylated domain originally compensating for the perturbed expression of *H19* caused by IVF procedures). The impact of vitrification and warming cycles on the histone modifications was evaluated in a mouse model, which showed that histone 3 lysine 9 (H3K9) methylation and histone 4 lysine 5 (H4K5) acetylation were significantly increased [116].

CURRENT TECHNIQUES FOR THE ASSESSMENT OF SPERMATOZOA EPIGENETICS

Spermatozoa have a variety of unique features that complicate experimental handling and treatment, including a small size and a robust and resistant surface. The goal of molecular andrology is therefore to establish effective techniques to enable more effective and accurate research on sperm epigenetics [109].

Isolation of chromatin, DNA, and RNA from a single spermatozoan is the most critical step. Unlike somatic cells, spermatozoa are too robust to be lysed by proteinase K; therefore, cell lysis must be carried out with mechanic or chemical treatment [117].

DNA Methylation Analysis

Currently, three protocols are available for the investigation of DNA methylation:

- Sodium bisulfite treatment of denatured DNA,
- Restriction enzymatic digestion of genomic DNA,
- Enrichment of methylated DNA using affinity matrices, antibodies, or methylated DNA-binding proteins.

A variety of downstream analytical and enzymatic methods exist, based on specific aims of the experiment. Sodium bisulfite causes conversion of unmethylated cytosines, but not methylated cytosines into uracil residues. The resulting difference between methylated and unmethylated CpGs can be easily detected by PCR.

If genome-wide CpG methylation assessment is required, DNA has to be pretreated with enzyme digestion or sonication, followed by antibodies or proteins binding methylated cytosines. The patterns are quantified using microarrays [109].

DNA–Histone Binding Analysis

A typical method to analyze DNA–histone binding is through chromatin immunoprecipitation (ChIP), which highlights the binding of specific DNA–protein interactions. Isolated DNA is broken into fragments, and after immunoprecipitation of protein–DNA complexes using specific antibodies, proteins are removed from the DNA, and DNA can be analyzed using quantitative PCR, sequencing, or microarrays [118].

Sperm RNA Analysis

This analysis requires the isolation of high-quality RNA, which is usually achieved by mechanic treatment. Two standard procedures are available for the examination of sperm RNA. These include the standard quantitative reverse transcription PCR or microarrays using either fluorescence labeling or linear amplification [109].

However, although being informative, these analyzes are unable to provide a more complex view of sperm transcriptomics as compared to the RNA sequencing. Besides being recognized as a more accurate assessment of the transcriptome, RNA sequencing has the advantage of being able to identify and quantify novel tissue isoforms and variants. Moreover, unrecognized coding and noncoding transcripts can be discovered and their potential functions assessed by RNA-Seq [119]. Sperm RNAs examined by RNA sequencing may provide an insight into the developmental and functional viability along with the identification of potential elements delivered by sperm that are likely of significance upon fertilization.

THE FUTURE OF EPIGENETICS

As opposed to the genome, which is a stable unit over a lifetime, the epigenome is constantly changing, and it emphasizes on the influence that external factors and the environment have on the genome.

Epigenetic research has just recently gained ground and new exciting research has led to:

- An accumulation of scientific knowledge about regulation of the gene expression regulation in biological homeostasis.
- Creation of methods to artificially regulate gene expression [120].

Therefore, future epigenetic research may contribute to the development of treatments for alterations specifically related to the male gamete, as the spermatozoa epigenetic program is unique and tailored to meet the requirements of this highly specialized cell. Also, new data from epigenetic research may be helpful in understanding how male fertility may be altered by the environment or stress and how to measure the damage accurately, which may be useful in the prevention and/or protection against future such damage.

Epigenetics has recently gained significant scientific attention since it has added a new dimension to genomic and proteomic research, and holds great promise in capturing the subtle yet very important regulatory elements that might drive normal and abnormal sperm function.

References

[1] Li ZX, Ma X, Ma, Wang ZG. A differentially methylated region of the *DAZ1* gene in spermatic and somatic cells. Asian J Androl 2006;8:61–7.

[2] Rajender S, Avery K, Agarwal A. Epigenetics, spermatogenesis, and male infertility. Mutat Res 2011;727:62–71.

[3] Frésard L, Morisson M, Brun JM, Collin A, Pain B, Minvielle F, et al. Epigenetics and phenotypic variability: some interesting insights from birds. Genet Select Evol 2013;45:16.

[4] Griffiths AJF, Miller JH, Suzuki DT, Lewontin RC, Gelbart WM, editors. Genetics and the organism: an introduction to genetic analysis. 7th ed. New York: W.H. Freeman; 2000. ISBN 0-7167-3520-2.

[5] Waddington CH. The epigenotype. Endeavour 1942;1:18–20.

[6] Reik W, Dean W. Back to the beginning. Nature 2002;420(6912):127.

[7] Krawetz SA. Paternal contribution: new insights and future challenges. Nat Rev Genet 2005;6(8):633–42.

[8] Carrell DT. Contributions of spermatozoa to embryogenesis: assays to evaluate their genetic and epigenetic fitness. Reprod Biomed Online 2008;16(4):474–84.

[9] Agarwal A, Said TM. Role of sperm chromatin abnormalities and DNA damage in male infertility. Hum Reprod Update 2003;9(4):331–45.

[10] Sutovsky P, Schatten G. Paternal contributions to the mammalian zygote: fertilization after sperm–egg fusion. Int Rev Cytol 2000;195:1–65.

[11] Kumar M, Kumar K, Jain S, Hassan T, Dada R. Novel insights into the genetic and epigenetic paternal contribution to the human embryo. Clinics (Sao Paulo) 2013;68(Suppl. 1):5–14.

[12] Ng HH, Bird A. DNA methylation and chromatin modification. Curr Opin Genet Dev 1999;9:158–63.

[13] Biermann K, Steger K. Epigenetics in male germ cells. J Androl 2007;28:466–80.

[14] Illingworth RS, Bird AP. CpG islands – "a rough guide". FEBS Lett 2009;583:1713–20.

[15] Jenkins TG, Carrell DT. The paternal epigenome and embryogenesis: poising mechanisms for development. Asian J Androl 2011;13:76–80.

[16] Weber M, Hellmann I, Stadler MB, Ramos L, Paabo S, Rebhan M, et al. Distribution, silencing potential, and evolutionary impact of promoter DNA methylation in the human genome. Nat Genet 2007;39:457–66.

[17] Ichiyanagi T, Ichiyanagi K, Miyake M, Sasaki H. Accumulation and loss of asymmetric nonCpG methylation during male germ-cell development. Nucleic Acids Res 2012;41(7):738–45.

[18] Eden S, Cedar H. Role of DNA methylation in the regulation of transcription. Curr Opin Genet Dev 1994;4:255–9.

[19] Okano M, Bell DW, Haber DA, Li E. DNA methyltransferases *Dnmt3a* and *Dnmt3b* are essential for *de novo* methylation and mammalian development. Cell 1999;99:247.

[20] Bestor TH. Activation of mammalian DNA methyltransferase by cleavage of a Zn binding regulatory domain. EMBO J 1992;11:2611–7.

[21] Li E, Bestor TH, Jaenisch R. Targeted mutation of the DNA methyltransferase gene results in embryonic lethality. Cell 1992;69:915–26.

[22] Carrell DT, Hammoud SS. The human sperm epigenome and its potential role in embryonic development. Mol Hum Reprod 2010;16:37–47.

[23] Hajkova P, Erhardt S, Lane N, Haaf T, El-Maarri O, Reik W, et al. Epigenetic reprogramming in mouse primordial germ cells. Mech Dev 2002;117:15–23.

[24] Hajkova P, Ancelin K, Waldmann T, Lacoste N, Lange UC, Cesari F, et al. Chromatin dynamics during epigenetic reprogramming in the mouse germ line. Nature 2008;452:877–81.

[25] Ueda T, Abe K, Miura A, Yuzuriha M, Zubair M, Noguchi M, et al. The paternal methylation imprint of the mouse *H19* locus is acquired in the gonocyte stage during foetal testis development. Genes Cells 2000;5:649–59.

[26] Li JY, Lees-Murdock DJ, Xu GL, Walsh CP. Timing of establishment of paternal methylation imprints in the mouse. Genomics 2004;84:952–60.

[27] Tilghman SM. The sins of the fathers and mothers: genomic imprinting in mammalian development. Cell 1999;96:185–93.

[28] Oakes CC, la Salle S, Smiraglia DJ, Robaire B, Trasler JM. Developmental acquisition of genome-wide DNA methylation occurs prior to meiosis in male germ cells. Dev Biol 2007;307:368–79.

[29] Arnaud P. Genomic imprinting in germ cells: imprints are under control. Reproduction 2010;140:411–23.

[30] Boissonnas CC, Jouannet P, Jammes H. Epigenetic disorders and male subfertility. Fertil Steril 2013;99:624–31.

[31] Yanagimachi R. Male gamete contributions to the embryo. Ann NY Acad Sci 2005;1061:203–7.

[32] Oliva R. Protamines and male infertility. Hum Reprod Update 2006;12:417–35.

[33] van Roijen HJ, Ooms MP, Spaargaren MC, Baarends WM, Weber RF, Grootegoed JA, et al. Immunoexpression of testis-specific histone 2b in human spermatozoa and testis tissue. Hum Reprod 1998;13:1559–66.

[34] Rousseaux S, Gaucher J, Thevenon J, Caron C, Vitte AL, Curtet S, et al. Spermiogenesis: histone acetylation triggers male genome reprogramming. Gynecol Obstet Fertil 2009;37:519–22.

[35] Nair M, Nagamori I, Sun P, Mishra DP, Rheaume C, Li B, et al. Nuclear regulator Pygo2 controls spermiogenesis and histone H3 acetylation. Dev Biol 2008;320:446–55.

[36] Meistrich ML, Mohapatra B, Shirley CR, Zhao M. Roles of transition nuclear proteins in spermiogenesis. Chromosoma 2003;111:483–8.

[37] Shirley CR, Hayashi S, Mounsey S, Yanagimachi R, Meistrich ML. Abnormalities and reduced reproductive potential of sperm from Tnp1- and Tnp2-null double mutant mice. Biol Reprod 2004;71:1220–9.

[38] Suganuma R, Yanagimachi R, Meistrich ML. Decline in fertility of mouse sperm with abnormal chromatin during epididymal passage as revealed by ICSI. Hum Reprod 2005;20:3101–8.

[39] Corzett M, Mazrimas J, Balhorn R. Protamine 1: protamine 2 stoichiometry in the sperm of eutherian mammals. Mol Reprod Dev 2002;61:519–27.

[40] Carrell DT, Emery BR, Hammoud S. Altered protamine expression and diminished spermatogenesis: what is the link? Hum Reprod Update 2007;13:313–27.

[41] Cree LH, Balhorn R, Brewer LR. Single molecule studies of DNA–protamine interactions. Protein Pept Lett 2011;18:802–10.

[42] Hammoud SS, Nix DA, Zhang H, Purwar J, Carrell DT, Cairns BR. Distinctive chromatin in human sperm packages genes for embryo development. Nature 2009;460:473–8.

[43] Carrell DT, Liu L. Altered protamine 2 expression is uncommon in donors of known fertility, but common among men with poor fertilizing capacity, and may reflect other abnormalities of spermiogenesis. J Androl 2001;22:604–10.

[44] Aoki VW, Liu L, Jones KP, Hatasaka HH, Gibson M, Peterson CM, et al. Sperm protamine 1/protamine 2 ratios are related to *in vitro* fertilization pregnancy rates and predictive of fertilization ability. Fertil Steril 2006;86:1408–15.

[45] Simon L, Castillo J, Oliva R, Lewis SE. Relationships between human sperm protamines, DNA damage, and assisted reproduction outcomes. Reprod Biomed Online 2011;23:724–34.

[46] Fenic I, Sonnack V, Failing K, Bergmann M, Steger K. *In vivo* effects of histone-deacetylase inhibitor trichostatin-A on murine spermatogenesis. J Androl 2004;25:811–8.

[47] Okada Y, Scott G, Ray MK, Mishina Y, Zhang Y. Histone demethylase JHDM2A is critical for *Tnp1* and *Prm1* transcription and spermatogenesis. Nature 2007;450:119–23.

[48] Baarends WM, Wassenaar E, van der Laan R, Hoogerbrugge J, Sleddens-Linkels E, Hoeijmakers JH, et al. Silencing of unpaired chromatin and histone H2A ubiquitination in mammalian meiosis. Mol Cell Biol 2005;25:1041–53.

[49] Khalil AM, Boyar FZ, Driscoll DJ. Dynamic histone modifications mark sex chromosome inactivation and reactivation during mammalian spermatogenesis. Proc Natl Acad Sci USA 2004;101:16583–7.

[50] Godmann M, Auger V, Ferraroni-Aguiar V, Di Sauro A, Sette C, Behr R, et al. Dynamic regulation of histone H3 methylation at lysine 4 in mammalian spermatogenesis. Biol Reprod 2007;77:754–64.

[51] Jenuwein T, Allis CD. Translating the histone code. Science 2001;293:1074–80.

[52] Filipponi D, Feil R. Perturbation of genomic imprinting in oligozoospermia. Epigenetics 2009;4:27–30.

[53] Tan M, Luo H, Lee S, Jin F, Yang JS, Montellier E, et al. Identification of 67 histone marks and histone lysine crotonylation as a new type of histone modification. Cell 2011;146:1016–28.

[54] Montellier E, Rousseaux S, Zhao Y, Khochbin S. Histone crotonylation specifically marks the haploid male germ cell gene expression program: post- meiotic male-specific gene expression. Bioessays 2012;34:187–93.

[55] Ostermeier GC, Dix DJ, Miller D, Khatri P, Krawetz SA. Spermatozoal RNA profiles of normal fertile men. Lancet 2002;360:772–7.

[56] Hamatani T. Human spermatozoal RNAs. Fertil Steril 2012;97:275–81.

[57] Dadoune JP. Spermatozoal RNAs: what about their functions? Microsc Res Tech 2009;72:536–51.

[58] Lalancette C, Miller D, Li Y, Krawetz SA. Paternal contributions: new functional insights for spermatozoal RNA. J Cell Biochem 2008;104:1570–9.

[59] Garrido N, Martinez-Conejero JA, Jauregui J, Horcajadas JA, Simon C, Remohi J, et al. Microarray analysis in sperm from fertile and infertile men without basic sperm analysis abnormalities reveals a significantly different transcriptome. Fertil Steril 2009;91:1307–10.

[60] World Health Organization. WHO Manual for the Standardised Investigation and Diagnosis of the Infertile Couple. Cambridge: Cambridge University Press; 2000.

[61] Johnson MD. Genetic risks of intracytoplasmic sperm injection in the treatment of male infertility: recommendations for genetic counseling and screening. Fertil Steril 1998;70(3):397–411.

[62] van Assche EV, Bonduelle M, Tournaye H, Joris H, Verheyen G, Devroey P, et al. Cytogenetics of infertile men. Hum Reprod 1996;11(Suppl. 4):1–24.

[63] Miyagawa Y, Tsujimura A, Matsumiya K, Takao T, Tohda A, Koga M, et al. Outcome of gonadotropin therapy for male hypogonadotropic hypogonadism at university affiliated male infertility centers: a 30-year retrospective study. J Urol 2005;173(6):2072–5.

[64] Pryor JL, Kent-First M, Muallem A, van Bergen AH, Nolten WE, Meisner L, et al. Microdeletions in the Y chromosome of infertile men. New Engl J Med 1997;336(8):534–9.

[65] Vogt P, Edelmann A, Kirsch S, Henegariu O, Hirschmann P, Kiesewetter F, et al. Human Y chromosome azoospermia factors (AZF) mapped to different subregions in Yq11. Hum Mol Genet 1996;5(7):933–43.

[66] Muslumanoglu MH, Turgut M, Cilingir O, Can C, Ozyurek Y, Artan S. Role of the AZFd locus in spermatogenesis. Fertil Steril 2005;84(2):519–22.

[67] Foresta C, Moro E, Rossi A, Rossato M, Garolla A, Ferlin A. Role of the AZFa candidate genes in male infertility. J Endocrinol Invest 2000;23(10):646–51.

[68] Kamp C, Huellen K, Fernandes S, Sousa M, Schlegel PN, Mielnik A, et al. High deletion frequency of the complete AZFa sequence in men with Sertoli-cell-only syndrome. Mol Hum Reprod 2001;7(10):987–94.

[69] Teng YN, Lin YM, Sun HF, Hsu PY, Chung CL, Kuo PL. Association of DAZL haplotypes with spermatogenic failure in infertile men. Fertil Steril 2006;86(1):129–35.

[70] Marques CJ, Carvalho F, Sousa M, Barros A. Genomic imprinting in disruptive spermatogenesis. Lancet 2004;363:1700–2.

[71] Marques CJ, Costa P, Vaz B, Carvalho F, Fernandes S, Barros A, et al. Abnormal methylation of imprinted genes in human sperm is associated with oligozoospermia. Mol Hum Reprod 2008;14:67–74.

[72] Kobayashi H, Hiura H, John RM, Sato A, Otsu E, Kobayashi N, et al. DNA methylation errors at imprinted loci after assisted conception originate in the parental sperm. Eur J Hum Genet 2009;17(2):1582–91.

[73] Hammoud S, Purwar J, Pflueger C, Cairns BR, Carrell DT. Alterations in sperm DNA methylation patterns at imprinted loci in two classes of infertility. Fertil Steril 2009;94(5):1728–33.

[74] Houshdaran S, Cortessis VK, Siegmund K, Yang A, Laird PW, Sokol RZ. Widespread epigenetic abnormalities suggest a broad DNA methylation erasure defect in abnormal human sperm. PLoS One 2007;2:e1289.

[75] Boissonnas CC, Abdalaoui HE, Haelewyn V, Fauque P, Dupont JM, Gut I, et al. Specific epigenetic alterations of IGF2-H19 locus in spermatozoa from infertile men. Eur J Hum Genet 2010;18:73–80.

[76] Anway MD, Cupp AS, Uzumcu M, Skinner MK. Epigenetic transgenerational actions of endocrine disruptors and male fertility. Science 2005;308(5727):1466–9.

[77] Ziglioli F, Maestroni U, Dinale F, Ciuffreda M, Cortellini P. Carcinoma *in situ* (CIS) of the testis. Acta Biomed 2011;82:162–9.

[78] Brait M, Sidransky D. Cancer epigenetics: above and beyond. Toxicol Mech Methods 2011;21(4):275–88.

[79] Godmann M, Lambrot R, Kimmins S. The dynamic epigenetic program in male germ cells: its role in spermatogenesis, testis cancer, and its response to the environment. Microsc Res Tech 2009;72:603–19.

[80] Omisanjo OA, Biermann K, Hartmann S, Heukamp LC, Sonnack V, Hild A, et al. *DNMT1* and *HDAC1* gene expression in impaired spermatogenesis and testicular cancer. Histochem Cell Biol 2007;127:175–81.

[81] Yamada S, Kohu K, Ishii T, Ishidoya S, Ishidoya S, Hiramatsu M, et al. Gene expression profiling identifies a set of transcripts that are up-regulated inhuman testicular seminoma. DNA Res 2004;11:335–44.

[82] Arai E, Nakagawa T, Wakai-Ushijima S, Fujimoto H, Kanai Y. DNA methyltransferase 3B expression is associated with poor outcome of stage I testicular seminoma. Histopathology 2012;60:E12–8.

[83] Minami K, Chano T, Kawakami T, Ushida H, Kushima R, Okabe H. DNMT3L is a novel marker and is essential for the growth of human embryonal carcinoma. Clin Cancer Res 2010;16:2751–9.

[84] Eckert D, Biermann K, Nettersheim D, Gillis AJM, Steger K, Jäck HM, et al. Expression of BLIMP1/PRMT5 and concurrent histone H2A/H4 arginine 3 dimethylation in fetal germ cells, CIS/IGCNU and germ cell tumors. BMC Dev Biol 2008;8:106.

[85] Dhar S, Thota A, Rao MRS. Insights into role of bromodomain, testis-specific (Brdt) in acetylated histone H4-dependent chromatin remodeling in mammalian spermiogenesis. J Biol Chem 2012;287:6387–405.

[86] Gryder B, Sodji QH, Oyelere AK. Targeted cancer therapy: giving histone deacetylase inhibitors all they need to succeed. Future Med Chem 2012;4:505–24.

[87] Waterland RA, Jirtle RL. Transposable elements: targets for early nutritional effects on epigenetic gene regulation. Mol Cell Biol 2003;23:5293–300.

[88] Zhang J, Zhang F, Didelot X, Bruce KD, Cagampang FR. Maternal high fat diet during pregnancy and lactation alters hepatic expression of insulin like growth factor-2 and key microRNAs in the adult offspring. BMC Genomics 2009;10:478–89.

[89] Heijmans B, Tobi E, Stein A, Putter H, Blauw G. Persistent epigenetic differences associated with prenatal exposure to famine in humans. PNAS 2008;105:17046–9.

[90] Tobi EW, Lumey LH, Talens RP, Kremer D, Putter H. DNA methylation differences after exposure to prenatal famine are common and timing and sex-specific. Hum Mol Genet 2009;18:4046–53.

[91] Wolff GL, Kodell RL, Moore SR, Cooney CA. Maternal epigenetics and methyl supplements affect agouti gene expression in Avy/a mice. FASEB J 1988;12:949–57.

[92] Yoon J, Baek S. Molecular targets of dietary polyphenols with anti-inflammatory properties. Yonsei Med J 2005;46:585–96.

[93] Yang C, Landau J, Huang M, Newmark H. Inhibition of carcinogenesis by dietary polyphenolic compounds. Annu Rev Nutr 2001;21:381–406.

[94] Kim Y. Folate and carcinogenesis: evidence, mechanisms, and implications. J Nutr Biochem 1999;10:66–88.

[95] Quintanilla-Vega B, Hoover D, Bal W, Silbergeld EK, Waalkes MP, Anderson LD. Lead effects on protamine-DNA binding. Am J Ind Med 2000;38:324–9.

[96] Singh KP, DuMond JW Jr. Genetic and epigenetic changes induced by chronic low dose exposure to arsenic of mouse testicular Leydig cells. Int J Oncol 2007;30(1):253–60.

[97] Perera F, Tang WY, Herbstman J, Tang D, Levin L, Miller R, et al. Relation of DNA methylation of 5′-CpG island of ACSL3 to transplacental exposure to airborne polycyclic aromatic hydrocarbons and childhood asthma. PLoS One 2009;4(2):e4488.

[98] Novikova SI, He F, Bai J, Cutrufello NJ, Lidow MS, Undieh AS. Maternal cocaine administration in mice alters DNA methylation and gene expression in hippocampal neurons of neonatal and prepubertal offspring. PLoS ONE 2008;3:1919.

[99] Dolinoy DC, Huang D, Jirtle RL. Maternal nutrient supplementation counteracts bisphenol A-induced DNA hypomethylation in early development. Proc Natl Acad Sci 2007;104:13056–61.

[100] Yaoi T, Itoh K, Nakamura K, Ogi H, Fujiwara Y, Fushiki S. Genome-wide analysis of epigenomic alterations in fetal mouse forebrain after exposure to low doses of bisphenol A. Biochem Biophys Res Commun 2008;376:563–7.

[101] Guerrero-Bosagna CM, Sabat P, Valdovinos FS, Valladares LE, Clark SJ. Epigenetic and phenotypic changes result from a continuous pre and post natal dietary exposure to phytoestrogens in an experimental population of mice. BMC Physiol 2008;8:17.

[102] Andersen HR, Schmidt IM, Grandjean P, Jensen TK, Budtz-Jorgensen E, Kjaerstad MB, et al. Impaired reproductive development in sons of women occupationally exposed to pesticides during pregnancy. Environ Health Perspect 2008;116:566–72.

[103] Yauk C, Polyzos A, Rowan-Carroll A, Somers CM, Godschalk RW, Van Schooten FJ, et al. Germ-line mutations, DNA damage, and global hypermethylation in mice exposed to particulate air pollution in an urban/industrial location. Proc Natl Acad Sci 2008;105:605–10.

[104] McLachlan JA. Environmental signaling: what embryos and evolution teach us about endocrine disrupting chemicals. Endocr Rev 2001;22:319–41.

[105] Li S, Hursting SD, Davis BJ, McLachlan JA, Barrett JC. Environmental exposure, DNA methylation, and gene regulation: lessons from diethylstilbesterol-induced cancers. Anne NY Acad Sci 2003;983:161–9.

[106] Lyn-Cook BD, Blann E, Payne PW, Bo J, Sheehan D, Medlock K. Methylation profile and amplification of proto-oncogenes in rat pancreas induced with phytoestrogens. Proc Soc Exp Biol Med 1995;208:116–9.

[107] Day JK, Bauer AM, DesBordes C, Zhuang Y, Kim BE, Newton LG, et al. Genistein alters methylation patterns in mice. J Nutr 2002;132:2419S–23S.

[108] Skinner MK, Manikkam M, Guerrero-Bosagna C. Epigenetic transgenerational actions of environmental factors in disease etiology. Trends Endocrinol Metab 2010;21(4):214–22.

[109] Schagdarsurengin U, Paradowska A, Steger K. Analysing the sperm epigenome: roles in early embryogenesis and assisted reproduction. Nat Rev Urol 2012;9(11):609–19.

[110] Fagundes NS, Michalczechen-Lacerda VA, Caixeta ES, Machado GM, Rodrigues FC, Melo EO, et al. Methylation status in the intragenic differentially methylated region of the *IGF2* locus in *Bos taurus indicus* oocytes with different developmental competencies. Mol Hum Reprod 2011;17:85–91.

[111] Miki H, Lee J, Inoue K, Ogonuki N, Noguchi Y, Mochida K, et al. Microinsemination with first-wave round spermatids from immature male mice. J Reprod Dev 2004;50:131–7.

[112] Anckaert E, Romero S, Adriaenssens T, Smitz J. Effects of low methyl donor levels in culture medium during mouse follicle culture on oocyte imprinting establishment. Biol Reprod 2010;83:377–86.

[113] Wright K, Brown L, Brown G, Casson P, Brown S. Microarray assessment of methylation in individual mouse blastocyst stage embryos shows that *in vitro* culture may have widespread genomic effects. Hum Reprod 2011;26:2576–85.

[114] Iliadou AN, Janson PC, Cnattingius S. Epigenetics and assisted reproductive technology. J Intern Med 2011;270(5):414–20.

[115] Wang Z, Xu L, He F. Embryo vitrification affects the methylation of the *H19/Igf2* differentially methylated domain and the expression of *H19* and *Igf2*. Fertil Steril 2010;93:2729–33.

[116] Yan LY, Yan J, Qiao J, Zhao PL, Liu P. Effects of oocyte vitrification on histone modifications. Reprod Fertil Devel 2010;22:920–5.

[117] Smith ZD. A unique regulatory phase of DNA methylation in the early mammalian embryo. Nature 2012;484:339–44.

[118] Dada R, Kumar M, Jesudasan R, Fernández JL, Gosálvez J, Agarwal A. Epigenetics and its role in male infertility. J Assist Reprod Genet 2012;29(3):213–23.

[119] Wang Z, Gerstein M, Snyder M. RNA-Seq: a revolutionary tool for transcriptomics. Nat Rev Genet 2009;10(1):57–63.

[120] Ito Y. Trends in research of epigenetics, a biological mechanism that regulates gene expression. Sci Technol Trends 2009;33:11–24.

Index

A

ACOG. *See* American College of Obstetricians and Gynecologists (ACOG)

Acquired immunodeficiency syndrome (AIDS), 151

Acrosome reaction (AR), 111, 390

ADD. *See* Attention deficit disorder (ADD)

Adult disease
developmental origins of, 241
developmental programming, 245

AGA. *See* Appropriate-for-gestational-age (AGA)

Alcohol consumption and fertility, 4, 149, 358, 401
effect on spermatogenesis, 358
effect on testosterone secretion, 358
fetal alcohol spectrum disorder, 150
hypothalamic–pituitary–testicular signaling impairment, 4
isoprostanes concentrations, 150
ovulatory disorders, 150
sertoli-only cell syndrome, 149
spermatogenic arrest, 149

Alpha-phenyl-*N*-*t*-butylnitrone, 52

Alzheimer's disease, 207, 248, 342

Amaranthus hybridusis, 213

American College of Obstetricians and Gynecologists (ACOG), 139, 175, 176

American Congress of Obstetrician and Gynecologists (ACOG), 124

American Society for Reproductive Medicine (ASRM), 11, 76

Amino acids, 164, 207, 214, 237, 243, 285
for body homeostasis, 237
exogeneous, 164
for health and reproduction performance of livestock, 164
NEFA concentrations, 40
for oxidative damage, 285
role of folic acid in synthesis, 214

Aminoglycoside antibiotics, 151

Anabolic-androgenic steroids, 151
effect on fertility, 151
erectile dysfunction (ED), 151
sperm count reduction, 151

Animal health, 159
and animal welfare, 159
effect of physical activity, 166
relationship with nutrition, 163
role of diet, 162
role of vitamin E, 165

Animal reproduction, 160
effect of physical activity, 166
role of vitamin E, 165
role of zinc and manganese, 166

Anorexia nervosa, 99

Antidepressants, 151

Antiestrogen treatment
clomiphene citrate, 93
effect on semen quality and male fertility, 93
tamoxifen, 93

Antioxidants, 148
DNA damage repair, role in, 148
oxidative stress reduction, role in, 148
reactive nitrogen species, 148
reactive oxygen species, 148
systematic review, 148
treatment, 94, 402
arginine, 403
inflammatory response, 403
side effects, 403
carnitine, 402
carotenoids, 403
coenzyme Q10, 404
folic acid, 405
glutathione and *N*-acetylcysteine, 404
omega fatty acids, 404
pentoxifylline, 404
phytoestrogens, 405
selenium, 403
deficiency of, 403
dietary sources of, 403
superoxide dismutase, 404
vitamin C, 402
vitamin E, 402
zinc, 402

Antiphospholipid antibody syndrome, 284

Antisperm antibodies (ASA), 391

Apoptosis, 23, 343
and apoptotic genes, 343
bcl-2 family of proteins, 343
correlation with different semen parameters, 329
death receptors, activation of, 343
germ cell, 347
kinase-signaling pathways, activation of, 343
and maternal obesity, 305
of sertoli cells, 150
and zinc, 341, 344

Appropriate-for-gestational-age (AGA), 267

AR. *See* Acrosome reaction (AR)

Arsenic (As), 52
antioxidant-related enzymes, inhibition of, 52
DNA repair enzymes, inhibition of, 52
effects on fertility in males, 360
effects on health, 52
oxidative stress, alteration of, 52
protein function enzyme inhibition, alteration of, 52

ART. *See* Assisted reproductive technologies (ART)

Artificial hormonal stimulation, 313

ASA. *See* Antisperm antibodies (ASA)

ASD. *See* Autism spectrum disorder (ASD)

ASRM. *See* American Society for Reproduction Medicine (ASRM)

Assisted reproductive technologies (ART), 3, 35, 75, 99, 195, 254, 303, 334, 355, 391, 397
application and availability of, 78
cycles per million inhabitants, 78
cycles per million women of a fertile age, 78
health policies, 78
body-mass index (BMI)-related metabolic composition
of human serum, 35
of preovulatory follicular fluid, 35
data collection systems, 75
Europe, 76
effect of obesity on assisted reproduction treatment, 303
intracytoplasmic sperm injection (ICSI) cycles, 79
number of babies born per country and per region, 80
number of cycles per million, 80
principal outcomes, 83
Australia data for 2011, 84
Canada data for 2011, 84
China, 84
EIM data for 2010, 83
ICMART data for 2006, 83
India data for 2009, 84
Japan data for 2010, 84
Latin America data for 2011, 83
New Zealand data for 2011, 84
South Africa data for 2010, 84
USA data for 2010, 83
procedures, 77, 85
cost of, 77
in developing countries, 77
in Europe, 77
oocyte, sperm and embryo donation, 77
preimplantation genetic diagnosis (PGD), 77
surrogacy, 77
in United States, 77
walking egg project, 77
religious perspective of, 78
Catholic religion, 78
Judaism, 78
Muslim religion, 78

Assisted reproductive technologies (ART) (*cont.*)
serum and follicular fluid composition, relation between, 35
single-embryo transfer (SET) percentage, 79
treatment cycles, 79
Australia, 82
Canada, 82
China, 82
Latin America, 82
New Zealand, 82
South Africa (SARA), 83
South Korea, 82
United States, 82
Astaxanthin, 94, 95
Attention deficit disorder (ADD), 321
Autism spectrum disorder (ASD), 309
Autoimmune disorders, 273, 342
Azoospermia, 119, 147, 237, 306, 314, 370
and chromosomal abnormality, 314
nonobstructive (NOA), 375
and paternal obesity, 306
in underweight men, 147
vitamin E levels, 370

B
Bacillus subtilis, 210
Bariatric surgery, 112, 137
adjustable gastric band (AGB) and gastric sleeve (GS), 112
effect on erectile function, 114
effect on *in vitro* fertilization, 114
effect on obesity-associated hypogonadotropic hypogonadism, 112
effect on sex quality, 114
effect on sex steroid hormone levels, 112
effect on sperm parameters, 114
estradiol levels, 113
gonadotropin amelioration, 112
hypogonadism, 113
intestinal bypass, 112
osteoblast-produced hormone (OCN) recovery, 112
Roux-en-Y gastric bypass (RYGB)/duodenal switch, 112
variations in androgens, 113
variations in body-mass index (BMI), 113
Barker's hypothesis, 241
Bayley scales of infant development (BSID), 231
Beckwith–Weidemann syndrome (BWS), 316
Behavioral change techniques, 178
action planning, 178
goal setting, 178
self monitoring, 178
Benign prostatic enlargement (BPE), 221
Benzo[*a*]pyrene, 21
in cigarette smoke, 20
and DNA damage, 21
Bioactive compounds, 205, 207
Biomarkers, for toxic metals, 46
meconium, 46
placenta, 46
Bladder cancer, 224
BMI. *See* Body mass index (BMI)

Body mass index (BMI), 133, 173, 227, 237, 317, 327, 355
relation with male fertility, 356
relation with semen, 356
and sex hormones, 110
BPD. *See* Bronchopulmonary dysplasia (BPD)
BPE. *See* Benign prostatic enlargement (BPE)
Brachytherapy, 223, 400
Brain development, 242
early life malnutrition and aging brain, 248
nutritional deficiencies and psychiatric disorders, 247
ADHD etiopathology, 248
malnutrition, 247
nutritional inadequacies and their neurological consequences, 246
amino acids, role of, 246
gestational vitamin D insufficiency, 246
malnutrition, moderate, 247
overnutrition, effects of, 247
perinatal protein malnutrition, 247
polyunsaturated fatty acids, role of, 246
trace elements, deficiencies of, 246
nutritional intervention, 248
challenges of nutritional neuroscience, 248
nutritional insult, mechanism of action of, 249
Brief sexual function inventory (BSFI), 112
Bronchopulmonary dysplasia (BPD), 48, 287
BSFI. *See* Brief Sexual Function Inventory (BSFI)
BSID. *See* Bayley scales of infant development (BSID)
Bulimia nervosa, 99
BWS. *See* Beckwith-Weidemann syndrome (BWS)

C
Cadmium (Cd), 51
alpha-phenyl-*N*-*t*-butylnitrone, 52
effect on health, 52
endoplasmic reticulum (ER) stress, 52
hematocrit, effect on, 52
hemoglobin content, effect on, 52
lipid peroxidation, 52
male reproductive effects, 360
mean corpuscular volume (MCV) levels, effect on, 52
oxidative stress, induction of, 52
uses of, 51
Zn-dependent enzymes, inhibition of, 52
Caffeine, effect on fertility, 5, 150
CAM. *See* Complementary/alternative/nonconventional medicine (CAM)
cAMP. *See* Cyclic adenosine monophosphate (cAMP)
Canadian Assisted Reproductive Technologies Register (CARTR), 76
Canadian Fertility and Andrology Society (CFAS), 76
Cannabinoids, 150, 358
apoptosis of Sertoli cells, 150
fertilizing capacity of sperm, effect on, 359
hypothalamic-pituitary-gonadal (HPG) axis, effect on, 150

male reproductive physiology, effect on, 358
reduction in luteinizing hormone (LH), 150
testosterone production in the Leydig cells, 150
Carbon monoxide, 22, 28, 49
Carcinogens, 20, 22, 28
Carnitine, 36, 370, 375, 402
acetyl, 94
palmitoyl transferase 1, 36
semen parameters, effect on, 375
spermatogenesis, effect on, 375
CARTR. *See* Canadian Assisted Reproductive Technologies Register (CARTR)
Caspases, 343
apoptotic executioners, 343
apoptotic initiators, 343
cytokine processors, 343
roles in amplification of apoptotic signals, 343
Centers for Disease Control and Prevention (CDC), 76, 146
Cerebrovascular accident (CVA), 212
Cesarean rate, 134
CFAS. *See* Canadian Fertility and Andrology Society (CFAS)
cGMP. *See* Cyclic guanosine monophosphate (cGMP)
CHD. *See* Coronary heart disease (CHD)
CHERG. *See* Child Health Epidemiology Research Group (CHERG)
Child Health Epidemiology Research Group (CHERG), 235
Chlamydia infections, 151
Cholesterol, 35, 36, 39, 285
and adverse metabolic health, 38
androgen synthesis, 384
cellular *de novo* synthesis from acetate, derived from, 39
concentrations in the follicular fluid, 37
within the follicle, 39
HDL, 36
LDL, 399
serum, 125, 176, 196
and zinc deficiency, 343
Chorus olitorius, 213
Chromatin packaging, hyperinsulinemic conditions on, 328
Chromosome 20p13, 273
Chromosome 10q26, 273
Chronic inflammation and infertility, 94
activation of the nuclear factor kappa B (Nf-kB), 94
cyclo-oxygenase enzymes, activation of, 94
elevated levels of interleukins, 94
Cigarette smoking, 20, 28, 146
carbon monoxide, 28
female fertility, effect on, 146
FSH levels, 146
genotoxin, 20
human carcinogen benzo[*a*]pyrene, 20
maternal erythrocyte folate levels, 30
microRNAs (miRNAs) functioning, effect on, 20
nicotine, 28
oxidative stress, 30

progesterone levels, 146
protamine 2 (PRM2) germ cell specific gene, upregulation of, 21
SALF germ cell specific transcription factor, upregulation of, 21
toxic metals, 46
TRIM26 zinc finger encoding gene, downregulation of, 21
Clomiphene citrate, 93, 123, 321, 371, 402
Coelomic metaplasia theory, 274
Cognitive-behavioral therapy, 187
Complementary/alternative/nonconventional medicine (CAM), 253–255
Congenital malformations, 61, 314, 397
  dietary intake of nitrate, effects of, 61
  nitrate exposure in drinking water, 61
  nitrosatable drugs, effects of, 61
  N-nitroso compounds, 61
Coriandrum sativum, 210
Coronary heart disease (CHD), 211
Corpus luteum (CL), 100
Corticotropin-releasing hormone (CRH), 47
C-reactive protein (CRP), 36, 136, 147
CRH. *See* Corticotropin-releasing hormone (CRH)
CRP. *See* C-reactive protein (CRP)
Cryopreservation, 313
  of embryos, 85
  of oocytes, 85
  policies, 85
  techniques, 85
CVA. *See* Cerebrovascular accident (CVA)
Cyclic adenosine monophosphate (cAMP), 391
Cyclic guanosine monophosphate (cGMP), 391
Cytochrome P450 1A1(CYP1A1), 48

D

Death inducing signaling complex (DISC), 343
Dehydroepiandrosterone sulfate (DHEAS), 319
Delivery rates (DRs), 83, 84
de novo mutations, 21, 23
Depression, 12, 123, 176, 248, 256, 294, 298
  adolescent, 248
  effect on sex quality of life, 114
  and herbal remedies, 254
  and infertility, 293, 294
  and nutritional deficiencies, 247
  treatment of, 256
    hyperforin, 256
    hypericin, 256
DET. *See* Double embryo transfer (DET)
Detoxification processes, 48
  cytochrome P450 1A1(CYP1A1), 48
  glutathione S-transferase theta 1 (GSTT1), 48
  phase I, 48
  phase II, 48
Developmental origins of adult disease hypothesis (DoAD), 241
DHEAS. *See* Dehydroepiandrosterone sulfate (DHEAS)

DIC. *See* Disseminated intravascular coagulation (DIC)
Diet
  effect on fertility, 149, 194
    animal protein, 149
    epidemiological information, 194
    intake of fat dairy food, 149
    intake of sugar-sweetened beverages, 149
    intake of trans fatty acids, 149
    ovulation, effect on, 149
    role of insulin resistance, 195
    sperm quality, effect on, 149
    total fat intake, 149
  effect on pregnancy and offspring development, 195
    implantation failures and miscarriages, 195
  effect on puberty, 194
  energy-restricted, 186
  influence on animal health, 162
Dietary practices and fertility, 148
Diet-induced obese (DIO) mouse models, 120
  control diet (CD) feeding, 120
  high fat diet (HFD) feeding, 120
Dioxins, 92, 93
DISC. *See* Death inducing signaling complex (DISC)
Disseminated intravascular coagulation (DIC), 284
DNA damage, 21, 23, 398
  in diabetic men, 306
  due to obesity, 124
  due to sexually transmitted infections, 5
  due to tobacco smoking, 4
  effect of antioxidants, 148
  effect of xenobiotics and toxic metals, 400
  in infertile men, 145
  oxidative stress, 398
  protective role of glutathione, 404
  role of tumor suppressor gene p53, 343
  sperm, 23, 307
  transgenerational, 23
    embryotrophic factors, 23
    oviductal apoptosis inducing factor, 23
    paternal genetic factors, 23
  transgenerational destabilization, 21
DNA fragmentation, 328, 397
  Comet assay, 371
  effect of bariatric surgery, 114
  effect of heat, 400
  effect of obesity, 401
  effect of vitamin E, 371, 402
  effect of Wi-Fi radiation, 400
  effect of xenobiotics and toxic metals, 400
  and oxidative stress, 400
  relation with cellular zinc, 347
  sperm, 111, 397
DNA-fragmented spermatozoa
  fertilization of, 21
  intracytoplasmic sperm injection (ICSI), use of, 21
DNA methylation, 6, 21, 22, 40, 246, 335, 375
DNA methyltransferases (DNMT), 22, 412
DNA packaging, 328, 333
DNA synthesis, 385, 393
  meiosis, initiation of, 385
  mitotic proliferation of cells, 385

DNA transcription, 402
DNMT. *See* DNA methyltransferases (DNMT)
DoAD. *See* Developmental origins of adult disease hypothesis (DoAD)
Docosahexaenoic acid (DHA), 95, 163, 374, 404
Double embryo transfer (DET), 315
Dysmenorrhea, 273, 279
Dyspareunia, 273

E

Early life programming, role of maternal nutrition in, 241
  nutrition, effects of, 242
  pseudo-experiments, performing of, 242
ECM. *See* Extracellular matrix (ECM)
Elective single embryo transfer (eSET), 315
Embryo development, 100
  conceptus, 100
  effect of undernutrition, 101
  insulin-like growth factors (IGF), 101
  during periconceptional period, 102
  during preimplantation period, 102
  undernutrition, 101
Embryonic rest theory, 274
Embryo transfer policies, 85
Endocrine disrupters, 5, 191, 360
  bisphenol A (BPA), 5
  chemicals mimicking endogenous estrogens, 5
  chlordecone, 5
  diethylstilbestrol (DES), 5
  human exposure, sources of, 360
  methoxychlor, 5
  nonylphenol, 5
  phytoestrogens genistein, 5
  polychlorinated biphenyls (PCBs), 5
  role in reproductive outcomes, 360
  targeting male reproductive tract, 360
  xenoestrogens, 5, 361
Endocrine-disrupting compounds (EDC). *See* Endocrine disrupters
Endometrial glands, 100, 273
Endometriosis, 95, 149, 273, 274
  cytokines, role of, 278
    interleukin 6 (IL-6), 278
    interleukin 1 (IL-1) for activation of T-cells, 278
    interleukin 8 (IL-8) for chemotaxis, 278
    TNF-α as proinflammatory cytokine, 278
  diagnosis of, 273
    laparoscopy, 273
    peritoneal markers, 273
    serum analysis, 273
  iron, role of, 277
    bilirubin production, 278
    therapeutic target, 278
  origin of, 274
    coelomic metaplasia theory, 274
    embryonic rest theory, 274
    Sampson's implantation theory, 274
    vascular and lymphatic metastasis theory, 275
  risk factors, 273
    genetic linkage on chromosome 20p13, 273
    genetic linkage on chromosome 10q26, 273

Endometriosis *(cont.)*
symptoms of, 273
abnormal bleeding, 273
dysmenorrhea, 273
dyspareunia, 273
infertility, 273
treatment of, 278
antioxidants, 279
catalase, 279
glutaredoxin, 279
glutathione reductase, 279
danazol therapy, 278
glutathione peroxidase, 279
vitamin E and vitamin C, 279
Endothelium-dependent vasodilatation, 213
Energy sources, 163
carbohydrates, 163
fats, 163
fatty acids, 163
microminerals, 165
protein, 164
vitamins, 164
Environmental contaminants, 7, 89
effect on sperm quality in Flanders, 92
human exposure to, 7
strategy of reduction, 92
in utero exposure to, 46
Environmental risks, 3
to fertility, mitigation of, 11
American Society for Reproductive
Medicine (ASRM) infertility
prevention campaign, 11
FertiSTAT, 11
health promotion, 11
risk communication, 11
perception, population-based, 9
perception to infertility, 7
models of, 7
social amplification of risk framework, 7, 8
EPA. *See* U.S. Environmental Protection
Agency (EPA)
Epididymis, structure and function, 387
cell types, 388
sperm maturation, 389
sperm transport, 390
Epigenetic reprogramming, 411
Epigenetics, 21, 22, 194, 411
and art, 418
chromatin reorganization, 413
current techniques for assessment of
spermatozoa, 418
DNA histone binding analysis, 419
DNA methylation analysis, 419
sperm RNA analysis, 419
DNA methylation, 21, 412
histone modifications, 21, 414
impact of environment on, 417
of the male gamete, 411
male infertility, 415
and nutrition, 417
sperm RNAs, 414
and testicular cancer, 416
EPO. *See* Evening primrose oil (EPO)
EQS. *See* Erection quality scale (EQS)
ERBA. *See* Estrogen receptor binding assay
(ERBA)

Erectile dysfunctions (ED), 112, 151
Erection quality scale (EQS), 112
Escherichia coli, 257
eSET. *See* Elective single embryo transfer
(eSET)
ESHRE. *See* European Society of Human
Reproduction and Embryology
(ESHRE)
Essential elements, 46
copper (Cu), 46
effect of acute stress, 46
effect of trauma, 46
iron (Fe), 46
manganese (Mn), 46
toxic levels of, 46
zinc (Zn), 46
Estradiol 17-beta, 93, 97, 209
Estrogen, 5, 125, 215, 405
biosynthesis of, 238
cytochrome P459 enzyme, role of, 238
effect of nitric oxide, 277
effect of obesity, 238
erythrocyte selenium and GPx, 264
metabolism, 96, 97
16-alfa-OH-estrogens, 96
heat shock protein 70, 96
soy isoflavones, 96
Estrogen receptor binding assay (ERBA), 90
application to environmental samples, 90
glutathion-S-transferase (GST) tag, 90
human estrogen receptor (TER), 90
principle of, 91
total estrogen activity, assessment of, 91
Estrone, 97
ESWL. *See* Extracorporeal shockwave
lithotripsy (ESWL)
Euphorbia hirta, 210
European IVF monitoring (EIM), 76
data for year 2010, 83
European Society of Human Reproduction
and Embryology (ESHRE), 75
Evening primrose oil (EPO), 258
Extracellular matrix (ECM) metalloproteinase
inducer, 23
basigin, 23
CD147, 23
Extracorporeal shockwave lithotripsy
(ESWL), 223

F
FAD. *See* Flavin adenine dinucleotide
(FAD)
Farm animals, 159, 166
aggression, 167
breeding values, 160
cattle, 160
feather pecking, 167
group size, 167
locomotion, 166
pig, 161
poultry, 161
rearing system, 167
reproductive efficiency, 160, 161, 162
reproductive traits, 160
sheep, 161
stocking density, 167

Fatty acids
changes through adverse metabolic health,
38
apoptosis, 38
cumulus–oocyte complex, 38
endoplasmic reticulum stress, 38
lipid accumulation, 38
NEFA concentrations, 38
polycystic ovarian syndrome (PCOS), 38
changes through diet, 37, 163
cumulus–oocyte complex, 38
immunomodulatory effects, 163
linoleic acid, 38
NEFA concentrations, 37
n-3 polyunsaturated fatty acids, 38
polyunsaturated fatty acid (PUFA)
concentrations, 37
ω-3 PUFAs, 163
in ovarian follicular fluid, 37
in ovarian follicular fluid, changes in
composition
through adverse metabolic health, 38
through diet, 37
Feeding strategy, 162
Female infertility, 95. *See also* Infertility
endometriosis, 95
homocysteine in blood, elevated level
of, 95
nutritional intake of PUFAs, 95
polycystic ovary syndrome, 95
Female obesity, 120. *See also* Obesity
Female reproductive disorders, 191
animal models, 192
pig, 193
Göttingen minipigs, 193
Yucatan, 193
rabbit, 193
rodents, 192
monogenic models, 192
sheep, 193
Fertility
and age, 145
testosterone levels, 145
effect of lifestyle factors, 145
of females
effect of age, 146
impact of obesity, 120
of males, 397
effect of bariatric surgery, 112
effect of obesity, 124
processes involved, 160
rates of, 145
and weight management, 124
Fertility Society of Australia (FSA), 76
Fetal development
complications in, 287
activation of polymorphonuclear
leukocytes, 287
effect of obesity, 122
role of folic acid, 137
Filial genome, transgenerational
destabilization of, 21
First-degree relatives, 273
Flavin adenine dinucleotide
(FAD), 276
Flavin mononucleotide (FMN), 276

FMN. *See* Flavin mononucleotide (FMN)
Folate, 27
  deficiency, 27
  low level, and neural tube defects, 29
  physiological functions, role in, 27
  risk of recurrent wheezing in children, 31
  sources of, 27
Folic acid, 27, 137, 214
  deficiency, 135
  supplementation, 30
    maternal, 30
    neonatal, 30
Follicle stimulating hormone (FSH), 321, 383
Follicular fluid, 35
  biochemical parameters in, 36
    C-reactive protein (CRP), 36
    total protein, 36
    urea, 36
  blood–follicular fluid barrier, modulation
      of, 35
  fatty acids, 37
  hormonal parameters, 37
    insulin, 37
    insulin-like growth factor 1 (IGF), 37
  model of the high-producing dairy cow, 35
    ketogenesis, 35
    lipolysis, 35
    peripheral insulin sensitivity, 35
  oocyte and embryo quality, 35
  parameters related to carbohydrate
      metabolism, 37
  parameters related to lipid metabolism, 36
    carnitine, 36
    cholestrol, 36
  relation with serum, 35
Free-radicals-related disease, 48
  bronchopulmonary dysplasia (BPD), 48
  intraventricular hemorrhage (IVH), 48
  necrotizing enterocolitis (NEC), 48
  reperfusion injury after hypoxia, 48
  retinopathy of prematurity (ROP), 48
Frozen embryo replacement (FER), 83
Frozen embryo transfers (FETs), 83
FSA. *See* Fertility Society of Australia (FSA)
FSH. *See* Follicle stimulating hormone (FSH)

G

GDM. *See* Gestational diabetes mellitus
      (GDM)
Generally recognized as safe (GRAS), 210
Genitourinary malformations, 314
Gestational and obesity adverse outcomes
  gestational weight gain (GWG), 138
  maternal, 133
  mechanism of, 135
  neonatal, 134
Gestational diabetes mellitus (GDM), 133,
      138, 267
Gestational weight gain (GWG), 138, 173
  achieving healthy, 178
    behavioral change techniques, 178
    bottom–up approach, 179
  guidelines, 173
  health care provider advice on, 174
  interventions, 177
    effectiveness, lack of, 178

fit for delivery study, randomized trial,
      178
  regular weight monitoring, 177
  systematic reviews, 177
postpartum weight retention, 174
recommendations, 173, 174
Glucose, 37
  lactate, 37
  pyruvate, 37
  uptake and metabolism, 327
Glucose transporters (GLUTS), 327
Glutathione-S-transferases (GST), 90
  mu 1 (GSTM1), 48
  theta 1 (GSTT1), 48
GLUTS. *See* Glucose transporters (GLUTS)
GLVs. *See* Green leafy vegetables (GLVs)
Gnetum africana, 213
GnRH. *See* Gonadotropin releasing hormone
      (GnRH)
Gonadotropin releasing hormone (GnRH),
      383
  androgen production (LH-testosterone), 383
  sperm production (FSH-inhibin), 383
Gonorrheal infections, 151
Granulosa cells
  and NEFA concentration, 39
  steroidogenesis, 39
GRAS. *See* Generally recognized as safe
      (GRAS)
Green leafy vegetables (GLVs), 205, 206
  cardiovascular disease and leafy vegetable,
      211
    carotenoids, 211
    flavonoids, 212
  dietary fiber, 213
  green leafy vegetables on mother and fetus
      health, effect of, 214
    fertility, 215
    folic acid, 214
  hypertension and leafy vegetables, 212
  lowering the risk of type 2 diabetes, 210
  phenolic compounds, properties of, 210
  role of vegetables as medicine, 207, 208
  source of phytochemicals, 207, 208
    classification of, 207, 209
  therapeutic value and health benefits of,
      207
    antidiabetic properties, 209
      hyperglycemia, 209
    antimicrobial and anti-inflammatory
        activity, 210
    antioxidant activity, 211
GWG. *See* Gestational weight gain (GWG)

H

Haber–Weiss reaction, 284
Halo cells, 388
HBA. *See* Hyaluronic binding assay (HBA)
hCG. *See* Human chorionic gonadotropin
      (hCG)
Heat shock proteins (HSP), 276
Heavy metals, 92, 93, 360
  arsenic (As), 46, 360
  cadmium (Cd), 46, 360
    effect on hypothalamic-pituitary-adrenal
        (HPA) axis, 47

effect on oxidative stress, 48
  effect on regulation of detoxification genes,
      48
  experimental studies on human
      reproductive systems, 360
  lead (Pb), 46, 360
  mercury (Hg), 46, 360
  pituitary hormones, receptor binding of,
      360
  and preterm mortality and morbidity, 46
  in utero exposure to, 46
Heme-oxygenase (HO), 49
HEPA. *See* High-efficiency air particulate
      (HEPA)
Herbal remedies (HR), 253, 254
  acupuncture, 255
  herbal therapy, 255
  infertility, 254
  pharmacological properties, 254
  during pregnancy, 255
    epidemiological studies, 255
Herbs, 257
  American cranberry, 257
  blue cohosh, 258
  castor oil, 258
  echinacea, 257
  evening primrose oil, 258
  garlic, 257
  ginger, antiemetic effects of, 256
  observational cohort study, 257
  raspberry leaf plant, 258
High-efficiency air particulate (HEPA), 417
Histone modifications, 21, 414
  acetylation, 414
  crotonylation, 414
  methylation, 414
  phosphorylation, 414
  ubiquitination, 414
HIV. *See* Human immunodeficiency virus
      (HIV)
Hormone disrupters, 89
  development of essays for detection, 90
  elimination from enviornment, 92
  elimination from living organisms, 92
  estrogen receptor binding assay (ERBA), 90
  historical aspects of, 89
  interference with sex hormones, 90
  in vitro assays, 90
  in vivo test systems, 90
HPA. *See* Hypothalamus-pituitary-adrenal
      (HPA)
HR. *See* Herbal remedies (HR)
HSP. *See* Heat shock proteins (HSP)
Human aortic endothelial cells (HAECS), 346
Human chorionic gonadotropin (hCG), 303
Human embryos, 100
  elongation of, 100
  histotroph, 100
  interferon tau (IFN- τ), secretion of, 100
  preimplantation period, 100
Human genome project, 411
Human germ cell toxicants/mutagens, 19
Human immunodeficiency virus (HIV), 151,
      268
Human papillomavirus (HPV) infection, 151
Human paraoxonases (PONs), 49

Hyaluronic binding assay (HBA), 328
8-Hydroxy-2′-deoxyguanosine (8-OH 2-dG), 51, 94, 95
Hyperglycemia and adverse pregnancy outcomes (HAPO)
study, 135
Hypothalamic gonadal (HPG) axis, 120
Hypothalamic-pituitary-adrenal (HPA) axis, 47, 245
Hypoxia-inducible factor (HIF)-1α, 22

I

IARC. *See* International Agency for Research on Cancer (IARC)
ICMART. *See* International Committee Monitoring Assisted Reproductive Technologies (ICMART)
ICSI. *See* Intracytoplasmic sperm injection (ICSI)
IGF. *See* Insulin-like growth factor (IGF)
IGFBP1. *See* Insulin-like growth factor binding protein 1 (IGFBP1)
IIEF standard. *See* International index of erectile function (IIEF) standard
Illicit drugs and male fertility, 150
Imprinting disorders, 316, 397
　Angelman syndrome of mental retardation, 316
　Beckwith–Weidemann syndrome (BWS), 316
Indian Society of Assisted Reproduction (ISAR), 77
Infertility, 75, 111, 145, 293. *See also various infertilities*
　consequences of, 294
　　depression, 294
　　posttraumatic stress disorder, 294
　　social support, 294
　cycle of, 293
　　negative pregnancy result, 293
　　symptoms of, 293
　definition of, 3
　effect of obesity, 3
　factors, 153
　gender and risk perception, 8
　　occupational reproductive risk assessments, 8
　　reproductive health, 8
　impact of adoption, 296
　and lifestyle, 3
　　alcohol, 4
　　body weight, 3
　　caffeine, 5
　　sexually transmitted infections (STI), 5
　　tobacco, 4
　and nutrition, 99
　and one's relationship, 294
　public health approaches, 12
　　health assessment, 12
　　health policy, 12
　　health promotion, 12
　　health services, 12
　　proposed framework, 12
　risk factors, 3
　risk perception of, 7
　　environmental, 7, 9, 10
　　fertility awareness, 9

　　gender, 8
　　general public, 10
　　lead exposure, 10
　　lifestyle habits, 9
　　occupational contaminant exposures, 11
　　population-based environmental risk perception, 9
　　preconception, 9
　　race/ethnicity, 8
　　smoking, 10
　　stress, 10
　and stress management, 293, 298
Influence of male hyperinsulinemia on IVF outcome, 332
　embryo quality and IVF outcome for groups A and B, 333, 334
　results, 333
Institute of Medicine (IOM) guidelines, 173
Insulin growth factors (IGF), 101, 102, 163, 245
　IGF-I, 37, 327
　　IGF binding proteins (IGFBPs), 37
　　oocyte's developmental competence, influence on, 37
　IGF-II, 327
Insulin-like growth factor binding protein 1 (IGFBP1), 304
Insulin receptor, 328
Interferon tau (IFN- τ), 100
International Agency for Research on Cancer (IARC), 20
International Committee for Monitoring Assisted Reproductive Technology (ICMART), 75, 76, 99
　Australia, 76
　Canada, 76
　data for year 2006, 83
　India, 77
　Israel, 77
　Japan, 76
　Latin America, 76
　Middle East Society, 77
　New Zealand, 76
　South Africa (SARA), 77
　total estimated number of cycles in 2006 in individual countries, 80
　total estimated number of cycles in 2008 in the region, 79
　United States, 76
International index of erectile function (IIEF) standard, 112
International Working Group for Registers on Assisted Reproduction (IWGROAR), 76
Intracytoplasmic sperm microinjection (ICSI), 21, 125, 313, 397
Intrauterine growth retardation (IUGR), 45, 193, 284
Intraventricular hemorrhage (IVH), 287
In utero environment, 241
　epigenetic mechanisms, 246
　genetic and environmental factors, 245
　glucocorticoids, effects of, 245
　human epidemiological studies, 245
　maternal calorie restriction models, 246
　nutritional programming mechanisms, 245

　predictive adaptive response, 245
　prenatal undernutrition, 246
　probable mechanism of action, 244
　thrifty gene hypothesis, 245
In vitro fertilization (IVF), 20, 75, 222, 313, 328, 356, 391
　cardiovascular and metabolic effects, 317, 318
　　blood pressure, 317
　　body mass index, 317
　　lipid metabolism regulators, 317
　　metabolic profile, 317, 318
　　vascular dysfunction, 317
　cerebral palsy, 319
　cognitive function, 320
　effect of bariatric surgery, 114
　endocrine, 318
　general health and well-being, 319
　growth and pubertal development, 318
　　growth, 318
　　puberty, 319
　imprinting disorders
　　Angelman syndrome of mental retardation, 316
　　animal models, 316
　　Beckwith–Weidemann syndrome (BWS), 316
　　epidemiological studies, 316
　　Silver–Russel syndrome, 316
　malignancy, 317
　　nonhereditary retinoblastoma, 317
　neuromotor development, 320
　ophthalmological and auditory disorders, 318
　respiratory and allergy disorder, 318
　school performance, 320
　socio-emotional development and psychological disorders, 320
　　attention deficit disorder, 321
　　autism, 321
　testicular function, 319
IOM. *See* Institute of Medicine (IOM) guidelines
Ipomoea involucrate, 210
ISAR. *See* Indian Society of Assisted Reproduction (ISAR)
IUGR. *See* Intrauterine growth retardation (IUGR)
IVF. *See* in vitro fertilization (IVF)
IVF data, summary of, 331, 332
IVH. *See* Intraventricular hemorrhage (IVH)
IWGROAR. *See* International Working Group for Registers on Assisted Reproduction (IWGROAR)

K

Kallmann's syndrome, 415
Klinefelter's syndrome, 415

L

Lactuca sativa, 210
Large-for-gestationalage (LGA), 195
Latin American Network for Reproductive Medicine (REDLARA), 76
LBW. *See* Low birth weight (LBW)

Lead (Pb), 47
  Ca-gated K channels, effect on, 49
  Ca-mediated cellular processes, alteration
      of, 47
  cholinergic receptors, effect on, 49
  and corticotropin-releasing hormone
      (CRH), 47
  cytochrome oxidase activity, 47
  effect on hypothalamic-pituitary-adrenal
      (HPA) axis, 47
  effect on oxidative stress, 48
  N-methyl-D-aspartic acid (NMDA)
      receptor, effect on, 49
  other effects, 49
  and placenta, 47
  syncytiotrophoblast cells, storage in, 47
  voltage-gated Ca channels, effect on, 49
  voltage-gated K channels, effect on, 49
Leptin receptor *(LEPR)*, 192, 194, 195
Leukemia inhibitory factor (LIF), 304, 384
Leydig cells, 384
LGA. *See* Large-for-gestational age (LGA)
LIF. *See* Leukemia inhibitory factor (LIF)
Lifestyle intervention
  implementation of, 187
    intrinsic resistance to changing behavior,
        187
    lifestyle counselors, 187
  for resumption of ovulation, 184
    abdominal fat, loss of, 184, 185, 187
      lifestyle intervention program, 184
    diet and dietary composition, 186
    lifestyle factors, 186
    physical activity, increased, 184
      hypothalamic suppression, 185
      insulin sensitivity improvement, 185
    weight reduction, 184, 186
    studies of, 184
Linoleic acid, 38, 165, 404
Lipid peroxidation, 211, 283
Livestock species
  effect of physical activity, 166
  influence of diet, 162
  production of, 162
  reproductive performance of, 160
    cattle, 160
    pig, 161
    poultry, 161
    sheep, 161
Lorcaserin, 137
Low birth weight (LBW), 45, 52, 229
Lower urinary tract symptoms (LUTS), 221
Low molecular weight heparin (LMWH), 139
Lubricants, 152
Luteolytic mechanism, 100
LUTS. *See* Lower urinary tract symptoms
      (LUTS)

M

Macrosomia, 124, 134, 136, 267
Male fecundity, 124, 152
Male germ cells, 237
Male infertility, 237, 238, 369, 397, 415. *See also*
      Infertility
  epigenetic factors, 415
  genetic causes of, 415

Male obesity, 109, 119, 124. *See also* Obesity
  cardiovascular diseases (CVD), 109
  genetic model of leptin deficiency ($Ob^{-/-}$),
      124
  gonadotropin levels, 109
  impaired reproductive health, 125
  intracytoplasmic sperm injection (ICSI), 125
  IVF treatment, 125
  metabolic syndrome (MetS), 109
  MetS, 111
  reproductive dysfunction, 124
  and reproductive dysfunction, 119
  reproductive functions, 111
  rodent studies, 124
  semen quality, 111
  sex hormone-binding globulin (SHBG)
      concentrations, 119
  sperm oxidative stress, 125
  and sperm quality, 119
  testosterone concentrations, 119
  type 2 diabetes mellitus (T2DM), 109
Male patients, 355
  caffeine, 355
    biological pathways for affecting on
        reproductivity, 361
    decrease in fertilization capacity, 361
    effects of, 361
  food intake, 361
    dairy food, 361
      environment estrogens, 361
      saturated fats causing low sperm
          counts, 361
      semen quality, low, 361
    red meat, 362
      positive effect of, 362
    sweet foods and beverages, 362
      insulin resistance factor, 362
    vegetables, legumes and fruits, 361
      Mediterranean dietary pattern
          adherence, 362
      presence of pesticides, 362
  reproductive health, 359
    occupational and environmental factors,
        359
      endocrine disruptor compounds, 360
      heavy metals, 360
      organic solvents, 359
  weight and exercise, 355
    excessive exercising, 357
      hypothalamic-pituitary-gonadal axis,
          affecting of, 357
    overweight, 355
      obesity, 355
    sedentary, 357
      anabolic environment, 357
      negative effects, 357
      positive effects, 357
      semen and hormone parameters,
          change of, 357
    underweight, 356
Malondialdehyde (MDA), 51
MAR. *See* Matrix attachment regions (MAR)
Marijuana, 150
Maternal calorie restriction models, 246
Maternal obesity, 122, 135. *See also* Obesity
  breastfeeding rates, 139

chemokine mRNA, 122
and fertility, relationship between, 303
  maternal obesity and assisted
      reproductive treatment, 303
  maternal obesity and embryo
      implantation, 304
    case-control study, 304
    endometrial receptivity impairement,
        305
    glandular LIF and female BMI,
        negative correlation between, 304
  maternal obesity and offspring health,
      305
    chronic inflammation, 305
    congenital anomalies, detection rate
        of, 306
    cytokines and adipokines, changes in
        homeostasis of, 305
  maternal obesity and the ovary, 304
    adipose tissue for steroid, 304
    embryo grading, 304
    embryo quality, role of, 304
    negative effects of metabolites, 304
    negative environmental exposures, 304
  obesity and pregnancy outcomes, 305
    spontaneous miscarriage, rate of, 305
metabolic inflammation, 136
and postpartum, 135
sheep studies, 122
vitamin D deficiency, 136
weight loss study, 122
Maternal smoking, 21, 27, 28
  case studies, 28
  congenital malformations, 21, 28
  correlation between serum homocysteine
      and serum folate levels, 30
  fetal mortality and morbidity, 21
  folate concentrations, 29
  growth retarding effects, 28
  harmful effects arising in newborn and
      nursling periods, 29
  harmful effects arising in prenatal period,
      28
  homocysteine concentrations, 28, 29
  low birth weight infant delivery, risk of, 28
  lower birth weight, 21
  MTHFR activity, 29
  and neonatal serum folate levels, 29
  placental abruption, 28
  placenta previa, 28
  risk of asthma or recurrent wheezing in
      children, 31
Matrix attachment regions (MARs), 411
Matrix metalloproteinases (MMP), 51, 276
Maximum contaminant level (MCL), 61
MCL. *See* Maximum contaminant level
      (MCL)
MDA. *See* Malondialdehyde (MDA)
Menstrual dysfunction, 122
Mentha piperita, 210
Mercury (Hg), 50
  chemical forms of, 50
  1955 disaster in Minamata Bay, Japan, 51
  elemental (Hg0), 50
  ethyl (EtHg), 50
  matrix metalloproteinases, 51

Mercury (Hg) *(cont.)*
  methyl (MeHg), 50
    neurotoxicity, 50
    NF-E2-related factor 2 (Nrf2), expression of, 50
    nonprotein thiols, interaction with, 50
    sulfhydryl-containing proteins, interaction with, 50
  methylmercury (MeHg), 50
  organic, 51
  uses of, 50
Mesoderm specific transcript (MEST), 416
MEST. *See* Mesoderm specific transcript (MEST)
Metabolic carcinogen stimulation, 211
Metabolic syndrome, 134, 241
Metformin therapy and dietary supplements on semen parameters in, 334
Methionine adenosyltransferase 2A, 22
Methionine synthase *(MTR)*, 375
N-methyl-D-aspartic acid (NMDA) receptor, 49
Methylenetetrahydrofolate reductase *(MTHFR)*, 375
Microminerals, 165
Micronutrients, 227, 242, 341
  Bhutta study 2009, 229
  Christian study 2003, 229
  criteria for considering studies, 228
  data collection and analysis, 228
    analysis issues, unit of, 228
    data extraction and management, 228
      review manager software, use of, 228
    study selection, 228
  deficiencies of, 227
  Dieckmann study 1943, 229
  Fawzi study 2007, 229
  identification of studies, search methods for, 228
  Kaestel study 2005, 230
  Osrin study 2005, 230
  Ramakrishnan study 2003, 230
  Roberfroid study 2008, 230
  Summit study 2008, 231
  Sunawang study 2009, 231
  Tofail study 2008, 231
  Zeng study 2008, 231
Middle East Fertility Society, 77
MIF. *See* Müllerian inhibitory factor (MIF)
Minocyclin, 151
Mitochondrial dysfunction, 121
Mitochondrial membrane potential (MMP), 307, 328
Mitochondrial outer membrane permeabilization (MOMP), 343
MMP. *See* Mitochondrial membrane potential (MMP)
MOET. *See* Multiple ovulation and embryo transfer (MOET) programs
MOMP. *See* Mitochondrial outer membrane permeabilization (MOMP)
Monozygotic twinning, 315
Mouse models, 120
  Db$^{-/-}$ mouse, 120
  diet-induced obese (DIO) mouse, 120
  nonleptin-deficient polygenic mouse, 120
  Ob$^{-/-}$ mouse, 120

Müllerian inhibitory factor (MIF), 343
Multiple ovulation and embryo transfer (MOET) programs, 102

N

NADPH. *See* Nicotinamide adenine dinucleotide phosphate (NADPH)
NAS. *See* National Academy of Sciences (NAS)
National Academy of Sciences (NAS), 61
National ART Registry of India (NARI), 77
National Birth Defects Prevention Study (NBDPS), 61
National Health Interview Survey (NHIS), 254
National Institute for Health and Clinical Excellence (NICE), 139
National Registry of Japan, 76
NBDPS. *See* National birth defects prevention study (NBDPS)
NEFA. *See* Nonesterified fatty acid (NEFA)
Neonatal mortality and morbidity, 45, 228
Neonatal serum folate levels
  and maternal smoking, 29
  in preterm infants, 30
Nervous system, development of, 242
  brain development, 242
    pattern of events, 242
  deficiency, timing of, 243
    effect of timing, 243
    maternal malnutrition, 243
  deficiency, types of, 243
    protein calorie malnutrition (PCM), 243
  dose or severity, 243
  duration of, 244
Neural tube defects (NTDs), 65
Neutraceutical treatment, 93
NF-E2-related factor 2 (Nrf2), 50
NF-kB. *See* Nuclear factor kappa B (NF-kB)
NHIS. *See* National Health Interview Survey (NHIS)
NICE. *See* National Institute for Health and Clinical Excellence (NICE)
Nicotinamide adenine dinucleotide phosphate (NADPH), 276
Nicotine, 20, 28, 400
Nigella sativa, 238
Nitarte
  birth defects in offspring, 65, 66, 68
  dietary intake, 62, 69
    Willett food frequency questionnaire, 69
  in drinking water
    and central nervous system (CNS) defects, 67
    and congenital heart defects, 68
    and late adverse pregnancy outcomes, 68
    and neural tube defects (NTDs), 68
  in maternal drinking water sources, 66
  maternal exposure to, 61
Nitric oxide (NO)
  detrimental action of, 277
  and endometriosis, 276
  production of, 276
Nitric oxide synthase (NOS), 276
  endothelial NOS (eNOS), 276
  inducible NOS (iNOS), 276
  neuronal NOS (nNOS), 276

Nitrites
  dietary intake and birth defects, 69
    Texas neural tube defect project, 69
    Willett food frequency questionnaire, 69
  maternal exposure to, 61
Nitrofurantoin, 151
Nitrosamines, 69
Nitrosatable drugs
  and birth defects, 62
    epidemiologic studies of, 64
  during first trimester of pregnancy, 63
  NBDPS control women, 1997–2005, 63
  neural tube defects (NTDs), studies of, 65
  and nitrate/nitrite
    joint exposure and birth defects, 69
      epidemiologic studies, 70
      NBDPS populations, 69
      Texas neural tube defect study, 69
  prenatal use and vitamin C, 71
N-nitroso compounds
  cell-mediated nitrosation, 62
  endogenous formation, 62
  exogenous sources, 62
  maternal exposure to, 61
Nonesterified fatty acid (NEFA), 35, 102
  concetrations, 39
    for cumulus, granulosa and theca cells, 39
    effect on oocyte and embryo quality, 39
    for oocyte, 40
    for resultant embryo, 40
  genes encoding enzymes, 40
Nonsteroidal anti-inflammatory drugs (NSAIDs), 278
Normozoospermic males, 330
NOS. *See* Nitric oxide synthase (NOS)
n-3 Polyunsaturated fatty acids, 38
NSAIDs. *See* Nonsteroidal anti-inflammatory drugs (NSAIDs)
NSFG. *See* National Survey of Family Growth (NSFG)
Nuclear factor kappa B (NF-kB), 276
Nutrition, 99, 242, 341
  and animal health, 163
  deficiencies, 242
    qualitative malnutrition, 242
    quantitative nutritional deviation, types of, 242
  effect on embryo development, 101
  effect on oxidative stress, 162
  energy status, 163
  factors, 224
  and fertility, 148
  relationship with reproduction, 162
  relation with reproductive axis, 99
  role in infertility, 99

O

Obesity, 3, 35, 119, 133, 183, 303, 355. *See also various obesities*
  and adipose tissue dysfunction, 184
  adverse maternal and neonatal outcomes, 133, 134
  alteration in molecular makeup of sperm, 124
  assisted reproductive technologies (ART), 3

and associated diseases, 303
and benign urological disease, 221
    calculi formation, 222
        extracorporeal shockwave lithotripsy, 223
        nephrolithotomy, 223
    fertility, 222
        inhibin B, role of, 222
    lower urinary tract symptoms, 221
    physical exercise and weight loss, 224
    renal disease, 223
    sexual dysfunction, 221
    stress urinary incontinence, 221
cardiovascular disease (CVD), 119
case-control study, 135
changes in serum composition, 35
contribution to anovulation, 183
definition of, 183
effect on ovary, 121
effect on pituitary, 121
endoplasmic reticulum (ER) stress, 121
fat accumulation, 183
and female fertility, 120
fetal development, 122
follicular microenvironment, 35
histopathological and molecular effects of, 304
hypothalamic gonadal (HPG) axis, 120
hypothalamus-pituitary-testis axis, dysregulation in, 109, 121
impact on male fecundity and fertility, 124
impaired glucose tolerance (IGT), 119
and infertility, studies related to, 237
    sperm count, 237
    sperm quality and function, 238
lipotoxicity, 121
metabolic syndrome, 119
monogenic leptin signaling defects, 120
and mouse models, 120
nonesterified fatty acid (NEFA) concentrations, 35
offspring programming, 122
oocyte and embryo quality, 35, 121
and overweight, 122, 146
    anovulatory infertility, 123
    assisted reproductive technologies (ART) outcomes, effect on, 123
    body mass index (BMI), effect on, 123
    C-reactive protein levels, 147
    early menarche, 122
    fecundity, effect on, 147
    menstrual dysfunction, 122
    meta-analysis, 147
    offspring health, 123
    oocyte development, 147
    ovulatory dysfunction, 122
    polycystic ovulatory syndrome (PCOS), 122, 147
    pregnancy complications, 123
    semen parameters and body-mass index (BMI), 147
    sex-hormone-binding globulin (SHBG) levels, 147
pregnancy complications, 3
prevalence in women, 183, 184
related pregnancy complications, 183

risk of reproductive failure, 35
serum cholesterol concentrations, 35
type 2 diabetes mellitus (DM2), 119
unfolded protein response (UPR), 121
and urological malignancy, possible mechanisms, 223
    bladder cancer, 224
    prostate cancer, 223
    renal cancer, 223
    testicular cancer, 224
Obesity-associated anovulation, 99
Obesity management, 137
    bariatric surgery, 137
    blood pressure monitoring, 139
    CONQUER study, 137
    dietary modifications, 137
    folate supplementation, 139
    gestational diabetes mellitus, 138
    lorcaserin, 137
    pedometers, use of, 137
    pharmacological therapy, 137
    postbariatric surgery dietary guidelines, 137
    during pregnancy, 138
    thromboembolism, 139
    Vitamin D supplementation, 139
    weight-loss maintenance, 137
Offspring health, 123
Offspring programming, 122
8-OH 2-dG. See 8-Hydroxy-2'-deoxyguanosine (8-OH 2-dG)
Oligospermic men, 370
Oligozoospermia, 415
Oocyte and embryo quality, 35, 121
Organic compounds, 205
Organic solvents, 359
    meta-analysis results, 359
    negative effect of, 359
OSI. See Oxidative stress index (OSI)
Ovarian hyperstimulation protocol, 313
Overall pregnancy rates (PRs), 83
Overnutrition
    embryo losses, 101
    progesterone level, increase in, 101
Over-the-counter supplements, 370
Ovine estrous cycle, 100
Ovis aries, 100
Ovulation rate, 102
Ovulatory dysfunction, 122
Oxidative stress, 48, 94, 148, 214, 238, 283, 399
    adenosine triphosphate (ATP) production, inhibition of, 94
    antioxidants, 48
    biomarkers of
        8-hydroxy-2'-deoxyguanosine (8-OH 2-dG), 51
        malondialdehyde (MDA), 51
    and DNA fragmentation prevention, 400
        alternative source of free radicals, 400
            brachytherapy and ionizing radiations, 400
            ethanol, 401
            heat, 400
            leucocytes, 400
            mobile phone radiation, 400
            obesity, 401

            smoke, 400
            varicocele, 400
            Wi-Fi radiation, 400
            xenobiotics and toxic metals, 400
    DNA impairment, 94
    and endometriosis, 275
    human paraoxonases (PONs), 49
    8-hydroxy-2'-deoxyguanosine (8-OH 2-dG), 94
    and male infertility, 369
        DNA replication, impairment of, 369
        electromagnetic radiation, 369
        lipid peroxidation, induction of, 369
        smoking, 369
        varicoceles, 370
    markers in endometriosis, 275
        heat shock protein 70, 276
        macrophages, 276
        matrix metalloproteinases, 276
        mesothelial cells, destruction of, 276
        transcription factor NF-kB, 276
        vascular endothelial growth factor, 276
    in preeclampsia, 287
        vitamin C, role of, 287
    reactive oxygen species (ROS), 48, 94
Oxidative stress index (OSI), 286

P
PAR. See Predictive adaptive response (PAR)
Paraoxonase 1 (PON1), 49
Paternal germ cells, 23
Paternal mutations, in offsprings, 21
    effect of cigarette smoke, 21
    father's age, effect of, 21
Paternal obesity, and fertility, 306. See also Obesity
    embryo development, and IVF outcome, 308
        embryonic genome activation, 308
        fetal heart rates, decrease in, 308
        paternal BMI on IVF outcomes, impact of, 308
        paternal obesity in isolation, 308
    hormone levels and seminal plasma, 306
        androgens aromatization, 306
        BMI, testosterone, and SHBG, negative relationship between, 306
    paternal obesity and offspring health, 309
        epidemiological studies, 309
        offspring phenotype programming, 309
    semen parameters, 306
Paternal smoking, 21, 22
    alteration in gene expression, 21
    chromosomal aberrations, 21
    early postnatal exposures of the newborn, 19
    genetic/epigenetic changes in fetus, 19
    mutations, 21
    oxidative DNA damage, 21
    postconception exposures to embryo and newborn, 19
    preconceptional exposure of maternal and paternal gametes, 19
    prenatal in utero exposure of the fetus, 19
    strand breaks, 21

PCM. *See* Protein calorie malnutrition (PCM)
PCOS. *See* Polycystic ovarian syndrome (PCOS)
Pedersen hypothesis on hyperglycemia during pregnancy, 135
Pedometers, 137
PEM. *See* Protein energy malnutrition (PEM)
Perinatal complications, 315
  birth weight, 316
    PAPP-A as a marker, 316
  multifetal gestation, 315
    elective single embryo transfer (eSET), 315
    multiple gestations, 315
    multiple pregnancies risks, 315
    multiple twinning, 315
    preterm birth, 315
Perinatal mortality, 45
  definition of, 228
  environmental pollutants, effect of, 45
  industrial pollution, effect of, 45
  risk factors, 45
Pesticides, 92, 93
Petroselinum crispum, 210
Phospholipids, 39
Physical activity
  in pregnancy, 176
    barriers faced, 177
    engaging women in physical activity, 177
    exercise benefits, 176
    intrapersonal barriers, 177
  and weight loss, 147
    effect on fertility, 147
Pituitary gland, obesity effect on, 121
Placenta, 46
  abruption of, 28
  as active transporter of essential elements, 46
  as biomarker for toxic metals, 46
  and lead (Pb), 47
  role in fetal development, 46
Placental growth factor (PlGF), 196
Placenta previa, 28
PlGF. *See* Placental growth factor (PlGF)
Polyaromatic hydrocarbons (PAHs), 92, 93
Polycystic ovarian syndrome (PCOS), 38, 95, 122, 123, 133, 147, 183, 194, 265
  chronic anovulation, 265
  hyperandrogenism, 265
  menstrual abnormalities, 265
  polycystic ovary appearance on ultrasound, 265
Polyunsaturated fatty acids (PUFA), 37, 94, 370, 404
PONs. *See* Human paraoxonases (PONs)
Portulaca oleracea, 210
Postconception, 19, 65, 138
Predictive adaptive response (PAR), 245
Preeclampsia, 133, 283
  antioxidant power, decrease of, 286
    superoxide dismutase, use of, 286
  cellular signaling, interactions with, 286
    redox-sensitive transcription factors, activation of, 286
    ROS-induced activation of apoptosis-regulating signal kinase 1, 286
  definition of, 283
  DNA oxidation, 286
    clinically diagnosed mitochondrial dysfunction, 286
    mitochondrial DNA, role of, 286
  lipid peroxidation, 285
  mother and baby, effects on, 284
    maternal complications, 284
    premature delivery, 284
  pathophysiology of, 283
    abnormal placentation, 283
  protein alteration, 285
  related oxidative stress, 284
    fetal placental vasculature, 285
  risk factors for, 284
    chronic kidney disease, 284
    history of chronic hypertension, 284
    hypertensive systems, 284
    renal associated symptoms, 284
  treatment of, 284
    delievery methods, 284
    goals of treatment, 284
    risk of recurrence, 284
Pregnancy, 254, 256
  and chronic low-grade inflammation, 136
  complications, 313
    effect of obesity, 123
  establishment of, 100
  herbal remedies
    and fertility, 254
    use of, 255
      herbal remedies, most investigated, 256, 257
      popular herbal remedies, 255
      rationale for, 255
  with multiple micronutrients, 227
  preimplantation period of, 100
Prepregnancy, 137
Prescription drugs, 151
  antibiotics, 151
  antidepressants, 151
  effect on fertility, 151
  psychotropics, 151
Preterm birth (PTB), 45, 315
Prostate cancer, 223
  cohort study, 223
  treatment of, 223
Protamine 2 (PRM2) germ cell specific gene, 21
Protein calorie malnutrition (PCM), 243
Protein coding genes, 411
Protein energy malnutrition (PEM), 244, 342
Proteins, 36, 164
  in birds, 164
  in cattle, 164
  dietary amino acids, 164
  reproductive performance, effect on, 164
  rumen undegradable protein, 164
  urea nitrogen concentrations, 164
Psychological stress
  alpha-amylase, 148
  and fertility, 148
  hypothalamus–pituitary–adrenal axis, effect on, 148
  menstrual cycle, effect on, 148
  plasma cortisol levels, 148
  salivary cortisol, 148
  sperm quantity, effect on, 148
Psychoneuroimmunology, 294
Psychotropics, 151
PTB. *See* Preterm birth (PTB)
PUFAs. *See* Polyunsaturated fatty acids (PUFAs)

Q

Qualisperm, 95, 96
Qualitative malnutrition, 242
  macronutrients, 242
  micronutrients, 242
Quality of life, sexual, 112

R

Race/ethnicity and risk perception to infertility, 8
Randomized controlled trials (RCT), 232, 254, 265, 371
  neonatal mortality, risk of, 232, 234, 235
  perinatal mortality, 232, 233
  stillbirth analysis, 232–234
Raphanus sativus, 210
RCIS. *See* Reactive chlorine species (RCIS)
RCT. *See* Randomized controlled trials (RCT)
RDA. *See* Recommended daily allowance (RDA)
Reactive chlorine species (RCIS), 285
Reactive nitrogen species (RNS), 276, 285
Reactive oxygen species (ROS), 94, 148, 274, 283, 327, 346, 358, 369
  acrosome reaction, role in, 148
  definition of, 284
  oocyte maturation, role in, 148
  primary generation of, 284
  sperm maturation, role in, 148
  sperm–oocyte fusion, role in, 148
  steroidogenesis, role in, 148
Recommended daily allowance (RDA), 205, 264
Recreational substances, and illicit drugs, 357
  alcohol consumption, 358
    decreasing sperm count, 358
    gynecomastia, 358
    negative effects, 358
    studies of, 358
    testicular atrophy, 358
  cigarette smoking
    seminal plasma antioxidants, 358
    sperm motility reduction, 358
  cocaine, 359
    animal studies of, 359
    sperm motility, reduction of, 359
    teratogenic effects, 359
  marijuana, 358
    acrosome reaction impairement, 359
    sperm motility, reduction of, 359
REDLARA. *See* Latin American Network for Reproductive Medicine (REDLARA)
Reference nutrient intakes (RNI), 265
Renal cancer, 223
  laparoscopic approach, 224
Reproductive and metabolic implications in women, 122

Reproductive dysfunction, 119, 125
  male obesity-induced, 125
  and obesity, 119
Reproductive health, 8
  DNA methylation, 6
  environmental determinants of, 6
  environmental disruption of, 5
  gene-environment interactions, 6
Respiratory distress syndrome (RDS), 45
Retinopathy of prematurity (ROP), 287
Retrograde menstruation, 277
Review Manager software, use of, 228
Risk mitigation, barriers to, 11
RNI. *See* Reference nutrient intakes (RNI)
RNS. *See* Reactive nitrogen species (RNS)
ROP. *See* Retinopathy of prematurity (ROP)
ROS. *See* Reactive oxygen species (ROS)
Rosiglitazone, 121

S
Sampson's implantation theory, 274
SART. *See* Society for Assisted Reproductive
    Technology (SART)
SCSA. *See* Sperm chromatin structure assay
    (SCSA)
Second mitochondrial activation of caspases
    (SMAC), 343
Selenium (Se), 261
  aging, role in, 264
  dietary selenium, 262, 263
  and early childhood, 267
  organic forms, 261
  and pregnancy, 265
    gestational diabetes, 267
    miscarriage, 266
      definition of, 266
      selenium deficiency, as a factor
        for, 266
    normal pregnancy, 265
    preeclampsia, 266
      definition of, 266
      genes for functional polymorphisms,
        266
    preterm birth, 267
      inflammation, 267
      small-for-gestational-age deliveries, 267
  and reproductive health, 264
    female fertility, 265
    male fertility, 264
      antioxidant capacity, 265
      spermatogenesis, 264, 414
      spermatozoa motility, 264
      testicular development, 264
  and vitamin E cancer prevention trial, 224
Selenoproteins, 261, 262, 372
  actions of, 261
  classes of, 261
  type 3 iodothyronine deiodinase (DiO3),
    266
Semen
  analysis, 111
  effect of bariatric surgery, 114
  parameters, 334
    and body-mass index (BMI), 111
    and functional assays in
      hyperinsulinemic patients, 328, 329

quality
  effect of antiestrogen treatment, 93
  xenoestrogen contamination, 93
samples, 328
Sertoli cells, 384, 386
  germinative cells, 385, 386
    arrangement of, 385, 388
  spermatogonia, types of, 385
Sex hormone binding globulin (SHBG), 147,
    183, 304, 328
Sex hormones, 109
  correlation with body-mass index (BMI),
    110
  effect of bariatric surgery, 112
  sex steroid-binding globulin, 109
Sex quality
  effect of bariatric surgery, 114
Sexually transmitted infections (STI), 5, 151
  chlamydia, 151
  gonorrhea, 151
  human papillomavirus (HPV) infection,
    151
  infection with *Chlamydia trachomatis*, 5
  infection with *Neisseria gonorrhoeae*, 5
  syphilis, 151
SGA. *See* Small-for-gestational-age (SGA)
SHBG. *See* Sex hormone binding globulin
    (SHBG)
Sheep as an animal model, 100
Silver–Russell syndrome (SRS), 316, 416
Single embryo transfer (SET), 83, 85
SMAC. *See* Second mitochondrial activation
    of caspases (SMAC)
Small-for-gestational-age (SGA), 45, 195, 267
Smoking, 22. *See also* Maternal smoking
  cigarette. *See* Cigarette smoking
  DNA damage in sperm, 19
  and DNA methylation, 22
  maternal. *See* Maternal smoking
  postChernobyl radioactive exposure, 19
  postconceptional, 19
  preconceptional, 19
  and pregnancy, 28
  tobacco. *See* Tobacco smoking
  transgenerational changes
    via epigenetic modifications, 19
    via stable chromosomal alterations, 19
Society for Assisted Reproductive Technology
    (SART), 76
Sperm
  cell proliferation, 327
  damage, 397
    antioxidants usage, 399, 401
      SUVIMAX, 399
    factors for, 397
    oxidative stress, 399
  impairment of, 125
    glycidamide formation, 125
    HFD components, 125
    reproductive dysfunction, 125
  maturation, 390
    chromatin condensation, 390
    sperm remodelling, 390
  molecular makeup of, 124
    HFD-feeding model, 124
    microRNA content, 124

motility
  correlation with year at semen
    analysis, 94
  relation with nutritional intake
    of long-chain PUFAs, 94
  relation with nutritional intake
    of short-chain PUFAs, 94
oxidative stress, 125
quality
  characteristics in infertility patients, 96
  decline of, 89
  indicators, 328
quantity
  effect of sleep, 152
Spermatid maturation, stages of, 386
Spermatocyte labeling curves, 393
Spermatogenesis, 23, 151, 237, 385, 403
  cell division stages, 385, 387
  demethylation–remethylation process, 413
  and germ cell apoptosis, 347
    abnormally accelerated apoptosis of
      germ cells, 347
    basic processes, 347
    regulation of gene expression in germ
      cells, 347
    spermatogenesis, maintenance of, 347
  hyperacetylation, 413
  sperm protamination, 413
  transition proteins, role of, 413
  in vivo kinetics, 392
    noninvasive direct kinematic
      measurement of spermatogenesis, 392
    seminiferous epithelial cycle,
      identification of, 392
Spermatozoa, epigenetic modifications of,
    411, 412
  DNA methylation, 411
  histone retention, 411
  histone tail modifications, 411
  protamine incorporation into chromatin,
    411
Spermatozoon, 22, 92, 358, 386, 389, 391, 398.
    See also Sperm
Sperm chromatin structure assay (SCSA), 307
Sperm donors
  hormonal and BMI values of, 331
  hyperinsulinemic
    in vitro fertilization, 330
      initial assessment of research subjects,
        330
      insulin and hormonal testing, 330
      results, 331
  insulin ranges of, 330
Sperm function, 390
  acrosome integrity, 391
  capacitation, 390
  hydrolytic enzymes, role of, 391
  incubation with blockers, 391
    nifedipine, 391
    trifluoperazine, 391
    verapamil, 391
  spermatozoa hyperactivation, 390
  spermatozoon and ZP, binding of, 391
  sperm–oocyte fusion, 391
  sperm penetration through zona pellucida
    (ZP), 390

Sperm function (*cont.*)
  toxis substances, 391
    β-lapachone, 391
    genistein, 391
    isoflavone, 391
    ZP glycoprotein-3 (ZP3) for AR initiation, 391
Sperm functional assay, 111
  acrosome reaction (AR), 111
  zona pellucida (ZP) binding, 111
Spermiogenesis, 386
Spousal involvement, 299
SRS. *See* Silver–Russell syndrome (SRS)
Stress, 295
  chronic stress, 297, 298
  and infertility, 293
  management, 298
    assertion training, 298
    cognitive challenging, 298
    psychoneuroimmunological factors, 298
    relaxation techniques, 298
  real and self induced, 295
  time urgency perfectionism, 295
Stressors, 148
Supplements, 370
  antioxidants, 370
  arginine, 376
  carnitines, 375
  coenzyme Q10, 374
    energy production, 374
  folic acid, 375
    cofactor for DNA synthesis, 375
    effects of, 375
  glutathione, 373
    free-radical scavenger, 373
    male factor infertility, 373
  lycopene, 375
    studies of, 375
  micronutrients, 370
  polyunsaturated fatty acids, 374
    types of, 374
  selenium, 372
    dietary selenium level, decrease of, 372
    seminal concentrations, increase of, 372
    studies helping in improving pregnancy rates, 372
  vitamin A, 371
  vitamin C, 371
    extracellular matrices creation, 371
    studies of, 371
  vitamin E, 370
    external stressors effect on spermatogenesis, 371
    germ cell integrity improvement function, 370
  zinc, 372
    abnormal semen parameters, balancing of, 372
    cofactor for embryogenesis, 372
    male infertility treatment, 373
    oral zinc supplementation, 373
Surface water samples
  ERBA test, 90, 91
  human breast cancer cell test, 90
  from Schelde river
    total estrogen activity, 91
    in vitro breast cancer cell test, 91
    yeast test, 90
Sus scrofa ferus, 194
Sus scrofa meridionalis, 194
Syphilis, 151

T

Talinum triangulare, 213
Tamoxifen, 93
TAR. *See* Total antioxidant response (TAR)
TAS. *See* Total antioxidant status (TAS)
TCM. *See* Traditional Chinese medicine (TCM)
Telifaria accidentlis, 213
Terminal uridine nick-end labeling (TUNEL) analysis, 307, 344
Testicular cancer, 5, 90, 224
Testicular dysgenesis syndrome, 5
Testis, structure and function, 383
  androgen production, 384
    leukemia inhibitory factor (LIF), 384
    platelet-derived growth factor-α (PDGF-α), 384
    steroidogenic pathways, 384, 385
    testosterone production, 384
  Leydig or interstitial cells, 384
  peritubular cells, 384
  seminiferous tubules, 383
  Sertoli cells layering, 384
  spermatozoon structure, 386, 389
  sperm production, 384
  weight and volume, 383
Tetramethyl rhodamine methyl ester (TMRM) oocytes, 121
Texas neural tube defect study, 69
TGF. *See* Transforming growth factor (TGF)
Thrifty gene hypothesis, 245
Thromboprophylaxis, 139
Time urgent perfectionist (TUP), 296
  example of, 297
  factors including, 296
TM. *See* Traditional medicine (TM)
Tobacco smoking, 4, 20
  adverse reproductive outcome, associated with, 21
  effect on women's health, 4
  epidemic, 19
  and infertility, 4
  nicotine, 20
  pre-conceptional maternal exposure, 20
  pre-conceptional paternal exposure, 20
  second-hand exposure, 4
  sperm DNA damage, 4
TOS. *See* Total oxidant status (TOS)
Total antioxidant response (TAR), 286
Total antioxidant status (TAS), 286
Total oxidant status (TOS), 286
Trace elements, 261
  copper (Cu), 166
  manganese (Mn), 166
  molybdenum (Mo), 166
  selenium (Se), 166
  zinc (Zn), 166
Traditional Chinese medicine (TCM), 253
Traditional medicine (TM), 253
Trans forming growth factor (TGF), 213
Transgenerational DNA alterations, 23
TUNEL. *See* Terminal uridine nick-end labeling (TUNEL) analysis
TUP. *See* Time urgent perfectionist (TUP)

U

Undernutrition, 101
  embryonic development, 101
  oocyte quality, 101
  preganancy rates, 101
  progesterone concentrations, 101
Underweight females, and fertility, 147
  ovarian disorder, 147
Urea, 36, 164
Urinary tract infection (UTI), 257
Urtica dioica, 210
U.S. Agency for Toxic Substances and Disease Registry (ATSDR), 46
U.S. Environmental Protection Agency (EPA), 61
UTI. *See* Urinary tract infection (UTI)

V

Vanishing twin phenomenon, 315
Varicocele, 400
Varicocelectomy, 370, 373
Vascular and lymphatic metastasis theory, 275
Vascular endothelial growth factor (VEGF), 196, 276
VEGF. *See* Vascular endothelial growth factor (VEGF)
Venous thrombosis, 134
Vitamins, 164
  reproductive performance, role in, 164
  vitamin A, 165
  vitamin C, 71, 165
    epidemiologic studies, 71
    inhibition of nitrosation, 71
    prenatal exposures to nitrosatable compounds and birth defects in offspring, 71
  vitamin D deficiency, 135, 136
  vitamin E, 165
Voltage-operated calcium channels (VOCCs), 391

W

Weight loss
  long term effects of, 222
  strategies, 126
Weight management, 124
Weight retention, long term, 177
WHO. *See* World Health Organization (WHO)
Wi-Fi radiation, 400
Women's reproductive health, 133
World Health Organization (WHO), 19, 45, 99, 227, 263, 306, 362
  report on obesity, 133
  sperm parameters, 125

X

Xenobiotics, 400
Xenoestrogens, 90, 94, 96
    effect on sperm quality, 93
    in flemish surface waters, 91
X-ray absorptiometry, 317

Z

Zinc (Zn)
    absorption, 341
        dependent proteins, 341
        factors affecting, 341
        intracellular, regulation of, 341
        transport proteins, 341
    and apoptosis, 344
        in culture medium, 344
        deficiency conditions, 344, 346
        function against oxidative stress, 346
        ICE/CED-3 proteases, activation of, 345
        low-zinc cultures increasing apoptosis, 344
        mitochondrial-dependent apoptosis, 345
        p^{53} causing cell cycle arrest, 346
        potential targets, 346
        proliferation of the prostatic cells, balancing of, 344
        with redox properties, 346
        for sulfhydryl groups, affinity for, 346
    biology of, 341
        anticancerous properties, 341
        role in antioxidant functions, 341
        role in enzyme functions, 341
        role in immune function, 341
        secretion of proteins, regulation of, 341
        structure stabilizing property, 341
    deficiency, 342
        apoptotic neuronal death, 345
        importance in human nutrition, 342
        plasma concentrations, 342
        related dysfunctions, 342
        respiratory epithelial cells, 345
    dietary supplementation, 238
    in male reproduction, 343
        content in adult testis, 343
        protection against degenerative changes, 343
        protection against toxic effects, 343
        structural integrity of DNA, maintaining of, 343
    for testicular function, 238
Zinc finger proteins, 341
Zingiber officinale, 256
Zona pellucida (ZP) binding, 111, 390
ZP. *See* Zona pellucida (ZP) binding